Adhesion Molecules

THE HANDBOOK OF IMMUNOPHARMACOLOGY

Series Editor: Clive Page
King's College London, UK

Titles in this series

Cells and Mediators

Immunopharmacology of Eosinophils
(edited by H. Smith and R. Cook)

Immunopharmacology of Mast Cells and Basophils
(edited by J.C. Foreman)

Adhesion Molecules
(edited by C.D. Wegner)

Lipid Mediators
(edited by F. Cunningham)

Immunopharmacology of Lymphocytes
(edited by M. Rola-Plezczynski)

Immunopharmacology of Neutrophils
(edited by P.G. Hellewell and T.J. Williams)

Immunopharmacology of Macrophages and other Antigen-Presenting Cells
(edited by C.A.F.M. Bruijnzeel-Koomen and E.C.M. Hoefsmit)

Immunopharmacology of Platelets
(edited by M. Joseph, forthcoming)

Systems

Immunopharmacology of the Gastrointestinal System
(edited by J.L. Wallace)

Immunopharmacology of Joints and Connective Tissue
(edited by J.T. Dingle and M.E. Davies)

Immunopharmacology of the Heart
(edited by M.J. Curtis)

Immunopharmacology of Epithelial Barriers
(edited by R. Goldie)

Immunopharmacology of the Renal System
(edited by C. Tetta)

Immunopharmacology of the Microcirculation
(edited by S. Brain)

Drugs

Immunotherapy for Immune-related Diseases
(edited by W.J. Metzger, forthcoming)

Immunopharmacology of Free Radical Species
(edited by D. Blake and P.G. Winyard, forthcoming)

Immunopharmacology of AIDS (forthcoming)

Immunosuppressive Drugs
(forthcoming)

Glucocorticosteroids
(forthcoming)

Angiogenesis (forthcoming)

Adhesion Molecules

edited by

C.D. Wegner
Abbott Labs., Abbott Park, Illinois, USA

ACADEMIC PRESS
Harcourt Brace & Company, Publishers
London San Diego New York
Boston Sydney Tokyo Toronto

ACADEMIC PRESS LIMITED
24/28 Oval Road
London NW1 7DX

United States Edition published by
ACADEMIC PRESS INC.
San Diego, CA 92101

A catalogue record for this book
is available from the British Library

ISBN: 9780127414409

Typeset by Mathematical Composition Setters Ltd, Salisbury, Wiltshire

Transferred to digital print 2007

Printed and bound by CPI Antony Rowe, Eastbourne

Contents

4. *Role of Cell Adhesion Molecules in Tumor Progression and Metastasis* 71

Steven M. Albelda

5. *Concepts in Adhesion Regulation* 89

Tamas Schweighoffer *and* Stephen Shaw

6. *Specificity and Expression of Carbohydrate Ligands* 111

John B. Lowe

7. *The Role of Cell Adhesion in the Evolution of Inflammatory Arthritis* 141

Nancy Oppenheimer-Marks *and* Peter E. Lipsky

8. *Pathogenesis of Ischemia/Reperfusion: Role of Neutrophil-endothelial Cell Adhesion* 163

Ronald J. Korthuis *and* D. Neil Granger

9. *Lung Inflammation* 191

Craig D. Wegner

10. *Leucocyte-endothelial Adhesion Molecules* in vivo: *A role in Physiological Functions and Pathophysiological Vasculidities* 209

Lawrence W. Argenbright *and* Thomas C. Noonan

11. *Molecular Interaction of Viruses with Host Cell Receptors* 223

Thomas L. Lentz

12. *Bacterial Adhesins* 253

Soman N. Abraham

13. *Cell Adhesion Molecules in Host-parasite Interactions* 277

Christian F. Ockenhouse

Contributors

S.N. Abraham
Departments of Pathology and Molecular Microbiology, Washington University School of Medicine,
and
Department of Pathology and Laboratory Medicine, Jewish Hospital of St Louis,
216 South Kingshighway,
St Louis, MO 63110
USA

S.M. Albelda
Assistant Professor of Medicine,
Pulmonary and Critical Care Division,
Department of Medicine,
University of Pennsylvania School of Medicine,
Philadelphia, Pennsylvania,
USA
and
Adjunct Assistant Professor
The Wistar Institute of Anatomy and Biology,
3601 Spruce Street,
Philadelphia, PA 19104–4268,
USA

L.W. Argenbright
Department of Pharmacology,
Boehringer Ingelheim Pharmaceuticals, Inc,
900 Ridgebury Road,
PO Box 368,
Ridgefield, CT 06877,
USA

D.N. Granger
Department of Physiology,
LSU Medical Center,
1501 Kings Highway,
Shreveport, LA 71130–3932,
USA

R.J. Korthuis
Department of Physiology,
Lousiana State University Medical Center,
School of Medicine in Shreveport,
Shreveport, LA 71130,
USA

T.L. Lentz
Department of Cell Biology,
Yale University School of Medicine,
333 Cedar St,
New Haven, CT 06510,
USA

P. Lipsky
The Harold C. Simmons Arthritis Center,
The University of Texas Southwestern Medical Center,
5223 Harry Hines Blvd.,
Dallas, TX 75235–8884,
USA

J.B. Lowe
Associate Professor of Pathology,
Assistant Investigator, Howard Hughes Medical Institute,
University of Michigan Medical School,
MSRBI, Room 3510,
1150 West Medical Center Drive,
Ann Arbor, MI 48109–0650,
USA

S.C. Meuer
Division of Applied Immunology,
German Cancer Research Center,
Im Neuenheimer Feld 280,
D-6900 Heidelberg,
Germany

T.C. Noonan
Department of Pharmacology,
Boehringer Ingelheim Pharmaceuticals, Inc,
900 Ridgebury Road,
PO Box 368,
Ridgefield, CT 06877,
USA

C.F. Ockenhouse
Department of Immunology,
Walter Reed Army Institute of Research,
Walter Reed Army Medical Center,
Washington, DC 20307,
USA

N. Oppenheimer-Marks
The Harold C Simmons Arthritis Center,
The University of Texas Southwestern Medical Center,
5323 Harry Hines Blvd,
Dallas, TX 75235–8884,
USA

R. Rothlein
Boehringer Ingelheim Pharmaceuticals Inc.
90 East Ridge,
Ridgefield, CT 06877,
USA

L.A. Scharschmidt
Boehringer Ingelheim Pharmaceuticals, Inc,
90 East Ridge,
Ridgefield, CT 06877,
USA

T. Schweighoffer
Experimental Immunology Branch,
National Cancer Institute,
NIH,
Bethesda, MD 20892,
USA

S.B. Shappell
Speros P Martel Laboratory,
Section of Leukocyte Biology,
Departments of Pediatrics and Microbiology and Immunology,
Baylor College of Medicine,
Houston, TX 77030,
USA

S. Shaw
Experimental Immunology Branch,
National Cancer Institute,
NIH
Bethesda, MD 20892,
USA

C.W. Smith
Speros P Martel Laboratory,
Section of Leukocyte Biology,
Departments of Pediatrics and Microbiology and Immunology,
Baylor College of Medicine,
Houston, TX 77030,
USA

C.D. Wegner
Abbott Labs.,
1 Abbott Park Road,
Department 46R, BldgAP9,
Abbott Park, IL 60064–3500,
USA

Series Preface

The consequences of diseases involving the immune system such as AIDS, and chronic inflammatory diseases such as bronchial asthma, rheumatoid arthritis and atherosclerosis, now account for a considerable economic burden to governments worldwide. In response to this, there has been a massive research effort investigating the basic mechanisms underlying such diseases, and a tremendous drive to identify noval therapeutic applications for the prevention and treatment of such diseases. Despite this effort, however, much of it within the pharmaceutical industries, this area of medical research has not gained the prominence of cardiovascular pharmacology or neuropharmacology. Over the last decade there has been a plethora of research papers and publications on immunology, but comparatively little written about the implications of such research for drug development. There is also no focal information source for pharmacologists with an interest in diseases affecting the immune system or the inflammatory response to consult, whether as a teaching aid or as a research reference. The main impetus behind the creation of this series was to provide such a source by commissioning a comprehensive collection of volumes on all aspects of immunopharmacology. It has been a deliberate policy to seek editors for each volume who are not only active in their respective areas of expertise, but who also have a distinctly ***pharmacological*** bias to their research. My hope is that *The Handbook of Immunopharmacology* will become indispensable to researchers and teachers for many years to come, with volumes being regularly updated.

The series follows three main themes, each theme represented by volumes on individual component topics. The first covers each of the major cell types and classes of inflammatory mediators. The second covers each of the major organ systems and the diseases involving the immune and inflammatory responses that can affect them. The series will thus include clinical aspects along with basic science. The third covers different classes of drugs that are currently being used to treat inflammatory disease or diseases involving the immune system, as well as novel classes of drugs under development for the treatment of such diseases.

To enhance the usefulness of the series as a reference and teaching aid, a standardized artwork policy has been adopted. A particular cell type, for instance, is represented identically throughout the series. An appendix of these standard drawings is published in each volume. Likewise, a standardized system of abbreviations of terms has been implemented and will be developed by the editors involved in individual volumes as the series grows. A glossary of abbreviated terms is also published in each volume. This should facilitate cross-referencing between volumes. In time, it is hoped that the glossary will be regarded as a source of standard terms.

While the series has been developed to be an integrated whole, each volume is complete in itself and may be used as an authoritative review of its designated topic.

I am extremely grateful to the officers of Academic Press, and in particular to Dr Carey Chapman, for their vision in agreeing to collaborate on such a venture, and greatly hope that the series does indeed prove to be invaluable to the medical and scientific community.

C.P. Page

Preface

In the 1970's and 1980's the terms "cell-cell interactions" and "cell-cell communications" became common to discussions on inflammatory responses and their resolution. Elaborate pathways depicting cytokines, growth factors, and feedback mechanisms interconnecting multiple cells were standard in attempts to comprehend and explain the pathogenesis of various diseases. More recently, there has been a renewed appreciation of the influence of intimate cell-cell contact, or "adhesion", the ultimate in cell-cell communication, in the regulation of almost every aspect of host defence and the inflammatory response.

Leukocytes must adhere to vascular endothelial cells, extracellular matrix components and tissue cells in order to migrate from the peripheral circulation to sites of inflammation. T-lymphocyte activation on adherence to class II positive antigen-presenting cells is an essential initial step in cascades leading to the production of antibodies, cytotoxic T-lymphocytes (CTL) and cell-mediated immune responses. Likewise, effector leukocytes [CTL, natural killer (NK) cells and granulocytes] must adhere to appropriate target cells (e.g. virally infected/foreign antigen-expressing airway epithelium, tumor cells or transplanted allogeneic tissue) to perform their targeted lytic function.

Beyond simply providing the traction for leukocyte trafficking and the mooring for cell docking, cell surface-adhesive glycoproteins perform an active role in directing and regulating inflammatory/immune responses. Their enhanced expression on inflamed endothelium and epithelium support the focal accumulation and retention of leukocytes at sites of inflammation. On leukocyte activation, some leukocyte adhesion molecules increase their binding avidity and/or expression, while others are shed presumably, to accommodate docking and migration, respectively. Finally, adhesion itself, under appropriate co-stimulatory/co-operative conditions, results in the signalling, priming or activation of leukocytes.

Similarly, cell surface adhesive molecules are utilized by viruses, bacteria and parasites to adhere to, replicate on the surface of, and gain entry into their "host" cells. Thus, an understanding of the expression, specificity, intracellular signalling and function of leukocyte, mesenchymal and epithelial cell surface adhesion molecules is essential to our comprehension and therapeutic interruption of the pathogenesis of inflammatory and infectious diseases as well as to the retention or protection of host defence mechanisms.

In contrast to several other books and chapters recently published on this topic, the objective of this volume is to provide a comprehensive overview of the adhesion molecules, processes and concepts that govern both inflammatory and infectious diseases, while at the same time driving deeply into the specific molecular and cellular mechanisms as well as tissue/disease specific *in vivo* pathways involved. We begin with an overview chapter to conceptually introduce some of the molecules that mediate leukocyte adhesion from their discovery utilizing monoclonal antibodies and a congenital adhesion deficiency to their antagonism in preliminary clinical trials as novel therapeutics. The next three chapters provide in-depth analysis of the structure, distribution and function of the cell surface glycoproteins that regulate lymphocyte (specific immune response), granulocyte (acute inflammatory response) and metastatic cell (malignant processes) adhesions, respectively. Two chapters detailing the molecular structure, intracellular pathways, specificity of carbohydrate interactions, and signalling of the molecules that regulate leukocyte-leukocyte and leukocyte-mesenchymal cell interactions follow. With this as background, the probable contributions of specific molecules to inflammatory diseases in various organs are explored in the next four chapters. Finally, and unique to this volume on adhesion molecules, the comparable (and in some cases same) cell surface molecules that mediate virus, bacteria and parasite interactions with host cells are reviewed.

While important specific details remain for future studies, the state-of-the-art background foundations contained in this volume seem firm. Thus, this volume will serve as a valuable resource for many years to come.

C.D. Wegner

1. Leukocyte Adhesion in Inflammation: From Discovery to the Clinic

Robert Rothlein and Linda A. Scharschmidt

1. Introduction

There is a multifactorial basis to explain this decade's explosion in research leading to our current understanding of the molecular mechanisms involved in cell adhesion as well as the central role that leukocyte adhesion plays in the proper functioning of a host defense response. First, there has been an exponential burst in our capabilities to dissect molecular mechanisms of cellular behavior. This has been facilitated with the sophisticated techniques of molecular biology and monoclonal antibody (mAb) production as well as the generation of new computer driven analytical techniques at both the molecular and cellular level. Second, leukocyte adhesion processes are of interest to a broad spectrum of academic and clinical disciplines, including immunology, cell biology, physiology, molecular biology, biochemistry, virology, and is just now extending into clinical studies. Third, there is an intense industrial interest in pursuing the identification and clinical applications of molecules that interfere with the leukocyte adhesion process. In this chapter, we will give an historical perspective into the development of an anti-ICAM-1 mAb as a potential therapeutic for inflammation. We will highlight the critical experiments that fuelled the enthusiasm for ICAM-1 directed research and motivated establishment of clinical trials with anti-ICAM-1 mAb for kidney transplantation and rheumatoid arthritis.

By way of background, it is clear that leukocyte-leukocyte and leukocyte-target cell adhesion are necessary events for the host defense system to function normally in inflammation. Specifically: (1) Leukocytes must first attach to the endothelial cells which line blood vessels prior to migrating from the circulation to sites of inflammation (Perry and Granger, 1991; Argenbright *et al.*, 1991); (2) Lymphocytes must first adhere to antigen presenting cells in order for such specific immunological processes such as antibody production or the generation and elaboration of antigen specific thymus derived lymphocyte cells (T cells) to occur (Lipsky and Rosenthal, 1975); (3) Finally, for leukocytes to perform effector functions such as lysis of target cells, they must first attach to these target cells. The lytic mediators secreted by the leukocytes achieve high local concentrations at the leukocyte-target cell junction, thus facilitating the destruction of the target cell (Bierer and Burakoff, 1988).

1.1 Preclinical Experimentation

Three separate lines of evidence obtained during the last 10 years have converged to implicate the CD18 family of

Adhesion Molecules
ISBN 0-12-741440-1

molecules as being partially responsible for mediating leukocyte adhesion to cellular substrates. The first line of evidence comes from the use of adhesion antagonists in *in vitro* experimental models. The second is a fascinating genetic experiment of nature in which humans are unable to express most or all of the CD18 family. The third line of evidence is evolving in clinical research with a specific adhesion antagonist in human kidney transplantation and rheumatoid arthritis.

The CD18 (or LFA-1) family of adhesion molecules consists of CD11a,CD18 (LFA-1); CD11b,CD18 (MAC-1) and CD11c,CD18 (p150,95) (reviewed in Springer *et al.*, 1987; Martz, 1987; Springer, 1990). These molecules, which are distributed in various combinations on all leukocytes, are each heterodimers with distinct α subunits that associate with a common β subunit. Structurally, the leukocyte adhesion molecules are members of the integrin superfamily and specifically they belong to the β-subgroup. Although the leukocyte adhesion molecules mediate adhesion processes, members of the CD18 family are unique in that they appear to recognize ligands that, for the most part, do not contain RGD sequences. Furthermore, their activity is dependent on Mg^{2+} rather than Ca^{2+} (Martz, 1980; Martz *et al.*, 1982; Rothlein and Springer, 1986).

Primary evidence to show the function of these molecules and, in fact, the importance of leukocyte adhesion in the generation and maintenance of inflammation has been elucidated in many experimental systems in which mAb to these molecules, used as antagonists, were found to inhibit multiple events associated with inflammation.

In vitro, mAb to CD18 or CD11a inhibit lymphocyte mediated lytic events such as cytotoxic T-cell activity and natural killer (NK) cell activity (Krensky *et al.*, 1983; Sanchez-Madrid *et al.*, 1983; Kohl *et al.*, 1984). These same mAbs also block lymphocyte trafficking processes, specifically attachment of lymphocytes to vascular endothelium and epidermal cells (Haskard *et al.*, 1986; Dustin *et al.*, 1988). Furthermore these mAb inhibit *in vitro* models of general and/or specific immunological responsiveness, such as antibody formation, mitogen- and antigen-induced T-cell proliferation, and the mixed lymphocyte response (MLR) (Davignon *et al.*, 1981; Dougherty and Hogg, 1987; Dougherty *et al.*, 1988; Boyd *et al.*, 1988). All of these processes have in common the requirement of cell-cell interaction. The mechanism by which mAb to the adhesion molecules inhibit the above processes is thought to be mediated by their ability to antagonize the interaction of the appropriate CD18 family member with its ligand(s) rather than cellular elimination, since these mAb have not been shown to be cytotoxic themselves, either in the presence of complement or in the presence of cells capable of mediating antibody-dependent cellular cytotoxicity (ADCC) (Kohl *et al.*, 1984).

MAb to CD18 and to CD11 also inhibit granulocyte function such as attachment to endothelium (Smith *et al.*, 1988), homotypic aggregation (Anderson *et al.*, 1986), binding to inactivated C3b (iC3b) coated particles (Beller *et al.*, 1982; Anderson *et al.*, 1986) and ADCC (Kohl *et al.*, 1984). These observations suggest that the CD18 family of adhesion molecules play a role in processes involved in acute inflammation as well as those involved in specific immunological processes.

The second and perhaps more compelling line of evidence, suggesting an important role of the CD18 family of adhesion molecules, came from the identification of a group of individuals, who due to genetic defects, were unable to express the normal number of leukocyte adhesion molecules on their cell surfaces (Anderson and Springer, 1987; Todd and Freyer, 1988; Kishimoto *et al.*, 1989). This hereditary defect is called leukocyte adhesion deficiency (LAD) syndrome. Patients were classified into a severely deficient group (expressing between 0–1% of the normal number of adhesion molecules on the leukocyte surface) and a moderately deficient group (expressing between 2–10% of the normal number of adhesion molecules on the leukocyte surface) (Anderson *et al.*, 1985). Both groups of LAD patients presented with the following clinical signs: delayed umbilical cord separation, frequent skin infections with little leukocyte infiltrate and the inability to form pus. The severely deficient patients, in addition to the above, also had depressed T-cell responses, presumably because the inability of their T-cells to attach to the antigen-presenting cell resulting in a defective antigen-presenting capacity.

In vitro, lymphocytes from these LAD patients behaved as did normal lymphocytes if the normal lymphocytes were in the presence of anti-LFA-1 mAb. Both demonstrated defective proliferative and effector responses in assays requiring cell-cell interactions (Mentzer *et al.*, 1986). Not only did these patients teach us that the CD18 family of adhesion molecules were involved in leukocyte adhesion, but also that if one could identify a specific antagonist of the CD18 family of adhesion molecules, one could attenuate an inflammatory response without having other unwanted side effects. This became apparent since the CD18 deficiency manifested itself clinically only as an inability to mount an inflammatory response, in that the LAD patients had no other apparent defects (e.g., failure to develop *in utero*, predisposition to malignancy, etc.).

About the time that the role of the CD18 family of adhesion molecules was being elucidated, a ligand of LFA-1, intercellular adhesion molecule-1 (ICAM-1) was discovered. It is the mAb to ICAM-1 which has been taken into the clinical trials. ICAM-1 was functionally identified by testing various antibodies from hybridomas generated from spleen cells of mice immunized against LFA-1 deficient lymphocytes in an LFA-1/ICAM-1 dependent homotypic aggregation assay (Rothlein and Springer, 1986; Rothlein *et al.*, 1986). One antibody was identified. Subsequently, the antigen that was affinity purified using this antibody, when inserted into planar membranes or immobilized on plastic, was able to mediate lymphocyte adhesion (Marlin and Springer, 1987; Makgoba *et al.*, 1988). This adhesion was inhabitable by mAb directed against either ICAM-1 or LFA-1.

Structurally, ICAM-1 is a member of the immunoglobulin supergene family with five Ig-like domains, a single membrane-spanning region and a short cytoplasmic tail (Staunton *et al.*, 1988; Simmons *et al.*, 1988). Further characterization of ICAM-1 revealed that there is little if any expressed on unperturbed tissue, but it is inducible *in vitro* on multiple cell types including hematopoietic cells, endothelial cells and fibroblasts with cytokines such as IL-1a, IL-1b (interleukins), TNFa, TNFb (tumor necrosis factor) and/or IFNg (interferon) depending on the cell type (Dustin *et al.*, 1986; Dustin *et al.*, 1988; Prober *et al.*, 1986; Frohman *et al.*, 1989). Functionally, anti-ICAM-1 mAb inhibits *in vitro* assays such as MLRs (Boyd *et al.*, 1988), antigen-induced proliferation (Dougherty *et al.*, 1988), cytotoxic T-cell activity (Mentzer *et al.*, 1988; Makgoba *et al.*, 1989), and granulocyte and lymphocyte attachment to endothelium (Smith *et al.*, 1988; Smith *et al.*, 1989; Dustin and Springer, 1988). In short, antibodies to either CD18 or ICAM-1 inhibit the same leukocyte functions. Further evidence suggesting the role of ICAM-1 in the immune response came when ICAM-1, transfected into L cells expressing low amounts of HLA-DR molecules, enhanced class II restricted proliferation when compared to L cells expressing the human histocompatability antigen (HLA-DR) alone (Altmann *et al.*, 1989). Also ICAM-1 defective mutants were unable to properly present antigen to mouse T cells (Dang *et al.*, 1990). These data suggested, that like the CD18 family of adhesion molecules, ICAM-1 expression on endothelial cells and other cell types were critical for the normal function of the host defense system. What came as somewhat of a surprise to us was the finding that ICAM-1, like CD18, also mediated neutrophil adhesion to endothelium. Initially, when ICAM-1 was identified as the ligand of leucocyte function-associated antigen-1 (LFA-1), LFA-1 was thought to be functional only in lymphocyte mediated processes. Smith and coworkers, using anti-ICAM-1 in a neutrophil/endothelial cell adhesion assay, found that ICAM-1 also inhibited neutrophil adhesion to, and migration through, endothelium (Smith *et al.*, 1988). Furthermore, he found that there was both an LFA-1 component and a membrane attack molecule-1 (MAC-1) component to this adhesion, depending on the activation state of the neutrophil (Smith *et al.*, 1989). This finding was later confirmed with a direct binding assay showing that MAC-1 recognizes domain 3 of ICAM-1 (Diamond *et al.*, 1990; Diamond *et al.*, 1991).

Equally compelling evidence for the role of ICAM-1 in inflammation was the finding that ICAM-1 was expressed on multiple cell types at sites of inflammation. As mentioned previously, there is a low constitutive expression of ICAM-1 on venular endothelial cells, however its expression is markedly increased at inflammatory sites (Dustin *et al.*, 1986). Furthermore, there is increased ICAM-1 expression not only on endothelial cells, but on multiple other cell types including keratinocytes in inflammatory skin lesions (Vejlsgaard *et al.*, 1989; Griffiths and Nickoloff, 1989), transplanted liver bile duct and perivenular hepatocytes during rejection (Adams *et al.*, 1989) or in native livers with primary biliary cirrhosis (Adams *et al.*, 1991). ICAM-1 expression is also enhanced on endothelium in the human brain surrounding multiple sclerosis (MS) plaques as well as experimental models of MS, termed experimental allergic encephalitis, in rodents (Sobel *et al.*, 1990; Frohman *et al.*, 1989; Raine *et al.*, 1990; Cannella *et al.*, 1990; Cannella *et al.*, 1991; Raine, 1991; Wilcox *et al.*, 1990), transplanted kidney glomerulii and tubules during rejection (Cosimi *et al.*, 1990) as well as lung epithelial cells following antigen provocation (Wegner *et al.*, 1990). Increased ICAM-1 is also found on melanoma cells following metastasis (Matsui *et al.*, 1987; Natali *et al.*, 1990).

With this compelling evidence, it became critical to test the *in vivo* activity of antibodies to ICAM-1 in animal models of neutrophil and/or lymphocyte mediated inflammatory disease. Some of the earliest experiments to be performed were those by Barton and colleagues demonstrating that anti-ICAM-1 mAb inhibited neutrophil influx into rabbit lungs following systemic activation with phorbol esters (Barton *et al.*, 1989), providing *in vivo* evidence that anti-ICAM-1 mAb did indeed block neutrophil function. More recently, it has been shown that anti-ICAM-1 mAb also mitigated eosinophil influx and airway hyperresponsiveness in a non-human primate model of antigen-induced airway hyperresponsiveness (Wegner *et al.*, 1990), a model that has both non-specific and specific inflammatory components. This suggested a potential clinical role for the use of anti-ICAM-1 mAb in both acute and chronic pulmonary diseases.

Subsequently, ICAM-1 antagonism was tested in a model of immunological kidney allograft rejection. Here anti-ICAM-1 mAb, when given as the sole form of immunosuppressive therapy for 12 days only, starting two days prior to transplant, increased the time to rejection of allogeneic kidneys in non-human primate recipients. The average time of allograft survival more than doubled from an average of less than 10 days in the untreated animals to greater than 22 days in the treated animals (Cosimi *et al.*, 1990). This provided *in vivo* evidence of the ability of a mAb to ICAM-1 to inhibit lymphocyte function. It was also noted that when the monkeys were given anti-ICAM-1 as the sole induction therapy, the initial deterioration in kidney function usually seen in the immediate post-transplant period and manifested by a rise in serum creatinine, was also mitigated. It is felt by some that this initial rise in creatinine occurs, in part, as a result of neutrophil-mediated damage, also termined "reperfusion injury". As will be explained later, the decision to first examine anti-ICAM-1 mAb clinically as an induction therapy in human renal transplantation stemmed from the desire to exploit both its anti-neutrophil as well as anti-lymphocyte potential.

A second set of experiments in this same model showed that anti-ICAM-1 mAb also reversed acute allograft rejection episodes. In these experiments, monkeys received kidney transplants with cyclosporin A (CyA) as the sole immunosuppressive therapeutic agent. CyA doses were then tapered until subtherapeutic levels were achieved. This resulted in an acute rejection episode detected by a rise in serum creatinine. At this point, anti-ICAM-1 was administered daily for 10 days while the subtherapeutic dose of CyA was maintained. Results of these experiments reveal that anti-ICAM-1 normalized the rise in serum creatinine in all animals tested and markedly prolonged kidney allograft survival time (Cosimi *et al.*, 1990). These data thus suggested a role for an ICAM-1 antagonist in mitigating specific immunological processes such as those involved in allogeneic solid organ transplantation.

Another model of disease in which anti-ICAM-1 had been found to be effective in mitigating the inflammatory response is a rabbit spinal cord ischemia/reperfusion model. This model, thought to be representative of stroke, tests the effects of an adhesion-antagonistic on neutrophil-mediated damage. In this model, antibodies to either CD18 or ICAM-1 allowed the rabbit to be subjected to longer ischemic times without any apparent subsequent deleterious effects (Clark *et al.*, 1991b; Clark *et al.*, 1991a).

Recently it was reported that anti-ICAM-1 mAb were effective in inhibiting a Shwartzman type reaction in rabbits. In this model a stimulus such as lipopolysaccharide (LPS) was administered subcutaneously to rabbits. Twenty-four hours later, the rabbit was injected with a neutrophil activating agent such as zymosan, along with technecium labelled red blood cells. Hemorrhage was measured 4 h after the zymosan injection by monitoring the amount of radioactivity that had accumulated at the site of the LPS injection. The results from these experiments suggest that anti-ICAM-1 or anti-CD18 mAb given prior to either the initial subcutaneous priming or the systemic challenge markedly reduced the amount of hemorrhage in these animals compared to saline or irrelevant antibody controls (Argenbright and Barton, 1992). Finally, it was recently reported that a mouse anti-rat ICAM-1 was able to inhibit the development of arthritic lesions in a rat model of antigen induced arthritis (Iigo *et al.*, 1991).

1.2 ANTI-ICAM-1: THE CLINICAL EXPERIENCE

The findings described in this chapter support a role for the use of anti-ICAM-1 in a number of diseases. The specific anti-ICAM-1 selected for clinical trials is an IgG2a which binds to domain 2 of ICAM-1. This antibody, called BIRR1 (also called R6.5), blocks both lymphocyte and neutrophil adhesion (Smith *et al.*, 1988; Diamond *et al.*, 1990). BIRR1 does not inhibit bactericidal or phagocytic functions of granulocytes, which would not be the case with an antagonist of CD18 (unpublished data).

The first clinical areas selected for study with BIRR1 have been solid organ transplantation and RA. The unavailability of a less immunogenic therapeutic agent, such as a humanized anti-ICAM-1 or a small molecular weight compound, limit the full exploration of the therapeutic potential of ICAM-1 antagonism at this time, but early evidence suggests that even the murine antibody is effective and relatively safe.

1.3 RENAL TRANSPLANT

The renal transplant program was the first clinical program to start. It is currently in phase II with a double blind, placebo controlled study ongoing. The other solid organ indications, liver and heart transplant, are all in early phase I–II trials.

The phase I–II studies were performed at the Massachusetts General Hospital by Dr C.B. Cosimi. This study involved cadaveric renal allograft recipients at "high" risk for post transplant complications such as delayed graft function (DGF) and acute rejection. These patients were given BIRR1 prophylactically as an add-on medication to conventional triple therapy immunosuppression (azathioprine, steroids, and cyclosporine A). The objectives of this study were to evaluate the safety and pharmacokinetics of BIRR1 and to hopefully get a clinical read on BIRR1's ability to reduce the incidence of delayed graft function (DGF) and acute rejection (Tolkoff-Rubin *et al.*, 1991). Both DGF and acute rejection increase the morbidity and cost of renal transplantation.

It was originally intended that only ten patients, five each on two dosing regimens, would be studied. These dosing regimens were direct extrapolations from those which were efficacious in the cynomolgus monkeys, as well as the investigator's experience with OKT3. It was anticipated that both of these regimens would provide BIRR1 levels in a potentially therapeutic range.

It turned out that the original dosing projections were far too low, and the treatment period (two weeks) impractically long. Thus, in order to fulfill the objective of the study (i.e. to find two dosing regimens which achieved the pharmacokinetic target), a total of five dosing regimens were studied in 18 patients (Table 1.1).

Of the five dosing regimens studied, the two highest regimens consistently met the pharmacokinetic target. Both were 6-day regimens: 160 mg pretransplant loading, followed by 40 mg/day × 5 days (abbreviated as 160/40 × 5), and a 160/80 × 5 regimen, for a total dose of 360 mg and 560 mg, respectively.

The reasons for the increased dosing requirement for BIRRl in renal allograft recipients may have to do with any one (or a combination) of the following:

(i) Expression of ICAM-1 in a chronically ill human versus healthy monkey. In this situation ICAM-1, which is normally present only in small amounts, is markedly up-regulated in the presence of inflammation. This could cause an "ICAM-1 sink", requiring more BIRRl in order to saturate all the ICAM-1 sites. Since the ethics of developing an immunogenic molecule such as a murine monoclonal antibody (mAb) dictate that phase I studies not be done in normal individuals, we will learn more about this as we investigate BIRRl in different diseases.
(ii) The presence of a circulating form of ICAM-1 (cICAM-1) in humans. Again, this was not detected in monkeys. Although not much is known yet about the significance of cICAM-1 in humans, its levels are increased in the presence of disease and could be binding to BIRRl, thus blocking its actions (Rothlein *et al.*, 1991).
(iii) Higher avidity of BIRRl for human ICAM-1 than non-human primate ICAM-1 with a corresponding slower rate of dissociation.
(iv) Additional factors as yet unknown.

As the study was designed originally, efficacy comparisons could only be made with an historical control. As the study evolved, however, 18 patients were enrolled who received five different regimens. Eleven patients received BIRRl regimens which achieved the pharmacokinetic target; seven did not (Table 1.1). The outcome of these two groups can be compared.

The incidence of both acute rejection (followed out to three months) and delayed graft function (DGF) (a shorter-term endpoint evaluable by the end of the first post transplant week) was lower in the 11 patients with adequate BIRRl levels. This is depicted below:

	Patients with low BIRRl levels N=7	Patients with target BIRRl levels N=11
Rejection	7/7 (100%)	3/11 (27%)
DGF	6/7 (86%)	6/11 (56%)

Sixteen of the 18 patients developed an immune response to BIRRl (HAMA), which is to be expected. A high incidence of BIRRl anti-idiotype antibodies was found, but only two of 10 human anti-mouse antibody (HAMA) responders tested had antibodies which cross-reacted with OKT3, another mouse IgG2a monoclonal antibody. The eight patients in this study who required OKT3, however, did not demonstrate any evidence of anaphylaxis nor an attenuated response to this additional monoclonal antibody. This provides early evidence suggesting that prophylactic treatment of a transplant patient with BIRRl does not limit the selection of subsequent anti-rejection therapy, should this be needed. This hypothesis will be further tested in controlled clinical trials.

1.4 RHEUMATOID ARTHRITIS (RA)

BIRRl is also being evaluated in an investigator's IND study in an open-label, dose escalating phase I/II study in patients with severe rheumatoid arthritis (RA) by Dr Peter Lipsky (University of Texas, Southwestern Medical Center, Dallas) (Kavanaugh *et al.*, 1992b; Kavanaugh *et al.*, 1992a). Although primarily a protein kinase (PK) and safety study in its focus, preliminary efficacy is also assessed using the Paulus criteria described below. The population under study includes patients with active disease of greater than four years duration who are functional class I–III and who have failed at least two disease-modifying drugs (DMARDS). The patient is allowed to be on non-steroidal anti-inflammatory drugs (NSAIDS) or ≤10 mg prednisone/day during active treatment with BIRRl. The BIRRl regimens used are depicted in Table 1.2. Of the 21 patients enrolled, safety and efficacy information is available on the first 17

Table 1.1 Dosing regimen of Anti-ICAM-1 in phase I/II kidney transplant trial

Patient	*Dosing Regimen*	*Total dose/duration of treatment*	*Achieved PK target*
1–3	20 mg/10 mg × 13 days	150 mg/14 days	0/3
4–7	40 mg/10 mg × 14 days	300 mg/13 days	1/4
8–13	80 mg/60 mg/40 × 4 days	300 mg/6 days	5/6
14–16	160 mg/80 mg × 5 days	560 mg/6 days	3/3
17–18	160 mg/40 mg × 5 days	360 mg/6 days	2/2
			11/18

Table 1.2 Summary of results of Anti-ICAM-1 in phase I/II kidney transplant trial

Patient #	*Dosing regimen (mg)*	*Total dose/days*	*Achieved target levels while on treatment*
1–2	60/20 × 4	140 mg/5 days	0/2
3–13	120/40 × 4	280 mg/5 days	11/11
14–16	240/80 × 4	560 mg/5 days	3/3
17–21	120/120	240 mg/2 days	3/3

patients. The mean duration of RA in this group is 20 years and the mean number of DMARDS failed is four, underscoring the severity of disease in these patients.

In all of the dosing regimens studied, there have been no serious adverse events. Both the 60/20 mg × 4 day and 120/40 mg × 4 day regimens were well tolerated. The pharmacokinetic target was achieved in all eleven patients at the latter regimen.

Efficacy is assessed in this study using the modified Paulus criteria, which examines the following six parameters: joint swelling, joint tenderness, morning stiffness, patient/physician global assessment, and erythrocyte sedimentation rate (ESR). In order to be classified as a "responder" a patient must have a ⩾20% reduction in 4/6 parameters. The results are shown below:

Degree of Response	*Day 8*	*Day 15*	*Day 30*	*Day 60*
Marked	4	4	4	2
Moderate	6	3	3	5
None	7	10	10	10

The seven responders at day 60 were all on the 120/40 × 4 regimen. Thus, in severe RA, BIRR1 has been tested at several dosing regimens. The 120/40 × 4 dosing regimen (280 mg total) is well tolerated, achieves the protein kinase (PK) target, and has resulted in a clinical response in more than half the patients.

To summarize the clinical experience to date, in open-label studies, 18 renal and 21 patients with RA have received BIRR1 using various dosing regimens. In each indication a different dosing regimen has been required to achieve the pharmacokinetic target, suggesting that these diseases have different amounts of ICAM-1 expression, cICAM-1, and/or other factors that may influence BIRR1 serum levels.

BIRR1 has been well tolerated at the dosing regimens which achieve the pharmacokinetic target. At these dosing regimens preliminary evidence of efficacy is also suggested in the two indications tested to date. Safety and efficacy are being objectively assessed in an ongoing placebo-controlled, double-blind, 200 patient renal transplant study.

In conclusion, the inhibition of CD18/ICA-1 interactions inhibits leukocyte trafficking, antigen presentation and leukocyte mediated cytolysis. The inhibition of which one of these three processes is most responsible for the dramatic reduction in the inflammatory responses described to date and summarized in this chapter is unclear. It will probably be a combination of all three. Regardless, it is clear that antagonists of these interactions offer an exciting new target to prevent inflammatory responses associated with transplantation, autoimmunity and acute inflammation.

2. *References*

Adams, D.H., Hubscher, S.G., Shaw, J., Rothlein, R. and Neuberger, J.M. (1989). Intercellular adhesion molecule 1 on liver allografts during rejection. Lancet 2, 1122–1125.

Adams, D.H., Hubscher, S.G., Shaw, J., Johnson, G.D., Babbs, C., Rothlein, R. and Neuberger, J.M. (1991). Increased expression of intercellular adhesion molecule 1 on bile ducts in primary biliary cirrhosis and primary sclerosing cholangitis. Hepatology 14, 426–431.

Altmann, D.M., Hogg, N., Trowsdale, J. and Wilkinson, D. (1989). Cotransfection of ICAM-1 and HLA-DR reconstitutes human antigen-presenting cell function in mouse L cells. Nature 338, 512–514.

Anderson, D.C., Schmalstieg, F.C. Finegold, M.J., Hughes, B.J., Rothlein, R., Miller, L.J., Kohl, S., Tosi, M.F., Jacobs, R.L., Waldrop, T.C., Goldman, A.S., Shearer, W.T. and Springer, T.A. (1985). The severe and moderate phenotypes of heritable Mac-1, LFA-1 deficiency: their quantitative definition and relation to leukocyte dysfunction and clinical features. J. Inf. Dis. 152, 668–689.

Anderson, D.C., Miller, L.J., Schmalstieg, F.C., Rothlein, R. and Springer, T.A. (1986). Contributions of the Mac-1 glycoprotein family to adherence-dependent granulocyte functions: Structure-function assessments employing subunit-specific monoclonal antibodies. J. Immunol. 137, 15–27.

Anderson, D.C. and Springer, T.A. (1987). Leukocyte adhesion deficiency: an inherited defect in the Mac-1, LFA-1, and p150,95 glycoproteins. Ann. Rev. Med. 38, 175–194.

Argenbright, L.W., Letts, L.G. and Rothlein, R. (1991). Monoclonal antibodies to the leukocyte membrane CD18 glycoprotein complex and to intercellular adhesion molecule 1 inhibit leukocyte-endothelial adhesion in rabbits. J. Leukocyte Biol. 49, 253–257.

Argenbright, L.W. and Barton, R.W. (1992). Interactions of leukocyte integrins with intercellular adhesion molecule 1 in the production of inflammatory vascular injury *in vivo*. The Shwartzman reaction revisited. J. Clin. Invest. 89, 259–272.

Barton, R.W., Rothlein, R., Ksiazek, J. and Kennedy, C. (1989). The effect of anti-intercellular adhesion molecule-1 on phorbol-ester-induced rabbit lung inflammation. J. Immunol. 143, 1278–1282.

Beller, D.I., Springer, T.A. and Schreiber, R.D. (1982). Anti-Mac-1 selectively inhibits the mouse and human type three complement receptor. J. Exp. Med. 156, 1000–1009.

Bierer, B.E. and Burakoff, S.J. (1988). T-cell adhesion molecules. FASEB 2, 2584–2590.

Boyd, A.W., Wawryk, S.O., Burns, G.F. and Fecondo, J.V. (1988). Intercellular adhesion molecule 1 (ICAM-1) has a central role in cell-cell contact-mediated immune mechanisms. Proc. Natl Acad. Sci. USA 85, 3095–3099.

Cannella, B., Cross, A.H. and Raine, C.S. (1990). Upregulation and coexpression of adhesion molecules correlate with relapsing autoimmune demyelination in the central nervous system. J. Exp. Med. 172, 1521–1524.

Cannella, B., Cross, A.H. and Raine, C.S. (1991). Adhesion-related molecules in the central nervous system: upregulation correlates with inflammatory cell influx during relapsing experimental autoimmune encephalomyelitis. Lab. Invest. 65, 23–31.

Clark, W.M., Madden, K.P., Rothlein, R. and Zivin, J.A. (1991a). Reduction of central nervous system ischemic injury in rabbits using leukocyte adhesion antibody treatment. Stroke 22, 877–883.

Clark, W.M., Madden, K.P., Rothlein, R. and Zivin, J.A. (1991b). Reduction of central nervous system ischemic injury by monoclonal antibody to intercellular adhesion molecule. J. Neurosurg. 75, 623–627.

Cosimi, A.B., Conti, D., Delmonico, F.L., Preffer, F.I., Wee, S.-L.., Rothlein, R., Faanes, R. and Colvin, R.B. (1990). *In vivo* effects of monoclonal antibody to ICAM-1 (CD54) in non-human primates with renal allografts. J. Immunol. 144, 4604–4612.

Dang, L.H., Michalek, M.T., Takei, F., Benaceraff, B. and Rock, K.L. (1990). Role of ICAM-1 in antigen presentation demonstrated by ICAM-1 defective mutants. J. Immunol. 144, 4082–4091.

Davignon, D., Martz, E., Reynolds, T., Kürzinger, K. and Springer, T.A. (1981). Monoclonal antibody to a novel lymphocyte function-associated antigen (LFA-1): mechanism of blocking of T lymphocyte-mediated killing and effects on other T and B lymphocyte functions. J. Immunol. 127, 590–595.

Diamond, M.S., Staunton, D.E., De Fougerolles, A.R., Stacker, S.A., Garcia-Aguilar, J., Hibbs, M.L. and Springer, T.A. (1990). ICAM-1 (CD54): a counter-receptor for Mac-1 (CD11b/CD18). J. Cell Biol. 111, 3129–3139.

Diamond, M.S., Staunton, D.E., Marlin, S.D. and Springer, T.A. (1991). Binding of the integrin Mac-1 (CD11b/CD18) to the third immunoglobulin-like domain of ICAM-1 (CD54) and its regulation by glycosylation. Cell 65, 961–971.

Dougherty, G.J. and Hogg, N. (1987). The role of monocyte lymphocyte function-associated antigen 1 (LFA-1) in accessory cell function. Eur. J. Immunol. 17, 943–947.

Dougherty, G.J., Murdoch, S. and Hogg, N. (1988). The function of human intercellular adhesion molecule-1 (ICAM-1) in the generation of an immune response. Eur. J. Immunol. 35–39.

Dustin, M.L., Rothlein, R., Bhan, A.K., Dinarello, C.A. and Springer, T.A. (1986). Induction by IL-1 and interferon, tissue distribution, biochemistry, and function of a natural adherence molecule (ICAM-1). J. Immunol. 137, 245–254.

Dustin, M.L. and Springer, T.A. (1988). Lymphocyte function associated antigen-1 (LFA-1) interaction with intercellular adhesion molecule-1 (ICAM-1) is one of at least three mechanisms for lymphocyte adhesion to cultured endothelial cells. J. Cell Biol. 107, 321–331.

Dustin, M.L., Singer, K.H., Tuck, D.T. and Springer, T.A. (1988). Adhesion of T lymphoblasts to epidermal keratinocytes is regulated by interferon gamma and is mediated by intercellular adhesion molecule-1 (ICAM-1) J. Exp. Med. 167, 1323–1340.

Frohman, E.M., Frohman, T.C., Dustin, M.L., Vayuvegula, B., Choi, B., Gupta, A., van den Noort S. and Gupta, S. (1989). Induction of ICAM-1 expression on human fetal astrocytes by interferon-gamma, tumor necrosis factor-alpha and interleukin-1: relevance to intracerebral antigen presentation. J. Neuroimmunol. 23, 117–124.

Griffiths, C.E.M. and Nickoloff, B.J. (1989). Keratinocyte intercellular adhesion molecule-1 (ICAM-1) expression precedes dermal T lymphocytic infiltration in allergic contact dermatitis (*Rhus dermatitis*). Am. J. Pathol. 135, 1045–1053.

Haskard, D., Cavender, D., Beatty, P., Springer, T. and Ziff, M. (1986). T lymphocyte adhesion to endothelial cells: mechanisms demonstrated by anti-LFA-1 monoclonal antibodies. J. Immunol. 137, 2901–2906.

Iigo, Y., Takashi, T., Tamatani, T., Miyasaka, M., Higashida, T., Yagita, H., Okumura, K. and Tsukada, W. (1991). ICAM-1-dependent pathway is critically involved in the pathogenesis of adjuvant arthritis in rats. J. Immunol. 147, 4167–4171.

Kavanaugh, A., Norris, S., Rothlein, R., Nicholls, L.A. and Lipsky, P.E. (1992a). Pharmacokinetic Analysis of Rheumatoid Arthritis Patients Treated with an Anti-CD54 (Intercellular Adhesion Molecule-1, ICAM-1) Monoclonal Antibody. Arthritis Rheum. 35, S106–S106. (Abstract)

Kavanaugh, A., Nicholls, L.A. and Lipsky, P.E. (1992b). Treatment of Refractory Rheumatoid Arthritis with an Anti-CD54 (Intercellular Adhesion Molecule-1, ICAM-1) Monoclonal Antibody. Arthritis Rheum. 35, S43–S43. (Abstract)

Kishimoto, T.K., O'Connor, K. and Springer, T.A. (1989). Leukocyte adhesion deficiency: aberrant splicing of a conserved integrin sequence causes a moderate deficiency phenotype. J. Biol. Chem. 264, 3588–3595.

Kohl, S., Springer, T.A., Schmalstieg, F.C., Loo L.S. and Anderson, D.C. (1984). Defective natural killer cytotoxicity and polymorphonuclear leukocyte antibody-dependent cellular cytotoxicity in patients with LFA-1/OKM-1 deficiency. J. Immunol. 133, 2972–2978.

Krensky, A.M., Sanchez-Madrid, F., Robbins, E., Nagy, J., Springer, T.A. and Burakoff, S.J. (1983). The functional significance, distribution, and structure of LFA-1, LFA-2, and LFA-3: cell surface antigens associated with CTL-target interactions. J. Immunol. 131, 611–616.

Lipsky, P.E. and Rosenthal, A.S. (1975). Macrophage-lymphocyte interaction. II. Antigen-mediated physical interactions between immune guinea pig lymph node lymphocytes and syngeneic macrophages. J. Exp. Med. 141, 138.

Makgoba, M.W., Sanders, M.E., Luce, G.E.G., Dustin, M.L., Springer, T.A., Clark, E.A., Mannoni, P. and Shaw, S. (1988). ICAM-1: definition by multiple antibodies of a ligand for LFA-1 dependent adhesion of B, T and myeloid cell. Nature 331, 86–88.

Makgoba, M.W., Sanders, M.E., Luce, G.E.G., Gugel, E.A., Dustin, M.L., Springer, T.A. and Shaw, S. (1989). Intercellular Adhesion Molecule-1 (ICAM-1) monoclonal antibody inhibits cytotoxic T lymphocyte recognition. Ann. NY Acad. Sci. 532, 427–428.

Marlin, S.D. and Springer, T.A. (1987). Purified intercellular adhesion molecule-1 (ICAM-1) is a ligand for lymphocyte function-associated antigen 1 (LFA-1). Cell 51, 813–819.

Martz, E. (1980). Immune lymphocyte to tumor cell adhesion: magnesium sufficient, calcium insufficient. J. Cell Biol. 84, 584–598.

Martz, E., Parker, W.L., Gately, M.K. and Tsoukas, C.D. (1982). The role of calcium in the lethal hit of T lymphocyte-mediated cytolysis. Adv. Exp. Biol. Med. 146, 121–143.

Martz, E. (1987). LFA-1 and other accessory molecules functioning in adhesions of T and B lymphocytes. Human Immunol. 18, 3–37.

Matsui, M., Temponi, M. and Ferrone, S. (1987). Characterization of a monoclonal antibody-defined human melanoma-associated antigen susceptible to induction by immune interferon. J. Immunol. 139, 2088–2095.

Mentzer, S.J., Bierer, B.E., Anderson, D.C., Springer, T.A. and Burakoff, S.J. (1986). Abnormal cytolytic activity of lymphocyte function-associated antigen-1-deficient human cytolytic T lymphocyte clones. J. Clin. Invest. 78, 1387–1391.

Mentzer, S.J., Rothlein, R., Springer, T.A. and Faller, D.V. (1988). Intercellular adhesion molecule-1 (ICAM-1) is involved in the cytolytic T lymphocyte interaction with human synovial cells. J. Cell. Physiol. 137, 173–178.

Natali, P., Nicotra, M.R., Cavaliere, R., Bigotti, A., Romano, G., Temponi, M. and Ferrone, S. (1990). Differential expression of intercellular adhesion molecule 1 in primary and metastatic melanoma lesions. Cancer Res. 50, 1271–1278.

Perry, M.A. and Granger, D.N. (1991). Role of CD11/CD18 in shear rate-dependent leukocyte-endothelial cell interactions in cat mesenteric venules. J. Clin. Invest. 87, 1798–1804.

Prober, J.S., Gimbrone Jr., M.A., Lapierre, L.A., Mendrick, D.L., Fiers, W., Rothlein, R. and Springer, T.A. (1986). Overlapping patterns of activation of human endothelial cells by interleukin 1, tumor necrosis factor and immune interferon. J. Immunol. 137, 1893–1896.

Raine, C.S., Lee, S.C., Scheinberg, L.C., Duijvestijn, A.M. and Cross, A.H. (1990). Adhesion molecules on endothelial cells in the central nervous system: an emerging area in the neuroimmunology of multiple sclerosis. Clin. Immunol. Immunopathol. 57, 173–187.

Raine, C.S. (1991). Multiple sclerosis: A pivotal role for the T cell in lesion development. Neuropathol. Appl. Neurobiol. 17, 265–274.

Rothlein, R. and Springer, T.A. (1986). The requirement for lymphocyte function-associated antigen 1 in homotypic leukocyte adhesion stimulated by phorbol ester. J. Exp. Med. 163, 1132–1149.

Rothlein, R., Dustin, M.L., Marlin, S.D. and Springer, T.A. (1986). A human intercellular adhesion molecule (ICAM-1) distinct from LFA-1. J. Immunol. 137, 1270–1274.

Rothlein, R., Mainolfi, E.A., Czajkowski, M. and Marlin, S.D. (1991). A form of circulating ICAM-1 in human serum. J. Immunol. 147, 3788–3793.

Sanchez-Madrid, F., Nagy, J., Robbins, E., Simon, P. and Springer, T.A. (1983). A human leukocyte differentiation antigen family with distinct α subunits and a common β subunit: the lymphocyte function-associated antigen (LFA-1), the C3bi complement receptor (OKM1/Mac-1), and the p150,95 molecule. J. Exp. Med. 158, 1785–1803.

Simmons, D., Makgoba, M.W. and Seed, B. (1988). ICAM, an adhesion ligand of LFA-1, is homologous to the neural cell adhesion molecule NCAM. Nature 331, 624–627.

Smith, C.W., Rothlein, R., Hughes, B.J., Mariscalco, M.M., Schmalstieg, F.C., and Anderson, D.C. (1988). Recognition of an endothelial determinant for CD18 – dependent neutrophil adherence and transendothelial migration. J. Clin. Invest. 82, 1746–1756.

Smith, C.W., Marlin, S.D., Rothlein, R., Toman, C. and Anderson, D.C. (1989). Cooperative Interactions of LFA-1 and Mac-1 with Intercellular Adhesion Molecule-1 in Facilitating Adherence and Transendothelial Migration of Human Neutrophils *in vitro*. J. Clin. Invest. 83, 2008–2017.

Sobel, R.A., Mitchell, M.E. and Fondren, G. (1990). Intercellular adhesion molecule-1 (ICAM-1) in cellular immune reactions in the human central nervous system. Am. J. Pathol. 136, 1309–1316.

Springer, T.A., Dustin, M.L., Kishimoto, T.K. and Marlin, S.D. (1987). The lymphocyte function-associated LFA-1, CD2, and LFA-3 molecules: cell adhesion receptors of the immune system. Ann. Rev. Immunol. 5, 223–252.

Springer, T.A. (1990). Adhesion receptors of the immune system. Nature 346, 425–434.

Staunton, D.E., Marlin, S.D., Stratowa, C., Dustin, M.L. and Springer, T.A. (1988). Primary structure of intercellular adhesion molecule 1 (ICAM-1) demonstrates interaction between members of the immunoglobulin and integrin supergene families. Cell 52, 925–933.

Todd, R.F. and Freyer, D.R. (1988). The CD11/CD18 leukocyte glycoprotein deficiency. Hem./Onc. Clinics. N. A. 2, 13–31.

Tolkoff-Rubin, N., Rothlein, R., Scharschmidt, L., Colvin, R., Auchincloss, H., Delmonico, F., Norris, S., Wood, C. and Cosimi, A.B. (1991). Immunosuppression with Anti-ICAM-1 (CD54) mAb in Renal Allograft Recipients. J. Am. Soc. Nephrology 2, 280. (Abstract)

Vejlsgaard.G.L., Ralfkiaer, E., Avnstorp, C., Czajkowski, M., Marlin, S.D., and Rothlein, R. (1989). Kinetics and characterization of intercellular adhesion molecule-1 (ICAM-1) expression on keratinocytes in various inflammatory skin lesions and malignant cutaneous lymphomas. J. Amer. Acad. Dermtol. 20, 782–790.

Wegner, C.D., Gundel, R.H., Reilly, P., Haynes, N., Letts, L.G. and Rothlein, R. (1990). Intercellular adhesion molecule 1 (ICAM-1) in the pathogenesis of asthma. Science 247, 456–459.

Wilcox, C.E., Ward, A.M.V., Evans, A., Baker, D., Rothlein, R. and Turk, J.L. (1990). Endothelial cell expression of the intercellular adhesion molecule-1 (ICAM-1) in the central nervous system of guinea pigs during acute and chronic relapsing experimental allergic encephalomyelitis. J. Neuroimmunol. 30, 43–51.

formation. These represent the firmest and most resistant components that an organism possesses. In contrast, intercellular contact formation with regard to the immune system – whose cellular components recirculate throughout the organism – has to be of transient nature, and therefore, low affinity.

Recent observations strongly support the notion that the various families of "adhesion molecules" of the immune system represent receptors capable of transmitting signals into the intracellular environment. Such signals regulate and "fine tune" the immune response at the level of the individual immunocompetent cell in its respective micro-environment. In this regard, a number of reports have demonstrated that these molecules are physically and/or functionally associated with intracellular signalling proteins such as protein kinases, phosphatases, nucleotide binding proteins and components which influence the distribution and organization of cytoskeletal determinants. One would, therefore, propose to call such molecules "accessory receptors" because this provides a more appropriate description of the roles that these molecules fulfil. Nevertheless, in this chapter, the terms "adhesion receptor" and "accessory receptor" will be employed interchangeably.

In order to explain, from a conceptual viewpoint, the impact of accessory receptors on the immune response, a number of earlier experiments have to be remembered. Once monoclonal antibodies existed which reacted with individual receptor molecules, it had become possible to investigate functional features of individual receptors employing tissue culture systems composed of highly purified cell populations (Meuer *et al.*, 1984; Meuer, Acuto *et al.*, 1984). This possibility was a major breakthrough in molecular immunology since until then, functional studies had to be performed with mixed cellular systems in which multiple molecular interactions occurred simultaneously. Thus, it had been virtually impossible to address individual receptor functions.

Monoclonal antibodies are unique in their capacity to bind to particular functional epitopes of an individual receptor structure and, depending on their mode and place of interaction, produce special functional effects (Bierer *et al.*, 1989). For example, those antibodies which are directed at domains involved in ligand binding can block the physical interaction of a receptor with its specific ligand and, thus, inhibit individual molecular functions (Mueller *et al.*, 1989). Conversely, those antibodies which react with active receptor sites can, under certain circumstances, mimic the action of the natural receptor ligand and, thus, be helpful in elucidating consequences of functional ligand binding. Such monoclonal antibodies have yielded tremendous insight into the roles of individual receptor structures expressed by particular immunocompetent cell types (Reinherz *et al.*, 1982; Meuer *et al.*, 1983; Meuer *et al.*, 1984).

One of the most important findings made in the course of such studies was that engagement of the antigen-specific receptor on T lymphocytes in the absence of engagement of additional accessory receptors is not sufficient to produce clonal expansion and/or immune activation (Reinherz *et al.*, 1982; Meuer *et al.*, 1983; Meuer, Acuto *et al.*, 1984; Meuer *et al.*, 1984; Bierer *et al.*, 1989; Knapp, 1989; Mueller *et al.*, 1989). Nevertheless, exclusive delivery of the primary signal of antigen recognition to the T cell receptor leads to modulation of this molecular complex followed by a state of lymphocyte unresponsiveness (anergy) – under certain experimental circumstances even cell deletion and death through apoptosis (Reinherz *et al.*, 1982; Meuer *et al.*, 1983; Meuer *et al.*, 1984; Mueller *et al.*, 1989). However, when the T cell receptor was triggered concomitantly with particular adhesion receptors, T-cell proliferation and differentiation into functional effector cells was achieved (Emmrich *et al.*, 1987). It seems justified to conclude that the recognition structures, namely T-cell receptor and cell surface immunoglobulin, determine the specificity of a given immune reaction, whereas accessory receptors are responsible for determining the functional type of the immune response. Therefore, immune intervention at the level of adhesion receptors provides potent perspectives for immunopharmacology, particularly if one understands the role of these molecules as regulators of the local immune response.

As mentioned above, besides expressing one type of antigen recognition structure, each lymphocyte carries a multitude of different adhesion receptors in its cell surface (Bierer and Burakoff, 1988; Springer, 1990). In the course of immune reactions initiated through recognition of specific antigen in appropriate Major Histocompatibility Complex (MHC) association, several or even many adhesion receptors are simultaneously engaged and produce a multiplicity of parallel signals towards the responding immunocompetent cell. However, the number of different signals transmitted to the lymphocyte is restricted by the fact that the various ligands of adhesion receptors are not equally distributed throughout the organism. Rather, during its journey through the body, a given lymphocyte will, depending on the respective local micro-environment, encounter a selective and restricted combination of ligand molecules expressed by "other" cells or extracellular matrix. It is this specific combination of ligand molecules which adjusts lymphocyte reactivity to specific local requirements. Therefore, besides the existence of several (genetically) predetermined and specialized lymphocyte subpopulations a dynamic concept must be formulated in which the respective functions of lymphocytes are modulated by the combination of accessory signals specific for a given micro-environment.

A further degree of complexity exists at the level of adhesion receptors in that these molecules appear to interact physically and functionally with each other

2. Accessory Receptors: Regulators of the Local Immune Response

Stefan C. Meuer

1. Introduction

To fulfil its multiple functions for protecting the integrity and physical individuality of an organism, the immune system has developed an abundant set of receptor molecules which are expressed in the plasma membranes of immunocompetent cells, i.e. lymphocytes and mononuclear phagocytes. Immune responses which are controlled by these receptors are featured by two principally different components, namely: (1) target specificities; and (2) functional activities such as help, suppression, cytotoxicity, inducer of help, inducer of suppression, contrasuppression etc. Accordingly, two different sets of receptor molecules can be distinguished from each other.

The specificity of immunological reactions is based on the expression of transmembrane receptor complexes which are composed of variable regions and constant domains. Antigen preceptors (i.e. T-cell receptor on T lymphocytes and surface immunoglobulin on B lymphocytes) result from somatic rearrangements of a large set of specialized genes which encode for clonally distributed recognition molecules (Meuer, Acuto *et al.*, 1984). These amount to a number of approx. 10^{10} possible different antigen receptors (this estimate is based on the numbers of variable, diversity, joining and constant region genes that have been cloned and sequenced so far). Given the exquisite fine specificity of immune reactions *in vivo*, one has to postulate that antigen receptor molecules are highest in a hierarchy of receptors involved in immunoregulation.

In addition to specific antigen receptors, a large set of non-variable transmembrane molecules have been defined by means of monoclonal antibodies (Knapp, 1989). The function of the latter was initially proposed to mediate contact between immunocompetent cells and other cells as well as with extracellular matrix constituents.

However, functional investigations employing monoclonal antibodies demonstrated that contact formation represents only one and perhaps not the most important function of these molecules. For several reasons, therefore, the term "adhesion molecules" has to be considered at least an incomplete description, if not even a misleading one. One would rather use such a term for components mediating constant adhesion via intercellular junctions, thus providing the basis for tissue/organ

Adhesion Molecules
ISBN 0-12-741440-1

(Schraven *et al.*, 1990). However, it seems clearly not to be the case that one adhesion receptor employs just one or the other particular molecule as an exclusive partner for interaction (Koretsky *et al.*, 1990). Rather, again depending on the ligand combination, each receptor at the lymphocyte surface would be free to interact with another one and, based on the molecular features of their interactions, may function differently.

A third level of complexity exists in the fact that the cell surface density of accessory molecules is highly variable and influenced by the differentiation state of the individual lymphocyte (Hemler, 1990; Springer, 1990). Thus, many adhesion receptors are present in much greater numbers on activated cells than on resting ones, and these differences can in some cases exceed one order of magnitude. Fig. 2.1 summarizes these considerations schematically.

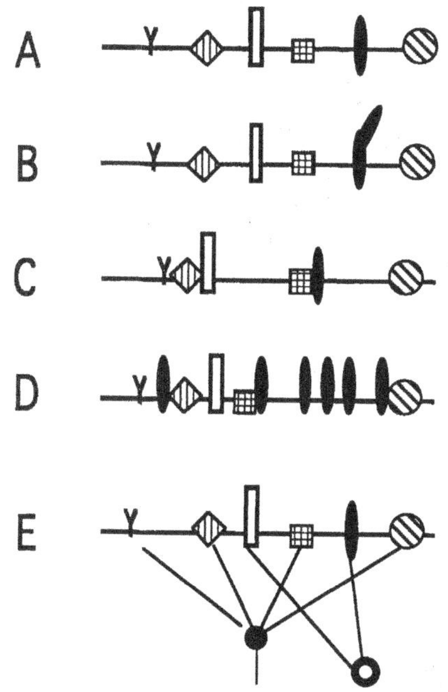

Figure 2.1 Regulatory principles in accessory receptor function. (A) A linearized segment of the outer cell membrane of an immunocompetent cell in which multiple accessory receptors are concomitantly expressed is schematically shown. (B) Lymphocyte triggering through the antigen receptor can induce a transient conformational change in one particular accessory molecule, thereby enhancing binding affinity for its natural ligand. (C) Depending on the respective local ligand expression, accessory receptors can physically and, more important, functionally interact with each other (receptor complexes). Such molecular interactions are not mutually exclusive. (D) Expression of accessory molecules varies tremendously depending on the differentiation and activation state of immunocompetent cells. Quantitative levels in the expression of accessory molecules influence their function in a dose/response fashion. (E) A considerable degree of redundancy exists at the cell surface level. Thus, several accessory molecules share common intracellular signalling pathways which are responsible for mediating critical signals towards the nucleus. Common intracellular signal transducing components represent most important targets for clinical immune intervention.

A considerable degree of redundancy appears to exist at the cell surface level. This means one would not anticipate that each adhesion receptor possesses its own intracellular signalling pathway or produces its own individual functional cellular response reaction. It seems justifiable to expect that several accessory molecules share one intracellular signalling pathway which is responsible for one or the other important signal relevant for gene regulation. Under such a premise, elucidation of the intracellular signalling machinery linked to adhesion receptors will be crucial for further considerations regarding immune intervention at this level. If it is true that several adhesion molecules, by means of acting on the same intracellular signalling pathway, can "replace" each other's function, then strategies for immunosuppressive therapies aimed at targeting one individual cell surface molecule may be too selective. In order to omit a critical signal one would then rather concentrate on common intracellular signalling components and/or reactions. Conversely, however, with regard to immunostimulation, efficient triggering of one cell surface receptor by an appropriate reagent might represent a relevant approach.

The functions of most adhesion receptors known so far have been investigated individually by monoclonal antibodies. Once all of their ligands are defined and the local distribution of the latter is better understood, one will have to establish more complex *in vitro* systems in order to evaluate the influences of a particular local environment on lymphocyte reactivity. Therefore, although the various adhesion receptor systems will be described individually in this chapter, it may be premature to draw conclusions regarding rational pharmacological immune intervention because the present knowledge of the biological functions of accessory molecules is highly simplistic, and given the above mentioned complexities, represents a very restricted view.

2. *Immunoglobulin Superfamily of Accessory Receptors*

A variety of adhesion receptors are genetically linked to immunoglobulin (Ig), the T-cell receptor complex and the MHC gene products. Because of their structural similarities, which are highlighted by their expression of variable numbers of 90–100 amino acid domains usually held together by disulfide bonds, a large superfamily of molecules was established (Alzari *et al.*, 1988; Williams and Barclay, 1988). More recently, structural analysis revealed that CD2, CD4, CD8, CD28, CD31, CD54, CD58, ICAM-2, VCAM-1 (Springer, 1990), and several additional molecules that were first identified on cells of the nervous system (e.g. NCAM) (Schad and Greenstein, 1989) belong to this Ig-superfamily as well (Fig. 2.2).

The adhesion receptors of the Ig-superfamily are, at present, by far the best characterized among the

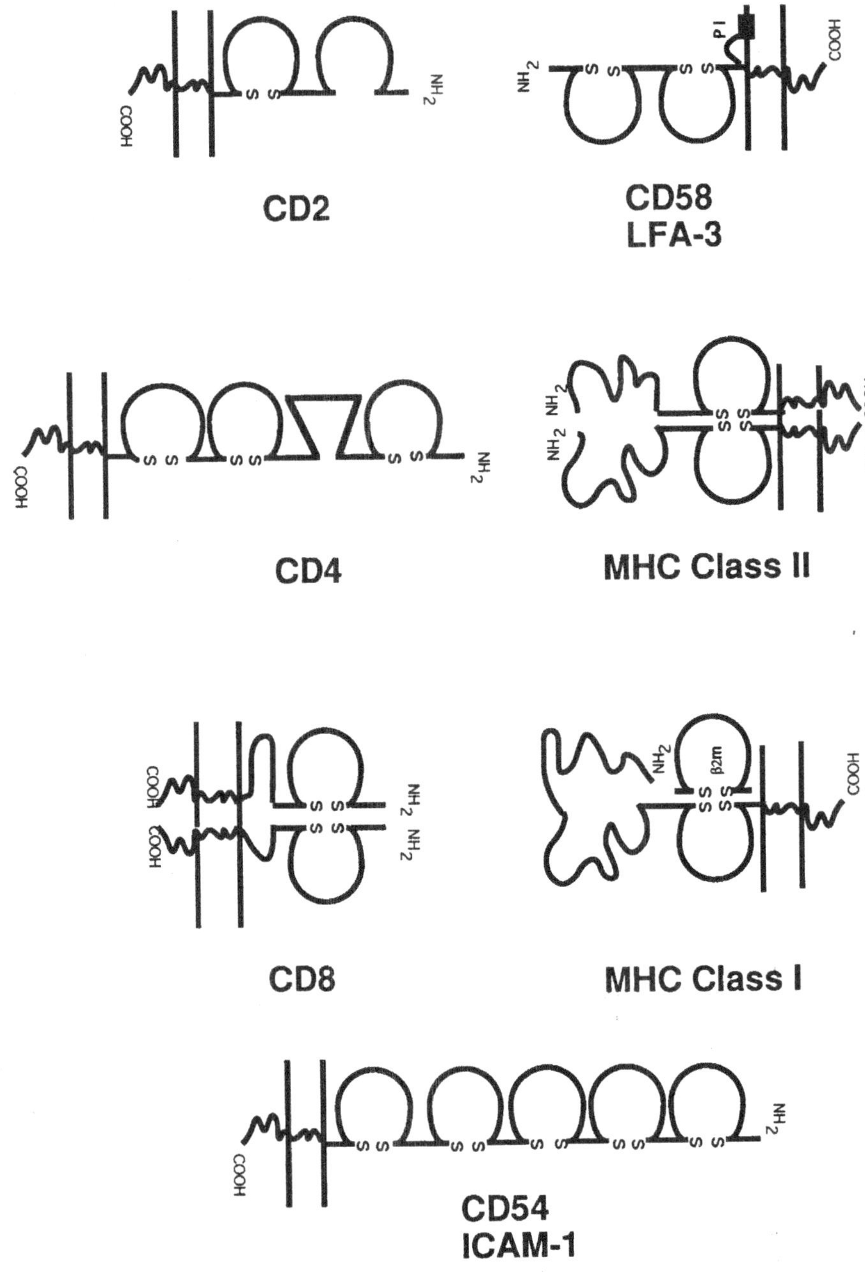

Figure 2.2 Accessory receptors of the immunoglobulin superfamily are characterized by variable numbers of 90–100 amino acid domains held together by disulfide bonds. Some members of this superfamily represent receptor/counter receptor pairs of reciprocal cell types (CD2/CD58; CD4/MHC class II; CD8/MHC class I). Interestingly, although receptors and counter-receptors can be co-expressed on the same cell in its activated state, no evidence exists for an interaction there. Thus, no biochemical evidence for e.g. ligand/receptor complexes between CD2 and CD58, CD4 and MHC class II or CD8 and MHC class I has been demonstrated, e.g. on activated T lymphocytes. Molecular interactions occur only with "other" cells expressing the appropriate ligand. This figure depicts a limited number of accessory receptors of the immunoglobulin superfamily.

molecules mediating cell-cell interactions in the immune response. This applies to structural features (genetic organization, functional epitopes, tertiary structures), the endogenous and exogenous (virus) ligands, and functional significance for lymphocyte activation including their signalling machineries. This article will, therefore, focus on this particular group of adhesion receptors which is now known to be intimately involved in lymphocyte differentiation, activation and recirculation (Bierer *et al.*, 1989; Springer, 1990). Moreover, since the majority of information on such structures has been collected from studies in the human system, experimental animal work will not be dealt with here. This seems particularly justified because the data available on adhesion molecules in murine and other species are largely in accordance with earlier findings made in the human system.

2.1 CD2

CD2, formerly known as the E-rosette receptor or T11, is a 50 kDa glycoprotein expressed by the vast majority of all human non-β-lymphocytes (i.e. thymocytes, T lymphocytes, natural killer (NK) cells) (Meuer *et al.*, 1984; Knapp, 1989; Beyers *et al.*, 1991). Only a very minor subpopulation of T lymphocytes was recently found not to be endowed with this adhesion receptor (Faure *et al.*, 1988). Although CD2 appears to be highly conserved in evolution, its expression on hematopoetic cells varies somewhat between man and murine species. Thus, unlike in human cells, CD2 is also found in the membrane of rat macrophages and mouse B lymphocytes (Clark *et al.*, 1988; Yagita *et al.*, 1989).

In man, CD2 is one of the earliest structures transcribed and expressed during lymphocyte ontogeny. It is believed that this glycoprotein plays a fundamental role in thymic differentiation (Moingeon *et al.*, 1989). An important feature of CD2 is its increase in cell surface density in the course of lymphocyte activation suggesting an important role in lymphocyte effector functions. A large number of monoclonal antibodies directed at CD2 (Knapp, 1989) are available and these have been shown to define several distinct functional epitopes which cluster into two major groups (Meuer *et al.*, 1984; Knapp, 1989; Moingeon *et al.*, 1989). "Conventional" CD2 monoclonal antibodies react with epitopes that are constitutively expressed in the resting cell state. A second group, designated as CD2R (Knapp, 1989) monoclonal antibodies, define epitopes which are barely detectable on resting cells but are rapidly exposed during activation due likely to a confirmational change in the CD2 molecule (Meuer *et al.*, 1984). The former have been shown by studies with deletion mutants and mutagenized plasmid constructs to bind to the N-terminal Ig-like domain 1 of CD2, whereas the binding sites for the latter map to the C-terminal domain of the extracellular portion (domain 2) (Peterson and Seed, 1987).

The N-terminal domain (domain 1) is responsible for binding of CD2 to its natural ligand (Peterson and Seed, 1987), CD58 (see below). At least two epitopes that are present in this region of the CD2 molecule interact with at least two regions of the CD58/LFA-3 counter-receptor. The functional significance of domain 2 is largely obscure at present. However, the fact that combinations of non-cross-blocking monoclonal antibodies directed at domains 1 plus 2, respectively, can induce T-cell proliferation and expression of the functional repertoire (Meuer *et al.*, 1984; Moingeon *et al.*, 1989) *in vitro* has prompted speculations on a second ligand of the CD2 molecule which, in combination with CD58, triggers an "alternative pathway of lymphocyte activation" (Meuer *et al.*, 1984).

A number of earlier studies indicated that the CD2 glycoprotein participates crucially in lymphocyte activation. Thus, it was initially found that monoclonal CD2 antibodies (directed at domain 1) exert inhibitory activities on T-cell activation (Howard *et al.*, 1981; Meuer *et al.*, 1984; Peterson and Seed, 1987; Knapp, 1989; Mueller *et al.*, 1989) induced by a number of different stimuli, and in addition, block lymphocyte effector function, including cell mediated lympholysis by cytotoxic T cells and NK cells (Huet *et al.*, 1986). The functional activity of such monoclonals could not simply be explained by inhibition of cell-cell contact since it was later demonstrated that combinations of CD2 monoclonal antibodies (see above) produce strong mitogenic effects (Meuer *et al.*, 1984), thus, indicating that CD2, besides mediating intercellular adhesion, functions as a signalling molecule. The combinations of monoclonal CD2 antibodies required for producing mitogenic effects towards human T lymphocytes do not relate to specialized epitopes of the CD2 molecule (Meuer *et al.*, 1984; Knapp, 1989). Rather, multiple combinations of at least two monoclonal antibodies, have been found to trigger lymphocyte activation, and such antibodies can be directed at a multiplicity of epitopes that exist on CD2. Therefore, it appears more likely that a combination of antibodies is required to produce a molecular change in the intracellular signalling apparatus of CD2 (which under physiological conditions is controlled by other cell surface receptors, such as the antigen receptor (Meuer *et al.*, 1984) in T lymphocytes and the Fc receptor in NK cells (Anderson *et al.*, 1990). Such a notion is further supported by the phenomenon that CD2-mediated – polyclonal – T-cell activation seems not to occur under physiological circumstances *in vivo* where immune responses are clearly dictated by clonal selection through the antigen-specific T cell receptor. The T cell receptor, therefore, must exert control over accessory signalling receptors, including CD2, in order to secure antigen specificity of an immune response. Evidence in support of such a view is described below.

The natural ligand of CD2 is CD58/LFA-3 (Plunkett *et al.*, 1987), which will be dealt with in depth below.

This has been substantiated by demonstrating that purified CD58 binds to CD2 specifically through interaction with domain 1 (Plunkett *et al.*, 1987). Particularly, studies with recombinant soluble CD58 (Albert-Wolf *et al.*, 1991), which inhibits in a competitive fashion multiple lymphocyte activities, have firmly established the role of CD2/CD58 in lymphocyte-dependent immune responses. Moreover, cells naturally expressing CD58, CD58-transfectants and purified or recombinant CD58 molecules, exert potent comitogenic effects on T-cell activation *in vitro* (Springer, 1990; Albert-Wolf *et al.*, 1991). With regard to recombinant CD58 molecules, their degree of oligomerization appears to be crucially relevant for binding affinity and functional activity (Albert-Wolf *et al.*, 1991). This is not surprising since the physiological interaction of CD2 with its ligand is of a multimeric nature.

It should be mentioned, in addition, that a CD58 homologue exists on sheep red blood cells (named T11/Ts=T11 target structure) (Hunig, 1985) and that this provides an explanation for the capacity of primate T-cells to form rosettes with red blood cells of the sheep.

A number of data suggest that the signals mediated through CD2 towards T cells are in part independent of T-cell receptor-mediated responses. Thus, although intracellular signalling events such as tyrosine phosphorylation of a number of proteins, calcium mobilization, activation of protein kinase C(PKC) and stimulation of $p21^{ras}$ appear to be shared by the T cell receptor and the CD2-dependent activation pathways (Yang *et al.*, 1986; Downward *et al.*, 1990), a set of T-cell receptor-independent discrete reactions occur selectively following activation through CD2. For example, induction of responsiveness to monokines (IL-1/IL-6) in T cells is dependent on a signal mediated through CD2 (Endler Jobst *et al.*, 1991). Moreover, dephosphorylation of a cytosolic 19 kDa phosphoprotein occurs in response to CD2 activation but not through the T cell receptor (Samstag *et al.*, 1991). Depletion of PKC activity by prolonged exposure to phorbol ester shuts off T-cell receptor-mediated activation, but rather upregulates the CD2-dependent alternative pathway (Cantrell *et al.*, 1988). Analogous functional activities are produced by inhibitors of PKC such as sphinganine (Qiao *et al.*, 1991). Moreover, in NK cells, which per definition are CD3/T cell receptor negative, CD2 can clearly function as a signalling molecule (Siliciano *et al.*, 1985; Scott *et al.*, 1989). Finally, the T-cell receptor/CD3 and the CD2-dependent pathways of activation segregate in some specialized compartments such as the *lamina propria* of the human intestine (Pirzer *et al.*, 1990; Qiao *et al.*, 1991) as well as in some immunodeficiency disorders (Meuer *et al.*, 1987; Meuer *et al.*, 1989; Roux *et al.*, 1991).

CD2 itself is not sufficient to explain its function as a triggering receptor in T lymphocytes and NK cells. This became evident when the human CD2 gene was transfected into non-lymphoid cell types. There, CD2 was not capable of mediating early activation events such as intracellular calcium mobilization (Moingeon *et al.*, 1989). One conclusion from these data was that, similar to the T-cell antigen receptor, which consists of a ligand binding portion (the α/β heterodimer) and a series of non-covalently associated signalling molecules (the CD3 molecular complex) (Meuer, Acuto *et al.*, 1984), CD2 must be associated in the membrane with a set of additional molecules responsible for its specific receptor/signalling function.

Based on functional effects on the CD2 pathway by monoclonal antibodies directed at the protein tyrosine phosphatase CD45, it was shown that the latter transmembrane glycoprotein which possesses enzymatic activity in its intracellular domain is physically associated with CD2 (Schraven *et al.*, 1989; Schraven *et al.*, 1990). More recently, a whole set of other cytoplasmic molecules were found to form a CD2 molecular signalling complex (Schraven *et al.*, 1991). These include – besides the tyrosine phosphatase CD45 – two tyrosine kinases of the *src* family, namely $p56^{lck}$ and $p59^{fyn}$ as well as a phosphoprotein complex, termed pp29–32 whose function is unclear at present (Schraven *et al.*, 1992). Pp29–32, however, which undergoes rapid intramolecular changes following lymphocyte activation is a potential substrate for both $p56^{lck}$ and CD45 (Brown *et al.*, 1989). Importantly, two additional phosphoproteins, namely the ζ and ε chains of the antigen receptor complex (Schraven *et al.*, 1992) co-precipitate with CD2. These two molecules which are also part of a functional T-cell receptor complex could provide a physical and functional link between these two receptor systems (Brown *et al.*, 1989). In fact, modulation experiments provided strong indications for the fact that the same ζ and ε chains are linked with CD2 and CD3-TCR (Schraven *et al.*, 1992). Such data also suggests for the first time a mechanism by which the T cell receptor could exert control over the CD2-dependent pathway. Thus, T cell receptor modulation which induces a state of desensitization in which T cells do not respond to activation through CD2 (Meuer *et al.*, 1984), is associated with uncoupling of the ζ and ε chains from the CD2 molecular complex (Schraven *et al.*, 1993). The kinetics of this modulation and uncoupling event may represent a crucial parameter with regard to immunomodulation strategies. These data are in line with the findings that the cytoplasmic domain of CD2 is involved in signal transduction (He *et al.*, 1988; Chang *et al.*, 1989; Bierer *et al.*, 1990).

Taken together, signals mediated through CD2 for T-cell activation appear to provide co-stimulation for T-cell activation and, therefore, represent potential targets for new immunosuppressive strategies. Recent experiments in murine species where co-injection of antigen plus blocking anti-CD2 antibody produced selective and long-lasting tolerance support this view (Guckel *et al.*, 1991).

2.2 CD4

CD4, a 50 kDa glycoprotein, is expressed by a subpopulation of thymocytes (common thymocyte), mature thymocytes and, under normal circumstances, approx. 60% of circulating mature T lymphocytes (Reinherz and Schlossman, 1980; Littman, 1985; Bierer *et al.*, 1989; Knapp, 1989). Moreover, expression of CD4 has been detected on cells of the monocyte lineage, not only in peripheral blood, but also in end stage differentiated cells, e.g. in skin and brain (Levy *et al.*, 1985). The human CD4 polypeptide carries four extracellular domains, the most distal of which has extensive homology to immunoglobulin light chain variable regions. Like CD2, CD4 fulfils the role of an adhesion structure as well as co-receptor (Barber *et al.*, 1989; Rudd, 1990). Signal transduction through CD4 occurs through its intracellular domain which is linked to the *src* family kinase $p56^{lck}$ (Barber *et al.*, 1989; Shaw *et al.*, 1989; Rudd, 1990; Glaichenhaus *et al.*, 1991).

T lymphocytes expressing CD4 molecules are usually specific for class II products of the MHC (Krensky *et al.*, 1982; Meuer *et al.*, 1982). Based on studies with human T-cell clones it was postulated that CD4 interacts with a non-polymorphic portion of MHC class II (Doyle and Strominger, 1987; Gay *et al.*, 1987) and cooperates with the T-cell receptor which also binds to MHC (plus antigenic peptide). Transfection studies demonstrated that binding of T cell receptor and CD4 to the same antigen presenting cell is the most efficient stimulus for activation. The interaction sites between CD4 and MHC class II have been mapped by means of deletion mutants (Clayton *et al.*, 1989; Lamarre *et al.*, 1989; Camerini and Seed, 1990). The elucidation of the cristal structure by 2.4 Å resolution demonstrated that binding site(s) for MHC class II molecules exist in domains 1 and 2 which both have the immunoglobulin fold (Kwong *et al.*, 1990; Wang *et al.*, 1990).

Besides mediating cell-cell interactions through binding to MHC class II molecules, CD4 has been found to represent the receptor for the HIV-1 retrovirus, the causative agent of AIDS (Maddon *et al.*, 1986; Clayton *et al.*, 1989; Lamarre *et al.*, 1989; Camerini and Seed, 1990; Kwong *et al.*, 1990; Wang *et al.*, 1990). The regions of the CD4 molecule mediating interactions with MHC class II or, alternatively, gp120 of HIV-1 appear to be similar if not identical (Kwong *et al.*, 1990; Wang *et al.*, 1990).

A structural interaction in the T-cell membrane of CD4 with the T-cell antigen receptor complex has been suggested from co-modulation studies as well as co-capping experiments (Saizawa *et al.*, 1987; Wegand *et al.*, 1987; Anderson *et al.*, 1988).

Signalling through CD4 appears to be related to its 38 amino acid cytoplasmic tail. Recently it was demonstrated that CD4 associates with the cytoplasmic protein tyrosine kinase $p56^{lck}$, and that this interaction occurs through a unique 32 amino acid, comprising N-terminal sequence of $p56^{lck}$ and the 38 C-terminal residues of CD4 (Littman, 1987; Barber *et al.*, 1989; Shaw *et al.*, 1989; Viellette *et al.*, 1989; Rudd, 1990; Abraham *et al.*, 1991; Glaichenhaus *et al.*, 1991). Moreover, crosslinking of CD4 molecules on the cell surface by monoclonal antibodies enhances the tyrosine kinase activity of $p56^{lck}$ enzyme. A physical and functional linkage between T-cell receptor and CD4 was also suggested from the fact that under certain experimental conditions CD3 ζ chains coprecipitate with the CD4/$p56^{lck}$ complex (Burgess *et al.*, 1991). All in all, the signalling machinery of CD4 appears to be quite similar to that described above for CD2 and, therefore, one might speculate that the intracellular signalling pathways engaged through these accessory receptors are closely related.

Most recently, a gene has been identified encoding for a product termed LAG3 which is selectively transcribed in activated NK cells and T lymphocytes (Triebel *et al.*, 1990). It codes for a membrane protein with 4 extracellular Ig-like domains. The sequence data, the compared exon/intron organization and the chromosomal localization revealed that LAG3 is closely related to CD4. However, its function is not known at present.

2.3 CD8

CD8 exists as a dimeric membrane glycoprotein composed of two disulfide linked subunits (Snow and Terhorst, 1983). CD8 is expressed on a subpopulation of thymocytes (common thymocyte where it is co-expressed with CD4), mature thymocytes, and approx. thirty per cent of circulating mature T lymphocytes (Reinherz and Schlossman, 1980; Littman, 1987; Bierer *et al.*, 1989). Furthermore, a subset of γ/δ cells and a minor population of NK cells express this differentiation antigen (Moebius *et al.*, 1991).

The human CD8 molecule was initially described as a homodimer consisting of two 34 kDa subunits since only one CD8 molecule was observed and since only the CD8 α-gene was known (Kavathas *et al.*, 1984; Littman *et al.*, 1985). However, following the identification of the human CD8 β-gene (Sukhatme *et al.*, 1985) it became evident that both genes, CD8 α and CD8 β, are expressed in CD8 positive T lymphocytes, suggesting a heterodimeric structure of this molecule (Norment and Littman, 1988; Shiue *et al.*, 1988).

Transfection studies employing mouse and human CD8 α- and CD8 β-cDNA respectively, have shown that the CD8 α-gene products can be expressed as homodimers on mouse L cells. In contrast, surface expression of CD8 β-proteins required co-transfection of the CD8 α-gene and was linked to the expression of CD8 α-molecules (DiSanto *et al.*, 1988; Gorman *et al.*, 1988). Most recent studies then demonstrated that "conventional" T cells, expressing the α/β T cell antigen receptor, co-express CD8 α/α homodimers and α/β heterodimers of different molecular masses (67 kDa and 75 kDa, respectively). In contrast, NK cells, T-cell receptor γ/δ cells and a rare population of mature CD4+/CD8+ T lymphocytes, express exclusively the CD8 α-gene products (α/α homodimer) (Moebius *et al.*, 1991). Differential functions of these two different isoforms of CD8 are unknown at present.

Human T cells, expressing CD8 (CD8 α/β heterodimer), are specific for human class I MHC molecules (Meuer *et al.* 1982). Subsequently, CD8 was found to bind to the α-3 domain of class I MHC molecules (Salter *et al.*, 1989). Similarly to that described above for CD4-MHC II, CD8 cooperates with the T cell receptor in MHC class I specific cellular immune reactions and (Samstag *et al.*, 1988), again, binding of both, T cell receptor and CD8 to the same cell, seems to be most effective.

The 28 amino acid comprising cytoplasmic region of human CD8 α plays a crucial role in its signal transducing functions. A short sequence motive containing two critical cystein residues mediates its interaction with the amino terminal domain of $p56^{lck}$. The cytoplasmic domain of human CD8 β does not contain such a sequence and, thus, should not interact with $p56^{lck}$ (Chalupny *et al.*, 1991; Miceli *et al.*, 1991). Phosphorylation of the CD8 α cytoplasmic tail on serine residues regulates its expression upon activation. Again, with regard to signalling, there seem to exist close similarities between the adhesion receptors CD2, CD8, and CD4 (Acres *et al.*, 1986; Blue *et al.*, 1987).

Immune intervention employing e.g. monoclonal antibodies directed at the adhesion receptors CD2, CD4, and CD8 was shown to provide a quite efficient approach to immunosuppression in experimental animal systems (Waldmann, 1989; Guckel *et al.*, 1991). However, if it holds true that these three molecules share the same intracellular activation pathway then one would not be surprised that immune intervention through one of these molecules is only partially sufficient *in vivo*.

2.4 CD58 (LFA-3)

CD58, a heavily glycosilated protein of 42–70 kDa is expressed on a broad range of hematopoetic and non-hematopoetic cells including endothelium, epithelium and various stromal tissues (Knapp, 1989). CD58 is a member of the immunoglobulin superfamily with two extracellular domains, the distal of which interacts with CD2 (Fig. 2.2). There is a considerable sequence homology between CD58 and CD2, the counter-receptor/ligand (Seed, 1987; Wallner *et al.*, 1987).

Two forms of CD58 are present on physiological cells, one containing a typical transmembrane domain and a second one which is linked through a phosphatidyl inositol (PI) anchor (Dustin *et al.*, 1987; Seed, 1987; Wallner *et al.*, 1987). Patients suffering from Paroxysmal nocturnal hemoglobinuria (PNH) lack functional PI (phospoltidylinositol)-anchored molecules and, therefore, do not express the PI-linked form of CD58 as well (Selvaraj *et al.*, 1987). However, no functional consequences for the function of CD58 in the immune response have been found upon investigation of cells from healthy versus PNH patients.

The interaction of CD2 and CD58 is involved in many essential immunological processes: both cell associated and soluble CD58 can activate T cells via an alternative pathway in conjunction with other mitogenic CD2 monoclonal antibodies, at least *in vitro*, (Meuer *et al.*, 1984; Hunig *et al.*, 1987; Selvaraj *et al.*, 1987; Takai *et al.*, 1987; Tiefenthaler *et al.*, 1987; Moingeon *et al.*, 1989; Springer, 1990). It participates in effector target cell binding, T-cell monocyte and T-cell/B cell interactions as well as in lymphocyte recirculation through interaction between endothelial cells (CD58) and T cells (CD2). With regard to the latter, it is now believed that while a number of additional cell-cell interaction molecules are essential for T cell-endothelial cell contact, CD2-CD58 attachment subserves a critical bidirectional signalling function, thus inducing endothelial cells to produce soluble mediators (IL-1, IL-6, TNF-α) and T cells to respond to them (Le *et al.*, 1990; Webb *et al.*, 1990; Endler Jobst *et al.*, 1991).

The molecular basis of the signalling function of CD58 is not known at present since no associated molecules have been demonstrated so far. The PI-linked form may share a common signalling subunit with other PI-linked molecules as recently suggested (Stefanova *et al.*, 1991).

2.5 CD54 (ICAM-1)

CD54 (intercellular adhesion molecule 1 = ICAM-1) is a cell surface adhesion protein which serves as a counter-receptor for the integrin LFA-1 (CD11a/18, see below). It is a heavily glycosilated 85–110 kDa single chain molecule expressing 5 tandem immunoglobulin-like domains (Fig. 2.2). CD54 is expressed on a variety of hematopoetic and non-hematopoetic cells. Like CD58, inflammatory mediators can strongly upregulate its expression in local areas (see below). ICAM-1 also serves as a receptor/ligand for Mac-1 (CD11b/18) and thus represents a component involved in activated neutrophil adhesion to endothelial cells and to ICAM-1 containing membranes (Dustin *et al.*, 1986; Rothlein *et al.*, 1986; Dustin *et al.*, 1988; Makgoba *et al.*, 1988; Simmons *et al.*, 1988; Staunton *et al.*, 1988; Smith *et al.*, 1989; Valent *et al.*, 1991).

Importantly, ICAM-1 is employed as a receptor by the major group of rhinoviruses, members of *Picorna* virus family which cause 40–50% of common colds (Staunton *et al.*, 1989; Staunton *et al.*, 1990). Over 100 immunologically non-cross-reactive rhinoviruses have been defined, of which 90% bind to ICAM-1. A putative receptor binding site has been localized in the viral capside protein and binds to domains 1 and 2 of CD54. This site is virtually identical with the binding site for LFA-1 and, thus, rhinoviruses appear to mimic LFA-1 in binding to these most membrane distal and most accessible sites of the ICAM-1 molecule (Staunton *et al.*, 1990). In contrast, the binding site on CD54 for Mac-1 is clearly distinct from that for LFA-1 and rhinovirus and maps to domain 3 of the ICAM-1 molecule (Diamond *et al.*, 1991).

ICAM-1 mediates co-stimulation in the activation of resting T lymphocytes by TcR/CD3 mAbs. The cytokines IL-1 β and IL-6 augment this adhesion ligand-mediated co-stimulatory effect (compare CD58/CD2) (Van Seventer *et al.*, 1991). However, the signalling apparatus of ICAM-1 has not been identified. Moreover, functional consequences of this ICAM-1 mediated signal are not known at present.

3. The Integrin Family of Accessory Receptors

A series of interrelated heterodimers, consisting of α- and β-subunits, respectively, forms the integrin family (Hynes, 1987; Berg *et al.*, 1989; Kishimoto *et al.*, 1989) (Fig. 2.3). Based on the nature of the β-subunits utilized to form the cell surface heterodimer, three distinct subfamilies have been defined. In the β1-subfamily (VLA-subfamily) one of six different α subunits, that have been identified so far, associate with the β1-glycoprotein. In the β2-subfamily, there are three distinct α subunits that associate with β2 and in the β3-subfamily two different α-subunits have been identified. In some cases, however, α6 was shown to associate with β4 instead of β1. αV which usually associates with β3 can associate with β5 (alternatively called βX) and, at least in murine species, α4 which is usually found in the β1-subfamily can also associate with a distinct subunit called βp (Berg *et al.*, 1989; Kishimoto *et al.*, 1989).

Members of the integrin family of adhesion receptors are expressed on nearly all cell types and are involved in their interactions with extracellular matrix (fibronectin, laminin, collagen) and transient cell to cell adhesion (Kishimoto *et al.*, 1989; Krissansen *et al.*, 1990; Savill *et al.*, 1990). Besides this, however, several more recent studies also suggest that signals can be transduced through integrin type receptors. For example, a monoclonal antibody directed at β1 (CD29) synergizes with CD2 or CD3 mediated T cell stimulation to increase c-AMP (Groux *et al.*, 1989). Moreover, monoclonal antibodies directed at β2 are comitogenic in the stimulation of T-cell proliferation by CD3 antibody (Isler *et al.*, 1988). Anti-VLA4 (β1/α4) monoclonal antibodies induce strong cell clustering and are comitogenic with CD2 monoclonal antibodies *in vitro* (author's unpublished observations). Integrin α subunits have three or four tandem repeats of a putative divalent action binding motif and require calcium or magnesium for function. This correlates with the requirement for magnesium in T-cell adhesion and in binding of purified CD58 to purified ICAM-1 (Dustin *et al.*, 1989; Dustin *et al.*, 1991).

A very important response, occurring in the form of inside-out signalling has been demonstrated with regard to the interaction of the β2 integrin LFA-1 (CD11a/18) and its ligand ICAM-1 (CD54) (Dustin *et al.*, 1989). There, triggering of the T-cell receptor induces a transient confirmational change in the LFA-1 molecule which depends on its intact intracellular domain (Dustin *et al.*, 1989; Dustin *et al.*, 1991). This change also alters (enhances) the affinity of LFA-1/ICAM-1 interaction, thus initially mediates stronger adhesion and subsequently allows for de-adhesion.

Before the family of integrins had been defined a number of studies were performed with the 4B4 monoclonal antibody which later turned out to be directed at the β1 subunit (CD29). Since expression of integrin molecules is highly variable and determined by the differentiation state (resting versus activated lymphocytes etc.) it was possible to separate $CD29^{high}$ from $CD29^{low}$ expressing cells and to investigate their functional properties (Morimoto *et al.*, 1985; Sanders *et al.*, 1988). The subset with higher CD29 levels also has increased levels of other adhesion receptors such as CD58, CD2, LFA-1, CD45RO, and CD44. That this subset contains the property of increased responsiveness to recall antigens is not surprising given the now known contribution of virtually all adhesion receptors in facilitating and augmenting immune reactions (Sanders *et al.*, 1988; Cerottini and McDonald, 1989).

In accordance with such a function, a series of monoclonal antibodies directed at integrin molecules when employed individually have been found to inhibit cellular immune responses such as cell mediated lympholysis and the mixed lymphocyte reaction (Sanchez Madrid *et al.*, 1982). In a reciprocal fashion, isolated matrix proteins which interact with integrin molecules through an RGD-containing domain were shown to boost proliferation of lymphocytes *in vitro* and/or secretion of enhanced amounts of e.g. Interferon-gamma (Matsuyama *et al.*, 1989).

Loss of leucocyte integrin expression which leads to profound defects in adhesion are features of a clinical syndrome called leucocyte adhesion deficiency (LAD) (Anderson and Springer, 1987). LAD is a recessive inherited disorder where single amino acid substitutions in a particular conserved region of β2 prevent association with the α subunit. As a consequence, patients experience progressive necrotic soft tissue infections, diminished pus formation, and impaired wound healing (Krensky *et al.*, 1985).

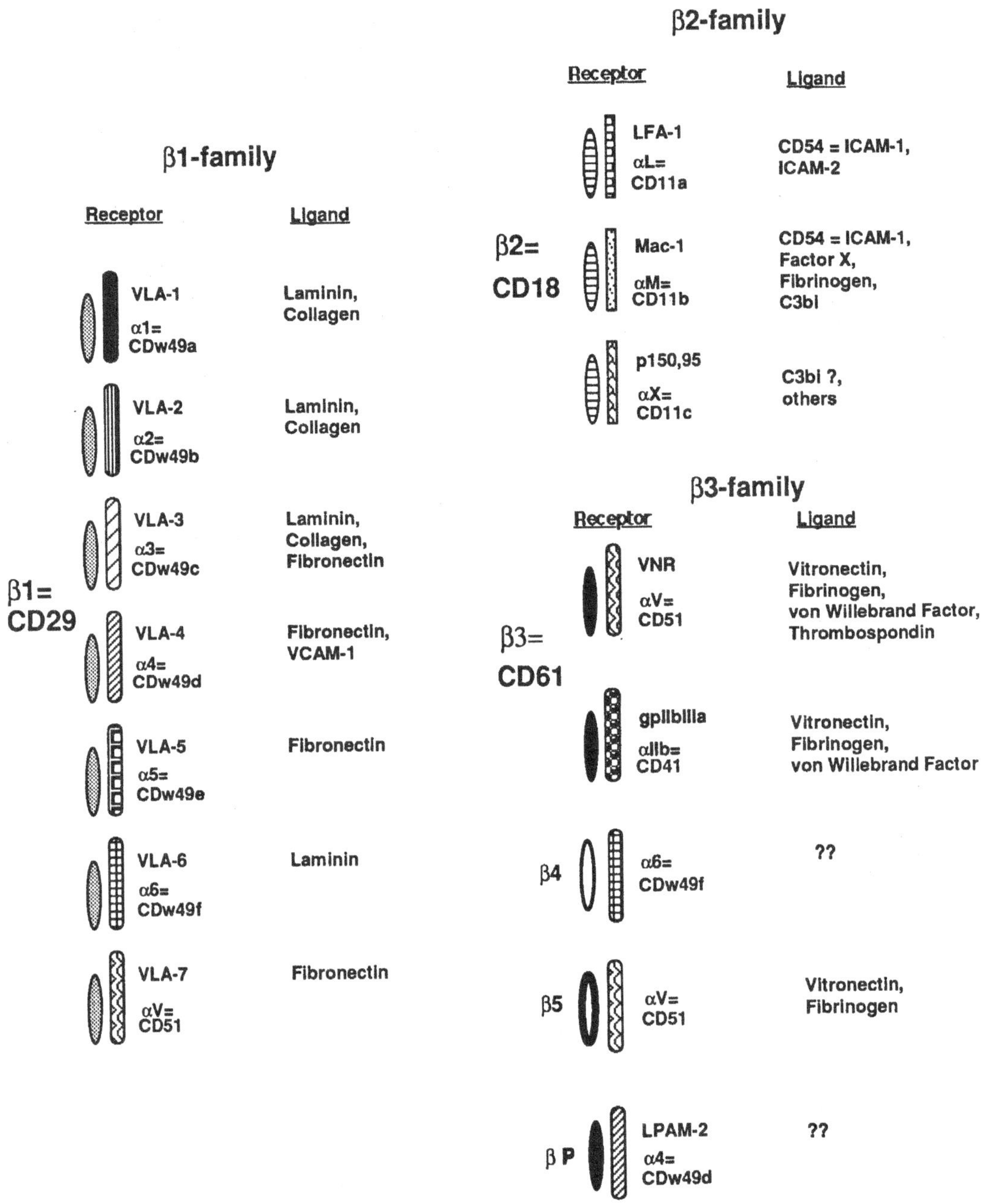

Figure 2.3 Accessory receptors of the integrin family are interrelated heterodimers consisting of α and β-subunits. Based on the nature of the β-subunit three distinct subfamilies have been defined so far. The β1- and β3-families interact predominantly with extracellular matrix components while the β2-family interacts with counter-receptors on other cell types. Although this figure delineates the present knowledge on the integrin families, one has to expect a multitude of additional members to be discovered in the near future.

4. *The Selectin Family of Accessory Receptors*

Three cell surface molecules which were originally investigated independently as lymphocyte homing receptors in lymphoid tissues (lymph nodes/pares patches) form the selectin family of adhesion receptors (L-Selectin, P-Selectin, E-Selectin) (Fig. 2.4). The structural highlights of selectins are N-terminal C-type lectin domains followed by a C-terminal growth factor like module, then multiple copies of the consensus sequence units characteristic of complement binding proteins, a transmembrane segment and a

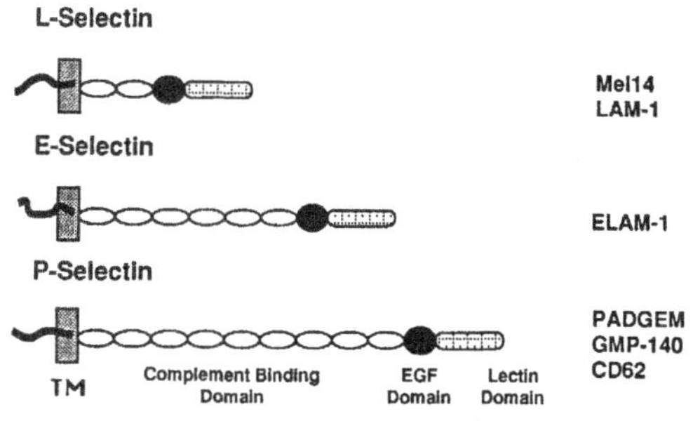

Figure 2.4 The selectin family of accessory receptors contains three members which are structurally related and encoded by closely linked genes. They function in a variety of cell interactions in the vasculature. (This figure also gives the former names of members of the selectin family).

short cytoplasmic domain. The three selectin proteins are structurally related and are encoded by closely linked genes which arose by gene duplication (Bevilacqua *et al.*, 1989; Camerini *et al.*, 1989; Goldstein *et al.*, 1989; Johnston *et al.*, 1989; Lasky *et al.*, 1989; Tedder *et al.*, 1990). Selectins function in a variety of cell interactions in the vasculature and are expressed both on leucocytes and endothelial cells.

Selectins are involved in cell to cell adhesion, particularly with regard to lymphocyte-endothelial cell interactions. Expression of selectins on the cell surface is highly variable. For example, L-selectin is lost from the surface of lymphocytes and granulocytes shortly after activation (Kishimoto *et al.*, 1989; Tedder *et al.*, 1990). P-selectin (CD62) and E-selectin are transiently expressed on the surface of monokine or thrombine activated endothelial cells (Bevilacqua *et al.*, 1989; Hattori *et al.*, 1989). Selectins apparently bind to phosphorylated oligosaccarides as ligands (Picker *et al.*, 1991) consistent with the presence of a lectin-homologous domain. For example, E-selectin recognizes the Sialyl-lewis X determinant, a granulocyte carbohydrate also found on the surface of some tumor cell lines (Phillips *et al.*, 1990; Walz *et al.*, 1990). The function of selectins at present is proposed to ly in an early step in leucocyte binding to endothelium. In contrast, integrins (see above) which are also increased on the leucocyte surface and then remain permanently upregulated are required for the subsequent event of trans-endothelial migration (Dustin and Springer, 1991).

5. *CD44*

CD44 is a polymorphic membrane protein recognizing hyaluronate (Aruffo *et al.*, 1990; Seth *et al.*, 1991; Stamenkovic *et al.*, 1991). Its proposed roles encompass lymphocyte activation, matrix adhesion and the attachment of lymphocytes to lymph node high endothelial venules. In addition to participating in T cell homing and adhesion, CD44 is suggested to play a major role in delivering a transmembrane signal to cytotoxic T lymphocytes which triggers the lytic machinery, even in situations where the T cell receptor is not occupied, suggesting that CD44 mediates a mechanism of non-specific tissue damage observed at sites of inflammation (Seth *et al.*, 1991). Like many other adhesion receptors, CD44 is upregulated in expression after activation through the T-cell receptor complex, indicating that CD44 represents a T-cell activation antigen. Monoclonal antibodies to CD44 enhance T-cell proliferation induced by anti-CD3 and anti-CD2. Antibodies to CD44, moreover, block lymphohemopoesis in long-term bone marrow cultures (Huet *et al.*, 1989; Denning *et al.*, 1990; Miyake *et al.*, 1990; Bochner *et al.*, 1991; von Andrian *et al.*, 1991).

Importantly, a variant of the CD44 glycoprotein was found on metastasizing subclones of some carcinoma cell lines, however, not on the non-metastasizing counterpart (Gunthert *et al.*, 1991). Expression by transfecting this variant into non-metastasizing cells conferred metastatic potential and suggests evidence for a causal role of this class of glycoproteins in the metastatic process.

6. *CD28*

CD28 was first identified as a 44 kDa homodimeric glycoprotein expressed on 80% of human peripheral blood cells (June *et al.*, 1990). CD28 exists in both, monomeric and homodimeric forms (Aruffo and Seed, 1987). Whether isolated subunits and disulfide bonded dimers coexist on the surface of T cells is unknown. CD28 consists of 202 amino acid residues. It represents a transmembrane protein with an extracellular domain containing 134 amino acids with 5 N-linked glycosylation sites (Aruffo and Seed, 1987). CD28 is homologous to immunoglobulin variable regions domains and now known to serve as a receptor for the human B lymphocyte activation antigen B7. The interaction between B7 and CD28 provides a potent co-stimulatory signal for human T cell activation via the T cell receptor (Freeman *et al.*, 1991; Linsley *et al.*, 1991). A second receptor for B7 is CTLA4, a predicted membrane receptor of cytotoxic T-cells which is homologous to CD28 and whose gene maps to the same chromosomal band as the gene for CD28 (Linsley *et al.*, 1991).

Monoclonal antibodies binding to the CD28 receptor were found to influence a number of immunologic functions *in vivo*, including the production of lymphokines and the mixed lymphocyte reaction (Bjorndahl *et al.*, 1989; Thompson *et al.*, 1989; June *et al.*, 1990). The degree of crosslinking of CD28 controls in part stipulatory versus inhibitory activity of anti-CD28 antibodies since monovalent fragments constantly inhibit while bivalent stimulation converts the inhibitory activity into stimulatory activity.

The primary mechanism by which stimulation through CD28 augments lymphokine production in mature T cells has been suggested by inhibition of degradation of lymphokine mRNAs. As a result of this stabilization, the steady state levels of lymphokine mRNAs increase, leading to enhanced translation and subsequent protein secretion (Thompson *et al.*, 1989). Importantly, while the major transcriptional stimulation of IL-2 mediated through the antigen receptor or CD2 is completely abolished by the immunosuppressant Cyclosporin-A, stimulation through CD28 is independent of intracellular calcium, activation of protein kinase C and Cyclosporin-resistant (Bjorndahl *et al.*, 1989; Thompson *et al.*, 1989; June *et al.*, 1990; Ledbetter *et al.*, 1990). Moreover, choleratoxin which induces irreversible activation of G proteins and, thus, increases intracellular cAMP, inhibits CD3 and CD2 dependent activation whereas CD28 stimulated T-cell proliferation is resistant to this agent. Moreover, unlike stimulation through the T-cell receptor, CD28 mediated activation does not induce tyrosine phosphorylation of the T-cell receptor ζ chain (Ledbetter *et al.*, 1990), suggesting that CD28 may not affect the tyrosine kinase/phosphatase signal transduction pathway.

7. *Immunopathology of Accessory Receptors*

As mentioned above, expression of most adhesion receptors is highly variable and, in general, strongly increased on activated, as opposed to resting, leucocytes. Since many adhesion molecules interact with other adhesion receptors whose kinetics of expression are quite similar, this means that the signals that are exchanged among these receptors and counter-receptors strongly upregulate local cytokine-driven inflammatory processes. In fact, tremendous upregulation of accessory receptors in inflammatory sites has been well documented by a number of investigators in chronic organ specific disorders (Bishop and Hall, 1989; Volpes *et al.*, 1990; Autschbach *et al.*, 1991).

As mentioned earlier, the cell surface expression of most adhesion receptors varies tremendously depending on the cellular differentiation state, the nature of signals that are delivered and the respective local environment. It is not known as yet how the cell surface density of adhesion molecules is regulated. In some cases, such as CD58 and CD54, virtually each cell appears to be capable of producing and expressing such molecules provided it is appropriately stimulated (Bishop and Hall, 1989). Particularly, certain inflammatory cytokines appear to control expression of genes encoding for some adhesion molecules and, therefore, these molecules are found abundantly in inflammatory sites. Given that their capacity to transmit signals depends to some extent on the cell surface density (the higher the degree of multimerization the higher the avidity with which they interact with their ligands), one has to assume that adhesion receptors and inflammatory mediators may even form autocatalytic cycles. There, signalling through adhesion receptors could lead to enhanced production of cytokines which, in turn, promote membrane expression of the same accessory molecules. Not unlikely, therefore, such mechanisms may be intimately involved in the perpetuation of chronic inflammation. Immune intervention at this level would be beneficial because of the interruption of uncontrolled autostimulatory mechanisms.

Conversely, a number of reports have demonstrated that malignant cells express unusually low levels of adhesion receptors (Gregory *et al.*, 1988; Johnson *et al.*, 1989; Nickoloff *et al.*, 1989; Rice and Bevilacqua, 1989; Rousset *et al.*, 1989; Schirren *et al.*, 1992). The extent of expression not surprisingly determines crucially tumor cell behaviour *in vivo* both with regard to metastatic spread and susceptibility to autologous immune effector mechanisms (Johnson *et al.*, 1989; Rice and Bevilacqua, 1989). For example, down-modulation of CD54 and CD58 was reported to exist in freshly isolated acute T-cell leukemias and to correlate with the white blood cell count (Schirren *et al.*, 1992). Moreover, freshly recovered tumor cell blasts could not be killed by autologous cytotoxic T lymphocytes. However, when malignant tumor targets were exposed to TNF-α, a mediator capable of inducing expression of these accessory molecules, then autologous killing was restored. This example illustrates that analysis of adhesion receptors and counter-receptors may represent an important means to monitor and interpret *in vivo* effects of therapies with biological response modifiers.

8. *Circulating Forms of Accessory Receptors*

So far the accessory receptors CD8, CD54, and CD58 have been demonstrated also to exist in circulating forms in human body fluids (Giblin *et al.*, 1989; Becker *et al.*, 1991; Rothlein *et al.*, 1991; Hoffmann *et al.*, 1992). Soluble natural variants of CD54 and CD58 have been purified and characterized biochemically, and in both cases were shown to bind to their cell surface ligands LFA-1 and CD2, respectively. Moreover, at higher concentrations, such molecules competitively inhibit cell to cell interactions. Not surprisingly and in line with their enhanced expression during inflammation on the cell surface, elevated levels of circulating CD54 and CD58 were detected in some inflammatory disorders (Becker *et al.*, 1991; Rothlein *et al.*, 1991; Hoffmann *et al.*, 1992).

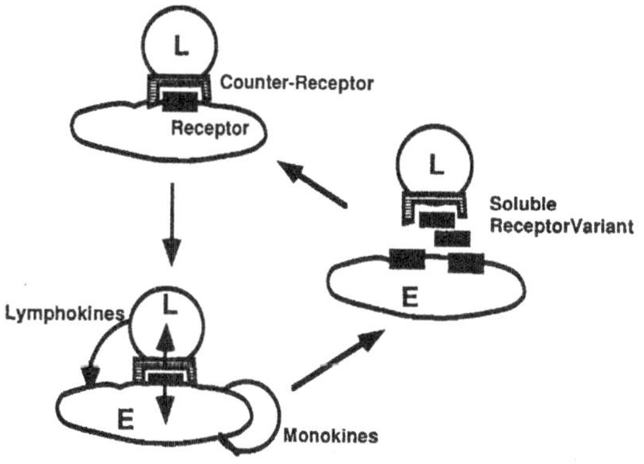

Figure 2.5 A potential function of natural soluble receptor variants may exist in regulating transient cell to cell attachment. Binding of receptor to its counter-receptor can induce the secretion of cytokines which, in turn, induce soluble receptor forms. The latter could act as competitive inhibitors of their membrane standing counterparts and thereby mediate de-adhesion. A disturbed balance of cell surface versus soluble receptor could critically determine the outcome of local immune responses during chronic inflammatory processes.

The physiological significance of low levels of circulating adhesion molecules is not entirely clear at present. However, given the fact that the binding of CD54 and CD58 to their respective ligand molecules produces signals which lead to secretion of mediators capable of inducing soluble CD54 and/or CD58, one could postulate that soluble forms of adhesion receptors serve a physiological function for de-adhesion. Such a mechanism would represent a second strategy (Fig. 2.5) in addition to transient conformational changes in binding affinity (e.g. in the case of LFA-1; Dustin and Springer, 1989) for transient interactions of immunocompetent cells with their multiple environments. On the other hand an impaired balance of cell surface versus soluble forms of adhesion receptors can be expected to represent a critical regulatory parameter in determining pathological local immune responses.

9. *Potential Approaches to Specific Immune Modulation*

The present tools for clinical immune intervention in immunodeficiencies, cancer, autoimmune disease, transplant rejection, and allergic disorders are highly unselective and clearly insufficient. The identification, biochemical characterization and molecular cloning of genes encoding for functionally important cell surface receptors involved in the regulation of local lymphocyte responsiveness (particularly the accessory molecules described herein) opens up a future area for immunomodulatory reagents.

Potential drugs at hand already are monoclonal antibodies (Waldmann, 1989). Their unique fine specificity and high binding affinity makes them first rate tools for immunopharmacology. Monoclonal antibodies (mAbs) can be applied as complete antibody molecules, Fab_2 or Fab (fast electron bombardment) fragments or even Fv subfragments composed of variable domains only. The present restrictions in the application of monoclonal antibodies or antibody fragments exists in a limited knowledge regarding their metabolism and distribution in the organism. For example, complete antibody molecules due to their Fc portions are usually trapped by the reticulo-endothelial system of the liver, bone marrow and spleen.

This non-specific uptake may to some extent differ, depending on the antibody isotope, but clearly represents a problem for the application of all antibodies that have been studied so far. For some special indications, however, non-specific local enrichment may be favorable, e.g. for targeting immune reactions that occur in these organs (liver transplant rejection, autoimmune liver disease etc). Conversely, removal of Fc portions strongly reduces anti-body half life *in vivo*. Another problem which is encountered during diagnostic or therapeutic application of monoclonal antibodies *in vivo* exists in their antigenicity, both with regard to the fact that most antibodies are foreign proteins (made from rodent cells) and the inevitable immune response against idiotypes that are delivered at high concentrations. While the former problem can be solved by producing chimeric, humanized antibodies employing recombinant techniques, one has to expect that anti-idiotypic responses cannot be avoided in general and therefore limit a prolonged application of immunomodulatory antibodies.

Another potential approach to immune intervention at the level of adhesion receptors exists in the recombinant production and delivery of soluble variants of naturally occurring membrane linked components as competitive inhibitors of cell-cell interactions (Marlin *et al.*, 1990; Albert-Wolf *et al.*, 1991; Dana *et al.*, 1991). While this looks like an elegant approach to specific immune modulation largely avoiding antigenicity, one has to consider that the affinity of variant receptor molecules can at best be as high as that of their natural counterparts. Therefore, one would anticipate that such reagents have to be administrated at tremendous concentrations particularly when attempting to influence local processes where high receptor/ligand densities occur. At present, only two examples for such soluble receptor variants exist which are active *in vitro* (Marlin *et al.*, 1990; Albert-Wolf *et al.*, 1991). As yet, nothing is known about their tissue distribution, half life or other important parameters under *in vivo* conditions.

Cloning of genes encoding for adhesion receptors now enables the elucidation of tertiary structures by molecular biophysics. Such knowledge which will yield information on functional domains could eventually result in the creation of low molecular weight organic compounds by means of molecular and computer-assisted drug design. This approach could well represent the strategy of the future. One advantage of such compounds clearly exists in the possibility to apply them orally.

Finally, given the problem of redundancy at the cell surface level as addressed above, approaches aimed at influencing intracellular signalling reactions may provide extremely potent means for immune intervention in the future. However, there the question of how to specifically target a particular molecule and/or molecular reaction (kinase, phosphatase) is unsolved at present. Intracellular enzymes appear to be of rather low substrate specificity (although there may be a few exceptions to this rule). The fact that particular kinases, for example, act on a limited number of molecular components involved in signal transduction may rather be determined by their physical interactions with particular substrate molecules. Again, the distinction between active sites and binding domains for their substrates may be important because influences at the former level could be much too unselective, whereas disruption of specific molecular interactions at the binding level (e.g. SH_2 domains) (Koch *et al.*, 1991; Saito *et al.*, 1991) could be more favorable. This consideration then also asks for the identification of the signalling cascade components, their molecular cloning and expression, their biophysical characterization, and subsequent design of organic compounds which can penetrate the plasma membrane and interfere with selective binding domains. Figs. 2.6 and 2.7 summarize potential strategies for immune intervention at the level of accessory receptors.

In conclusion, the fact that accessory receptors are the functional determinators of immune reactivity at particular local sites of an organism makes them the most favorable targets for immune intervention. At present, however, clinical approaches are limited by the complexity inherent to the action of specific sets of receptor/ligand combinations engaged in the multiple micro-environments, the complexity of molecular interactions among adhesion receptors themselves, and the unknown basis of their regulation and tremendously variable expression at the cell surface. Moreover, too little information exists regarding their intracellular signalling machineries. Nevertheless, given the enormous progress that was made in this field during a very short period of time and the powerful technical procedures that are now available to address structures and functions at the molecular level, it seems reasonable to suppose that in the near future powerful approaches can be delineated for a more rational and specific area of immunopharmacology.

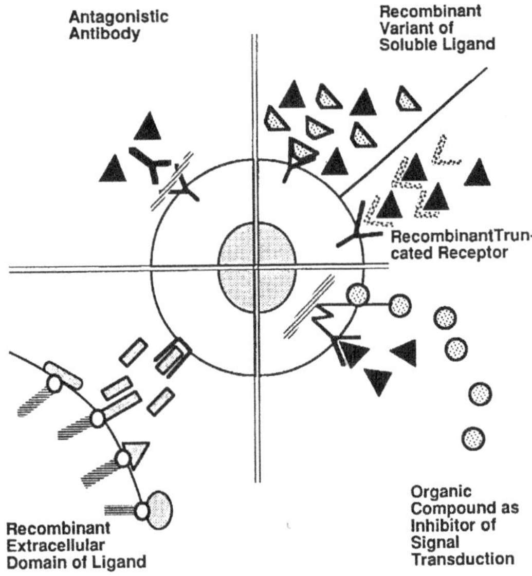

Figure 2.7 A multitude of strategies for immune intervention arise from structural informations on accessory receptors. Antagonistic antibodies with their high binding activity to functional sites of accessory receptors represent potent competitive inhibitors of natural ligands. Recombinant technology enables to produce soluble variants of extracellular ligand domains, inactive variants of soluble ligands, and recombinant truncated receptors domains. In addition, compounds capable of penetrating the plasma membrane can interfere with the signalling process (e.g. Cyclosporin-A).

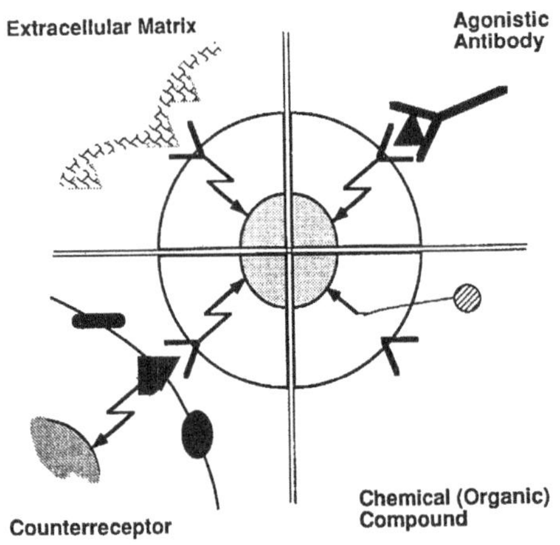

Figure 2.6 Accessory receptor triggering naturally occurs through a counter-receptor or, alternatively, extracellular matrix constituents. A powerful tool to study the function of accessory receptors represent agonistic antibodies which mimic the functions of natural ligands. In addition, some chemical compounds shortcircuit receptor/ligand interaction and activate the intracellular signalling machinery (e.g. phorbol esters).

10. *References*

Abraham, N., Miceli, M.C., Parnes, J.R. and Veillette, A. (1991). Enhancement of T-cell responsiveness by the lymphocyte-specific tyrosine protein kinase $p56^{lck}$. Nature 350, 62–66.

Acres, R.B., Conlon, P.J., Mochizuki, D.Y. and Gallis, B. (1986). Rapid phosphorylation and modulation of the T4 antigen on cloned helper T-cells induced by phorbol myristate acetate or antigen. J. Biol. Chem. 261, 16 210–16 214.

Albert-Wolf, M., Meuer, S.C., and Wallich, R. (1991). Dual function of recombinant human CD58: inhibition of T-cell adhesion and activation via the CD2 pathway. Intern. Immunol. 3, 1335–1347.

Alzari, P.M., Lascombe, M.B. and Poljak, R.J. (1988). Three-dimensional structure of antibodies. Annu. Rev. Immunol. 6, 555–580.

Anderson, D.C. and Springer, T.A. (1987). Leukocyte adhesion deficiency: an inherited defect in the Mac-1, LFA-1, and p150,95 glycoproteins. Annu. Rev. Med. 38, 175–194.

Anderson, P., Blue, M.L. and Schlossman, S.F. (1988). Comodulation of CD3 and CD4. Evidence for a specific association between CD4 and approximately 5% of the CD3:T-cell receptor complexes on helper T lymphocytes. J. Immunol. 140, 1732–1737.

Anderson, P., Caligiuri, M., O'Brien, C., Manley, T., Ritz, J. and Schlossman, S.F. (1990). Fc gamma receptor type III (CD16) is included in the zeta NK receptor complex expressed by human natural killer cells. Proc. Natl Acad. Sci. USA 87, 2274–2278.

Aruffo, A. and Seed, B. (1987). Molecular cloning of a CD28 cDNA by a high-efficiency COS cell expression system. Proc. Natl Acad. Sci. USA 84, 8573–8577.

Aruffo, A., Stamenkovic, I., Melnick, M., Underhill, C.B. and Seed, B. (1990). CD44 is the principal cell surface receptor for hyaluronate. Cell 61, 1303–1313.

Autschbach, F., Meuer, S.C., Moebius, U., Manns, M., Hess, G., Meyer zum Buschenfelde, K.-H., Thoenes, W., and Dienes, H.-P. (1991) Hepatocellular expression of lymphocyte function-associated antigen 3 in chronic hepatitis. Hepatology 14, 223–230.

Barber, E.K., Dasgupta, J.D., Schlossman, S.F., Trevillyan, J.M., and Rudd, C.E. (1989). The CD4 and CD8 antigens are coupled to a protein-tyrosine kinase ($p56^{lck}$ that phosphorylates the CD3 complex. Proc. Natl Acad. Sci. USA 86, 3277–3281.

Becker, J.C., Dummer, R., Hartmann, A.A., Burg, G., and Schmidt, R.E. (1991). Shedding of ICAM-1 from human melanoma cell lines induced by IFN-gamma and tumor necrosis factor-alpha. J. Immunol. 147, 4398–4401.

Berg, E.L., Goldstein, L.A., Jutila, M.A., Nakache, M., Picker, L.J., Streeter, P.R., Wu, N.W., Zhou, D. and Butcher, E.C. (1989). Homing receptors and vascular addressins: cell adhesion molecules that direct lymphocyte traffic. Immun. Rev. 108, 1–18.

Bevilacqua, M.P., Stengelin, S., Gimbrone, M.A., Jr. and Seed, B. (1989). Endothelial leukocyte adhesion molecule 1: an inducible receptor for neutrophils related to complement regulatory proteins and lectins. Science 243, 1160–1165.

Beyers, A.D., Simon, J.D., Cantrell, D.A., Iszquierdo, M. and Williams, A.F. (1991) Autonomous roles for the cytoplasmic domains of the CD2 and CD4 T-cell surface antigens. EMBO J. 10, 377–385.

Bierer, B.E., and Burakoff, S.J. (1988) T-cell adhesion molecules. FASEB Journal 2, 2584–2590.

Bierer, B.E., Sleckman, B.P., Ratnofsky, S.E. and Burakoff, S.J. (1989). The biological roles of CD2, CD4, and CD8 in T-cell activation. Annu. Rev. Immunol. 7, 579–599.

Bierer, B.E., Bogart, R.E. and Burakoff, S.J. (1990). Partial deletions of the cytoplasmic domain of CD2 result in a partial defect in signal transduction. J. Immunol. 144, 785–789.

Bishop, G.A., and Hall, B.M. (1989) Expression of leucocyte and lymphocyte adhesion molecules in the human kidney. Kidney Intern. 36, 1078–1085.

Bjorndahl, J.M., Sung, S.S., Hansen, J.A. and Fu, S.M. (1989). Human T-cell activation: differential response to anti-CD28 as compared to anti-CD3 monoclonal antibodies. Eur. J. Immunol. 19, 881–887.

Blue, M.L., Hafler, D.A., Craig, K.A., Levine, H. and Schlossman, S.F. (1987). Phosphorylation of CD4 and CD8 molecules following T-cell triggering. J. Immunol. 139, 3949–3954.

Bochner, B.S., Luscinskas, F.W., Gimbrone, M.A., Jr., Newman, W., Sterbinsky, S.A., Derse Anthony, C.P., Klunk, D. and Schleimer, R.P. (1991). Adhesion of human basophils, eosinophils, and neutrophils to interleukin 1-activated human vascular endothelial cells: contributions of endothelial cell adhesion molecules. J. Exp. Med. 173, 1553–1557.

Brown, M.H., Cantrell, D.A., Brattsand, G., Crumpton, M.J., and Gullberg, M. (1989) The CD2 antigen associates with the T-cell antigen receptor CD3 antigen complex on the surface of human T lymphocytes. Nature 339, 551–553.

Burgess, K.E., Odysseos, A.D., Zalvan, C., Druler, B.J., Anderson, P., Schlossman, S.F., and Rudd, C.E. (1991). Biochemical identification of a direct physical interaction between the CD4:$p56^{lck}$ and Ti/TcR/CD3 complexes. Eur. J. Immunol. 21, 1663–1668.

Camerini, D., James, S.P., Stamenkovic, I. and Seed, B. (1989). Leu-8/TQ1 is the human equivalent of the Mel-14 lymph node homing receptor. Nature 342, 78–82.

Camerini, D. and Seed, B. (1990). A CD4 domain important for HIV-mediated syncytium formation lies outside the virus binding site. Cell 60, 747–754.

Cantrell, D.A., Verbi, W., Davies, A., Parker, P. and Crumpton, M.J. (1988). Evidence that protein kinase C differentially regulates the human T lymphocyte CD2 and CD3 surface antigens. Eur. J. Immunol. 18, 1391–1396.

Cerottini, J.C. and MacDonald, H.R. (1989). The cellular basis of T-cell memory. Annu. Rev. Immunol. 7, 77–89.

Chalupny, N.J., Ledbetter, J.A. and Kavathas, P. (1991). Association of CD8 with $p56^{lck}$ is required for early T-cell signalling events. EMBO J. 10, 1201–1207.

Chang, H.C., Moingeon, P., Lopez, P., Krasnow, H., Stebbins, C. and Reinherz, E.L. (1989). Dissection of the human CD2 intracellular domain. Identification of a segment required for signal transduction and interleukin 2 production. J. Exp. Med. 169, 2073–2083.

Clark, S.J., Law, D.A., Paterson, D.J., Puklavec, M. and Williams, A.F. (1988). Activation of rat T lymphocytes by anti-CD2 monoclonal antibodies. J. Exp. Med. 167, 1861–1872.

Clayton, L.K., Sieh, M., Pious, D.A. and Reinherz, E.L. (1989). Identification of human CD4 residues affecting class II MHC versus HIV-1 gp120 binding. Nature 339, 548–551.

Dana, N., Fathallah, D.M. and Arnaout, M.A. (1991). Expression of a soluble and functional form of the human beta 2 integrin CD11b/CD18. Proc. Natl Acad. Sci. USA 88, 3106–3110.

Denning, S.M., Le, P.T., Singer, K.H. and Haynes, B.F. (1990). Antibodies against the CD44 p80, lymphocyte homing receptor molecule augment human peripheral blood T-cell activation. J. Immunol. 144, 7–15.

Diamond, M.S., Staunton, D.E., Marlin, S.D. and Springer, T.A. (1991). Binding of the integrin Mac-1 (CD11b/CD18) to the third immunoglobulin-1 ke domain of ICAM-1 (CD54) and its regulation by glycosylation. Cell 65, 961–971.

DiSanto, J.P., Knowles, R.W. and Flomenberg, N. (1988). The human Lyt-3 molecule requires CD8 for cell surface expression. EMBO J. 7, 3465–3470.

Downward, J., Graves, J.D., Warne, P.H., Rayter, S. and Cantrell, D.A. (1990). Stimulation of p21ras upon T-cell activation. Nature 346, 719–723.

Doyle, C. and Strominger, J.L. (1987). Interaction between CD4 and class II MHC molecules mediates cell adhesion. Nature 19–25; 330, 256–259.

Dustin, M.L., Rothlein, R., Bhan, A.K., Dinarello, C.A. and Springer, T.A. (1986). Induction by IL-1 and interferon-gamma: tissue distribution, biochemistry, and function of a natural adherence molecule (ICAM-1). J. Immunol. 137, 245–254.

Dustin, M.L., Sanders, M.E., Shaw, S. and Springer, T.A. (1987). Purified lymphocyte function-associated antigen 3 binds to CD2 and mediates T lymphocyte adhesion. J. Exp. Med. 165, 677–692.

Dustin, M.L. and Springer, T.A. (1988). Lymphocyte function-associated antigen-1 (LFA-1) interaction with intercellular adhesion molecule-1 (ICAM-1) is one of at least three mechanisms for lymphocyte adhesion to cultured endothelial cells. J. Cell Biol. 107, 321–331.

Dustin, M.L. and Springer, T.A. (1989). T-cell receptor cross-linking transiently stimulates adhesiveness through LFA-1. Nature 19; 341, 619–624.

Dustin, M.L. and Springer, T.A. (1991). Role of lymphocyte adhesion receptors in transient interactions and cell locomotion. Annu. Rev. Immunol. 9, 27–66.

Emmrich, F., Kanz, L. and Eichmann, K. (1987). Crosslinking of the T-cell receptor complex with the subset-specific differentiation antigen stimulates interleukin 2 receptor expression in human CD4 and CD8 T-cells. Eur. J. Immunol. 17, 529–534.

Endler Jobst, B., Schraven, B., Hutmacher, B. and Meuer, S.C. (1991). Human T-cell responses to IL-1 and IL-6 are dependent on signals mediated through CD2. J. Immunol. 146, 1736–1742.

Faure, F., Jitsukawa, S., Meuer, S., Bohuon, C., Triebel, F. and Hercend, T. (1988). Identification of a $CD2^-$ $CD3^+$ T-cell receptor-$gamma^+$ peripheral blood lymphocyte subpopulation. J. Immunol. 140, 2128–2132.

Freeman, G.J., Gray, G.S., Gimmi, C.D., Lombard, D.B., Zhou, L.J., White, M., Fingeroth, J.D., Gribben, J.G. and Nadler, L.M. (1991). Structure, expression, and T-cell co-stimulatory activity of the murine homologue of the human B lymphocyte activation antigen B7. J. Exp. Med. 174, 625–631.

Gay, D., Maddon, P., Sekaly, R., Talle, M.A., Godfrey, M., Long, E., Goldstein, G., Chess, L., Axel, R., Kappler, J. *et al.* (1987). Functional interaction between human T-cell protein CD4 and the major histocompatibility complex HLA-DR antigen. Nature 328, 626–629.

Giblin, P., Ledbetter, J.A. and Kavathas, P. (1989). A secreted form of the human lymphocyte cell surface molecule CD8 arises from alternative splicing. Proc. Natl Acad. Sci. USA 86, 998–1002.

Glaichenhaus, N., Shastri, N., Littman, D.R. and Turner, J.M. (1991). Requirement for association of $p56^{lck}$ with CD4 in antigen-specific signal transduction in T-cells. Cell 64, 511–520.

Goldstein, L.A., Zhou, D.F., Picker, L.J., Minty, C.N., Bargatze, R.F., Ding, J.F. and Butcher, E.C. (1989). A human lymphocyte homing receptor, the hermes antigen, is related to cartilage proteoglycan core and link proteins. Cell 56, 1063–1072.

Gorman, S.D., Sun, Y.H., Zamoyska, R. and Parnes, J.R. (1988). Molecular linkage of the Ly-3 and Ly-2 genes. Requirement of Ly-2 for Ly-3 surface expression. J. Immunol. 140, 3646–3653.

Gregory, C.D., Murray, R.J., Edwards, C.F. and Rickinson, A.B. (1988). Downregulation of cell adhesion molecules LFA-3 and ICAM-1 in Epstein-Barr virus-positive Burkitt's lymphoma underlies tumor cell escape from virus-specific T cell surveillance. J. Exp. Med. 167, 1811–1824.

Greve, J.M., Davis, G., Meyer, A.M., Forte, C.P., Yost, S.C., Marlor, C.W., Kamarck, M.E. and McClelland, A. (1989). The major human rhinovirus receptor is ICAM-1. Cell 56, 839–847.

Groux, H., Huet, S., Valentin, H., Pham, D. and Bernard, A. (1989). Suppressor effects and cyclic AMP accumulation by the CD29 molecule of CD4+ lymphocytes. Nature 339, 152–154.

Guckel, B., Berek, C., Lutz, M., Altevogt, P., Schirrmacher, V. and Kyewski, B.A. (1991). Anti-CD2 antibodies induce T-cell unresponsiveness *in vivo*. J. Exp. Med. 174, 957–967.

Gunthert, U., Hofmann, M., Rudy, W., Reber, S., Zoller, M., Haussmann, I., Matzku, S., Wenzel, A., Ponta, H. and Herrlich, P. (1991). A new variant of glycoprotein CD44 confers metastatic potential to rat carcinoma cells. Cell 65, 13–24.

Hattori, R., Hamilton, K.K., Fugate, R.D., McEver, R.P. and Sims, P.J. (1989). Stimulated secretion of endothelial von Willebrand factor is accompanied by rapid redistribution to the cell surface of the intracellular granule membrane protein GMP-140. J. Biol. Chem. 264, 7768–7771.

He, Q., Beyers, A.D., Barclay, A.N. and Williams, A.F. (1988). A role in transmembrane signaling for the cytoplasmic domain of the CD2 T lymphocyte surface antigen. Cell 54, 979–984.

Hemler, M.E. (1990). VLA proteins in the integrin family: structures, functions, and their role on leukocytes. Annu. Rev. Immunol. 8, 365–400.

Hoffmann, J.C., Dengler, T.J., Knolle, P.A., Albert-Wolf, M., Roux, M., Wallich, R. and Meuer, S.C. (1993). A soluble form of the adhesion receptor CD58 (LFA-3) is present in human body fluids. Eur. J. Immunol. 23, 3003–3010.

Howard, F.D., Ledbetter, J.A., Wong, J., Bieber, C.P., Stinson, E.B., and Herzenberg, L.A. (1981) A human T-lymphocyte differentiation marker defined by monoclonal antibodies that block E-rosette formation. J. Immunol. 126, 2117–2122.

Huet, S., Wakasubi, H., Sterkers, G., Gilmour, J., Tursz, T., Boumsell, L. and Bernard A. (1986) T-cell activation via CD2

(T, gp50): the role of accessory cells in activating resting T cells via CD2. J. Immunol. 137, 1420–1428.

Huet, S., Groux, H., Caillou, B., Valentin, H., Prieur, A.M. and Bernard, A. (1989). CD44 contributes to T-cell activation. J. Immunol. 143, 798–801.

Hughes, C.C., Savage, C.O. and Pober, J.S. (1990). Endothelial cells augment T-cell interleukin 2 production by a contact-dependent mechanism involving CD2/LFA-3 interaction. J. Exp. Med. 171, 1453–1467.

Hunig, T. (1985). The cell surface molecule recognized by the erythrocyte receptor of T lymphocytes. Identification and partial characterization using a monoclonal antibody. J. Exp. Med. 162, 890–901.

Hunig, T., Tiefenthaler, G., Meyer sum Buschenfelde, K.H. and Meuer, S.C. (1987). Alternative pathway activation of T cells by binding of CD2 to its cell-surface ligand. Nature 19–25; 326, 298–301.

Hynes, R.O. (1987) Integrins: a Family of Cell Surface Receptors. Cell 48, 549–554.

Isler, P., Salvi, S., Rapin, C., Giuffre, L., Cerottini, J.C. and Carrel, S. (1988). Antigen-independent activation of T cells mediated by a novel cell surface heterodimer (Tp135-145). Eur. J. Immunol. 18, 1491–1498.

Johnson, J.P., Stade, B.G., Holzmann, B., Schwable, W. and Riethmuller, G. (1989). *De novo* expression of intercellular-adhesion molecule 1 in melanoma correlates with increased risk of metastasis. Proc. Natl Acad. Sci. USA 86, 641–644.

Johnston, G.I., Cook, R.G. and McEver, R.P. (1989). Cloning of GMP-140, a granule membrane protein of platelets and endothelium: sequence similarity to proteins involved in cell adhesion and inflammation. Cell 56, 1033–1044.

June, C.H., Ledbetter, J.A., Linsley, P.S. and Thompson, C.B. (1990). Role of the CD28 receptor in T-cell activation. Immunol. Today 11, 211–216.

Kavathas, P., Sukhatme, V.P., Herzenberg, L.A. and Parnes, J.R. (1984). Isolation of the gene encoding the human T-lymphocyte differentiation antigen Leu-2 (T8) by gene transfer and cDNA subtraction. Proc. Natl Acad. Sci. USA 81, 7688–7692.

Kishimoto, T.K., Jutila, M.A., Berg, E.L. and Butcher, E.C. (1989). Neutrophil Mac-1 and MEL-14 adhesion proteins inversely regulated by chemotactic factors. Science 245, 1238–1241.

Kishimoto, T.K., Larson, R.S., Corbi, A.L. Dustin, M.L., Staunton, D.E. and Springer, T.A. (1989) The leuocyte integrins. Adv. Immunol. 46, 149–182.

Knapp, W. (1989). "Leucocyte Typing IV White Cell Differentiation Antigens". Oxford University Press.

Koch, C.A., Anderson, D., Moran, M., Ellis, C. and Pawson, T. (1991) SH2 and SH3 domains: elements that control interactions of cytoplasmic signalling proteins. Science 252, 668–674.

Koretsky, G.A., Picus, J., Thomas, M.L. and Weiss, A. (1990). Tyrosine phosphatase CD45 is essential for coupling T-cell antigen receptor to the phosphatidyl inositol pathway. Nature 346, 66–68.

Krensky, A.M., Reiss, C.S., Mier, J.W., Strominger, J.L. and Burakoff, S.J. (1982). Long-term human cytolytic T-cell lines allospecific for HLA-DR6 antigen are OKT4$^+$. Proc. Natl Acad. Sci. USA 79, 2365–2369.

Krensky, A.M., Mentzer, S.J., Clayberger, C., Anderson, D.C., Schmalstieg, F.C., Burakoff, S.J. and Springer, T.A. (1985). Heritable lymphocyte function-associated antigen-1 deficiency: abnormalities of cytotoxicity and proliferation associated with abnormal expression of LFA-1. J. Immunol. 135, 3102–3108.

Krissansen, G.W., Elliott, M.J., Lucas, C.M., Stomski, F.C., Berndt, M.C., Cheresh, D.A., Lopez, A.F. and Burns, G.F. (1990). Identification of a novel integrin β subunit expressed on cultured monocytes (macrophages). Evidence that one alpha subunit can associate with multiple β subunits. J. Biol. Chem. 265, 823–830.

Kwong, P.D., Ryu, S.E., Hendrickson, W.A., Axel, R., Sweet, R.M., Folena Wasserman, J. Hensley, P. and Sweet, R.W. (1990). Molecular characteristics of recombinant human CD4 as deduced from polymorphic crystals. Proc. Natl Acad. Sci. USA 87, 6423–6427.

Lamarre, D., Capon, D.J., Karp, D.R., Gregory, T., Long, E.O. and Sekaly, R.P. (1989). Class II MHC molecules and the HIV gp 120 envelope protein interact with functionally distinct regions of the CD4 molecule. EMBO J. 8, 3271–3277.

Lasky, L.A., Singer, M.S., Yednock, T.A., Dowbenko, D., Fennie, C., Rodriguez, H., Nguyen, T., Stachel, S. and Rosen, S.D. (1989). Cloning of a lymphocyte homing receptor reveals a lectin domain. Cell 56, 1045–1055.

Le, P.T., Vollger, L.W., Haynes, B.F. and Singer, K.H. (1990). Ligand binding to the LFA-3 cell adhesion molecule induces IL-1 production by human thymic epithelial cells. J. Immunol. 144, 4541–4547.

Ledbetter, J.A., Martin, P.J., Spooner, C.E., Wofsy, D., Tsu, T.T., Beatty, P.G. and Gladstone, P. (1985). Antibodies to Tp67 and Tp44 augment and sustain proliferative responses of activated T cells. J. Immunol. 135, 2331–2336.

Ledbetter, J.A., Imboden, J.B., Schieven, G.L., Grosmaire, L.S., Rabinovitch, P.S., Lindsten, T., Thompson, C.B. and June, C.H. (1990). CD28 ligation in T-cell activation: evidence for two signal transduction pathways. Blood 75, 1531–1539.

Levy, J.A., Shimabukuro, J., McHugh, T., Casavant, C., Stites, D. and Oshiro, L. (1985). AIDS-associated retroviruses (ARV) can productively infect other cells besides human T helper cells. Virology 147, 4441–4484.

Linsley, P.S., Brady, W., Urnes, M., Grosmaire, L.S., Damle, N.K. and Ledbetter, J.A. (1991). CTLA-4 is a second receptor for the B cell activation antigen B7. J. Exp. Med. 174, 561–569.

Littman, D.R., Thomas, Y., Maddon, P.J., Chess, L. and Axel, R. (1985) The isolation and sequence of the gene encoding T8: a molecule defining functional classes of T lymphocytes. Cell 40, 237–246.

Littman, D.R. (1987). The structure of the CD4 and CD8 genes. Annu. Rev. Immunol. 5, 561–584.

Maddon, P.J., Dalgleish, A.G., McDougall, J.S., Clapham, P.R., Weiss, R.A. and Axel, R. (1986). The T4 gene encodes the AIDS virus receptor and is expressed in the immune system and the brain. Cell 47, 333–348.

Makgoba, M.W., Sanders, M.E., Ginther Luce, G.E., Dustin, M.L., Springer, T.A., Clark, E.A., Mannoni, P. and Shaw, S. (1988). ICAM-1 a ligand for LFA-1-dependent adhesion of B, T and myeloid cells. Nature 331, 86–88.

Marlin, S.D., Staunton, D.E., Springer, T.A., Stratowa, C., Sommergruber, W. and Merluzzi, V.J. (1990). A soluble form of intercellular adhesion molecule-1 inhibits rhinovirus infection. Nature 344: 70.

Matsuyama, T., Yamada, A., Kay, J., Yamada, K.M., Akiyama, S.K., Schlossman, S.F. and Morimoto, C. (1989). Activation of CD4 cells by fibronectin and anti-CD3 antibody. A synergistic effect mediated by the VLA-5 fibronectin receptor complex. J. Exp. Med. 170, 1133–1148.

Meuer, S.C., Schlossman, S.F. and Reinherz, E.L. (1982). Clonal analysis of human cytotoxic T lymphocytes: T4+ and T8+ effector T cells recognize products of different major histocompatibility complex regions. Proc. Natl Acad. Sci. USA 79, 4395–4399.

Meuer, S.C., Hodgdon, J.C., Hussey, R.E., Protentis, J.P., Schlossman, S.F. and Reinherz, E.L. (1983). Antigen-like effects of monoclonal antibodies directed at receptors on human T-cell clones. J. Exp. Med. 158, 988–993.

Meuer, S.C., Acuto, O., Hercend, T., Schlossman, S.F. and Reinherz, E.L. (1984). The human T-cell receptor. Annu. Rev. Immunol. 2, 23–50.

Meuer, S.C., Hussey, R.E., Fabbi, M., Fox, D., Acuto, O., Fitzgerald, K.A., Hodgdon, J.C., Protentis, J.P., Schlossman, S.F. and Reinherz, E.L..(1984). An alternative pathway of T-cell activation: a functional role for the 50 kDa T11 sheep erythrocyte receptor protein. Cell 36, 897–906.

Meuer, S.C., Hauer, M., Kurz, P., Meyer zum Buschenfelde, K.H. and Kohler, H. (1987). Selective blockade of the antigen-receptor-mediated pathway of T-cell activation in patients with impaired primary immune responses. J. Clin. Invest. 80, 743–749.

Meuer, S.C., Roux, M.M. and Schraven, B. (1989). The alternative pathway of T-cell activation: biology, pathophysiology, and perspectives for immunopharmacology. Clin. Immunol. Immunopathol. 50, S133–S139.

Miceli, M.C., von Hoegen, P. and Parnes, J.R. (1991). Adhesion versus coreceptor function of CD4 and CD8: role of the cytoplasmic tail in coreceptor activity. Proc. Natl Acad. Sci. USA 88, 2623–2627.

Miyake, K., Underhill, C.B., Lesley, J. and Kincade, P.W. (1990). Hyaluronate can function as a cell adhesion molecule and CD44 participates in hyaluronate recognition. J. Exp. Med. 172, 69–75.

Moebius, U., Kober, G., Griscelli, A.L., Hercend, T. and Meuer, S.C. (1991). Expression of different CD8 isoforms on distinct human lymphocyte subpopulations. Eur. J. Immunol. 21, 1793–1800.

Moingeon, P., Chang, H.C., Wallner, B.P., Stebbins, C., Frey, A.Z. and Reinherz, E.L. (1989). CD2-mediated adhesion facilitates T lymphocyte antigen recognition function. Nature 339, 312–314.

Moingeon, P., Chang, H.C., Sayre, P.H., Clayton, L.K., Alcover, A., Gardner, P. and Reinherz, E.L. (1989). The structural biology of CD2. Immunol. Rev. 111, 111–144.

Morimoto, C., Letvin, N.L., Boyd, A.W., Hagan, M., Brown, H.M., Kornacki, M.M. and Schlossman, S.F. (1985). The isolation and characterization of the human helper inducer T cell subset. J. Immunol. 134, 3762–3769.

Mueller, D.L. Jenkins, M.K., and Schwartz, R.H. (1989). Clonal expansion versus functional clonal inactivation: a co-stimulatory signalling pathway determines the outcome of T cell antigen receptor occupancy. Annu. Rev. Immunol. 7, 445–480.

Nickoloff, B.J., Griffiths, C.E., Baadsgaard, O., Voorhees, J.J., Hanson, C.A. and Cooper, K.D. (1989). Markedly diminished epidermal keratinocyte expression of intercellular adhesion molecule-1 (ICAM-1) in Sezary syndrome. JAMA 261, 2217–2221.

Norment, A.M. and Littman, D.R. (1988). A second subunit of CD8 is expressed in human T cells. EMBO J. 7, 3433–3439.

Peterson, A. and Seed, B. (1987). Monoclonal antibody and ligand binding sites of the T-cell erythrocyte receptor (CD2). Nature 329, 842–846.

Picker, L.J., Warnock, R.A., Burns, A.R., Doerschuk, C.M., Berg, E.L. and Butcher, E.C. (1991). The neutrophil selectin LECAM-1 presents carbohydrate ligands to the vascular selectins ELAM-1 and GMP-140. Cell 66, 921–933.

Pirzer, U.C., Schrmann, G., Post, S., Betzler, M., and Meuer, S.C. (1990) Differential responsiveness to CD3-Ti versus CD2-dependent activation of human intestinal T lymphocytes. Eur. J. Immunol. 20, 2339–2342.

Phillips, M.L., Nudelman, E., Gaeta, F.C., Perez, M., Singhal, A.K., Hakomori, S. and Paulson, J.C. (1990). ELAM-1 mediates cell adhesion by recognition of a carbohydrate ligand, sialyl-Lex. Science 250, 1130–1132.

Plunkett, M.L., Sanders, M.E., Selvaraj, P., Dustin, M.L. and Springer, T.A. (1987). Rosetting of activated human T lymphocytes with autologous erythrocytes. Definition of the receptor and ligand molecules as CD2 and lymphocyte function-associated antigen 3 (LFA-3). J. Exp. Med. 165, 664–676.

Qiao, L., Schurmann, G., Betzler, M. and Meuer, S.C. (1991). Down-regulation of protein kinase C activation in human lamina propria T lymphocytes: influence of intestinal mucosa on T-cell reactivity. Eur. J. Immunol. 21, 2385–2389.

Qiao, L., Schrmann, G., Betzler, M., and Meuer, S.C. (1991) Activation and Signalling Status of Human Lamina Propria T Lymphocytes. Gastroent. 101, 1529–1536.

Reinherz, E.L. and Schlossman, S.F. (1980) The differentiation and function of human T lymphocytes. Cell 19, 621–827.

Reinherz, E.L., Meuer, S., Fitzgerald, K.A., Hussey, R.E., Levine, H. and Schlossman, S.F. (1982). Antigen recognition by human T lymphocytes is linked to surface expression of the T3 molecular complex. Cell 30, 735–743.

Rice, G.E. and Bevilacqua, M.P. (1989). An inducible endothelial cell surface glycoprotein mediates melanoma adhesion. Science 246, 1303–1306.

Rothlein, R., Dustin, M.L., Marlin, S.D. and Springer, T.A. (1986). A human intercellular adhesion molecule (ICAM-1) distinct from LFA-1. J. Immunol. 137, 1270–1274.

Rothlein, R., Mainolfi, E.A., Czajkowski, M. and Marlin, S.D. (1991). A form of circulating ICAM-1 in human serum. J. Immunol. 147, 3788–3793.

Rousset, F., Billaud, M., Blanchard, D., Figdor, C., Lenoir, G.M., Spits, H. and De Vries, J.E. (1989). IL-4 induces LFA-1 and LFA-3 expression on Burkitt's lymphoma cell lines. Requirement of additional activation by phorbol myristate acetate for induction of homotypic cell adhesions. J. Immunol. 143, 1490–1498.

Roux, M., Schraven, B., Roux, A., Gamm, H., Mertelsmann, R. and Meuer, S. (1991). Natural inhibitors of T-cell activation in Hodgkin's disease. Blood 78, 2365–2371.

Rudd, C.E. (1990) CD4, CD8 and the TcR-CD3 complex: a novel class of protein-tyrosine kinase receptor. Immunol. Today 11, 400–406.

Saito, H. and Streuli, M. (1991). Molecular characterization of protein tyrosine phosphatases. Cell Growth and Differentiation 2, 59–65.

Saizawa, K., Rojo, J. and Janeway, C.A., Jr. (1987). Evidence for a physical association of CD4 and the CD3:alpha: beta T-cell receptor. Nature 328, 260–263.

Salter, R.D., Norment, A.M., Chen, B.P., Clayberger, C., Krensky, A.M., Littman, D.R. and Parham, P. (1989). Polymorphism in the alpha 3 domain of HLA-A molecules affects binding to CD8. Nature 338, 345–347.

Samstag, Y., Emmrich, F. and Staehelin, T. (1988). Activation of human T lymphocytes: differential effects of CD3- and CD8-mediated signals [published erratum appears in Proc. Natl Acad. Sci. USA 1990 Jan;87(2):856. Proc. Natl Acad. Sci. USA 85, 9689–9693.

Samstag, Y., Bader, A. and Meuer, S.C. (1991). A serine phosphatase is involved in CD2-mediated activation of human T lymphocytes and natural killer cells. J. Immunol. 147, 788–794.

Sanchez Madrid, F., Krensky, A.M., Ware, C.F., Robbins, E., Strominger, J.L., Burakoff, S.J. and Springer, T.A. (1982). Three distinct antigens associated with human T-lymphocyte-mediated cytolysis: LFA-1, LFA-2, and LFA-3. Proc. Natl Acad. Sci. USA 79, 7489–7493.

Sanders, M.E., Makgoba, M.W., Sharrow, S.O., Stephany, D., Springer, T.A., Young, H.A. and Shaw, S. (1988). Human memory T lymphocytes express increased levels of three cell adhesion molecules (LFA-3, CD2, and LFA-1) and three other molecules (UCHL1, CDw29, and Pgp-1) and have enhanced IFN-gamma production. J. Immunol. 140, 1401–1407.

Savill, J., Dransfield, I., Hogg, N. and Haslett, C. (1990). Vitronectin receptor-mediated phagocytosis of cells undergoing apoptosis. Nature 343, 170–173.

Schad, V.C. and Greenstein J.L. (1989) T-cell accessory molecules mediating cell adhesion and signal transduction. Curr. Op. Immunol. 2, 123–128.

Schirren, A., Vlpel, H. and Meuer, S.C. (1992). Adhesion molecules on freshly recovered T leukemias promote tumor-directed lympholysis. Blood 79, 138–143.

Schraven, B., Roux, M., Hutmacher, B. and Meuer, S.C. (1989). Triggering of the alternative pathway of human T-cell activation involves members of the T 200 family of glycoproteins. Eur. J. Immunol. 19, 397–403.

Schraven, B., Samstag, Y., Altevogt, P. and Meuer, S.C. (1990). Association of CD2 and CD45 on human T lymphocytes. Nature 345, 71–74.

Schraven, B., Kirchgessner, H., Gaber, B., Samstag, Y. and Meuer, S. (1991). A functional complex is formed in human T lymphocytes between the protein tyrosine phosphatase CD45, the protein tyrosine kinase p56lck and pp32, a possible common substrate. Eur. J. Immunol. 21, 2469–2477.

Schraven, B., Schirren, A., Kirchgessner, H., Siebert, B. and Meuer, S.C. (1992). Four CD45/P56^{lck}-associated phosphoproteins (pp29–pp32) undergo alterations in human T-cell activation. Eur. J. Immunol. 22, 1857–1863.

Schraven, B., Schirren, A., Kirchgessner, H., Siebert, B., Wallich, R., Henning, S., Samstag, Y. and Meuer, S.C. (1993). Alterations of CD2 association with T cell receptor signalling molecules in 'CD2 unresponsive' human T lymphocytes. Eur. J. Immunol. 23, 119–123.

Scott, C.F., Jr., Bolender, S., McIntyre, G.D., Holldack, J., Lambert, J.M., Venkatesh, Y.P., Morimoto, C., Ritz, J. and Schlossman, S.F. (1989). Activation of human cytolytic cells through CD2/T11. Comparison of the requirements for the induction and direction of lysis of tumor targets by T cells and NK cells. J. Immunol. 142, 4105–4112.

Seed, B. (1987). An LFA-3 cDNA encodes a phospholipid-linked membrane protein homologous to its receptor CD2. Nature 329, 840–842.

Selvaraj, P., Dustin, M.L., Silber, R., Low, M.G. and Springer, T.A. (1987). Deficiency of lymphocyte function-associated antigen 3 (LFA-3) in paroxysmal nocturnal hemoglobinuria. Functional correlates and evidence for a phosphatidylinositol membrane anchor. J. Exp. Med. 166, 1011–1025.

Selvaraj, P., Plunkett, M.L., Dustin, M., Sanders, M.E., Shaw, S. and Springer, T.A. (1987). The T lymphocyte glycoprotein CD2 binds the cell surface ligand LFA-3. Nature 326, 400–403.

Seth, A., Gote, L., Nagarkatti, M. and Nagarkatti, P.S. (1991). T-cell-receptor-independent activation of cytolytic activity of cytotoxic T lymphocytes mediated through CD44 and gp90MEL-14. Proc. Natl Acad. Sci. USA 88, 7877–7881.

Shaw, A.S., Amrein, K.E., Hammond, C., Stern, D.F., Sefton, B.M. and Rose, J.K. (1939). The lck tyrosine protein kinase interacts with the cytoplasmic tail of the CD4 glycoprotein through its unique amino-terminal domain. Cell 59, 627–636.

Shiue, L., Gorman, S.D. and Parnes, J.R. (1988). A second chain of human CD8 is expressed on peripheral blood lymphocytes. J. Exp. Med. 168, 1993–2005.

Siliciano, R.F., Pratt, J.C., Schmidt, R.E., Ritz, J. and Reinherz, E.L. (1985). Activation of cytolytic T lymphocyte and natural killer cell function through the T11 sheep erythrocyte binding protein. Nature 317, 428–430.

Simmons, D., Makgoba, M.W. and Seed, B. (1988). ICAM, an adhesion ligand of LFA-1, is homologous to the neural cell adhesion molecule NCAM. Nature 331, 624–627.

Smith, C.W., Marlin, S.D., Rothlein, R., Toman, C. and Anderson, D.C. (1989). Co-operative interactions of LFA-1 and Mac-1 with intercellular adhesion molecule-1 in facilitating adherence and transendothelial migration of human neutrophils *in vitro*. J. Clin. Invest. 83, 2008–2017.

Snow, P.M. and Terhorst, C. (1983). The T8 antigen is a multimeric complex of two distinct subunits on human thymocytes but consists of homomultimeric forms on peripheral blood T lymphocytes. J. Biol. Chem. 258, 14 675–14 681.

Springer, T.A. (1990). Adhesion receptors of the immune system. Nature 346, 425–434.

Stamenkovic, I., Aruffo, A., Amiot, M. and Seed, B. (1991). The hematopoietic and epithelial forms of CD44 are distinct polypeptides with different adhesion potentials for hyaluronate-bearing cells. EMBO J. 10, 343–348.

Staunton, D.E., Marlin, S.D., Stratowa, C., Dustin, M.L. and Springer, T.A. (1988). Primary structure of ICAM-1 demonstrates interaction between members of the immunoglobulin and integrin supergene families. Cell 52, 925–933.

Staunton, D.E., Merluzzi, V.J., Rothlein, R., Barton, R., Marlin, S.D. and Springer, T.A. (1989). A cell adhesion molecule, ICAM-1, is the major surface receptor for rhinoviruses. Cell 56, 849–853.

Staunton, D.E., Dustin, M.L., Erickson, H.P. and Springer, T.A. (1990). The arrangement of the immunoglobulin-like domains of ICAM-1 and the binding sites for LFA-1 and rhinovirus [published erratum appears in Cell 1990 Jun 15;61(2):1157]. Cell 61, 243–254.

Stefanova, I., Horejsi, V., Ansotegui, I.J., Knapp, W., Stockinger, H. (1991) GPI-anchored cell surface molecules complexed to protein tyrosine kinases. Science 254, 1016–1019.

Sukhatme, V.P., Sizer, K.C., Vollmer, A.C., Hunkapiller, T., Parnes, J.R. (1985). The T-cell differentiation antigen Leu-2/T8 is homologous to immunoglobulin and T-cell receptor variable regions. Cell 40, 591–597.

Takai, Y., Reed, M.L., Burakoff, S.J. and Herrmann, S.H. (1987). Direct evidence for a receptor-ligand interaction between the T-cell surface antigen CD2 and lymphocyte-function-associated antigen 3. Proc. Natl Acad. Sci. USA 84, 6864–6868.

Tedder, T.F., Penta, A.C., Levine, H.B. and Freedman, A.S. (1990). Expression of the human leukocyte adhesion molecule, LAM1. Identity with the TQ1 and Leu-8 differentiation antigens. J. Immunol. 144, 532–540.

Thompson, C.B., Lindsten, T., Ledbetter, J.A., Kunkel, S.L., Young, H.A., Emerson, S.G., Leiden, J.M. and June, C.H. (1989). CD28 activation pathway regulates the production of multiple T-cell-derived lymphokines/cytokines. Proc. Natl Acad. Sci. USA 86, 1333–1337.

Tiefenthaler, G., Hunig, T., Dustin, M.L., Springer, T.A. and Meuer, S.C. (1987). Purified lymphocyte function-associated antigen-3 and T11 target structure are active in CD2-mediated T-cell stimulation. Eur. J. Immunol. 17, 1847–1850.

Triebel, F., Jitsukawa, S., Baixeras, E., Roman, S., Genevee, C., Viegas Pequignot, E. and Hercend, T. (1990). LAG-3, a novel lymphocyte activation gene closely related to CD4. J. Exp. Med. 171, 1393–1405.

Valent, P., Bevec, D., Maurer, D., Besemer, J., Di Padova, F., Butterfield, J.H., Speiser, W., Majdic, O., Lechner, K. and Bettelheim, P. (1991). Interleukin 4 promotes expression of mast cell ICAM-1 antigen. Proc. Natl Acad. Sci. USA 88, 3339–3342.

Van Seventer, G.A., Newman, W., Shimizu, Y., Nutman, T.B., Tanaka, Y., Horgan, K.J., Gopal, T.V., Ennis, E., O'Sullivan, D., Grey, H. and Shaw, S. (1991). Analysis of T-cell stimulation by superantigen plus major histocompatibility complex class II molecules or by CD3 monoclonal antibody co-stimulation by purified adhesion ligands VCAM-1, ICAM-1, but not ELAM-1. J. Exp. Med. 1174, 901–913.

Veillette, A., Bookman, M.A., Horak, E.M., Samelson, L.E. and Bolen, J.B. (1989). Signal transduction through the CD4 receptor involves the activation of the internal membrane tyrosine-protein kinase $p56^{lck}$. Nature 338, 257–259.

Volpes, R., van den Oord, J.J., and Desmet V.J. (1990) Immunohistochemical study of adhesion molecules in liver inflammation. Hepatology 12, 59–65.

von Andrian, U.H., Chambers, J.D., McEvoy, L.M., Bargatze, R.F., Arfors, K.E. and Butcher, E.C. (1991). Two-step model of leukocyte-endothelial cell interaction in inflammation: distinct roles for LECAM-1 and the leukocyte beta 2 integrins *in vivo*. Proc. Natl Acad. Sci. USA 88, 7538–7542.

Waldmann, H. (1989). Manipulation of T Cell Responses with Monoclonal Antibodies. Annu. Rev. Immunol. 7, 407–444.

Wallner, B.P., Frey, A.Z., Tizard, R., Mattaliano, R.J., Hession, C., Sanders, M.E., Dustin, M.L. and Springer, T.A. (1987). Primary structure of lymphocyte function-associated antigen 3 (LFA-3). The ligand of the T lymphocyte CD2 glycoprotein. J. Exp. Med. 166, 923–932.

Walz, G., Aruffo, A., Kolanus, W., Bevilacqua, M. and Seed, B. (1990). Recognition by ELAM-1 of the sialyl-Lex determinant on myeloid and tumor cells. Science 250, 1132–1135.

Wang, J.H., Yan, Y.W., Garrett, T.P., Liu, J.H., Rodgers, D.W., Garlick, R.L., Tarr, G.E., Husain, Y., Reinherz, E.L. and Harrison, S.C. (1990). Atomic structure of a fragment of human CD4 containing two immunoglobulin-like domains. Nature 348, 411–418.

Webb, D.S., Shimizu, Y., Van Seventer, G.A., Shaw, S. and Gerrard, T.L. (1990). LFA-3, CD44, and CD45: physiologic triggers of human monocyte TNF and IL-1 release. Science 249, 1295–1297.

Weyand, C.M., Goronzy, J. and Fathman, C.G. (1987). Modulation of CD4 by antigenic activation. J. Immunol. 138, 1351–1354.

Williams, A.F. and Barclay, A.N. (1988). The immunoglobulin superfamily-domains for cell surface recognition. Annu. Rev. Immunol. 6, 381–405.

Yagita, H., Nakamura, T., Karasuyama, H., and Okumura, K. (1989) Monoclonal antibodies specific for murine CD2 reveal its presence on B as well as T-cells. Proc. Natl Acad. Sci. USA 86, 645.

Yang, S.Y., Chouaib, S. and Dupont, B. (1986). A common pathway for T-lymphocyte activation involving both the CD3-Ti complex and CD2 sheep erythrocyte receptor determinants. J. Immunol. 137, 1097–1100.

3. Acute Inflammatory Response: Granulocyte Migration and Activation

Scott B. Shappell and C. Wayne Smith

1. Introduction

A major component of the acute inflammatory process is migration of circulating neutrophils from the vascular compartment through the endothelium and perivascular structures into tissues, the steps of which have been thoroughly documented morphologically in a variety of models *in vivo* (Marchesi and Gowans, 1964; Marchesi, 1964; Grant, 1974; Hurley, 1963; Shaw, 1980). This review will address the contributions of specific adhesion molecules to these processes and the activation of mechanisms used by neutrophils to damage tissue.

Adhesion Molecules
ISBN 0-12-741440-1

2. *Adhesion Molecules and Neutrophil Margination at Inflammatory Sites*

Intravital microscopy has clearly revealed that neutrophils can roll along the vessel wall in post-capillary venules (Atherton and Born, 1972; Atherton and Born, 1973; Ley *et al.*, 1989; Fiebig *et al.*, 1991; Ley *et al.*, 1991a), and that this phenomenon is markedly increased in the initial stages of acute inflammation (House and Lipowsky, 1987; Zimmerman and Granger, 1990; Kubes *et al.*, 1990; Hernandez *et al.*, 1987). The rolling cells often stop, change shape and emigrate into the surrounding tissue. Initially, it was Arfors *et al.*, 1987 who pointed out the likelihood that the molecules involved in the rolling phenomenon were different from those involved in the stopping and transmigration phenomena. In a study on the effects of anti-CD18 monoclonal antibody 60.3, in a rabbit model of inflammation, systemic administration of this antibody almost completely inhibited stationary adhesion and emigration of neutrophils, but the rolling phenomenon still occurred. Smith *et al.*, (1989a) and Lawrence *et al.*, (1990) used a parallel plate flow chamber to assess the ability of isolated neutrophils to adhere while flowing past a confluent monolayer of human umbilical vein endothelial cells (HUVEC). Without stimulation, few neutrophils were found to interact with the endothelial monolayer, but when the endothelial cells were stimulated with IL-1 for 3–4 h prior to the flow experiment, a significant increase in adhesion was seen at wall shear stresses below 3 dynes/cm^2. Consistent with observations *in vivo*, two types of adhesion were observed, rolling and stationary. At a wall shear stress of approximately 2 dynes/cm^2, rolling cells were usually moving at $\leq 10\ \mu$m/sec, in contrast to cells that did not interact with the endothelial surface where the rate of movement was approximately 2000 μm/sec. Stationary cells often changed shape and migrated beneath the monolayer as previously seen (Smith *et al.*, 1988). Abbassi *et al.* (1991) extended these observations using canine neutrophils and canine jugular vein endothelial cell monolayers. In both the Lawrence *et al.* and Abbassi *et al.* studies, anti-CD18 monoclonal antibodies (mAbs) were without effect on the number of neutrophils associating with the cytokine-stimulated monolayers at wall shear stresses around 2 dynes/cm^2, but there was a pronounced effect on the behaviour of the neutrophils. The rate of rolling was significantly increased, trans-endothelial migration was markedly reduced, and few neutrophils stopped on the apical surface of the endothelium. It was found by Lawrence *et al.* (1990) and Smith *et al.* (1989a) that CD18-deficient neutrophils associated with the monolayers as well as normal cells, but their rate of rolling was increased and transmigration did not occur. Monoclonal antibody R6.5, recognizing a functional epitope on intercellular adhesion molecule-1 (ICAM-1), a ligand for CD18 integrins expressed on cytokine-activated endothelial cells (Smith *et al.*, 1988), also did not reduce the number of monolayer-associated neutrophils. Perry and Granger (1991) demonstrated the relevance of these issues to conditions *in vivo* by showing that the rolling velocity of neutrophils in the cat mesenteric venules was increased by administration of anti-CD18 monoclonal antibody IB4. These studies indicate that both *in vitro* and *in vivo*, the adherence mechanisms of CD18–ICAM-1 do not account for the initial rolling (i.e. margination), but they apparently play critical roles in events subsequent to margination. Nonetheless, the phenomenon of rolling is a form of neutrophil–endothelial adherence, and recent experiments have shown that members of the selectin family of adhesion molecules are critical to this event.

2.1 L-SELECTIN

This member of the selectin family, also known as leucocyte adhesion molecule-1 (LAM-1) (Tedder *et al.*, 1990) was originally identified in the mouse mAb MEL-14, and numerous mAbs are now available against the human homologue (Kishimoto *et al.*, 1990; Spertini *et al.*, 1991). On lymphocytes, L-selectin functions as the peripheral lymph node homing receptor and binds through its lectin domain to a carbohydrate-containing ligand on high endothelial venules in peripheral lymph nodes (Yednock and Rosen, 1989). Smith *et al.* (1991c) and Abbassi *et al.* (1991) found that anti-L-selectin mAb would reduce by greater than 60% the number of neutrophils associating with cytokine-stimulated endothelial monolayers under conditions of flow described above (Lawrence *et al.*, 1990; Smith *et al.*, 1989a). Additionally, antibodies to L-selectin did not reduce the ability of adherent neutrophils to migrate through cytokine-stimulated endothelial monolayers *in vitro* (Smith *et al.*, 1991c; Abbassi *et al.*, 1991; Anderson *et al.*, 1991), suggesting that the principal role for this glycoprotein on neutrophils is in margination. The topography of L-selectin on the neutrophil may be critical to this function. A recent report shows L-selectin to be concentrated on the tips of membrane folds extending from the surface of unstimulated neutrophils (Picker *et al.*, 1991b), sites on the cell surface most likely to initially contact the apical surface of the endothelial cells. In contrast, CD18 appeared on flat areas of the cell surface as well as on the surface of these projections, and was abundant on the membranes of cytoplasmic granules (Bainton *et al.*, 1987).

Von Andrian *et al.* (1991) reported that systemic administration of anti-L-selectin mAb, DREG-200, in rabbits greatly reduced the number of rolling cells *in vivo*, and Ley *et al.* (1991b) reported that systemic administration in rats of a soluble recombinant form of L-selectin, or a polyclonal anti-L-selectin antiserum, also reduced by greater than 80% the number of rolling cells *in vivo*. The soluble form of L-selectin presumably blocked a ligand on activated endothelial cells. Thus, it appears that both *in vitro* and *in vivo*, the constitutively expressed neutrophil glycoprotein, L-selectin, interacts with a newly up-regulated structure on the surface of endothelial cells, and accounts for a significant portion of the initial margination. Studies in the flow chamber *in vitro* (Smith *et al.*, 1991c; Abbassi *et al.*, 1991; Anderson *et al.*, 1991) demonstrated that reducing the number of neutrophils marginating on endothelial monolayers (e.g. with anti-L-selectin mAb) reduced trans-endothelial migration simply by limiting the number of neutrophils available to migrate. This mechanism may explain the anti-inflammatory effects of anti-L-selectin monoclonal antibody, MEL-14, seen by Lewinsohn *et al.* (1987) and Jutila *et al.* (1989) in mice, and the anti-inflammatory effects of the soluble L-selectin (Watson *et al.*, 1991).

L-selectin-dependent neutrophil-endothelial adhesion is lost following chemotactic stimulation of the neutrophil (Hallmann *et al.*, 1991; Smith *et al.*, 1991c). This can be most clearly seen under conditions of flow where the contributions of the newly activated CD18 heterodimers do not obscure the loss of L-selectin-dependent adherence (Smith *et al.*, 1991c; Abbassi *et al.*, 1991; Anderson *et al.*, 1991). The most likely explanation for this phenomenon is that stimulation results in shedding of L-selectin from the cell surface (Jutila *et al.*, 1989; Kishimoto *et al.*, 1989a; Griffin *et al.*, 1990; Jutila *et al.*, 1990; Smith *et al.*, 1991c; Abbassi *et al.*, 1991). There is a high degree of correlation between the amount of L-selectin on the neutrophil surface and adhesion to cytokine-stimulated endothelial monolayers under conditions of flow *in vitro* (Smith *et al.*, 1991c; Anderson *et al.*, 1991). Jutila *et al.* (1989) demonstrated that isolated murine neutrophils had reduced ability to migrate into thioglycollate-induced peritonitis if they had been activated *in vitro* to shed L-selectin. When activated cells were labelled and returned to the animal intravenously, a high percentage continued to circulate throughout the time course of the inflammation, and few localized in the inflamed peritoneum. Thus, the loss of L-selectin function (either by shedding or inhibition with mAb) may reduce the ability of neutrophils to localize at sites of inflammation, apparently by reducing margination on inflamed endothelium.

2.2 E-SELECTIN

Margination *in vitro* does not occur at venous wall shear stresses unless the endothelial cells have been stimulated (Lawrence *et al.*, 1990). Conditions that up-regulate E-selectin cause pronounced margination, thus supporting speculation that E-selectin could play an important role. However, anti-L-selectin mAbs produce between 60–80% inhibition of neutrophil margination both *in vitro* and *in vivo* (Smith *et al.*, 1991c; Anderson *et al.*, 1991; von Andrian *et al.*, 1991), and ⩾60% inhibition of neutrophil accumulation in peritoneal inflammation, leaving the impression that E-selectin may not be sufficient to sustain but a minor part, if any, of the margination phenomenon. However, E-selectin may function as a necessary accessory molecule along with L-selectin. Recent studies by Mulligan *et al.* (1991) have shown that anti-E-selectin mAb, CL3 (originally made against human E-selectin (Kishimoto *et al.*, 1991; Picker *et al.*, 1991a)) cross-reacts with E-selectin in the rat. When administered systemically, CL3 significantly inhibited neutrophil influx into glycogen-induced peritonitis, and immune complex-induced inflammation in the skin and lungs. Gundel *et al.* (1991) found that anti-human E-selectin mAb CL2 (Kishimoto *et al.*, 1991; Picker *et al.*, 1991a) was cross-reactive with cynomolgus E-selectin. Monkeys sensitive to *Ascaris suum* extract expressed E-selectin on lung vascular endothelium within 6 h following a single inhalation exposure to the antigen. This correlated with accumulation of neutrophils in the lungs, and late-phase airway obstruction. Systemic administration of CL2 blocked both influx of neutrophils and the airway obstruction. Though these observations do not reveal whether E-selectin is primarily involved in margination or transmigration, they indicate an important role for E-selectin in acute inflammation at these sites.

Recent work supports the idea that E-selectin is involved in the same adherence pathway as neutrophil L-selectin. Kishimoto *et al.* (1991) initially observed that the inhibitory effects of anti-L-selectin and anti-E-selectin mAbs were not additive, but each produced additive inhibition when combined with anti-ICAM-1. Neither anti-L-selectin nor anti-E-selectin were inhibitory when neutrophil adhesion to endothelial monolayers was primarily determined by ICAM-1, i.e. at more than 8 h after stimulation of the endothelial cells with IL-1 (Smith *et al.*, 1988; Luscinskas *et al.*, 1991). Chemotactic stimulation of neutrophils under conditions shown to reduce neutrophil L-selectin to very low levels, markedly inhibited E-selectin-dependent neutrophil adhesion. Anti-L-selectin mAbs significantly inhibited neutrophil adhesion to COS (fibroblast-like kidney cell-line established from simian cells) and L cells transfected with E-selectin. Since these studies were carried out under static conditions, more recent work has assessed neutrophil adhesion under conditions of flow (Abbassi, O., Kishimoto, T.K., McIntyre, L.V. and Smith, C.W., Baylor College of Medicine, Houston, unpublished observations). Using E-selectin transfected L-cell monolayers in the flow chamber as previously described (Smith *et al.*, 1991c), neutrophils were found to exhibit a rolling type adhesion that was almost completely inhibited by anti-E-selectin mAb, CL3, and was inhibited by greater than 65% by anti-L-selectin antibodies. Chemotactic stimulation shown to markedly deplete neutrophil surface L-selectin reduced adhesion to the same extent as anti-L-selectin.

The hypothesis that E-selectin and neutrophil L-selectin may utilize the same adhesion pathway during margination was strengthened by the recent studies of Picker *et al.* (1991b). They found that the carbohydrate moiety recognized by the pectin domain of E-selectin, sialyl Lewis X (Lowe *et al.*, 1990; Phillips *et al.*, 1990; Walz *et al.*, 1990), is present on L-selectin isolated from neutrophils, but not on L-selectin isolated from lymphocytes. Additionally, mouse pre-B cells, L1-2, transfected with E-selectin adhered to surfaces with adsorbed neutrophil L-selectin but not surfaces with lymphocyte L-selectin. This adherence was inhibited by antibodies to E-selectin, sialyl Lewis X or L-selectin. Thus, it appears that neutrophil L-selectin is modified with E-selectin-binding oligosaccharide structures that are not present on lymphocyte L-selectin, and these structures support E-selectin-dependent neutrophil margination on IL-1-stimulated endothelial cells under conditions of flow.

2.3 P-SELECTIN (CD62)

A third member of the selectin family appears to be involved in margination and possibly activation of the neutrophil at the endothelial surface. P-selectin is constitutively expressed in normal, non-inflamed venular endothelium (McEver *et al.*, 1989). It is found within the Weibel–Palade bodies (Hattori *et al.*, 1989; McEver *et al.*, 1989; Bonfanti *et al.*, 1989), and is rapidly translocated to the apical surface of the cell following stimulation with such agents as thrombin or histamine (Hattori *et al.*, 1989; McEver *et al.*, 1989; Geng *et al.*, 1990). Following stimulation with these agents, P-selectin remains on the cell surface for less than 1 h (Hattori *et al.*, 1989). Patel *et al.* (1991) found that its stay was markedly prolonged when low level oxidants were the stimuli for translocation from the Weibel–Palade bodies. Geng *et al.* (1990) and Patel *et al.* (1991) demonstrated that P-selectin can under static conditions attach neutrophils to endothelial cells. Lawrence and Springer (1991) demonstrated that P-selectin in artificial planar membranes will cause neutrophils to roll under conditions of flow at wall shear stresses similar to those demonstrated by Smith *et al.* (1991c) and Abbassi *et al.* (1991), to cause L-selectin-dependent rolling on endothelial cells. Thus, it appears that each member of the selectin family has a high enough rate of reaction with its carbohydrate epitope to catch neutrophils as they flow by the endothelium.

In addition to the possible role of P-selectin in promoting margination, Lorant *et al.* (1991) have obtained evidence that binding of P-selectin with its ligand on neutrophils potentiates CD11/CD18 adhesive responses stimulated by agonists such as chemotactic factors. This would result in strengthening of the adhesion, and possibly promote more effective leucocyte migration.

3. *Three Stages in the Process of Leucocyte Extravasation*

Fig. 3.1 illustrates schematically, stages in the extravasation of neutrophils from post-capillary venules in which adhesion molecules appear to play important roles. The rolling adhesion often seen early during inflammation, as discussed above, involves members of the selectin family. P-selectin may be in position to catch previously unstimulated neutrophils at very early stages since it is constitutively expressed and rapidly mobilized. As a

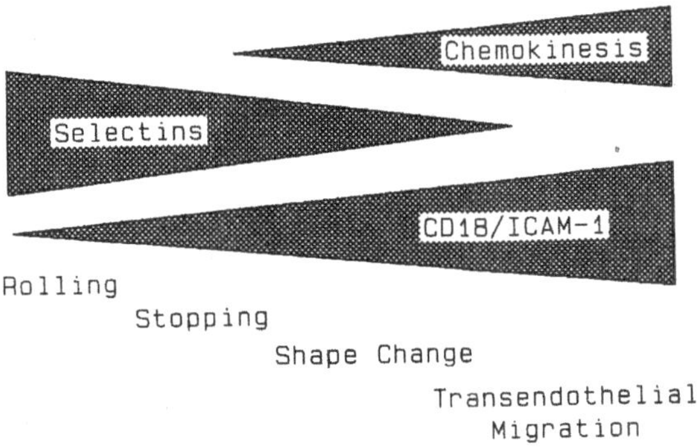

Figure 3.1 Schematic representation of the most likely contributions of adhesion molecules to the emigration of neutrophils in post-capillary venules at sites of inflammation. The triangles represent the relative importance of the selectins, CD18/ICAM-1 and chemokinetic agents. Each member of the selectin family has been shown to support rolling adhesion of previously unactivated neutrophils, and the selectins are therefore thought to play a necessary and predominant role in the earliest stages of neutrophil localization. Their contribution is subsequently limited, though, by an increasing contribution of the CD18 integrins (CD11a/CD18 and CD11b/CD18) and ICAM-1 acting as accessory adhesion molecules required for the stopping of rolling leucocytes. Shape change of the stationary neutrophils is necessary for cell locomotion and is thought to be stimulated by chemotactic factors acting at the interface of the neutrophil and the endothelial cell. Such factors as IL-8 and PAF are thought to be produced by the endothelial cells. They act as chemokinetic stimuli that not only promote cell shape change, but they bring about a marked alteration in the relative contributions of L-selectin and CD18 integrins – L-selectin is shed and the function of CD18 integrins is markedly augmented. It is now clear that CD18 integrins are necessary for effective trans-endothelial migration under most circumstances. This model predicts that blocking either the selectin side (rolling) or the CD18/ICAM-1 side (stopping and transmigration) would greatly reduce the number of neutrophils that localize at sites of inflammation. See Tables 3.1 through 3.4 for a list of published *in vivo* studies that support this conclusion.

result of cytokine production tumor necrosis factor, (TNF), up-regulation of E-selectin and possibly ligands for L-selectin would serve to sustain the phenomenon of margination. While the velocity of rolling appears to be influenced by CD18 integrins (Lawrence *et al.*, 1990; Abbassi *et al.*, 1991; Lawrence and Springer, 1991; Perry and Granger, 1991), they do not appear to be capable of independently sustaining the rolling phenomenon (Lawrence *et al.*, 1990; Abbassi *et al.*, 1991; Lawrence and Springer, 1991; Smith *et al.*, 1991c) even at venous wall shear stresses.

Many of the rolling neutrophils stop on the activated endothelium. The cells change shape, becoming ruffled, flattened and bipolar. The mechanisms that account for this shape change are not well defined, but locally generated chemotactic factors provide the stimulus, and the adhesive mechanisms that initially catch the neutrophils may potentiate the neutrophils response (Lo *et al.*, 1991; Lorant *et al.*, 1991). During this stage an important transition occurs in the neutrophil, not only from an inactivated to a motile cell, but from a cell that is primarily dependent on selectins for adhesion to one that is predominately dependent on CD18 integrins (Kishimoto *et al.*, 1989a; Jutila *et al.*, 1989). Down-regulation of neutrophil L-selectin and up-regulation of neutrophil CD11b/CD18 occur upon contact of neutrophils with cytokine-stimulated endothelial cells *in vitro* (Smith *et al.*, 1991c), and neutrophils have markedly reduced surface L-selectin after diapedesis *in vivo* (Kishimoto *et al.*, 1989a; Jutila *et al.*, 1989). Trans-endothelial migration of neutrophils in most of the settings evaluated requires CD18 integrins. Though some evidence indicates that CD18-independent neutrophil extravasation can occur *in vivo* (Doerschuk *et al.*, 1990), morphological and molecular mechanisms remain obscure.

This model is consistent with the bulk of the evidence obtained *in vitro*. It is important to note, however, that the experimental settings *in vitro* do not fully reproduce the complex conditions *in vivo*, and the relative contributions of each adhesive mechanism are likely to vary with different vascular beds and inflammatory etiologies. If, for example, neutrophil influx occurs through smaller vessels than post-capillary venules, then rolling-type margination may not occur. Rather, physical factors may determine the initial stationary phase of localization prior to neutrophil activation, or systemically activated neutrophils may lodge in small diameter vessels (Worthen *et al.*, 1989). Evidence regarding the role of selectins *in vivo* is much more limited than for CD18 (see Tables 3.1–3.4), and at present there is no published evidence supporting a role for P-selectin in neutrophil localization *in vivo*.

4. Trans-endothelial Migration

The migration of neutrophils through confluent monolayers of endothelial cells *in vitro* has been investigated as a possible model for the phenomenon of leucocyte emigration at sites of acute inflammation. Beesley *et al.*, (1979) examined the interaction of porcine neutrophils with visually confluent porcine endothelial monolayers on glass coverslips (Beesley *et al.*, 1978). A portion of the neutrophils contacting the apical endothelial cell surface were observed to migrate between the endothelial cell junctions and crawl beneath the monolayer. The transmigrated cells could be easily distinguished from the non-migrating cells under phase contrast microscopy, by their markedly flattened appearance in the focal plane of the basal surface of the monolayer. The migration was quite rapid, and individual cells moved from above to

Table 3.1 Inhibitory effect of anti-CD18 mAbs on neutrophil emigration

Adhesion Molecule	Target Organ	Species	Investigators
CD18, β_2	Skin	Rabbit	(Arfors *et al.*, 1987)
			(Rampart and Williams, 1988)
			(Nourshargh *et al.*, 1989)
			(Price *et al.*, 1987)
	Heart	Primate	(Winquist *et al.*, 1990)
	Heart	Dog	(Dreyer *et al.*, 1991)
	Heart	Cat	(Ma *et al.*, 1991)
	Lung	Rabbit	(Doerschuk *et al.*, 1990)
			(Horgan *et al.*, 1992)
			(Barton *et al.*, 1989b)
	Synovium	Rabbit	(Jasin *et al.*, 1990)
	Intestine	Cat	(Granger *et al.*, 1991b)
	Skeletal muscle	Dog	(Jerome *et al.*, 1992)
	Kidney	Rat	(Mulligan *et al.*, 1992a)
	Lung	Rat	(Mulligan *et al.*, 1992c)

Table 3.2 Inhibitory effect of anti-CD11b and anti-CD11a mAbs on neutrophil emigration

Adhesion Molecule	*Target Organ*	*Species*	*Investigators*
CD11b, Mac-1	Foot pad	Mouse	(Rosen *et al.*, 1989)
	Peritoneum	Mouse	(Rosen and Gordon, 1987)
	Lung	Mouse	(Rosen, 1990)
	Liver	Mouse	(Jaeschke *et al.*, 1991)
	Intestine	Rabbit	(Argenbright *et al.*, 1991)
	Liver	Rat	(Smith *et al.*, 1992)
	Peritoneum	Rat	(Mulligan *et al.*, 1992b)
CD11a, LFA-1	Peritoneum	Mouse	(Jutila *et al.*, 1989)
	Intestine	Rabbit	(Argenbright *et al.*, 1991)

Table 3.3 Inhibitory effect of anti-CD54 (ICAM-1) mAbs on neutrophil emigration

Adhesion Molecule	*Target Organ*	*Species*	*Investigators*
CD54, ICAM-1	Lung	Rabbit	(Barton *et al.*, 1989a)
	Peritoneum	Mouse	(Doerschuk *et al.*, 1991)
	Intestine	Cat	(Granger *et al.*, 1991a)
	Intestine	Rabbit	(Argenbright *et al.*, 1991)
	Skin	Rabbit	(Argenbright and Barton, 1992)
	Kidney	Rat	(Mulligan *et al.*, 1992a)

Table 3.4 Inhibitory effect of anti-L-selectin and anti-E-selectin mAbs on neutrophil emigration

Adhesion Molecule	*Target Organ*	*Species*	*Investigators*
L-selectin (Mel-14, LAM-1, LECAM-1)	Skin	Mouse	(Lewinsohn *et al.*, 1987)
	Peritoneal	Mouse	(Jutila *et al.*, 1989) (Watson *et al.*, 1991)
	Intestine	Rat	(Ley *et al.*, 1991b)
	Intestine	Rabbit	(von Andrian *et al.*, 1991)
E-selectin (ELAM-1)	Lung	Rat	(Mulligan *et al.*, 1991)
	Skin	Rat	(Mulligan *et al.*, 1991)
	Lung	Primate	(Gundel *et al.*, 1991)
	Eye	Rat	(Till *et al.*, 1992)

below the monolayer within 30 sec. The sequence of events was very similar to that of migrating neutrophils *in vivo* (Grant, 1974). The initial contact between neutrophils and endothelium was apparently mediated by microvillous-like extensions of organelle-free neutrophil cytoplasm, while the main body of the cell remained roughly spherical and out of contact. Subsequent granulocyte contact involved the close apposition of areas of the blood cell to the endothelial surface, creating small indentations, and at points of closest contact bundles of filaments were often found in the endothelial cytoplasm just beneath the plasma membrane. Neutrophils located above the area where two endothelial cells abutted were often seen with pseudopodia extruded, separating the two endothelial cells. Examination of leucocytes in various stages of migration led the authors to conclude that the cytoplasm including granules flowed beneath the monolayer first, and the portion of the cell containing the nucleus followed last. The flattening of the granulocyte beneath the endothelium was maintained after migration was complete. These observations supported earlier work (Lackie and Debono, 1977) showing that neutrophil migration *in vitro* is not inhibited by intercellular contact, and that neutrophils adhere sufficiently well to endothelial cells to crawl across their surface.

4.1 TRANSMIGRATION IN RESPONSE TO CHEMOTACTIC GRADIENTS

Taylor *et al.* (1981) modified the culture conditions so that a chemotactic gradient could be established across the endothelial monolayer. They grew confluent bovine aortic endothelial cell monolayers on gelatin treated polycarbonate filters, inserted them into chemotaxis chambers, and demonstrated that ^{51}Cr-labelled human neutrophils would migrate into the lower compartment of the chamber containing the chemotactic tripeptide, f-Met-Leu-Phe (10 nM). Migration could be detected within 30 min and continued throughout the 90 min observation period. Very few neutrophils were found in the lower compartment if f-Met-Leu-Phe was omitted, suggesting that under these culture conditions trans-endothelial migration was not a spontaneous event.

Since these techniques involved the culture of endothelial cells on artificial surfaces, there was concern that some important features of the vessel wall that might physically impede or otherwise alter the migration of neutrophils were not reproduced in these models. Furie *et al.*, (1984) prepared confluent monolayers of bovine microvascular endothelial cells on human amniotic membrane. These monolayers restricted the passage of macromolecules and electrical current. When a concentration gradient of f-Met-Leu-Phe or leukotriene B_4 (LTB_4) was established across these membranes, trans-endothelial migration of isolated human neutrophils was markedly increased (Furie *et al.*, 1987). Since the evaluation of neutrophil migration was made visually on cross-sections of monolayer preparations, it was possible to determine that migration was very rare in the absence of chemotactic factors. These investigators found that in response to f-Met-Leu-Phe (10 nM), neutrophils adhered to the endothelium within 2 min and began to migrate within 5 min. At 10 min, the majority of the migrating cells were beneath the endothelium. In this model, between 20–40% of the neutrophils added to the chamber eventually migrated across the endothelial monolayer, and those neutrophils that did not migrate either failed to adhere or they detached from the apical surface of the endothelial cells. Transmission electron microscopic studies revealed some of the same features seen in the studies of Beesley *et al.* (1979). Pseudopodia of the adhering neutrophils appeared to indent the apical plasma membranes of the endothelial cells, but migration began by insertion of a pseudopod between two endothelial cells with the neutrophils maintaining close contact with the adjacent endothelial cells as they traversed the monolayer. In contrast to the previous studies, endothelial cells grown on the stromal surface of the amniotic membrane produced an underlying basement membrane, and following emigration, the neutrophils were observed lying between the endothelium and this basement membrane tightly opposed to the basal surface of the endothelial cells, a phenomenon repeatedly seen in studies *in vivo* (Williamson and Grisham, 1961; Hurley, 1963; Marchesi, 1964; Shaw, 1980; Granstein *et al.*, 1986). With time, the neutrophils moved past the basement membrane into the amniotic tissue.

Huber and Weiss (1989) prepared confluent monolayers of HUVEC on type I collagen gels on polycarbonate filters, and demonstrated that after 21 days in culture these monolayers produced a continuous basement membrane with biochemical and structural features of subendothelial basement membranes *in vivo*. As observed in the Furie *et al.* study (1987), transmigration of neutrophils did not occur unless a chemotactic gradient was established across the monolayer, and then, approximately 40% of the neutrophils contacting the apical surface of the monolayer migrated past the endothelial cells into the collagen gel. As observed *in vivo*, neutrophil migration appeared to be transiently impeded by the basement membrane.

4.2 TRANSMIGRATION INDUCED BY CYTOKINE STIMULATION OF THE ENDOTHELIUM

Numerous studies *in vivo* indicate that interleukin-1 (IL-1) and tumor necrosis factor-α (TNFα) may play an important role in the extravasation of neutrophils (Beck *et al.*, 1986; Movat *et al.*, 1987; Cybulsky *et al.*, 1988; Henderson and Pettipher, 1988; Rampart and Williams, 1988; Sayers *et al.*, 1988; Sharp *et al.*, 1988; Wankowicz *et al.*, 1988). While these studies deal with the accumulation of neutrophils at the site of inflammation, they do not distinguish effects on leucocyte adhesion to the luminal surface of the vascular endothelium from effects on subsequent leucocyte migration. Accumulation of leucocytes at a site of inflammation could be due to either function, and these cytokines are well known to increase the adhesiveness of cultured endothelial cells for previously unstimulated neutrophils (Pober and Cotran, 1990). This question has been investigated using variations of the *in vitro* models described above. Smith *et al.* (1988), used a variation of the Beesley *et al.*, model (Beesley *et al.*, 1979). They prepared visually confluent monolayers of HUVEC on fibronectin-coated glass coverslips, and allowed isolated human neutrophils to settle onto these monolayers. Direct visualization using phase contrast microscopy revealed that in contrast to the observations of Beesley and colleagues, the isolated neutrophils were spherical, and remained spherical after contacting monolayers of unstimulated HUVEC over a 30 min observation period. However, if the HUVEC monolayers had been exposed previously to IL-1 (0.3–3.0 U/ml) or endotoxin (LPS) (1–10 ng/ml) for 3 h, a high percentage of the contacting neutrophils would rapidly change shape and migrate: beneath the monolayer, there assuming the flattened morphology observed by several other investigators (Armstrong and Lackie, 1975; Beesley *et al.*, 1979; Buchanan *et al.*, 1982; Gimbrone *et al.*, 1984; Tonnesen *et al.*, 1984). Under basal conditions, less than 1% of the neutrophils exhibited this behavior, but between 35–78% of the cells migrated following cytokine stimulation of the endothelial cells.

Several investigators have confirmed that cytokine stimulation of endothelial monolayers *in vitro* provokes trans-endothelial migration of previously unstimulated neutrophils. As neutrophils contact the apical surface of the stimulated endothelium, they rapidly assume a bipolar shape, and usually greater than 40% of the cells migrate through the monolayer. These events occur with endothelial cells grown on polycarbonate filters (Moser *et al.*, 1989) human amniotic membrane (Furie and McHugh, 1989) or type I collagen gels (Luscinskas *et al.*, 1991). IL-1, TNFα, or lipopolysaccharide (LPS) stimulation of the endothelium for 4 h provides the optimum set of conditions (Smith *et al.*, 1988; Luscinskas *et al.*, 1991), and migration is almost complete within 10 min after the neutrophils contact the endothelium (Smith *et al.*, 1988; Furie and McHugh, 1989; Luscinskas *et al.*, 1991). The reason that only a portion of the neutrophils contacting the monolayer migrate is not clear. In most studies, the non-migrating cells either do not attach to the apical surface of the endothelial cells, or only transiently attach. In fact, this population could not be recruited to migrate if a chemotactic gradient was established across cytokine stimulated monolayers (Furie and McHugh, 1989; Moser *et al.*, 1989). The endothelial monolayer does not appear to limit penetration under these experimental conditions since the number of neutrophils migrating per unit area of endothelium is directly proportional to the number added to the chamber over a wide range for both chemotactic and endothelial cell induced migration.

4.3 CYTOKINE-STIMULATED ENDOTHELIAL CELLS ACTIVATE NEUTROPHIL MOTILITY

The apparent activation of neutrophils following contact with cytokine-stimulated neutrophils is unlikely to be a direct effect of IL-1 or TNFα since in most experimental protocols the endothelial cells are washed prior to introduction of the neutrophils, and current evidence indicates that IL-1 is incapable of directly stimulating neutrophil locomotion (Yoshimura *et al.*, 1987; Kharazmi *et al.*, 1988). Endothelial cells are capable of producing such chemotactic factors following a variety of stimuli (O'Brien *et al.*, 1984; Mercandetti *et al.*, 1984; Gudewicz *et al.*, 1988; Farber *et al.*, 1985; Farber *et al.*, 1986; Breviario *et al.*, 1988; Bussolino *et al.*, 1988; Schroder and Christophers, 1989; Gimbrone *et al.*, 1989; Strieter *et al.*, 1989; Griffin *et al.*, 1990; Zimmerman *et al.* 1990). Two are of particular interest since they are produced by cytokine stimulated endothelial cells within the time during which neutrophil transmigration is provoked by the monolayers. Platelet activating factor (PAF) may be produced by HUVEC following IL-1 and TNFα stimulation (Bussolino *et al.*, 1988) [though this is not altogether clear (Zavoico *et al.*, 1989)], and is well known to be chemotactic for neutrophils (Goetzl and Pickett, 1980). It has recently been implicated in the extravasation of neutrophils in mesenteric vessels following ischemia and reperfusion (Kubes *et al.*, 1990). IL-8 is produced by HUVEC for an extended time following cytokine stimulation (Gimbrone *et al.*, 1989), and is well known as a chemotactic factor for neutrophils (Matsushima and Oppenheim, 1989; Baggiolini *et al.*, 1989; Detmers *et al.*, 1990; Hechtman *et al.*, 1991). There is no direct evidence to indicate that the stimulated endothelial cells create a chemotactic gradient across the monolayer toward which the neutrophils migrate. Moser *et al.* (1989) found that when C5a was added with the neutrophils above the endothelial monolayer (i.e. in a position to create a negative gradient), IL-1-stimulated monolayers were still able to induce significant increases in transmigration. Such results are difficult to interpret and may reflect simply an effect of the chemokinetic influence of the endothelial factors. Several investigators have found that once the neutrophils pass through the endothelial junctions, they continue to migrate into the matrix supporting the monolayer, and thus away from the monolayer. This suggests chemokinetic rather than directed migration (Luscinskas *et al.*, 1991; Furie and McHugh, 1989; Huber and Weiss, 1989; Moser *et al.*, 1989).

Two recent studies support the conclusion that endothelial derived chemotactic factors are directly responsible for trans-endothelial migration. Huber *et al.* (1991) used the *in vitro* model developed by Huber and Weiss (1989) in evaluating the contribution of IL-8 to the transmigration of neutrophils induced by IL-1β, TNFα or LPS stimulation of HUVEC. They found a correlation between the ability of the endothelial monolayers to synthesize IL-8 and their ability to promote neutrophil transmigration. Immunolocalization of IL-8 revealed that it was in association with both the endothelial cells and the underlying structures supporting the monolayer (basement membrane and type 1 collagen gel). Furthermore, antiserum to IL-8 inhibited neutrophil transmigration by greater than 60%. Kuijpers *et al.* (1992) used a platelet-activating factor (PAF) receptor antagonist as well as antiserum to IL-8 to provide evidence that both PAF and IL-8 contribute to neutrophil activation and the subsequent trans-endothelial migration upon contact with the cytokine stimulated HUVEC monolayers.

5. *Contributions of Adherence to Trans-endothelial Migration*

In most of the studies discussed above, the investigators found that approximately 50% of the neutrophils contacting the apical surface of endothelial cell monolayers transmigrated under maximal stimulus conditions. The non-migrating cells either do not adhere to or eventually detached from the monolayer, and adhesion of the migrating population is evident within a few minutes following their contact with the monolayer (Furie *et al.*, 1987; Furie and McHugh, 1989). Stimuli such as chemotactic factors lead to increased adhesion of neutrophils to the endothelium, and in a concentration gradient across the monolayer induce trans-endothelial migration. Gimbrone *et al.* (1984) found that LTB_4 at concentrations as low as 0.1 nM increased the attachment of ^{111}In-labelled human neutrophils to monolayers of bovine aortic endothelial cells on glass or plastic coverslips when added for 10 min to the cultures. This effect appeared not to be species specific since LTB_4 increased human neutrophil attachment to HUVEC, bovine vena cava, human and baboon saphenous vein and SV-40 transformed human umbilical vein endothelial cells (HUVEC). Tonnesen *et al.* (1984) demonstrated that f-Met-Leu-Phe, C5a and C5a des arg were each effective in increasing human neutrophil adherence to unstimulated HUVEC monolayers grown on plastic when added to the cell cultures at 10 nM. This increase was not specific for endothelium since neutrophil adhesion to human skin fibroblasts and smooth muscle cells was also increased as well. These investigators' adhesion assay was designed to study low affinity adhesions, and used static conditions and immediate glutaraldehyde fixation of adherent cells. Under these conditions the onset of stimulated adhesion occurred within 30 s of adding the stimulus, was maximal at 2 min, and persisted for up to 45 min. Since preincubated and washed neutrophils showed augmented adhesion while preincubated and washed endothelium did not, they concluded that the chemotactic factors acted primarily on the neutrophils. Others have confirmed this adherence-enhancing effect of chemotactic factors (Gamble *et al.*, 1985; Charo *et al.*, 1986; Pohlman *et al.*, 1986; Zimmerman and McIntyre, 1988; Smith *et al.*, 1989b; Tonnesen *et al.*, 1989), and of particular interest are the observations that two of the factors produced by stimulated endothelial cells, PAF (Tonnesen *et al.*, 1984; Entman *et al.*, 1990) and IL-8 (Carveth *et al.*, 1989; Detmers *et al.*, 1990), are both active in this regard.

While such studies clearly show the potential of chemotactic factors to augment neutrophil adhesion to endothelial cells, other studies indicate that the role of chemotactic factors may be rather complex. Charo *et al.* (1986) confirmed that C5a could increase neutrophil adherence, but noted a reciprocal relationship between chemotactic migration and adherence to endothelial cells. Concentrations of C5a (e.g. 1 nM) that induced optimal chemotaxis (i.e. migration through micropore filters toward a higher concentration of C5a) reduce neutrophil adhesion to endothelial monolayers, while concentrations of C5a that inhibited migration (e.g, 10 nM) significantly increased adherence. Tonnesen (1984) found that C5a did not increase adherence at concentrations less than 10 nM. Gimbrone *et al.* (1989), Hectman *et al.* (1990) and Smith *et al.* (1991c); Kishimoto *et al.* (1991); Abbassi *et al.* (1991) have also found that chemotactic factors such as IL-8, C5a, f-Met-Leu-Phe and PAF significantly reduced neutrophil adherence to endothelial monolayers. Thus it is evident that the relationship between chemotactic stimulation and neutrophil-endothelial adhesion or trans-endothelial migration is complex and may lead to increases or decreases in adhesion by as yet poorly defined mechanisms.

An additional issue may be pertinent to this discussion. One of the possible endothelial-derived chemotactic factors, PAF, is largely retained by the endothelial cells (Bussolino *et al.*, 1988; Zimmerman *et al.*, 1990) and presented on the plasma membrane (Zimmerman *et al.*, 1990). There it may act as a surface-bound stimulus (Zimmerman *et al.*, 1990) in the microenvironment of the vessel wall where fluid flow may preclude establishment of a stable gradient of soluble chemotactic factors. Zigmond *et al.* (1977, 1978) has clearly shown that a stable gradient of soluble chemotactic factor is required for directional migration *in vitro*. However, at the interface between neutrophils and endothelial cells a surface bound stimulus may provide the necessary chemokinetic migration to promote trans-endothelial migration of adherent leucocytes.

5.1 SPECIFIC ADHESION MOLECULES AND MIGRATION TOWARD CHEMOTACTIC GRADIENTS

5.1.1 CD18 Integrins

Case reports from several laboratories have described children with recurrent bacterial infections, marked leucocytosis, and profound reductions in the migration of neutrophils into inflamed tissue. Neutrophils from these patients are almost completely deficient in the ability to adhere to protein-coated glass *in vitro*, and they show markedly reduced adhesion to endothelial cells (Hayward *et al.*, 1979; Crowley *et al.*, 1980; Abramson *et al.*, 1981; Bissenden *et al.*, 1981; Arnaout *et al.*, 1982; Bowen *et al.*, 1982; Davies *et al.*, 1982; Fischer *et al.*, 1983; Anderson *et al.*, 1984; Kobayashi *et al.*, 1984; Harlan *et al.*, 1985). This syndrome is caused by a deficiency of CD18 (Anderson and Springer, 1987; Todd and Freyer, 1988) which prevents or profoundly reduces the surface expression of functional β_2 integrins, LFA-1 (CD11a/CD18), Mac-1 (CD11b/CD18), p150,95 (CD11c/CD18) (Springer *et al.*, 1984). Harlan *et al.* (1985) used the *in vitro* model of trans-endothelial migration described by Taylor *et al.* (1981) to investigate the ability of neutrophils from two of these patients to migrate toward a chemotactic gradient. While up to 46% of normal neutrophils transmigrated toward f-Met-Leu-Phe in this study, 9% or less of the patient neutrophils migrated. They also demonstrated that an mAb to CD18 (60.3) reduced the trans-endothelial migration of normal neutrophils to the same low level as that of the CD18-deficient neutrophils, and postulated that the absence of the CD18 glycoproteins prevented neutrophil adherence to endothelium following chemotactic stimuli, thereby accounting for the failure of these patients' neutrophils to emigrate at sites of inflammation.

5.1.2 CD11b/CD18 (Mac-1, Mo1, CR3)

Chemotactic migration of CD18-deficient neutrophils across a protein-coated plastic surface *in vitro* is less than 20% of that of normal cells (Schmalstieg *et al.*, 1986). Chemotactic migration of normal neutrophils on artificial surfaces is greatly diminished by mAbs against CD18 (Anderson *et al.*, 1986; Dana *et al.*, 1986), and the specific heterodimer of greatest importance in this experimental setting is CD11b/CD18. Monoclonal antibodies against CD11b are markedly inhibitory while those against CD11a are not. Stimuli such as chemotactic factors and phorbol esters increase the adhesive functions of constitutive surface CD111b/CD18 (Detmers *et al.*, 1987; Philips *et al.*, 1988), and they promote rapid (within minutes) translocation of CD11b/CD18 to the cell surface from intracellular stores in the secondary granules (Todd *et al.*, 1984; O'Shea *et al.*, 1985; Bainton *et al.*, 1987; Petty *et al.*, 1987; Stevenson *et al.*, 1987; Jones *et al.*, 1989; Jones *et al.*, 1990). Adhesion to endothelial cells *in vitro* following chemotactic stimulation of neutrophils is also greatly inhibited by anti-CD11b mAbs (Gamble *et al.*, 1985; Pohlman *et al.*, 1986; Zimmerman and McIntyre, 1988; Smith *et al.*, 1989b; Lo *et al.*, 1989b).

Several investigators have proposed that chemotactic stimulation of neutrophils influences the cell surface distribution of adhesion sites, promoting not only the adherence of the advancing lamellipodium to the substratum, but the transport of the adhesion sites to the uropod (Ryan *et al.*, 1974; Stossel, 1978; Smith and Hollers, 1980; Springer and Anderson, 1986; Francis *et al.*, 1989). Francis *et al.* (1989) proposed a model whereby sequential expression of cell surface CD11b/CD18 would contribute to the locomotion of adherent neutrophils. They evaluated the surface distribution of CD11b/CD18 on migrating neutrophils using fluorescein- and rhodamine-labelled anti-CD11b, applied sequentially to chemotactically stimulated cells, and provided evidence that CD11b/CD18 is up-regulated during locomotion as CD11b/CD18-positive granules fuse with the plasma membrane near the lamellipodia. The newly expressed CD11b/CD18 is then swept to the uropod as the cell migrates, and accumulates on the retraction fibers. The observations that CD11b/CD18-dependent adhesion sites exhibit the same pattern of appearance and redistribution (Smith and Hollers, 1980; Anderson *et al.*, 1986) supports the hypothesis that such a mechanism may contribute to neutrophil locomotion.

Until recently, published evidence did not support the view that newly up-regulated CD11b/CD18 participates in adhesion (Lopez *et al.*, 1986; Buyon *et al.*, 1988a; Buyon *et al.*, 1988b; Vedder and Harlan, 1988; Lo *et al.*, 1989a; Schleiffenbaum *et al.*, 1989; Schwartz and Harlan, 1989). Rather, it seemed that constitutive CD11b/CD18 undergoes a qualitative change following stimulation of the neutrophil which promotes its participation in adhesion. Dransfield and Hogg (Dransfield and Hogg, 1989) have obtained evidence for a conformational change in the heterodimer recognized by an mAb that defines a Mg^{2+}-dependent and temperature-sensitive epitope, and Detmers *et al.* (Detmers *et al.*, 1987) have provided evidence that phorbol ester stimulation of adherent neutrophils results in transient clustering of CD11b/CD18, a phenomenon that would likely increase the avidity of CD11b/CD18-dependent binding. Potential cellular mechanisms regulating qualitative changes in CD11/CD18 heterodimers are considered further below in Section 6. While both qualitative and quantitative changes in surface CD11b/CD18 have been documented following exposure of neutrophils to chemotactic factors, the role of the newly mobilized CD11b/CD18 was not known. Hughes *et al.* (1990) have recently shown that the newly mobilized CD11b/CD18 can function in adhesion to protein-coated glass and plastic if the neutrophils are subsequently exposed to an increase in the concentration of the chemotactic stimulus. This appears to be involved in the continued ability of the neutrophil to migrate as concentrations of the chemotactic stimulus increase.

5.1.3 CD11a/CD18 (LFA-1)

The contribution of CD11b/CD18 to trans-endothelial migration in a chemotactic gradient was evaluated by Furie *et al.* (1991b) using the model with HUVEC grown on amniotic membrane (Furie *et al.*, 1987). They evaluated the effects of anti-CD11b mAbs previously shown to markedly inhibit adhesion to and migration of neutrophils across protein-coated surfaces, and found that they inhibited trans-endothelial migration through unstimulated endothelial monolayers by less than 40%. In addition, they found that anti-CD11a mAb, previously shown to be ineffective in reducing adhesion to and migration across protein-coated artificial surfaces, inhibited trans-endothelial migration by greater than 50%. The combination of anti-CD11a and anti-CD11b antibodies was almost as inhibitory as anti-CD18. Thus, it appears that in contrast to migration on artificial surfaces, CD11a/CD18 appears to be sufficient for substantial chemotactic migration of neutrophils through confluent endothelial monolayers, though both CD11a/CD18 and CD11b/CD18 are needed for optimum migration.

5.1.4 ICAM-1 (CD54)

The known ligands for CD11a/CD18, intercellular adhesion molecules 1 and 2 (ICAM-1, ICAM-2) are constitutively expressed on unstimulated endothelial cells both in culture and *in vivo* (Dustin *et al.*, 1986; Prober *et al.*, 1986; Munro *et al.*, 1989; Munro *et al.*, 1991; Staunton *et al.*, 1989; de Fougerolles *et al.*, 1991). Isolated, unstimulated neutrophils are capable of using CD11a/CD18 to adhere to isolated ICAM-1 densely inserted in a planar artificial membrane, but they adhere very little to resting endothelial cell monolayers unless stimulated (e.g. with phorbol ester or chemotactic factors). Furie *et al.* (1991b) also studied anti-ICAM-1 mAbs, R6.5, and showed that it inhibited migration to the same degree as anti-CD11a, and inhibition was no greater when anti-ICAM-1 was combined with anti-CD11a. In contrast, when anti-ICAM-1 was combined with anti-CD11b, the inhibition was almost additive, suggesting that under chemotactic conditions, transmigration across unstimulated endothelial monolayers also involves CD11a/CD18-ICAM-1 interactions. A ligand for CD11b/CD18 under these conditions remains to be defined. Stimulation of neutrophils with phorbol esters also promotes adhesion dependent on CD11a/CD18 (Lo *et al.*, 1989a) interacting with constitutive ICAM-1 on the endothelial cell (Lo *et al.*, 1989b; Diamond *et al.*, 1990). Since, there is no intracellular storage pool for CD11a/CD18 in the neutrophil, and since stimulation does not alter cell surface levels of this heterodimer (Kishimoto *et al.*, 1989b), qualitative changes in the ability of CD11a/CD18 to interact with endothelial ICAM-1 account for enhanced adhesion. The resulting adhesion is transient (i.e. peaking within 15 min and returning to baseline within 30 min), and the qualitative change appears to be reversible (Lo *et al.*, 1989a). As described below in Section 6.1, Lymphocyte CD11a/CD18-ICAM-1 interactions can be rapidly modulated by cell stimulation without changes in the density of CD11a/CD18 on the cell surface. The time course of this change in avidity is similar in stimulated neutrophils and lymphocytes, peaking within 15 min and returning to baseline within 30 min (Dustin and Springer, 1989; Lo *et al.*, 1989a), and is recognized in both cell types (as is the case for CD11b/CD18 on neutrophils) by the appearance of a Mg^{2+}-dependent, temperature sensitive epitope (Dransfield and Hogg, 1989). Such a modulation in CD11a/CD18 may allow neutrophils to adhere to and detach from endothelial ICAMs in the process of transmigration. Recent evidence indicates that trans-endothelial migration of lymphocytes *in vitro* involves CD11a/CD18-ICAM-1 interactions as well (Kavanaugh *et al.*, 1991; Oppenheimer-Marks *et al.*, 1991).

Chemotactically stimulated neutrophils can use both CD11a/CD18 and CD11b/CD18 for locomotion as discussed above. CD11b/CD18 provides the neutrophil with greater flexibility, allowing adhesion-dependent migration across artificial surfaces as well as tissue cells. It binds to a number of different proteins thereby adhering neutrophils to a wide variety of surfaces (Wright *et al.*, 1983; Ross *et al.*, 1985; Wright and Jong, 1986; Altieri and Edgington, 1988; Russell and Wright, 1988; Lo *et al.*, 1989b; Wright *et al.*, 1989; Relman *et al.*, 1990; Shappell *et al.*, 1990c). The intracellular pool appears to provide replacement for CD11b/CD18 lost from the cell surface during locomotion (Francis *et al.*, 1989). In contrast, CD11a/CD18 plays a more limited role, apparently allowing adhesion to and migration over surfaces bearing ICAM-1, and possibly ICAM-2.

5.2 SPECIFIC ADHESION MOLECULES AND MIGRATION INDUCED BY CYTOKINE STIMULATION OF THE ENDOTHELIAL CELLS

Emigration of neutrophils is unlikely to result simply from a chemotactic gradient across unstimulated endothelium *in vivo*, though intradermal injection of C5a des arg, f-Met-Leu-Phe and LTB_4 in rabbits results in rapid emigration of neutrophils that is maximum within 30 min (Cybulsky *et al.*, 1989), a time frame that would not allow for *de novo* synthesis of inflammatory determinants. Intradermal injection of IL-1 or TNFα produces leucocyte emigration that is delayed by 30–60 min in onset and peaks within 2–3 h (Cybulsky *et al.*, 1989). Inhibition of protein synthesis greatly reduces cytokine-induced extravasation, but has no effect on the emigration that follows injection of chemotactic agents (Rampart and Williams, 1988; Cybulsky *et al.*, 1989). Protein synthesis-independent and -dependent mechanisms may contribute to different phases of leucocyte localization (e.g. rapid margination after injury (Allison *et al.*, 1955) followed by more sustained efflux of leucocytes). The experimental models that provide stable chemotactic gradients *in vitro* reveal some leucocyte and endothelial functions in isolation, but they fail to take into account newly up-regulated endothelial adhesion molecules and the effects of flow on the interactions of leucocytes and endothelial cells. This point is clearly illustrated by the fact that chemotactic stimulation does not increase adhesion of neutrophils to unstimulated endothelial cells under conditions of flow until the wall shear stress is reduced to below that observed in post-capillary venules (Worthen *et al.*, 1987; Lawrence *et al.*, 1990), the site where most leucocyte emigration occurs. However, as discussed earlier, adhesion of neutrophils to thrombin- or cytokine-stimulated endothelial cell monolayers *in vitro* does occur at wall shear stresses within the low venous range (Lawrence *et al.*, 1990; Abbassi *et al.*, 1991; Anderson *et al.*, 1991; Smith *et al.*, 1991c). Chemotactic stimulation of neutrophils alone may be insufficient to localize these cells at an inflammatory site without coincident expression of some rapidly mobilized or newly synthesized endothelial molecules. Beyond promoting margination of neutrophils, the cytokine-stimulated endothelium activates trans-endothelial migration of neutrophils.

5.2.1 β_2 Integrins

The trans-endothelial migration of neutrophils provoked by cytokine-stimulated endothelial monolayers *in vitro* (Smith *et al.*, 1988; Furie and McHugh, 1989; Moser *et al.*, 1989; Luscinskas *et al.*, 1991) essentially fails to occur when neutrophils from patients with CD18 deficiency are used (Smith *et al.*, 1988; Smith *et al.*, 1989b; Anderson *et al.*, 1990; Smith *et al.*, 1991c), and anti-CD18 mAbs almost completely inhibit the transmigration of normal neutrophils (Smith *et al.*, 1988; Smith *et al.*, 1989b; Lawrence *et al.*, 1990; Abbassi *et al.*, 1991; Anderson *et al.*, 1991; Luscinskas *et al.*, 1991; Smith *et al.*, 1991c). There are important contributions of both CD11b/CD18 and CD11a/CD18. Anti-CD11b mAbs produce less than 40% inhibition of migration, anti-CD11a mAbs produce greater than 50% inhibition (Smith *et al.*, 1989b; Anderson *et al.*, 1990), and combining the antibodies gives an additive inhibitory effect equivalent to that of anti-CD18 mAbs. Jutila *et al.* (1989) found that systemic administration of either anti-CD11a or anti-CD11b mAbs *in vivo* significantly inhibited (by greater than 60%) the influx of neutrophils into the thioglycollate-inflamed peritoneal cavity of mice.

5.2.2 CD11c/CD18 (p150,95)

Relatively little is known about the contribution of this β_2 integrin to the adhesion of neutrophils to endothelial cells. Stacker and Springer (1991) isolated the functional heterodimer, and demonstrated that when adsorbed to a plastic surface, CD11c/CD18 would support the adhesion of endothelial cells. Unstimulated endothelial cells bound poorly, but stimulation of these cells with IL-1β or endotoxin for 18 h caused a significant increase in the level of binding to CD11c/CD18. This binding could be specifically inhibited with antibodies to CD11c. The endothelial ligand remains undefined. Both neutrophils and monocytes have this integrin on their surfaces, and it is reasonable to conclude that it may participate in the adhesion and migration of these leucocytes through activated endothelium, though published evidence indicates that its contribution to adhesion is relatively small (Keizer *et al.*, 1987; Arnaout *et al.*, 1988; Luscinskas *et al.*, 1989; Beekhuizen *et al.*, 1990).

5.2.3 ICAM-1 (CD54)

Among the numerous changes in endothelial cells that occur following cytokine stimulation (Pober and Cotran, 1990), the increase in adhesiveness for unstimulated neutrophils seen in the first few hours is largely attributable to synthesis and surface expression of ICAM-1 (Smith *et al.*, 1988) and E-selectin (Bevilacqua *et al.*, 1987; Bevilacqua *et al.*, 1989; Luscinskas *et al.*, 1989). When evaluated in a static system at the peak time for neutrophil adhesion following IL-1 or TNFα stimulation (3–4 h), anti-ICAM-1 mAbs R6.5 (Smith *et al.*, 1988) and Hu5/3 (Luscinskas *et al.*, 1991) each produces partial inhibition (30–50%), and anti-E-selectin mAbs CL2 and CL3 (Kishimoto *et al.*, 1991; Picker *et al.*, 1991a), and H18/7 (Bevilacqua *et al.*, 1987; Luscinskas *et al.*, 1991) each produced partial inhibition (30–50%). Combined use of anti-ICAM-1 and anti-E-selectin antibodies causes essentially additive inhibition of adhesion (Kishimoto *et al.*, 1991; Luscinskas *et al.*, 1991) (70–80%), indicating that ICAM-1 and E-selectin are components of distinct adherence pathways.

Since trans-endothelial migration requires the β_2 integrins, principally CD11a/CD18 and CD11b/CD18, endothelial ICAM-1 should also play an important role. As discussed above, CD11a/CD18-ICAM-1 interactions apparently account for a portion of the transmigration of neutrophils in a chemotactic gradient (Furie *et al.*, 1991b), but the ligand for CD11b/CD18 under those conditions has not been defined. Experimental protocols using isolated ICAM-1 densely inserted in planar artificial membranes have shown that isolated neutrophils exposed to a chemotactic factor exhibit a high degree of binding, and that anti-CD11b mAbs reduce this adherence by approximately 50% (Smith *et al.*, 1989b; Anderson *et al.*, 1990). Diamond *et al.* (1991) more directly addressed the question of CD11b/CD18 binding to ICAM-1 by demonstrating that COS cells transfected with both cDNAs for CD18 and CD11b bound to isolated ICAM-1, and that anti-ICAM-1 mAb, R6.5 (Smith *et al.*, 1988), was almost completely inhibitory. The reciprocal experiment was also performed showing that COS cells and HUVEC expressing ICAM-1 bound to planar membranes containing isolated CD11b/CD18 heterodimer. The interaction of CD11b/CD18 and CD11a/CD18 with ICAM-1 is not identical, though. Monoclonal antibodies recognizing different domains of ICAM-1 exhibit differential ability to inhibit these interactions (e.g. R6.5 (Smith *et al.*, 1988) binding to domain 2 inhibited CD11b/CD18 and LB2 (Clark *et al.*, 1986) binding to domain 1 inhibited CD11a/CD18), and the binding of CD11b/CD18 with ICAM-1 was less shear resistant than bound CD11a/CD18. Using domain-deleted and chimerical forms of ICAM-1, Staunton *et al.* (1990) demonstrated that CD11a/CD18 bound to the first domain of ICAM-1, and Diamond *et al.* (1991) demonstrated that CD11b/CD18 bound to the third domain of ICAM-1. Diamond *et al.* (1991) suggested that CD11b/CD18 and CD11a/CD18 on the same leucocyte may be able to bind to a single ICAM-1 molecule, and they provided information on the structural feasibility of such an interaction. In addition, they found that the size of the N-linked oligosaccharide chain on domain 3 of ICAM-1 affects the interaction between CD11b/CD18 and ICAM-1. Adhesion depends primarily on CD11b/CD18 if ICAM-1 has a smaller N-linked side chain, but on CD11a/CD18 if ICAM-1 has a complex carbohydrate side chain.

ICAM-1 from different cell types has been shown to differ in the level and type of N-linked carbohydrates (Dustin *et al.*, 1986), and thereby may regulate biological interactions of different cells with neutrophils. For example, Entman and colleagues have found that canine neutrophils adhere to isolated, cytokine-stimulated canine cardiac myocytes (Entman *et al.*, 1990) primarily through a mechanism dependent on CD11b/CD18 and ICAM-1 (Entman *et al.*, 1991b), and anti-CD11a mAbs have little effect on this adhesion. In contrast, ICAM-1-dependent adhesion of canine neutrophils to cytokine-stimulated canine jugular vein endothelial cells is largely dependent on CD11a/CD18 with some contribution of CD11b/CD18. Furthermore, the biological consequences of these interactions appear to be distinct. Upon contacting cytokine-stimulated endothelial monolayers, a high percentage of canine neutrophils rapidly migrate through the monolayers (Abbassi *et al.*, 1991), but upon contacting the cytokine-stimulated myocytes, neutrophils adhere, show little tendency to migrate, and within minutes begin to secrete reactive oxygen species (Entman *et al.*, 1991b). Adherence dependent hydrogen peroxide production (Nathan, 1987; Nathan *et al.*, 1989) can be blocked by anti-CD11b and anti-CD18 mAbs (Shappell *et al.*, 1990c), but evidence is lacking that anti-CD11a mAbs can inhibit the oxidative burst in neutrophils (Shappell *et al.*, 1990c), (see below, Section 7.1.2).

5.2.4 E-selectin

The adhesiveness of HUVEC monolayers for CD18-deficient neutrophils peaks at 4 h after IL-1 stimulation and declines thereafter to near baseline values within 8 h (Smith *et al.*, 1988). This time course is consistent with that of endothelial surface expression of E-selectin (Pober *et al.*, 1986): and Luscinskas *et al.* (1989, 1991) published evidence from mAb blocking studies that E-selectin-dependent and CD18-ICAM-1-dependent adherence mechanisms are distinct. However, trans-endothelial migration of CD18-deficient neutrophils through cytokine stimulated endothelial monolayers at the peak time of adhesion is extremely low (Smith *et al.*, 1988; Smith *et al.*, 1989b), in marked contrast to normal neutrophils. These results indicate that CD18 integrins are necessary for transmigration, and E-selectin-dependent adhesion is not sufficient for migration. Supporting this conclusion is the fact that patients with CD18-deficiency have a profound deficit in migration of neutrophils into sites of inflammation. There is, however, an important exception. Hawkins and Anderson (1991) observed in one deceased CD18-deficient patient accumulation of neutrophils in the alveolar spaces in association with gram positive infection of the lungs. Doerschuk *et al.*, (1990) found that systemic administration of anti-CD18 mAb, 60.3, almost completely inhibited emigration of neutrophils into the lungs of rabbits in response to phorbol ester, *E. coli* organisms and *E. coli* endotoxin, but was without effect on the response to *S. pneumoniae*. In contrast to the results in lung, 60.3 largely prevented the emigration of neutrophils into the peritoneal cavity of rabbits in response to *S. pneumoniae*, indicating that mechanisms of extravasation may vary with different vascular beds. These studies show that trans-endothelial migration *in vivo* can be CD18-independent, but whether E-selectin is the predominant determinant under these conditions has not been assessed. No published studies *in vitro* have revealed an E-selectin-dependent, CD18-independent mechanism for transmigration of neutrophils.

Studies of the participation of E-selectin in neutrophil migration *in vitro* have provided inconsistent results. Luscinskas *et al.* (1991) found that trans-endothelial migration peaked following IL-1 stimulation of HUVEC at a time when surface expression of E-selectin was highest (i.e. 4 h), and that anti-E-selectin mAb, H18/7, almost completely inhibited migration. Anti-ICAM-1 mAb, Hu5/3, was also markedly inhibitory in their model, supporting the conclusion that both ICAM-1 and E-selectin were necessary for migration. Kishimoto *et al.* (1991), using the *in vitro* model of trans-endothelial migration reported by Smith *et al.* (1988), found that anti-E-selectin mAb, CL2 and CL3, had little effect on migration. In both investigations, adhesion of isolated, unstimulated neutrophils to cytokine stimulated HUVEC was partially inhibited by either anti-ICAM-1 or anti-E-selectin mAb, and when these antibodies were combined in the same assay, inhibition was additive. In light of the observations *in vivo* suggestive of varying mechanisms for neutrophil emigration in different vascular beds, the apparent inconsistency *in vitro* may reflect differing requirements for E-selectin in neutrophil transmigration. One technical difference between these studies that may be significant was that in the Luscinskas and colleagues study, endothelial cell monolayers were prepared on type I collagen gels, and in the Kishimoto and colleagues study, they were prepared on fibronectin-coated glass. Since the activation of neutrophil motility in these *in vitro* models is endothelial-dependent, investigation of possible differential influences of the substrates on the relevant endothelial functions may prove important. Furie *et al.* (1991a) have recently evaluated the effects of anti-E-selectin mAbs BB11 (Benjamin *et al.*, 1990) and H18/7 (Luscinskas *et al.*, 1991) on neutrophil migration through endothelial monolayers grown on amnion (Furie and McHugh, 1989). They found that when the endothelium was stimulated with low concentrations of IL-1β (0.1–0.15 U/ml), anti-E-selectin antibodies were partially effective, but when the IL-1 level was optimal for provoking trans-endothelial migration, these antibodies were ineffective in contrast to anti-CD18 which profoundly inhibited migration. They concluded that E-selectin was not necessary for migration in this assay.

As discussed above, the mechanism(s) by which neutrophil motility is activated by cytokine-stimulated endothelial cells is unknown, but may involve endothelial derived chemotactic factors such as PAF and IL-8. Lo *et al.* (1991) have presented evidence for a potential additional determinant. They found that co-incubating neutrophils, iC3b-coated erythrocytes and cytokine-stimulated HUVEC on fibronectin-coated plastic for 20 min at 37°C resulted in a high degree of erythrocyte binding to the neutrophils that was inhibited by preincubation of the HUVEC with anti-E-selectin mAbs, BB11 or H18/7, but not by anti-ICAM-1 mAb LB-2 (an antibody that inhibits only CD11a/CD18-ICAM-1 interactions (Diamond *et al.*, 1991)). Activation of iC3b binding also occurred when neutrophils were allowed to contact for 1 h at 37°C soluble recombinant E-selectin adsorbed to a plastic surface. Anti-E-selectin antibodies reduced the number of erythrocytes bound per neutrophil to baseline levels (i.e. when neutrophils contacted glycophorin or human serum albumin-coated plastic). In addition, the soluble recombinant E-selectin stimulated neutrophil motility. These authors speculated that E-selectin may provoke motility of interacting neutrophils, and increase CD18-dependent adhesion to endothelial cells. If such a mechanism primarily determined the transmigration seen in the Luscinskas *et al.* study (1991), then both E-selectin and ICAM-1 would be necessary for migration. However, since there appears to be redundancy in the endothelial mechanisms that stimulate neutrophil motility (e.g. IL-8 and PAF also functioning in this role (Huber *et al.*, 1991; Kuijpers *et al.*, 1992)), then anti-E-selectin antibodies may not inhibit migration of adherent neutrophils.

6. *Cellular and Molecular Mechanisms Regulating Neutrophil β2-Integrin Dependent Adhesion*

6.1 CD11a/CD18

Activation of CD11a/CD18 mediated adhesion may involve conformational changes of the heterodimer, transient reversible phosphorylation of the cytoplasmic domain of the CD18 subunit, or transient changes in cytoskeletal relationships. In lymphocyte studies, a recently described monoclonal antibody NKI-L16, appears to recognize an "activation" epitope on CD11a/CD18 in a Ca^{2+}-dependent manner (Keizer *et al.*, 1988; Van Kooyk *et al.*, 1991). As opposed to other CD11a/CD18 mAbs that inhibit lymphocyte adhesive interactions, this antibody is capable of stimulating CD11a/CD18-dependent lymphocyte aggregation. Although the ability of lymphocytes and lymphocyte cell lines to aggregate appears to correlate with the expression of this epitope, its expression alone is not sufficient for cell aggregation. Presumably, NKI-L16 mAb binding induces a conformational change that correlates with enhanced adhesiveness. Lymphocyte aggregation can be induced by stimuli such as phorbol esters or antibodies to T-cell antigen receptor-(TcR)-associated CD3 (Rothlein and Springer, 1986; Rothlein *et al.*, 1986; Dustin and Springer, 1989). It is likely that these stimuli, acting through intracellular signalling mechanisms, induce a similar conformational change or other activational responses yielding the adhesive state of CD11a/CD18. PMA (phorbol myristate acetate) induces not only CD11a/CD18/ICAM-1 dependent leucocyte homotypic aggregation, but also stimulates adhesion of isolated peripheral blood lymphocytes and CD11a/CD18 bearing lymphocyte cell lines to purified ICAM-1 (Dustin and Springer, 1989). PMA likely exerts its regulatory effects via activation of protein kinase C (PKC), perhaps resulting in either direct phosphorylation of the CD11a or CD18 cytoplasmic domains or regulating altered association with the cytoskeleton. The PMA induced activation of CD11a/CD18 mediated binding to ICAM-1 is blocked by the PKC inhibitor staurosporine (Dustin and Springer, 1989). In addition to PMA-induced activation of CD11a/CD18 avidity, Dustin and Springer demonstrated that crosslinking (T-cell receptor) TCR with anti-CD3 mAbs causes transient adhesion through CD11a/CD18 (Dustin and Springer, 1989). This activated state of CD11a/CD18 is less prolonged than that induced by phorbol esters, and in contrast to PMA-induced adhesion, is only partially reduced by staurosporine (Dustin and Springer, 1989), suggesting an activation mechanism potentially independent of PKC. Additionally, whereas the adhesion induced by TCR crosslinking was inhibited by elevation of cAMP (dibutyryl cAMP or forskolin plus phosphodiesterase inhibition), the PMA-stimulated adhesion to ICAM-1 substrate was not affected (Dustin and Springer, 1989). Similar differential mechanisms for activating prolonged or transient adhesion by phorbol esters vs. chemotactic factors may exist for neutrophil CD11b/CD18 dependent adhesion as outlined below.

Employing COS cells transiently expressing wild type or serially truncated cytoplasmic domain mutants of the α or β subunits of CD11a/CD18, Hibbs *et al.* (1991b) recently demonstrated that regulation of CD11a/CD18 adhesion to ICAM-1 is mediated by the β subunit. Interestingly, the β subunit mutants (although to a lesser degree than wild type expressing cells) could be induced to adhere with the NKI-L16 mAb described above, indicating further that conformational changes induced by this mAb may function analogously to the physiological regulatory mechanisms provided by the cytoplasmic domain of the β subunit. CD11a and CD18-expressing stably transfected (Epstein-Barr virus) EBV-transformed B lymphoblastoid cells established from a patient with leucocyte adhesion deficiency exhibited augmented adhesion to immobilized ICAM-1 with PMA stimulation. In contrast, lymphoblastoid cells expressing truncated or altered CD18 cytoplasmic domains demonstrated little or no adhesion to ICAM-1 with or without PMA, indicating that the β subunit cytoplasmic domain is required for the PMA (likely PKC mediated)-regulated increased adhesiveness.

Studies with intact neutrophils indicate that PKC-mediated phosphorylation of the cytoplasmic domain of CD18 correlates with increased adhesiveness to CD11b/CD18. No data exists *per se* regarding mechanisms of regulation of CD11a/CD18 binding in neutrophils. As CD11b/CD18 mediates neutrophil binding to protein-coated surfaces, unstimulated HUVEC, and other defined ligands such as iC3b and fibrinogen, most studies regarding regulation of neutrophil integrin-mediated binding have focused on CD11b/CD18. However, CD11a/CD18 clearly participates in binding of neutrophils to ICAM expressing endothelium and appears to recognize ICAM-1 during neutrophil migration through both unstimulated (Furie *et al.*, 1991b) and cytokine-stimulated (Smith *et al.*, 1989b) endothelium. Regulation of neutrophil CD11a/CD18 binding to ICAM-1 likely occurs through similar mechanisms as in lymphocytes. Neither is associated with any quantitative change in CD11a/CD18 expression (Kishimoto *et al.*, 1989b). As the β subunit mediates regulation of CD11a/CD18 adhesion (at least in lymphocytes) and CD11b/CD18 and CD11a/CD18 share the same β subunit (CD18), it is likely that the same intracellular molecular mechanisms participate in regulation of adhesion through both leucocyte integrins (as well as CD11c/CD18), with perhaps the specificity of the substrate (as well as relative surface expression) determining which integrin mediates a specific adhesive interaction. For example, in contrast to other substrates, it appears that both CD11b/CD18 and CD11a/CD18 can bind to purified ICAM-1 (Marlin and Springer, 1987; Diamond *et al.*, 1990; Diamond *et al.*, 1991; Staunton *et al.*, 1990). We have recently observed that whereas unstimulated neutrophils (expressing roughly similar levels of CD11a/CD18 and CD11b/CD18) exhibit little (≤5%) adhesion to ICAM-1 expressing stably transfected L-cells, 10 nM fMLP stimulation causes approximately 40% of neutrophils to adhere (absence of shear stress), and this augmented adhesion is mediated almost exactly half by CD11a/CD18 and half by CD11b/CD18 (Shappell and Wayne Smith, unpublished observations). These results indicate that in neutrophils, the intracellular signals induced by low levels of a chemotactic stimulus are sufficient to activate a similar level of adhesion by both CD11a/CD18 and CD11b/CD18 to a physiologically relevant substrate.

6.2 CD11b/CD18

In contrast to CD11a/CD18, which is constitutively expressed in the neutrophil plasma membrane and does not show any increased surface expression as a result of cell activation, CD11b/CD18 is not only constitutively expressed, but has a large intracellular storage pool. CD11b/CD18 appears to be contained largely in a subset of specific granules which are rich in gelatinase and low in vitamin B12 binding protein content (Arnaout *et al.*, 1984; Todd *et al.*, 1984; Bainton *et al.*, 1987; Jones *et al.*, 1988). Stimulation of neutrophils with phorbol esters or chemotactic factors results in a two- to ten-fold increase in the surface expression of CD11b/CD18 as a consequence of granule membrane and plasma membrane fusion during degranulation (Arnaout *et al.*, 1984; Berger *et al.*, 1984; Todd *et al.*, 1984). Buyon *et al.* (1988a) have demonstrated that newly up-regulated CD11b/CD18 is not required for the homotypic aggregation of neutrophils and that up-regulation is not sufficient to promote aggregation. Pretreatment of neutrophils with anti-CD11b/CD18 mAbs and subsequent washing is sufficient to inhibit homotypic aggregation despite the increased surface expression of CD11b/CD18 not bound by monoclonal antibody (Buyon *et al.*, 1988a). Pharmacological studies have also allowed dissociation of up-regulation and aggregation. Whereas some agents inhibit both up-regulation and aggregation in response to chemotactic stimulation (Shappell *et al.*, 1990a), treatment of neutrophils with relatively high concentrations of the non-steroidal anti-inflammatory agents sodium salicylate, piroxicam, and indomethacin inhibits neutrophil aggregation in response to 100 nM *N*-formyl-methionyl-leucyl-phenylalanine (fMLP) without reducing the increased surface expression of CD11b/CD18 at the time point (1 min) at which aggregation is maximal (Buyon *et al.*, 1988b). In reciprocal experiments, it was observed that 1 mM of the local anesthetic agent tetracaine was able to substantially inhibit the increased surface expression of CD11b/CD18 without inhibiting CD11b/CD18 dependent homotypic aggregation in response to the same level of fMLP stimulation (Buyon *et al.*, 1988b). Similar types of pharmacological experiments have demonstrated that increased surface expression of CD11b/CD18 is not necessary for chemotactic factor-stimulated adhesion to unstimulated endothelium (Vedder and Harlan, 1988).

It is important to recognize that studies demonstrating that it is possible to pharmacologically dissociate CD11b/CD18 dependent adhesion from CD11b/CD18 up-regulation under certain experimental conditions do not rule out a contribution of newly up-regulated CD11b/CD18 to adhesive interactions occurring under conditions in which increased surface expression does occur. Newly up-regulated CD11b/CD18 may become subject to the same regulatory mechanisms responsible for modulating avidity of constitutively expressed heterodimers, either by continued presence of a prolonged stimulatory factor, such as phorbol esters, or in response to successive exposures to chemotactic stimuli. The latter would be particularly important in mediating transient adhesive interactions and subsequent migration in response to progressively higher concentrations of a soluble chemotactic agent. The studies of Francis *et al.* (1989) and Hughes *et al.* (1990) discussed above are consistent with this concept.

The above data clearly indicate that CD11b/CD18 in the plasma membrane undergoes some modification resulting in increased avidity for its various substrates. Furthermore, both constitutively expressed and newly up-regulated CD11b/CD18 are capable of being regulated in this manner. The potential molecular basis for activation of CD11b/CD18 binding has been most extensively studied using phorbol esters as a stimulus. Whereas unstimulated human neutrophils show little capacity to bind iC3b-coated erythrocytes (E-C3bi), neutrophils stimulated with PMA have a transient increase in the capacity to bind and ingest E-C3bi (Wright and Meyer, 1986). This response appears to be maximal at approximately 15 min and returns to baseline at about 50 min. The loss of enhanced receptor activity appears to require the continued presence of phorbol stimulus and to be an active process that was temperature dependent. This transient increased affinity of CD11b/CD18 for iC3b in response to phorbol esters may be related to the clustering of CD11b/CD18 in the plasma membrane. Using immunoelectron microscopy, Detmers *et al.* (1987) demonstrated that neutrophil CD11b/CD18 clusters in the plasma membrane in response to PMA. This response was also transient, being evident at 10 min, greater at 25 min, and returning towards baseline at 50 min, paralleling the time course of augmented binding and phagocytosis of E-C3bi under the same conditions. In contrast, stimulation of neutrophils with fMLP (10–50 nM) did not result in any detectable CD11b/CD18 clustering at 10 or 20 min, and consistent with the correlation of clustering and augmented ligand binding, did not augment the capacity of cells to bind or ingest E-C3bi. In theory, integrin clustering in the plasma membrane would enhance the normally low affinity of integrin mediated interactions by decreasing the off rate during binding to surface bound ligand. Such receptor clustering likely involves interaction with the neutrophil cytoskeleton. That this response is induced by PMA would suggest that it involves a PKC-dependent phosphorylation event, either by direct phosphorylation of CD11b/CD18 subunits, a β_2-integrin associated protein, or a cytoskeletal associated protein, resulting in enhanced integrin-cytoskeleton interaction. That a PKC-dependent phosphorylation event is involved is supported by the observations of Wright *et al.* (1986) that pretreatment of neutrophils with thiophosphate exaggerated the PMA-induced increased binding of E-C3bi. The augmented attachment of E-Cbi was also prolonged (still elevated at 80 min), presumably due to the relatively stable phosphorylation (less susceptible to cellular phosphatases) of whatever the responsible involved proteins are.

Two separate groups of investigators have subsequently shown that CD18 is itself phosphorylated in response to phorbol ester stimulation of neutrophils (Chatila *et al.*, 1989; Buyon *et al.*, 1990). In unstimulated mononuclear cells and neutrophils, the three α subunits associated with it CD18 (CD11a, CD11b, and CD11c) are all constitutively phosphorylated, with little change evident with phorbol ester or fMLP stimulation. In contrast, there was little constitutive phosphorylation of CD18, but in response to PMA, a majority of CD18 is phosphorylated in both mononuclear cells and neutrophils (Chatila *et al.*, 1989; Buyon *et al.*, 1990). This response is rapid (essentially maximal at one minute) and prolonged, being sustained for at least 30 min, consistent with the ability of PMA to induce sustained activation of PKC in leucocytes, as described further below regarding respiratorary burst activation. In contrast, Chatila *et al.* (1989) observed that fMLP-induced phosphorylation of CD18 was very transient in both mononuclear cells and neutrophils, occurring at 30 and 60 s and returning to essentially baseline values at 5 min. This is consistent with the very transient nature of membrane associated PKC activation in response to chemotactic factors in neutrophils (detailed further below) and correlates well with the time course of reversible CD11b/CD18 mediated homotypic aggregation induced with the same (100 nM) concentration of fMLP (maximal at 1 min and essentially over by 5 min vs. the sustained aggregation induced with phorbol esters). The majority of the phosphorylation of CD18 in PMA or fMLP-stimulated monocytes represented phosphoserine, with only trace amounts of phosphothreonine or phosphotyrosine (Chatila *et al.*, 1989).

Although these studies clearly suggest a potential regulatory role for CD18 cytoplasmic domain phosphorylation in regulating adhesiveness of leucocyte integrins (particularly as the kinetics of phosphorylation parallel the time course of homotypic aggregation in neutrophils), such results do not establish a cause and effect relationship. As PMA induces a prolonged activation of PKC in neutrophils, sustained phosphorylation of CD18 may simply reflect the sustained activation of PKC in the cell membranes under such conditions and the availability of potential phosphorylation sites in CD18. However, adhesive regulation could still be mediated through a distinct, unrelated phosphorylation event, perhaps resulting in a conformational change in CD11/CD18 or membrane clustering as described above, both of which could be more directly mediated through associated membrane and/or cytoskeletal proteins.

In a follow up to their elegant studies demonstrating that the cytoplasmic domain of CD18 regulates CD11a/CD18 avidity Hibbs *et al.*, (1991b), Hibbs *et al.* (1991a) have provided strong evidence for this latter possibility for CD11a/CD18 mediated adhesion to ICAM-1. Using CD18 subunit mutants generated by oligonucleotide-directed mutagenesis and transiently or stably expressed with wild-type CD11a in COS cells or EBV transformed leucocyte adhesion deficiency (LAD) lymphoblast cells, respectively, these authors demonstrated that a serine residue at the 756 amino acid position in the CD18 cytoplasmic domain is the predominantly phosphorylated amino acid in PMA stimulated cells. However, serine phosphorylation could be dissociated from PMA-induced binding of cells to immobililized ICAM-1. B-cells expressing a mutant β subunit in which the serine at 756 was replaced with a cysteine did not demonstrate augmented CD18 phosphorylation in response to PMA, but still bound ICAM-1 in a manner stimulated by PMA to the same degree as wild-type CD18 expressing cells (Hibbs *et al.*, 1991a). Sites identified as crucial for regulated binding to ICAM-1 were a phenylalanine at 766 (four amino acids proximal to the COOH-terminus) and a series of three contiguous threonine residues at 758–760, regions highly homologous to the corresponding regions in the cytoplasmic domains of β_1 and β_3 integrin subunits as well. Phosphorylation of the identified crucial threonine residues does not appear to be required for augmented adhesion to ICAM-1 and does not appear to account for the role of these amino acids in PMA augmented adhesion of CD11a/CD18 to ICAM-1, although a partial contribution of this mechanism was not excluded (Hibbs *et al.*, 1991a).

It appears that CD11a/CD18 and CD11b/CD18 may potentially be regulated in a similar fashion through their identical β subunits. However, these studies addressing CD11a/CD18-ICAM-1 interactions cannot necessarily be translated directly to mechanisms for regulation of neutrophil CD11b/CD18-dependent interactions. No rigid data exists as yet clearly demonstrating specific cytoskeletal associations of CD11a/CD18 or CD11b/CD18 in intact neutrophils, so that it is possible that differential associations do exist. Furthermore, it is not clear if the β subunits associated with a specific α chain are preferentially phosphorylated in intact cells. Immunoprecipitation studies in PMA-stimulated neutrophils offer some hint as to this possibility. In the studies of Buyon *et al.* (1990) immunoprecipitation with either CD11b or CD11c mAb allowed detection of inducible phosphorylation of CD18, but immunoprecipitations of either resting or stimulated neutrophils with the anti-CD11a mAb did not detect any phosphorylated β chain. It is conceivable that the ability of integrins to activate specific signal transduction mechanisms in leucocytes (see below in Section 7.1.3) will also be specifically regulated, perhaps in a manner different than the regulation of adhesion *per se*, as association with different cellular elements may be involved in these two different functions of adhesion molecules. Expression systems demonstrating the appropriate biochemical regulatory steps in modulating adhesion, and participating in signal transduction of adhesion-related events and mutagenesis studies as employed by Hibbs *et al.* (1991b, 1991a) will likely provide a very powerful approach to further clarifying these complex integrin interactions in leucocytes.

Finally, evidence has been generated recently indicating that upon neutrophil stimulation, transient changes in membrane lipid composition might occur such that CD11b/CD18 (and CD11a/CD18) adhesiveness is modulated in an allosteric manner by an acidic amphophilic lipid (Hermanowski-Vosatka *et al.*, 1992). Addition of partially purified acidic lipids from PMA, TNF, or fNLP-stimulated neutrophils cause an increase in CD11b/CD18 dependent-binding of iC3b-coated erythrocytes to intact neutrophils. Similar lipid extracts from unstimulated neutrophils were without similar effect. The formation of this adhesion-promoting factor was apparently transient, paralleling the transient increase in CD11b/CD18 avidity of neutrophils stimulated under similar conditions. This lipid factor, termed integrin modulating factor-1 (IMF-1) by the authors, was also able to promote neutrophil adhesion to fibrinogen and unstimulated endothelial monolayers (Hermanowski-Vosatka *et al.*, 1992). Monoclonal antibody studies revealed that adhesion to HUVEC under these conditions was mediated partly by CD11b/CD18 and partly by CD11a/CD18, indicating that this lipid factor may be capable of modulating adhesion involving both these leucocyte integrins and supporting further the notion presented above, that CD11b/CD18 and CD11a/CD18 dependent adhesion of neutrophils may be regulated through molecularly similar or identical mechanisms. The adhesion promoting effect of this lipid moiety appeared to be due to a direct effect on the integrin molecules rather than a consequence of cell activation. Addition of IMF-1 did not result in cell activation as assessed by degranulation, and most importantly, addition of IMF-1 to immobilized purified CD11b/CD18 heterodimer was able to induce binding of iC3b-coated erythrocytes (Herrnanowski-Vosatka *et al.*, 1992). The chemical nature of this lipid moiety was not specifically characterized by traditional physico-chemical means. Its molecular weight by size exclusion chromatography was 340 + or −16 Da (Hermanowski-Vosatka *et al.*, 1992) and its chemical properties are consistent with its possible identity as an unsaturated fatty acid. Its specific identification will allow for the consideration of its novel adhesion-modulating biological function in the context of well-defined pathways of phospholipid turnover in stimulated neutrophils.

In summary, modulation of neutrophil CD11b/CD18 and CD11a/CD18 dependent adhesion appears to be a complexly, perhaps multifactorally, regulated process that might involve covalent modification of the CD18 cytoplasmic domain, transient changes in cytoskeletal relationships, and/or allosteric modulation by specific membrane lipids. These processes might translate to changes in the spatial configuration of the extracellular regions of the heterodimers, either by producing actual physical clustering of heterodimers in the plasma membrane or by exposing certain adhesion-mediating "activation" epitopes.

7. *Integrin-mediated Neutrophil Activation*

7.1 ADHESION-DEPENDENT RESPIRATORY BURST

Neutrophil β_2 integrins are capable of mediating a prolonged activation of reactive oxygen production by adherent neutrophils. The molecular mechanisms whereby this response is generated are still incompletely understood. However, much is known about the structure and potential regulation of the reduced nicotinamide adenine dinucleotide phosphate (NADPH) oxidase complex by which neutrophils produce superoxide and hydrogen peroxide. Consideration of the structure and function of the NADPH oxidase complex is thus useful in examining the manner in which β_2 integrins may affect neutrophil reactive oxygen production.

7.1.1 Structure and Regulation of the NADPH Oxidase of Neutrophils

Neutrophils produce superoxide anion upon stimulation with a variety of soluble and particulate stimuli by virtue of a multicomponent, transmembrane electron transport system termed the NADPH oxidase (Klebanoff, 1988). This enzyme complex is inactive in resting cells, but undergoes assembly and activation under appropriate conditions and utilizes NADPH on the cytoplasmic side as the electron donor for the one electron reduction of molecular oxygen in the extracellular fluid, or in the interior of invaginated phagosomes during phagocytosis. Likely participants in electron transfer include a flavoprotein, which may accept electrons from NADPH, and a b-cytochrome, which may represent the terminal oxidase capable of directly reducing molecular oxygen (Klebanoff, 1988). In addition, studies on NADPH oxidase activation in cell-free systems (Bromberg and Pick, 1985; Curnutte *et al.*, 1987b; Curnutte *et al.*, 1987a) have revealed the requirement for multiple cytosolic components (Curnutte *et al.*, 1989). Two of these factors are proteins, p47 and p67, which appear to undergo translocation to the plasma membrane upon stimulation of the neutrophil respiratory burst (Ambruso *et al.*, 1990; Clark *et al.*, 1990) and which are absent in distinct forms of autosomal recessively inherited chronic granulomatous disease (Volpp *et al.*, 1988; Clark *et al.*, 1989; Curnutte *et al.*, 1989; Lomax *et al.*, 1989).

The cytochrome b_{558} is a heterodimer with an α subunit of 22 kDa and a β subunit of 91 kDa (Parkos *et al.*, 1987). The b-cytochrome spectrum is absent in the classic X-linked form of chronic granulomatous disease (CGD) (Segal *et al.*, 1978; Segal, 1987). CGD is a rare disorder of neutrophil and monocyte oxidative metabolism, in which the patient's cells fail to undergo a respiratory burst upon stimulation. The most common form of CGD is the X-linked cytochrome b_{558}-negative form (approximately two-thirds of CGD patients (Ohno *et al.*, 1986; Clark *et al.*, 1989)), in which the 91 kDa β subunit of the cytochrome is defective, resulting in the absence of both the α and β subunits (Segal, 1987). In contrast, CGD patients who lack the b-cytochrome but who have an autosomal pattern of inheritance (about 5% of patients), likely possess a defect in the synthesis of the α subunit (Ohno *et al.*, 1986; Clark *et al.*, 1989).

Detailed analysis of autosomal recessive, cytochrome positive forms of CGD in cell-free oxidase systems has allowed for identification of multiple cytosolic factors which may function in regulation of neutrophil superoxide generation. The cell free assay utilizes membrane and cytosol fractions from unstimulated neutrophils, with superoxide generation occurring in the presence of Mg^{2+} and NADPH with normal cytosol and membranes upon addition of a membrane perturbing detergent such as sodium dodecyl sulfate (Bromberg and Pick, 1985) or the sodium salts of long-chain unsaturated fatty acids such as arachidonate (Bromberg and Pick, 1984; McPhail *et al.*, 1985). In complementation studies using mixtures of patient and normal cell fractions, it can be shown that membrane fractions from cells from patients with cytochrome-negative CGD are abnormal (i.e. do not support superoxide generation when mixed with normal neutrophil cytosol), consistent with the membrane location of the cytochrome (Weening *et al.*, 1985; Curnutte *et al.*, 1987b). In contrast, studies with cell fractions from patients with autosomal recessive CGD revealed deficient activity of the cytosol (Nunoi *et al.*, 1988; Curnutte *et al.*, 1989). Studies with fractionated normal cytosol demonstrated that multiple cytosolic components are required for superoxide generation and that two different cytosolic factors could be missing in distinct and complementing (with regard to restoring cell-free oxidase activity) autosomal recessive forms of CGD (Nunoi *et al.*, 1988; Curnutte *et al.*, 1989). Use of a polyclonal antiserum prepared against a guanosine triphosphate agarose-binding cytosolic fraction active in the cell-free system revealed that such patients lack either a 47 or 67 kDa protein (Volpp *et al.*, 1988). Recent examination of a large number of CGD patients by immunoblot analysis for these factors demonstrated that approximately one-third of all CGD patients are lacking the 47 kDa cytosolic factor, whereas lack of p67 is rarer, accounting for approximately 5% of all CGD cases (Clark *et al.*, 1989).

The cDNAs for both the p47 and p67 cytosolic oxidase factors have been cloned and sequenced (Lomax *et al.*, 1989; Volpp *et al.*, 1989; Leto *et al.*, 1990). Both p47 and p67 appear to have two segments of significant similarity to a non-catalytic domain (SH3) of a number of *src*-related oncogene products of the non-receptor tyrosine kinase family (Leto *et al.*, 1990). This motif is found in a variety of other proteins, including phospholipase C-γ, non-erythroid alpha-spectrin, and GAP (GTPase activating protein), which are located at the cortical cytoplasm in a variety of cell types, and hence this domain may function in membrane skeleton recognition and have functional significance with regard to membrane translocation of p47 and p67 upon cell activation (Lomax *et al.*, 1989; Volpp *et al.*, 1989; Leto *et al.*, 1990) (and references therein). In addition, the COOH-terminus of p47 has multiple serine residues that conform to consensus PKC phosphorylation sites (Lomax *et al.*, 1989; Volpp *et al.*, 1989), of significance with regard to oxidase activation as indicated below.

Evidence from studies with non-hydrolyzable analogs of guanosine triphosphate (GTP) in the cell-free system indicate the possible role of a GTP-binding protein in the activation of the NADPH oxidase (Gabig *et al.*, 1987; Seifert and Schultz, 1987). A potential candidate protein is a *ras*-related protein that has recently been demonstrated to be associated with cytochrome b558 in stimulated human neutrophils (Quinn *et al.*, 1989). Partial purification of cytochrome b from stimulated neutrophils resulted in co-isolation of a 22 kDa protein with an amino terminal sequence similar to the human RAS proteins and which was recognized by a mAb directed against the GTP-binding domain of RAS proteins (Quinn *et al.*, 1989). Screening a cDNA library with oligonucleotides based on the partial amino acid sequence, resulted in the isolation of a cDNA clone identical to *ras*-related rap-1 (Quinn *et al.*, 1989). Furthermore, immuno-affinity purification experiments revealed co-purification of both cytochrome subunits and the 22 kDa ras-related protein with either anti-cytochrome antibody or anti-ras antibody conjugated matrices (Quinn *et al.*, 1989). However, other investigators found that the $p21^{rap1}$ protein completely dissociates from the cytochrome during its progressive purification (to a form still completely functional as the sole membrane component in the cell-free assay system) (Abo *et al.*, 1991). These authors proposed that the GTP-binding cytosolic component required for oxidase stimulation in the cell-free assay is a heterodimer composed of the small GTP-binding protein $p21^{rac1}$ and the GDP-dissociation inhibitor rhoGDI (Abo *et al.*, 1991). Partially purified recombinant $p21^{rac1}$ stimulated superoxide production in the cell-free assay when first preincubated with GTP-gamma-S (Abo *et al.*, 1991). $p21^{rac1}$ is a member of the *rho* subfamily of *ras* proteins, which may function to regulate actin polymerization in different cell types. Hence, the $p21^{rac1}$/GDI complex may function in promoting or regulating the assembly of the oxidase complex at the plasma membrane of activated neutrophils (Abo *et al.*, 1991).

In addition to the possible regulatory function of guanine nucleotides, substantial evidence exists for a role for PKC in the activation of the respiratory burst in neutrophils. Phorbol esters, such as phorbol myristate acetate (PMA), are potent activators of neutrophil reactive oxygen secretion (Tauber, 1987), although they are obviously non-physiological. Activation of PKC may contribute to stimulation or the assembly of the neutrophil oxidase by the phosphorylation of certain components of the NADPH oxidase-complex (Tauber, 1987). As mentioned above, the predicted amino acid sequence of the 47 kDa oxidase component contains several potential PKC phosphorylation sites (Lomax *et al.*, 1989; Volpp *et al.*, 1989). Stimulation of human neutrophils with PMA, opsonized latex beads, calcium ionophore A23187, or high concentrations of fMLP results in the phosphorylation of a 47 kDa protein (Heyworth and Segal, 1986). Phosphorylation of the 47 kDa protein is absent in neutrophils or neutrophil cytoplasts prepared from patients with the autosomal recessive form of CGD (Heyworth and Segal, 1986; Kramer *et al.*, 1988), indicating the identity of this PKC substrate to the p47 factor described above. The existence of several pI forms of the 47 kDa phosphoprotein after polymorphonuclear neutrophil's (PMNs) are activated (6.8–9.5), compared to the basic pI in unstimulated cells (approx. 10) and its overall basic charge predicted by amino acid sequence, is consistent with multiple phosphorylations of the same protein (Okamura *et al.*, 1988; Curnutte *et al.*, 1989; Lomax *et al.*, 1989; Volpp *et al.*, 1989). Stimulation of intact neutrophils with PMA results in a subsequent deficiency of the cytosolic component (vs. that of resting cells) in the cell-free oxidase system (Ambruso *et al.*, 1990), and it appears that there is a sequential relationship between the phosphorylation of p47 and its appearance in the membrane (Heyworth *et al.*, 1989). Nauseef and colleagues (Nauseef *et al.*, 1991) recently demonstrated the PKC-dependent cytoskeletal and membrane association of the p47 and p67 oxidase components. Whereas p67 was constitutively present in a triton-insoluble "cytoskeletal" fraction, p47 was localized in the triton soluble fraction of unstimulated neutrophils. PMA stimulation of neutrophils (at concentrations optimal for respiratory burst activation) resulted in the cytoskeletal association of a major portion ($\geqslant$50%) of p47 and the membrane translocation of both p47 and p67. Staurosporine inhibited the PMA-induced secretion of superoxide and concomitantly inhibited p47 phosphorylation and the cellular redistribution of both p47 and p67 (Nauseef *et al.*, 1991). Hence, the data are consistent with a PKC-dependent assembly of the oxidase at the membrane-cytoskeleton interface.

The physiological regulation of PKC is presumed to be mediated by the formation of diacylglycerol (DAG) (Nishizuka, 1986; Berridge, 1987; Nishizuka, 1988; Rando, 1988; Kikkawa *et al.*, 1989). The neutrophil respiratory burst is activated by synthetic membrane permeable diacylglycerols, such as 1-oleoyl-2-acetyl-glycerol (OAG) (Tauber, 1987), and the superoxide generation achieved in response to PMA or OAG is blocked by inhibitors of PKC, such as the isoquinoline sulfonamide derivatives H7 and C1 (Hidaka *et al.*, 1984; Gerard *et al.*, 1986; Berkow *et al.*, 1987). However, in contrast to agents like PMA and OAG, chemotactic factors, such as fMLP, LTB_4, interleukin-8, and complement fragment C5a, for which neutrophils have specific membrane receptors, elicit little reactive oxygen from neutrophils stimulated in suspension, even at concentrations 2–3 orders of magnitude higher than those required for the stimulation of maximal chemotaxis and adhesion (Snyderman and Pike, 1984; Nathan, 1987). Furthermore, the small respiratory burst elicited by these receptor-mediated stimuli has been largely found to not be inhibited by PKC inhibitors (Gerard *et al.*, 1986; Berkow *et al.*, 1987; Gaudy *et al.*, 1988). These types of results have been taken to indicate the existence of multiple pathways for activating the neutrophil respiratory burst. Recently, Kramer *et al.* (1989) used a 1,2-*O*-dialkylglycerol inhibitor of PKC (1-*O*-hexadecyl-2-*O*-methyl-*rac*-glycerol or AMG-C_{16}) to investigate the role of PKC and specific phosphorylation of p47 in the neutrophil superoxide generation in response to phorbol esters vs. chemotactic factors. AMG-C_{16} inhibited superoxide generation by suboptimally phorbol dibutyrate (PDBu)-stimulated neutrophils in a dose-dependent manner (that was diminished by increasing doses of stimulus). 125 μM AMG-C_{16} inhibited the superoxide production when neutrophils were stimulated with low doses of fMLP (e.g. 30–100 nM), but had little to modest effect when μM levels of fMLP were employed (Kramer *et al.*, 1989). Effects on the respiratory burst were correlated with PKC activity in intact cells by examining phosphorylation of p47 in cytoplasts. Cytoplasts contain only one phosphoprotein at 47 kDa, the 47 kDa protein associated with oxidase activity as described above (Kramer *et al.*, 1988). With either 4α-phorbol 12,13-dibutyrate (PDBu) or fMLP stimulation of cytoplasts, there was a dose-dependent phosphorylation of p47. However, with similar levels of superoxide production, there was always a lesser extent of p47 phosphorylation with fMLP than observed with PDBu (Kramer *et al.*, 1989). When AMG-C_{16} inhibition of the respiratory burst was overcome by increasing the concentration of the stimulus, there was good correlation between the extent of p47 phosphorylation and activity recovery with PDBu as a stimulus, but with fMLP, recovery of respiratory burst activity occurred initially without the simultaneous recovery of observable p47 phosphorylation (Kramer *et al.*, 1989). These data add further support to the notion that besides PKC activation, additional signals (elicited by Ca^{2+}-mobilizing agonists such as fMLP) exist for the activation of the oxidase complex in intact cells. However, it should be borne in mind that most investigators find phorbol esters to be much more potent activators of neutrophil superoxide and hydrogen peroxide generation than very high levels of agents such as fMLP. Most importantly, the respiratory burst in response to phorbol esters is sustained (e.g. greater than 1 h), whereas that initiated by chemotactic factors is extremely transient (e.g. 2–5 min). Therefore, even if examination at relatively early time points (e.g. typically 15 min) reveals similar quantities of superoxide under the different stimulation conditions, the absolute total quantities (particularly relevant for oxidative injury of other cell types) achieved with phorbol esters are 50–100 fold or so greater.

The neutrophil plasma membrane receptors for chemotactic factors such as fMLP, C5a, and LTB_4 appear to be coupled by a pertussis toxin-sensitive G (guanine nucleotide binding heterotrimeric) protein to a phospholipase C enzyme (Snyderman and Uhing, 1988). Stimulation of these receptors results in the hydrolysis of polyphosphoinositides (particularly phosphatidylinositol 4,5-bisphosphate or PIP_2) with formation of inositol 1,4,5-triphosphate (IP_3), which mobilizes intracellular calcium, and sn-1,2-diacylglycerol (DAG) (reviewed in Berridge, 1987). However, the formation of diacylglycerol under such circumstances is extremely transient (Honeycutt and Niedel, 1986) and may be insufficient for sustained stimulation of the NADPH complex and production of substantial superoxide or hydrogen peroxide.

Further evidence for the differential ability of phorbol esters vs. chemotactic factors to stimulate a sustained PKC-dependent respiratory burst comes from studies addressing the membrane translocation of PKC. PKC is a family of isoenzymes, all composed of a single polypeptide chain (Kikkawa *et al.*, 1989). The cDNAs for numerous isozymes have been cloned and sequenced, and reveal an overall conserved structure consisting of an N-terminal, relatively lipophilic regulatory domain and a C-terminal catalytic domain (Kikkawa *et al.*, 1989). These domains are joined by a "hinge region", which represents a site for proteolytic cleavage (e.g. calpain, trypsin), yielding the free catalytic domain which is enzymatically active (in the absence of Ca^{2+} and phospholipid) (Nishizuka, 1986; Berridge, 1987; Kikkawa *et al.*, 1989). Stimulation of cells with phorbol esters (which bind to the hydrophobic N-terminal domain) results in membrane translocation of PKC and enzyme activation (Nishihira and O'Flaherty, 1985; Wolfson *et al.*, 1985; Nishizuka, 1986; Berridge, 1987;

Nishizuka, 1988; Kikkawa *et al.*, 1989). Presumably, formation of diacylglycerol (DAG) in the plasma membrane (accompanied by a rise in Ca^{2+} in the case of stimulated neutrophils) results in a Ca^{2+}-dependent PKC translocation and activation. However, as expected from the above comments, the accumulation of Ca^{2+} and DAG in neutrophils stimulated with ligands binding to (PIP_2-specific) phospholipase C-coupled receptors is extremely transient, so that one cannot demonstrate a sustained translocation of PKC to neutrophil membranes following chemotactic-factor stimulation. Using [^{3}H]-phorbol binding and enzyme activity assays, Nishihira and colleagues showed that whereas PKC was translocated to the membrane following 15 min incubations with PMA or A23187, similar translocation was not observed with fMLP, C5a, LTB_4, or PAF (absence of cytochalasin pretreatment) (Nishihira *et al.*, 1986). We have made similar observations by immunoblot analysis using a mAb that recognizes the hinge region (site of limited proteolysis) of PKC (Young *et al.*, 1988) and that cross-reacts with the human α and β PKC isozymes, the predominant forms in human neutrophils (Pontremoli *et al.*, 1990; Obel *et al.*, 1991). In contrast to the increased membrane translocation seen with PMA stimulation, cytosol and membrane distribution of PKC in cells stimulated with even 1 μM fMLP was essentially the same as in unstimulated neutrophils when analysed at 15 min (Shappell, unpublished results). Although it appears that stimulation with such agonists might induce a transient association of PKC with the membrane (O'Flaherty *et al.*, 1990), the brief nature of the DAG and/or Ca^{2+} increase responsible for stimulating translocation and a potential simultaneous calcium-dependent translocation and activation of calpain (resulting in cleavage of PKC as described above) (Melloni *et al.*, 1985) might dictate that PKC membrane association is highly transient in response to these physiological type stimuli. In the case of potential proteolysis, the kinase (catalytic) domain is no longer capable (sufficiently hydrophobic) to remain membrane associated, yet may be responsible for phosphorylating cytosolic components (such as proteins in the cytoskeleton) (Kishimoto *et al.*, 1983; Pontremoli *et al.*, 1986; Pontremoli *et al.*, 1988). Given the membrane location of the NADPH oxidase complex as described above, sustained activation in a PKC-dependent manner might require localization of active enzyme in the membrane (Pontremoli *et al.*, 1988), a situation that could perhaps be achieved by a sustained accumulation of diacylglycerol (analogous to the conditions achieved with cell stimulation by the membrane-associating, metabolically inert phorbol esters). Consistent with this notion is the fact that pretreatment of neutrophils with the fungal metabolite cytochalasin B allows for marked and sustained production of superoxide and concomitant diacylglycerol accumulation upon subsequent exposure to chemotactic factors (Honeycutt and Niedel, 1986). In contrast to the results obtained when cells are stimulated with chemotactic factors alone, numerous investigators have been readily able to demonstrate translocation of PKC activity to membranes (at later time points, such as 15 min) when neutrophils are pretreated with cytochalasin B and then stimulated with chemotactic factors (Nishihira *et al.*, 1986; Pike *et al.*, 1986; Christiansen *et al.*, 1988).

7.1.2 Integrin Mediated Adhesion and Activation of the Neutrophil Respiratory Burst

In contrast to the inability of chemotactic factors or cytokines, such as TNF, to induce a sustained respiratory burst from neutrophils in suspension, Nathan (1987) reported that recombinant TNFα and recombinant TNFβ caused a marked production of H_2O_2 by neutrophils simultaneously exposed to polystyrene surfaces coated with serum, extracellular matrix proteins, or HUVEC. These studies and subsequent studies by Nathan's laboratory and ours on the role of integrins in this response are performed using 96-well microtiter tissue culture plates with the quantitation of H_2O_2 based on the oxidation of the fluorescent compound scopoletin to a non-fluorescent product. This reaction is followed in a plate reading fluorometer and is catalysed by horseradish peroxidase in the incubation buffer, which also includes sodium azide (to inhibit myeloperoxidase and catalase in order to facilitate H_2O_2 accumulation and quantitation). Studies with neutrophils from CGD patients and with catalase in the media demonstrate that fluorescence changes are specific for oxidase activation and H_2O_2 generation (Nathan, 1987). As previously observed, neutrophils exposed to uncoated polystyrene undergo an immediate respiratory burst. Nathan demonstrated that coating the polystyrene plates with serum, the extracellular matrix proteins fibronectin, vitronectin, or laminin, or HUVEC prevented this plastic induced response, with the duration of suppression (time until spontaneous release of H_2O from otherwise unstimulated neutrophils) varying with the specific substrate (e.g. >3–4 h with laminin and HUVEC) (Nathan, 1987). Neutrophils on such coated polystyrene plates underwent a massive and prolonged production of H_2O_2 in response to levels of TNF and fMLP that were insufficient to stimulate a detectable respiratory burst from cells in suspension (but which are sufficient to promote adhesion). The production of H_2O_2 was prolonged (e.g. 1–3.5 h, depending on substrate), with final quantities achieved being essentially of the same magnitude as with PMA. H_2O_2 production with chemotactic or cytokine stimulation under these conditions occurred following a distinct lag period (e.g. 15–90 min depending on the substrate). Importantly, under the same conditions, the respiratory burst induced by PMA does not exhibit a lag phase. The lag phase with TNF-stimulated neutrophils on serum coated surfaces was not affected by pretreatment of neutrophils with TNFα in suspension prior to exposure to the coated surface or by allowing neutrophils to contact the surface before adding TNF (Nathan, 1987). Other soluble stimuli have been tested to promote the neutrophil respiratory burst under these or similar conditions (see below), and in general the ability of agents to promote such a response agrees with the agent's capacity to promote neutrophil adhesion. For example, all chemotactic factors examined thus far, including C5a, PAF, and LTB_4 will promote the adhesion-dependent respiratory burst (Shappell *et al.*, 1990c); (Shappell, unpublished observations; C.F. Nathan, personal communication). The colony stimulating factor GM-CSF, which promote CD18-dependent adhesion and increase surface expression of CD11b/CD18 (Arnaout *et al.*, 1986), is also capable of stimulating the production of H_2O_2 by neutrophils simultaneously exposed to serum or laminin-coated surfaces (Nathan, 1989). In contrast, IFNα, IFNβ, IFNγ, PDGF, IL-1β, IL-3, M-CSF, or endotoxin do not appear capable of inducing the neutrophil respiratory burst under these conditions. Importantly, in contrast to the marked production of H_2O_2 by neutrophils, adherent monocytes (from the same blood samples) do not demonstrate H_2O_2 production in response to TNFα, TNFβ, G-CSF, or GM-CSF over the same time course (Nathan, 1987; Nathan, 1989).

Using mAbs to integrin subunits and neutrophils from patients with leucocyte adhesion deficiency (LAD), we have investigated the role of specific CD11/CD18 integrins in mediating H_2O_2 production by chemotactic-factor stimulated neutrophils on protein-coated surfaces, endothelial monolayers, and during neutrophil adhesion to isolated cardiac myocytes. As with the extracellular matrix proteins described above, collagen I prevented the plastic induced production of H_2O_2 in a microtiter plate assay. fMLP-stimulated neutrophils did not adhere to collagen type I-coated glass coverslips (visual adherence assay as described above for neutrophil-endothelial interactions), and consistent with the role of adhesion in mediating respiratory burst activation, no detectable H_2O_2 production occurred with neutrophils exposed to collagen I-coated polystyrene and a range of fMLP concentrations (Shappell *et al.*, 1990c). In contrast, PMA induces a marked respiratory burst from neutrophils on collagen I, similar to that described on other protein surfaces. In contrast to this substrate, we observed that whereas unstimulated neutrophils exhibited little adhesion to coverslips coated with keyhole limpet hemocyanin (KLH), 10 nM fMLP induced adhesion of a majority of neutrophils ($\geqslant$60%). CD18-deficient neutrophils did not demonstrate stimulated adhesion to this substrate, and the adhesion of stimulated normal neutrophils was essentially completely inhibited by the anti-CD18 mAbs R15.7 and the anti-CD11b mAb 904, indicating that KLH is a CD11b/CD18 dependent adhesive substrate (Shappell *et al.*, 1990c). As with extracellular matrix proteins, KLH-coating prevented the spontaneous respiratory burst induced by plastic and supported a dose-dependent production of H_2O_2 by fMLP-stimulated cells that was prolonged and occurred with a lag phase of approximately 50 min. In contrast to the essentially normal production of H_2O_2 by CD18-deficient neutrophils on uncoated plastic (Nathan *et al.*, 1989; Shappell *et al.*, 1990c) and in response to phorbol esters in suspension, CD18-deficient neutrophils did not exhibit H_2O_2 production with fMLP stimulation on KLH (Shappell *et al.*, 1990c). The production of H_2O_2 by 10 nM fMLP-stimulated normal neutrophils on KLH was blocked by anti-CD11b and anti-CD18 mAbs (e.g. 5 μg/ml, added at time zero of incubation), indicating that CD11b/CD18 can mediate respiratory burst activation (Shappell *et al.*, 1990c). Similar results were obtained using platelet activating factor (PAF)-stimulated canine neutrophils on collagen I and KLH (Shappell *et al.*, 1990c). In addition, as described by Nathan for TNF, fMLP also stimulated a delayed and prolonged production of H_2O_2 by human neutrophils exposed to HUVEC monolayers (Shappell *et al.*, 1990c). PAF induced a similar response with canine neutrophils exposed to cultured jugular vein endothelial cells (Shappell *et al.*, 1990c). CD18-deficient human neutrophils did not demonstrate any detectable H_2O_2 production when stimulated with fMLP on HUVEC, and the respiratory burst of fMLP-stimulated human and PAF-stimulated canine neutrophils was blocked by both anti-CD11b or anti-CD18 mAbs. Anti-CD11a/CD18 mAbs were without effect under any of the above conditions and none of the mAbs tested had any effect on the respiratory burst induced with artificial stimuli in suspension (Shappell *et al.*, 1990c). Hence, it appears that CD11b/CD18 is responsible for the activation of the neutrophil NADPH oxidase during adhesion to endothelial cells.

Nathan, Wright, and colleagues presented evidence for a role of CD18 in the production of H_2O_2 by TNF-stimulated neutrophils on extracellular matrix proteins (Nathan *et al.*, 1989). In contrast to normal neutrophils, CD18-deficient cells did not undergo a respiratory burst in response to TNF on plastic coated with serum, fibronectin, vitronectin, fibrinogen, thrombospondin, or laminin (Nathan *et al.*, 1989). Ninety minutes of pre-exposure at 37°C and continuous exposure throughout the assay to 30 μg/ml of the anti-CD18 mAb IB4 completely inhibited the H_2O_2 production by stimulated normal neutrophils on surfaces coated with FBS, fibrinogen, laminin, and thrombospondin, but not at all on fibronectin or vitronectin (or uncoated polystyrene) (Nathan *et al.*, 1989). Interestingly, pre-exposure to mAb was required (in contrast to results described above) and preincubation at 4°C was not sufficient to induce inhibition, despite continued presence of mAb throughout the assay period (conducted at 37°C). In addition, numerous other integrin mAbs were without effect when used under the same conditions as necessary to obtain inhibition with IB4 (Nathan *et al.*, 1989).

As the magnitude of the respiratory burst under such adhesion-related conditions is similar to that induced by phorbol esters (typically employed for cytotoxicity assays), this has important implications for neutrophil mediated target cell injury during adhesive interactions. In fact, conditions of CD11b/CD18-dependent adhesion and respiratory burst activation of neutrophils exposed to unstimulated endothelial monolayers result in endothelial injury that is reduced by anti-CD18 and anti-CD11b mAbs (Von Asmuth *et al.*, 1991). TNF pretreatment of neutrophils caused approximately 20% neutrophil adhesion to unstimulated HUVEC monolayers (10 min incubation and a visual adhesion assay similar to that described by Smith *et al.*, 1988; Smith *et al.*, 1989b and as detailed above). This adhesion was reduced substantially (more than 50%) by anti-CD18 mAbs and about 30% by various anti-CD11b mAbs (Von Asmuth *et al.*, 1991). Incubation of TNF-pretreated neutrophils with HUVEC resulted in substantial H_2O_2 production and endothelial cell detachment ($\geqslant$60%) during 3 h protocols. Similar to our observations with chemotactic-factor induced adhesion to HUVEC (Shappell *et al.*, 1990c), H_2O_2 production was similarly (approximately 80%) inhibited by anti-CD18 mAbs and the anti-CD11b mAb 904 (Von Asmuth *et al.*, 1991). These same mAbs also reduced endothelial cell detachment (with greater inhibition observed with the anti-CD18 mAbs). Anti-CD11a/CD18 mAbs were without effect. Interestingly, two other anti-CD11b mAbs (44 and PDV11.2) reduced adhesion to a similar degree as anti-CD11b mAb 904, but in contrast to 904, did not reduce H_2O_2 production or reduce endothelial detachment. TNF pretreatment of HUVEC (4 h) increased ELAM-1 and ICAM-1 expression and increased neutrophil adhesion (to 62%). Under the static conditions employed, a substantial amount (about 40%) of this augmented adhesion was reduced by the anti-ELAM-1 monoclonal antibody ENA2, but only a minor inhibition was observed with the anti-ICAM1 mAb RR1/1. Cytokine pretreatment of HUVEC did not appear to augment the magnitude of H_2O_2 production of TNF-stimulated neutrophils adherent to HUVEC, nor did it result in any increase in endothelial detachment. With cytokine-pretreated HUVEC, neither the TNF-induced neutrophil respiratory burst nor cytotoxicity (detachment) was reduced by the anti-ELAM-1 or anti-ICAM-1 mAbs employed. However, the injury and respiratory burst under these conditions appeared to be CD11b/CD18 dependent. This is similar to experiments performed in our laboratories with HUVEC (Shappell *et al.*, 1990c; and as described above) and cardiac myocytes (Entman *et al.*, 1991b; and as described below). Hence, elucidating the possible involvement of ICAM-1 in supporting a CD11b/CD18-ICAM-1 interaction in neutrophil activation is critically dependent upon utilization of anti-ICAM-1 mAbs that block portions of the protein involved in CD11b/CD18 dependent [domain 3 (Diamond *et al.*, 1991)] vs. CD11a/CD18 dependent [domains 1 and 2 (Staunton *et al.*, 1990)] adhesion. Monoclonal antibody (mAb) RR1/1, for example, appears to block CD11a/CD18-ICAM-1 interactions, but not CD11b/CD18-ICAM-1 interactions (Diamond *et al.*, 1990), and hence would not address the issue of involvement of ICAM-1 in a CD11b/CD18 dependent neutrophil activation response. As described below, we have observed a potential involvement of a CD11b/CD18-ICAM-1 interaction in the activation of the respiratory burst and cytotoxic response of neutrophils adherent to cytokine-stimulated cardiac myocytes (Entman *et al.*, 1991b).

Substantial evidence indicates that neutrophils contribute to myocardial injury during ischemia and reflow (reviewed in Entman *et al.*, 1991a; Smith *et al.*, 1991a). In addition to leucocyte plugging of the microvasculature and direct endothelial injury, neutrophils migrating into the inflamed myocardium may directly induce myocyte injury through release of reactive oxygen species and proteolytic enzymes. As neutrophil adhesion may be a significant stimulus for these cytotoxic responses, we have investigated the capacity of neutrophils to directly adhere and destroy isolated cardiac myocytes. Canine neutrophils stimulated with chemotactic factors adhere to canine cardiac myocytes that have been stimulated with phorbol esters or cytokines, such as IL-1, TNFα, and IL-6 (Entman *et al.*, 1990; Smith *et al.*, 1991b; Youker *et al.*, 1992). The capacity of stimulated myocytes to support neutrophil adhesion is dependent upon protein synthesis (Entman *et al.*, 1990). This adhesion is CD18-dependent as evidenced by its inhibition by anti-CD18 mAb R15.7 (Entman *et al.*, 1990). As with CD18-dependent adhesion to protein-coated surfaces and endothelium, this neutrophil-myocyte adhesion results in the prolonged elaboration of H_2O_2 that is also blocked by anti-CD18 mAbs (Entman *et al.*, 1990). When analyzed with the extracellular detection system for H_2O_2 based on the horseradish peroxidase catalyzed oxidation of scopoletin as described above, H_2O_2 during neutrophil-myocyte adhesion was detected following a pronounced lag period (Entman *et al.*, 1990). More recently, we have demonstrated that the adhesion of chemotactic factor-stimulated canine neutrophils to cytokine-stimulated canine myocytes is dependent upon expression of myocyte ICAM-1 (Smith *et al.*, 1991b). Cytokine stimulation of myocytes results in induction of ICAM-1 mRNA and expression of ICAM-1 on the cell surface, and the neutrophil-myocyte adhesion is blocked by a mAb to canine ICAM-1 (Smith *et al.*, 1991b). Furthermore, this adhesion appears to be mediated by canine neutrophil CD11b/CD18 and not CD11a/CD18, as it is blocked by a mAb that reacts with canine CD11b/CD18 but is not affected by a mAb to canine CD11a/CD18 (that is capable of inhibiting neutrophil-endothelial interactions dependent upon CD11a/CD18

(Smith *et al.*, 1991b). We have further investigated the oxidative nature of this cell-cell interaction using myocytes loaded with the intracellular probe dihydrodichlorofluorescein (DCFH), which is oxidized by H_2O_2 hydroxyl radical, or lipid peroxides to a fluorescent derivative. Using this detection system, neutrophil adhesion to cytokine stimulated myocytes results in a rapid (within minutes) myocyte oxidation (increased DCF fluorescence) that is accompanied by myocyte toxicity (contracture). Myocyte oxidation and injury is blocked by mAb that block adhesion (anti-CD18, anti-ICAM-1, anti-CD11b) and unaffected by anti-CD11a/CD18 or control mAbs (Entman *et al.*, 1991b). This cellular interaction may not be associated with the extracellular elaboration of neutrophil derived oxidants as indicated by the lack of superoxide detection using extracellular cytochrome c reduction or spin-trapping with non-cell-permeable traps and the lack of effect on myocyte oxidation by exogenously added superoxide dismutase and catalase (Entman *et al.*, 1991b). In contrast, pretreatment of myocytes with iron-chelators and hydroxyl radical scavengers (capable of remaining cell-associated) prevents myocyte oxidative injury although there is no apparent reduction of neutrophil adhesion (Entman *et al.*, 1991b). These experiments offer further evidence for the capacity of CD11b/CD18 to initiate activation of the neutrophil NADPH oxidase.

7.1.3 Possible Signal Transduction Mechanisms in β_2-integrin Mediated Neutrophil Oxidase Activation

Nathan and colleagues recently presented evidence for a role of changes in cAMP and actin reorganization in the β_2 integrin-dependent sustained respiratory burst of TNFα-stimulated neutrophils on serum (FBS) or fibrinogen (FGN)-coated plastic (Nathan and Sanchez, 1990). Neutrophils stimulated with TNF on FBS or FGN underwent a sustained drop in intracellular cAMP. Whereas unstimulated cells on coated plastic had a rapid drop in cAMP that was quickly restored to control levels and maintained for the 2 h experiment duration, TNF-stimulated cells demonstrated a sustained depression of cAMP that did not return to baseline values until the termination of the respiratory burst. Pretreatment with the anti-CD18 mAb IB4 under conditions that inhibit H_2O_2 production by TNF-stimulated cells on FBS prevented the sustained drop in cAMP. Furthermore, whereas CD18-deficient neutrophils demonstrated a sustained depression of cAMP on uncoated plastic (a response that does not require cytokine or chemotactic factor stimulation), there was no such response with TNF-stimulation of cells exposed to serum-coated plastic. That this fall in cAMP under conditions of the integrin-dependent respiratory burst might be essential for subsequent NADPH oxidase activation was further evidenced by the fact that agents that elevate cAMP blocked the respiratory burst if added before, but not after, its onset. Cytochalasins also blocked H_2O_2 production if added before, but not after, its onset. Cytochalasins also blocked the TNF-induced fall in cAMP of cells on FBS, and agents that elevated cAMP prevented cell spreading and actin filament assembly if added before cell spreading began. The potential involvement of cytoskeletal reorganization as a consequence of or a contributor to integrin-mediated neutrophil activation is considered further below.

Substantial evidence exists supporting a role for protein kinase C (PKC) in the regulation of the NADPH oxidase in neutrophils under certain activation conditions. We have recently demonstrated the requirement for PKC activity in the integrin-mediated (CD11b/CD18 dependent) respiratory burst of fMLP-stimulate neutrophils on keyhole limpet haemocyanin (KLH). In contrast to the inability of the isoquinoline sulfonamide protein kinase inhibitors to block reactive oxygen production by neutrophils stimulated with high levels of chemotactic factors in suspension (Gerard *et al.*, 1986; Berkow *et al.*, 1987), H7 exhibited a concentration-dependent inhibition of H_2O_2 production by 10 nM fMLP-stimulated human neutrophils on the CD11b/CD18 substrate KLH (Shappell *et al.*, 1990b). HA1004, which inhibits cyclic nucleotide dependent-protein kinases with similar potency have indicated cytoskeletal linkage of CD11b/CD18 in neutrophils stimulated by exposure to immobilized C3 or phorbol myristate acetate (PMA), respectively. Further studies are needed to more definitively demonstrate cytoskeletal association of CD11b/CD18 in neutrophils and to elucidate possible mechanisms that might regulate such associations.

Hence, the neutrophil β_2-integrin mediated respiratory burst may be accompanied by a qualitative change in the nature of adhesion, resulting in cytoskeletal reorganization and NADPH oxidase activation. Consistent with this is the fact that under conditions of a CD18-dependent respiratory burst (TNF stimulation of neutrophils on FBS or fibrinogen), actin polymerization and cell spreading occur with a general temporal correlation with the generation of H_2O_2, and the adhesion-dependent respiratory burst is inhibited by cytochalasin if it is added prior to but not after the onset of H_2O_2 generation (Nathan and Sanchez, 1990). In addition, as detailed above regarding cytoskeletal association of specific oxidase components, the cytoskeleton may have a more "mechanical" function in the assembly of an active NADPH oxidase. Babior and colleagues recently reported that upon PMA activation of human neutrophils, the more acidic (more phosphorylated) forms of the oxidase associated family of 48 kDa (p47) phosphoproteins were associated with the cytoskeleton and that 20% of the cytochrome b558 and most of the respiratory burst oxidase activity was cytoskeletal associated (Babior *et al.*, 1990; Woodman *et al.*, 1991). Conceivably, simultaneous cytoskeletal reorganization and formation of lipid mediators activating PKC may function in assembling and activating the NADPH oxidase of neutrophils at sites of CD18-mediated contact. In addition to the large quantities of oxidants generated, formation of H_2O_2 or other oxidizing species at the cellular interface during neutrophil integrin-mediated adhesion has obvious important consequences for escaping extracellular and overwhelming intracellular antioxidant defense mechanisms in cytotoxic oxidative injury such as that postulated to occur during myocardial ischemia reflow injury (Lucchesi and Mullane, 1986; Mullane and Smith, 1990; Entman *et al.*, 1991a) or other inflammatory disorders.

Our studies on activation mechanisms of the integrin-mediated respiratory burst have focused on conditions in which adhesion of neutrophils is clearly CD11b/CD18 dependent. Given the specific nature of the adhesion-dependent phenomenon being studied, it is clear that antibodies that block adhesion to a particular surface should inhibit transduction mechanisms and cellular responses which are adhesion-dependent. However, the relationship between CD11b/CD18 and other protein surfaces which have been utilized in studying neutrophil adhesion-dependent activation (Nathan *et al.*, 1989) is less certain. CD11b/CD18 appears to be the major neutrophil protein participating in the binding of cells to fibrinogen (Wright *et al.*, 1988), for example, although a recent report suggests that CD11c/CD18 may bind fibrinogen, depending on stimulation conditions (Loike *et al.*, 1990). In addition, although CD11b/CD18 may participate in the binding of neutrophils to other extracellular matrix or plasma proteins, such as fibronectin, von Willebrand factor, and laminin, other mechanisms, including separate integrin molecules, may also contribute. For example, Gresham *et al.* (1989) have reported the existence of an integrin receptor on neutrophils, distinct from members of the CD11/CD18 family, that mediates augmented neutrophil phagocytosis in response to arginine-glycine-asparagine (RGD)-containing proteins. (See below, section on Enhanced Phagocytosis). In addition, a recent report demonstrated the presence of a β_1 integrin on neutrophils capable of mediating CD18-independent binding to laminin (Bohnsack *et al.*, 1990). Although multiple mechanisms, perhaps even including non-integrin constituents, may participate in the actual physical association of neutrophils to complicated protein or other cell surfaces, CD18- and perhaps CD11b/CD18-dependent adhesion, in particular, may be unique in the ability to transduce certain signals to adherent cells.

The manner in which CD11b/CD18-dependent adhesion activates phospholipid hydrolysis and the specific nature of the phospholipases activated during integrin-mediated adhesion to result in the formation of diacylglycerol remain unknown (Fig. 3.2). Whether similar biochemical activation processes are initiated at the membrane during CD11b/CD18 dependent adhesion vs. phagocytosis is also uncertain. CD11b/CD18 (CR3) is well known to mediate the binding and ingestion of iC3b-coated particles (Wright *et al.*, 1983; Wright *et al.*, 1987). Fallman and colleagues demonstrated that phagocytosis of iC3b- (or IgG-) opsonized yeast particles is associated with the increased formation of inositol phosphates and diacylglycerol (Fallman *et al.*, 1989). PTX pretreatment had little if any effect on iC3b-mediated phagocytosis, the associated formation of IP_3, or DAG formation (Fallman *et al.*, 1989). Similarly, the component of the adhesion-dependent respiratory burst due to β_2-integrin mediated adhesion *per se* (that is, in contrast to the increased surface expression of CD11b/CD18 and induction of CD18-dependent adhesion in response to soluble stimuli) appears to be insensitive to PTX-pretreatment. PTX-pretreatment of neutrophils abolished the subsequent respiratory burst of fMLP-stimulated human neutrophils on fetal calf serum coated-polystyrene, but had no effect on that induced by TNFα (Meurer and MacIntyre, 1989). If the mechanism of activation of the delayed respiratory burst is the same regardless of the initial stimulus used to promote adhesion (as indicated by the similar kinetics and dependence on β_2 integrins for H_2O_2 production induced by either chemotactic factors or cytokines (Nathan, 1987; Nathan *et al.*, 1989; Meuerer and MacIntyre, 1989; Shappell *et al.*, 1990c)), inhibition by PTX with fMLP may be due to the well-known prevention of PIP_2 hydrolysis, Ca^{2+} mobilization, and subsequent cellular responses, such as CD11b/CD18 up-regulation and adhesion. In contrast, although TNF is well known to promote CD11b/CD18 up-regulation and CD11b/CD18-dependent adhesion (Lo *et al.*, 1989a; Detmers *et al.*, 1990), such processes are not likely to be mediated in a manner analogous to the G-protein-coupled chemotactic factor receptors, as the molecular structure of the TNF-receptor is quite distinct (e.g. single transmembrane domain) (Loetscher *et al.*, 1990; Schall *et al.*, 1990) from the predicted structure of G-protein-coupled receptors. Such results would indicate that in contrast to the transient accumulation of diacylglycerol (DAG) and the brief respiratory burst induced by chemotactic factors with neutrophils in suspension, the transduction mechanisms responsible for the integrin-mediated respiratory burst may not depend on a PTX-sensitive G-protein.

In addition, in contrast to the transient formation of DAG that may occur in response to chemotactic factor stimulation (Snyderman and Uhing, 1988), the accumulation of DAG upon phagocytosis of iC3b-opsonized particles was much more sustained (Fallman *et al.*, 1989), similar to that observed under conditions of CD11b/CD18 dependent adhesion (although not preceded by a lag period) and with cytochalasin pretreatment of fMLP-stimulated neutrophils (Honeycutt and Niedel, 1986; Rider and Niedel, 1987; Shappell *et al.*, 1991). This accumulation of DAG associated with phagocytosis (and phagocytosis itself) was decreased by preincubation of neutrophils with forskolin under conditions that induced a three-fold rise in cAMP levels that was maintained for almost an hour (Fallman *et al.*, 1989). In light of the data summarized above (Nathan and Sanchez, 1990) that the adhesion-dependent respiratory burst induced by TNFα on FBS and fibrinogen is associated with a sustained decrease in cAMP during the lag phase, it is tempting to speculate that these two signalling pathways interact in the regulation of the adhesion-dependent production of hydrogen peroxide. The sustained accumulation of diacylglycerol during CD11b/CD18 dependent adhesion, its rough temporal correlation with the time course of prolonged H_2O_2 production, and the ability of PKC inhibitors to terminate the respiratory burst while it is in progress would indicate that continued PKC modulation of the NADPH oxidase complex, possible by virtue of sustained DAG presence in the membrane, is necessary for the prolonged generation of neutrophil reactive oxygen species. In contrast, agents that elevate cAMP in adherent neutrophils block the generation of H_2O_2 only if added during the lag period, being ineffective once the prolonged respiratory burst has ensued (Nathan and Sanchez, 1990). Owing to previous demonstrations of elevated cAMP impairing DAG formation under other stimulus conditions (Fallman *et al.*, 1989) and the fact that agents elevating cAMP do not inhibit the respiratory burst of cells on FBS in response to PMA (Nathan and Sanchez, 1990), it is tempting to speculate that such agents prevent the formation or accumulation of diacylglycerol (Della Bianca *et al.*, 1986; Agwu *et al.*, 1991a; Tyagi *et al.*, 1991) rather than acting at subsequent steps in oxidase regulation.

The specific phospholipid pools or species from which the sustained DAG accumulation derives under conditions of CD11b/CD18 mediated phagocytosis or adhesion are still undefined. Calcium depleted neutrophils retained the majority of their phagocytic properties for iC3b-coated particles. Phagocytosis was still accompanied by accumulation of diacylglycerol, although no formation of IP_3 could be detected under these conditions, perhaps indicating a significant contribution of a non-PIP_2 phospholipid source, such as phosphatidylcholine (PC), in the process of sustained DAG accumulation. In addition to the well-defined phospholipase C hydrolysis of polyphosphatidylinositides leading to the transient formation of inositol phosphates and DAG upon binding of chemotactic factors to their specific membrane receptors (Snyderman and Uhing, 1988), phosphatidylcholine as a source of diacylglycerol is receiving increasing attention in the mechanisms of activation of a variety of cell types including neutrophils. Furthermore, phosphatidylcholine breakdown may involve both phospholipase C and D (Exton, 1988). Phospholipase D has recently begun receiving increasing attention in the analysis of neutrophil NADPH oxidase regulation (Agwu *et al.*, 1989; Billah *et al.*, 1989; Bonser *et al.*, 1989; Gelas *et al.*, 1989; Rossi *et al.*, 1990; Thompson *et al.*, 1990), either with regard to potential biological activities of phosphatidic acid, *per se*, or via subsequent formation of DAG (via PA phosphatase). These studies have emphasized the potential pathway of phospholipase D generation of PA and subsequent phosphatidate phosphohydrolase mediated formation of diradylglycerols (DRG) in neutrophils stimulated with cytochalasin B and fMLP (Agwu *et al.*, 1989; Billah *et al.*, 1989; Bonser *et al.*, 1989; Gelas *et al.*, 1989; Rossi *et al.*, 1990; Thompson *et al.*, 1990). Furthermore, the formation of phosphatidic acid, *per se* has been causally linked to the respiratory burst, based on inhibitor studies. The role of phosphatidic acid in regulating the neutrophil respiratory burst and its potential mechanisms is a subject of ongoing investigation (Agwu *et al.*, 1991b). Whether DAG accumulation during integrin-mediated neutrophil activation occurs via activation of a PLC or a PLD remains to be determined. The possible relationships between integrin (CD11b/CD18)-mediated adhesion, cytoskeletal reorganization, oxidase assembly, and PLC or PLD formation of regulatory second messengers are summarized in Fig. 3.2.

7.2 INTEGRIN-MEDIATED ADHESION AND NEUTROPHIL DEGRANULATION

In addition to the potential role of oxidants in neutrophil-mediated target cell injury during adhesion, neutrophils contain in their primary and secondary granules a variety of proteolytic enzymes capable, in theory, of contributing to tissue injury during neutrophil activation (Henson *et al.*, 1988). These enzymes, including the serine protease elastase, are released into phagosomes during the ingestion and destruction of microorganisms and may also be released into the extracellular space during inflammation (Henson and Johnston, 1987; Henson *et al.*, 1988). Elastase has been implicated in certain aspects of endothelial damage in some *in vitro* experiments, including in the non-lytic detachment of endothelial cells caused by activated neutrophils (Harlan *et al.*, 1981). In addition, proteolytic enzymes have been demonstrated to possibly act synergistically with oxidants in mediating endothelial injury under certain experimental conditions (Varani *et al.*, 1989), and elastase was implicated in lytic injury of human microvascular endothelial cells by chemotactic factor stimulated neutrophils following pretreatment with low levels of lipopolysaccharide (LPS) (Smedly *et al.*, 1986).

The potential ability of integrins to specifically regulate degranulation (i.e. in a manner analogous to the ability of CD11b/CD18 to activate the respiratory burst) is thus an important consideration given the participation of CD11/CD18 heterodimers in mediating adhesion to endothelium, the supporting extracellular matrix, and specific tissue parenchymal cells. Fehr and colleagues (Schleiffenbaum *et al.*, 1989) reported that when neutrophils were preincubated in suspension with control, anti-CD11b, or anti-CD18 monoclonal antibodies for 30 min at 37°C, then stimulated in suspension with 2.5 μM fMLP for an additional 30 min, and then allowed to adhere to uncoated plastic for an additional 30 min, substantial release of vitamin B_{12} binding protein occurred. This degranulation was inhibited by preincubation with mAb 60.3 (anti-CD18) or 60.1 (anti-CD11b) and appeared to correlate with the inhibition of adhesion. Similar results were obtained when elastase was measured in the supernatant or when endotoxin-stimulated neutrophils were subsequently allowed to adhere (Schleiffenbaum *et al.*, 1989).

Using a hemolytic plaque assay in which plastic plates are treated with poly-L-lysine, protein A-coated sheep erythrocytes, antibodies to enzyme marker proteins such as lactoferrin, and guinea pig complement prior to addition of stimulus and neutrophils, Richter and coworkers demonstrated that TNF and GM-CSF (as well as fMLP) cause secretion of the secondary granule protein lactoferrin (Richter *et al.*, 1989). This semi-quantitative assay is thus able to detect degranulation by single adherent cells (by subsequent plaque counting), although extent of degranulation cannot be precisely assessed. In contrast to lactoferrin release during 2 h incubations of such adherent cells, secretion of the primary granule enzyme myeloperoxidase (MPO) did not occur with fMLP, TNF, or GM-CSF unless cytochalasin B was also included (Richter *et al.*, 1989). Similar to the adhesion-induced respiratory burst by fMLP vs. TNF stimulated neutrophils on serum as described above (Meurer and McIntyre, 1989), PTX-pretreatment of neutrophils inhibited lactoferrin plaque formation induced by fMLP (and GM-CSF) stimulation, but not that induced by TNF (or PMA) stimulation. In addition, chelating intracellular Ca^{2+} by pretreating neutrophils with increasing concentrations of quin-2 caused a progressive inhibition of TNF-induced lactoferrin secretion (Richter *et al.*, 1989). These same investigators have demonstrated that under conditions of adhesion, neutrophils exhibit spontaneous oscillations of intracellular Ca^{2+} concentration, and such Ca^{2+} changes appear to be required for TNF to induce degranulation from cells stimulated under these conditions (Richter *et al.*, 1990b). Furthermore, pretreatment with mAbs to either CD18 or CD11b inhibit the oscillations of cytosolic free Ca^{2+} on coverslips coated with human serum albumin (HSA) and these same mAbs inhibit TNF and fMLP (but not PMA)-induced lactoferrin release as assessed by the same hemolytic plaque assay (Richter *et al.*, 1990a). The effect of the mAbs in the microspectrofluorometry assay for intracellular Ca^{2+} appeared to be possibly related to the inhibition of cell spreading rather than preventing initial contact (Richter *et al.*, 1990a).

Other investigators recently described the release of specific (secondary) and tertiary granule markers by TNFα-stimulated neutrophils on fibrinogen (Hanlon *et al.*, 1991). Adhesion under these conditions is CD11b/CD18-dependent and was demonstrated by the authors to steadily increase until a maximum at about 40 min and essentially to be blocked completely by the anti-CD18 mAb 1B4 (same). Neutrophils thus stimulated released substantial lactoferrin and gelatinase into the incubation medium, with an apparent lag period of 20–30 min (and an overall time course somewhat similar to H_2O_2 production under the same conditions in other laboratories). Although adhesion under these conditions is clearly CD18-dependent, no data was presented on the effects of the CD18-mAb on degranulation. In contrast to lactoferrin and gelatinase, no release of MPO into the media was observed under similar conditions, nor was there any detectable elastase activity when assessed using either a fluorogenic peptide substrate or a radioimmunoassay for an elastase generated cleavage product of FGN (the first 21 N-terminal amino acids of the Aα chain). It would be of great interest to determine if such a degradation product were formed on the adhesion-substrate *per se*, as elastase release at the site of integrin-mediated adhesion would have great implications for neutrophil-mediated matrix digestion or cell injury.

7.3 ENHANCED PHAGOCYTOSIS

One of the most clearly established ligands for CD11b/CD18 (CR3) is the iC3b component of complement (Wright *et al.*, 1983; Wright *et al.*, 1987) which serves to opsonize particles during complement activation. Hence, CD11b/CD18 clearly participates in the phagocytosis of iC3b-coated particles. As also mentioned previously, phagocytosis of iC3b- (or IgG) opsonized yeast particles is associated with the increased accumulation of diacylglycerol that may be derived largely from choline-containing phospholipids (Fallman *et al.*, 1989). In addition, because of as yet ill-defined relationships to other cellular constituents such as certain cytoskeletal components, CD11b/CD18 may participate in augmented cellular responses upon binding to surfaces or opsonized particles that are not CD11b/CD18 ligands. For example, Gresham *et al.*, (1991) have also demonstrated that CD18 integrins are essential for the augmented phagocytosis of IgG or C4b opsonized erythrocytes. Such enhanced phagocytosis can be elicited by a variety of cytokines, phorbol esters, or RGDS-containing proteins (Gresham *et al.*, 1988; Gresham *et al.*, 1989). Phorbol ester stimulated ingestion of IgG or C4b-opsonized particles appears to depend on the activation of the respiratory burst (Gresham *et al.*, 1988; Gresham *et al.*, 1991). It is deficient in neutrophils from patients with CGD and is blocked in normal neutrophils by inclusion of catalase (Gresham *et al.*, 1988). In contrast, RGD enhanced phagocytosis is apparently inactivated by oxidizing species derived from the H_2O_2-MPO system, is normal in CGD cells, and requires the inclusion of catalase for demonstration in normal neutrophils (Gresham *et al.*, 1989). CD18-heterodimers (specifically CD11b/CD18) are required for both pathways of stimulated ingestion, despite the fact that the ligands are CD11b/CD18 independent. LAD neutrophils, which exhibit essentially normal baseline phagocytosis (Gresham *et al.*, 1991), fail to amplify ingestion of either IgG- or C4b-coated erythrocytes in response to either phorbol esters or RGD-containing stimuli (Gresham *et al.*, 1991). Interestingly, certain CD18 and CD11b mAbs can induce similar defects in normal neutrophils (Arnaout *et al.*, 1983; Graham *et al.*, 1989; Gresham *et al.*, 1991). Furthermore, both pathways of enhanced phagocytosis, dependent on the presence of CD11b/CD18 and in particular, perhaps a microfilament-maintained immobile subset of CD11b/CD18 (Graham *et al.*, 1989), are apparently mediated by PKC activation (Gresham *et al.*, 1991). This further illustrates a connection between CD11b/CD18 dependent neutrophil responses and PKC activation, as illustrated by our studies for the adhesion-dependent production of H_2O_2 via DAG formation.

7.4 ARACHIDONIC ACID RELEASE AND METABOLISM

Release of free arachidonic acid (AA) may serve multiple functions in activated neutrophils. In addition to its well defined role as the substrate for 5-lipoxygenase metabolism and hence the precursor for the biologically active leukotrienes, AA has been postulated to have a more direct effect in activation of neutrophil responses, particularly modulation of the respiratory burst. AA may function in such a signal transduction capacity through one or more proposed mechanisms, including activation of PKC (McPhail *et al.*, 1984), modulation of GTP binding to cytosolic or membrane regulatory proteins (Gabig *et al.*, 1990; Abramson *et al.*, 1991), or a more

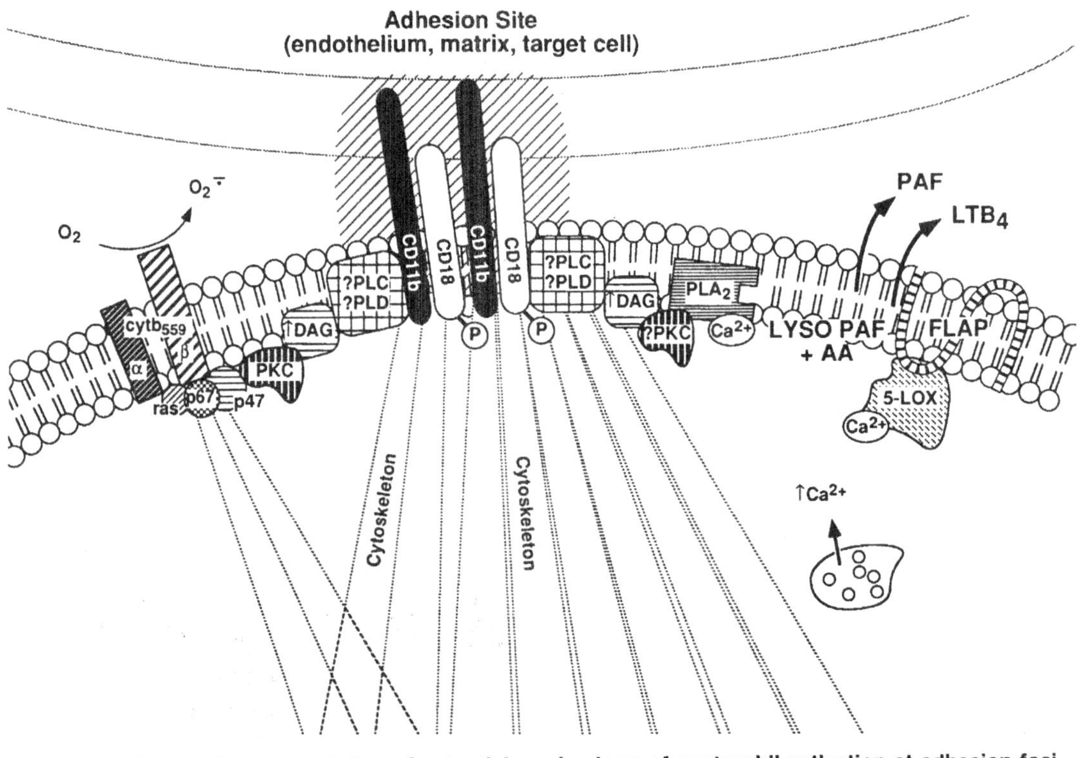

Figure 3.2 Schematic representation of potential mechanisms of neutrophil activation at adhesion foci. Cytoskeletal association of CD11b/CD18 and NADPH oxidase components and simultaneous generation of oxidase regulating intracellular mediators such as cAMP and DAG may result in the assembly and activation of the NADPH oxidase at sites of integrin-mediated adhesion. Integrin-mediated adhesion may activate other cytotoxic and pro-inflammatory neutrophil responses, such as degranulation and formation of lipid-derived inflammatory mediators. Integrin-mediated regulation of arachidonate metabolism has not been specifically demonstrated, but DAG accumulation due to CD11b/CD18-dependent adhesion may contribute to PLA_2 regulation in PKC-dependent or independent mechanisms, providing for amplification of the inflammatory response through formation of the chemotactic lipid mediators PAF and LTB_4. (Abbreviations: PLC, phospholipase C; PLD, phospholipase D; PLA_2, phospholipase A_2; DAG, diacylglycerol; PKC, protein kinase C; AA, arachidonic acid; PAF, platelet activating factor; 5-LOX, 5-lipoxygenase; FLAP, 5-lipoxygenase activating protein. P indicates regulatable phosphorylation of CD18.

direct effect on the membrane localized NADPH oxidase. However, the high concentrations required and potential lack of specificity leave the physiological relevance of these functions currently uncertain.

In contrast to the uncertainty of its role as a signal transduction molecule in neutrophils, AA clearly serves as the substrate for the production of the biologically active 5-lipoxygenase products, including the neutrophil derived chemoattractant leukotriene B_4 (LTB_4) (Lewis and Austen, 1988; Borgeat, 1989; Samuelsson and Funk, 1989) and the cysteinyl-leukotrienes LTC_4 and LTD_4 that can be generated during neutrophil-platelet and neutrophil-endothelial interactions (Feinmark and Cannon, 1986; Feinmark and Cannon, 1987; Maclouf and Murphy, 1988; Maclouf *et al.*, 1989; Maclouf *et al.*, 1990). In addition, the enrichment of AA in the *sn*-2 position of 1-O-alkyl-2-acyl-*sn*-glycero-3-phosphocholine in neutrophil membranes means that a phospholipase A_2 mediated AA liberation for LOX metabolizm is simultaneous with the formation of 1-O-alkyl-2-lyso-GPC, the immediate precursor for the formation of platelet activating factor (PAF) (1-O-alkyl-2-acetyl-GPC) (Chilton *et al.*, 1984; Chilton and Connell, 1988; Chilton, 1989) (See Fig. 3.2).

Production of lipoxygenase products by neutrophils appears to require both the activation of the 5-LOX enzyme and the availability of substrate AA (Clancy *et al.*, 1983). 5-LOX is a Ca^{2+} and ATP dependent enzyme (Rouzer and Samuelsson, 1985), that appears to undergo Ca^{2+}-dependent translocation to an 18 kDa plasma membrane protein (5-lipoxygenase-activating protein or FLAP) as a key step in its activation in intact cells (Rowzer and Kargman, 1985; Dixon *et al.*, 1990; Miller *et al.*, 1990). In contrast to early reports (Jubiz *et al.*, 1982; Salari *et al.*, 1985), most investigators observe that stimulation of neutrophils in suspension with chemotactic factors alone does not result in the production of detectable LTB_4 or its metabolites (Clancy *et al.*, 1983; Palmer and Salmon, 1983; Haines *et al.*, 1987; Dahinden *et al.*, 1988; Osaki *et al.*, 1989; Shappell *et al.*, 1990a), in contrast, for example, to the marked production of LTB_4 that occurs with ionophore stimulation of neutrophils. That chemoattractant stimulation (likely by increased intracellar Ca^{2+}) is sufficient to stimulate the 5-LOX enzyme is evidenced by the ability of neutrophils stimulated with these agents to produce leukotrienes in the presence of exogenous AA (Clancy *et al.*, 1983; Palmer and Salmon, 1983; Haines *et al.*, 1987; Osaka *et al.*, 1989). The inability of neutrophils to produce appreciable leukotrienes from endogenous AA in response to even high (e.g. μM) levels of chemotactic factors alone appears to be due to the inability of such stimulation conditions to mobilize AA in human neutrophils (Clancy *et al.*, 1983).

The mechanisms controlling AA release in intact neutrophils are still not completely characterized. Studies using phorbol ester potentiation of AA release in submaximally stimulated neutrophils (Liles *et al.*, 1987) and inhibitors of PKC with ionophore stimulated PMNs (McIntyre *et al.*, 1987; Gailbraith, 1988) have indicated a potential role for PKC in regulating AA hydrolysis. Other studies have indicated that PLA_2 may be more directly modulated by DAG itself. DAG incorporation into membranes makes substrate more accessible to PLA_2 and stimulates PLA_2 activity *in vitro* (Dawson *et al.*, 1984; Kramer *et al.*, 1987). This PLA_2 undergoes translocation to the plasma membrane in a Ca^{2+}-dependent manner and its activity is stimulated by DAG and PIP_2 (Channon and Leslie, 1990). As integrin mediated adhesion of neutrophils can cause prolonged accumulation of diacylglycerol (and PKC-dependent activation of the respiratory burst), adhesion may represent a physiological mechanism for stimulating neutrophil AA release and leukotriene production (Fig. 3.2).

Note added in proof

Fuortes *et al.* have recently shown that integrin-dependent adhesion of TNF-stimulated PMNs to protein-coated surfaces is associated with tyrosine phosphorylation of several proteins. Tyrosine kinase inhibitors blocked cell spreading and the adhesion-dependent production of hydrogen peroxide (J. Cell Bio. 120: 777, 1993). Possible relationships between adhesion-dependent tyrosine kinase activation and other signalling pathways described herein remain to be elucidated.

8. References

Abbassi, O., Lane, C.L., Krater, S.S., Kishimoto, T.K., Anderson, D.C., McIntire, L.V. and Smith, C.W. (1991). Canine neutrophil margination mediated by pectin adhesion molecule-1 (LECAM-1) *in vitro*. J. Immunol. 147, 2107–2115.

Abo, A., Pick, E., Hall, A., Totty, N., Teahan, C.G. and Segal, A.W. (1991). Activation of the NADPH oxidase involves the small GTP-binding protein p21rac1. Nature 353, 668–670.

Abramson, J.S., Mills, E.L., Sawyer, M.K., Regelmann, W.R., Nelson, J.D. and Quie, P.G. (1981). Recurrent infections and delayed separation of the umbilical cord in an infant with abnormal phagocytic cell locomotion and oxidative response during particle phagocytosis. J. Pediatr. 99, 887–894.

Abramson, S.B., Leszczynska-Piziak, J. and Weissmann, G. (1991). Arachidonic acid as a second messenger: interactions with a GTP-binding protein of human neutrophils. J. Immunol. 147, 231–236.

Agwu, D.E., McPhail, L.C., Chabot, M.C., Daniel, L.W., Wykle, R.L., and McCall, C.E. (1989). Choline-linked phosphoglycerides: a source of phosphatidic acid and diglycerides in stimulated neutrophils. J. Biol. Chem. 264, 1405–1413.

Agwu, D.E., McCall, C.E. and McPhail, L.C. (1991a). Regulation of phospholipase D-induced hydrolysis of choline-containing phosphoglycerides by cyclic AMP in human neutrophils. J. Immunol. 146, 3895–3903.

Agwu, D.E., McPhail, L.C., Sozzani, S., Bass, D.A. and McCall, C.E. (1991b). Phosphatidic acid as a second messenger in human polymorphonuclear leukocytes. Effects on activation of NADPH oxidase. J. Clin. Invest. 88, 531–539.

Allison, F., Smith, M.R. and Wood, W.B. (1955). Studies on the pathogenesis of acute inflammation. I. The inflammatory reaction to thermal injury observed in the rabbit ear chamber. J. Exp. Med. 102, 655–668.

Altieri, D.C. and Edgington, T.S. (1988). The saturable high affinity association of Factor X to ADP-stimulated monocytes defines a novel function of the Mac-1 receptor. J. Biol. Chem. 263, 7007–7015.

Ambruso, D.R., Bolscher, B.G.J.M., Stokman, P.M., Verhoeven, A.J., and Roos, D. (1990). Assembly and activation of the NADPH:O_2 oxidoreductase in human neutrophils after stimulation with phorbol myristate acetate. J. Biol. Chem. 265, 924–930.

Anderson, D.C. and Springer, T.A. (1987). Leukocyte adhesion deficiency: an inherited defect in the Mac-1, LFA-1 and p150,95 glycoproteins. Ann. Rev. Med. 38, 175–194.

Anderson, D.C., Schmalstieg, F.C., Kohl, S., Arnaout, M.A., Hughes, B.J., Tosi, M.F., Buffone, G.J., Brinkley, B.R., Dickey, W.D., Abramson, J.S., Springer, T.A., Boxer, L.A., Hollers, J.M. and Smith, C.W. (1984). Abnormalities of polymorphonuclear leukocyte function associated with a heritable deficiency of high molecular weight surface glycoproteins (GP138): common relationship to diminished cell adherence. J. Clin. Invest. 74, 536–551.

Anderson, D.C., Miller, L.J., Schmalstieg, F.C., Rothlein, R. and Springer, T.A. (1986). Contributions of the Mac-1 glycoprotein family to adherence-dependent granulocyte functions: structure-function assessments employing subunit-specific monoclonal antibodies. J. Immunol. 137, 15–27.

Anderson, D.C., Rothlein, R., Marlin, S.D., Krater, S.S. and Smith, C.W. (1990). Impaired transendothelial migration by neonatal neutrophils: abnormalities of Mac-1 (CD11b/CD18)-dependent Adherence Reactions. Blood 78, 2613–2621.

Anderson, D.C., Abbassi, O., Kishimoto, T.K., Koenig, J.M., McIntire, L.V. and Smith, C.W. (1991). Diminished lectin-, epidermal growth factor-, complement binding domain-cell adhesion molecule-1 on neonatat neutrophils underlies their impaired CD18-independent adhesion to endothelial cells *in vitro*. J. Immunol. 146, 3372–3379.

Arfors, K.E., Lundberg, C., Lindbom, L., Lundberg, K., Beatty, P.G., and Harlan, J.M. (1987). A monoclonal antibody to the membrane glycoprotein complex CD18 inhibits polymorphonuclear leukocyte accumulation and plasma leakage *in vivo*. Blood 69, 338–340.

Argenbright, L.W., Letts, L.G. and Rothlein, R. (1991). Monoclonal antibodies to the leukocyte membrane CD18 glycoprotein complex and to intercellular adhesion molecule-1 inhibit leukocyte-endothelial adhesion in rabbits. J. Leuk. Biol. 49, 253–257.

Argenbright, L.W. and Barton, R.W. (1992). Interactions of leukocyte integrins with intercellular adhesion molecule 1 in the production of inflammatory vascular injury *in vivo*. The Shwartzman reaction revisited. J. Clin. Invest. 89, 259–272.

Armstrong, P.B. and Lackie, J.M. (1975). Studies on intercellular invasion *in vitro* using rabbit peritoneal neutrophil granulocytes (PMNs) (I) Role of contact inhibition in locomotion. J. Cell Biol. 65, 439–462.

Arnaout, M.A. (1990). Leukocyte Adhesion Molecule Deficiency: its structural basis, pathophysiology and implications for modulating the inflammatory response. Immunol. Revs. 114, 145–180.

Arnaout, M.A., Pitt, J., Cohen, H.J., Melamed, J., Rosen, F.S. and Colten, H.R. (1982). Deficiency of a granulocyte-membrane glycoprotein (gp150) in a boy with recurrent bacterial infections. N. Eng. J. Med. 306, 693–699.

Arnaout, M.A., Todd, R.F. III, Dana, N., Melamed, J., Schlossman, S.F. and Colten, H.R. (1983). Inhibition of phagocytosis of complement C3 – or immunoglobulin G – coated particles and of C3bi binding by monoclonal antibodies to a monocyte-granulocyte membrane glycoprotein (Mo1). J. Clin. Invest. 72, 171–179.

Arnaout, M.A., Spits, H., Terhorst, C., Pitt, J. and Todd, R.F., III (1984). Deficiency of a leukocyte surface glycoprotein (LFA-1) in two patients with Mo1 deficiency. J. Clin. Invest. 74, 1291–1300.

Arnaout, M.A., Wang, E.A., Clark, S.C. and Sieff, C.A. (1986). Human recombinant granulocyte-macrophage colony-stimulating factor increases cell-to-cell adhesion and surface expression of adhesion-promoting surface glycoproteins on mature granulocytes. J. Clin. Invest. 78, 597–601.

Arnaout, M.A., Lanier, L.L. and Faller, D.V. (1988). Relative contribution of the leukocyte molecules Mo1, LFA-1 and p150/95 (LeuM5) in adhesion of granulocytes and monocytes to vascular endothelium is tissue and stimulus-specific. J. Cell. Physiol. 137, 305–313.

Atherton, A. and Born, G.V.R. (1972). Quantitative investigations of the adhesiveness of circulating polymorphonuclear leukocytes to blood vessel walls. J. Physiol. 222, 447–474.

Atherton, A. and Born, G.V.R. (1973). Relationship between the velocity of rolling granulocytes and that of the blood flow in venules. J. Physiol 233, 157–165.

Babior, B.M., Curnutte, J.T. and Okamura, N. (1990). The respiratory burst oxidase and certain members of the 48K phosphoprotein family are associated with the neutrophil cytoskeleton. Blood 72 (Suppl. 1), 141a.

Baggiolini, M., Walz, A. and Kunkel, S.L. (1989). Neutrophil-activating peptide-1/interleukin 8, a novel cytokine that activates neutrophils. J. Clin. Invest. 84, 1045–1049.

Bainton, D.F., Miller, L.J., Kishimoto, T.K. and Springer, T.A. (1987). Leukocyte adhesion receptors are stored in peroxidase-negative granules of human neutrophils. J. Exp. Med. 166, 1641–1653.

Barton, R.W., Rothlein, R., Ksiazek, J. and Kennedy, C. (1989a). The effect of anti-intercellular adhesion molecule-1 on phorbol-ester-induced rabbit lung inflammation. J. Immunol. 143, 1278–1282.

Barton, R.W., Rothlein, R., Ksiazek, J. and Kennedy, C. (1989b). In "Leukocyte Adhesion Molecules: Structure, Function, and Regulations", (eds T.A. Springer, D.C. Anderson, A.S. Rosenthal and R. Rothlein), pp 149–156. New York, Springer-Verlag.

Beck, G., Habicht, G.S., Benach, J.L. and Miller, F. (1986). Interleukin-1: a common endogenous mediator of inflammation and the local Shwartzman reaction. J. Immunol. 136, 3025–3032.

Beekhuizen, H., Tilburg, J.C. and VanFurth, R. (1990). Characterization of monocyte adherence to human macrovascular and microvascular endothelial cells. J. Immunol. 145, 510–518.

Beesley, J.E., Pearson, J.D., Carleton, J.S., Hutchings, S.A. and Gordon, J.L. (1978). Interactions of leukocytes with vascular cells in culture. J. Cell Sci. 33, 85–101.

Beesley, J.E., Pearson, J.D., Hutchings, A., Carleton, J.S. and Gordon, J.L. (1979). Granulocyte migration through endothelium in culture. J. Cell Sci. 38, 237–248.

Benjamin, C., Dougas, I., Chi-Rosso, G., Luhowskyj, S., Rosa, M., Newman, B., Osborn, L., Vassallo, C., Hession, C., Goelz, S., McCarthy, K. and Lobb, R. (1990). A blocking monoclonal antibody to endothelial-leukocyte adhesion molecule-1 (ELAM-1). Biochem. Biophy. Res. Comm. 171, 348–353.

Berger, M., O'Shea, J.J., Cross, A.S., Folks, T.M., Chused, T.M., Brown, E.J. and Frank, M.M. (1984). Human neutrophils increase expression of C3bi as well as C3b receptors upon activation. J. Clin. Invest. 74, 1566–1571.

Berkow, R.L., Dodson, R.W. and Kraft, A.S. (1987). The effect of a protein kinase C inhibitor, H-7, on human neutrophil oxidative burst and degranulation. J. Leukocyte Biol. 41, 441–446.

Berridge, M.J. (1987). Inositol trisphosphate and diacylglycerol: two interacting second messengers. Ann. Rev. Biochem. 56, 159–193.

Bevilacqua, M.P., Pober, J.S., Mendrick, D.L., Cotran, R.S. and Gimbrone, Jr., M.A. (1987). Identification of an inducible endothelial-leukocyte adhesion molecule. Proc. Natl Acad. Sci. USA 84, 9238–9242.

Bevilacqua, M.P., Stengelin, S., Gimbrone, Jr., M.A. and Seed, B. (1989). Endothelial leukocyte adhesion molecule 1: an inducible receptor for neutrophils related to complement regulatory proteins and pectins. Science 243, 1160–1165.

Billah, M.M., Eckel, S., Mullmann, T.J., Egan, R.W. and Siegel, M.I. (1989). Phosphatidylcholine hydrolysis by phospholipase D determines phosphatidate and diglyceride levels in chemotactic peptide-stimulated human neutrophils. Involvement of phosphatidate phosphohydrolase in signal transduction. J. Biol. Chem. 264, 17 069–17 077.

Bissenden, J.G., Haeney, M.R., Tarlow, M.J. and Thompson, R.A. (1981). Delayed separation of the umbilical cord severe widespread infections and immunodeficiency. Arch. Dis. Child. 56, 397–399.

Bohnsack, J.F., Akiyama, S.K., Damsky, C.H., Knape, W.A. and Zimmerman, G.A. (1990). Human neutrophil adherence to laminin *in vitro*. Evidence for a distinct neutrophil integrin receptor for laminin. J. Exp. Med. 171, 1221–1237.

Bonfanti, R., Furie, B.C., Furie, B. and Wagner, D.D. (1989). PADGEM (GMP140) is a component of Weibel-Palade bodies of human endothelial cells. Blood 73, 1109–1112.

Bonser, R.W., Thompson, N.T., Randall, R.W. and Garland, L.G. (1989). Phospholipase D activation is functionally linked to superoxide generation in the human neutrophil. Biochem. J. 264, 617–620.

Borgeat, P. (1989). Biochemistry of the lipoxygenase pathways in neutrophils. Can. J. Physiol. Pharmacol. 67, 936–942.

Bowen, T.J., Ochs, H.D., Altman, L.C., Price, T.H., Van Epps, D.E., Brautigan, D.L., Rosin, R.E., Perkins, W.D., Babior, B.M., Klebanoff, S.J. and Wedgwood, R.J. (1982). Severe recurrent bacterial infections associated with defective adherence and chemotaxis in two patients with neutrophils deficient in a cell-associated glycoprotein. J. Pediatr. 101, 932–940.

Breviario, F., Bertocchi, F., Dejana, E. and Bussolino, F. (1988). IL-1 induced adhesion of polymorphonuclear leukocytes to cultured human endothelial cells. J. Immunol. 141, 3391–3397.

Bromberg, Y. and Pick, E. (1984). Unsaturated fatty acids stimulate NADPH-dependent superoxide production by cell-free system derived from macrophages. Cell. Immunol. 88, 213–221.

Bromberg, Y. and Pick, E. (1985). Activation of NADPH-dependent superoxide production in a cell-free system by sodium dodecyl sulfate. J. Biol. Chem. 260, 13 539–13 545.

Brown, E.J., Hooper, L., Ho, T. and Gresham, H.D. (1990). Integrin-associated protein: a 50 kDa plasma membrane antigen physically and functionally associated with integrins. J. Cell Biol. 111, 2785–2794.

Buchanan, M.R., Crowley, C.A., Rosin, R.E., Gimbrone, Jr., M.A. and Babior, B.M. (1982). Studies on the interaction between GP-180 deficient neutrophils and vascular endothelium. Blood 60, 160–165.

Burn, P., Kupter, A. and Singer, S.J. (1988). Dynamic membrane-cytoskeleton interactions: specific association of integrin and talin arises *in vivo* after phorbol ester treatment of peripheral blood lymphocytes. Proc. Natl Acad. Sci. USA 85, 497–501.

Bussolino, F., Camussi, G. and Baglioni, C. (1988). Synthesis and release of platelet-activating factor by human vascular endothelial cells treated with tumor necrosis factor or interleukin 1 alpha. J. Biol. Chem. 263, 11 856–11 861.

Buyon, J.P., Abramson, S.B., Philips, M.R., Slade, S.G., Ross, G.D., Weissmann, G. and Winchester, R.J. (1988a). Dissociation between increased surface expression of Gp165/95 and homotypic neutrophil aggregation. J. Immunol. 140, 3156–3160.

Buyon, J.P., Philips, M.R., Abramson, S.B., Slade, S.G., Weissmann, G. and Winchester, R. (1988b). In "Leukocyte Adhesion Molecules: Structure, Function, and Regulation" (eds T.A. Springer, D.C. Anderson, A.S. Rosenthal and R. Rothlein), pp. 72–83. New York, Springer-Verlag.

Buyon, J.P., Slade, S.G., Reibman, J., Abramson, S.B., Philips, M.R., Weissman, G. and Winchester, R. (1990). Constitutive and induced phosphorylation of the alpha and beta-chains of the CD11/CD18 leukocyte integrin family. J. Immunology 144, 191–197.

Carveth, H.J., Bohnsack, J.F., McIntyre, T.M., Baggiolini, M., Prescott, S.M. and Zimmerman, G.A. (1989). Neutrophil activating factor (NAF) induces polymorphonuclear leukocyte adherence to endothelial cells and to subendothelial matrix proteins. Biochem. Biophys. Res. Commun. 162, 387–393.

Channon, J.Y. and Leslie, C.C. (1990). A calcium-dependent mechanism for associating a soluble arachidonyl-hydrolyzing phospholipase A2 with membrane in the macrophage cell line RAW 264.7. J. Biol. Chem. 265, 5409–5413.

Charo, I.F., Yuen, C., Perez, H.D. and Goldstein, I.M. (1986). Chemotactic peptides modulate adherence of human polymorphonuclear leukocytes to monolayer of cultured endothelial cells. J. Immunol. 136, 3412–3419.

Chatila, T.A., Geha, R.S. and Arnaout, M.A. (1989). Constitutive and stimulus-induced phosphorylation of CD11/CD18 leukocyte adhesion. J. Cell Biol. 109, 3435–3444.

Chilton, F.H. (1989). Potential phospholipid source(s) of arachidonate used for the synthesis of leukotrienes by the human neutrophil. Biochem. J. 258, 327–333.

Chilton, F.H. and Connell, T.R. (1988). 1-ether-linked phosphoglycerides: major endogenous sources of arachidonate in the human neutrophil. J. Biol. Chem. 263, 5260–5265.

Chilton, F.H., Ellis, J.M., Olson, S.C. and Wylde, R.L. (1984). 1-O-alkyl-2-arachidonyl-*sn*-glycero-3-phosphocholine, a common source of platelet-activating factor and arachidonate in human polymorphonuclear leukocytes. J. Biol. Chem. 259, 12 014–12 019.

Christiansen, N.O., Larsen, C.S., Juhl, H. and Esmann, V. (1988). Membrane-associated protein kinase C activity in superoxide-producing human polymorphonuclear leukocytes. J. Leukocyte Biol. 44, 33–40.

Clancy, R.M., Dahinden, C.A. and Hugli, T.E. (1983). Arachidonate metabolism of human polymorphonuclear leukocytes stimulated by N-formyl-Met-Leu-Phe or complement component C5a is independent of phospholipase activation. Proc. Natl Acad. Sci. USA 80, 7200–7204.

Clark, E.A., Ledbetter, J.A., Holly, R.C., Dinndorf, P.A. and Shu, G. (1986). Polypeptides on human beta lymphocytes associated with cell activation. Human Immunol. 16, 100–113.

Clark, R.A., Malech, H.L., Gallin, J.I., Nunoi, H., Volpp, B.D., Pearson, D.W., Nauseef, W.M. and Curnutte, J.T. (1989). Genetic variants of chronic granulomatous disease: Prevalence of deficiencies of two cytosolic components of the NADPH oxidase system. N. Engl. J. Med. 321, 647–652.

Clark, R.A., Volpp, B.D., Leidal, K.G. and Nauseef, W.M. (1990). Two cytosolic components of the human neutrophil respiratory burst oxidase translocate to the plasma membrane during cell activation. J. Clin. Invest. 85, 714–721.

Corbi, A.L., Kishimoto, T.K., Miller, L.J. and Springer, T.A. (1988). The human leukocyte adhesion glycoprotein Mac-1 (complement receptor type 3, CD11b) alpha subunit. J. Biol. Chem. 263, 12 403–12 411.

Crowley, C.A., Curnutte, J.T., Rosin, R.E., Andre-Schwartz, J., Gallin, J.I., Klempner, M., Snyderman, R., Southwick, F.S., Stossel, T.P. and Babior, B.M. (1980). An inherited abnormality of neutrophil adhesions: its genetic transmission and its association with a missing protein. N. Eng. J. Med. 302, 1163–1168.

Curnutte, J.T., Kuver, R. and Babior, B.M. (1987a). Activation of the respiratory burst oxidase in a fully soluble system from human neutrophils. J. Biol. Chem. 262, 6450–6452.

Curnutte, J.T., Kuver, R. and Scott, P.J. (1987b). Activation of neutrophil NADPH oxidase in a cell-free system. J. Biol. Chem. 262, 5563–5567.

Curnutte, J.T., Scott, P.J. and Mayo, L.A. (1989). Cytosolic components of the respiratory burst oxidase: Resolution of four components, two of which are missing in complementing types of chronic granulomatous disease. Proc. Natl Acad. Sci. USA 86, 825–829.

Currie, M.S., Padmanabhan, J., Cohen, H.J. and Rao, K.M.K. (1990). Modulation of neutrophil LAM1 and CR3 cytoskeletal attachment by protein kinases. J. Leukoc. Biol. Suppl, 21a (Abst).

Cybulsky, M.I., McComb, D.J. and Movat, H.Z. (1988). Neutrophil leukocyte emigration induced by endotoxin-mediator roles of interleukin 1 and tumor necrosis factor alpha 1. J. Immunol. 140, 3144–3149.

Cybulsky, M.I., McComb, D.J. and Movat, H.Z. (1989). Protein synthesis dependent and independent mechanisms of neutrophil emigration: different mechanisms of inflammation in rabbits induced by interleukin-1, tumor necrosis factor alpha or endotoxin versus leukocyte chemoattractants. Am. J. Pathol. 135, 227–235.

Dahinden, C.A., Zingg, J., Maly, F.E. and De Weck, A.L. (1988). Leukotreine production in human neutrophils primed by recombinant human granulocyte/macrophage colony-stimulating factor and stimulated with the complement component C5a and FMLP as second signals. J. Exp. Med. 167, 1281–1295.

Dana, N., Styrt, B., Griffin, J.D., Todd, III, R.F., Klempner, M.S. and Arnaout, M.A. (1986). Two functional domains in the phagocyte membrane glycoproteins Mol identified with monoclonal antibodies. J. Immunol. 137, 3259–3263.

Davies, E.G., Isaacs, D. and Levinsky, R.J. (1982). Defective immune interferon production and natural killer activity associated with poor neutrophil mobility and delayed umbilical cord separation. Clin. Exp. Im. 50, 454–460.

Dawson, R.C.M., Irvine, R.F., Bray, J. and Quinn, P.J. (1984). Long-chain unsaturated diacylglycerols cause a perturbation in the structure of phospholipid bilayers rendering them susceptible to phospholipase attack. Biochem. Biophys. Res. Commun. 125, 836–842.

de Fougerolles, A.R., Stacker, S.A., Schwarting, R. and Springer, T.A. (1991). Characterization of ICAM-2 and evidence for a third counter-receptor for LFA-1. J. Exp. Med. 174, 253–267.

Della Bianca, V., De Togni, P., Grzeskowiak, M., Vicentini, L.M. and Di Virgilio, F. (1986). Cyclic AMP inhibition of phosphoinositide turnover in human neutrophils. Biochim. Biophys. Acta 886, 441–447.

Detmers, P.A., Wright, S.D., Olsen, E., Kimball, B. and Cohn, Z.A. (1987). Aggregation of complement receptors on human neutrophils in the absence of ligand. J. Cell Biol. 105, 1137–1145.

Detmers, P.A., Lo, S.K., Olsen-Egbert, E., Walz, A., Baggiolini, M., and Cohn, Z.A. (1990). Neutrophil-activating protein 1/interleukin 8 stimulates the binding activity of the leukocyte adhesion receptor CD11b/CD18 on human neutrophils. J. Exp. Med. 171, 1155–1162.

Diamond, M.S., Staunton, D.E., deFougerolles, A.R., Stacker, S.A., Garcia-Aguilar, J., Hibbs, M.L. and Springa, T.A. (1990). ICAM-1 (CD54): a counter-receptor for Mac-1 (CD11b/CD18). J. Cell Biol. 111, 3129–3139.

Diamond, M.S., Staunton, D.E., Marlin, S.D. and Springer, T.A. (1991). Binding of the integrin Mac-1 (CD11b/CD18) to the third immunoglobulin-like domain of ICAM-1 (CD54) and its regulation by glycosylation. Cell 65, 961–971.

Dixon, R.A.F., Diehl, R.E., Opas, E., Rands, E., Vickers, P.J., Evans, J.F., Gillard, J.W. and Miller, D.K. (1990). Requirement of a 5-lipoxygenase-activating protein for leukotriene synthesis. Nature 343, 282–284.

Doerschuk, C.M., Winn, R.K., Coxson, H.O. and Harlan, J.M. (1990). CD18-dependent and independent mechanisms of neutrophil emigration in the pulmonary and systemic microcirculation of rabbits. J. Immunol. 144(6), 2327–2333.

Doerschuk, C.M., Burns, A.R., English, D. and Takei, F. (1991). The role of ICAM-1 in neutrophil emigration induced by peritoneal instillates of *S. pneumonia* in mice. FASEB J. 5, 548 (Abstr).

Dransfield, I. and Hogg, N. (1989). Regulated expression of Mg^{2+} binding epitope on leukocyte integrin alpha subunits. EMBO J. 8(12), 3759–3765.

Dreyer, W.J., Michael, L.H., West, M.S., Smith, C.W., Rothlein, R., Rossen, R.D., Anderson, D.C. and Entman, M.L. (1991). Neutrophil accumulation in ischemic canine myocardium: insights into the time course, distribution, and mechanism of localization during early reperfusion. Circulation 84, 400–411.

Dustin, M.L. and Springer, T.A. (1989). T-Cell receptor cross-linking transiently stimulates adhesiveness through LFA-1. Nature, Lond. 341, 619–624.

Dustin, M.L., Rothlein, R., Bhan, A.K., Dinarello, C.A. and Springer, T.A. (1986). Induction by IL-1 and interferon-gamma: tissue distribution, biochemistry, and function of a natural adherence molecule (ICAM-1). J. Immunol. 137, 245–254.

Entman, M.L., Youker, K., Shappell, S.B., Siegel, C., Rothlein, R., Dreyer, W.J., Schmalstieg, F.C. and Smith, C.W. (1990). Neutrophil adherence to isolated adult canine myocytes: evidence for a CD18-dependent mechanism. J. Clin. Invest. 85, 1497–1506.

Entman, M.L., Michael, L.H., Rossen, R.D., Dreyer, W.J., Anderson, D.C., Taylor, A.A. and Smith, C.W. (1991a). Inflammation in the course of early myocardial ischemia. FASEB J. 5, 2529–2537.

Entman, M.L., Youker, K., Shoji, T., Kukielka, G.L., Shappell, S.B., Taylor, A.A. and Smith, C.W. (1992). Neutrophil induced oxidative injury of cardiac myocytes: A compartmented system requiring CD11b/CD18-ICAM-1 adherence. J. Clin. Invest. 90, 1335–1345.

Exton, J.H. (1988). Mechanisms of action of calcium-mobilizing agonists: some variations on a young theme. FASEB 2, 2670–2676.

Fallman, M., Lew, D.P., Stendahl, O. and Andersson, T. (1989). Receptor-mediated phagocytosis in human neutrophils is associated with increased formation of inositol phosphates and diacylglycerol. Elevation in cytosolic free calcium and formation of inositol phosphates can be dissociated from accumulation of diacylglycerol. J. Clin. Invest. 84, 886–891.

Farber, H.W., Center, D.M. and Rounds, S. (1985). Bovine and human endothelial cell production of neutrophil chemoattractant activity in response to components of the angiotensin system. Circ. Res. 57, 898–903.

Farber, H.W., Weller, P.F., Rounds, S., Beer, D.J. and Center, D.M. (1986). Generation of lipid neutrophil chemo-attractant activity by histamine-stimulated cultured endothelial cells. J. Immunol. 137, 2918–2923.

Feinmark, S.J. and Cannon, P.J. (1986). Endothelial cell leukotriene C4 synthesis results from intercellular transfer of leukotriene A4 synthesized by polymorphonuclear leukocytes. J. Biol. Chem. 261, 16 466–16 472.

Feinmark, S.J. and Cannon, P.J. (1987). Vascular smooth muscle cell leukotriene C4 synthesis: requirement for transcellular leukotriene A4 metabolism. Biochim. Biophys. Acta 922, 125–135.

Fiebig, E., Ley, K. and Arfors, K.-E. (1991). Rapid leukocyte accumulation by "spontaneous" rolling and adhesion in the exteriorized rabbit mesentery. Int. J. Microcirc. Clin. Exp. 10, 127–144.

Fischer, A., Descamps-Latscha, B., Gerota, I., Scheinmetzler, C. and Virelizier, J.L. (1983). Bone marrow transplantation for inborn error of phagocytic cells associated with defective adherence, chemotaxis, and oxidative response during opsonized particle phagocytosis. Lancet 2, 473–476.

Francis, J.W., Todd, R.F., Boxer, L.A. and Petty, H.R. (1989). Sequential expression of cell surface C3bi receptors during neutrophil locomotion. J. Cell. Physiol. 140, 519–523.

Furie, M.B. and McHugh, D.D. (1989). Migration of neutrophils across endothelial monolayers is stimulated by treatment of the monolayers with interleukin-1 or tumor necrosis factor-alpha. J. Immunology 143, 3309–3317.

Furie, M.B., Cramer, E.V., Naprstek, B.L. and Silverstein, S.C. (1984). Cultured endothelial cell monolayers that restrict the transendothelial passage of macromolecules and electrical current. J. Cell. Biol. 98, 1033–1041.

Furie, M.B., Naprstek, B.L. and Silverstein, S.C. (1987). Migration of neutrophils across monolayers of cultured microvascular endothelial cells. J. Cell Science 88, 161–175.

Furie, M.B., Burns, M.J., Tancinco, M.C.A., Benjamin, C.D. and Lobb, R.R. (1991a). Endothelial-leukocyte adhesion molecule-1 is not required for the migration of neutrophils across IL-1-stimulated endothelium *in vitro*. J. Immunol. (In press).

Furie, M.B., Tancinco, M.C.A. and Smith, C.W. (1991b). Monoclonal antibodies to leukocyte integrins CD11a/CD18 and CD11b/CD18 or intercellular adhesion molecule-1 (ICAM-1) inhibit chemo-attractant-stimulated neutrophil transendothelial migration *in vitro*. Blood 78, 2089–2097.

Gabig, T.G., English, D., Akard, L.P. and Schell, M.J. (1987). Regulation of neutrophil NADPH oxidase activation in a cell-free system by guanine nucleotides and fluoride. J. Biol. Chem. 262, 1685–1690.

Gabig, T.G., Eklund, E.A., Potter, G.B. and Dykes, J.R. III, (1990). A neutrophil GTP-binding protein that regulates cell-free NADPH oxidase activation is located in the cytosolic fraction. J. Immunol. 145, 945–951.

Gailbraith, G.M.P. (1988). Effect of protein kinase C inhibitors on calcium ionophore-induced arachidonic acid mobilization in human leukocytes. Immunopharmacology 16, 63–69.

Gamble, J.R., Harlan, J.M., Klebanoff, S.J. and Vadas, M.A. (1985). Stimulation of the adherence of neutrophils to umbilical vein endothelium by human recombinant tumor necrosis factor. Proc. Natl Acad. Sci. USA 82, 8667–8674.

Gaudry, M., Perianin, A., Marquetty, C. and Hakim, J. (1988). Negative effect of a protein kinase C inhibitor (H-7) on human polymorphonuclear neutrophil locomotion. Immunology 63, 715–719.

Gelas, P., Ribbes, G., Record, M., Terce, F. and Chap, H. (1989). Differential activation by fMet-Leu-Phe and phorbol ester of a plasma membrane phosphatidylcholine-specific phospholipase D in human neutrophil. FEBS Lett. 251, 213–218.

Geng, J.G., Bevilacqua, M.P., Moore, K.L., McIntyre, T.M., Prescott, S.M., Kim, J.M., Bliss, G.A., Zimmerman, G.A. and McEver, R.P. (1990). Rapid neutrophil adhesion to activated endothelium mediated by GMP-140. Nature 343, 757–760.

Gerard, C., McPhail, L.C., Marfat, A., Stimler-Gerard, N.P., Bass, D.A. and McCall, C.E. (1986). Role of protein kinases in stimulation of human polymorphonuclear leukocyte oxidative metabolism by various agonists. J. Clin. Invest. 77, 61–65.

Gimbrone, Jr., M.A., Brock, A.F. and Schafer, A.I. (1984). Leukotiene B_4 stimulates polymorphonuclear leukocyte adhesion to cultured vascular endothelial cells. J. Clin. Invest. 74, 1552–1555.

Gimbrone, Jr., M.A., Obin, M.S., Brock, A.F., Luis, E.A., Hass, P.E., Hebert, C.A., Yip, Y.K., Leung, D.W., Lowe, D.G., Kohr, W.J., Darbonne, W.C., Bechtol, K.B. and Baker, J.B. (1989). Endothelial interleukin-8: a novel inhibitor of leukocyte-endothelial interactions. Science 246, 1601–1603.

Ginis, I. and Tauber, A.I. (1990). Activation mechanisms of adherent human neutrophils. Blood 76, 1233–1239.

Goetzl, E.J. and Pickett, W.C. (1980). The human PMN leukocyte chemotactic activity of complex hydroxy-eicosatetraenoic acids (HETES). J. Immunol. 125, 1789–1794.

Graham, I.L., Gresham, H.D. and Brown, E.J. (1989). An immobile subset of plasma membrane CD11b/CD18 (Mac-1) is involved in phagocytosis of targets recognized by multiple receptors. J. Immunol. 142, 2352–2358.

Granger, D.N., Russell, J., Arfors, K.E., Rothlein, R. and Anderson, D.C. (1991a). Role of CD11/CD18 and ICAM-1 in ischemia-reperfusion induced leukocyte adherence and emigration in mesenteric venules. FASEB J. 5, 1753a (Abstr).

Granger, D.N., Russell, J., Arfors, K.E., Rothlein, R. and Anderson, D.C. (1991b). Role of CD11/CD18 and ICAM-1 in ischemia-reperfusion induced leukocyte adherence and emigration in mesenteric venules. FASEB J. 5, A1753 (Abstr).

Granstein, R.D., Margolis, R.J., Mizel, S.D. and Sauder, D.N. (1986). *In vivo* inflammatory activity of epidermal cell-derived lymphocyte activating factor and recombinant interleukin-1 in the mouse. J. Clin. Invest. 77, 1020–1027.

Grant, L. (1974). The sticking and emigration of white blood cells in inflammation. In "The Inflammatory Process" (eds B.W. Zweifach, L. Grant, and R.T. McCluskey), pp. 205–221. New York, Academic Press.

Gresham, H.D., McGarr, J.A., Shackelford, P.G. and Brown, E.J. (1988). Studies on the molecular mechanisms of human Fc receptor-mediated phagocytosis. J. Clin. Invest. 82, 1192–1201.

Gresham, H.D., Goodwin, J.L., Allen, P.M., Anderson, D.C. and Brown, E.J. (1989). A novel member of the integrin receptor family mediates Arg-Gly-Asp (RGD)-stimulated neutrophil phagocytosis. J. Cell Biol. 108, 1935–1943.

Gresham, H.D., Graham, I.L., Anderson, D.C. and Brown, E.J. (1991). Leukocyte adhesion deficient (LAD) neutrophils fail to amplify phagocytic function in response to stimulation: evidence for CD11b/CD18-dependent and -independent mechanisms of phagocytosis. J. Clin. Invest. 88, 588–597.

Griffin, J.D., Spertini, O., Ernst, T.J., Belvin, M.P., Levine, H.B., Kanakura, Y. and Tedder, T.F. (1990). Granulocyte-macrophage colony-stimulating factor and other cytokines regulate surface expression of the leukocyte adhesion molecule-1 on human neutrophils, monocytes, and their precursors. J. Immunol. 145, 576–584.

Gudewicz, P.W., Odekon, L.E., DelVecchio, P.J. and Saba, T.M. (1988). Generation of neutrophil chemotactic activity by phorbol ester-stimulated calf pulmonary artery endothelial cells. J. Leuk. Biol. 440, 1–7.

Gundel, R.H., Wegner, C.D., Torcellini, C.A., Clarke, C.C., Haynes, N., Rothlein, R., Smith, C.W. and Letts, L.G. (1991). Endothelial-leukocyte adhesion molecule-1 mediates antigen-induced acute airway inflammation and late-phase airway obstruction in monkeys. J. Clin. Invest. 88, 1407–1411.

Haines, K.A., Giedd, K.N., Rich, A.M., Korchak, H.M. and Weissmann, G. (1987). The leukotriene B4 paradox: neutrophils can, but will not, respond to ligand-receptor interactions by forming leukotriene B4 or its omega-metabolites. Biochem. J. 241, 55–62.

Hallmann, R., Jutila, M.A., Smith, C.W., Anderson, D.C., Kishimoto, T.K. and Butcher, E.C. (1991). The peripheral lymph node homing receptor, LECAM-1, is involved in CD-18-independent adhesion of human neutrophils to endothelium. Biochem. Biophys. Res. Commun. 174, 236–243.

Hanlon, W.A., Stolk, J., Davies, P., Humes, J.L., Mumford, R. and Bonney, R.J. (1991). rTNFalpha facilitates human polymorphonuclear leukocyte adherence to fibrinogen matrices with mobilization of specific and tertiary but not azurophilic granule markers. J. Leuk. Biol. 50, 43–48.

Harlan, J.M., Killen, P.D., Harker, L.A., Striker, G.E. and Wright, D.G. (1981). Neutrophil-mediated endothelial injury *in vitro*: mechanisms of cell detachment. J. Clin. Invest. 68, 1394–1403.

Harlan, J.M., Killen, P.D., Senecal, F.M., Schwartz, B.R., Yee, E.K., Taylor, R.F., Beatty, P.G., Price, T.H. and Ochs, H.D. (1985). The role of neutrophil membrane glycoprotein GP-150 in neutrophil adherent to endothelium *in vitro*. Blood 66, 167–178.

Hattori, R., Hamilton, K.K., Fugates, R.D., McEver, R.P. and Sims, P.J. (1989). Stimulated secretion of endothelial von Willebrand factor is accompanied by rapid redistribution to the cell surface of the intracellular granule membrane protein GMP-140. J. Biol. Chem. 264, 7768–7771.

Hawkins, H.K., Heffelfinger, S. and Anderson, D.C. (1991). Leukocyte adhesion deficiency: clinical and post-mortem observations. Pediat. Pathol. 12, 119–130.

Hayward, A.R., Leonard, J., Wood, C.B.S., Harvey, B.A.M., Greenwood, M.C. and Soothill, J.F. (1979). Delayed separation of the umbilical cord, widespread infections, and defective neutrophil mobility. Lancet 1, 1099–1101.

Hechtman, D.H., Cybulsky, M.I., Fuchs, H.J., Baker, J.B. and Gimbrone, Jr., M.A. (1991). Intravascular IL-8. Inhibitor of polymorphonuclear leukocyte accumulation at sites of acute inflammation. J. Immunol. 147, 883–892.

Hectman, D.H., Cybulsky, M.I., Baker, J.B. and Gimbrone, M.A. (1990). Intravenous endothelial interleukin-8 reduces neutrophil accumulation at intradermal sites of inflammation. FASEB J. 4, A890 (Abstr).

Henderson, B. and Pettipher, E.R. (1988). Comparison of the *in vivo* inflammatory activities after intra-acticular injection of natural and recombinant IL-1alpha and IL-1beta in the rabbit. Biochem. Pharm. 37, 4171.

Henson, P.M. and Johnston, R.B. (1987). Tissue injury in inflammation. Oxidants, proteinases, and cationic proteins. J. Clin. Invest. 79, 669–674.

Henson, P.M., Henson, J.E., Fittschen, C., Kimani, G., Bratton, D.L. and Riches, D.W.H. (1988). In "Inflammation: Basic Principles and Clinical Correlates" (eds J.I. Gallin, I.M. Goldstein and R. Snyderman), pp. 363–390. New York, Raven Press Ltd.

Hermanowski-Vosatka, A., Van Strijp, J.A.G., Swiggard, W.J. and Wright, S.D. (1992). Integrin modulating factor-1: a lipid that alters the function of leukocyte integrins. Cell 68, 341–352.

Hernandez, L.A., Grisham, M.B., Twohig, B., Arfors, K.E., Harlan, J.M. and Granger, D.N. (1987). Role of neutrophils in ischemia-reperfusion-induced microvascular injury. Am. J. Physiol. 238, H699-H703.

Heyworth, P.G. and Segal, A.W. (1986). Further evidence for the involvement of a phosphoprotein in the respiratory burst oxidase of human neutrophils. Biochem. J. 239, 723–731.

Heyworth, P.G., Shrimpton, C.F. and Segal, A.W. (1989). Localization of the 47 kDa phosphoprotein involved in the respiratory-burst NADPH oxidase of phagocytic cells. Biochem. J. 260, 243–248.

Hobbs, M.L., Jakes, S., Stacker, S.A., Wallace, R.W. and Springer, T.A. (1991a). The cytoplasmic domain of the integrin lymphocyte function-associated antigen 1-beta subunit: sites required for binding to intercellular adhesion molecule 1 and the phorbol ester-stimulated phosphorylation site. J. Exp. Med. 174, 1227–1238 (Abstr).

Hobbs, M.L., Xu, H., Stacker, S.A. and Springer, T.A. (1991b). Regulation of adhesion to ICAM-1 by the cytoplasmic domain of LFA-1 integrin beta subunit. Science 251, 1537–1642.

Hidaka, H., Inagaki, M., Kawamoto, S. and Sasaki, Y. (1984). Isoquinolinesulfonamides, novel and potent inhibitors of cyclic nucleotide dependent protein kinase and protein kinase C. Biochemistry 23, 5036–5041.

Honeycutt, P.J. and Niedel, J.E. (1986). Cytochalasin B enhancement of the diacylglycerol response in formyl peptide-stimulated neutrophils. J. Biol. Chem. 261, 15 900–15 905.

Horgan, M.J., Ge, M., Gu, J., Rothlein, R. and Malik, A.B. (1992). Role of ICAM-1 in neutrophil-mediated lung vascular injury after occlusion and reperfusion. Am. J. Physiol. 261, H1578–H1584.

Horowitz, A., Duggan, K., Buck, C.A., Beckerle, M.C. and Burridge, K. (1986). Interaction of plasma membrane fibronectin receptor with talin – a transmembrane linkage. Nature 320, 531–533.

House, S.D. and Lipowsky, H.H. (1987). Leukocyte-endothelium adhesion: microhemodynamics in mesentery of the cat. Microvas. Res. 34, 363–379.

Huber, A.R. and Weiss, S.J. (1989). Disruption of the subendothelial basement membrane during neutrophil diapedesis in an *in vitro* construct of a blood vessel wall. J. Clin. Invest. 83, 1122–1136.

Huber, A.R., Kunkel, S.L., Todd, III, R.F. and Weiss, S.J. (1991). Regulation of transendothelial neutrophil migration by endogenous interleukin-8. Science 254, 99–105.

Hughes, H., Gentry, D.L., McGuire, G.M. and Taylor, A.A. (1990). Lipoxygenase product formation following myocardial ischemia. FASEB J. 4, A1145 (Abstr).

Hurley, J.V. (1963). An electron microscopic study of leucocytic emigration and vascular permeability in rat skin. Austral. J. Exp. Biol. Med. Sci. 41, 171–186.

Hynes, R.O. (1987). Integrins: a family of cell surface receptors. Cell 48, 549–554.

Jaeschke, H., Farhood, A.I. and Smith, C.W. (1991). Neutrophil-induced liver cell injury in endotoxin shock is a CD11b/CD18-dependent mechanism. Am. J. Physiol. 261, c-1051–c-1056.

Jasin, H.E., Lightfoot, E., Kavanaugh, A., Rothlein, R., Faanes, R.B., and Lipsky, P.E. (1990). Successful treatment of chronic antigen-induced arthritis in rabbits with monoclonal antibodies to leukocyte adhesion molecules. Arthritis Rheum. 33, s34 (Abstr).

Jerome, S.N., Smith, C.W. and Korthuis, R.J. (1993). CD18-dependent adherence reactions play an important role in the development of the no-reflow phenomenon. Am. J. Physiol. 264, H479-H483.

Jones, D.H., Anderson, D.C., Burr, B.L., Rudloff, H.E., Smith, C.W., Krater, S.S. and Schmalstieg, F.C. (1988). Quantitiation of intracellular Mac-1 (CD11b/CD18) pools in human neutrophils. J. Leuk. Biol. 44, 535–544.

Jones, D.H., Schmalstieg, F.C., Hawkins, H.K., Burr, B.L., Rudloff, H.E., Krater, S.S., Smith, C.W. and Anderson, D.C. (1989). In "Leukocyte Adhesion Molecules: Structure, Function, and Regulation" (eds T.A. Springer, D.C. Anderson, A.S. Rosenthal and R. Rothlein), pp. 106–124. New York, Springer-Verlag.

Jones, D.H., Schmalstieg, F.C., Dempsey, K., Krater, S.S., Nannen, D.D., Smith, C.W. and Anderson, D.C. (1990). Subcellular distribution and mobilization of Mac-1 (CD11b/CD18) in neonatal neutrophils. Blood 75, 488–498.

Jubiz, W., Radmark, O., Malmsten, C., Hansson, G., Lindgren, J.A., Palmblad, J., Uden, A.M. and Samuelsson, B. (1982). A novel leukotriene produced by stimulation of leukocytes with formylmethionylleucylphenylalanine. J. Biol. Chem. 257, 6106–6110.

Jutila, M.A., Rott, L., Berg, E.L. and Butcher, E.C. (1989). Function and regulation of the neutrophil MEL-14 antigen *in vivo*: comparison with LFA-1 and MAC-1. J. Immunol. 143, 3318–3324.

Jutila, M.A., Kishimoto, T.K. and Butcher, E.C. (1990). Regulation and lectin activity of the human neutrophil peripheral lymph node homing receptor. Blood 75, 1–5.

Kavanaugh, A.F., Lightfoot, E., Lipsky, P.E. and Oppenheimer-Marks, N. (1991). Role of CD11/CD18 in adhesion and transendothelial migration of T-cells. Analysis utilizing CD18-deficient T-cell clones. J. Immunol. 146, 4149–4156.

Keizer, G.D., te Velde, A.A., Schwarting, R., Figdor, C.G. and De Vries, J.E. (1987). Role of p150,95 in adhesion, migration, chemotaxis and phagocytosis of human monocytes. Eur. J. Immunol. 17, 1317–1322.

Keizer, G.D., Visser, W., Vliem, M. and Figdor, C.G. (1988). A monoclonal antibody (NKL-L16) directed against a unique epitope on the alpha-chain of human leukocyte function-associated antigen 1 induces homotypic cell-cell interactions. J. Immunol. 140, 1393–1400.

Kharazmi, A., Nielsen, H. and Bendzen, K. (1988). Recombinant interleukin alpha and beta prime human monocyte superoxide production but have no effect on chemotaxis and oxidative burst response of neutrophils. Immunobiology 177, 32–37.

Kikkawa, U., Kishimoto, A. and Nishizuka, Y. (1989). The protein kinase C family: heterogeneity and its implications. Annu. Rev. Biochem. 58, 31–44.

Kishimoto, A., Kajikawa, N., Shiota, M. and Nishizuka, Y. (1983). Proteolytic activation of calcium-activated, phospholipid-dependent protein kinase by calcium-dependent neutral protease. J. Biol. Chem. 258, 1156–1164.

Kishimoto, T.K., O'Connor, K., Lee, A., Roberts, T.M. and Springer, T.A. (1987). Cloning of the beta subunit of the leukocyte adhesion proteins: homology to an extracellular matrix receptor defines a novel supergene family. Cell 48, 681–690.

Kishimoto, T.K., Jutila, M.A., Berg, E.L. and Butcher, E.C. (1989a). Neutrophil Mac-1 and MEL-14 adhesion proteins inversely regulated by chemotactic factors. Science 245, 1238–1241.

Kishimoto, T.K., Larson, R.S., Corbi, A.L., Dustin, M.L., Staunton, D.E. and Springer, T.A. (1989b). The leukocyte integrins. Advances in Immunol. 46, 149–182.

Kishimoto, T.K., Larson, R.S., Corbi, A.L., Dustin, M.L., Staunton, D.E. and Springer, T.A. (1989c). In "Leukocyte Adhesion Molecules: Structure, Function, and Regulation" (eds T.A. Springer, D.C. Anderson, A.S. Rosenthal and R. Rothlein) pp. 7–43. New York, Springer-Verlag.

Kishimoto, T.K., Jutila, M.A. and Butcher, E.C. (1990). Identification of a human peripheral lymph node homing receptor: a rapidly down-regulated adhesion molecule. Proc. Natl Acad. Sci. USA 87, 2244–2248.

Kishimoto, T.K., Warnock, R.A., Jutila, M.A., Butcher, E.C., Lane, C.L., Anderson, D.C. and Smith, C.W. (1991). Antibodies against human neutrophil LECAM-1 (LAM-1/Leu-8/DREG-56 antigen) and endothelial cell ELAM-1 inhibit a common CD18-independent adhesion pathway *in vitro*. Blood 78, 805–811.

Klebanoff, S.J. (1988). In "Inflammation: Basic Principles and Clinical Correlates" (eds J.I. Gallin, I.M. Goldstein and R. Snyderman) pp. 391–444. New York, Raven Press, Ltd.

Kobayashi, K., Fujita, K. and Okino, F. (1984). An abnormality of neutrophil adhesion: autosomal recessive inheritance associated with missing neutrophil glycoproteins. Pediatrics 73, 606–610.

Kramer, I.M., Verhoeven, A.J., Van der Bend, R.L., Weening, R.S., and Roos, D. (1988). Purified protein kinase C phosphorylates a 47-kDa protein in control neutrophil cytoplasts but not in neutrophil cytoplasts from patients with the autosomal form of chronic granulomatous disease. J. Biol. Chem. 263, 2352–2357.

Kramer, I.M., Van der Bend, R.L., Tool, A.T.J., Van Blitterswijk, W.J., Roos, D. and Verhoeven, A.J. (1989). 1-O-hexadecyl-2-O-methylglycerol, a novel inhibitor of protein kinase C, inhibits the respiratory burst in human neutrophils. J. Biol. Chem. 264, 5876–5884.

Kramer, K.M., Checani, G.C. and Deykin, D. (1987). Stimulation of Ca^{2+}-activated human platelet phospholipase A2 by diacylglycerol. Biochem. J. 248, 779–783.

Kubes, P., Ibbotson, G., Russell, J., Wallace, J.L. and Granger, D.N. (1990). Role of platelet-activating factor in ischemia/reperfusion-induced leukocyte adherence. Am. J. Physiol. 259, G300–G305.

Kuijpers, T.W., Haldert, B.C., Hart, M.H.L. and Roos, D. (1992). Neutrophil migration across monolayers of cytokine-prestimulated endothelial cells: a role for platelet-activating factor and IL-8. J. Cell Biol. 117, 565–572.

Lackie, J.M. and Debono, D. (1977). Interactions of neutrophil granulocytes and endothelium *in vitro*. Microvasc. Res. 13, 107–112.

Lawrence, M.B. and Springer, T.A. (1991). Leukocytes roll on a selectin at physiological flow rates: distinction from and prerequisite for adhesion through integrins. Cell 65, 1–20.

Lawrence, M.B., Smith, C.W., Eskin, S.G. and McIntire, L.V. (1990). Effect of venous shear stress on CD18-mediated neutrophil adhesion to cultured endothelium. Blood 75, 227–237.

Leto, T.L., Lomax, K.J., Volpp, B.D., Nunoi, H., Sechler, J.M.G., Nauseef, W.M., Clark, R.A., Gallin, J.I. and Malech, H.L. (1990). Cloning of a 67-kDa neutrophil oxidase factor with similarity to a noncatalytic region of p60c-src. Science 248, 727–730.

Lewinsohn, D.M., Bargatze, R.F. and Butcher, E.C. (1987). Leukocyte-endothelial cell recognition: evidence of common molecular mechanism shared by neutrophils, lymphocytes, and other leukocytes. J. Immunol. 138, 4313–4321.

Lewis, R.A. and Austen, K.F. (1988). In "Inflammation: Basic Principles and Clinical Correlates" (eds J.I. Gallin, I.M. Goldstein and R. Snyderman) pp. 121–128. New York, Raven Press, Ltd.

Ley, K., Lundgren, E., Berger, E. and Arfors, K.E. (1989). Shear-dependent inhibition of granulocyte adhesion to cultured endothelium by dextran sulfate. Blood 73, 1324–1330.

Ley, K., Cerrito, M. and Arfors, K.-E. (1991a). Sulfated polysaccharides inhibit leukocyte rolling in rabbit mesentery venules. Am. J. Physiol. 26, H1667–H1673.

Ley, K., Gaehtgens, P., Fennie, C., Singer, M.S., Lasky, L.A. and Rosen, S.D. (1991b). Lectin-like cell adhesion molecule 1 mediates leukocyte rolling in mesenteric venules *in vivo*. Blood 77, 2553–2555.

Liles, W.C., Meier, K.E. and Henderson, W.R. (1987). Phorbol myristate acetate and the calcium ionophore A23187 synergistically induce release of LTB4 by human neutrophils: involvement of protein kinase C activation in regulation of the 5-lipoxygenase pathway. J. Immunol. 138, 3396–3402.

Lo, S.K., Detmer, P.A., Levin, S.M. and Wright, S.D. (1989a). Transient adhesion of neutrophils to endothelium. J. Exp. Med. 169, 1779–1793.

Lo, S.K., Van Seventer, G.A., Levin, S.M. and Wright, S.D. (1989b). Two leukocyte receptors (CD11a/CD18) mediate transient adhesion to endothelium by binding to different ligands. J. Immunol. 143, 3325–3329.

Lo, S.K., Lee, S., Ramos, R.A., Lobb, R., Rosa, M., Chi-Rosso, G., and Wright, S.D. (1991). Endothelial-leukocyte adhesion molecule 1 stimulates the adhesive activity of leukocyte integrin CR3 (CD11b/CD18, Mac-1, alpha m beta 2) on human neutrophils. J. Exp. Med. 173, 1493–1500.

Loetscher, H., Pan, Y.E., Lahm, H.W., Gentz, R., Brockhaus, M., Tabuchi, H. and Lesslauer, W. (1990). Molecular cloning and expression of the human 55 kDa tumor necrosis factor receptor. Cell 61, 351–359.

Loike, J.D., Sodeik, B., Weitz, J., Wright, S.D. and Silverstein, S.C. (1990). P150/95 on neutrophils is a receptor for fibrinogen. FASEB J. 4, 2185a (Abstr).

Lomax, K.J., Leto, T.L., Nunoi, H., Gallin, J.I. and Malech, H.I. (1989). Recombinant 47 kDa cytosol factor restores NADPH oxidase in chronic granulomatous disease. Science 245, 409–413.

Lopez, A.F., Williamson, D.J., Gamble, J.R., Begley, C.G., Harlan, J.M., Klebanoff, S.J., Waltersdorph, A., Wong, G., Clark, S.C. and Vadas, M.A. (1986). Recombinant human granulocyte-macrophage colony-stimulating factor stimulates *in vitro* mature human neutrophil and eosinophil function, surface receptor expression, and survival. J. Clin. Invest. 78, 1220–1228.

Lorant, D.E., Patel, K.D., McIntyre, T.M., McEver, R.P., Prescott, S.M. and Zimmerman, G.A. (1991). Co-expression of GMP-140 and PAF by endothelium stimulated by histamine or thrombin: a juxtacrine system for adhesion and activation of neutrophils. J. Cell Biol. 115, 223–234.

Lowe, J.B., Stoolman, L.M., Nair, R.P., Larsen, R.D., Berhend, T.L., and Marks, R.M. (1990). ELAM-1-dependent cell adhesion to vascular endothelium determined by a transfected human fucosyltransferase cDNA. Cell 63, 475–484.

Lucchesi, B.R. and Mullane, K.M. (1986). Leukocytes and ischemia induced myocardial injury. Ann. Rev. Pharm. Tox. 26, 201–224.

Luscinskas, F.W., Brock, A.F., Arnaout, M.A. and Gimbrone, Jr., M.A. (1989). Endothelial-leukocyte adhesion molecule-1-dependent and leukocyte (CD11/CD18)-dependent mechanisms contribute to polymorphonuclear leukocyte adhesion to cytokine-activated human vascular endothelium. J. Immunol. 142, 2257–2263.

Luscinskas, F.W., Cybulsky, M.I., Kiely, J-M., Peckins, C.S., Davis, V.M. and Gimbrone, M.A. (1991). Cytokine-activated human endothelial monolayer support enhanced neutrophil transmigration via a mechanism involving both endothelial-leukocyte adhesion molecule-1 and intercellular adhesion molecule-1. J. Immunol. 146, 1617–1625.

Ma, X.-L., Tsao, P.S. and Lefer, A.M. (1991). Antibody to CD-18 exerts endothelial and cardiac protective effects in myocardial ischemia and reperfusion. J. Clin. Invest. 88, 1237–1243.

Maclouf, J.A. and Murphy, R.C. (1988). Transcellular metabolism of neutrophil-derived leukotriene A4 by human platelets. J. Biol. Chem. 263, 174–181.

Maclouf, J.A., Murphy, R.C. and Henson, P.M. (1989). Transcellular sulfidopeptide leukotriene biosynthetic capacity of vascular cells. Blood 74, 703–707.

Maclouf, J.A., Murphy, R.C. and Henson, P.M. (1990). Transcellular biosynthesis of sulfidopeptide leukotrienes during receptor-mediated stimulation of human neutrophil/platelet mixtures. Blood 76, 1838–1844.

Marchesi, V.T. (1964). Some electron microscopic observations on interactions between leukocytes, platelets, and endothelial cells in acute inflammation. Ann. NY Acad. Sci. 116, 774–788.

Marchesi, V.T. and Gowans, J.L. (1964). The migration of lymphocytes through the endothelium of venules in lymph nodes: an electron microscope study. Proc. Roy. Soc. Med. 159, 283–290.

Marlin, S.D. and Springer, T.A. (1987). Purified intercellular adhesion molecule-1 (ICAM-1) is a ligand for lymphocyte function-associated antigen 1 (LFA-1). Cell 51, 813–819.

Matsushima, K. and Oppenheim, J.J. (1989). Review article: interleukin 8 and MCAF: novel inflammatory cytokines inducible by IL-1 and TNF. Cytokine 1, 2–10.

McEver, R.P., Beckstead, J.H., Moore, K.L., Marshall-Carlson, L. and Bainton, D.F. (1989). GMP-140, a platelet alpha-granule membrane protein, is also synthesized by vascular endothelial cells and is localized in Weibel-palade bodies. J. Clin. Invest. 84, 92–99.

McIntire, T.M., Reinhold, S.L., Prescott, S.M. and Zimmerman, G.A. (1987). Protein kinase C activity appears to be required for the synthesis of platelet activating factor and leukotriene B4 by human neutrophils. J. Biol. Chem. 262, 15 370–15 376.

McPhail, L.C., Clayton, C.C. and Snyderman, R. (1984). A potential second messenger role for unsaturated fatty acids: activation of Ca^{2+}-dependent protein kinase. Science 224, 622–625.

McPhail, L.C., Shirley, P.S., Clayton, C.C. and Snyderman, R. (1985). Activation of the respiratory burst enzyme from human neutrophils in a cell-free system. J. Clin. Invest. 75, 1735–1739.

Melloni, E., Pontremoli, S., Michetti, M., Sacco, O., Sparatore, B., Salamino, F. and Horecker, B.L. (1985). Binding of protein kinase C to neutrophil membranes in the presence of Ca^{2+} and its activation by a Ca^{2+}-requiring proteinase. Proc. Natl Acad. Sci. USA 82, 6435–6439.

Mercandetti, A.J., Lane, T.A. and Colmerauer, M.E.M. (1984). Cultured human endothelial cells elaborate neutrophil chemoattractants. J. Lab. Clin. Med. 104, 370–377.

Meurer, R. and McIntyre, D.E. (1989). Lack of effect of pertussis toxin on TNF-α-induced formation of reactive oxygen intermediates by human neutrophils. Biochem. Biophys. Res. Commun. 159, 763–769.

Miller, D.K., Gillard, J.W., Vickers, P.J., Sadowski, S., Léveillé, C., Mancini, J.A., Charleson, P., Dixon, R.A.F., Ford-Hutchinson, A.W., Fortin, R., Gauthier, J.Y., Rodkey, J., Rosen, R., Rouzer, C., Sigal, I.S., Strader, C.D. and Evans, J.F. (1990). Identification and isolation of a membrane protein necessary for leukotriene production. Nature 343, 278–281.

Moser, R., Schleiffenbaum, B., Groscurth, P. and Fehr, J. (1989). Interleukin 1 and tumor necrosis factor stimulate human vascular endothelial cells to promote transendothelial neutrophil passage. J. Clin. Invest. 83, 444–455.

Movat, H.Z., Burrowes, C.E., Cybulsky, M.I. and Dinarello, C.A. (1987). Acute inflammation and a shwartzman-like reaction induced by interleukin-1 and tumor necrosis factor. Am. J. Pathol. 129, 463–476.

Mullane, K.M. and Smith, C.W. (1990). The role of leukocytes in ischemic damage, reperfusion injury and repair of the myocardium. In "Pathophysiology of Severe Ischemic Myocardial Injury" (ed H.M. Piper), pp. 239–267. Dordrecht, Kluwer Academic Publishers.

Mulligan, M.S., Varani, J., Dame, M.K., Lane, C.L., Smith, C.W., Anderson, D.C. and Ward, P.A. (1991). Role of endothelial-leukocyte adhesion molecule 1 (ELAM-1) in neutrophil-mediated lung injury in rats. J. Clin. Invest. 88, 1396–1406.

Mulligan, M.S., Johnson, K.J., Todd, III, R.F., Issekutz, T.B., Miyasaka, M.J., Tamatani, T., Smith, C.W., Anderson, D.C. and Ward, P.A. (1993). Requirements for leukocyte adhesion molecules in nephrotoxic nephritis. J. Clin. Invest. 91, 577–587.

Mulligan, M.S., Varani, J., Warren, J.S., Till, G.O., Smith, C.W., Anderson, D.C., Todd, III, R.F. and Ward, P.A. (1992b). Roles of beta-2 integrins of rat neutrophils in complement- and oxygen radical-mediated acute inflammatory injury. J. Immunol. 148, 1847–1857.

Mulligan, M.S., Warren, J.S., Smith, C.W., Anderson, D.C., Yeh, C.G., Rudolph, A.R. and Ward, P.A. (1992c). Lung injury after deposition of IgA immune complexes. Requirements for CD18 and L-arginine. J. Immunol. 148, 3086–3092.

Munro, J.M., Pober, J.S. and Cotran, R.S. (1989). Tumor necrosis factor and interferon-gamma induce distinct patterns of endothelial activation and associated leukocyte accumulation in skin of papio anubis. Am. J. Pathol. 135, 121–133.

Munro, J.M., Pober, J.S. and Cotran, R.S. (1991). Recruitment of neutrophils in the local endotoxin response: association with *de novo* endothelial expression of endothelial leukocyte adhesion molecule-1. Lab. Invest. 64, 295–299.

Nathan, C. and Sanchez, E. (1990). Tumor necrosis factor and CD11/CD18 (beta2) integrins act synergistically to lower cAMP in human neutrophils. J. Cell Biol. 111, 2171–2181.

Nathan, C.F. (1987). Neutrophil activation on biological surfaces: massive secretion of hydrogen peroxide in response to products of macrophages and lymphocytes. J. Clin. Invest. 80, 1550–1560.

Nathan, C.F. (1989). Respiratory burst in adherent human neutrophils: Triggering by colony-stimulating factors CSF-GM and CSF-G. Blood 73, 301–306.

Nathan, C.F., Srimal, S., Farber, C., Sanchez, E., Kabbash, L., Asch, A., Gailit, J. and Wright, S.D. (1989). Cytokine-induced respiratory burst of human neutrophils: dependence on extracellular matrix proteins and CD11/CD18 integrins. JCB 109, 1341–1349.

Nauseef, W.M., Volpp, B.D., McCormick, S., Leidal, K.G. and Clark, R.A. (1991). Assembly of the neutrophil respiratory burst oxidase. Protein kinase C promotes cytoskeletal and membrane association of cytosolic oxidase components. J. Biol. Chem. 266, 5911–5917.

Nishihira, J. and O'Flaherty, J.T. (1985). Phorbol myristate acetate receptors in human polymorphonuclear neutrophils. J. Immunol. 135, 3439–3443.

Nishihira, J., McPhail, L.C. and O'Flaherty, J.T. (1986). Stimulus-dependent mobilization of protein kinase C. Biochem. Biophys. Res. Commun. 134, 587–594.

Nishizuka, Y. (1986). Studies and perspectives of protein kinase C. Science 233, 305–312.

Nishizuka, Y. (1988). Review article. The molecular heterogeneity of protein kinase C and its implications for cellular regulation. Nature 334, 661–665.

Nourshargh, S., Rampart, M., Hellewell, P.G., Jose, P.J., Harlan, J.M. and Williams, T.J. (1989). Accumation of IIIIn-neutrophils in rabbit skin in allergic and non-allergic reactions *in vivo*. Inhibition by neutrophil pretreatment *in vitro* with reconizing the CD18 antigen. J. Immunol. 142, 3193–3198.

Nunoi, H., Rotrosen, D., Gallin, J.I. and Malech, H.L. (1988). Two forms of autosomal chronic granulomatous disease lack distinct neutrophil cytosol factors. Science 242, 1298–1301.

O'Brien, R.F., Seton, M.P., Makarski, J.S., Center, D.M. and Rounds, S. (1984). Thiourea causes endothelial cells in tissue culture to produce neutrophil chemoattractant activity. Am. Rev. Respir. Dis. 130, 103–109.

O'Flaherty, J.T., Jacobson, D.P., Redman, J.F. and Rossi, A.G. (1990). Translocation of protein kinase C in human polymorphonuclear neutrophils. Regulation by cytosolic Ca^{2+}-independent and Ca^{2+}-dependent mechanisms. J. Biol. Chem. 265, 9146–9152.

O'Shea, J.J., Brown, E.J., Seligmann, B.E., Metcalf, J.A., Frank, M.M. and Gallin, J.I. (1985). Evidence for distinct intracellular pools of receptors for C3b and C3bi in human neutrophils. J. Immunol. 134, 2580–2587.

Obel, D., Rasmussen, L.H. and Christiansen, N.O. (1991). Protein kinase C subtypes in human neutrophils. Scand. J. Clin. Lab. Invest. 51, 299–302.

Ohno, Y., Buescher, E.S., Roberts, R., Metcalf, J.A. and Gallin, J.I. (1986). Re-evaluation of cytochrome b and flavin adenine dinucleotide in neutrophils from patients with chronic granulomatous disease and description of a family with probable autosomal recessive inheritance of cytochrome b deficiency. Blood 67, 1132–1138.

Okamura, N., Curnutte, J.T., Roberts, R.L. and Babior, B.M. (1988). Relationship of protein phosphorylation to the activation of the respiratory burst in human neutrophils. Defects in the phosphorylation of a group of closely related 48-kDa proteins in two forms of chronic granulomatous disease. J. Biol. Chem. 263, 6777–6782.

Oppenheimer-Marks, N., Davis, L.S., Bogue, D.T., Ramberg, J. and Lipsky, P.E. (1991). Differential utilization of ICAM-1 and VCAM-1 during the adhesion and trans-endothelial migration of human T lymphocytes. J. Immunol. 147, 2913–2921.

Osaki, M., Sumimoto, H., Takeshige, K., Cragoe, E.J., Jr., Hori, Y. and Minakami, S. (1989). Na^+/H^+ exchange modulates the production of leukotriene B4 by human neutrophils. Biochem. J. 257, 751–758.

Otey, C.A., Pavalko, F.M. and Burridge, K. (1990). An interaction between alpha-actinin and the beta 1 integrin subunit *in vitro*. J. Cell Biol. 111, 721–729.

Palmer, R.M.J. and Salmon, J.A. (1983). Release of leukotriene B4 from neutrophils and its relationship to degranulation induced by N-formyl-methionyl-leucyl-phenylalanine, serum-treated zymosan and the ionophore A23187. Immunology 50, 65–73.

Pardi, R., Inverardi, L., Rugarli, C. and Bender, J.R. (1992). Antigen-receptor complex stimulation triggers protein kinase C-dependent CD11a/CD18-cytoskeleton association in T lymphocytes. J. Cell Biol. 116, 1211–1220.

Parkos, C.A., Allen, R.A., Cochrane, C.G. and Jesaitis, A.J. (1987). Purified cytochrome b from human granulocyte plasma membrane is comprised of two polypeptides with relative molecular weights of 91 000 and 22 000. J. Clin. Invest. 80, 732–742.

Patel, K.D., Zimmerman, G.A., Prescott, S.M., McEver, R.P. and McIntyre, T.M. (1991). Oxygen radicals induce human endothelial cells to express GMP-140 and bind neutrophils. J. Cell Biol. 112, 749–759.

Perry, M.A. and Granger, D.N. (1991). Role of CD11/CD18 in shear rate-dependent leukocyte-endothelial cell interactions in cat mesenteric venules. J. Clin. Invest. 87, 1798–1804.

Petty, H.R., Francis, J.W., Todd, III, R.F., Petrequin, P.R. and Boxer, L.A. (1987). Neutrophil C3bi receptors: formation of membrane clusters during cell triggering requires intracellular granules. J. Cell. Physiol. 133, 235–242.

Philips, M., Buyon, J.P., Winchester, R., Weissmann, G. and Abramson, S. (1988). Upregulation of iC3b receptors (CR3) is neither necessary nor sufficient to promote neutrophil aggregation. J. Clin. Invest. 82, 495–501.

Phillips, M.L., Nudelman, E., Gaeta, F.C.A., Perez, M., Singhal, A.K., Hakomori, S. and Paulson, J.C. (1990). ELAM-1 mediates cell adhesion by recognition of a carbohydrate ligand, Sialyl-Le^x. Science 250, 1130–1132.

Picker, L.J., Kishimoto, T.K., Smith, C.W., Warnock, R.A. and Butcher, E.C. (1991a). ELAM-1 is an adhesion molecule for skin-homing T-cells. Nature 349, 796–799.

Picker, L.J., Warnock, R.A., Burns, A.R., Doerschuk, C.M., Berg, E.L. and Butcher, E.C. (1991b). The neutrophil selectin LECAM-1 presents carbohydrate ligands to the vascular selectins ELAM-1 and GMP-140. Cell 66, 921–933.

Pike, M.C., Jakoi, L., McPhail, L.C. and Snyderman, R. (1986). Chemoattractant-mediated stimulation of the respiratory burst in human polymorphonuclear leukocytes may require appearance of protein kinase activity in the cell's particulate fraction. Blood 67, 909–913.

Pober, J.S. and Cotran, R.S. (1990). The role of endothelial cells in inflammation. Transplantation 50, 537–544.

Pober, J.S., Gimbrone, Jr., M.A., Lapierre, L.A., Mendrick, D.L., Fiers, W., Rothlein, R. and Springer, T.A. (1986). Overlapping patterns of antigenic modulation by interleukin 1, tumor necrosis factor and immune interferon. J. Immunol. 137, 1893–1896.

Pohlman, T.H., Stanness, K.A., Beatty, P.G., Ochs, H.D. and Harlan, J.M. (1986). An endothelial cell surface factors(s) induced *in vitro* by lipopolysaccharide, interleukin-1, and tumor necrosis factor increases neutrophil adherence by a CDw18 (LFA)-dependent mechanism. J. Immunol. 136, 4548–4553.

Pontremoli, S., Melloni, E., Michetti, M., Salamino, F., Sparatore, B., Sacco, O. and Horecker, B.L. (1986). Differential mechanisms of translocation of protein kinase C to plasma membranes in activated human neutrophils. Biochem. Biophys. Res. Commun. 136, 228–234.

Pontremoli, S., Melloni, E., Damiani, G., Salamino, F., Sparatore, B., Michetti, M. and Horecker, B.L. (1988). Effects of a monoclonal anti-calpain antibody on responses of stimulated human neutrophils. J. Biol. Chem. 263, 1915–1919.

Pontremoli, S., Melloni, E., Sparatore, B., Michetti, M., Salamino, F., and Horecker, B.L. (1990). Isozymes of protein kinase C in human neutrophils and their modification by two endogenous proteinases. J. Biol. Chem. 265, 706–712.

Price, T.H., Beatty, P.G. and Corpuz, S.R. (1987). *In vivo* inhibition of neutrophil function in the rabbit using monoclonal antibody to CD18. J. Immunol. 139, 4174–4177.

Quinn, M.T., Parkos, C.A., Walker, L., Orkin, S.H., Dinauer, M.C. and Jesaitis, A.J. (1989). Association of a *Ras*-related protein with cytochrome b of human neutrophils. Nature 342, 198–200.

Rampart, M. and Williams, T.J. (1988). Evidence that neutrophil accumulation induced by interleukin-1 requires both local protein biosynthesis and neutrophil CD18 antigen expression *in vivo*. Br. J. Pharmacol. 94, 1143–1151.

Rando, R.R. (1988). Regulation of protein kinase C activity by lipids. FASEB J. 2, 2348–2355.

Relman, D., Tuomanen, E., Falkow, S., Golenbock, D.T., Saukkonen, K. and Wright, S.D. (1990). Recognition of a bacterial adhesin by an integrin: macrophage CR3 (alpha m beta2, CD11b/CD18) binds filamentous hemagglutinin of bordetella pertussis. Cell 61, 1375–1382.

Richter, J., Andersson, T. and Olsson, I. (1989). Effect of tumor necrosis factor and granulocyte/macrophage colony-stimulating factor on neutrophil degranulation. J. Immunol. 142, 3199–3205.

Richter, J., Ng-Sikorski, J., Olsson, I. and Andersson, T. (1990a). Tumor necrosis factor-induced degranulation in adherent human neutrophils is dependent on CD11b/CD18-integrin-triggered oscillations of cytosolic free Ca^{2+}. Proc. Natl Acad. Sci. USA 87, 9472–9476.

Richter, J., Olsson, I. and Andersson, T. (1990b). Correlation between spontaneous oscillations of cytosolic free Ca^{2+} and tumor necrosis factor-induced degranulation in adherent human neutrophils. J. Biol. Chem. 265, 14 358–14 363.

Rider, L.G. and Niedel, J.E. (1987). Diacylglycerol accumulation and superoxide anion production in stimulated human neutrophils. J. Biol. Chem. 262, 5603–5608.

Rosen, H. (1990). Role of CR3 in induced myelomonocytic recruitment: insights from *in vivo* monoclonal antibody studies in the mouse. J. Leuk. Biol. 48, 465–469.

Rosen, H. and Gordon, S. (1987). Monoclonal antibody to the murine type 3 complement receptor inhibits adhesion of myelomonocytic cells *in vitro* and inflammatory cell recruitment *in vivo*. J. Exp. Med. 166, 1685–1701.

Rosen, H., Gordon, S. and North, R.J. (1989). Exacerbation of murine listeriosis by a monoclonal antibody specific for the type 3 complement receptor of myelomonocytic cells. J. Exp. Med. 170, 27–37.

Ross, G.D., Thompson, R.A., Walport, M.J., Springer, T.A., Watson, J.V., Ward, R.H.R., Lida, J., Newman, S.L., Harrison, R.A. and Lachmann, P.J. (1985). Characterization of patients with an increased susceptibility to bacterial infections and a genetic deficiency of leukocyte membrane complement receptor type 3 and the related membrane antigen LFA-1. Blood 66, 882–890.

Rossi, F., Grzeskowiak, M., Della Bianca, V., Calzetti, F. and Gandini, G. (1990). Phosphatidic acid and not diacylglycerol generated by phospholipase D is functionally linked to the activation of the NADPH oxidase by FMLP in human neutrophils. Biochem. Biophys. Res. Commun. 168, 320–327.

Rothlein, R. and Springer, T.A. (1986). The requirement for lymphocyte function-associated antigen 1 in homotypic stimulated by phorbol ester. J. Exp. Med. 163, 1132–1149.

Rothlein, R., Dustin, M.L., Marlin, S.D. and Springer, T.A. (1986). A human intercellular adhesion molecule (ICAM-1) distinct from LFA-1. J. Immunol. 137, 1270–1274.

Rouzer, C.A. and Samuelsson, B. (1985). On the nature of the 5-lipoxygenase reaction in human leukocytes: enzyme purification and requirement for multiple stimulatory factors. Proc. Natl Acad. Sci. USA 82, 6040–6044.

Rowzer, C.A. and Kargman, S.A. (1985). Translocation of 5-lipoxygenase to the membrane in human leukocytes challenged with ionophore A23187. J. Biol. Chem. 263, 10 980–10 988.

Ruoslahti, E. and Pierschbacher, M.D. (1987). New perspectives in cell adhesion: RGD and integrins. Science 238, 491–497.

Russell, D.G. and Wright, S.D. (1988). Complement receptor type 3 (CR3) binds to an Arg-Gly-Asp-containing region of the major surface glycoprotein, gp63, of leishmania promastigotes. J. Exp. Med. 168, 279–292.

Ryan, G.B., Borysenko, J.Z. and Karnovski, M.J. (1974). Factors affecting the redistribution of surface-bound concanavalin A on human polymorphonuclear leukocytes. J. Cell Biol. 62, 351–365.

Salari, H., Braquet, P., Naccache, P. and Borgeat, P. (1985). Characterization of effect of N-formyl-methionyl-leucyl-phenylalanine on leukotriene synthesis in human polymorphonuclear leukocytes. Inflammation 9, 127–137.

Samuelsson, B. and Funk, C.D. (1989). Enzymes involved in the biosynthesis of leukotriene B_4. J. Biol. Chem. 264, 19 469–19 472.

Sayers, T.J., Wiltrout, T.A., Bull, C.A., Denn, A.C., Pilaro, A.M. and Lokesh, B. (1988). Effect of cytokines on polymorphonuclear neutrophil infiltration in the mouse: prostaglandia- and leukotriene-independent induction of infiltration by IL-1 and tumor necrosis factor. J. Immunol. 141, 1670–1677.

Schall, T.J., Lewis, M., Koller, K.J., Lee, A., Rice, G.C., Wong, G.H.W., Gatanaga, T., Granger, G.A., Lentz, R., Raab, H., Kohr, W.J. and Goeddel, D.V. (1990). Molecular cloning and expression of a receptor for human tumor necrosis factor. Cell 61, 361–370.

Schleiffenbaum, B., Moser, R., Patarroyo, M. and Fehr, J. (1989). The cell surface glycoprotein Mac-1 (CD11b/CD18) mediates neutrophil adhesion and modulates degranulation independently of its quantitative cell surface expression. J. Immunol. 142, 3537–3542.

Schmalstieg, F.C., Rudloff, H.E., Hillman, G.R. and Anderson, D.C. (1986). Two dimensional and three dimensional movement of human polymorphonuclear leukocytes: two fundamentally different mechanisms of location. J. Leuk. Biol. 40, 677–691.

Schroder, J-M. and Christophers, E. (1989). Secretion of novel and homologous neutrophil-activating peptides by LPS-stimulated human endothelial cells. J. Immunol. 142, 244–251.

Schwartz, B.R. and Harlan, J.M. (1989). Sulfhydryl reducing agents promote neutrophil adherence without increasing surface expression of CD11b/CD18 (Mac-1, Mo-1). Biochem. Biopsy. Res. Comm. 165, 51–57.

Segal, A.W. (1987). Absence of both cytochrome b-245 subunits from neutrophils in X-linked chronic granulomatous disease. Nature 326, 88–90.

Segal, A.W., Jones, O.T.G., Webster, D. and Allison, A.C. (1978). Absence of a newly described cytochrome b from neutrophils of patients with chronic granulomatous disease. Lancet 2, 446–449.

Seifert, R. and Schultz, G. (1987). Fatty-acid-induced activation of NADPH oxidase in plasma membranes of human neutrophils depends on neutrophil cytosol and is potentiated by stable guanine nucleotides. Eur. J. Biochem. 162, 563–569.

Shappell, S.B., Taylor, A.A., Hughes, H., Mitchell, J.R., Anderson, D.C. and Smith, C.W. (1990a). Comparison of antioxidant and nonantioxidant lipoxygenase inhibitors on neutrophil function. Implications for pathogenesis of myocardial reperfusion injury. J. Pharmacol. Exp. Ther. 252, 531–538.

Shappell, S.B., Taylor, A.A. and Smith, C.W. (1990b). Mac-1 (CD11/CD18) mediates adherence-dependent H_2O_2 production by neutrophils through protein kinase C activation. FASEB J. 4, 2179a (Abstr).

Shappell, S.B., Toman, C., Anderson, D.C., Taylor, A.A., Entman, M.L. and Smith, C.W. (1990c). Mac-1 (CD11b/CD18) mediates adherence-dependent hydrogen peroxide production by human and canine neutrophils. J. Immunol. 144, 2702–2711.

Shappell, S.B., Taylor, A.A., Hughes, H. and Smith, C.W. (1991). Neutrophil (PMN) CD18-mediated adherence-dependent H_2O_2 production is accompanied by diacylglycerol (DAG) accumulation. J. Cell Biochem. (Suppl) 15C, 179a (Abstr).

Sharpe, R.J., Margolis, R.J., Askari, M., Amento, E.P. and Granstein, R.D. (1988). Induction of dermal and subcutaneous inflammation by recombinant cachectin/tumor necrosis factor (TNF-alpha) in the mouse. J. Invest. Dermatol. 91, 353–357.

Shaw, J.O (1980). Leukocytes in chemotactic-fragment-induced lung inflammation: vascular emigration and alveolar surface migration. Am. J. Pathol. 101, 283–302.

Smedly, L.A., Tonnesen, M.G., Sandhaus, R.A., Haslett, C., Guthrie, L.A., Johnston, R.B., Henson, P.M. and Worthen, G.S. (1986). Neutrophil mediated injury to endothelial cells, enhancement by endotoxin and essential role of neutrophil elastase. J. Clin. Invest. 77, 1233–1243.

Smith, C.W. and Hollers, J.C. (1980). Motility and adhesiveness in human neutrophils. Redistribution of chemotactic factor induced adhesion sites. J. Clin. Invest. 65, 804–812.

Smith, C.W., Rothlein, R., Hughes, B.J., Mariscalco, M.M., Schmalstieg, F.C. and Anderson, D.C. (1988). Recognition of an endothelial determinant for CD18-dependent human neutrophil adherence and transendothelial migration. J. Clin. Invest. 82, 1746–1756.

Smith, C.W., Marlin, S.D., Rothlein, R., Lawrence, M.B., McIntire, L.V. and Anderson, D.C. (1989a). In "Leukocyte Adhesion Molecules: Structure, Function, and Regulation" (eds T.A. Springer, D.C. Anderson, A.S. Rosenthal and R. Rothlein), pp. 170–189. New York, Springer-Verlag.

Smith, C.W., Marlin, S.D., Rothlein, R., Toman, C. and Anderson, D.C. (1989b). Cooperative interactions of LFA-1 and Mac-1 with intercellular adhesion molecule-1 in facilitating adherence and transendothelial migration of human neutrophils *in vitro*. J. Clin. Invest. 83, 2008–2017.

Smith, C.W., Anderson, D.C., Taylor, A.A., Rossen, R.D. and Entman, M.L. (1991a). Leukocyte adhesion molecules and myocardial ischemia. TCM 1, 167–170.

Smith, C.W., Entman, M.L., Lane, C.L., Beaudet, A.L., Ty, T.I., Youker, K., Hawkins, H.K. and Anderson, D.C. (1991b). Adherence of neutrophils to canine cardiac myocytes *in vitro* is dependent on intercellular adhesion molecule-1. J. Clin. Invest. 88, 1216–1223.

Smith, C.W., Kishimoto, T.K., Abbassi, O., Hughes, B.J., Rothlein, R., McIntire, L.V., Butcher, E. and Anderson, D.C. (1991c). Chemotactic factors regulate lectin adhesion molecule 1 (LECAM-1)-dependent neutrophil adhesion to cytokine-stimulated endothelial cells *in vitro*. J. Clin. Invest. 87, 609–618.

Smith, C.W., Farhood, A. and Jaeschke, H. (1992). Monoclonal antibody against CD11b/CD18 (Mac-1) integrin protects in a model of hepatic ischemia-reperfusion injury. FASEB J. 6, 1303a (Abstr).

Snyderman, R. and Pike, M.C. (1984). Chemo-attractant receptors on phagocytic cells. Ann. Rev. Immunol. 2, 257–281.

Snyderman, R. and Uhing, R. (1988). In "Inflammation: Basic Principles and Clinical Correlates" (eds J.I. Gallin, I.M. Goldstein and R. Snyderman), New York, Raven Press.

Southwick, F.S., Dabiri, G.A., Paschetto, M. and Zigmond, S.H. (1989). Polymorphonuclear leukocyte adherence induces acti polymerization by a transduction pathway which differs from that used by chemo-attractants. J. Cell Biol. 109, 1561–1569.

Spertini, O., Kansas, G.S., Reimann, K.A., Mackay, C.R. and Tedder, T.F. (1991). Function and evolutionary conservation of distinct epitopes on the leukocyte adhesion molecule-1 (TQ-1, Leu-8) that regulate leukocyte migration. J. Immunol. 147, 942–949.

Springer, T.A. and Anderson, D.C. (1986). In "Biochemistry of Macrophages", pp. 102–126. London, Ciba Foundation Symposium.

Springer, T.A., Thompson, W.S., Miller, L.J., Schmalstieg, F.C. and Anderson, D.C. (1984). Inherited deficiency of the Mac-1, LFA-1, p150,95 glycoprotein family and its molecular basis. J. Exp. Med. 160, 1901–1918.

Stacker, S.A. and Springer, T.A. (1991). Leukocyte integrin p150,95 (CD11c/CD18) functions as an adhesion molecule binding to a counter-receptor on stimulated endothelium. J. Immunol. 146, 648–655.

Staunton, D.E., Dustin, M.L. and Springer, T.A. (1989). Functional cloning of ICAM-2, a cell adhesion ligand for LFA-1 homologous. Nature 339, 61–64.

Staunton, D.E., Dustin, M.L., Erickson, H.P. and Springer, T.A. (1990). The arrangement of the immunoglobulin-like domains of ICAM-1 and the binding sites for LFA-1 and rhinovirus. Cell 61, 243–254.

Stevenson, K.B., Nauseef, W.M. and Clark, R.A. (1987). The neutrophil glycoprotein Mo1 is an integral membrane protein of plasma membranes and specific granules. J. Immunol. 139, 3759–3763.

Stossel, T.P. (1978). In "Leukocyte Chemotaxis" (eds J.I. Gallin and P.G. Quie), pp. 143–147. New York, Raven Press.

Strieter, R.M., Kunkel, S.L., Showell, H.J., Remick, D.G., Phan, S.H., Ward, P.A. and Marks, R.M. (1989). Endothelial cell gene expression of a neutrophil chemotactic factor by TNF-alpha, LPS, and IL-1-beta. Science 243, 1467–1469.

Tauber, A.I. (1987). Review: Protein kinase C and the activation of the human neutrophil NADPH-oxidase. Blood 69, 711–720.

Taylor, R.F., Price, T.H., Schwartz, S.M. and Dale, D.C. (1981). Neutrophil-endothelial cell interactions on endothelial monolayers grown on micropore filters. J. Clin. Invest. 67, 584–587.

Tedder, T.F., Penta, A.C., Levine, H.B. and Freedman, A.S. (1990). Expression of the human leukocyte adhesion molecule, LAM1 identity with the TQ1 and Leu-8 differentiation antigens. J. Immunology 144, 532–540.

Thompson, N.T., Tateson, J.E., Randall, R.W., Spacey, G.D., Bonser, R.W. and Garland, L.G. (1990). The temporal relationship between phospholipase activation, diradylglycerol formation and superoxide production in the human neutrophil. Biochem. J. 271, 209–213.

Till, G.O., Mulligan, M.S., Lee, S., Reimer, W., Ward, P.A. and Marak, G.E. (1992). Adhesion molecules in experimental phacoanaphylactic endophthalmitis. Invest. Ophthal. Vis. Sci. 33, 795a (Abstr).

Todd, III, R.F. and Freyer, D.R. (1988). The CD11/CD18 leukocyte glycoprotein deficiency. Hematology/Oncology Clinics of North America 2, 13–28.

Todd, III, R.F., Arnaout, M.A., Rosin, R.E., Crowley, C.A., Peters, W.A. and Babior, B.M. (1984). Subcellular localization of the large subunit of Mo1 ($Mo1_1$; formerly gp110), a surface glycoprotein associated with neutrophil adhesion. J. Clin. Invest. 74, 1280–1290.

Tonnesen, M.G., Smedly, L.A. and Henson, P.M. (1984). Neutrophil-endothelial cell interactions: modulation of neutrophil adhesiveness induced by complement fragments C5a and $C5a_{desArg}$, and formyl-methionyl-leucyl-phenylalanine *in vitro*. J. Clin. Invest. 74, 1581–1592.

Tonnesen, M.G., Anderson, D.C., Springer, T.A., Knedler, A., Avdi, N. and Henson, P.M. (1989). Adherence of neutrophils to cultured human microvascular endothelial cells: stimulation by chemotactic peptides and lipid mediators and dependence upon the Mac-1, LFA-1, p150,95 glycoprotein family. J. Clin. Invest. 83, 637–646.

Tyagi, S.R., Olson, S.C., Burnham, D.N. and Lambeth, J.D. (1991). Cyclic AMP-elevating agents block chemoattractant activation of diradylglycerol generation by inhibiting phospholipase D activation. J. Biol. Chem. 266, 3498–3504.

Van Kooyk, Y., Weder, P., Hogervorst, F., Verhoeven, A.J., Van Seventer, G.A., te Velde, A.A., Borst, J., Keizer, G.D. and Figdor, C.G. (1991). Activation of LFA-1 through a Ca^{2+}-dependent epitope stimulates lymphocyte adhesion. J. Cell Biol. 112, 345–354.

van Kuppevelt, T.H., Languino, L.R., Gailit, J.O., Suzuki, S. and Ruoslahti, E. (1986). An alternative cytoplasmic domain of the integrin $Beta_3$. Proc. Natl Acad. Sci. USA 86, 5415–5418.

Varani, J., Ginsburg, I., Schuger, L., Gibbs, D.F., Bromberg, J., Johnson, K.J., Ryan, U.S. and Ward, P.A. (1989). Endothelial cell killing by neutrophils: synergistic interactions of oxygen products and proteases. Am. J. Pathol. 135, 435–438.

Vedder, N.B. and Harlan, J.M. (1988). Increased surface expression of CD11b/CD18 (Mac-1) is not required for stimulated neutrophil adherence to cultured endothelium. J. Clin. Invest. 81, 676–682.

Volpp, B.D., Nauseef, W.M. and Clark, R.A. (1988). Two cytosolic neutrophil oxidase components absent in autosomal chronic granulomatous disease. Science 242, 1295–1297.

Volpp, B.D., Nauseef, W.M., Donelson, J.E., Moser, D.R. and Clark, R.A. (1989). Cloning of the cDNA and functional expression of the 47-kilodalton cytosolic component of human neutrophil respiratory burst oxidase. Proc. Natl Acad. Sci. USA 86, 7195–7199.

von Andrian, U.H., Chambers, J.D., McEvoy, L.M., Bargatze, R.F., Arfors, K.-E. and Butcher, E.C. (1991). Two step model of leukocyte-endothelial cell interaction in inflammation: distinct roles for LECAM-1 and the leukocyte beta-2 integrins *in vivo*. Proc. Natl Acad. Sci. USA 88, 7538–7542.

Von Asmuth, E.J.U., van der Linden, C.J., Leeuwenberg, J.F.M. and Buurman, W.A. (1991). Involvement of the CD11b/CD18 integrin, but not of the endothelial cell adhesion molecules ELAM-1 and ICAM-1 in tumor necrosis factor-alpha-induced neutrophil toxicity. J. Immunol. 147, 3869–3875.

Walz, G., Aruffo, A., Kolanus, W., Bevilacqua, M.P. and Seed, B. (1990). Recognition by ELAM-1 of the Sialyl-Lewis X determinant on myeloid and tumor cells. Science 250, 1132–1135.

Wankowicz, Z., Megyeri, P. and Issekutz, A. (1988). Synergy between tumour necrosis factor alpha and interleukin-1 in the induction of polymorphonuclear leukocyte migration during inflammation. J. Leuk. Biol. 43, 349–356.

Watson, S.R., Fennie, C. and Lasky, L.A. (1991). Neutrophil influx into an inflammatory site inhibited by a soluble homing receptor-IgG chimaera. Nature 349, 164–167.

Weening, R.S., Corbeel, L., de Boer, M., Lutter, R., van Zwieten, R., Hamers, M.N. and Roos, D. (1985). Cytochrome b deficiency in an autosomal form of chronic granulomatous disease. J. Clin. Invest. 75, 915–920.

Williamson, J.R. and Grisham, J.W. (1961). Electron microscopy of leukocytic margination and emigration in acute inflammation in dog pancreas. Am. J. Pathol. 39, 239–256.

Winquist, R., Frei, P., Harrison, P., McFarland, M., Letts, G., Van, G., Andrews, L., Rothlein, R. and Hintze, T. (1990). An anti-CD18 mAb limits infarct size in primates following myocardial ischemia and reperfusion. Circulation 82, III-701.

Wolfson, M., McPhail, L.C., Nasrallah, V.N. and Snyderman, R. (1985). Phorbol myristate acetate mediates redistribution of protein kinase C in human neutrophils: potential role in the activation of the respiratory burst enzyme. J. Immunol. 135, 2057–2062.

Woodman, R.C., Ruedi, J.M., Jesaitis, A.J., Okamura, N., Quinn, M.T., Smith, R.M., Curnutte, J.T. and Babior, B.M. (1991). Respiratory burst oxidase and three of four oxidase-related polypeptides are associated with the cytoskeleton of human neutrophils. J. Clin. Invest. 87, 1345–1351.

Worthen, G.S., Smedly, L.A., Tonnesen, M.G., Ellis, D., Voelkel, N.F., Reeves, J.T. and Henson, P.M. (1987). Effects of shear stress on adhesive interaction between neutrophils and cultured endothelial cells. J. Appl. Physiol. 63(5), 2031–2041.

Worthen, G.S., Schwab III, B., Elson, E.L. and Downey, G.P. (1989). Mechanics of stimulated neutrophils: cell stiffening induces retention of capillaries. Science 245, 183–186.

Wright, S.D. and Jong, M.T.C. (1986). Adhesion-promoting receptors on human macrophages recognize *Escherichia coli* by binding to lipopolysaccharide. J. Exp. Med. 164, 1876–1888.

Wright, S.D. and Meyer, B.C. (1986). Phorbol esters cause sequential activation and deactivation of complement receptors on polymorphonuclear leukocytes. J. Immunol. 136, 1759–1764.

Wright, S.D., Rao, P.E., VanVoorhis, W.C., Craigmyle, L.S., Iida, K., Talle, M.A., Westbery, E.F., Goldstein, G. and Silverstein, S.C. (1983). Identification of the C3bi receptor of human monocytes and macrophages with monoclonal antibodies. Proc. Natl Acad. Sci. USA 80, 5699–5703.

Wright, S.D., Reddy, P.A., Jong, M.T.C. and Erickson, B.W. (1987). C3bi receptor (complement receptor type 3) recognizes a region of complement protein C3 containing the sequence Arg-Gly-Asp. Proc. Natl Acad. Sci. USA 84, 1965–1968.

Wright, S.D., Weitz, J.I., Huang, A.J., Levin, S.M., Silverstein, S.C., and Loike, J.D. (1988). Complement receptor type three (CD11b/CD18) of human polymorphonuclear leukocytes recognizes fibrinogen. Proc. Natl Acad. Sci. USA 85(2), 7734–7738.

Wright, S.D., Lo, S.K. and Detmers, P.A. (1989). In "Leukocyte Adhesion Molecules: Structure, Function and Regulation" (eds T.A. Springer, D.C. Anderson, R. Rothlein and A.S. Rosenthal). pp 190–207. New York, Springer-Verlag.

Yednock, T.A. and Rosen, S.D. (1989). Lymphocyte homing. Advances in Immunol. 44, 313–378.

Yoshimura, T., Matsushima, K., Oppenheim, J.J. and Leonard, E.J. (1987). Neutrophil chemotactic factor produced by lipopolysaccharide (LPS)-stimulated human blood mononuclear leukocytes: partial characterization and separation from interleukin 1 (IL-1). J. Immunol. 139, 788–793.

Youker, K., Smith, C.W., Anderson, D.C., Miller, D., Michael, L.H., Rossen, R.D. and Entman, M.L. (1992). Neutrophil adherence to isolated adult cardiac myocytes: induction by cardiac lymph collected during ischemia and reperfusion. J. Clin. Invest. 89, 602–609.

Young, S., Rothbard, J. and Parker, P.J. (1988). A monoclonal antibody recognising the site of limited proteolysis of protein kinase C. Eur. J. Biochem. 173, 247–252.

Zavoico, G.B., Ewenstein, B.M., Schafer, A.I. and Pober, J.S. (1989). IL-1 and related cytokines enhance thrombin-stimulated PGI2 production in cultured endothelial cells without affecting thrombin-stimulated von Willebrand Factor secretion or platelet-activating factor biosynthesis. J. Immunol. 142, 3993–3999.

Zhou, M.J., Todd, R.F., Anderson, C.L. and Petty, H.R. (1990). Linkage of CR3, FcRII and FcRIII to the neutrophil's cortical microfilaments. FASEB J. 4, 2056a (Abstr).

Zigmond, S.H. (1977). Ability of polymorphonuclear leukocytes to orient in gradients of chemotactic factors. J. Cell Biol. 75, 606–616.

Zigmond, S.H. (1978). Chemotaxis by polymorphonuclear leukocytes. J. Cell Biol. 77, 269–287.

Zimmerman, B.J. and Granger, D.N. (1990). Reperfusion-induced leukocyte infiltration: role of elastase. Am. J. Physiol. 259, H390–H394.

Zimmerman, G.A. and McIntyre, T.M. (1988). Neutrophil adherence to human endothelium *in vitro* occurs by CDw18 (Mo1, MAC-1/LFA-1/GP150,95) glycoprotein-dependent and independent mechanisms. J. Clin. Invest. 81, 531–537.

Zimmerman, G.A., McIntyre, T.M., Mehra, M. and Prescott, S.M. (1990). Endothelial cell-associated platelet-activating factor: a novel mechanism for signaling intercellular adhesion. J. Cell Biol. 110, 529–540.

4. Role of Cell Adhesion Molecules in Tumor Progression and Metastasis

Steven M. Albelda

1. Cell Adhesion Molecules involved in Tumor Progression and Metastasis

Cell-cell and cell-substratum interactions are mediated by specific cell adhesion molecules. These cell adhesion molecules play a central role in many basic cellular functions including maintenance of normal tissue organization, cell migration, and regulation of proliferation and differentiation. Although a great deal of information has been obtained about the molecular structure, biochemical properties and cellular distribution of cell adhesion molecules (Buck and Horwitz, 1987; Hynes, 1987; Hemler, 1990; Albelda and Buck, 1990; Springer, 1990; Edelman and Crossin, 1991; Takeichi, 1991), much less is known about how they contribute to complex physiological events. The purpose of this chapter is to summarize the information available about the roles played by these molecules in tumor growth and metastasis.

Adhesion Molecules
ISBN 0-12-741440-1

Over the past fifteen years, dozens of specific cell surface molecules responsible for cell-cell and cell-substratum interactions have been identified and characterized to varying degrees (reviewed in Albelda and Buck, 1990). It has become increasingly apparent that many of these proteins can be placed into well-defined molecular families or superfamilies. Among the best characterized of these groupings are the integrins, cadherins, and the immunoglobulin superfamily.

1.1 INTEGRINS

1.1.1 Structure and Classification

The integrins are heterodimers consisting of non-covalently associated α and β subunits (Buck and Horwitz, 1987; Hynes, 1987; Ruoslahti and Pierschbacher, 1987; Hemler, 1990) that mediate cell adhesion. Certain of the integrins function as extracellular matrix receptors binding to such proteins as collagen laminin, fibrinogen, and fibronectin. In some cases (i.e. the fibronectin receptor), specific peptide regions within extracellular matrix proteins have been identified as the integrin binding sites. The best described of these regions is the arginine-glycine-aspartic acid (RGD) sequence found in fibronectin, fibrinogen, vitronectin, type I collagen and laminin (Ruoslahti and Pierschbacher, 1987). Peptides containing this sequence have the ability to inhibit cell adhesion to certain of these extracellular proteins. Other integrins participate in cell-cell adhesion, and some (i.e. the VLA-4 integrin or Mac-1) can serve in both capacities.

As originally described (Hynes, 1987), integrins were divided into three subfamilies, each with a common β subunit capable of associating with a specific group of α subunits. More recent work has shown that there are at least eight different β subunits and that certain α subunits can combine with more than one β subunit (Cheresh *et al.*, 1989; Krissansen *et al.*, 1990; Vogel *et al.*, 1990). In addition, most of the integrins bind to more than one ligand. Because of the complexity of the integrin family, we have found it useful to group integrins according to their cell-binding activity (Table 4.1). In this simplified organization, integrins can be segregated into one of three groups: those that function as cell-cell adhesion molecules (primarily found on leucocytes), those that bind primarily to the major constituents of the basement membrane (i.e. collagen and laminin), and those that bind primarily to the extracellular matrix proteins found during early development, inflammation, and wound healing (i.e. fibronectin, fibrinogen, vitronectin, thrombospondin). Only the $\alpha_3\beta_1$ integrin has been shown to bind to both types of matrix proteins.

Table 4.1 Simplified classification of integrins based on binding characteristics

Subunit	Ligands
1. Integrins that function as cell-cell adhesion molecules	
Subunit	*Ligands*
α_L/β_2 (LFA-1)	ICAM-1
α_M/β_2 (MAC-1)	ICAM-1, C3bi, Endotoxin
α_X/β_2 (gp150,95)	?
α_4/β_1	VCAM-1
II. Integrins that bind primarily to basement membrane proteins	
Subunit	*Ligands*
α_1/β_1	laminin/collagen
α_2/β_1	collagen/laminin
α_3/β_1	laminin/collagen/fibronectin
α_6/β_1	laminin
α_6/β_4	laminin
III. Integrins that bind primarily to matrix proteins of inflammation, wound healing, and development	
Subunit	*Ligands*
α_4/β_1	fibronectin (CSII site)
α_5/β_1	fibronectin (RGD site)
α_V/β_1	fibronectin
α_V/β_3	vitronectin, fibrinogen, thrombospondin, von Willebrand's Factor
α_V/β_5	vitronectin

1.1.2 Tissue Distribution

With the availability of monoclonal antibodies (mAbs) against specific α and β subunits, it has become possible to determine the integrin repertoire of cultured cells and tissues (Albelda *et al.*, 1989; Buck *et al.*, 1990; Hemler, 1990; Virtanen *et al.*, 1990). A number of generalizations can be made. Integrin expression appears to be universal; at least one, and generally several, integrins have been found on almost every cell or tissue studied to date (adult red blood cells being an exception). Furthermore, integrin expression is both tissue and cell type specific and has been shown to change during normal physiological processes (e.g. the menstrual cycle, Lessey *et al.*, 1992). The expression of integrins on cultured cells is not always identical to that expressed by the same cells in tissues, however, suggesting that there are regulatory constraints placed upon integrin expression *in situ* that are different from those in culture (Albelda *et al.*, 1990).

The cells of most tissues express $\alpha_1\beta_1$, $\alpha_2\beta_1$, $\alpha_3\beta_1$ and $\alpha_6\beta_1$; integrins primarily required for adhesion to collagen and laminin. In contrast, the $\alpha_5\beta_1$ fibronectin receptor is poorly expressed in most tissues. Although there is very limited, if any, expression of the classical vitronectin receptor, $\alpha_V\beta_3$, in epithelial cells, the α_V subunit is expressed, in association with alternative β subunits (β_1, β_5, and β_6) (Cherish *et al.*, 1989, Sheppard *et al.*, 1990, Vogel *et al.*, 1990). Leucocytes generally contain integrins of the β_2 subfamily, along with $\alpha_4\beta_1$, $\alpha_5\beta_1$, and $\alpha_6\beta_1$ (Hemler, 1990). The α_{IIb}/β_3 integrin is expressed primarily by platelets.

1.2 CADHERINS

1.2.1 Structure and Classification

Cadherins are calcium dependent transmembrane cell-cell adhesion molecules that bind to one another by means of homophilic interactions (reviewed in Takeichi, 1991). Cadherins are involved in establishing and maintaining intercellular connections and are probably one of the most important groups of CAM's in the formation of cell-cell associations. Four of the cadherins have been cloned; all have a similar primary structure consisting of 723–748 amino acids (Hatta *et al.*, 1988). Each contains a putative signal peptide, an extracellular region with 3 repeated domains, a highly hydrophobic region that functions as a transmembrane anchor, and a relatively long cytoplasmic tail. The overall homology between cadherins from different tissues and different species is between 50–60%. Information from gene cloning, along with immunological comparisons have allowed a tentative classification of the cadherin family into three subclasses (Takeichi, 1991). These are: (1) the E-cadherins found in adult epithelial cells (also known as uvomorulin, Cell CAM 120/80, Arc-1, or L-CAM); (2) the N-cadherins found in adult neural tissues and muscle (also termed A-CAM or N-Cal-CAM); and (3) the P-cadherins found primarily in placenta and epithelium, but transiently expressed in other tissue during development. This initial classification is rapidly expanding as other subclasses of cadherins are identified. For example, an endothelial cell cadherin, closely related to N-cadherin has been recognized (Heimark *et al.*, 1990; Liaw *et al.*, 1990). In addition, molecules that share protein sequence similarities with the cadherins, but have different overall structures [e.g. desmoglein, an intracellular component of desmosomes (Koch *et al.*, 1990)] have recently been described.

Cadherins appear to function by connecting cells to each other by homophilic interactions, in which they bind selectively to identical cadherin types (Nose *et al.*, 1990). For example, E-cadherin binds selectively to E-cadherin. Cadherin function is also dependent on interactions between the cytoplasmic domain and the elements of the cellular cytoskeleton. This interaction is not direct, however. The intracellular portion of the molecule binds to three cytoplasmic proteins termed catenin α, β, and γ (Ozawa *et al.*, 1989). These molecules, along with the cadherin molecule, localize to the zonula adherins of the cell and participate in junction formation and stabilization.

1.2.2 Tissue Distribution

Expression of cadherins is developmentally regulated (see below), thus implicating these molecules as important morphogenetic molecules (Takeichi, 1988). In the adult, almost all epithelia express both E-cadherin and P-cadherin (Eidelman *et al.* 1989; Shimoyama *et al.*, 1989). E-cadherin tends to be distributed uniformly throughout the epithelial layer while P-cadherin is concentrated at the more basal layers. The presence of cadherins is important for the maintenance of epithelial and endothelial structure, and alterations in their expression may thus be expected to be associated with loss of normal cell-cell integrity. N-cadherin is found primarily on neural tissues and in the lens of the eye (Takeichi, 1988). It is also present, however, in cardiac and skeletal muscles, where it plays an important role in maintaining the structure of the intercalated discs between adjacent muscle cells (Volk and Geiger, 1986).

1.3 IMMUNOGLOBULIN SUPERFAMILY

The immunoglobulin superfamily (Ig-SF) encompasses a wide variety of molecules that share a common structural feature, the immunoglobulin homology unit. This structure consists of 70–110 amino acids organized into seven to nine β-pleated sheets. Each unit is stabilized by a constant disulfide bridge formed between two of the β strands (Hunkapillar and Hood, 1989). Most members of the Ig-SF are involved in cell-cell recognition (Williams and Barclay, 1988) and include molecules that function in cellular immunity (i.e. major histocompatibility complex [MHC] antigens, CD4, CD8, and the T-cell receptor), neural development (N-[cell adhesion molecule] CAM, L1), leucocyte trafficking (ICAM-1, VCAM-1, PECAM-1), and signal transduction (colony stimulating factor-1 receptor and the platelet-derived growth factor receptor). The function of some of these molecules, including many associated with malignant tissues (i.e. carcinoembryonic antigen [CEA]), has not yet been discovered.

1.3.1 Tissue Distribution

The evolutionary importance of the members of the Ig-SF is demonstrated by their presence on virtually every cell of the body (e.g. MHC antigens). Expression of growth factor receptors (i.e. [colony-stimulating factor-1] CSF-1 receptor and [platelet-derived growth factor] PDGF-receptor) is also widespread. Cells of the immune system are particularly well endowed with Ig-SF molecules. These include leucocyte specific molecules such as immunoglobulin, CD4, CD8, and the T-cell receptor, as well as Ig-SF members involved in leucocyte trafficking that are present on both leucocytes and endothelial cells (ICAM-1, VCAM-1, PECAM-1) (Albelda, 1991; Edelman and Crossin, 1991).

Other Ig-SF members have more restricted distributions. CEA is present primarily on cells of the gastrointestinal tract. There are a group of CAM's that play an important role in neural development (see below). The prototype of this subfamily is the neural cell adhesion molecule (N-CAM), which appears very early in development and directs neural histogenesis. Expression of

N-CAM is followed somewhat later in development by other "secondary" CAM's such as Ng-CAM (neural-glial CAM).

1.4 OTHERS

A number of other cell adhesion molecules, with distinctly different molecular structures have also been characterized to varying degrees. These include: (1) a 67–69 kDa laminin/elastin binding protein that has been identified on normal and tumor cells (Mecham, 1991); (2) the selectins, a group of three cell surface proteins found on endothelial cells, platelets, and leucocytes that are involved in vascular cell interactions (Laskey, 1991) and appear to bind to specific carbohydrate structures via an N-terminal region with lectin binding activity (Brandly *et al.*, 1990); and (3) CD44 (or H-CAM), a highly glycosylated surface protein with similarities to proteoglycan core proteins, that functions as an adhesion receptor for hyaluronic acid, as well as functioning in lymphocyte homing (Picker *et al.*, 1989).

2. *The Metastatic Cascade*

It is now apparent that the process of tumor metastasis is a complex cascade of events (Juliano, 1987), which includes the following series of steps (Fig. 4.1):
(1) Tumor growth, invasion, and release of neoplastic cells from the primary tumor;
(2) Movement of tumor cells into the lymphatics or vasculature;
(3) Survival of the cells in the circulation and interaction of cells with platelets and with the clotting system;
(4) Arrest of the tumor cells in distant sites via interactions with vascular or lymphatic endothelium and/or subendothelial basement membrane;
(5) Migration of tumor cells into tissue parenchyma;
(6) Growth of the tumor at the metastatic site.

Many, if not all of these steps, involve changes (either increases or decreases) in the ability of tumor cells to adhere to adjacent cells or to surrounding extracellular matrix. With the recent identification and characterization of specific cell surface molecules that mediate cell-cell and cell-substratum adhesion, the roles that these molecules play in the process of metastasis is being actively investigated. Receptors from the integrin, cadherin, immunoglobulin superfamily, and the 69 kDa laminin receptor families (Table 4.2) all probably play important roles in metastasis. In addition, a number of other cell surface molecules likely to be involved in cell adhesion (such as CD44, carbohydrates, and endothelial cell-tumor addressins) have been implicated in the metastatic process. The remainder of this chapter will summarize recent data describing how these cell adhesion molecules may fit into the metastatic cascade. This discussion will, of course, be limited in its scope. For additional information, the reader is referred to a number of excellent reviews that discuss the role of cell adhesion molecules in tumor growth and metastasis (Roos, 1984;

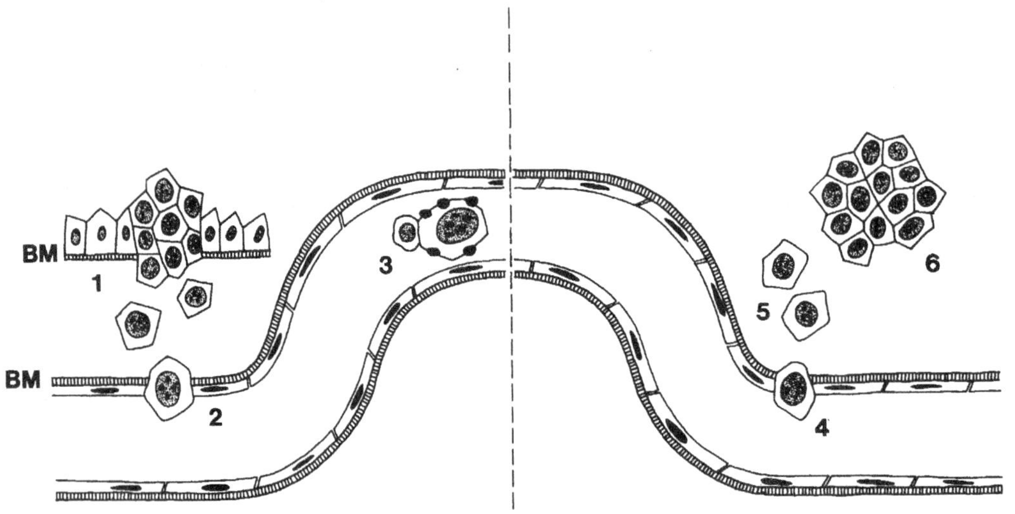

Figure 4.1 The process of tumor invasion and metastasis is a complex cascade of events that includes the following series of steps: (1) Tumor growth, invasion, and release of neoplastic cells from the primary tumor; (2) Movement of tumor cells into the lymphatics or vasculature; (3) Survival of the cells in the circulation and interaction of tumor cells with platelets and leucocytes with the vasculature; (4) Arrest of the tumor cells in distant sites via interactions with vascular or lymphatic endothelium and/or subendothelial basement membrane; (5) Migration of tumor cells into tissue parenchyma; and (6) Growth of the tumor at the metastatic site.

Table 4.2 Cell adhesion molecules important in tumor growth and metastasis

Tumor cell-tumor cell	*Tumor cell-substratum*	*Tumor cell-endothelium*
Integrins	Integrins	Integrins
Ig-SF*	Ig-SF (?)	Ig-SF
Cadherins	67–69 kDa laminin receptor	Selectins
	CD44	Carbohydrates
		Vascular addressins

* Ig-SF = Immunoglobulin superfamily

Nicolson, 1988; Ruoslahti and Giancotti, 1989, Zetter, 1990).

3. The Role of Specific Cell Adhesion Molecules in Tumor Progression and Metastasis

3.1 INTEGRINS

3.1.1 Expression of Integrins on Transformed Cell Lines

3.1.1.1 Fibroblasts

In order to determine the importance of integrins in tumor invasion and metastasis, a number of investigators have compared the expression of integrins on normal and transformed fibroblasts. In these studies, decreased amounts of fibronectin receptors (Chakrabarty *et al.*, 1989; Plantefarber and Hynes, 1989) or alterations in the distribution of fibronectin receptors (i.e. more diffuse and lacking clustering) were noted (Chen *et al.*, 1986; Akiyama *et al.*, 1990). A few reports have described tyrosine-specific phosphorylation of integrins in virally transformed chicken fibroblasts (Tapley *et al.*, 1989; Horvath *et al.*, 1990), however the significance of these findings remain unclear.

3.1.1.2 Melanocytes

Similar comparisons have been made between normal melanocytes or nevus cells and melanoma cells derived from primary or metastatic lesions (Albelda *et al.*, 1990; Klein *et al.*, 1991; Mortarini *et al.*, 1991). In general, cells of melanocytic lineage contain both collagen/laminin binding and fibronectin/fibrinogen binding integrins, however expression tends to be quite heterogeneous. Although there are no truly consistent differences between the integrin expression on melanocytes versus primary or metastatic melanoma cells in culture, cells derived from the more malignant melanomas tend to show expression of more integrins, particularly the α_2/β_1 and α_4/β_1 integrins. Expression of α_4/β_1 ([very late activation antigen-4] VLA-4) on melanoma cells may be particularly important. In addition to binding to an alternative binding site (non-RGD) on fibronectin, VLA-4 is the counter-receptor for the endothelial cell adhesion molecule VCAM-1 (also known as INCAM110) that has been shown to mediate melanoma-endothelial cell interactions *in vitro* (Rice and Bevilacqua, 1989). Expression of this receptor may thus be an important factor in the ability of a cell to metastasize.

3.1.1.3 Epithelial cells

Relatively few studies have been performed examining the integrin distribution on epithelial cell tumor lines. We have recently completed a comparison of normal bronchial epithelial cells to eight lung cancer lines (Mette, Albelda and Buck, in preparation). The normal epithelial cells contained the α_2, α_3, α_6, and α_v subunits with very little α_1, α_5, and β_3. Integrin expression on tumor cells was quite heterogeneous, with some tumor lines showing expression of additional types of integrins than normal bronchial epithelial cells and some tumor lines showing expression of fewer types of integrins.

In summary, although careful, comparative information is rather limited, the pattern of integrin expression on transformed, cultured cells appears to be complex and dependent on the cell type. Transformed fibroblasts tend to express lower amounts of integrins, especially the fibronectin receptor. Melanoma cells tend to have increased expression of certain integrin subunits (α_2 and α_4). No consistent changes were observed in a variety of lung cancer epithelial cell lines.

3.1.2 Expression of Integrins on Tumors *In Situ*

Although an analysis of CAM's expressed on cells in tissue culture is useful, the integrin repertoire of cultured cells does not always exactly mimic that found *in situ* (Albelda *et al.*, 1990; Albelda, 1991). In order to understand the role of integrins in tumor growth and metastasis, it is therefore important to study their distribution in a variety of human tumors at different stages of malignancy. Using immunohistochemical techniques, relatively complete studies have been performed examining the distribution of integrins in the progression of melanomas, neuroblastomas and a variety of epithelial cell carcinomas.

In contrast to most of the cell culture work, a relatively clear cut relationship relating integrin expression to tumor progression has been observed for malignant melanoma *in situ*. Two groups have independently observed an up-regulation of the β_3 subunit in invasive and metastatic melanomas when compared to melanocytes or non-invasive melanomas (McGregor *et al.*, 1989; Albelda *et al.*, 1990). A similar up-regulation of the $\alpha_v\beta_3$ vitronectin receptor has also been described in glioblastoma multiform, a malignant tumor of astrocytes (Gladson and

Cherish, 1990). Expression of the VLA-4 integrin was observed in about 40% of invasive and metastatic melanomas, whereas it was not expressed in non-malignant melanocytes (Albelda *et al.*, 1990). Unlike the consistent up-regulation of the vitronectin receptor, Natali *et al.*, (1991) have reported a down-regulation of the $\alpha 6$ subunit during melanoma tumor progression.

One group has recently examined integrin expression on neuroblastoma, another type of neuroectodermal tumor, and compared these results to melanoma (Favrot *et al.*, 1991). Forty-three tumors of various grades and states of differentiation were examined. Again, heterogeneity of integrin expression was observed; however, the appearance of three integrins, $\alpha_v\beta 3$, $\alpha 4\beta 1$, and $\alpha 6\beta 4$ was

Extracellular Matrix Distribution

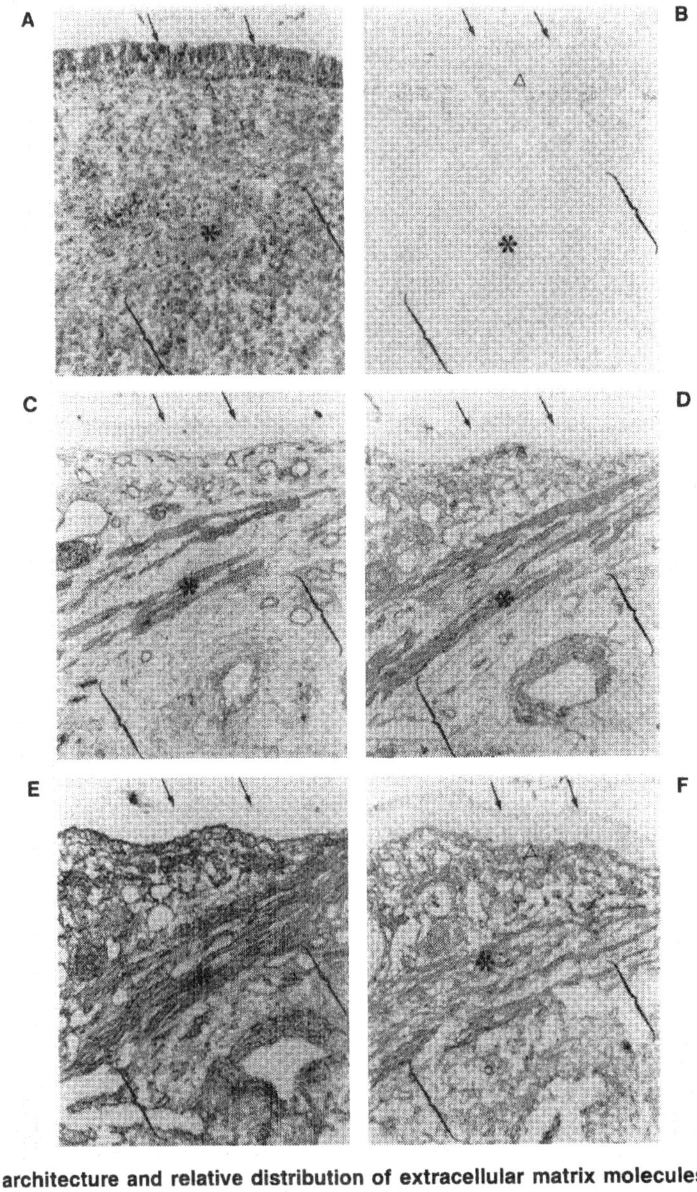

Figure 4.2 Tissue architecture and relative distribution of extracellular matrix molecules in a large cell lung cancer. A hematoxalin and eosin section (A) demonstrates the tumor mass (brackets) with the overlying histologically normal epithelium (arrows). The basement membrane and a piece of bronchial muscle, partly destroyed by the invading tumor are indicated with open arrowhead and asterisk respectively. To determine the non-specific background intensity of the immunoreactions an irrelevant monoclonal antibody was used (B). The staining with anti-laminin (C) and anti-collagen IV (D) mAb was confined to the basement membranes, whereas Coll III (E) and FN (F) reactivity could be detected throughout the tumor mass. No positivity could be observed in the epithelium.

restricted to n-myc negative neuroblastomas, a subgroup of tumors with a better prognosis.

The relationship of integrin expression and the progression of epithelial carcinomas of the lung, breast, skin, and gastrointestinal (GI) tract appears to be complex. In general, integrin expression in these tumors is heterogeneous, however there seems to be a trend toward expression of fewer types and amounts of integrins than on normal epithelial tissues.

We have recently described the integrin repertoire on normal bronchial epithelium and 11 epithelial cell derived lung cancers (Damjanovich *et al.* 1992). Frozen sections from human tumors were stained with monoclonal antibodies (mAbs) specific for extracellular matrix proteins and individual integrin subunits. Fig. 4.2A shows the histological appearance of a large cell carcinoma (brackets) with overlying normal epithelium (arrows). Note that this tumor has invaded airway smooth muscle (asterisk). Fig. 4.2 also shows the distribution of laminin (Fig. 4.2C), type IV collagen (Fig. 4.2D), type III collagen (Fig. 4.2E) and fibronectin (Fig. 4.2F) within the tumor. This particular carcinoma (Fig. 4.3) shows decreased expression of the α_1 (Fig. 4.3A) and α_3 (Fig. 4.3C) collagen/laminin receptors when compared to the normal epithelium (arrows). Expression of α_2 (Fig. 4.3B) and α_6 (Fig. 4.3D) are

Collagen-Laminin Receptors

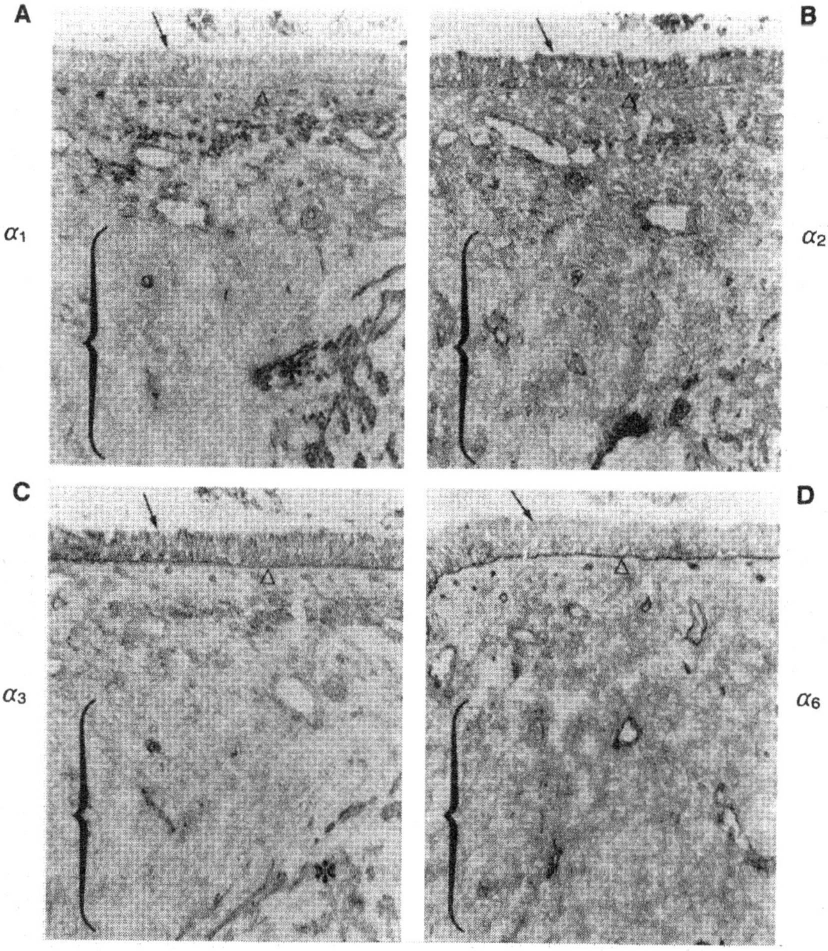

Figure 4.3 Distribution of collagen-laminin binding integrins in a large cell lung carcinoma. Serial frozen sections were immunostained with mAbs directed against the α_1 (A), α_2 (B), α_3 (C) and α_6 (D) subunits of the β_1 subfamily of integrins. The normal epithelium (arrows) serves as an internal control to the tumor (brackets) in every section. The α_2 (B) subunit was strongly expressed on both the normal and tumor cells, while α_3 (C) was weaker on the cancer cells than epithelium. The reactivity of the anti-α_1 (A) and the anti-α_6 (D) mAb's with the tumor was similar to that of the normal epithelium, but the cancer, lacking a well defined basement membrane, did not display basolateral intensification with anti-α_6. A piece of disrupted bronchial muscle is labelled by asterisks, the basement membrane of the normal epithelium is indicated by open arrowheads. Magnification ×125.

preserved. When the fibronectin/fibrinogen binding subunits were analyzed (Fig. 4.4), it was seen that the tumor, like normal epithelium, was negative for the $\alpha 4$ subunit (Fig. 4.4A) and was positive for α_V (Fig. 4.4C). However, both the $\alpha 5$ (Fig. 4.4B) and the $\beta 3$ subunit (Fig. 4.4D) were expressed in the tumor, unlike the overlying epithelium. Analysis of ten additional lung cancers revealed that expression of the $\alpha 2$ and $\alpha 3$ subunit tended to be maintained in most of the cancers. Eight of 11 tumors expressed fewer types of integrins than normal bronchial epithelium. For example, the $\alpha 6$ laminin receptor, strongly expressed on the normal epithelium, was present in only four of the 11 tumors. Similarly, the α_V integrin subunit was seen in only five of 11 tumors. On the other hand, at least two of the cancers (one illustrated above) expressed subunits ($\alpha 5$ and $\beta 3$ not normally seen on bronchial epithelium.

Similar types of analyses have been performed on carcinomas of the breast (Zutter *et al.*, 1990; Koukoulis *et al.*, 1991; Pignatelli *et al.*, 1991a, 1991b), colon (Pignatelli *et al.*, 1990; Koretz *et al.*, 1991), pancreas (Hall *et al.*, 1991) and skin (Peltonin *et al.*, 1989; Stamp *et al.*, 1991). Most of these studies have reported similar results. Each tumor type examined had a great deal of heterogeneity in integrin expression, but there was a general down-regulation of some integrin subunits (specifically the collagen/laminin receptor binding subunits $\alpha 2$, $\alpha 3$, and $\alpha 6$) in the more poorly differentiated carcinomas. In many of these studies, there was a relationship between the integrity of the basement

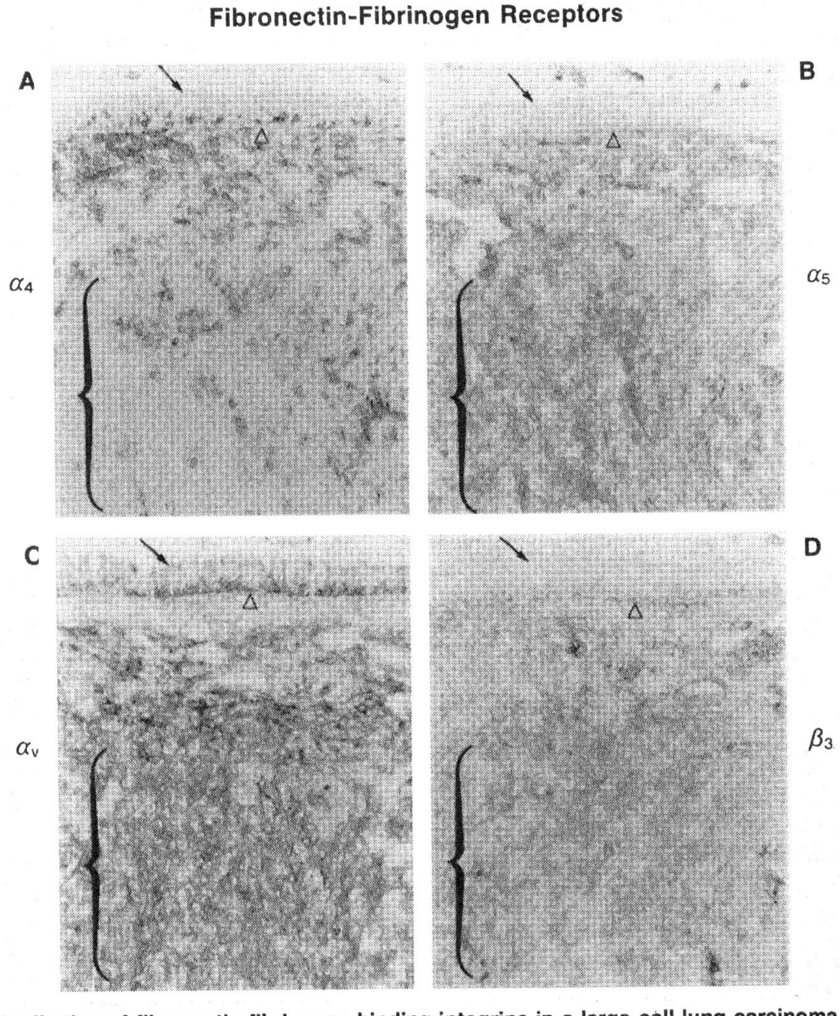

Figure 4.4 Distribution of fibronectin-fibrinogen binding integrins in a large cell lung carcinoma. Serial frozen sections were immunostained with mAbs directed against the α_4 (A); α_5 (B); α_v (C); and β_3 (D) subunits of integrins. No α_4 (A) staining could be detected in the normal epithelium (arrows) or the tumor (brackets), The positive cells are infiltrating lymphocytes. The α_5 (B) receptor subunit as well as the β_3 subunit (D) are up-regulated on the cancer cells. The α_v (C) subunit could be detected on both normal epithelium and tumor cells. Magnification ×125.

membranes and expression of integrins. Those tumors that were more invasive and lacked intact basement membranes tended not to express basement-membrane protein binding integrins such as α_6. Those cancer cells at the periphery of tumor nodules or within well-differentiated glands tended to resemble normal epithelial cells.

In summary, the pattern of integrin expression on tumor cells *in situ*, like that in cultured cells, appears to be complex and dependent on the tumor type. Malignant melanomas have a consistent up-regulation of specific integrins during tumor progression. Epithelial carcinomas, on the other hand, have an inconsistent alteration in their integrin expression, although in general, tend to express fewer integrins than normal epithelium. This down-regulation of integrins appears to be associated with loss of normal contact with underlying basement membrane. Whether this alteration in integrin expression is a cause or an effect of the changes in the interaction of tumor cells with their surrounding extracellular matrix remains to be determined.

3.1.3 Experimental Studies Linking Integrins to Tumor Invasion and Metastasis

Although important and useful, descriptive studies of integrins on tumor cells in culture and *in situ*, have told us little about the actual importance of integrins in tumor progression. To answer these questions, experimental studies are required. To date, two lines of evidence support an important role for integrins in tumor growth and metastasis. The first set of experiments have shown that inhibition of integrin function at the time of intravenous tumor injection into mice can reduce the number of metastasis formed. The second set of studies have examined the effect of altering the number or type of integrins on tumor cells and examining their ability to grow and metastasize.

As mentioned above, many integrins bind to their respective extracellular ligands at a specific three amino acid sequence, arginine-glycine-aspartic acid (RGD) (Ruoslahti and Pierschbacher, 1987). RGD-containing peptides can inhibit adhesion of many cell types to matrix proteins such as fibronectin and vitronectin. Humphries *et al.* (1986) took advantage of this observation and injected a peptide (gly-arg-gly-asp-ser: GRGDS) containing this sequence, along with the appropriate control peptides, into mice simultaneously with B16-F10 melanoma cells. They found a marked reduction in the number of lung metastases formed after two weeks in the GRGDS-injected mice but not in control mice. The authors postulated that the GRGDS peptide had interfered with the initial binding of the tumor cells during the attachment phase or interfered with cell migration during invasion. Work by Gehlsen and colleagues supported this hypothesis by showing inhibition of *in vitro* tumor cell invasion of an artificial basement membrane by GRD-containing synthetic peptides (Gehlsen *et al.*, 1988). Other explanations, such as interference of the peptide with natural killer (NK) cells or an inhibition of platelet-tumor interactions were suggested, but do not appear to be operative (Humphries *et al.*, 1988; Soszka *et al.*, 1991). These results have been duplicated by others with the observation that cyclic peptides containing the RGD sequence are even more effective (Kumagai *et al.*, 1991). Interestingly, a very similar inhibition of lung metastases has been reported with the injection of certain peptides obtained from snake venoms, the so-called "dysintegrins" (Soszka *et al.*, 1991).

An alternative approach to studying the importance of integrins in tumor progression is to compare the biological activity of tumor cells different only in the amount or types of integrins they express. Felding-Habermann *et al.* (1992) selected melanoma cells for low expression of the $\alpha_v\beta_3$ integrin. When these cells were injected into immunodeficient mice, they grew much more slowly than the parental cell line expressing normal levels of vitronectin receptor, despite the fact that both cell lines grew equally well in culture. The *in vivo* growth inhibition was reversed in the low-expressing cells by transfection of α_v cDNA (deoxyribonucleic acid). This study supports the importance of the vitronectin receptor in the progression of malignant melanoma suggested by the tissue distribution studies (Albelda *et al.* 1990).

The effect of changes in the $\alpha_5\beta_1$ fibronectin receptor on the ability of Chinese hamster ovary (CHO) cells to grow in nude mice was examined by two groups (Giancotti and Ruoslahti, 1990; Schreiner *et al.*, 1991). Schreiner *et al.* isolated naturally occurring clonal lines of CHO cells which showed decreased or increased amounts of the α_5/β_1 fibronectin receptor. The tumors produced by injection of the under-expressing clones were larger than those produced by injection of wild-type cells. On the other hand, the clones that were over-expressing the fibronectin receptor grew less rapidly than control cells suggesting an inverse correlation between level of expression of the fibronectin receptor and rate of tumor growth. Giancotti and Ruoslahti transfected human α_5 and β_1 cDNA's into transformed CHO cells and selected clonal lines over-expressing this fibronectin receptor. These over-expressor cells, unlike control cells, lost the ability to grow in soft agar and were non-tumorigenic when injected subcutaneously into nude mice.

A similar approach was used by Chan *et al.* (1991) who cloned the α_2 integrin subunit into a line of human rhabdomyosarcoma cells not normally expressing the functional $\alpha_2\beta_1$ collagen/laminin binding integrin. There was no difference in tumorigenicity between control and transfected cells, however injection of the α_2-transfected cells resulted in substantially more metastatic tumor colonies in nude mice compared to injection of non-transfected cells.

3.1.4 Summary and Conclusions

Despite a great deal of data suggesting that integrins are

important in the metastatic cascade, their actual roles in this process still remain unclear. The role of integrins on tumor progression and metastasis is cell-type specific. Distinctly different patterns of alterations in the expression of integrins in malignant mesenchymal cells (e.g. fibroblasts), melanomas, and epithelial tumors have been observed. The information available to date supports the following generalizations: (1) Loss of expression of the fibronectin receptor ($\alpha 5\beta 1$) seems to be an important alteration in malignant transformation of mesenchymal (fibroblast-like) cells. This is supported by the consistent loss of this receptor on transformed fibroblasts and clear-cut experimental evidence showing that expression of the fibronectin receptor is inversely correlated with the rate of tumor growth. Increased expression of the $\alpha 2\beta 1$ collagen/laminin binding integrin has been associated with an enhanced ability to metastasize in at least one type of mesenchymal cell (rhabdomyosarcoma); (2) Up-regulation of the vitronectin receptor seems to be associated with increased tumorigenicity of malignant melanoma and inhibition of this receptor during times of tumor circulation inhibit metastasis formation; (3) The role of integrins on epithelial carcinomas is complex. There tends to be a down-regulation of their expression during tumor progression although no uniform changes have been identified. It is not clear whether the down-regulation of integrins is a cause or an effect of the changes in basement membrane structure associated with carcinoma progression.

3.2 CADHERINS

3.2.1 Expression of Cadherins on Cell Lines

Although the cadherins were originally described in a developmental context, their importance in the maintenance of adult epithelial structures (Gumbiner and Simons, 1986) suggested that they may also play a role in epithelial cell tumor invasion. This hypothesis is consistent with the observation that MDCK cells transformed with Harvey and Maloney sarcoma viruses (which are constitutively invasive) were found not to express E-cadherin at their cell surfaces. Frixen *et al.* (1991) showed that carcinoma cell lines with an epithelioid phenotype were non-invasive and expressed E-cadherin while carcinoma lines with a fibroblastoid appearance were invasive and had lost E-cadherin expression. Navarro *et al.* (1991) also observed an inverse relationship between the amount of E-cadherin proteins produced on seven mouse epidermal keratinocyte lines representative of different stages of carcinogenesis and the tumorigenicity of the cell lines.

3.2.2 Expression of Cadherins in Tissues and Tumors

With evidence from cell culture to suggest that loss of cadherins plays a role in the invasive behavior of epithelial cells, a number of investigators have begun to examine the distribution of these proteins in human tumors using immunohistochemical techniques. The initial report of the distribution of cadherins in cancers was from Shimoyama *et al.* (1989). These authors stained 44 lung carcinomas using antibodies directed against both E-cadherin and P-cadherin. They found that the staining intensity of each cadherin varied considerably from case to case, but when present, staining was always restricted to cell-cell borders. No clear down-regulation of cadherins in the primary tumors or in metastases was noted. Shimoyama and Hirohashi (1991) also stained 54 gastric cancers with antibodies to E- and P-cadherin and found that P-cadherin was expressed in approximately half of the tumors examined, but in a rather heterogeneous fashion. E-cadherin was uniformly present in the well differentiated adenocarcinomas and tumors showing an "adherent" (i.e. tight intercellular adhesion between cells) phenotype. When tumors showing the "scattered" (i.e. lacking tight intercellular adhesion between cancer cells) phenotype were examined, only four completely lacked E-cadherin staining. Nineteen of 28 showing strong cadherin staining. In addition, E-cadherin was also expressed on 20 metastases examined. To make sure that this immunoreactive cadherin was intact, immunoblotting was performed and showed normal sized protein. Eidelman *et al.* (1989) studied 60 tumors of varying origins for their expression of E-cadherin. These authors could not identify a single case of epithelial malignancy lacking E-cadherin, although they did note a loss of the normal cell-cell staining in the more undifferentiated tumors.

In contrast, two groups have have observed more consistent changes in cadherin expression associated with malignancy. Shiozaki *et al.* (1991) examined E-cadherin expression in a selection of 55 human carcinomas. All normal epithelial tissue strongly expressed E-cadherin, however, they found a heterogeneous and reduced expression of E-cadherin in 73% of esophageal tumors, 25% of gastric carcinomas, and 45% of breast cancers. Some of the positively staining tumors had an abnormal distribution (cytoplasmic rather than at cell-cell borders) of cadherin. In general, no correlation was seen between the degree of differentiation and the loss of cadherin expression except for the esophageal tumors. The strongest correlation of loss of E-cadherin and tumor de-differentiation has been reported by Schipper *et al.* (1991). Thirty-two squamous cell carcinomas of the head and neck were analyzed. Their results showed that all twelve of the well differentiated carcinomas expressed E-cadherin normally, the moderately well differentiated carcinomas expressed "intermediate" amounts of cadherin or were heterogeneous, and that all of the poorly-differentiated tumors (five of five) and seven of eight tumors that had invaded lymph nodes were E-cadherin negative.

Finally, Mahoney *et al.* (1991) have identified a gene (the *fat* locus of *Drosophila*) that encodes a novel member of the cadherin gene superfamily and functions as a tumor suppressor gene. Mutation of this gene results in hyperplastic, tumor-like over-growth of larval imaginal discs, defects in differentiation and morphogenesis, and death during the pupal stage.

3.2.3 Experimental Studies Linking Cadherins to Tumor Invasion

The actual importance of normal cadherin expression in regulating the invasive behavior of epithelial cells was first suggested by Behrens *et al.* (1989) who found that when intercellular adhesion of a canine epithelial cell line (MDCK cells) was disrupted by the addition of an anti-E-cadherin antibody, the normally non-invasive MDCK cells acquired the ability to invade collagen gels and embryonal heart tissue. As mentioned above, Frixen *et al.* (1991) showed that invasive carcinoma cell lines with a fibroblastoid appearance had lost E-cadherin expression. A direct role for cadherins was demonstrated when invasiveness of these cells was prevented by transfection with E-cadherin cDNA. Similar observations and experiments were reported by two other groups. Vleminckx *et al.* (1991) transfected E-cadherin cDNA into highly invasive tumor lines and caused a decrease in *in vitro* invasion assays and production of *in vivo* tumors that were more differentiated than those produced by non-transfected cells. Introduction of a plasmid encoding E-cadherin-specific *anti*sense RNA into cells with high endogenous expression of E-cadherin, that resulted in down-regulation of E-cadherin, rendered the cells invasive. Navarro *et al.* (1991) also reported that transfection of E-cadherin into a highly malignant cell line with low endogenous cadherin expression suppressed its tumorigenicity. Chen *et al.* (1991) transfected E-cadherin cDNA into a mouse fibroblast line (L cells) and demonstrated inhibition of migration and of invasive behavior.

3.2.4 Summary and Conclusions

Good experimental evidence exists to support the hypothesis that loss of cadherin expression or function is associated with increased invasiveness in transformed epithelial cells. Unfortunately, these observations have not been consistently confirmed in immunohistochemical analysis of human tumors. It is possible that unknown host factors could play important roles in the heterogeneous nature of cadherin expression seen in some systems (Mareel *et al.*, 1991). Additional studies are needed. A more careful examination of specific tumor subpopulations (i.e squamous cell carcinomas of the head and neck) may reveal a more clearly defined relationship between tumor invasion and cadherin expression.

3.3 IMMUNOGLOBULIN SUPERFAMILY

Since a number of the members of the Ig-SF can function as cell-cell adhesion molecules, it is reasonable to postulate that these molecules may be involved in tumor invasion and metastasis. In some cases, a clear functional role of Ig-SF members has been identified. In others, only an association with malignancy has been established. Indeed, a number of lg-SF members were first identified by their consistent association with certain tumors. Because of the widely divergent nature of the Ig-SF, it is more difficult to make generalizations than with the integrins or cadherin families. For this reason, individual members of the Ig-SF will be discussed separately.

3.3.1 Carcinoembyronic Antigen

One of the best examples of an Ig-SF member associated with tumor is carcinoembyronic antigen (CEA). CEA was first described in 1965 as an 180 kDa oncofetal glycoprotein found in the serum of patients colon carcinoma. It was subsequently found to be associated with other tumors and is used clinically to monitor recurrences of colon and lung carcinomas. In 1987, CEA was cloned and found to be a member of the immunoglobulin superfamily (reviewed in Benchimol, 1989). Unlike many members of the Ig-SF, CEA is attached to the cell via a phosphatidylinositol bond and lacks a true cytoplasmic domain. It now appears that CEA is part of a family of related molecules that are expressed in a highly regulated developmental fashion and include non-specific cross-reacting antigen (NCA), biliary glycoprotein I (BGPI), and normal fetal antigen (NFA-1) (Hunkapillar and Hood, 1989). Expression of cloned cDNA has demonstrated that CEA and other members of this family function as true cell adhesion molecules in both calcium-dependent (CEA) and independent fashion (BGPI) (Benchimol *et al.*, 1989; Rojas *et al.*, 1990). It still remains unclear whether the increased expression of CEA on tumor cells results in any functional consequence in the development of malignancy. Benchimol *et al.* (1989) have suggested that the presence of excess CEA could interfere with the function of the normally present cell-cell adhesion molecules. It is also possible that there is no important functional role for CEA on tumor cells, but that its expression represents a true oncofetal marker.

3.3.2 N-CAM

Neural cell adhesion molecule (N-CAM) is a homophilic calcium dependent cell-cell adhesion molecule that is expressed primarily in the nervous system. The expression of N-CAM is altered developmentally. The findings that it is down-regulated in migrating cells and mesenchyme suggest that N-CAM might also be involved in malignancy (reviewed in Edelman and Crossin, 1991).

In contrast to the cadherins, however, the data to support this relationship is quite speculative. Decreased expression of N-CAM on chicken retinal cells transformed by Rous sarcoma virus has been reported (Brackenbury *et al.*, 1984), however, the loss of N-CAM expression on malignant cells, *in vivo*, has not been described. To the contrary, N-CAM has been visualized immunohistochemically on many neural and neuroendocrine tumors including Wilms' tumor, pituitary adenomas, pheochromocytomas, and small cell carcinomas of the lung (Carbone *et al.*, 1991; Edelman and Crossin, 1991, Jin *et al.*, 1991). For the most part, the form of N-CAM present on these tumors is the embryonic form, containing high levels of polysialic acid. This isoform of N-CAM is less adhesive than the adult form and may be relevant to the metastatic potential of the cells. One observation that supports this idea is that lines of small cell lung carcinomas that grow as adherent cells tend to express much less N-CAM than those lines that grow as floating aggregates (Doyle *et al.*, 1990).

A possible mechanism for N-CAM in the malignant process was suggested in a recent report that showed that N-CAM was involved in the contact-dependent inhibition of growth of a line of mouse fibroblasts (Aoki *et al.*, 1991). Viral transformation of this N-CAM expressing line resulted in loss of contact inhibition and a rapid and dramatic loss of N-CAM on the cell surface. Immobilized antibodies against N-CAM were able to inhibit growth. The authors suggest that crosslinking or engagement of N-CAM results in intracellular signals that modulate cell growth. Loss of these signals could conceivably contribute to neoplastic growth.

3.3.3 VCAM-1 (INCAM110)

Some tumor lines appear to adhere to cytokine stimulated endothelial cells to a higher degree than resting cells (Dejana *et al.*, 1988; Rice *et al.*, 1988). Two independent lines of investigation have identified a cytokine-inducible Ig-SF member (termed vascular cell adhesion molecule-1 [VCAM-1] or Inducible cell adhesion molecule-1 [INCAM-110]) found primarily on vascular endothelial cells that serves as a counter-receptor for the VLA-4 ($\alpha_4\beta_1$) integrin (Osborne *et al.*, 1989; Rice and Bevilacqua, 1989; Elices *et al.*, 1990). Since VLA-4 is primarily found on leucocytes, the most important function of VCAM-1 is probably in white blood cell-endothelial cell interactions, however, certain malignant cells, notably melanomas, also express VLA-4 both *in vitro* (Rice and Bevilacqua, 1989; Albelda *et al.*, 1990) and *in situ* (Albelda *et al.*, 1990). It thus seems likely that VCAM-1 serves as a tumor adhesion receptor for those melanoma cells expressing VLA-4, as well as for malignancies of the myeloid line that also express VLA-4.

3.3.4 Others

A number of other members of the IG-SF have been implicated in malignancy, although their roles remain uncertain. Increased expression of intercellular adhesion molecule-1 (ICAM-1) has been shown to be correlated with melanoma tumor progression and an increased risk of metastasis (Johnson *et al.* 1989). Circulating ICAM-1 also appears to be correlated with malignant disease (Tsujisaki *et al.*, 1991). Fearon *et al.* (1990) cloned a gene from chromosome 18q that was regularly *d*eleted in *c*olorectal *c*arcinomas (thus termed "DCC") and found that it encoded for a molecule in the Ig-SF with homology to N-CAM. It was postulated that loss of this gene could lead to an abnormality in cell-cell adhesion predisposing to the malignant phenotype. Finally, MUC-18 is a melanoma associated antigen associated with metastasis and poor prognosis. Cloning experiments have revealed that it is a member of the Is-SF with homology to N-CAM (Lehmann *et al.*, 1989). The ability of this molecule to function as an adhesion molecule and its possible role in tumor progression are currently unknown.

3.4 CD44

CD44, also known as the Hermes antigen, H-CAM, pgp-1 antigen, and extracellular matrix receptor ECM-III, is a widely distributed integral membrane protein that exists in a variety of forms with different molecular sizes ranging from 85 kDa to 160 kDa (Picker *et al.*, 1989). The molecular structure of CD44 reveals that it is highly glycosylated and related to cartilage link proteins and a related segment of proteoglycan core protein (Goldstein *et al.*, 1989; Stamenkovic *et al.*, 1989). The 90 kDa protein is the predominant form in leucocytes and is thought to be important in lymphocyte homing. The larger (150–160 kDa) isoform is found on epithelial and mesechymal cells and appears to function as a receptor for hyaluronic acid (Aruffo *et al.*, 1990; Culty *et al.*, 1990).

A number of lines of evidence implicate CD44 as a cell adhesion molecule with a possible role in tumor progression. The epithelial cell form seems to be expressed more highly in carcinomas than normal epithelium (Stamenkovic *et al.*, 1989). Birch *et al.* (1991) have isolated human melanoma cell variants based on expression of CD44. After i.v. (intravenous) injection into nude mice, the cells expressing low amounts of CD44 gave significantly fewer lung nodules than injection of the same number of cells expressing high levels of CD44. Gunthert *et al.* (1991) have isolated an antibody specific for a line of rat pancreatic carcinoma cells with high metastatic potential. Using this antibody, they have identified a cDNA (pMeta-1) that encoded for a close homologue of human and murine CD44. Interestingly, this variant molecule contained an additional extracellular domain of 162 amino acids, probably introduced by alternative splicing. This new variant of CD44 was expressed only on metastatic cell lines of two rat tumors and not on most normal rat tissues. Transfection and overexpression of pMeta-1 into non-metastasizing cells conferred full metastatic behavior.

How CD44 contributes to the malignant phenotype is not yet known. Evidence is accumulating that the higher molecular weight form of CD44 is a distinct form with different adhesion potential (Brown *et al.*, 1991; Stamenkovic *et al.* 1991). It is possible that this higher weight form confers metastatic potential by allowing cells to more firmly adhere to hyaluronic acid, thus aiding tumor implantation. Alternatively, it has been demonstrated that CD44, via its cytoplasmic domain, can serve in signal transduction (Shimizu *et al.* 1989; Koopman *et al.*, 1990). Its effect on metastasis could thus be mediated by certain intracellular signals induced by binding to hyaluronic acid.

3.5 THE 67–69 kDa LAMININ-ELASTIN BINDING PROTEIN

As one of the major components of basement membranes, laminin has been thought to play an important role in metastasis (Liotta, 1986). A number of laminin binding proteins (including several members of the integrin family) have recently been identified and characterized to varying degrees (Mecham, 1991). One of these proteins is a 67–69 kDa peripheral membrane protein that binds with high affinity to laminin on the surface of many tumor cells. (Terranova *et al.*, 1982; Liotta, 1986). This protein is found on many cell types, but appears to be altered in number and in the degree of occupancy in human carcinomas. Cells selected for their ability to bind laminin produced 10-fold more metastases following i.v. injection (Terranova *et al.*, 1982). Addition of intact laminin to the cell surface enhanced metastasis, while treating the cells with the receptor binding fragment inhibited metastasis (Barsky, 1984). One of the binding sites of the 67–69 kDa receptor is a region of laminin in the B1 chain containing the amino acids YIGSR (Graf *et al.*, 1987). Injection of this peptide into mice, simultaneously with injected melanoma cells markedly inhibited the formation of lung metastases (Iwamoto *et al.*, 1987).

More complete understanding of the 67–69 kDa laminin binding protein has been difficult because its molecular structure is not yet known for certain. Although a cDNA clone was isolated using mAbs and a λgt11 expression library, the physical and functional properties of the protein encoded by the clone (notably a predicted size of 33, not 67–69 kDa, the lack of a signal sequence, and no N-linked carbohydrate attachment sites) make it unclear if this sequence encodes for the correct molecule (reviewed in Mecham, 1991). Adding to the confusion about the 67 kDa laminin binding protein is the recent realization that it also binds to a hydrophobic sequence in the elastin molecule (VGVAPG) and is probably identical to a 67–69 kDa elastin receptor identified by Hinek *et al.* (1988) and Mecham *et al.* (1989). This hexapeptide (VGVAPG) has been shown to be chemotactic for certain carcinomas (Blood *et al.*, 1988).

It now appears that the 67–69 kDa receptor functions as a binding protein for both laminin and elastin. Either or both of these functions could have a role in the behavior of malignant cells. Interestingly, this binding is regulated by carbohydrates (notably galactoside sugars) (Hinek *et al.*, 1988). Since the receptor behaves as a peripheral membrane protein, it is postulated that the 67–69 kDa protein binds to another, heretofore unidentified, transmembrane protein which then interacts with the cytoskeleton (Mecham, 1991).

3.6 OTHERS

3.6.1 Carbohydrates and Lectins

It has been recognized for many years that cell surface carbohydrates play a role in metastasis (Lotan and Raz, 1988). Evidence to support this concept includes the facts that carbohydrate profiles change during malignant transformation, that lectins (sugar binding proteins) have been used to select for tumor cells with varying metastatic capacities, and that removal of cell-surface carbohydrates alters the ability of metastatic cells to attach to various matrix components (reviewed in Stanford *et al.*, 1986). The exact identity of these carbohydrate moieties and the endogenous lectins to which they bind is still an area of active investigation, however, at least one potential receptor-counter receptor pair has been identified.

A family of three transmembrane glycoproteins important in leucocyte-endothelial cell interactions with lectin-like activity has recently been identified (Laskey, 1991). These molecules called "LEC-CAM's" or "selectins" have a common structure characterized by an N-terminal lectin-like domain, an area homologous to the (epidermal growth factor) EGF receptor, a variable number of complement-like repeats, and a short cytoplasmic domain. It is now becoming evident that the lectin-like region of these molecules binds to specific carbohydrate structures on adjacent cells which are fucosylated and sialylated. The most important of these structures is defined as the sialylated Lewis X antigen (Brandly *et al.*, 1990). Since at least two of the selectins, P-selectin (also known as [GMP140] or PADGEM) and E-selectin (also known as ELAM-1), are present on activated endothelial cells, they provide potential receptors for cells bearing the sialylated Lewis X antigen. Colorectal carcinoma cells, especially metastatic cells, express the sialylated Lewis X antigen (Matsushita *et al.*, 1990). This fact helps to explain the observations of Lauri *et al.* (1991) that the adhesion of seven different colon carcinoma cell lines to cytokine-activated endothelial cells was inhibited by antibodies against E-selectin. This inhibition was not seen with other types of carcinoma cells, suggesting that adhesion of colonic carcinoma cells via E-selectin may be an important mechanism of metastasis. Undoubtedly other carbohydrate-endogenous lectin combinations will be discovered.

3.6.2 Vascular Addressins

One of the characteristic features of any metastatic tumor is its relatively unique pattern of organ distribution (Zetter, 1990). This organ preference is partially mediated by selective attachment of tumor cells to organ-specific microvascular endothelium (Pauli and Lee, 1986; Auerbach *et al.*, 1987). It is thought that this specificity is mediated through mechanisms that are similar to those seen with lymphocyte homing where specific lymphocyte adhesion receptors home to specific vascular addressins. The identity of one such addressin has been recently described by Zhu *et al.* (1991). These investigators identified a 90 kDa protein that was selectively expressed on distinct branches of lung vessels or on endothelial cells cultured on lung extracellular matrix. Antibodies directed against this lung-specific endothelial cell adhesion molecule (Lu-ECAM-1) markedly inhibited the ability of B16-F10 melanoma cells to metastasize to lung. Further characterization of Lu-ECAM-1 and discovery of other site-specific addressins will be important additions to our understanding of the metastatic process.

3.6.3 Intravascular Interactions

After entry into the blood stream, a metastatic cell undergoes certain adhesive interactions with circulating leucocytes and with platelets (Weiss *et al.*, 1988). It is likely, that in most cases, the attachment of polymorphonuclear phagocytes or NK-lymphocytes serve to inhibit metastasis. However, the arrest and metastasis of circulating cells can be promoted in locally inflamed areas (Orr *et al.*, 1985; Adamson *et al.*, 1987). This effect may be due to local release of cytokines which then activate endothelial cell adhesion receptors such as VCAM-1 or E-selectin.

The interactions of platelets with tumor cells is also an important variable in metastasis, especially in the development of pulmonary metastases (Weiss *et al.*, 1988; Nierodzik *et al.*, 1991). Induction of thrombocytopenia is associated with a decrease in the number of metastases. Reconstitution after platelet repletion reverses this effect (see Nierodizik *et al.*, 1991). Tumor cell-platelet adhesion may occur in the circulation or after the initial arrest has occurred. In either case, it appears that platelets may enhance the adhesion of the tumor cells to the extracellular matrix. Evidence has been provided to show that the platelet integrin gpIIb/IIIa is important in this process (Chopra *et al.*, 1988). Metastasis was markedly inhibited by mAbs to the IIb/IIIa integrin, as well as by an antibody against the gpIIb/IIIa ligand, von Willebrand's factor. Although controversial, the mechanism of the inhibition of metastasis by the RGD peptide (Humphries *et al.*, 1986) has been interpreted by some to be due to an inhibition of tumor cell-platelet inhibition. The importance of this platelet tumor interaction was emphasized by the recent demonstration that activation of platelets by thrombin could enhance tumor metastasis four to 413-fold with two different tumor lines (Nierodizik *et al.*, 1991).

4. *Summary and Conclusions*

Despite rapid advances in our understanding of the biology of cell adhesion, it is difficult to propose one simple scheme in which cell adhesion molecules can be related to tumor growth and metastasis. One major consideration is the very complexity of the metastatic process itself.

Another major obstacle in understanding the role of CAM's in metastasis is the well known problem of tumor heterogeneity and the phenotypic instability of metastatic cells over relatively short periods of time (Nicolson *et al.*, 1987). The cells that make up a metastatic focus may thus be quite different from the tumor cells that originally formed the lesion. It is quite possible that the selective pressures that initially enabled a cancer cell to form a metastatic lesion may be quite different than those which later favor rapid tissue growth.

With these caveats in mind, a few generalizations can be made about the relationship of cell adhesion molecules and cancer. Using the schema shown in Fig. 4.1, successful metastasis initially requires that normal cell-cell and cell-substratum adhesion be disrupted causing release of neoplastic cells from the primary tumor (Step 1). For epithelial tumors, down-regulation of cadherins and, perhaps, integrins, appear to be involved. This loss of cell adhesion must be followed by migration of tumor cells into the vascular system (Step 2), a step requiring efficient cell-substratum interactions. In melanomas, this step seems to require expression of the vitronectin (VN) receptor. After movement into the vascular system, firm cell-cell contacts must be made in order for the tumor cells to arrest in the circulation (Step 4). Evidence exists to implicate many CAM's in this process including the integrins, the 69 kDa laminin binding proteins, Ig-SF members such as VCAM-1, CD44, carbohydrates, selectins, and specific vascular addressins. Cell-cell adhesion must then be broken, as the tumor cells separate away from the endothelium and into the tissue parenchyma, where they again need effective cell-substratum adhesion mechanisms for migration (Step 5) and finally growth (Step 6).

In addition to alterations in cell adhesion, successful tumor invasion and spread involves many other changes such as uncontrolled cell growth, induction of angiogenesis, production of proteases, and dedifferentiation (Schirrmacher, 1985; Fidler, 1989; Herlyn and Malkowicz, 1991). It is perhaps overly simplistic to expect to find a change in any one cell adhesion molecule that will be consistently found in all types of malignancies. There appear to be specific situations, however, where a particular cell adhesion molecule plays a crucial role in the metastatic cascade. The goal of future research will be to identify such instances using specific mAbs in a variety of animal species and with the use of molecular biological and transgenic techniques. A more complete understanding of the involvement of cell adhesion molecules in tumor progression has already suggested a number of interesting possible novel therapeutic approaches (i.e. bioactive peptides or mAbs). It is highly likely that this knowledge, along with future discoveries, will soon be used to successfully treat patients with cancer.

5. *References*

Adamson, I.Y.R., Young, L. and Orr, F.W. (1987). Tumor metastasis after hyperoxic injury and repair of the pulmonary endothelium. Lab. Invest., 57, 71–77.

Akiyama, S.K., Larjava, H. and Yamada, K.M. (1990). Differences in the biosynthesis and localization of the fibronectin receptor in normal and transformed cultured human cells. Cancer Res. 50, 1601–1607.

Albelda, S.A. and Buck, C.A. (1990). Integrins and other cell adhesion molecules. FASEB J. 4, 2868–2880.

Albelda, S.M. (1991). Endothelial and epithelial cell adhesion molecules. Am. J. Respir. Cell Mol. Biol. 41, 195–203.

Albelda, S.M., Daise, M., Levine, E.N. and Buck, C.A. (1989). Identification and characterization of cell-substratum adhesion receptors on cultured human adult large vessel endothelial cells. J. Clin. Invest. 83, 1992–2002.

Albelda, S.M., Mette, S.A., Elder, D.E., Stewart, R., Damjanovich, L., Herlyn, M. and Buck, C.A. (1990). Integrin distribution in malignant melanoma: association of the β_3 subunit with tumor progression. Cancer Res. 50, 6757–6764.

Aoki, J., Umeda, M., Takio, K., Titani, K., Utsumi, H., Sasaki, M. and Inoue, K. (1991). Neural cell adhesion molecule mediated contact-dependent inhibition of growth of near-diploid mouse fibroblast cell line m5S/1M. J. Cell. Biol. 115, 1751–1761.

Aruffo, A., Stamenkovic, E., Melnick, M., Underhill, C.B. and Seed, B. (1990). CD44 is the principal cell surface receptor for hyaluronate. Cell. 61, 1303–1313.

Auerbach, R., Lu, W.C., Pardon, E., Gumkowski, F., Kaminska, G. and Kaminski, M. (1987). Specificity of adhesion between murine tumor cells and capillary endothelium: an *in vitro* correlate of preferential metastasis *in vivo*. Cancer Res. 47, 1492–1496.

Barsky, S.H., Rao, C.N., Williams, J.E. and Liotta, L.A. (1984). Laminin molecular domains which alter metastasis in a murine model. J. Clin. Invest. 74, 843–848.

Behrens, J., Mareel, M.M., VanRoy, F.M., and Birchmeier, W. (1989). Dissecting tumor cell invasion: epithelial cells acquire invasive properties after the loss of uvomorulin-mediated cell-cell adhesion. J. Cell Biol. 108, 2435–2447.

Benchimol, S., Fuks, A., Jothy, S., Beauchemin, N., Shirota, K. and Stanners, C.P. (1989). Carcinoembryonic antigen, a human tumor marker, functions as an intercellular adhesion molecule. Cell. 57, 327–334.

Birch, M., Mitchell, S. and Hart, I.R. (1991). Isolation and characterization of human melanoma cell variants expressing high and low levels of CD44. Cancer Res. 51, 6660–6667.

Blood, C.H., Sasse, J., Brodt, P. and Zetter, B.R. (1988). Identification of a tumor cell receptor for VGVAPG, an elastin-derived chemotactic peptide. J. Cell Biol. 107, 1987–1993.

Brackenbury, R., Greenberg, M.E. and Edelman, M. (1984). Phenotypic changes and loss of N-CAM-mediated adhesion in transformed embryonic chicken retinal cells. J. Cell Biol. 99, 1944–1954.

Brandley, B.K., Swiedler, S.J. and Robbins, P. (1990). Carbohydrate ligands of the LEC cell adhesion molecules. Cell. 63, 861–863.

Brown, T.A., Bouchard, T., St. John, T., Wayner, E.A. and Carter, W.G. (1991). Human keratinocytes express a new CD44 core protein (CD44E) as a heparan-sulfate intrinsic membrane proteoglycan with additional exons. J. Cell Biol. 113, 207–221.

Buck, C.A. and Horwitz, A.F. (1987). Cell surface receptors for extracellular matrix molecules. Annu. Rev. Cell Biol. 3, 179–205.

Buck, C., Albelda, S., Damjanovich, L., Edelman, J., Shin, D.T. and Solowska, J. (1990). Immunohistochemical and molecular analysis of β_1 and β_3 integrins. Cell Differentiation and Development. 32, 189–202.

Carbone, D.P., Koros, A.M.C., Linnoila, I., Jewett, P. and Gazdar, A.F. (1991). Neural cell adhesion molecule expression and messenger RNA splicing patterns in lung cancer cell lines are correlated with neuroendocrine phenotype and growth morphology. Cancer Res. 51, 6142–6149.

Chakrabarty, S., Jan, Y., Levine, A., McClenic, B. and Varani, J. (1989). Fibronectin/laminin and their receptors in aberrant growth control in FR3T3 cells transformed by *Ha-ras* oncogene and epidermal growth factor gene. Int. J. Cancer. 44, 325–331.

Chan, B.M.C., Matsuura, N., Takada, Y., Zetter, B.R. and Hemler, M.E. (1991). *In vitro* and *in vivo* consequences of VLA-2 expression on rhabdomyosarcoma cells. Science. 251, 1600–1602.

Chen, W. and Obrink, B. (1991). Cell-cell contacts mediated by E-cadherin (Uvomorulin) restrict invasive behavior of L cells. J. Cell Biol. 114, 319–327.

Chen, W., Wang, J., Hasegawa, T., Yamada, S.S. and Yamada, K.M. (1986). Regulation of fibronectin receptor distribution by transformation, exogenous fibronectin, and synthetic peptides. J. Cell Biol. 103, 1649–1661.

Cheresh, D., Smith, J., Cooper, H. and Quaranta, V. (1989). A novel vitronectin receptor integrin (α_v/β_x) responsible for distinct adhesive properties of carcinoma cells. Cell. 57, 59–69.

Chopra, H., Hatfield, J.S., Chang, Y.S., Grossi, I.M., Fitzgerald, L.A., O'Gara, C.Y., Marnett, L.J., Diglio, C.A., Taylor, J.D. and Honn, K.V. (1988). Role of tumor cell cytoskeleton and membrane glycoprotein IRGpIIb/IIIa in platelet adhesion to tumor cell membrane and tumor cell-induced platelet aggregation. Cancer Res. 48, 3787–3800.

Culty, M., Miyake, K., Kincade, P.W., Silorski, E., Butcher, E.C. and Underhill, C. (1990). The hyaluronate receptor is a member of the CD44 (H-CAM) family of cell surface glycoproteins. J. Cell Biol., 111, 2765–2774.

Damjanovich, L., Albelda, S.M., Mette, S.A. and Buck, C.A. (1992). Distribution of integrin cell adhesion receptors in normal and malignant lung tissue. Am. J. Resp. Cell Molec. Biol. 6, 197–206.

Dejana, E., Bertocchi, F., Bortolami, M.C., Regonesi, A., Tonta, A., Breviario, F. and Giavazzi, R. (1988). Interleukin 1 promotes tumor cell adhesion to cultured human endothelial cells. J. Clin. Invest. 82, 1466–1470.

Doyle, L.A., Borges, M., Hussain, A., Elias, A. and Tomiyasu, T. (1990). An adherent subline of a unique small-cell lung cancer cell line down-regulates antigens, of the neural cell adhesion molecule. J. Clin. Invest. 86, 1848–1854.

Edelman, G.M. and Crossin, K.L. (1991). Cell adhesion molecules: implications for a molecular histology. Annu. Rev. Biochem. 60, 155–190.

Eidelman, S., Damsky, C.H., Wheelock, M.J. and Damjanov, I. (1989). Expression of the cell-cell adhesion glycoprotein cell-CAM 120/80 in normal human tissues and tumors. Am. J. Pathol. 135, 101–110.

Elices, M.J., Osborn, L., Takada, Y., Crouse, C., Luhowskyj, S., Hemler, M.E. and Lobb, R.R. (1990). VCAM-1 on activated endothelium interacts with the leucocyte integrin VLA-4 at a site distinct from the VLA-4/fibronectin binding site. Cell. 60, 577–584.

Favrot, M.C., Combaret, V., Goillot, E., Lutz, P., Frappaz, D., Thiesse, P., Thyss, A., Dolbeau, D., Bouffet, E., Tabone, E. and Philip, T. (1991). Expression of integrin receptors on 45 clinical neuroblastoma specimens. Int. J. Cancer. 49, 347–355.

Fearon, E.R., Cho, K.R., Nigro, J.M., Kern, S.E., Simons, J.W., Ruppert, J.M., Hamilton, S.R., Preisinger, A.C., Thomas, G., Kinzler, K.W. and Vogelstein, B. (1990). Identification of a chromosome 18q gene that is altered in colorectal cancer. Science. 247, 49–56.

Felding-Habermann, B., Mueller, B.M., Romerdahl, C.A. and Cheresh, D.A. (1992). Involvement of integrin α_V gene expression in human melanoma tumorigenicity. J. Clin. Invest. (In press).

Fidler, I.J. (1989). Origin and biology of cancer metastasis. Cytometry. 10, 673–680.

Frixen, U.H., Behrens, J., Sachs, M., Eberle, G., Voss, B., Warda, A., Lochner, D. and Birchmeier, W. (1991). E-cadherin-mediated cell-cell adhesion prevents invasiveness of human carcinoma cells. J. Cell Biol. 113, 173–185.

Gehlsen, K.R., Argraves, W.S., Pierschbacher, M.D. and Ruoslahti, E. (1988). Inhibition of *in vitro* tumor cell invasion by Arg-Gly-Asp-containing synthetic peptides. J. Cell Biol. 106, 925–930.

Giancotti, F.G. and Ruoslahti, E. (1990). Elevated levels of the $\alpha_5\beta 1$ fibronectin receptor suppress the transformed phenotype of Chinese hamster ovary cells. Cell. 60, 849–859.

Gladson, C.L. and Cheresh, D.A. (1991). Glioblastoma expression of vitronectin and the $\alpha v/\beta 3$ integrin. J. Clin. Invest., 88: 1924–1932.

Goldstein, L.A., Zhou, D.F.H., Picker, L.J., Minty, C.N., Bargatze, R.F., Ding, J.F. and Butcher, E.C. (1989). A human lymphoctye homing receptor, the hermes antigen, is related to cartilage proteoglycan core and link proteins. Cell. 56, 1063–1072.

Graf, J., Iwamoto, M., Martin, G.R., Kleinman, H.K., Robey, F.A. and Yamada, Y. (1987). Identification of an amino acid sequence in laminin mediating cell attachment, chemotaxis, and receptor binding. Cell. 48, 989–996.

Gumbiner, B. and Simons, K. (1986). A functional assay for proteins involved in establishing an epithelial occluding barrier: identification of a uvomorulin-like polypeptide. J. Cell Biol. 102, 457–468.

Gunthert, U., Hormann, M., Rudy, W., Reber, S., Zoller, M., Baubmann, I., Matzku, S., Wenzel, A., Ponta, H. and Herrlich, P. (1991). A new variant of glycoprotein CD44 confers metastatic potential to rat carcinoma cells. Cell. 65, 13–24.

Hall, P.A., Coates, P., Lemoine, N.R. and Horton, M.A. (1991). Characterization of integrin chains in normal and neoplastic human pancreas. J. Pathol. 165, 33–41.

Hatta, K., Nose, A., Nagafuchi, A. and Takeichi, M. (1988). Cloning and expression of cDNA encoding a neural calcium-dependent cell adhesion molecule: its identity in the cadherin gene family. J. Cell Biol. 106, 873–881.

Heimark, R.L., Degner, M. and Schwartz, S.M. (1990). Identification of a Ca^{2+} -dependent cell-cell adhesion molecule in endothelial cells. J. Cell Biol. 110, 1745–1756.

Hemler, M.E. (1990). VLA proteins in the integrin family: structures, functions and their role on leucocytes. Annu. Rev. Immunol. 8, 365–400.

Herlyn, M. and Malkowicz, S .B . (1991). Regulatory pathways in tumor growth and invasion. Lab. Invest. 65, 262–271.

Hinek, A., Wrenn, D.S., Mecham, R.P. and Barondes, S.H. (1988). The elastin receptor: a galactoside-binding protein. Science. 239, 1539–1541.

Horvath, A.R., Elmore, M.A. and Kellie, S. (1990). Differential tyrosine-specific phosphorylation of integrin in Rous sarcoma virus transformed cells with differing transformed phenotypes. Oncogene. 5, 1349–1357.

Humphries, M.J., Olden, K. and Yamada, K.M. (1986). A synthetic peptide from fibronectin inhibits experimental metastasis of marine melanoma cells. Science. 233, 467–469.

Humphries, M.J., Yamada, K.M. and Olden, K. (1988). Investigation of the biological effects of anti-cell adhesive synthetic peptides that inhibit experimental metastasis to B16-F10 murine melanoma cells. J. Clin. Invest.

Hunkapiller, T.H. and Hood, L. (1989). Diversity of the immunoglobulin gene superfamily. Adv. Immunol. 44, 1–63.

Hynes, R.O. (1987). Integrins, a family of cell surface receptors. Cell. 48, 549–555.

Iwamoto, Y., Robey, F.A., Graf, J., Sasaki, M., Kleinman, H.K., Yamada, Y. and Martin, G.R. (1987). YIGSR, a synthetic laminin pentapeptide, inhibits experimental metastasis formation. Science, 238, 1132–1138.

Jin, L., Hemperly, J.J. and Lloyd, R.V. (1991). Expression of neural cell adhesion molecule in normal and neoplastic human neuroendocrine tissues. Am. J. Pathol. 138, 961–969.

Johnson, J.P., Stade, B.G., Holzmann, B., Schwable, W. and Riethmuller, G. (1989). *De novo* expression of intercellular-adhesion molecule in melanoma correlates with increased risk of metastasis. Proc. Natl Acad. Sci. USA. 86, 641–644.

Juliano, R.L. (1987). Membrane receptors for extracellular matrix macromolecules: relationship to cell adhesion and tumor metastasis. Biochimica et Biophysica Acta. 907, 261–278.

Klein, C.E., Steinmayer, T., Kaufmann, L. and Brocker, E. (1991). Identification of a melanoma progression antigen as integrin VLA-2. J. Invest. Dermatol. 96, 281–284.

Koch, P.J., Walsh, M.J., Schmelz, M., Goldschmidt, M.D., Zimbelmann, R., and Franke, W.W. (1990). Identification of desmoglein, a constitutive desmosomal glycoprotein, as a member of the cadherin family of cell adhesion molecules. Eur. J. Cell Biol. 53, 1–12.

Koopman, G., van Kooyk, Y., deGraaff, M., Meyer, C.J.L.M., Figdor, C.G. and Pals, S.T. (1992). Triggering of the CD44 antigen on T lymphocytes promotes T cell adhesion through the LFA-1 pathway. J. Biol. Chem. 145, 3589–3593.

Koretz, K., Schlag, P., Boumsell, L. and Moller, P. (1991). Expression of VLA-α_2, VLA-α_6, and VLA-β_1 chains in normal mucosa and adenomas of the colon, and in colon carcinomas and their liver metastases. Am. J. Pathol., 138, 741–750.

Koukoulis, G.K., Virtanen, I., Korhonen, M., Laitinen, L., Quaranta, V. and Gould, V.E. (1991). Immunohistochemical localization of integrins in the normal, hyperplastic, and neoplastic breast. Am. J. Pathol. 139, 787–799.

Krissansen, G.W., Elliot, M.J., Lucas, C.M., Stomski, F.C., Berndt, M.C., Cheresh, D.A., Lopez, A.F. and Burns, G.F. (1990). Identification of a novel integrin β subunit expressed on cultured monocytes (macrophages). J. Biol. Chem. 265, 823–830.

Kumagai, H., Tajima, M., Ueno, Y., Giga-Hama, Y. and Ohba, M. (1991). Effect of cyclic RGD peptide on cell adhesion and tumor metastasis. Biochem. Biophys. Res. Commun. 177, 1, 74–82.

Lasky, L.A. (1991). Lectin cell adhesion molecules (LEC-CAMs): a new family of cell adhesion proteins involved with inflammation. J. Cell. Biochem. 45, 139–146.

Laurie, D., Needham, L., Martin-Padura, I. and Dejana, E. (1991). Tumor cell adhesion to endothelial cells: Endothelial leukocyte adhesion molecule-1 as an inducible adhesive receptor specific for colon carcinoma cells. J. Natl Cancer Inst., 83, 1321–1324.

Lehmann, J.M., Riethmullar, G. and Johnson, J.P. (1989). MUC18, a marker of tumor progression in human melanoma, shows sequence similarity to the neural cell adhesion molecules of the immunoglobulin superfamily. Proc. Natl Acad. Sci. USA 86, 9891–9895.

Lessey, B.A., Damjanovich, L., Coutifaris, C., Castelbaum, A., Albelda, S.M. and Buck, C.A. Integrin adhesion molecules in the human endometrium: correlation with the normal and abnormal menstrual cycle. J. Clin. Invest., 1992 (In press).

Liaw, C.W., Cannon, C., Power, M.D., Kiboneka, P.K. and Rubin, L.L. (1990). Identification and cloning of two species of cadherins in bovine endothelial cells. EMBO Journal. 9, 2701–2708.

Liotta, L. (1991). Tumor invasion and metatases-role of the extracellular matrix. Cancer Res. 46, 1–7.

Lotan, R. and Raz, A. (1988). Endogenous lectins as mediators of tumor cell adhesion. J. Cell. Biochem. 37, 107–117.

Mahoney, P.A., Weber, U., Onofrechuk, P., Biessmann, H., Bryant, P.J. and Goodman, C.S. (1991). The *fat* tumor suppressor gene in *Drosophila* encodes a novel member of the cadherin gene superfamily. Cell. 67, 853–868.

Mareel, M.M., Behrens, J., Birchmeier, W., DeBruyne, G., Vleminckx, K., Hoogewus, A., Fiers, W.C. and VanRoy, F.M. (1991). Down-regulation of E-cadherin expression in madin darby canine kidney (MDCK) cells inside tumors of nude mice. Int. J. Cancer. 47, 922–928.

Matsushita, Y., Cleary, K.R., Ota, D.M., Hoff, S.D. and Irimura, T. (1990). Sialyl-demeric Lewis X antigen expressed on mucin-like glycoproteins in colorectal cancer metastases. Lab. Invest. 63, 780–791.

McGregor, B.C., McGregor, J.L., Weiss, L.M., Wood, G.S., Hu, C., Boukerchem, H. and Warnke, R.A. (1989). Presence of cytoadhesins (IIb-IIIa-like glycoproteins) on human metastatic melanomas but not on benign melanocytes. Am. J. Clin. Pathol. 92, 495–499.

Mecham, R.P. (1991). Receptors for laminin on mammalian cells. FASEB J. 5, 2538–2546.

Mecham, R.P., Hinek, A., Entwistle, R., Wrenn, D.S., Griffin, G.L. and Senior, R.M. (1989). Elastin binds to a multifunctional 67-kDa peripheral membrane protein. Biochemistry. 28, 3716–3722.

Mortarini, R., Anichini, A. and Parmiani, G. (1991). Heterogeneity for integrin expression and cytokine-mediated VLA modulation can influence the adhesion of human melanoma cells to extracellular matrix proteins. Int. J. Cancer. 47, 551–559.

Natali, P.G., Nicotra, M.R., Cavaliere, R., Giannarelli, D. and Bigotti, A. (1991). Tumor progression in human malignant melanoma is associated with changes in α_6/β_1 laminin receptor. Int. J. Cancer. 49, 168–172.

Navarro, P., Gomez, M., Pizarro, A., Gamallo, C., Quintanilla, M. and Cano, A. (1991). A role for the E-cadherin cell-cell adhesion molecule during tumor progression of mouse epidermal carcinogenesis. J. Cell Biol. 115, 517–533.

Nicholson, G.L. (1987). Tumor cell instability, diversification, and progression to the metastatic phenotype: from oncogene to oncofetal expression. Cancer Res. 47, 1473–1487.

Nicolson, G.L. (1988). Cancer metastasis: tumor cell and host organ properties important in metastasis to specific secondary sites. Biochemica et Biophysica Acta. 948, 175–224.

Nierodzik, M.L., Plotkin, A., Kajumo, F. and Karpatkin, S. (1991). Thrombin stimulates tumor-platelet adhesion *in vitro* and metastasis *in vivo*. J. Clin. Invest. 87, 229–236.

Nose, A., Tsuji, K. and Takeichi, M. (1990). Localization of specificity determining sites in cadherin cell adhesion molecules. Cell. 61, 147–155.

Orr, W., Adamson, I.Y.R. and Young, L. (1985). Pulmonary inflammation generates chemotactic activity for tumor cells and promotes lung metastasis. Am. Rev. Respir. Dis. 131, 607–611.

Osborn, L., Hession, C., Tizard, R., Vassallo, C., Luhowskyj, S., Chi-Rosso, G. and Lobb, R. (1989). Direct expression cloning of vascular cell adhesion molecule 1, a cytokine-induced endothelial protein that binds to lymphocytes. Cell. 59, 1203–1211.

Ozawa, M., Baribault, H. and Kemler, R. (1989). The cytoplasmic domain of the cell adhesion molecule uvomorulin associates with three independent proteins structurally related in different species. EMBO J. 8, 1711–1717.

Pauli, B.U. and Lee, C.L. (1988). Organ preference of metastasis: the role of organ-specifically modulated endothelial cells. Lab. Invest. 58, 379–387.

Peltonen, J., Larjava, H., Jaakkola, D., Grainick, H., Akiyama, S.K., Yamada, S.S., Yamada, K.M. and Uitto, J. (1989). Localization of integrin receptors for fibronectin, collagen, and laminin in human skin. J. Clin. Invest. 84, 1916–1923.

Picker, L.J., Nakache, M. and Butcher, E.C. (1989). Monoclonal antibodies to human lymphocyte homing receptors define a novel class of adhesion molecules on diverse cell types. J. Cell Biol. 109, 927–937.

Pignatelli, M. (1990). Integrin cell adhesion molecules and colorectal cancer. J. Pathol. 162, 95–97.

Pignatelli, M., Hanby, A.M. and Stamp, G.W.H. (1991a). Low expression of β_1, α_2 and α_3 subunits of VLA integrins in malignant mammary tumors. J. Pathol. 165, 25–32.

Pignatelli, M., Smith, M.E.F. and Bodmer, W.F. (1991b). Low expression of collagen receptors in moderate and poorly differentiated colorectal adenocarcinomas. Br. J. Cancer, 61, 636–638.

Plantefaber, L.C. and Hynes, R.O. (1989). Changes in integrin receptors on oncogenically transformed cells. Cell. 56, 281–290.

Rice, G.E., Gimbrone, M.A., Jr. and Bevilacqua, M.P. (1988). Tumor cell-endothelial interactions. Am. J. Pathol. 133, 204–210.

Rice, G.E. and Bevilacqua, M.P. (1989). An inducible endothelial cell surface glycoprotein mediates melanoma adhesion. Science. 246, 1303–1306.

Rojas, M., Fuks, A. and Slanners, C.P. (1990). Biliary glycoprotein, a member of the immunoglobulin supergene family, function *in vitro* as a Ca^{2+}-dependent intercellular adhesion molecule. Cell Growth and Differentiation. 1, 527–533.

Roos, E. (1984). Cellular adhesion, invasion, and metastasis. Biochemica et Biophysica Acta. 738, 263–284.

Ruoslahti, E. and Pierschbacher, M.D. (1987). New perspectives in cell adhesion: RGD and integrins. Science. 238, 491–497.

Ruoslahti, E. and Giancotti, F. (1989). Integrins and tumor cell dissemination. Cancer Cells. 1, 119–126.

Schipper, J.H., Frixen, U.H., Behrens, J., Unger, A., Jahnke, K. and Birchmeier, W. (1991). E-cadherin expression in squamous cell carcinomas of head and neck: inverse correlation with tumor dedifferentiation and lymph node metastasis. Cancer Res. 51, 6328–6337.

Schirrmacher, V. (1985). Cancer metastasis: experimental approaches, theoretical concepts, and impacts for treatment strategies. Adv. Cancer Res. 43, 1–73.

Schreiner, C., Fisher, M., Hussein, S. and Juliano, R.L. (1991). Increased tumorigenicity of fibronectin receptor deficient chinese hamster ovary cell variants. Cancer Res. 51, 1738–1740.

Sheppard, D., Rozzo, C., Starr, L., Quaranta, V., Erle, D.J. and Pytela, R. (1990). Complete amino acid sequence of a novel integrin β subunit (β_6) identified in epithelial cells using the polymerase chain reaction. J. Biol. Chem. 265, 11 502–11 507.

Shimizu, Y., Van Seventer, G.A., Siraganian, R., Wahl, L. and Shaw, S. (1989). Dual role of the CD44 molecule in T-cell adhesion and activation. J. Immunol., 143, 2457–2463.

Shimoyama, Y. and Hirohashi, S. (1991). Expression of E- and P-cadherin in gastric carcinomas. Cancer Res. 51, 2185–2192.

Shimoyama, Y., Hirohashi, S., Hirano, S., Noguchi, M., Schimosato, Y., Takeichi, M. and Abe, O. (1989). Cadherin cell-adhesion molecules in human epithelial tissues and carcinomas. Cancer Res. 49, 2128–2133.

Shiozaki, H., Tahara, H., Oka, H., Miyata, M., Kobayashi, K., Tamura, S., Iihara, K., Doki, Y., Hirano, S., Tadeichi, M. and Mori, T. (1991). Expression of immunoreactive E-cadherin adhesion molecules in human cancers. Am. J. Pathol. 139, 17–23.

Soszka, T., Kundsen, K.A., Bevilia, L., Rossi, C., Poggi, A. and Niewiarowski, S. (1991). Inhibition of murine melanoma cell-matrix adhesion and experimental metastasis by albolabrin, an RGD-containing peptide isolated from the venom of *Trimeresurus albolabris*. Exp. Cell Res. 196, 6–12.

Springer, T.A. (1990). Adhesion receptors of the immune system. Nature. 346, 425–434.

Stamenkovic, I., Amiot, M., Pesando, J.M. and Seed, B. (1989). A lymphocyte molecule implicated in lymph node homing is a member of the cartilage like protein family. Cell. 56, 1057–1062.

Stamenkovic, I., Aruffo, A., Amiot, M. and Seed, B. (1991). The hematopoietic and epithelial forms of CD44 are distinct polypeptides with different adhesion potentials for hyaluronate-bearing cells. EMBO. 10, 343–348.

Stamp, G.W.H. and Pignatelli, M. (1991). Distribution of β_1, α_1, α_2 and α_3 integrin chains in basal cell carcinomas. J. Patrol. 163, 307–313.

Stanford, D.R., Starkey, J.R. and Magnuson, J.A. (1986). The role of tumor-cell surface carbohydrate in experimental metastasis. Int. J. Cancer. 37, 435–444.

Takeichi, M. (1988). The cadherins: cell-cell adhesion molecules controlling animal morphogenesis. Development. 102, 639–655.

Takeichi, M. (1991). Cadherin cell adhesion receptors as a morphogenetic regulator. Science. 251, 1451–1455.

Tapley, P., Horwitz, A., Buck, C., Duggan, K. and Rohrschneider, L. (1989). Integrins isolated from Rous sarcoma virus-transformed chicken embryo fibroblasts. Oncogene. 4, 325–333.

Terranova, V.P., Liotta, L.A., Russo, R.G. and Martin, G.R. (1982). Role of laminin in the attachment and metastasis of murine tumor cells. Cancer Res. 42, 2265–2269.

Tsujisaki, M., Imai, K., Hirata, H., Hanzawa, Y., Masuya, J., Nakano, T., Sugiyama, T., Matsui, M., Hinoda, Y. and Yachi, A. (1991). Detection of circulating intercellular adhesion molecule-1 antigen in malignant diseases. Clin. Exp. Immunol. 85, 3–8.

Virtanen, I., Korhonen, M., Kariniemi, A.L., Gould, V.E., Laitinen, L. and Ylanne, J. (1990). Integrins in human cells and tumors. Cell Differentiation and Development, 32, 215–228.

Vleminckx, K., Vakaet, L., Jr., Mareel, M., Fiers, W. and VanRoy, F. (1991). Genetic manipulation of E-cadherin expression by epithelial tumor cells reveals an invasion suppressor role. Cell. 66, 107–119.

Vogel, B., Tarone, G., Giancotti, F.G., Gailit, J. and Ruoslahti, E. (1990). A novel fibronectin receptor with an unexpected subunit composition (α_v/β_1). J. Biol. Chem. 265, 2934–2937.

Volk, T. and Geiger, B. (1986). A-CAM: A 135 kDa receptor of intercellular adherens junction. I. Immunoelectron microscopic localization and biochemical studies. J. Cell Biol. 103, 1441–1450.

Weiss, L., Orr, F.W. and Honn, K.V. (1988). Interactions of cancer cells with the microvasculature during metastasis. FASEB J. 2, 12–21.

Williams, A.F. and Barclay, A.N. (1988). The immunoglobulin superfamily-domains for cell surface recognition. Ann. Rev. Immunol. 6, 381–405.

Zetter, B.R. (1990). The cellular basis of site-specific tumor metastasis. N. Eng. J. Med. 322, 605–612.

Zhu, D., Cheng, C.F. and Pauli, B.U. (1991). Mediation of lung metastasis of murine melanomas by a lung-specific endothelial cell adhesion molecule. Proc. Natl Acad. Sci. USA 88, 9568–9572.

Zutter, M.M., Mazoujian, G. and Santoro, S.A. (1990). Decreased expression of integrin adhesive protein receptors in adenocarcinoma of the breast. Am. J. Pathol. 137, 863–870.

5. *Concepts in Adhesion Regulation*

Tamas Schweighoffer and Stephen Shaw

1. *Introduction*

Adhesion is critical to the co-ordinated function of all multicellular organisms. However, its importance to immunological function has only been widely recognized in the last decade. Since adhesion is enormously complicated, simplifying principles need to be sought to make the complexity manageable. At the outset, it seemed that it would be sufficient to describe the adhesion molecules on the surface of a cell. That in itself has become formidable because of the large number of adhesion molecules which we now know can be expressed by individual leucocytes. However, an entirely new dimension of complexity has been added by a wave of recent information that the functional capacity of adhesion receptors is often dynamically regulated without changes in level of expression. Therefore, in this review we deal with these two broad categories by which adhesion can be controlled: regulation of the level of expression of the adhesion molecules (Section 2) and regulation of the functional adhesive capacity of those receptors (Section 3).

Adhesion Molecules
ISBN 0-12-741440-1

Given the breadth of the topic of regulation of adhesion, this review is necessarily illustrative not comprehensive. In addition, given our interest in T cell adhesion, illustrations are often drawn from T cells. The structure of this review is as follows. First, we introduce the concept of adhesion cascades and illustrate the concepts by outlining the steps in prototypic adhesion cascades: leucocyte interactions with endothelium (Butcher, 1991; Shimizu *et al.*, 1992a; Zimmerman *et al.*, 1992b). Next, we explore the regulation of the adhesion processes from a molecular point of view, reviewing step-by-step the process of regulation of expression, regulation of adhesive function.

Our expectation is that the fundamental features of leucocyte adhesion will be similar to those of adhesion by other cell types. However, mature leucocytes differ from virtually all other mature cells in the rate of their mobility. For example, cytotoxic T cells can repeatedly adhere to target cells, deliver the lethal hit and then disengage with a cycle time about 1–5 min (Poenie *et al.*, 1987). This confers a requirement that many of the adhesive properties of leucocytes must be subject to rapid regulation. Fig. 5.1 illustrates crudely the relationship between two broad kinds of issues relevant to regulation: regulation of expression vs. function and kinetics of regulation. Because of the rapid regulation of leucocyte adhesion, virtually all of the fast mechanisms have been defined in immunological systems.

1.1 CONCEPT OF ADHESION CASCADE

Initiation of adhesion is never a single event but rather a dynamic process sequentially involving multiple molecules and interactions. We therefore designate the process an "adhesion cascade", by analogy to enzymatic cascades of clotting and complement activation. Since characteristics of the cascade depend on the cells types or extracellular matrix components interacting, this in turn means that there are many different adhesion cascades.

KINETICS	"SLOW": >1 HOUR	"FAST": <1 HOUR
REGULATION OF EXPRESSION	TRANSCRIPTION ALTERNATIVE SPLICING TRANSLATION GLYCOSYLATION	ENZYMATIC CLEAVAGE (SHEDDING) GRANULAR STORAGE (SECRETION)
REGULATION OF FUNCTION		ENZYMATIC SYNTHESIS OR DEGRADATION OF REGULATOR MOLECULES PHOSPHORYLATION CYTOSKELETAL ASSOCIATION

Figure 5.1 Simplified grouping of the wide variety of elements which can be used in adhesion regulation.

Because the details of each cascade will differ, the overall picture is extremely complicated. However, the fundamental governing principles are rather uniform and shared between many cascades. It is those overall principles we seek to explore and illustrate in this review.

It is also probably useful to think about families of adhesion cascades. First (1) leucocyte interactions with endothelial cells is undoubtedly one of the most important families of adhesion cascades, in which there is intense current interest and much recent progress in understanding. A second (2), for the T cell or B cell, would be antigen-specific interactions with antigen presenting cells. Other broadly investigated and characterized areas are: (3) interactions of different leucocytes with extracellular matrix components; (4) allospecific cytotoxic T lymphocyte (CTL)-target recognition; (5) aggregation of platelets; and (6) migration of tumor cells.

Each step of an adhesion cascade involves a receptor-ligand interaction, involving at least one molecule on each interacting surface. Each step has multiple functions: besides providing physical force for association it also changes the state of the participants: this makes occurrence of steps in an ordered sequence possible, resulting in a cascade.

1.2 PROTOTYPIC ADHESION CASCADES: LEUCOCYTE INTERACTIONS WITH ENDOTHELIUM

To illustrate the complexities of the adhesion processes, we present a model of lymphoid-endothelial interaction (Butcher, 1991; Shimizu *et al.*, 1992; Zimmerman *et al.*, 1992b). This is a useful example because: (1) it demonstrates the cascade principle, which we believe is a crucial foundation to all adhesion processes; and (2) elements on which endothelial adhesion rely are recurrently used by other adhesion cascades.

The objective of leucocyte/endothelial adhesion cascades is for the endothelium to capture leucocytes moving through the vascular lumen and enable them to migrate into the underlying tissue. This is a formidable task because one starts with relatively non-adhesive leucocytes which are racing by, asks them to make an informed decision as to whether it is appropriate to migrate into this particular tissue, and if so brings them to a screeching halt. We view this process as a sequence of at least four steps, as illustrated in Fig. 5.2 (Shimizu *et al.*, 1992a): (1) tether; (2) trigger; (3) glue; (4) migration. Some critical features of the molecules which mediate these steps are given in Table 5.1.

For the sake of discussion, we start with "glue", the strong adhesion step which is an obvious requirement in this adhesion. In this and many other adhesion cascades this strong adhesion is mediated by a family of molecules called integrins. These are a large family of heterodimers

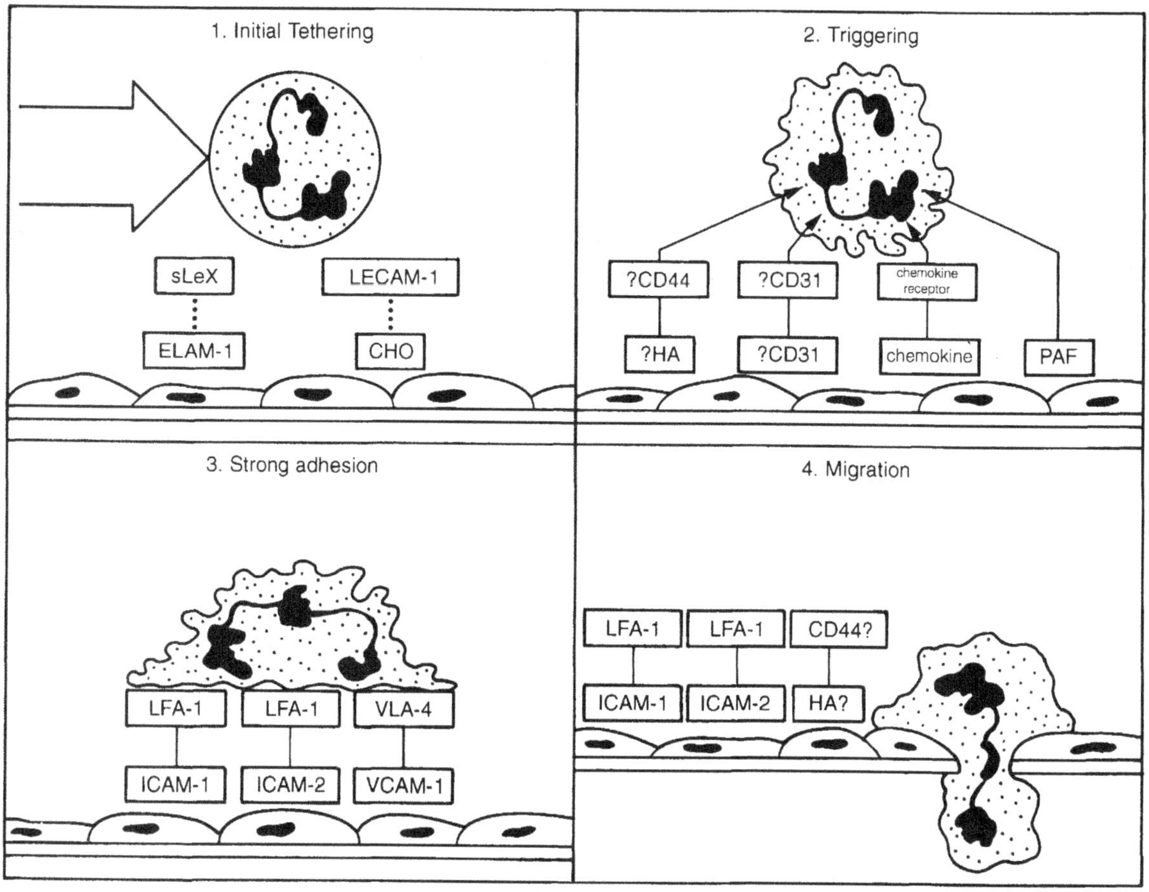

Figure 5.2 Sequential steps in simplified model of leucocyte-endothelial adhesion. The cascade starts with (1) tethering passing cells via carbohydrate-selectin interaction. Subsequently, several interaction pathways have the potential to produce a trigger signal (2), which results in strong adhesion (3) mediated by functionally up-regulated leucuocyte receptors. Adhering cells then may migrate through the endothelium into the surrounding tissue (4).

Table 5.1 Characteristics of tether, trigger and glue

Class	*Prototypic stucture*	*Characteristics*
Tether	Selectin	Adhesion during flow adhesion without activation
Trigger	Ig family Cytokine receptors	Adhesive strength not critical Mediates rapid adhesion induction
Glue	Integrin	Regulatable adhesion. Little adhesion on resting T cell. Strong adhesion when induced.

(see Section 2.5.1) represented on all cell types. Their utility in adhesion cascades is is not only due to their strength in adhesion, but also their capacity to undergo rapid changes in adhesive capacity (see Section 3). Which integrin will mediate leucocyte endothelial interaction depends on the leucocyte and the endothelium. T cells use primarily interaction of integrin LFA-1 with endothelial ligands intercellular adhesion molecules-1 and 2 (ICAM-1 and ICAM-2) and integrin VLA-4 (very late activation antigen-4) with endothelial ligand VCAM-1. Granulocytes use primarily CD11b/CD18 (Mac-1) interacting with ICAM-1.

Although such potent adhesive molecules as the integrins would seem sufficient to mediate adhesion in a single step interaction, they are not. One critical reason is that the integrins expressed on circulating leucocytes are virtually non-functional; they are "turned off". The step which precedes them must be a signalling step, which we designate "trigger", which mediates their rapid conversion from an inactive state to an active state. The mechanisms for such conversion are discussed in Section 3. In leucocyte-endothelial cascades, there is great diversity in the triggers (see Section 3.2). The triggers are perhaps best defined for granulocytes, whose adhesive capacity can be triggered by cytokines such as IL-1 and tumor necrosis factor-α (TNF-α) or by cell surface PAF (platelet activating factor) on activated endothelium (Zimmerman *et al.*, 1992b). For T cells, there are a number of potential triggers including cell surface molecules CD31 (Tanaka *et al.*, 1992), VLA-4 (Bednarczyk and McIntyre, 1990; Campanero *et al.*, 1990), CD44 (Koopman *et al.*, 1990) and cytokine receptors (Butcher, 1991; Tanaka *et al.*, 1993). Although cytokines can trigger integrin function, it is our understanding that they do so not by their presence in solution, but by their immobilization on the endothelial cells, as a decoration which is the functional equivalent to a cell surface ligand on the endothelium.

The preceding two steps, "trigger" and "glue", are still inadequate for mediating leucocyte endothelial interactions under the physiological conditions of flow. It appears that the fluid dynamics of cell contact with endothelium does not provide sufficient duration of contact (or intimacy of contact) to enable the trigger step. Accordingly, there is a requisite step of loose contact, which we designate "tether", which precedes the trigger. This rapid, transient, reversible adhesion is provided by a family of adhesion molecules referred to as selectins (Siegelman, 1991). Their interaction with ligands must have rapid kinetics which is uniquely suited to this function (Williams, 1991). For many leucocytes, the tethering is by leucocyte L-selectin which interacts with specific endothelial carbohydrates. However, the E-selectin (ELAM-1) and P-selectin (GMP-140) expressed with different kinetics on activated endothelium can also mediate this role by binding to specific carbohydrates on granulocytes or subsets of T cells (Walz *et al.*, 1990; Picker *et al.*, 1991a; Shimizu *et al.*, 1991).

The final step envisioned in this simplified cascade is "migration". After the leucocytes are captured from circulation by strong adhesion, that adhesion must be moderated to allow the cell to move into tissue. Undoubtedly multiple mechanisms are involved here. Two mechanisms are known for release. First, the L-selectin molecule is shed from the cell surface, apparently by proteolytic cleavage. Second, the strong integrin adhesion must be moderated, in part by the intrinsic transience of adhesion capacity induced by the known triggering signals (Dustin and Springer, 1989; Tanaka *et al.*, 1992).

This cascade concept fits intuitively with findings from simplified model systems (von Andrian *et al.*, 1991) and *ex vivo* videomicroscopy. Consider for example results from studies of T cell binding to gut (high endothelial venule [HEV]) (Bjerknes *et al.*, 1986). Many of the lymphocytes bouncing against the endothelial wall bind very transiently and come off. This process can be repeated. A fraction of these transient encounters (which we imagine are mediated by tethers) result in strong binding (which we imagine results from their encounter with appropriate triggers with resultant recruitment of integrin binding). Those bound cells subsequently migrate into tissue.

2. *Regulation of Expression of Adhesion Receptors and their Regulators*

Obviously adhesion molecules need to be expressed at the cell surface in order to mediate adhesion. There are multiple modes by which the level of expression and the structure of the expressed receptor are regulated.

2.1 INTRODUCTION – REGULATION AT THE CELLULAR LEVEL

Constitutive (long-term) expression of adhesion molecules is highly specific for cell-type. Most of the molecules are, however, not lineage specific, but the uniqueness consists of the combinations expressed. The combinatorial rules seem more *ad hoc* than rational; whatever works is used during evolution. That the $\alpha_6\beta_4$ hemidesmosome-associated integrin might be restricted to cells with epithelial origin is rather a rare exception (Kajiji *et al.*, 1989). Other exclusive markers (e.g. CD3 on T lymphocytes) have other primary function than adhesion itself.

Hematopoietic cells in general and T cells in particular may have a larger complement of adhesion molecules than mature sessile cells because they confer on the mobile cell the ability to function in many different micro-environments.

Within the leucocytes, in turn, there are again lineage-determined patterns of expression of adhesion molecules: VLA-4 is present on lymphocytes and eosinophils but not polymorphonuclear neutrophils (PMNs). Among lymphocytes, there are still substantial differences between T and B cells. Among T cells, naive cells preferentially express L-selectin while memory cells have more leucocyte function-associated antigen-3 (LFA-3). Finally, and among memory cells, only a small subset expresses the adhesion molecule implicated in skin-homing.

In addition, within the lineages or subsets there are also reversible (induced/response/short-term) changes in the expression of adhesion molecules; the easiest category to conceptualize are the inflammatory ones. On endothelial cells, TNF-α, IL-1 or bacterial lipopolysaccharide (LPS) rapidly induce expression of adhesion molecules: VCAM-1 (Osborn, 1990), ICAM-1 (Rothlein *et al.*, 1991), E-selectin (Bevilacqua *et al.*, 1989). Other molecules are also induced, which do not directly participate in adhesion, but have the potential to modify the interaction, like adenosine, nitric oxide (NO) granulocyte-macrophage colony-stimulating factor (GM-CSF), PAF (Kishimoto *et al.*, 1989; Zimmerman *et al.*, 1992b). These and other complexities in the regulation of expression of most adhesion molecules indicates that there must be multiple modes of regulation; it is most consistent with multiple upstream DNA regulatory elements which are now being appreciated in virtually every upstream sequence being analyzed (MHC-I, IL-2, VLA-4).

2.2 GENOMIC ORGANIZATION AND MOLECULAR EVOLUTION

Of the many adhesion molecules which have been cloned, some have been mapped to chromosomes. The most interesting feature is that structurally and functionally homologous molecules often map close to each other. Selectins constitute a gene cluster on human and mouse chromosome 1, where they map within 300 kb; this suggests that the selectin family arose by gene duplication before divergence of mouse and human. They are physically close to certain complement or receptor genes with which they show sequence homology (Watson *et al.*, 1990). Similarly, β_2-related α chains also have been reported to map to a single region. The short arm of human chromosome 1 (Sewell *et al.*, 1988) offers intriguing complexities: it not only codes members of the immunoglobulin-superfamily, the VCAM-1 gene (Cybulsky *et al.*, 1991) and the receptor-ligand pair CD2 and CD58 but also molecules with potential regulatory properties: the TCL5 locus may specify a helix-loop-helix type DNA-binding protein (Finger *et al.*, 1989), and the nearby lck is a lymphocyte-specific tyrosine kinase (Telfer and Rudd, 1991). At this moment we do not know whether the clustering has any significance, but in view of the recent developments regarding the chromosomal organization of the major histocompatibility complex/human leucocyte antigen (MHC/HLA) genes this is certainly an area where we can expect surprises.

Adhesion molecules are often mosaics of structural elements with adhesive function. For example, the selectins have incorporated structural domains from lectins, epidermal growth factor (EGF) and complement regulatory repeats (Bevilacqua *et al.*, 1989; Hession *et al.*, 1990). β_7 looks like a structural composite of β_2 and β_1 [Yuan *et al.*, 1990; Erle *et al.*, 1991]). Thus evolution has exploited existing structural components, recombined them, and optimized them for new adhesive interactions.

2.3 TRANSCRIPTION/TRANSLATION

Complexities of expression suggest that numerous different promoter and transcriptional regulator regions must be involved in the control of expression of adhesion molecules. In the upstream region of the VLA-α_4 gene several relevant elements have been identified (Rosen *et al.*, 1991): (1) tissue-specific expression is determined by a PU-box which is a myeloid/lymphoid specific enhancer and also by muscle-specific MyoD-binding site, which may explain why α_4 is present on rhabdomyosarcomas: (2) Transforming growth factor 3 (TGF-β) and serum are inferred to mediate transcriptional stimulation through an AP-1 site: (3) An upstream ATG codon was also found which is frequently seen in proto-oncogenes and is likely involved in translational control. VCAM−1 upstream region contains an AP−1 site, a sequence recognized by the GATA box family of zinc finger transcription factors and also one or two NF-kB binding site. Cytokines like IL-1 or TNF-α activate nuclear factor-kB (NF-kB), so the latter element may explain cytokine-induced expression of VCAM-1. (Cybulsky *et al.*, 1991). NF-kB and the AP-1 site are also involved in the regulation of E-selectin (Montgomery *et al.*, 1991)

Activation, differentiation and transformation all involve ensembles of changes, producing signals which drive transcription. Often a single inducer is not responsible, or has not yet been identified. During T cell activation, expression of more than 70 genes has been described, among them several adhesion molecules and molecules which have the capacity to modulate the adhesion process (Crabtree, 1989). Relevant transcription factors, like NF-kB are induced within 30 min; cytokines, like IFN (interferon), TGF-β or IL-4 and GM-CSF take 1–20 h. Some adhesion molecules, for example ICAM-1, is induced within hours. Members of the VLA-family and the cytokine RANTES are induced 4–14 days after initiation of the activation (Crabtree, 1989). As a sign of *in vivo* chronic activation VLA-1 is expressed by T cells from synovial fluid and tissue of rheumatoid arthritis (RA) patients (Hemler *et al.*, 1986), and a VLA-2+CD8+ subset was detected in autoimmune disease (AIDS) patients (Pantaleo *et al.*, 1990).

Differentiation is highly dependent on regulated expression of adhesion molecules during embryonic and early development: they direct and then keep cells in their specific micro-environment. For example, $\alpha_1\beta_1$ is expressed on microvascular endothelium but is absent from large vessels such as the aorta or umbilical cord vessels (Defilippi *et al.*, 1991).

Likewise, transformation results in an ensemble of changes in adhesion molecules which contribute to the tumorogenesis. Ras-transformed rodent cells have reduced expression of the fibronectin receptor $\alpha_5\beta_1$ (Plantefaber and Hynes, 1989). Experimental over-expression of $\alpha_5\beta_1$ can reverse the transformed phenotype of Chinese hamster ovary (CHO) cells (Giancotti and Ruoslahti, 1990), resulting in increased FN-matrix deposition, loss of anchorage-independent cell growth and loss of tumorigenicity in nude mice. Glioblastoma multiforme expresses $\alpha_v\beta_3$ and produces vitronectin (VN) in late stage tumors, when it grows as adherent, invasive mass, but both the receptor and the matrix protein are absent on normal or reactive or low-grade tissue (Gladson and Cheresh, 1991). Similarly, melanoma cells with vertical growth type and metastases express

$\alpha_v\beta_3$, which is not found on horizontal-growth melanoma or on normal skin melanocytes (Albelda *et al.*, 1990). Rhabdomyosarcoma cells with a transfected functional VLA-2 integrin showed enhanced metastatic capability; the *in vivo* tumorigenicity and *in vitro* growth rates remained, however, unaffected (Chan *et al.*, 1991a). Viral infection also acts in a complex way, which affects adhesion molecules; HIV-1 (human immunodeficiency virus) infection of human T lymphocytes results in enhanced $\alpha_5\beta_1$ integrin expression (Weeks *et al.*, 1991).

Soluble mediators are examples of singular factors which direct expression of adhesion molecules. Most members of the integrin family can be up-regulated by TGF-β; the regulated component seems to be the α chain (Heino *et al.*, 1989; Ignotz *et al.*, 1989). Under the influence of TGF-β *in vitro* the integrin VLA-α4 is replaced by a newly expressed α chain, recognized by monoclonal antibody (mAb) M290 in the mouse (Kilshaw and Murant, 1991) and HML-1 in human (Cerf-Bensussan *et al.*, 1992). TGF-β treatment of young (2–5 days) but not old TNF/IFN-induced HUVEC (human umbilical vein endothelial cell) inhibits adhesion of neutrophils and T cells (Gamble and Vadas, 1988). Since ICAM-1 and VCAM-1 expression does not change substantially, the change may involve an unidentified pathway. On HUVEC, there are dramatic changes in expression of adhesion molecules induced by cytokines. IL-1, TNF-α or LPS induce rapid expression of adhesion molecules: VCAM-1, E-selectin, ICAM-1. The induction is specific for cytokine and for target tissue. For example, all three are induced on HUVEC by IL-1 and TNF-α, but only VCAM-1 is induced by IL-4 (Graber *et al.*, 1990; Rice *et al.*, 1990; Thornhill *et al.*, 1990). VCAM-1 regulation also differs between different kinds of endothelium; on HUVEC it can be induced by IL-1 and TNF-α but on cultured microvascular endothelium it is induced by TNF-α but not IL-1 (Swerlick *et al.*, 1992). VLA-1 is induced on HUVEC by TNF-α, retinoic acid or phorbol myristate acebate (PMA) more slowly, with a peak at 48 h (Defilippi *et al.*, 1991).

2.4 ALTERNATE SPLICING

Alternate splicing is an extremely powerful strategy which generates diversity several ways. First, in some cases the intracytoplasmic portion of a molecule is changed, which means potentially different interaction with the cytoskeleton and altered capacity in signal-transduction (see Section 3.4.1). Examples are integrins β_1, β_3 (van Kuppevelt *et al.*, 1989), α_6, α_3 (Tamura *et al.*, 1991). The latter have two different (A and B) intracytoplasmic variants whose expression is somewhat cell-type dependent; mouse embryonic fibroblasts have only α_{6A}, ES cells have only α_{6B}, but most normal tissues and transformed lines have both. Similarly, α_{3B} was found only in mouse brain and heart, while other tissues have both α_3 forms.

Differential splicing is not restricted to cytoplasmic domains but can influence both transmembrane and extracellular portions. LFA-3 is anchored to the membrane in two different ways: one is a transmembrane form and the other a GPI-anchored form. Since there is only a single LFA-3 gene, most likely alternative splicing is responsible for the difference. Two different VCAM-1 molecules were found which are generated by alternative mRNA splicing (Cybulsky *et al.*, 1991). In one form one Ig-domain is missing from the middle of the extracellular portion; this may be a functionally distinct or tissue specific variant.

Findings with CD44 illustrate how many aspects of a single molecule can be regulated by differential splicing. Molecular cloning data suggest that there may be at least four forms of CD44: CD44E is characteristic for epithelial cells and CD44H is the predominant variant on hematopoietic cells. CD44H has an isoform with truncated cytoplasmic tail. (Brown *et al.*, 1991; Stamenkovic *et al.*, 1991; Lesley *et al.*, 1992). Accompanying the structural differences, the variants also differ in their functional capacities, e.g. hyaluronic acid-binding.

2.5 POST-TRANSLATIONAL MODIFICATION AND PROCESSING

Multiple steps of post-translational modification are required for correct function of adhesion molecules. We discuss here three major areas which are crucially important for the successful function of one or another class of molecules. Because of their importance, they also might represent prime targets for pharmacological intervention.

2.5.1 Combinatorial Association

For the integrins, combinatorial assembly of α and β chains is an effective tool for generation of diversity of surface receptors. With a set of the known eight β chains and more than 15 α chains, as many as 20 pairs have already been identified. It has been known for several years that individual β chains can associate with multiple different α chains, for example β with α1, α2, α3, α4, α5, α6; this gave rise to nomenclature wherein the molecules containing a shared β chain were given a family name – such as VLA for molecules including β_1 However, it is now apparent that a single α chain can pair with a number of different β chains: e.g. α_v which pairs with five different β chains (Cheresh, 1991a). The simple interpretation based on limited data is that the receptor specificity may be determined substantially by the α chain, but the β chain is critical for regulation and for effector function. For example, in α_v-containing integrins, vitronectin specificity is contributed by the α_v but cytoskeletal interactions by the β chain.

Successful association, besides being a combinatorial strategy, is also critical for surface expression of both chains. In leucocyte adhesion deficiency (LAD) patients, where the production of the β_2 chain is defective, the corresponding α chains are also absent from the cell surface; α chains remain in precursor stage until degraded (Springer *et al.*, 1984). In most cases, β chains appear to be synthesized in excess, (Heino *et al.*, 1989; Cherish, 1991b) and availability of α is rate-limiting. In transfection experiments, when α chain genes were transfected, additional β_1 was endogenously available and β_1-surface expression was increased (Teixido *et al.*, 1992). However, as a counter-example, on megakaryocytes αIIb was found to be in excess over β_3.

2.5.2 Glycosylation

Glycosylation is important in multiple ways: (1) carbohydrates may themselves act as adhesion structures; (2) carbohydrates may determine or modify ligand or receptor function, as in the case of ICAM-1 or $\alpha_6\beta_4$; (3) on proteoglycans glycosylation dictates their ability to accept decoration by cytokines, and thus regulates the adhesion cascade; (4) carbohydrates may influence susceptibility of proteins to degradation.

Carbohydrates are known to mediate cell adhesion, but they are probably still under-rated in importance because they are harder to define biochemically than proteins. The most prominent example for carbohydrates functioning as autonomous ligands in immunological systems are the lectin-mediated adhesion pathways. The selectin family of adhesion molecules, described briefly above in the model of leucocyte endothelial adhesion, is the best understood family. The specificity of lectin binding can be as precise as in protein/protein interactions. The specificity of the selectins includes both sialic acid and fucose in characteristic linkages to an oligosaccharide backbone (Trude *et al.*, 1990; Walz *et al.*, 1990; Aruffo *et al.*, 1991). The presence of the relevant carbohydrate on a cell is determined by the ensemble of glycosyltransferases expressed by the cell (Lowe *et al.*, 1990); in addition, only a subset of the surface glycoproteins express the distinctive carbohydrate sequences seen by each selectins (Berg *et al.*, 1991). Undoubtedly lectins are widely used in other adhesion systems including tumor metastasis (Lotan and Raz, 1988), embryogenesis (Grabel, 1984), and sperm/egg binding (Glabe *et al.*, 1982; Huang *et al.*, 1982).

Integrin α and β subunits are co-translationally modified with Asn-linked glycosyl side chains (Hemler, 1991), and then assemble slowly ($> = 4$ h) in the endoplasmic reticulum (ER), where there is further trimming of three glucosyl residues from the core oligosaccharides. This procedure influences the intracellular fate of the newly produced chains dramatically, since glucosidase inhibitors cause at least delay in assembly/cleavage/surface appearance. Glucosidase inhibitors also block myoblast adhesion to fibronectin and laminin, because integrins cannot get to the surface at all (Trudel and Holland, 1989). Later both chains are processed from "high mannose form" to mature, complex glycoproteins. This step is not essential from the point of view of surface expression, since mannosidase I and II inhibitors did not inhibit assembly and transport of $\alpha_v\beta_3$ to the cell surface. Even under natural circumstances, a proportion of β_3, α_X, α_M stays in the high mannose form. This process, however, may have functional importance; mannosidase-treated $\alpha_5\beta_1$ was assembled and inserted into the membrane but was not functional (Akiyama *et al.*, 1989). A mutant melanoma line with glycosylation-deficient β_1 was also deficient in adhesion and spreading on laminin and fibronectin, which suggests a relationship between glycosylation and metastatic potential (Oz *et al.*, 1989).

Carbohydrates on an adhesion ligand can influence the adhesive function of that ligand. There is an N-linked carbohydrate on the third domain of ICAM-1, near the site of ICAM-1 interaction with CD11b/CD18 (Diamond *et al.*, 1991). Destruction of that glycosylation site facilitates binding by CD11b/CD18 but does not influence binding by CD11a/CD18 at a site on the first domain. It remains to be proven if this particular glycosylation is regulated physiologically, but it seems likely that the concept will be correct. Another example, whose physiological relevance is better understood is N-CAM. The degree of sialylation varies dramatically between embryonic and adult forms in the brain; the more heavily sialylated embryonic form is less adhesive (Nybroe and Bock, 1990).

An interesting combinatorial possibility is when an adhesive protein itself is modified with an adhesive carbohydrate group; on developing mouse brain integrin β_1 but not on adult may be modified with the L2/HNK-1 epitope. Since L2/HNK-1 itself may be involved in adhesion (Kunemund *et al.*, 1988), it could contribute or modify integrin adhesion. The L2/HNK-1 common carbohydrate epitope is shared by other adhesion molecules: N-CAM, L1, myelin-associated glycoprotein, J1 (Kruse *et al.*, 1985) and a structure on NK cells (Abo *et al.*, 1982). Another example is the presence of the sialylated E-selectin ligands on the L-selectin (Picker *et al.*, 1991b). Finally, there is a carbohydrate epitope found on $\alpha_6\beta_4$ on epithelial cells, which is involved in NK and cytotoxic T cell cytotoxicity (Phillips *et al.*, 1991).

2.5.3 Others

Other post-translational modifications also occur before adhesion molecules reach the cell surface. For example, mature VLA-α4 undergoes cleavage after having associated with the β_1, within the Golgi or post-Golgi (Teixido *et al.*, 1992). It is increased following T cell activation and depends on cell (line) type. A number of other integrin α chains are also cleaved. Cleavage is not known to alter function (Teixido *et al.*, 1992) but might be expected to do so. β_3 on platelets may undergo palmitylation (Cierniewski *et al.*, 1989); this was not seen for other integrin subunits and the significance is unknown.

2.6 SECRETORY MECHANISMS

Regulated export of stored adhesion molecules is critical for cells with extensive granules (including platelets, endothelial cells, granulocytes, probably monocytes) but apparently not relatively agranular cells (like resting T and B cells). Two kinds of adhesion molecules are temporarily stored in granules: (1) integral membrane proteins which become cell surface proteins when the granules fuse with the plasma membrane. Among the best studied are P-selectin in platelets, CD11b/CD18 in granulocytes and E-selectin in endothelial cells; (2) secreted molecules which will become soluble glue. Thrombospondin and von Willebrandt factor are two such well-characterized proteins in platelets. The appeal of a secretory mechanism is the ability to very rapidly mobilize large amounts of these adhesion proteins. One limitation in this strategy is that it is easily exhausted; it is particularly appropriate to granulocytes and platelets which are essentially "disposable" cell types.

On endothelial cells or platelets P-selectin is translocated from secretory granules (Weibel–Palade bodies) to the plasma membrane within seconds after thrombin or histamine activation. On histamine-induced endothelial cell (EC) it is then rapidly internalized, so the surface expression is transient, with a maximum at 3–10 min, while on platelets it shows sustained expression. The biochemical regulation of the release was investigated on CD11b/CD18 in granulocytes: it is apparently a complex procedure, related to but distinct from exocytosis (Brown *et al.*, 1991).

3. *Regulation of the Function of Adhesion Molecules*

3.1 INTRODUCTION

As outlined above, it used to be safe to assume that determination of the expression of adhesion molecules of a cell would allow good predictions regarding the adhesive properties of that cell. Transcriptional and translational changes take usually 4–24 h and so influence adhesive interactions with a considerable latency. However, as mentioned above, immune cells undergo changes in adhesion much more rapidly than can be explained by such mechanisms and therefore must result from fast qualitative changes in the existing set of adhesion molecules.

Multiple lines of evidence have converged to convincingly establish the pervasiveness and importance of such functional regulation of adhesion. It has been appreciated for some time that the αIIb/β_3 integrin on platelets is functionally inactive until the platelet has been activated (Coller, 1986; Phillips *et al.*, 1991; Roth, 1992). Likewise the adhesive function of CD11b/CD18 on granulocytes is minimal until they are activated (Wright and Meyer, 1986). Finally, lymphocytes which are relatively non-adhesive become very adhesive following treatment with PMA (Patarroyo and Jondal, 1985; Skoglund *et al.*, 1988). Appreciation of the importance and elegance of these phenomena has emerged from studies which place these changes in a broader biological context. For T cells, the key observation which provided a basis for the rapid affinity cycles was that T cell receptor (TCR) crosslinking on resting T lymphocytes triggered prompt and transient increase in the avidity of LFA-1 for ICAM-1 (Dustin and Springer, 1989). Dustin and Springer referred to LFA-1 as a "servo-motor", by which they meant that its strong adhesion was governed by the recognition event itself. This mechanism explains why antigen-specific T cells interact strongly only with those cells that bear the antigen for which they are primed. Similar activation-dependent binding to extracellular matrix proteins was shown for members of the VLA family (Shimizu *et al.*, 1990c) and other integrins like VN receptor.

The integrin molecules are the prototype of regulatable effectors of adhesion. Most of the current knowledge regarding acute adhesion regulation arose from, and is based on, the integrins. However, functional regulation of adhesion also pertains to immunoglobulin supergene family molecules, selectins and presumably other families. For example, TCR crosslinking dramatically augments adhesion/interaction of CD8 with MHC class I (O'Rourke *et al.*, 1990). In addition, on either granulocytes or T cells, the binding of L-selectin to a carbohydrate ligand is rapidly augmented by activation of the cell (Spertini *et al.*, 1991).

This relatively new area is rich in biological diversity; it provides as much specificity and combinatorial depth as the more traditional modes of regulation by transcription and chain assembly. Acute regulation of adhesive function can be conceptualized as composed of three components illustrated in Fig. 5.3: (1) trigger/amplifier/transducer molecules; (2) biochemical coupling mechanisms; (3) regulatable effectors of adhesion.

Such systems would be easiest to understand and conceptualize if individual trigger molecules produced unique biochemical signals which acted exclusively on individual adhesion molecules. The available data indicate that there is indeed a measure of specificity; but there is also redundancy and cross-talk resulting in individual triggers modifying function of multiple adhesion molecules, and multiple triggers modifying function of individual adhesion molecule.

3.2 TRIGGER

Triggering the activation process relies on a receptor-ligand interaction. Participating ligand structures are: (1) integral membrane proteins, carbohydrates, lipids; (2) soluble molecules; or (3) soluble molecules decorating

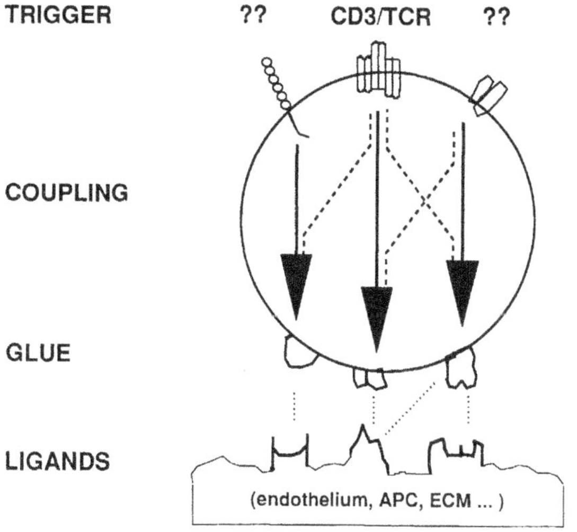

Figure 5.3 Simplified model of possible relationships between trigger molecules, biochemical second signals and the adhesion molecules whose function is regulated. CD3/TCR is shown as a particularly well-defined trigger molecule for T cells.

ECM or cell surface. Diverse families of surface molecules can serve as receptors, most characteristic are members of the immunoglobulin superfamily and the cytokine receptors. Probably, any signal transducing molecule can be pressed into service to regulate adhesion as much as it can regulate cellular metabolism or differentiation.

3.2.1 Cell-contact Mediated

Cell-contact mediated triggering is the major regulation pathway for CTL, and is a contributor in T cell-APC interactions and in endothelial adhesion. Crosslinking the TCR triggers increased LFA-1 avidity which peaks at 10–20 min and decreases to basal level by 40 min (Dustin and Springer, 1989). Antigen-specific stimulation of a murine T cell clone also induces transient VLA-mediated binding to the extracellular matrix (Chan *et al.*, 1991b). Crosslinking CD2 on T cells with a pair of mAbs results in a longer-term increase in LFA-1 avidity.

With the number of structures with identified trigger capacity continuously growing, some underlying principles begin to emerge:

1. There is redundancy at the cellular level: signalling through several different trigger structures may produce similar results. T-cell adhesion is triggered by mAbs to several surface structures besides CD3/TCR and CD2 on T cells: CD7, CD28, CD44, CD31 (Koopman *et al.*, 1990; Shimizu *et al.*, 1992b; Tanaka *et al.*, 1992). CD19, CD20, CD39, CD40, CD43, Leu-13 and HLA-DR on B cells all induce strong LFA-1 dependent, tyrosine kinase dependent adhesion (Barrett *et al.*, 1991; Kansas and Tedder, 1991). On macrophages, effects of anti-CD14 are similar (Lauener *et al.*, 1990).
2. There is conserved structure-function relationship in the trigger molecules; the very same structures can function as trigger molecules in different cell types. For example, anti-CD9 mAbs are potent triggers of platelet adhesion (Slupsky *et al.*, 1989), and anti-CD9 antibodies also augment neutrophil adherence to endothelium (Forsyth, 1991).
3. Within the redundancy there is nevertheless specificity. Although trigger molecules on T cells have in concept similar capacity in induction of adhesiveness, there are also examples of fine differences: in contrast to TCR/CD3, triggering CD31 preferentially induces adhesion through VLA-4/VCAM-1 and not so well through LFA-1/ICAM-1 (Tanaka *et al.*, 1992). This is most likely accomplished by using separate biochemical coupling pathways.
4. Adhesion-inducing molecules often are molecules known for their capacity to induce other activation and differentiation events (see Section 4.2).
5. Molecules which mediate adhesion can also trigger changes in adhesion. For example, the weak tethering of granulocytes via binding to ELAM-1 activates β_2 (CD18) and thus induces strong adhesion (Lo *et al.*, 1991; Kuijpers *et al.*, 1991).

One of the most striking examples of a triggering pathway is that of PAF on activated endothelium interacting with a PAF receptor on granulocytes to induce function of the CD11b integrin. PAF is a phospholipid not constituitively expressed on resting endothelial cells, but it is synthesized within a minute after stimulation of endothelium by thrombin, histamine, or LTC4. The enzymes involved in the synthesis, Ca^{2+}-dependent phospholipase A_2 and acetyltransferase are regulated by phosphorylation (Zimmerman *et al.*, 1991). PAF induces adhesion by binding to a G-protein-linked 7 domain transmembrane receptor which up-regulates β_2-integrins (reviewed in Zimmerman *et al.*, 1992b). This complex process was termed juxtacrine activation (Lorant *et al.*, 1991).

3.2.2 Cytokines and Soluble Factors

The second major activation pathway relies on signals transmitted by chemo-attractant factors or cytokines. Although its discussion is often in the context of cell-endothelial interactions, undoubtedly its relevance is far greater (Nathan and Sporn, 1991). Two different modes of action have been characterized:

1. the mediators are present in soluble form, and form a gradient which may also play a key role in orienting cells;
2. after their release the mediators are recaptured and presented by cell surface structures (predominantly proteoglycans) or extracellular matrix, called "decoration". This can result in relatively high local concentrations localized directly to the site of interaction, which facilitates efficient signal transduction. This

mode of cytokine function also establishes gradients which can be used for haptotaxis rather than chemotaxis.

The number of substances identified as activating factors is rapidly growing. In addition to the broadly expressed cytokines (e.g. IL-1, IL-6, TNF, IFN), it also includes members of the chemokine/RANTES/C-C/C-X-C family (e.g. IL-8, MIP-1β) (Oppenheim *et al.*, 1991; Schall, 1991; Tanaka *et al.*, 1993), colony-stimulating factors (G-CSF, GM-CSF (Bussolino *et al.*, 1989)), chemo-attractant peptides (C5a, FMLP [N-formyl-methionyl-leucyl-phenylalanine]), soluble fragments of ECM proteins [(fragments of fibronectin (Clark *et al.*, 1988), factor XIII, glycosaminoglycan (GAG) (Lanir *et al.*, 1988))] and neuropeptides (bombesin, gastrin-releasing peptide and neuromedin C (De la Fuente *et al.*, 1991).

For many of these cytokines, the nature of the receptor remains to be defined. Based on the striking lineage-specificity of action of many of the cytokines, one would predict that there is lineage specificity of a single receptor, or combinatorial specificity may be conferred by co-operation of multiple receptors. Some chemo-attractant factors target a remarkably narrow population of cells: RANTES acts primarily on memory T cells, MIP-1β on virgin T cells, MIP-1α on CD8+ cells and B cells, IL-5 on eosinophils (Coeffier *et al.*, 1991), C5a and the hMGSA/GROα family on neutrophils. Most "classic" cytokines (e.g. IL-1, IL-4, TGF-β) exert adhesion modulating effects on a variety of cells, but they may show different effects on different cells: TGF-β induces T and monocyte migration but not granulocytes (Adams *et al.*, 1991); IL-4 induces adhesion molecules on HUVEC but inhibits monocyte adhesion (Elliott *et al.*, 1991); IL-8 attracts granulocytes but not monocytes.

3.3 COUPLING

Our understanding is very primitive of the biochemical pathways which couple the trigger molecule to the adhesion molecule. We know a number of second messengers which can be generated by engagement of trigger molecules. A number of pharmacological inhibitors have been reported which induce adhesion or interfere with the coupling of the trigger to adhesion-induction. However, this information falls far short of a comprehensive understanding of the sequence of biochemical events which account for the regulatory coupling.

Various lines of evidence implicate protein phosphorylation in the coupling (as well as in the terminal regulatory events discussed below). The agent PMA, which is best know for its capacity to activate PKC, is a very effective inducer of integrin-mediated adhesion on a variety of cell types (Patarroyo and Jondal, 1985). Equally important, inhibitors of protein tyrosine phosphorylation are potent inhibitors of adhesion induced by a variety of trigger molecules (Rothman *et al.*, 1991). This fits with emerging concepts of the central role that protein phosphorylation plays in biochemical regulation. It also fits with findings that a number of the trigger molecules can induce protein phosphorylation (Klausner and Samelson, 1991; Kornberg *et al.*, 1991; Nojima *et al.*, 1992).

Intracellular calcium fluxes are also candidates for coupling trigger molecules to the regulated adhesion molecules. Calcium ionophores can induce augmentation of integrin-mediated adhesion (Shimizu *et al.*, 1992b). These findings complement data that the same triggering structures which induce Ca^{2+} influx are often those which mediate induction of adhesion (e.g. CD3, CD2, CD7, CD28 on T cells). The exact role and contribution of intracellular Ca^{2+} changes is still hard to appreciate, partly because Ca^{2+} interacts with other pathways, like PKC and IP/DAG, which regulate Ca^{2+} up or down, depending on the system. There are also some data about the spatial orientation of Ca^{2+} flux: it is lowest in leading edge of migrating cells (Brundage *et al.*, 1991), so together with kinases it might be important for migration in macrophages (Sozzani *et al.*, 1991). In turn, cell shape changes may be transduced by Ca^{2+} (Watson, 1991).

As depicted in Fig. 5.3, we envision multiple biochemical coupling pathways, with redundancy and co-operativity. This is consistent with the modest available data. For example in studies of the induction of T cell binding to fibronectin, the data suggest at least two pathways which co-operate in adhesion-induction: one involving Ca^{2+} and the other involving PKC activation. Analysis with inhibitors suggests that physiological triggers such as CD3 and CD7 use both. Furthermore, increases in cAMP inhibit adhesion induction (Dustin and Springer, 1989), potentially through the Ca^{2+}-dependent pathway (Shimizu *et al.*, 1992b) or through other as yet undefined mechanisms.

The general concepts outlined above for the biochemical second signals involved in adhesion induction parallels that known about biochemical second signals for leucocyte activation in general. This illustrates how closely adhesion induction is integrated with the other signalling processes within the cell. It also requires that intracellular integration of these signals must give rise to very specific consequences from these superficially chaotic and redundant second messages.

3.4 MECHANISMS FOR FUNCTIONAL REGULATION OF ADHESION

There are multiple mechanisms by which the adhesion capacity of expressed molecules can be regulated. Much remains to be definitively demonstrated with regard to mechanism, but the following areas are of particular interest.

3.4.1 Receptor Clustering and Cytokeletal Association

There are multiple lines of evidence for the importance of cytoskeleton. First, cytoskeletal reorganization almost invariably accompanies adhesion. In mAb-induced caps, talin colocalizes with $\beta 1$ integrins (Burn *et al.*, 1988). LFA-1 and talin also co-localize in activated but not in resting lymphocytes, and redistribute to the site of adhesion. Binding of fibrinogen to the platelet surface through activated $\alpha IIb\beta 3$ likely requires organization of platelet cytoskeleton, which is important for clot retraction [(Phillips *et al.*, 1989). Vitronectin receptor $\alpha_v\beta 3$ also forms focal contacts, and associates with the cytoskeleton, co-localizing with vinculin, talin, and actin-ends. (In contrast, $\alpha_v\beta_5$ shows distinct, non-focal contact distribution (Wayner *et al.*, 1991) emphasizing again the decisive role of β chains in intracellular communication.)] Second, agents like cytochalasin which interfere with cytoskeletal assembly interfere with adhesion. For example, LFA-1 or VLA-4 and VLA-5-dependent adhesion is blocked by cytochalasin B (Rothlein and Springer, 1986; Shimizu *et al.*, 1992b). Third, studies of transfectants with mutant adhesion molecules demonstrate alterations in adhesion and its regulation. Specific C-terminal deletions of the cytoplasmic tail of β_2 decrease ICAM-binding (Hibbs *et al.*, 1991a, 1991b). Similarly, affinity of $\alpha IIb\beta 3$ is controlled by the cytoplasmic domain of αIIb (O'Toole *et al.*, 1991). In contrast, cytoplasmic deletion mutants of CD2 which have lost their signal transduction capacity were still able to bind to LFA-3 comparably to the wild-type CD2 (Bierer *et al.*, 1990). Thus, cytoskeletal involvement is critical for regulation of adhesion.

Undoubtedly there are also multiple biophysical mechanisms that regulate cytoskeletal involvement and thus contribute to adhesion. One important mechanism in adhesion regulation is receptor clustering which can be mediated, at least in part, by cytoskeletal associations. This concept is woven from three kinds of evidence: (1) Theoretically, stability is favoured by multivalent interactions (Dower *et al.*, 1982); (2) Strong adhesion is often accompanied by specialized regions of contact, such as focal contacts, involving receptors and ligand co-cluster with heavy involvement of cytoskeletal elements (Turner and Burridge, 1991); (3) The functional activation of integrins can be associated with changes in their micro-clustering. This has been analyzed most carefully with respect to changes in function of the CD11b/CD18 integrin (MAC-1, CR3) on granulocytes. Inactive CD11b/CD18 is randomly distributed, while activation is associated with its movement into small clusters (Detmers *et al.*, 1987). Micro-clustering of CD11b/CD18 correlates very closely with functional activity of CD11b/CD18 with respect to many parameters including kinetics, and pharmacological inhibitors. Interestingly, its interaction with its ligand C3bi was most efficient when that ligand pre-existed in similarly sized clusters, containing 6–10 molecules (Hermanowski-Vosatka *et al.*, 1988).

3.4.2 Conformational Changes

Conformational flexibility is common in receptor molecules and changes in conformation of adhesion molecules appears to be a common mechanism for regulation of ligand binding. Changes which accompany the activated, binding state of adhesion molecules were detected by mAb binding; epitopes disappeared on ligand-occupied receptors (Frelinger *et al.*, 1988), or new induced epitopes appeared (Dransfield and Hogg, 1989). A mAb can differentially activate a $\beta 1$-type integrin expressed by developing retinal neurons: it promotes cell attachment to laminin but blocks it on vitronectin (Neugebauer and Reichardt, 1991).

Some of the newly induced epitopes, like the PAC1 on $\alpha IIb\beta_3$, are located at or near the binding pocket, because binding of the mAb also competed with ligands like glycine-arginine-glycine-aspartic acid serine-proline (GRGDSP) (Shattil *et al.*, 1985). The model is that a highly specific conformational change is occurring to facilitate ligand binding. In support of this model, it is observed that antigen binding Fab fragments which bind to this new conformation can also induce function (Frelinger *et al.*, 1991); apparently they shift the conformational equilibrium of $\alpha IIb\beta 3$ from the resting to the activated state, which results in fibrinogen binding and platelet aggregation.

Ligands themselves influence the binding capacity of the receptor. Arginine-glycine-asparagine (RGD) containing peptides stimulate neutrophil phagocytosis through the leucocyte response integrin (Gresham *et al.*, 1989). RGD-peptides can activate platelet integrin $\alpha IIb\beta_3$ to become a high affinity fibrinogen binding receptor (Duo *et al.*, 1991).

3.4.3 Regulatory Association with other Molecules

Regulation of adhesion can also occur by changes in the association of integrins with other molecules. In theory, these could be new associations of molecules already present, or associations with molecules generated during the regulatory event. A striking example is regulation of β_2 integrin function by integrin modulating factor (IMF) (Hermanowski-Vosatka *et al.*, 1992). IMF is an acidic, amphiphilic lipid, probably an unsaturated fatty acid or isoprenoid acid, distinct from PAF. Addition of IMF to cells or to purified CD11b/CD18 enhances binding. IMF is not present in resting PMN but is synthesised within one minute after stimulation with PMA, TNF, formyl-peptides, PAF, and it is eliminated within one hour; its appearance and disappearance from granulocytes parallel their transient adhesion. IMF thus appears to be an allosteric activator of leucocyte integrins.

This finding of lipid contribution to integrin function is consistent with an older literature regarding physical association between and functional inter-relationships between lipids and integrin. Gangliosides were shown to promote cell attachment by activating integrin? (Cheresh and Klier, 1986; Cheresh *et al.*, 1986) or $\beta_1\alpha_5$ (Stallcup, 1988). GD2 associates with the integrin in the presence of Ca^{2+}; chelation of the cation causes the dissociation of GD2 from the integrin and terminates binding. In addition, the Lipid A region of LPS and bacterial endotoxin is bound at a site distinct from RGD-binding on CD11b/CD18 and might induce a functionally active conformation (Wright and Detmers, 1988). Lipids have the necessary properties to effectively activate or modify function of adhesion molecules: they can be rapidly synthesized and eliminated and they are found physically associated with members of the integrin family (Prescott *et al.*, 1990) .

3.4.4 Divalent Cations

Divalent cations are required for many adhesive interactions, including those mediated by cadherins, lectins, some Ig family members and particularly those mediated by integrins. Our basic understanding of integrin-mediated adhesion has been advanced by studies of cation requirements outlined below. It remains to be seen whether local regulation of divalent cations at the adhesive site are used in adhesion regulation under physiological circumstances.

Integrin α-subunits contain 3–4 binding domains for divalent cations, which have homology to helix-loop-helix Ca-binding proteins (Cheresh, 1991b; Hemler, 1991). Mutations in this region abolishes cation-dependent conformational change (Loftus *et al.*, 1990), and as a result, mutant $\alpha IIB\beta_3$ is unable to bind to its fibrinogen ligand. The type of cation influences avidity dramatically: VLA-2 is a functional collagen-receptor only in the presence of Mg^{2+} (Staatz *et al.*, 1989), while binding is less avid in the presence of Ca^{2+} (Kirchhofer *et al.*, 1991). With the addition of different cations specificity of function can also be shifted: VLA-5 functions as fibronectin-receptor in the presence of Ca^{2+} or Mg^{2+}; substituting Mn for these augments affinity of VLA-5, but changes the relative specificity for short synthetic RGD-peptide compared to longer FN-fragments (Gailit and Ruoslahti, 1988). Recent findings (Dransfield *et al.*, 1992) suggest that avidly bound Ca^{2+} is responsible for maintaining the inactive state of LFA-1; functional activity is accompanied by the displacement of Ca^{2+}.

3.5 LINKS BETWEEN SIGNALS AND MECHANISMS

The foregoing sections describe signals (Section 3.3) and mechanisms (Section 3.4) involved in functional regulation of adhesion. So far it has been possible only to establish conceptual links between the two, not a detailed map of step by step biochemical events. For example, it is likely that the protein phosphorylation which is generated by the triggering molecules includes phosphorylation of either integrins or cytoskeletal elements, which thereby promotes the assembly of integrins and cytoskeleton. This plausible mechanism has not yet been fleshed out experimentally. Similarly, lateral association of molecules in the membrane, and their association with intracytoplasmic components are undoubtedly regulated by protein phosphorylation, by novel lipids and by Ca^{2+} fluxes, but a detailed picture of their role in adhesion regulation has yet to emerge.

4. *Signal Exchange as an Integral part of Adhesion*

Adhesion molecules are ideally suited to perceive the extracellular milieu and to transduce that information to the cell for determining its differentiation programme. In this section we deal briefly with the concept that engagement of adhesion molecules shapes several other cellular events. Although sometimes it is helpful to distinguish adhesion molecules from signal-transducing molecules (including the triggers or amplifiers described above), this distinction breaks down for at least two reasons. Since most receptor-ligand interactions can mediate adhesion under extreme circumstances [for example enzymes (Carter *et al.*, 1981)], it becomes somewhat arbitrary which are called adhesion molecules. Second, since most molecules involved in adhesion clearly have capacity to transduce signals, most molecules fall into both categories.

4.1 CELLULAR RESPONSES INDUCED BY ENGAGEMENT OF ADHESION MOLECULES

Cell interaction with its environment (ECM and other cells) via adhesion molecules is obviously critical in differentiation of all cells. We restrict our cells to findings of particular relevance to our immunological understanding. Engagement of natural ligands (B7 for CD28) or mAb occupancy/crosslinking (CD2, CD7, CD31) of trigger structures provide "second signals" for co-stimulation and cell proliferation for T cells when applied simultaneously with signalling through CD3/TCR (T cell receptor). Similarly, ligands of the integrin family (ICAM-1, laminin fibronectin, VCAM) also can facilitate TCR-dependent activation of T cells (Shimizu *et al.*, 1990a; van Seventer *et al.*, 1990; Damle and Aruffo, 1991). In fact, proliferation provides in some cases a more sensitive readout of receptor-ligand interaction than does adhesion.

Adhesion-induction (discussed in Section 3) is one of the functional responses to engagement of certain adhesion molecules; semantics then becomes interesting because these molecules are both triggers and effectors of adhesion. As mentioned in that section, ELAM-1 binding to its carbohydrate ligand on granulocytes induces CD11b/CD18 function on those cells (Lo *et al.*, 1991). Similarly, engagement of von Willebrandt factor (for example on exposed subendothelial collagen) by gpIb (CD42) results both in binding of the platelet and functional activation of the $\alpha IIb/\beta_3$ integrin (Roth, 1992).

Cytokine synthesis and/or release is likewise regulated by adhesive interactions. Anti-ICAM-1 mAb modulates the release of TNF-α, IFNγ and IL-1 on PHA-activated PBMC and T cells (Geissler *et al.*, 1990). On monocytes, immobilized anti-β_2 mAbs trigger the synthesis of cell-associated IL-1 (Couturier *et al.*, 1990) and engagement of monocyte LFA-3, CD44 or CD45 induces secretion of IL-1 and TNF (Webb *et al.*, 1990). VLA-2 mediated adhesion of monocytes to collagen induces release of IL-1 and this response is potentiated by VLA-5-dependent adhesion to fibronectin (Pacifici *et al.*, 1991; Pacifici *et al.*, 1992). Furthermore, occupancy and clustering of VLA-5 induces collagenase and stromelysin which facilitate migration (Werb *et al.*, 1989).

Adhesion may restrict the oxidative burst to specific sites. TNF and *N*-formyl-methionyl-leucyl-phenylalinine (FMLP) cause oxidative burst only when granulocytes are allowed to adhere, but not in suspension (Nathan, 1987). Naturally, PMNs from β_2 deficient LAD patients lack this regulative pathway (Nathan *et al.*, 1989). P-selectin can also interfere with superoxide anion release. (Wong *et al.*, 1991). Oxidative burst in eosinophils is also markedly influenced by the nature of ECM or cell to which the cells adhere (Dri *et al.*, 1991).

4.2 BROAD RANGE OF BIOCHEMICAL CHANGES AS A RESULT OF ADHESION

Given the diverse functional responses to adhesion, it is not surprising that the biochemical changes are correspondingly diverse. Interaction of VLA-5 on CD4+T cells with fibronectin induces the transcription factor AP-1, which, among other functions, influences expression of other adhesion molecules (Yamada *et al.*, 1991). Macrophage adhesion induces the level of mRNA for a large number of proteins; some are specifically induced by macrophage binding to collagen, others specifically by binding to fibronectin and others by both (Sporn *et al.*, 1990).

Protein tyrosine phosphorylation is also induced by ligation of adhesion molecules. Ligation of VLA-4 (either β_1 or α_4 chain) with mAbs or the CS1 ligand fragment induces specific phosphorylation of a 105 kDa protein. As would be expected, based on the differences between CD3 and VLA-4 functions, a different set of proteins is phosphorylated following VLA-4 crosslinking than following CD3 crosslinking (Nojima *et al.*, 1992). A 120 kDa protein was phosphorylated upon engagement with an integrin ligand (Guan *et al.*, 1991), and also a 130 kDa protein was found to be preferentially phosphorylated after clustering β_1 chains with antibody treatment (Kornberg *et al.*, 1991).

CD11b/CD18-integrin-occupancy triggers oscillations of cytosolic free Ca^{2+} in adherent human neutrophils, which is essential for subsequent TNF-induced degranulation (Richter *et al.*, 1990). Clustering of $\alpha_5\beta_1$ by insoluble fibronectin or engagement of LFA-1 on lymphocytes changes intracellular pH through activation of the Na^+/H^+ anti-porter (Schwartz *et al.*, 1991); cell anchorage also changes pH (Schwartz *et al.*, 1989; Ingber *et al.*, 1990; Schwartz *et al.*, 1990a; Schwartz *et al.*, 1990b). Anti-β_1-mAb (together with anti-CD2 or anti-CD3 mAbs) increases cAMP in T cells within 10–30 min and thus has suppressor effects on the helper function of CD4+T cells (Groux *et al.*, 1989).

5. *Termination of Adhesion*

Transiency is a key feature of the regulated adhesion for mobile leucocytes. Adhesiveness of a cell can be abolished either by physically loosing or changing the adhesion receptors or by returning the effector molecules to their inactive state.

5.1 MECHANISMS WHICH PHYSICALLY ALTER THE ADHESION MOLECULES

Multiple mechanisms can provide for removal/alteration of the adhesion molecules itself:

1. Internalization: clearing the receptor-ligand anchor can be achieved through internalization and subsequent degradation. MAbs to the fibronectin receptor were endocytosed and rapidly associated with coated pits (Raub and Kuentzel, 1989).
2. Recycling: 25–30% of hamster β_1 (Raub and Kuentzel, 1989) and $\alpha_5\beta_2$ complex (Sczekan and Juliano, 1990) appeared to be recycled. Both endo- and exocytosis have been demonstrated for VLA-5 from hamster, with 10–14% of the total inside during recycling. This process may transport receptors to the leading edge (Bretscher, 1989), and thereby sustain migration. That $\alpha_3\beta_1$, $\alpha_4\beta_1$ recycle much more slowly than $\alpha_5\beta_1$ (or not at all) indicates the decisive role of the α-chains (Bretscher, 1992).
3. Shedding: (limited) proteolysis has the advantage and disadvantage that it is less rapidly reversible. It also creates soluble product which may be important. L-selectin

(MEL-14) was lost acutely with granulocyte activation (Kishimoto *et al.*, 1989) paralleled by increased function of CD11b/CD18. PAF in the fluid phase also causes shedding of L-selectin. (Smith *et al.*, 1991) TNF-α-induced down-regulation of CD43 and CD44 on neutrophils was also explained by proteolysis (Campanero *et al.*, 1992). Fibronectin-degrading proteases are localized to the membrane of transformed cells, suggesting that targeted proteolysis may disrupt contacts between integrins and ECM proteins (Chen *et al.*, 1984; Chen and Chen, 1987). The susceptibility to proteolysis may change during the interaction itself. In the presence of RGD peptide, aIIbβ_3 undergoes an increased Stokes radius and thus its susceptibility to protease digestion increases (Phillips *et al.*, 1989).

5.2 TERMINATION OF BIOCHEMICAL SIGNALS

Termination of the active period of adhesion can result from decline in the triggering signals and in the activating biochemical pathways which are responsible for maintaining the functionally active state. Many signalling pathways are endowed with the capacity for desensitization of the signalling molecule. For example the CD3-soluble ligand pathway may be terminated by internalization of the whole complex. Most of the intracellular signals outlined above are self-limited; Ca^{2+} is taken up into intracellular depots and protein phosphorylation is removed by endogenous phosphatases. Thus, de-adhesion may not depend on a specific "termination signal"; on the contrary, sustained adhesion for leucocytes may often require additional signals.

Active signalling for termination of adhesion may also occur. For example, we wonder if there is a negative feedback loop since integrin engagement can induce cAMP (Groux *et al.*, 1989) and dbcAMP diminishes the induction of integrin adhesion (Dustin and Springer, 1989; Shimizu *et al.*, 1992b). In a different vein, IFNγ activation of macrophages enhances their lytic potential but decreases the binding activity of CR1 and CR3 (Wright *et al.*, 1986).

5.3 DE-ADHESION ASSOCIATED WITH CHANGES IN DIFFERENTIATION STATE

In contrast to the foregoing processes, which are critical to rapid de-adhesion, it is apparent that reduced adhesion also accompanies slower process of differentiation. Several examples will suffice:

1. Thymocyte differentiation: mature thymocytes lose their capacity to bind to fibronectin and may thereby be allowed to escape into the periphery (Utsumi *et al.*, 1991).
2. Environmental stress: ultraviolet B radiation selectively modulates ICAM-1 expression on murine epidermal Langerhans cells, and inhibits their function even at a low dose (Tang and Udey, 1991).
3. Malignant transformation: certain tumors, particularly Burkitt's lymphoma, have decreased expression of adhesion molecules which may facilitate their evasion of immune surveillance (Gregory, *et al.*, 1988; Gregory and Rickinson, 1988; Makgoba *et al.*, 1989). Down-regulation of E-cadherin in transformed epithelial cells results in increased invasiveness (Vleminckx *et al.*, 1991).

6. *Summary*

The specificity of adhesion is determined in a combinatorial fashion at many levels, combinations of: DNA-binding proteins acting on individual genes, exons into mature message, chains into molecules, molecules into complexes, and complexes into co-ordinated sequences of steps. Each cell-type (subset) has it own characteristic repertoire of adhesion molecules (none of them has the entire spectrum), which enables limited but still significant redundancy in available pathways.

The "personality" of individual leucocytes derives substantially from their adhesion molecules and how they regulate them. Granulocytes are virtually dependent on the β_2 family of integrins, as documented by the clinical deficiencies in the LAD (Anderson and Springer, 1987). In contrast, T cell and B cell function in this disease is relatively normal, reflecting alternative utilization of VLA-4 and presumably other molecules. Within peripheral T cells, expression of adhesion molecules is differentially regulated in different subsets, as are the trigger molecules which regulate adhesive function (Shimizu *et al.*, 1990b; Shimizu *et al.*, 1992b).

It important to understand that adhesion is not an "all or none" phenomenon, but is temporally and spatially (during migration) quite dynamic (Dustin and Springer, 1991). Likewise, the adhesion mechanisms at the single cell level must be continuously changing. This can be achieved by on/off regulation of individual molecules and individual associations between molecules. The degree of complexity in regulation of adhesion no doubt reflects the importance of adhesion in immunological function. Although simplifying generalization will continue to emerge, the overall complexity will be great. It is an area with great potential for pharmacological intervention.

7. *Acknowledgements*

We thank Y. Shimizu and D. Roberts for critical review. We thank A. Bernard, E. Butcher, D. Cheresh, M. Hemler, T. Kishimoto, C. Morimoto, M. Patarroyo, F. Sanchez-Madrid, S. Wright, K. Yamada, G. Zimmerman for reprints. TS is supported by a Fogarty Visiting Fellowship.

8. *References*

Abo, T., Cooper, M.D. and Balch, C.M. (1982). Characterization of HNK-1+ (Leu-7) human lymphocytes. I. Two distinct phenotypes of human NK cells with different cytotoxic capability. J. Immunol. 129, 1752–1757.

Adams, D.H., Hathaway, M., Shaw, J., Burnett, D., Elias, E. and Strain, A.J. (1991). Transforming growth factor-β induces human T lymphocyte migration *in vitro*. J. Immunol. 147, 609–612.

Akiyama, S.K., Yamada, S.S. and Yamada, K.M. (1989). Analysis of the role of glycosylation of the human fibronectin receptor. J. Biol. Chem. 264, 18 011–18 018.

Albelda, S.M., Mette, S.A., Elder, D.E., Stewart, R., Damjanovich, L., Herlyn, M. and Buck, C.A. (1990). Integrin distribution in malignant melanoma: association of the beta 3 subunit with tumor progression. Cancer Res. 50, 6757–6764.

Anderson, D.C. and Springer, T.A. (1987). Leucocyte adhesion deficiency: an inherited defect in the Mac-1, LFA-1, and p150,95 glycoproteins. Ann. Rev. Med. 38, 175–194.

Aruffo, A., Kolanus, W., Walz, G., Fredman, P. and Seed, B. (1991). CD62/P-selectin recognition of myeloid and tumor cell sulfatides. Cell 67, 35–44.

Barrett, T.B., Shu, G. and Clark, E.A. (1991). CD40 signalling activates CD11a/CD18 (LFA-1)-mediated adhesion in B cells. J. Immunol. 146, 1722–1729.

Bednarczyk, J.L. and McIntyre, B.W. (1990). A monoclonal antibody to VLA-4 α-chain (Cdw49d) induces homotypic lymphocyte aggregation. J. Immunol. 144(3), 777–784.

Berg, E.L., Yoshino, T., Rott, L.S., Robinson, M.K., Warnock, R.A., Kishimoto, T.K., Picker, L.J. and Butcher, E.C. (1991). The cutaneous lymphocyte antigen is a skin lymphocyte homing receptor for the vascular-lectin endothelial cell-leucocyte adhesion molecule. J. Exp. Med. 174.

Bevilacqua, M.P., Stengelin, S., Gimbrone, M.A., Jr. and Seed, B. (1989). Endothelial leucocyte adhesion molecule 1: an inducible receptor for neutrophils related to complement regulatory proteins and lectins. Science 243, 1160–1164.

Bierer, B.E., Bogart, R.E. and Burakoff, S.J. (1990). Partial deletions of the cytoplasmic domain of CD2 result in a partial defect in signal transduction. J. Immunol. 144(3), 785–789.

Bjerknes, M., Cheng, H. and Ottaway, C.A. (1986). Dynamics of lymphocyte-endothelial interactions *in vivo*. Science 231, 402–405.

Bretscher, M.S. (1989). Endocytosis and recycling of the fibronectin receptor in CHO cells. EMBO J. 8, 1341–1348.

Bretscher, M.S. (1992). Circulating integrins: $\alpha_5\beta_1$, $\alpha_6\beta_4$ and Mac-1, but $\alpha_3\beta_1$, $\alpha_4\beta_1$ or LFA-1. EMBO J. 11, 405–410.

Brown, G.E., Reed, E.B. and Lanser, M.E. (1991). Neutrophil CR3 expression and specific granule exocytosis are controlled by different signal transduction pathways. J. Immunol. 147, 965–971.

Brown, T.A., Bouchard, T., St John, T., Wayner, E. and Carter, W.G. (1991). Human keratinocytes express a new CD44 co-reprotein (CD44E) as a heparan-sulfate intrinsic membrane proteoglycan with additional exons. J. Cell Biol. 113, 207–211.

Brundage, R.A., Fogarty, K.E., Tuft, R.A. and Fay, F.S. (1991). Calcium gradients underlying polarization and chemotaxis of eosinophils. Science 254, 703–706.

Burn, P., Kupfer, A. and Singer, S.J. (1988). Dynamic membrane-cytoskeletal interactions: specific association of integrin and talin arises *in vivo* after phorbol ester treatment of peripheral blood lymphocytes. Proc. Natl Acad. Sci. USA 85, 497–501.

Bussolino, F., Wang, J.M., Defilippi, P., Turrini, F., Sanavio, F., Edgell, C.-J.S., Aglietta, M., Arese, P. and Mantovani, A. (1989). Granulocyte- and granulocyte-macrophage-colony stimulating factors induce human endothelial cells to migrate and proliferate. Nature 337, 471–473.

Butcher, E.C. (1991). Leucocyte-endothelial cell recognition: three (or more) steps to specificity and diversity. Cell 67, 1033–1036.

Campanero, M.R., Pulido, R., Ursa, M.A., Rodriguez-Moya, M., De Landazuri, M.O. and Sanchez-Madrid, F. (1990). An alternative leucocyte homotypic adhesion mechanism, LFA-1/ICAM-1 independent, triggered through the human VLA-4 integrin. J. Cell Biol. 110, 2157–2165.

Campanero, M.R., Pulido, R., Alonso, J.L., Pivel, J.P., Pimentel-Munioz, F.X., Fresno, M. and Sanchez-Madrid, F. (1991). Down-regulation by tumor necrosis factor-alpha of neutrophil cell surface expression of the sialophorin CD43 and the hyaluronate receptor CD44 through a proteolytic mechanism. Eur. J. Immunol. 21, 3045–3048.

Carter, W.G., Rauvala, H. and Hakomori, S.I. (1981). Studies on cell adhesion and recognition. (II) The kinetics of cell adhesion and cell spreading on surfaces coated with carbohydrate-reactive proteins (glycosidases and lectins) and fibronectin. J. Cell Biol. 88, 138–148.

Cerf-Bensussan, N., Begue, B., Gagnon, J. and Meo, T. (1992). The human intraepithelial lymphocyte marker HML-1 is an integrin consisting of a beta7 subunit associated with a distinctive alpha chain. Eur. J. Immunol. 22, 273–277.

Chan, B.M.C., Matsuura, N., Takada, Y., Zetter, B.R. and Hemler, M.E. (1991a). *In vitro* and *in vivo* consequences of VLA-2 expression on rhabdomyosarcoma cells. Science 251, 1600–1602.

Chan, B.M.C., Wong, J.G.P., Rao, A. and Hemler, M.E. (1991b). T cell receptor-dependent, antigen-specific stimulation of a murine T cell clone induces a transient, VLA protein-mediated binding to extracellular matrix. J. Immunol. 147, 398–404.

Chen, J.M. and Chen, W.T. (1987). Fibronectin-degrading proteases from the membranes of transformed cells. Cell 48, 193–203.

Chen, W.T., Olden, K., Bernard, B.A. and Chu, F.F. (1984). Expression of transformation-associated protease(s) that degrade fibronectin at cell contact sites. J. Cell Biol. 98, 1546–1555.

Cheresh, D.A. (1991a). Structure, function and biological properties of integrin $\alpha_v\beta_3$ on human melanoma cells. Cancer Metastasis Rev. 10, 3–10.

Cheresh, D.A. (1991b). In "*Advances in Molecular and Cell Biology, Vol. 7*".

Cheresh, D.A. and Klier, F.G. (1986). Disialoganglioside GD2 distributes preferentially into substrate-associated microprocesses on human melanoma cells during their attachment to fibronectin. J. Cell Biol. 102, 1887–1897.

Cheresh, D.A., Pierschbacher, M.D., Herzig, M.A. and Mujoo, K. (1986). Disialogangliosides GD2 and GD3 are involved in the attachment of human melanoma and neuroblastoma cells to extracellular matrix proteins. J. Cell Biol. 102, 688–696.

Cierniewski, C.S., Krzeslowska, J., Pawlowska, Z., Witas, H. and Meyer, M. (1989). Palmitylation of the glycoprotein IIb–IIIa complex in human blood platelets. J. Biol. Chem. 264, 12 158–12 164.

Clark, R.A., Wikner, N.E., Doherty, D.E. and Norris, D.A. (1988). Cryptic chemotactic activity of fibronectin for human monocytes resides in the 120 kDa fibroblastic cell-binding fragment. J. Biol. Chem. 263, 12 115–12 123.

Coeffier, E., Joseph, D. and Vargraftig, B.B. (1991). Activation of Guinea pig eosinophils by human recombinant IL-5. J. Immunol. 147, 2595–2602.

Coller, B.S. (1986). Activation affects access to the platelet receptor for adhesive glycoproteins. J. Cell Biol. 103, 451–456.

Couturier, C., Haeffner-Cavaillon, N., Weiss, L., Fischer, E. and Kazatchkine, M.D. (1990). Induction of cell-associated interleukin 1 through stimulation of the adhesion-promoting proteins LFA-1 (CD11a/CD18) and CR3 (CD11b/CD18) of human monocytes. Eur. J. Immunol. 20, 999–1005.

Crabtree, G.R. (1989). Contingent genetic regulatory events in T lymphocyte activation. Science 243, 355–361.

Cybulsky, M.I., Fries, J.W.U., Williams, A.J., Sultan, P., Eddy, R., Byers, M., Shows, T., Gimbrone, M.A. and Collins, T. (1991). Gene structure, chromosomal location, and basis for alternative mRNA splicing of the human VCAM1 gene. Proc. Natl Acad. Sci. USA 8, 7859–7863.

Damle, N.K. and Aruffo, A. (1991). Vascular cell adhesion molecule 1 induces T-cell antigen receptor-dependent activation of CD4+T lymphocytes. Proc. Natl Acad. Sci. USA 88, 6403–6407.

De la Fuente, M., Del Rio, M., Ferrandez, M.D. and Hernanz, A. (1991). Modulation of phagocytic function in murine peritoneal macrophages by bombesin, gastrin-releasing peptide and neuromedin C. Immunology 73, 205–211.

Defilippi, P., van Hinsberg, V., Bertolotto, A., Rossino, P., Silengo, L. and Tarone, G. (1991). Differential distribution and modulation of expression of $\alpha_1\beta_1$ integrin on human endothelial cells. J. Cell Biol. 114, 855–863.

Detmers, P.A., Wright, S.D., Olsen, E., Kimball, B. and Cohn, Z.A. (1987). Aggregation of complement receptors on human neutrophils in the absence of ligand. J. Cell Biol. 105, 1137–1145.

Diamond, M.S., Staunton, D.E., Marlin, S.D. and Springer, T.A. (1991). Binding of the integrin Mac-1 CD11b/CD18) to the third immunoglobulin-like domain of ICAM-1 (CD54) and its regulation by glycosylation. Cell 65, 961–971.

Dower, S.K., Titus, J.A. and Segal, D.M. (1982). The binding of multivalent ligands to cell surface receptors. In "*Cell surface dynamics: Concepts and models*" (eds A. S. Perelson, C. Delisi, and F. W. Wiegel), pp. 277–328. Marcel Dekker, Inc., New York and Basel.

Dransfield, I. and Hogg, N. (1989). Regulated expression of Mg^{2+} binding epitope on leukocyte integrin alpha subunits. EMBO J. 8, 3759–3765.

Dransfield, I., Cabanas, C., Craig, A. and Hogg, N. (1992). Divalent cation regulation of the function of the leukocyte integrin LFA-1. J.Cell Biol. 116, 219–226.

Dri, P., Cramer, R., Spessotto, P., Romano, M. and Patriarca, P. (1991). Eosinophil activation on biologic surfaces: production of O^-_2 in response to physiological soluble stimuli is differentially modulated by extracellular matrix components and endothelial cells. J. Immunol. 147, 613–620.

Duo, X., Plow, E.F., Frelinger, III, A.L., O'Toole, T.E., Loftus, J.C. and Ginsberg, M.H. (1991). Ligands "activate" integrin αIIb-β3 (platelet GPIIa-IIa). Cell 65, 409–416.

Dustin, M.L. and Springer, T.A. (1989). T-cell receptor cross-linking transiently stimulates adhesiveness through LFA-1. Nature 341, 619–624.

Dustin, M.L. and Springer, T.A. (1991). Role of lymphocyte adhesion receptors in transient interactions and cell locomotion. Ann. Rev. Immunol. 9, 27–66.

Elliott, M., Gamble, J.R., Park, L.S., Vadas, M.A. and Lopez, A.F. (1991). Inhibition of human monocyte adhesion by interleukin-4. Blood 77, 2739–2745.

Erle, D.J., Ruegg, C., Sheppard, D. and Pytela, R. (1991). Complete amino acid sequence of an integrin beta subunit (b7) identified in leucocytes. J. Biol. Chem. 266(17), 11 009–11 016.

Finger, L.R., Kagan, J., Christopher, G., Kurtzberg, J., Hershfield, M.S., Nowell, P.C. and Croce, C.M. (1989). Involvement of the TCL5 gene on human chromosome 1 in T cell leukemia and melanoma. Proc. Natl Acad. Sci. USA 86, 5039–5043.

Forsyth, K.D. (1991). Anti-CD-9 antibodies augment neutrophil adherence to endothelium. Immunology 72, 292–296.

Frelinger, A.L., III, Lam, S.C.-T., Plow, E.F., Smith, M.A., Loftus, J.C. and Ginsberg, M.H. (1988). Occupancy of an adhesive glycoprotein receptor modulates expression of an antigenic site involved in cell adhesion. J. Biol. Chem. 263, 12 397–12 402.

Frelinger, A.L., Du, X., Plow, E.F. and Ginsberg, M.H. (1991). Monoclonal antibodies to ligand-occupied conformers of integrin $\alpha II\beta_3$ (glycoprotein IIb–IIIa) alter receptor affinity, specificity and function. J. Biol. Chem. 266, 17 106–17 111.

Gailit, J. and Ruoslahti, E. (1988). Regulation of the fibronectin receptor affinity by divalent cations. J. Biol. Chem. 263, 12 927–12 932.

Gamble, J.R. and Vadas, M.A. (1988). Endothelial adhesiveness for blood neutrophils is inhibited by transforming growth factor-beta. Science 242, 97–99.

Geissler, D., Gaggl, S., Most, J., Greil, R., Herold, M. and Dierich, M. (1990). A monoclonal antibody directed against the human intercellular adhesion molecule (ICAM-1) modulates the release of tumor necrosis factor alpha, interferon gamma and interleukin 1. Eur. J. Immunol. 20, 2591–2596.

Giancotti, F.G. and Ruoslahti, E. (1990). Elevated levels of the alpha5beta1 fibronectin receptor suppress the transformed phenotype of chinese hamster ovary cells. Cell 60, 849–859.

Glabe, C.G., Grabel, L.B., Vacquier, V.D. and Rosen, S.D. (1982). Carbohydrate specificity of sea urchin sperm bindin: a cell surface lectin mediating sperm-egg adhesion. J. Cell Biol. 94, 123–128.

Gladson, C.L. and Cheresh, D.A. (1991). Glioblastoma expression of vitronectin and the $\alpha_v\beta_3$ integrin: adhesion mechanism for transformed glial cells. J. Clin. Invest. 88: 1924–1932 .

Grabel, L.B. (1984). Isolation of a putative cell adhesion mediating lectin from teratocarcinoma stem cells and its possible role in differentiation. Cell Differ. 15, 121–124.

Graber, N., Gopal, T.V., Wilson, D., Beall, L.D., Polte, T. and Newman, W. (1990). T-cells bind to cytokine-activated endothelial cells via a novel, inducible sialoglycoprotein and ELAM-1. J. Immunol. 145, 819–830.

Gregory, C.D. and Rickinson, A.B. (1988). Adhesion molecules, immunosurveillance, and B-cell malignancies. Lancet 1/30/88, 248.

Gregory, C.D., Murray, R.J., Edwards, C.F. and Rickinson, A.B. (1988). Downregulation of cell adhesion molecules LFA-3 and ICAM-1 in Epstein-Barr virus-positive Burkitt's lymphoma underlies tumour cell escape from virus-specific T cell surveillance. J. Exp. Med. 167, 1811–1824.

Gresham, H.D., Goodwin, J.L., Allen, P.M., Anderson, D.C. and Brown, E.J. (1989). A novel member of the integrin receptor family mediates arg-gly-asp-stimulated neutrophil phagocytosis. J. Cell Biol. 108, 1935–1943.

Groux, H., Huet, S., Valentin, H., Pham, D. and Bernard, A. (1989). Suppressor effects and cyclic AMP accumulation by the CD29 molecule of CD4+ lymphocytes. Nature 339, 152–154.

Guan, J.L., Trevitchick, J.E. and Hynes, R.O. (1991). Fibronectin/integrin interaction induces tyrosine phosphorylation of a 120 kDa protein. Cell Reg. 2, 951–964.

Heino, J., Ignotz, R.A., Hemler, M.E., Crouse, C. and Massague, J. (1989). Regulation of cell adhesion receptors by transforming growth factor-beta. Concomitant regulation of integrins that share a common β_1 subunit. J. Biol. Chem. 264, 380–388.

Hemler, M.E. (1991). In "*Receptors for Extracellular Matrix*" (eds R. P. Mecham and J. A. McDonald), pp 255–299. Academic Press, San Diego.

Hemler, M.E., Glass, D., Coblyn, J.S. and Jacobson, J.S. (1986). Very late activation antigens on rheumatoid synovial fluid T lymphocytes. Association with stage of T cell activation. J. Clin. Invest. 78, 696–702.

Hermanowski-Vosatka, A., Detmers, P.A., Gotze, O., Silverstein, S.C. and Wright, S.D. (1988). Clustering of ligand on the surface of a particle enhances adhesion to receptor-bearing cells. J. Biol. Chem. 263, 17 822–17 827.

Hermanowski-Vosatka, A., van Strijp, J.A.G., Swiggard, W.J. and Wright, S.D. (1992). Integrin modulating factor-1: a lipid that alters the function of leukocyte integrins. Cell 68, 341–352.

Hession, C., Osborn, L., Goff, D., Chi-Rosso, G., Vassallo, C., Pasek, M., Pittack, C., Tizard, R., Goelz, S., McCarthy, K., Hopple, S. and Lobb, R. (1990). Endothelial leukocyte adhesion molecule 1: Direct expression cloning and functional interactions. Proc. Natl Acad. Sci. USA 87, 1673–1677.

Hibbs, M.L., Xu, H., Stackler, S.A. and Springer, T.A. (1991a). Regulation of adhesion to ICAM-1 by the cytoplasmic domain of LFA-1 integrin beta subunit. Science 251, 1611–1613.

Hibbs, M.L., Jakes, S., Stacker, S.A., Wallace, R.W. and Springer, T.A. (1991b). The cytoplasmic domain of the integrin lymphocyte function-associated antigen 1 β subunit: sites required for binding to intercellular adhesion molecule 1 and the phorbol ester-stimulated phosphorylation site. J. Exp. Med. 174, 1227–1238.

Huang, T.T.F., Jr., Ohzu, E. and Yanagimachi, R. (1982). Evidence suggesting that L-fucose is part of a recognition signal for sperm-zona pellucida attachment in mammals. Gam. Res. 5, 355–361.

Ignotz, R.A., Heino, J. and Massague, J. (1989). Regulation of cell adhesion receptors by transforming growth factor-beta. Regulation of vitronectin receptor and LFA-1. J. Biol.Chem. 264, 389–392.

Ingber, D.E., Prusty, D., Frangioni, J.V., Cragoe, E.J., Jr., Lechene, C. and Schwartz, M.A. (1990). Control of intracellular pH and growth by fibronectin in capillary endothelial cells. J. Cell Biol. 110, 1803–1811.

Kajiji, S., Tamura, R.N. and Quaranta, V. (1989). A novel integrin (ōEβ4) from human epithelial cells suggests a fourth family of integrin adhesion receptors. EMBO J. 8, 673–680.

Kansas, G.S. and Tedder, T.F. (1991). Transmembrane signals generated through MHC class II, CD19, CD20, CD39, and CD40 antigens induce LFA-1-dependent and independent adhesion in human B cells through a tyrosine kinase-dependent pathway. J. Immunol. 147, 4094–4102.

Kilshaw, P.J. and Murant, S.J. (1991). Expression and regulation of beta (βpi) integrins on mouse lymphocytes: relevance to the mucosal immune system. Eur. J. Immunol. 21, 2591–2597.

Kirchhofer, D., Grzesiak, J. and Pierschbacher, M.D. (1991). Calcium as potential physiological regulator of integrin-mediated cell adhesion. J. Biol. Chem. 266, 4471–4477.

Kishimoto, T.K., Jutila, M.A., Berg, E.L. and Butcher, E.C. (1989). Neutrophil Mac-1 and MEL-14 adhesion proteins inversely regulated by chemotactic factors. Science 245, 1238–1241.

Klausner, R.D. and Samelson, L.E. (1991). T cell antigen receptor activation pathways: the tyrosine kinase connection. Cell 64, 875–878.

Koopman, G., van Kooyk, Y., De Graaff, M., Meyer, C.J.L.M., Figdor, C.G. and Pals, S.T. (1990). Triggering of the CD44 antigen on T lymphocytes promotes T cell adhesion through the LFA-1 pathway. J. Immunol. 145, 3589–3593.

Kornberg, L.J., Earp, H.S., Turner, C.E., Prockop, C. and Juliano, R.L. (1991). Signal transduction by integrins: increased protein tyrosine phosphorylation caused by clustering of beta1 integrins. Proc. Natl Acad. Sci. USA 88, 8392–8396.

Kruse, J., Keilhauer, G., Faissner, A., Timpi, R. and Schachner, M. (1985). The J1 glycoprotein – a novel nervous system cell adhesion molecule of the L2/HNK-1 family. Nature 316, 146.

Kuijpers, T.W., Hakkert, B.C., Hoogerwerf, M., Leeuwenberg, J.F. and Roos, D. (1991). Role of endothelial leucocyte adhesion molecule-1 and platelet-activating factor in neutrophil adherence to IL-1-prestimulated endothelial cells: endothelial leucocyte adhesion molecule-1-mediated CD18 activation. J. Immunol. 147, 1369–1376.

Kunemund, V., Jungalwala, F.B., Fischer, G., Chou, D.K.H., Keilhauer, G. and Schachner, M. (1988). The L2/HNK-1 carbohydrate of neural cell adhesion molecules is involved in cell interactions. J. Cell Biol. 106, 213–223.

Lanir, N., Ciano, P.S., Van de, Waver L., McDonagh, J., Dvorak, A.M. and Dvorak, H.F. (1988). Macrophage migration in fibrin gel matrices. (II). Effects of clotting factor XIII, fibronectin, and glycosaminoglycan content on cell migration. J. Immunol. 140, 2340–2349.

Lauener, R.P., Geha, R.S. and Vercelli, D. (1990). Engagement of the monocyte surface antigen CD14 induces lymphocyte function-associated antigen-1/intercellular adhesion molecule-1-dependent homotypic adhesion. J. Immunol. 145, 1390–1394.

Lesley, J., He, Q., Miyake, K., Hamann, A., Hyman, R. and Kincade, P.W. (1992). Requirements for hyaluronic acid binding by CD44: a role for the cytoplasmic domain and activation by antibody. J. Exp. Med. 175, 257–266.

Lo, S.K., Lee, S., Ramos, R.A., Lobb, R., Rosa, M., Chi-Rosso, G. and Wright, S.D. (1991). Endothelial-leucocyte adhesion molecule 1 stimulates the adhesive activity of leucocyte integrin CR3 (CD11b/CD18, Mac-1, $\alpha_m\beta_2$) on human neutrophils. J. Exp. Med. 173, 1493–1500.

Loftus, J.C., O'Toole, T.E., Plow, E.F., Glass, A., Frelinger, A.L. and Ginsberg, M.H. (1990). A β_3 integrin mutation abolishes ligand binding and alters divalent cation dependent conformation. Science 249, 915–918.

Lorant, D.E., Patel, K.D., McIntyre, T.M., McEver, R.P., Prescott, S.M. and Zimmerman, G.A. (1991). Co-expression of GMP-140 and PAF by endothelium stimulated by histamine and thrombin: a juxtacrine system for adhesion and activation of neutrophils. J. Cell Biol. 115, 223–234.

Lotan, R. and Raz, A. (1988). Endogenous lectins as mediators of tumor cell adhesion. J. Cell Biochem. 37, 107–117.

Lowe, J.B., Stoolman, L.M., Nair, R.P., Larsen, R.D., Berhend, T.L. and Marks, R.M. (1990). ELAM-1-dependent cell adhesion to vascular endothelium determined by a transfected human fucosyltransferase cDNA. Cell 63, 475–484.

Makgoba, M.W., Sanders, M.E. and Shaw, S. (1989). The CD2-LFA-3 and LFA-1-ICAM-1 pathways: relevance to T-cell recognition. Immunol Today 10, 417–422.

Montgomery, K.F., Osborn, L., Hession, C., Tizard, R., Goff, D., Vassallo, C., Tarr, P.I., Bomszstyk, K., Lobb, R., Harlan, J.M. and Pohlman, T.H. (1991). Activation of endothelial-leukocyte adhesion molecule 1 (ELAM-1) gene transcription. Proc. Natl Acad. Sci. USA 88, 6523–6527.

Nathan, C.F. (1987). Neutrophil activation on biological surfaces. J. Clin. Invest. 80, 1550–1560.

Nathan, C. and Sporn, M. (1991). Cytokines in context. J. Cell Biol. 113, 981–986.

Nathan, C., Srimal, S., Farber, C., Sanchez, E., Kabbash, L., Asch, A., Gailit, J. and Wright, S.D. (1989). Cytokine-induced respiratory burst of human neutrophils: dependence on extracellular matrix proteins and CD11/CD18 integrins. J. Cell Biol. 109, 1341–1349.

Neugebauer, K.M. and Reichardt, L.F. (1991). Cell-surface regulation of β_1-integrin activity on developing retinal neurons. Nature 350, 68–71.

Nojima, Y., Rothstein, D.M., Sugita, K., Schlossman, S.F. and Morimoto, C. (1992). Ligation of VLA-4 on T cells stimulates tyrosine phosphorylation of a 105 kDa protein. J. Exp. Med.

Nybroe, O. and Bock, E. (1990). Structure and function of the neural cell adhesion molecules NCAM and L1. Adv. Exp. Med. Biol. 265, 185–196.

O'Rourke, A.M., Rogers, J. and Mescher, M.F. (1990. Activated CD8 binding to class I protein mediated by the T-cell receptor results in signalling. Nature 346, 187–189.

O'Toole, T.E., Mandelman, D., Forsyth, J., Shattil, S.J., Plow, E.F. and Ginsberg, M.H. (1991). Modulation of the affinity of integrin alpha IIb beta3 (GPiib-iiia) by the cytoplasmic domain of alpha IIb. Science 254, 845–847.

Oppenheim, J.J., Zachariae, C.O.C., Mukaida, N. and Matsushima, K. (1991). Properties of the novel pro inflammatory supergene "intercrine" cytokine family. Ann. Rev. Immunol. 9, 617–648.

Osborn, L. (1990). Leucocyte adhesion to endothelium in inflammation. Cell 62, 3–6.

Oz, O.K., Campbell, A. and Tao, T.W. (1989). Reduced cell adhesion to fibronectin and laminin is associated with altered glycosylation of beta 1 integrins in a weakly metastatic glycosylation mutant. Int. J. Cancer 44, 343–347.

Pacifici, R., Carano, A., Santoro, S.A., Rifas, L., Jeffrey, J.J., Malone, J.D., McCracken, R. and Avioli, L.V. (1991). Bone matrix constituents stimulate interleukin-1 release from human blood mononuclear cells. J. Clin. Invest. 87, 221–228.

Pacifici, R., Basilico, C., Roman, J., Zutter, M.M., Santoro, S.A. and McCracken, R. (1992). Collagen-induced release of interleukin 1 from human blood mononuclear cells. J. Clin. Invest. 89, 61–67.

Pantaleo, G., Koenig, S., Baseler, M., Lane, H.C. and Fauci, A.S. (1990). Defective clonogenic potential of CD8+ T lymphocytes in patients with AIDS. Expansion *in vivo* of a non-clonogenic CD3+CD8+DR+CD25− T cell population. J. Immunol. 144, 1696–1704.

Patarroyo, M. and Jondal, M. (1985). Phorbol ester-induced adhesion (binding) among human mononuclear leucocytes requires extracellular Mg^{2+} and is sensitive to protein kinase C, lipoxygenase, and ATPase inhibitors. Immunobiology 170, 305–319.

Phillips, D.R., Charo, I.F., Parise, L.V. and Fitzgerald, L.A. (1989). The platelet membrane glycoprotein IIb–IIIa complex. Blood 71, 831–843.

Phillips, D.R., Charo, I.F. and Scarborough, R.M. (1991). GPIIb-IIIa: The responsive integrin. Cell 65, 359–362.

Phillips, J.H., McKinley, L., Azuma, M., Spits, H. and Lanier, L.L. (1991). A novel β_4,õ6 integrin-associated epithelial cell antigen involved in natural killer cell and antigen-specific cytotoxic T lymphocyte cytotoxicity. J. Exp. Med. 174, 1571–1581.

Picker, L.J., Kishimoto, T.K., Smith, C.W., Warnock, R.A. and Butcher, E.C. (1991a). ELAM-1 is an adhesion molecule for skin-homing T cells. Nature 349, 796–799.

Picker, L.J., Warnock, R.A., Burns, A.R., Doerschuk, C.M., Berg, E.L. and Butcher, E.C. (1991b). The neutrophil selectin LECAM-1 presents carbohydrate ligands to the vascular selectins ELAM-1 and GMP-140. Cell 66, 921–933.

Plantefaber, L.C. and Hynes, R.O. (1989). Changes in integrin receptors on oncogenically transformed cells. Cell 56, 281–290.

Poenie, M., Tsien, R.Y. and Schmitt-Verhulst, A. (1987). Sequential activation and lethal hit measured by (Ca^{2+})i in individual cytolytic T cells and targets. EMBO J. 6, 2223–2232.

Prescott, S.M., McIntyre, T.M. and Zimmerman, G.A. (1990). The role of platelet-activating factor in endothelial cells. Thromb. Haemost. 64, 99–103.

Raub, T.J. and Kuentzel, S.L. (1989). Kinetic and morphological evidence for endocytosis of mammalian cell integrin receptors by using an anti-fibronectin receptor beta subunit monoclonal antibody. Exp. Cell Res. 184, 407–426.

Rice, G.E., Munro, J.M. and Bevilacqua, M.P. (1990). Inducible cell adhesion molecule 110 (INCAM-110) is an endothelial receptor for lymphocytes. A CD11/CD18-independent adhesion mechanism. J. Exp. Med. 171, 1369–1374.

Richter, J., Ng-Sikorski, J., Olsson, I. and Andersson, T. (1990). Tumor neurosis factor induced degranulation in adherent human neutrophils is dependent on CD11b/CD18-integrin triggered oscillations of cytosolic free Ca^{2+}. Proc. Natl Acad. Sci. USA 87, 9472–9476.

Rosen, G.D., Birkenmeier, T.M. and Dean, D.C. (1991). Characterization of the alpha-4 integrin gene promoter. Proc. Natl Acad. Sci. USA 88, 4094–4098.

Roth, G.J. (1992). Platelets and blood vessels: the adhesion event. Immunol. Today 13, 100–105.

Rothlein, R. and Springer, T.A. (1986). The requirement for LFA-1 in homotypic leukocyte adhesion stimulated by phorbol ester. J. Exp. Med. 163, 1132–1149.

Rothlein, R., Czajkowski, M. and Kishimoto, T.K. (1991). Intercellular adhesion molecule-1 (ICAM-1) in the inflammatory response. Chem. Immunol. 50: 135–142.

Rothman, B.L., Blue, M.-L., Kelley, K.A., Wunderlich, D., Mierz, D.V. and Aune, T.M. (1991). Human T cell activation by OKT3 is inhibited by a monoclonal antibody to CD44. J. Immunol. 147, 2493–2499.

Schall, T.J. (1991). Biology of the RANTES/SIS cytokine family. Cytokine 3, 165–183.

Schwartz, M.A., Both, G. and Lechene, C. (1989). Effect of cell spreading on cytoplasmic pH in normal and transformed fibroblasts. Proc. Natl Acad. Sci. USA 86, 4525–4529.

Schwartz, M.A., Cragoe, E.J., Jr. and Lechene, C.P. (1990a). pH regulation in spread cells and round cells. J. Biol. Chem. 265, 1327–1332.

Schwartz, M.A., Rupp, E.E., Frangioni, J.V. and Lechene, C.P. (1990b). Cytoplasmic pH and anchorage-independent growth induced by v-Ki-ras, v-src or polyoma middle T. Oncogene 5, 55–58.

Schwartz, M.A., Lechene, C. and Ingber, D.E. (1991). Insoluble fibronectin activates the Na/H anti-porter by clustering and immobilizing integrin $\alpha_5\beta_1$ independent of cell shape. Proc. Natl Acad. Sci. USA 88, 7849–7853.

Sczekan, M.M. and Juliano, R.L. (1990). Protease resistance of the β subunit of the hamster fibronectin receptor. Evidence for differential cleavage of membrane-bound and soluble receptor. Eur. J. Biochem. 189, 83–87.

Swell, W.A., Palmer, R.W., Spurr, N.K., Sheer, D., Brown, M.H., Bell, Y. and Crumpton, M.J. (1988). The human LFA-3 gene is located on the same chromosome band as the gene for its receptor CD2. Immunogenetics 28, 278–282.

Shattil, S.J., Hoxie, J.A., Cunnigham, M. and Brass, L.F. (1985). Changes in the platelet membrane glycoprotein IIbIIIa complex during platelet activation. J. Biol. Chem. 260, 11 107–11 114.

Shimizu, Y., van Seventer, G.A., Horgan, K.J. and Shaw, S. (1990a). Co-stimulation of proliferative responses of resting CD4+ T cells by the interaction of VLA-4 and VLA-5 with fibronectin or VLA-6 with laminin. J. Immunol. 145, 59–67.

Shimizu, Y., van Seventer, G.A., Horgan, K.J. and Shaw, S. (1990b). Roles of adhesion molecules in T cell recognition: fundamental similarities between four integrins on resting human T cells (LFA-1, VLA-4, VLA-5, VLA-6) in expression, binding, and co-stimulation. Immunol. Rev. 114, 109–143.

Shimizu, Y., van Seventer, G.A., Horgan, K.J. and Shaw, S. (1990c). Regulated expression and function of three VLA (β_1) integrin receptors on T cells. Nature 345, 250–253.

Shimizu, Y., Shaw, S., Graber, N., Gopal, T.V., Horgan, K.J., van Seventer, G.A. and Newman, W. (1991). Activation-independent binding of human memory T cells to ELAM-1. Nature 349, 799–802.

Shimizu, Y., Newman, W., Tanaka, Y. and Shaw, S. (1992a). Lymphocyte/endothelial interactions. Immunol. Today 13, 106–112.

Shimizu, Y., van Seventer, G.A., Ennis, E., Newman, W., Horgan, K.J. and Shaw, S. (1992b). Crosslinking of the T cell-specific accessory molecules CD7 and CD28 modulates T cell adhesion. J. Exp. Med. 175, 577–582.

Siegelman, M. (1991). Sweetening the selectin pot. Current Biology 1, 125–128.

Skoglund, G., Patarroyo, M., Forsbeck, K., Nilsson, K. and Ingelman-Sundberg, M. (1988). Evidence for separate control by phorbol esters of CD18-dependent adhesion and translocation of protein kinase C in U-937 cells. Cancer Res. 48, 3168–3172.

Slupsky, J.R., Seehafer, J.G., Tang, S.C., Masellis-Smith, A. and Shaw, A.R. (1989). Evidence that monoclonal antibodies against CD9 antigen induce specific association between CD9 and the platelet glycoprotein IIb–IIIa complex. J. Biol. Chem. 21, 12 289–12 293.

Smith, C.W., Kishimoto, T.K., Abbas, O., Hughes, B., Rothlein, R., McIntire, L.V., Butcher, E.C. and Anderson, D.C. (1991). Chemotactic factors regulate lectin adhesion molecule 1 (LECAM-1)-dependent neutrophil adhesion to cytokine stimulated endothelial cells *in vitro*. J. Clin. Invest. 87, 609–618.

Sozzani, S., Luini, W., Molino, M., Jilek, P., Botazzi, B., Cerletti, C., Matsushima, K. and Mantovani, A. (1991). The signal transduction pathway involved in the migration induced by a monocyte chemotactic cytokine. J. Immunol. 147, 2215–2221.

Spertini, O., Kansas, G.S., Munro, J.M., Griffin, T.F. and Tedder, T.F. (1991). Regulation of leucocyte migration by activation of the leucocyte adhesion molecule-1 (LAM-1) selectin. Nature 349, 691–694.

Sporn, S.A., Eierman, D.F., Johnson, C.E., Morris, J., Martin, G., Ladner, M. and Haskill, S. (1990). Monocyte adherence results in selective induction of novel genes sharing homology with mediators of inflammation and tissue repair. J. Immunol. 144, 4434–4441.

Springer, T.A., Thompson, W.S., Miller, L.J., Schmalstieg, F.C. and Anderson, D.C. (1984). Inherited deficiency of the Mac-1, LFA-1, p 150,95 glycoprotein family and its molecular basis. J. Exp. Med. 160, 1901–1918.

Staatz, W.D., Rajpara, S.M., Wayner, E.A., Carter, W.G. and Santoro, S.A. (1989). The membrane glycoprotein Ia–IIa (VIA-2) complex mediates the Mg^{2+}-dependent adhesion of platelets to collagen. J. Cell Biol. 108, 1917–1924.

Stallcup, W.R. (1988). Involvement of gangliosides and glycoprotein fibronectin receptors in cellular adhesion to fibronectin. Exp. Cell Res. 177, 90–102.

Stamenkovic, I., Aruffo, A., Amiot, M. and Seed, B. (1991). The hematopoietic and epithelial forms of CD44 are distinct polypeptides with different adhesion potentials for hyaluronate-bearing cells. EMBO J. 10, 343–348.

Swerlick, R.A., Lee, K.H. and Lawley, T.J. (1992). Regulation of VCAM-1 on human dermal microvascular endothelial cells. J. Cell Biochem. 16A, 54. (Abstract)

Tamura, R.N., Cooper, H.M., Collo, G. and Quaranta, V. (1991). Cell type-specific integrin variants with alternative alpha chain cytoplasmic domains. Proc. Natl Acad. Sci. USA 88, 10 183–10 187.

Tanaka, Y., Albelda, S.M., Horgan, K.J., van Seventer, G.A., Shimizu, Y., Newman, W., Hallam, J., Newman, P.J., Buck, C.A. and Shaw, S. (1992). CD31/PECAM-1 is a preferential amplifier of VLA-integrin-mediated adhesion for distinctive T cell subsets. J. Exp. Med. 176: 245–253.

Tanaka, Y., Adams, D.H. and Shaw, S. (1993). Protoglycan on endothelial cells present adhesion-inducing cytokines to leukocytes. Immunol. Today 14: 111–115.

Tang, A. and Udey, M.C. (1991). Inhibition of epidermal langerhans cell function by low dose UVB radiation: UVB radiation selectively modulates ICAM-1 (CD54) expression. J. Immunol. 146: 3347–3355.

Teixido, J., Parker, C.M., Kassner, P.D. and Hemler, M.E. (1992). Functional and structural analysis of VLA-4 integrin subunit cleavage. J. Biol. Chem. 267 (In press).

Teller, J.C. and Rudd, C.E. (1991). A 32 kDa GTP-binding protein associated with the CD4-p56lck and CD8-p56lck T cell receptor complexes. Science 254, 439–441.

Thornhill, M.H., Kyan-Aung, U. and Haskard, D.O. (1990). IL-4 increases human endothelial cell adhesiveness for T cells but not for neutrophils. J. Immunol. 144, 3060–3065.

Trude, D.D., Singer, M.S., Lasky, L.A. and Rosen, S.D. (1990). Requirement for syalic acis on the endothelial ligand of a lymphocyte homing receptor. J. Cell Biol. 111, 2757–2764.

Trudel, G.C. and Holland, P.C. (1989). Effect of inhibitors of glycoprotein processing on integrin and the adhesion of myoblasts to extracellular matrix proteins. Biochem. Biophys. Res. Commun. 163, 1338–1343.

Turner, C.E. and Burridge, K. (1991). Transmembrane molecular assemblies in cell-extracellular matrix interactions. Curr. Opin. Cell Biol. 3, 849–853.

Utsumi, K., Sawada, M., Narumiya, S., Nagamine, J., Sakta, T., Iwagami, S., Kita, Y., Teraoka, H., Hirano, H., Ogata, M., Hamaoka, T. and Fujiwara, H. (1991). Adhesion of immature thymocytes to thymic stromal cells through fibronectin molecules and its significance for the induction of thymocyte differentiation. Proc. Natl Acad. Sci USA 88, 5685–5689.

van Kuppevelt, T.H., Languino, L.R., Gailit, J.O., Suzuki, S. and Ruoslahti, E. (1989). An alternative cytoplasmic domain of the integrin beta 3 subunit. Proc. Natl Acad. Sci. USA 86, 5415–5418.

van Seventer, G.A., Shimizu, Y., Horgan, K.J. and Shaw, S. (1990). The LFA-1 ligand ICAM-1 provides an important costimulatory signal for T cell receptor-mediated activation of resting T cells. J. Immunol. 144(12), 4579–4586.

Vleminckx, K., Vakaet, L., Mareel, M., Fiers, W. and Van Roy, F. (1991). Genetic manipulation of E-cadherin expression by epithelial tumor cells reveals an invasion suppressor role. Cell 66, 107–119.

von Andrian, U.H., Chambers, J.D., McEvoy, L.M., Bargatze, R.F., Arfors, K-E. and Butcher, E.C. (1991). Two-step model of leucocyte-endothelial cell interaction in inflammation: distinct roles for LECAM-1 and the leucocyte integrins *in vivo*. Proc. Natl Acad. Sci. USA 88, 7538–7542.

Walz, G., Aruffo, A., Kolanus, W., Bevilacqua, M. and Seed, B. (1990). Recognition by ELAM-1 of the Sialyl-Le^{x} determinant on myeloid and tumor cells. Science 250, 1132–1135.

Watson, M.L., Kingsmore, S.F., Johnston, G.I., Siegelman, M.H., Le Beau, M.M., Lemons, R.S., Bora, N.S., Howard, T.A., Weissman, I.L., McEver, R.P. and Seldin, M.F. (1990). Genomic organization of the selectin family of leukocyte adhesion molecules on human and mouse chromosome 1. J. Exp. Med. 172, 263–272.

Watson, P.A. (1991). Function follows form: generation of intracellular signals by cell deformation. FASEB J. 5, 2013–2019.

Wayner, E.A. *et al.*, (1991). Integrins $\alpha_v\beta_3$ and $\alpha_v\beta_5$ contribute to cell attachment to vitronectin but differentially distribute on the cell surface. J. Cell Biol. 113, 919–929.

Webb, D.S, Shimizu, Y., van Seventer, G.A., Shaw, S. and Gerrard, T.L. (1990). LFA-3, CD44, and CD45: Physiologic triggers of human monocyte TNF and IL-I release. Science 249, 1295–1297.

Weeks, B.S., Klotman, M.E., Dhawan, S., Kibbey, M., Rappaport, J., Kleinman, H.K., Yamada, K.M. and Klotman, P.E. (1991). HIV-1 infection of human T lymphocytes results in enhanced $\alpha_5\beta_1$ integrin expression. J. Cell Biol. 114, 847–853.

Werb, Z., Tremble, P.M., Behrendtsen, O., Crowley, E. and Damsky, C.H. (1989). Signal transduction through the fibronectin receptor induces collagenase and stromelysin gene expression. J. Cell Biol. 109, 877–889.

Williams, A.F. (1991). Out of equilibrium. Nature 352, 473–474.

Wong, C.S., Gamble, J.R., Skinner, M.P., Lucas, C.M., Berndt, M.C. and Vadas, M.A. (1991). Adhesion protein GMP140 inhibits superoxide anion release by human neutrophils. Proc. Natl Acad. Sci. USA 88, 2397–2401.

Wright, S.D. and Meyer, B.C. (1986). Phorbol esters cause sequential activation and deactivation of complement receptors on polymorphonuclear leucocytes. J. Immunol. 136, 1759–1764.

Wright, S.D., Detmers, P.A., Jong, M.T.C. and Meyer, B.C. (1986). Interferon Gamma depresses binding of ligand by C3b and C3bi receptors on cultured human monocytes, an effect reversed by fibronectin. J. Exp. Med. 163, 1245.

Wright, S.D. and Detmers, P.A. (1988). Adhesion-promoting receptors on phagocytes. J. Cell Sci. Suppl. 9, 99–120.

Yamada, A., Nikaido, T., Nojima, Y., Schlossman, S.F. and Morimoto, C. (1991). Activation of human CD4 T lymphocytes. Interaction of fibronectin with VLA-5 receptor on CD4 cells induces the AP-1 transcription factor. J. Immunol. 146, 53–56.

Yuan, Q., Jiang, W., Krissansen, G.W. and Watson, J.D. (1990). Cloning and sequence analysis of a novel beta2-related integrin transcript from T lymphocytes: homology of integrin cysteine-rich repeats to domain III of laminin B chains. Int. Immunol. 2, 1097–1108.

Zimmerman, G.A., Prescott, S.M. and McIntyre, T.M. (1992a). In "*Inflammation. Basic principles and clinical correlates*" (eds Gallin, Goldstein and Snyderman).

Zimmerman, G.A., Prescott, S.M. and McIntyre, T.M. (1992b). Endothelial cell interactions with granulocytes: tethering and signalling molecules. Immunol. Today 13, 93–99.

6. *Specificity and Expression of Carbohydrate Ligands*

John B. Lowe

1. *Introduction*

The surfaces of mammalian cells display a structurally diverse class of molecules known as complex carbohydrates, oligosaccharides, or glyco-conjugates. A substantial amount of experimental effort has been devoted to delineating the chemical nature and structure of these molecules. These efforts have been driven largely by the notion that an understanding of the function(s) of such molecules will be facilitated by a clear delineation of their structure. It has become clear from these analyses that cell surface oligosaccharides exist in many distinct structural forms, that in turn are a function of the lineage of the tissue or cell type that display them. Dynamic changes in cell surface oligosaccharide structure are observed in mammalian developmental contexts, and the structures of cell surface glyco-conjugates in fully differentiated tissues may also vary dramatically. These observations would imply that complex regulatory mechanisms exist for controlling oligosaccharide expression patterns and that such patterns have important functional correlates. Discovery and analysis of a group of mammalian proteins called lectins, with the capacity to specifically bind distinct oligosaccharide molecules, has implicated cell surface complex carbohydrates in adhesive processes (Drickamer, 1988). Recent years have seen a somewhat limited confirmation of this latter hypothesis; oligosaccharides have been shown to play important or essential roles in lysosomal enzyme targeting (via the mannose-6-phosphate receptor; Kornfeld, 1987), in clearances of desialylated serum glycoproteins (via the hepatic asialoglycoprotein receptor; Ashwell and Harford, 1982), and in the specific clearance of some pituitary glycoprotein hormones (Fiete *et al.*, 1991; Baenziger *et al.*, 1992). Increasingly direct evidence for oligosaccharide-dependent immune cell adhesion has also accumulated over the past several years. This chapter will provide an overview and summary of this experimental data, of the

Adhesion Molecules
ISBN 0-12-741440-1

oligosaccharide molecules that function in these processes, and of enzymes and mechanisms that regulate expression of these carbohydrate ligands.

2. *Oligosaccharides in Lymphocyte Homing*

In man, rodents, and other mammalian species, lymphocytes are known to circulate both in the vascular compartment and in the lymphatic system. In this pathway, circulating blood–born lymphocytes exit the vascular tree within the lymph nodes. They then travel through the parenchyma of lymphoid tissues, and return to the vascular tree via the lymphatics (Yednock and Rosen, 1989). This system functions to maximize opportunities for lymphocytes to interact with foreign antigens in lymph nodes and in other lymphoid tissues. Post-capillary venules of lymphoid organs are the sites of lymphocyte egress from the vasculature. The first event in the process is believed to be a specific adhesive interaction between molecules on the surfaces of the lymphocytes, and molecules expressed at the surface of the post-capillary venular endothelium.

Post-capillary venular endothelium in lymph nodes is both functionally and morphologically distinct from other post-capillary endothelia. Specifically, lymph node capillary endothelium is composed of plump, cuboidal cells, and the post-capillary venules lined by these cells are termed HEV, for high endothelial venules. The HEVs in different lymphoid organs are functionally distinct, in that they mediate adherence of specific subsets of circulating immune cells. Cell trafficking, biochemical and immunological analyses have defined several such distinct specificities for lymphocyte-HEV interactions (reviewed in Kansas *et al.*, 1991). These include lymphocytic adhesive interactions specific for HEV within lymphoid organs that drain lung tissue, for HEV within chronically inflamed synovia, and for HEV found in peripheral lymph nodes (PNHEV) and in lymphoid aggregates within the gut (i.e. Peyer's patches). It is believed that distinct adhesive molecules are expressed by these different types of HEV, and that these receptors mediate lymphoid organ-specific "homing" of lymphocytes bearing cognate counter-receptors. The nature of these organ-specific receptor-counter-receptor pairs has in some cases been defined by molecular cloning studies. Molecules that mediate lymphocyte homing to PNHEV have been shown to interact via carbohydrate-dependent mechanisms, and will be the focus of this section.

Many of the *in vitro* analyses of lymphocyte homing to PNHEV have employed the Stamper–Woodruff assay (Stamper and Woodruff, 1977). With this approach, viable lymphocytes are first applied to HEVs exposed in frozen sections of peripheral lymph nodes. Bound cells are then quantitated after washing, and are a measure of the adhesive interactions between the applied lymphocytes and the exposed HEVs. The Stamper–Woodruff assay has been used in the mouse to define a lymphocyte epitope, that mediates lymphocyte attachment to PNHEV. This epitope is recognized by a monoclonal antibody (mAb) termed MEL-14 (Gallatin *et al.*, 1983). The MEL-14 antibody was generated with an approach that sought to identify antibodies reactive against surface epitopes on cultured lymphoma cell lines with binding specificity for PNHEV (Gallatin *et al.*, 1983). MEL-14 reacts with all murine peripheral lymphocytes, and blocks adhesion of these cells to PNHEV, as assessed by the Stamper–Woodruff assay. Moreover, incubation of lymphocytes with saturating amounts of MEL-14 will prevent their homing to peripheral lymph nodes *in vivo*. By contrast, in Peyer's patches homing is not significantly perturbed by this maneuver, nor does the MEL-14 antibody prevent lymphocyte binding to Peyer's patches HEV.

3. *Oligosaccharide-dependent Inhibition of Lymphocyte-PNHEV interactions*

Immunoprecipitation experiments demonstrated that MEL-14 binds to a 90 kDa lymphocyte surface protein (termed $gp90^{MEL}$) (Gallatin *et al.*, 1983). By contrast, neutrophils express a slightly larger form of this molecule (Lewinsohn *et al.*, 1987). Molecular cloning studies have since indicated that the apparent discrepancy between the sizes of the lymphocyte and neutrophil molecules may be accounted for by differences in post-translational modification by glycosylation and/or ubiquitination (Siegelman *et al.*, 1986; St. John *et al.*, 1986). Initial functional analyses proceeded without any primary sequence information about the MEL-14-reactive molecule, however. In these initial studies (Stoolman and Rosen, 1983; Stoolman *et al.*, 1984), Stoolman and Rosen tested the possibility that lymphocyte-PNHEV binding interactions involved carbohydrate determinants. These investigations were prompted by a consideration of others' microscopic evaluation of PNHEV and related observations concerning the possible chemical nature of PNHEV adhesive molecules. The morphological analyses indicated that the surfaces of PNHEV are covered with a rather prominent glycocalyx, that could present the first entity encountered by homing lymphocytes. Others had also previously demonstrated that glutaraldehyde fixation of exposed PNHEV does not significantly diminish the capacity of these structures to bind lymphocytes (Stamper and Woodruff, 1976; Stamper and Woodruff, 1977). Glutaraldehyde fixation typically leaves oligosaccharide epitopes relatively intact, but often eliminates peptide determinants. Taken together, these observations were consistent with the possibility that lymphocytes recognized carbohydrate determinants expressed on the glycocalyx of the PNHEV surface (Stoolman and Rosen, 1983).

Stoolman and Rosen tested this hypothesis by analyzing a number of different monosaccharides and polysaccharides for their ability to inhibit attachment of lymphocytes to PNHEV (Stoolman and Rosen, 1983). They found that D-mannose and L-fucose, at relatively high concentrations (≥75 mM) and under conditions of high ionic strength, significantly inhibit lymphocyte-PNHEV interactions. By contrast, several other monosaccharides were ineffective. These investigators also surveyed several polysaccharides for inhibitory capacity. These substances included heparin, dextran sulfate, chondroitin sulfate, and fucoidan, a sulfated high molecular weight polymer containing $\alpha(1 \rightarrow 2)$- and $\alpha(1 \rightarrow 3)$-linked fucose residues (Mian and Percival, 1973). Of these molecules, only fucoidan potently inhibited binding, with half maximal inhibition at low nanomolar concentrations.

Stoolman and colleagues extended these observations in studies that tested several phosphorylated carbohydrates (Stoolman *et al.* 1984). This work demonstrated that mannose-6-phosphate and fructose-1-phosphate were competent to block lymphocyte adherence to PNHEV, when tested at physiological ionic strength. Half maximal inhibition was achieved at concentrations ranging between 2 and 5 mM; complete inhibition was obtained when the sugars were tested at 10 mM. By contrast, a series of other phosphorylated monosaccharides did not inhibit binding. A polymeric polysaccharide rich in mannose-6-phosphate moieties was also tested. This substance, known as PPME, is a mannan-based substance derived from the yeast *Hansunela holstii* (San Blas and Cunningham, 1974). PPME potently inhibited lymphocyte adherence, exhibiting half maximal activity at low nanomolar concentrations. By contrast, non-phosphorylated yeast mannans did not block binding. PPME-dependent binding inhibition occurred when lymphocytes were pre-treated with this compound, whereas PNHEV pretreatment was without effect. These observations yield the conclusion that PPME participates in an adhesive interaction with a lymphocyte surface receptor responsible for lymphocyte-PNHEV interactions. When considered together with the results of the initial study, these observations further suggest that this phosphorylated polymer, and also fucoidan, maintains a tertiary configuration that mimics one displayed by counter-receptors expressed on the surface of PNHEVs, and that the inhibitory monosaccharides may exhibit conformations similar to components of these molecules.

These observations were extended in experiments using fluorescent beads derivatized with PPME. Such beads were shown to bind to lymphoid cells that are capable of adhering to PNHEV, but did not bind to non-adherent cell types (Yednock *et al.*, 1987a; Stoolman and Ebling, 1989). Carbohydrate-containing compounds that inhibited lymphocyte-PNHEV attachment were equipotent in their relative abilities to interfere with PPME binding to lymphocytes. Moreover, PPME binding, like lymphocyte attachment, was shown to be calcium-dependent. Rank order inhibitory potencies found in the rodent system were recapitulated with human reagents, suggesting an evolutionary conservation of the receptor's carbohydrate binding specificity (Stoolman *et al.*, 1987). It was subsequently demonstrated directly that the lymphocyte receptor that recognizes PPME is in fact the MEL-14 epitope (Yednock *et al.*, 1987b). In these experiments, the density of MEL-14 expression was shown to correlate directly with PPME binding efficiency, and pretreatment of lymphocytes with the MEL-14 antibody blocked PPME binding. In aggregate, the results outlined above suggested that MEL-14 represented a cell surface glycoprotein capable of mediating lymphocyte-PNHEV adhesion via an interaction with a PNHEV-specific carbohydrate epitope. These results were consistent with the notion that the MEL-14 molecule represents a mammalian carbohydrate binding protein, or lectin.

Further work to confirm this hypothesis, and to determine the nature of endogenous carbohydrate molecule(s) recognized by this receptor, has been hampered by difficulties associated with isolating the ligand(s) in quantities sufficient for structural analysis. PNHEV are a relatively minor component of lymph node cell mass, making large-scale isolation problematic, and it has not yet been possible to isolate continuously growing cultured cell lines that express the ligand(s). Further characterization of natural ligands has therefore relied largely on methods that utilize glycohydrolases for elimination of ligand reactivity on exposed PNHEV, or in animal experiments. In the first studies employing glycohydrolases (Rosen *et al.*, 1985), ligand destruction was attempted using frozen sections of PNHEV (Rosen *et al.*, 1985). It was found that sialidase treatment of paraformaldehyde-fixed PNHEV abrogated lymphocyte binding. Equivalent levels of destruction were achieved when using either *Vibrio cholera* sialidase, or affinity-purified sialidase prepared from *Clostridium perfringens*. Neither enzyme preparation was detectably contaminated with other glycosidases, and ligand destruction was prevented if sialidase inhibitors (sialic acid and sialyllactose) were included in the reactions. These control experiments thus strongly suggest a specificity for sialic acid moieties. Furthermore, neuraminidase-mediated ligand destruction could be achieved only partially when treating mesenteric node HEV, and no inhibition was detected after treating Peyer's patch HEV. These results are entirely consistent with previous work indicating that the HEV in mesenteric nodes expresses both PNHEV and Peyer's patch ligands, and that Peyer's patches HEV expresses a recognition determinant distinct from that found on Peyer's patches HEV (Stevens *et al.*, 1982).

An *in vivo* correlate of these observations was subsequently obtained by assessing the effect of of intravenous sialidase injection on short-term accumulation of lymphocytes in experimental mice (Rosen *et al.*, 1989). Injection of *Vibrio or Clostridium sialidases* substantially diminished accumulation of lymphocytes into peripheral lymph nodes, but lymphocyte accumulation in other organs (Peyer's patches, spleen, kidney, and liver) was indistinguishable from that observed in control animals. Histological examination using a lectin that detects sialic acid residues (*Limax flavus*) indicated that intravenous sialidase effectively removed sialic acid residues from both PNHEV and Peyer's patches HEV. PNHEV isolated from sialidase-treated animals did not support lymphocyte adherence, whereas Peyer's patches HEV from these animals maintained normal binding parameters. This work, and observations derived from *in vitro* sialidase treatment, implied that sialic acid moieties are essential components of the endogenous PNHEV ligand for the MEL-14 receptor. When considered together with the carbohydrate blocking experiments outlined above, this work provided additional evidence that the MEL-14 molecule functions as a mammalian lectin.

4. *Molecular Cloning Identifies a Carbohydrate Recognition Domain in MEL-14*

The mammalian lectin hypothesis was subsequently confirmed via molecular cloning of cDNAs encoding the MEL-14 molecule, and its human counterpart leucocyte adhesion molecule-1 (LAM-1). MEL-14 cDNAs were isolated using peptide sequence information (Lasky *et al.*, 1989; Siegèlman *et al.*, 1989), whereas the human molecules were cloned by expression cloning techniques (Camerini *et al.*, 1989; Tedder *et al.*, 1989) or by cross-hybridization with the murine probe (Bowen *et al.*, 1989; Siegelman and Weissman, 1989). Table 6.1 summarizes nomenclature used to described these molecules, now known by the consensus term L-selectin (Bevilacqua *et al.*, 1991); Sequence analysis of the L-selectin molecules identified a set of distinct domains (Fig. 6.1). An amino-terminal signal sequence yields a type I transmembrane protein, oriented with its amino terminus and most of its mass on the outside of the cell. Evidence has been presented to suggest that alternatively spliced human L-selectin transcripts also yield a form of L-selectin that lacks the hydrophobic membrane-spanning domain, but which is membrane-associated via a phospholipid anchor (Camerini *et al.*, 1989).

The 116 residues at the amino terminus of the mature polypeptide are of major interest with respect to the functional attributes of L-selectin. Sequence analysis of this segment identified striking sequence similarity to the

Table 6.1 Nomenclature used to describe the L-selectin molecule (Ref. 30)

Previous names	*Ref.*
LEC-CAM-1 (former consensus)	Bevilacqua *et al.*, 1991
LAM-1 (human)	Tedder *et al.*, 1989
Leu8 (human)	Tedder *et al.*, 1990
TQ1 (human)	Tedder *et al.*, 1990
DREG-56 (human)	Kishimoto *et al.*, 1990
Mel-14 (mouse)	Gallatin *et al.*, 1983
Lymphocyte homing receptor (mouse)	Siegelman *et al.*, 1989
Ly-22 (mouse)	Siegelman *et al.*, 1989

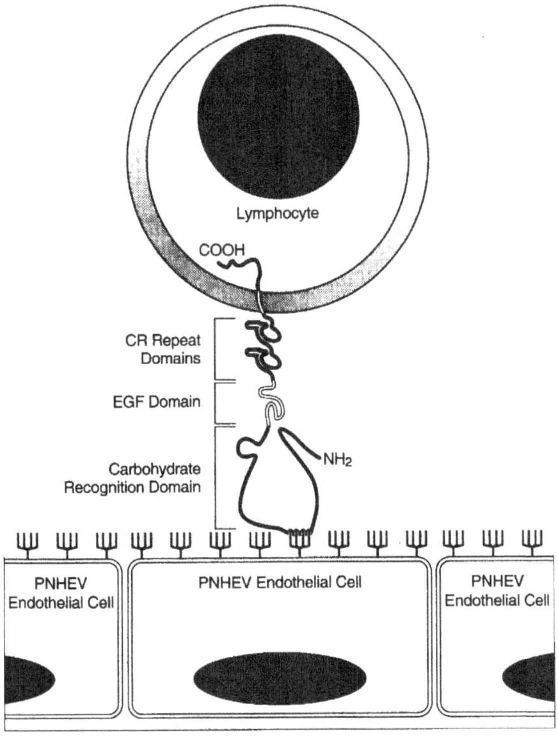

Figure 6.1 L-selectin domain structure. The structure predicted for the human L-selectin molecule is schematically illustrated, drawn to a scale that approximates the lengths of each protein domain. The carbohydrate recognition domain, CR repeats, and the EGF domain are shown folded in an arrangement proposed by Johnston *et al.* (1989) for the P-selectin molecule, based upon positions predicted for intrachain disulfide bonds. L-selectin is expressed by lymphocytes, as shown here, but also by neutrophils and other leucocytes as detailed in the text. The carbohydrate recognition of L-selectin functions to mediate adhesion between circulating lymphocytes and oligosaccharide ligand(s) (indicated by ψ) expressed along the surface of peripheral node high endothelium (PNHEV).

calcium-dependent carbohydrate binding proteins classified by Drickamer (1988). These molecules, also known as C-type lectins, are part of a family of proteins that include the asialoglycoprotein receptor, the fucose binding protein of Kupffer cells, a flesh fly lectin, and a variety of mannose binding proteins. Their CRDs, or carbohydrate recognition domains, are roughly 25–30% identical to the corresponding segment in L-selectin. Drickamer has identified four conserved cysteine residues amongst these molecules (Drickamer, 1988), which are also maintained in the L-selectin protein. By contrast, the amino acid sequences between these conserved residues varies substantially. These observations suggest that the conserved cysteines may serve to constrain the tertiary conformation of these lectins, whereas carbohydrate binding specificity may be dictated by intervening amino acids. The CRDs of the murine and human L-selectins share 83% sequence identity, consistent with data suggesting similar or identical ligand recognition specificities (Stoolman *et al.*, 1987).

The L-selectin CRD is located in a position adjacent to a peptide segment with primary sequence similarity to the epidermal growth factor precursor (EGF; Cooke *et al.*, 1987). This domain is 33 to 40 residues long, and is a motif found in many proteins, including Factor X (Enfield *et al.*, 1980), proteins encoded by the *Notch* locus of *Drosophila melanogaster* (Kidd *et al.*, 1986), and the low-density liproprotein (LDL) receptor (Davis *et al.*, 1987). The EGF domain is in turn adjacent to two tandemly repeated polypeptide segments, each with a unit length of 62 residues. The polypeptide unit maintains primary sequence similarity to a peptide motif found in many other proteins (Hourcade *et al.*, 1989), including factor XIII (Ichinose *et al.*, 1986), decay accelerating factor (DAF, Caras *et al.*, 1987), the human IL-2 receptor (Leonard *et al.*, 1984), complement receptor (CR) 1, and the receptor for C3d (Klickstein *et al.*, 1987). These repeating units are known as CR repeats, for their similarity to these latter complement receptor molecules. These surface-localized domains are displayed via a hydrophobic membrane-spanning segment, which is in turn linked to a short cytosolic segment.

The genomic structures of human (Ord *et al.*, 1990) and murine (Dowbenko *et al.*, 1991) L-selectin have been defined. The genes from the two species are organized similarly; single axons correspond to domains defined by the signal peptide sequence, the CRD, the EGF domain, the unit CR repeats, and transmembrane segments. Most of the cytosolic domain in each molecule is encoded by two exons. *In situ* hybridization data indicate that the human L-selectin gene resides on chromosome 1 between bands q22 and q25 (Tedder *et al.*, 1989 and unpublished data cited therein, and Watson *et al.*, 1990), at a position distinct from a site corresponding to other genes encoding the CR-repeat-containing proteins localized to band 132 (Weis *et al.*, 1987). As will be discussed below, genes encoding the other selectin family members (E-selectin and P-selectin) are structurally similar to L-selectin, and co-localize with the L-selectin gene to a position between 1q22 and 1q25. These considerations have suggested the possibility that the three selectin genes may represent an "adhesion molecule locus" (Tedder *et al.*, 1989) of very closely linked sequences that may have evolved by gene duplication.

5. *Functional Correlates of the L-selectin Domain Structure*

Antibody blocking approaches and recombinant chimeric L-selectin molecules have been used to delineate the functional importance of L-selectin domains. These studies have defined L-selectin segments that participate in PNHEV attachment, that bind PPME and fucoidan, and that react with various L-selectin mAbs (Bowen *et al.*, 1990; Siegelman *et al.*, 1990; Walz *et al.*, 1990; Kansas *et al.*, 1991). Bowen and colleagues investigated the role of the EGF domain in maintaining CRD conformation by testing recombinant murine L-selectin molecules assembled from its CRD, EGF and CR segments, and linked to the hinge, CH2 and CH3 regions of human IgG1 (Bowen *et al.*, 1990). This control protein was fully reactive with the MEL-14 antibody whereas a chimera lacking the EGF segment does not react with this antibody. To further define the position of the MEL-14 epitope within the murine L-selectin CRD, chimeras were constructed by replacing segments of the murine CRD with corresponding segments taken from the human L-selectin CRD, and these were tested for MEL-14 reactivity. The results indicate that the amino terminal 53 residues of murine L-selectin correspond to at least part of the MEL-14 epitope. Thus, since the MEL-14 antibody also blocks PPME binding (Yednock, 1987b), it follows that the CRD is indeed largely responsible for direct interaction with its carbohydrate ligand(s). These studies also imply an important role for the EGF segment in L-selectin carbohydrate binding function, operating most probably via a mechanism involving maintainence of a proper functional conformation of the CRD.

Kansas *et al.* (1991a; and 1991b) have subsequently analyzed the carbohydrate binding properties of a series of chimeric molecules comprised of segments taken from L-selectin and P-selectin. Chimeras containing the L-selectin CRD fused to the P-selectin EGF segment exhibit PPME and fucoidan binding properties virtually identical to the intact L-selectin molecule. Other chimeras were used to demonstrate that sequences in the COOH-terminal half of the L-selectin CRD are required to maintain its carbohydrate binding specificity. These experiments strongly suggest that the L-selectin EGF segment operates to maintain the CRD in a functional configuration, but does not directly interact with the L-selectin ligand or directly influence ligand binding specificity. They further indicate that these latter properties are determined primarily, if not completely, by the L-selectin CRD.

These results have been further extended by Spertini *et al.* (1991b), who have analyzed a large panel of monoclonal anti-L-selectin antibodies to determine epitope location and allied function. Functional endpoints examined included assay of effect of antibody binding on PPME binding and PNHEV attachment. Results from these studies and related ones (Kansas *et al.*, 1991a; Kansas *et al.*, 1991b) indicate that the L-selectin CRD maintains distinct subdomains. For example, antibodies LAM1-3, MEL-14, DREG-56, and TQ1 each block PPME and fucoidan binding and also inhibit attachment of lymphocytes to PNHEV. These reagents reconize the L-selectin CRD. By contrast, antibody LAM1-6, which also recognizes the CRD, blocks PNHEV attachment yet does not perturb binding of PPME, or fucoidan. Finally, the mAbs Leu8 and Lam1-10 also react with CRD determinant(s) yet neither diminishes PNHEV attachment, nor binding of PPME, or fucoidan. Considered together, these observations further support the prediction that the L-selectin lectin domain directly interacts with carbohydrate ligands.

Similar mAbs studies suggest that the EGF domain has a significant influence on the ability of L-selectin to mediate lymphocyte-PNHEV interaction. MAbs LAM1-1 and LAM-5, directed against the EGF domain of human L-selectin, apparently increase its affinity for PPME (Kansas *et al.*, 1991a; Spertini *et al.*, 1991a). Another antibody directed against the EGF domain of L-selectin (Ly-22, Siegelman *et al.*, 1990) may also enhance PPME affinity. Lymphocyte activation may also transiently up-regulate the affinity of L-selectin for carbohydrate ligands (Spertini *et al.*, 1991c). Considered together, these observations suggest the possibility that the EGF domain represents a mechanism used to modulate the binding affinity of the adjacent CRD, via interactions with endogenous molecules in activated leucocytes, or with its counter-receptor molecule(s).

The roles of the L-selectin CR repeats have been examined using recombinant chimeric murine L-selectin molecules, varying in their CR and EGF domains (Watson *et al.*, 1991). These reagents were tested for MEL-14 reactivity, for their ability to bind PPME, and for their ability to interfere with lymphocyte-PNHEV adhesion. These molecules were also used with histochemical and biochemical methods to assess their ability to interact with the endogenous L-selectin ligands on PNHEV. The control chimera consisted of the lectin, EGF, and CR domains of the murine L-selectin molecule, linked to the hinge, CH2 and CH3 domains of human IgG1 heavy chain Fc region. This MEL-14-reactive chimera was fully capable of binding PPME, blocked lymphocyte-PNHEV adhesion, stained sectioned PNHEV in histochemical analyses, and was capable of immunoprecipitating a 50 kDa radiolabelled molecule from [^{35}S]sulfate-labelled PNHEV. Deletion of the 2 CR domains from this chimera eliminated MEL-14 reactivity, PPME binding, and reactivity with PNHEV ligand(s). These observations imply that the CR repeats are essential for maintenance of CRD conformation (MEL-14 reactivity) and function (PPME binding and PNHEV ligand recognition).

Chimeric L-selectin molecules have been used more recently to further define the endogenous PNHEV ligand(s). Previous observations concerning the requirement for sialic acid residues have been confirmed (True. *et al.*, 1990) with the recombinant murine L selectin-IgG chimera containing the CRD-EGF-CR domains (Watson *et al.*, 1991). Binding of this molecule to PNHEV is effectively eliminated if the HEV are first treated with sialidases from either *Clostridium perfringens* or *Vibrio cholera*. Inclusion of sialic acid in the incubation mixture prevents destruction of HEV-expressed molecules bound by the chimera, again suggesting specificity for sialic acid residues. Binding of the chimera was also effectively inhibited by pre-treating the HEV with *Limax flavus* agglutinin, a lectin that recognizes terminal sialic acid determinants. However, binding of the chimera was not diminished on PNHEV sections pre-treated with *Limulus polyphemus* agglutinin, that also specifically recognizes sialic acid residues. While it is not possible to reconcile this apparent discrepancy from the experiments presented, it seems possible that it may be accounted for by variation in the specificities of each lectin for different sialic acid linkages (Liener *et al.*, 1986), and in the variety of sialic acid linkages (Schauer, 1985) and sialic acid modifications (O-acetylation, Cheresh *et al.*, 1984; N-glycolylation, Muchmore *et al.*, 1989) found on mammalian cell surface glyco-conjugates. In this context, while it does seem certain that sialic acid residues are essential for L-selectin-dependent PNHEV-lymphocyte interaction, it is not yet known if all, some, or only a fraction of these participate in this interaction at the surface of PNHEV. Likewise, the structural nature of the glycosidic bonds ($\alpha2 \rightarrow 3, \alpha2 \rightarrow 6$, or $\alpha2 \rightarrow 8$, for example) that link the critical sialic acid moieties to underlying molecules are not yet defined.

Nonetheless, recent experiments suggest that the scaffold that displays these sialic acid moieties may represent an O-linked glycoprotein (Imai *et al.*, 1991). In this work, and as shown previously (Watson *et al.*, 1991), the murine L-selectin-IgG chimera can immunoprecipitate a 50 kDa radiolabelled molecule from [^{35}S]sulfate-labelled peripheral lymph nodes. A lesser, and variable, amount of a similarly diffuse, but larger (90 kDa) radiolabelled molecule is also immunoprecipitated. Both molecules are also detected by immunoprecipitation with mAb MECA-79. This reagent has been shown previously to block lymphocyte adhesion to PNHEV (Streeter *et al.*, 1988; Butcher, 1990), and thus had been thought to recognize the L-selectin counter-receptor.

Immunoprecipitation with the L-selectin chimera also identifies the two molecules in mesenteric node HEV, but not in Peyer's patches HEV nor in several other lymphoid and non-lymphoid organs (Imai *et al.*, 1991). Both molecules are trypsin-sensitive; immunoprecipitation of these proteins is sialidase-sensitive, calcium-dependent, and is inhibitable by fucoidan and PPME, but not by the control polysaccharides chondroitin sulfates A and B, keratan sulfate, and yeast mannan. Radiolabelled fucose is also incorporated into the immunoprecipitated glycoproteins; as with the sulfate moieties, however, it remains to be determined if this substance functions as part of an oligosaccharide determinant recognized by the L-selectin CRD. By contrast, these radiolabelled molecules could not be identified by immunoprecipitation of lymph nodes radiolabelled with mannose or glucosamine. Since mannose is an abundant component of N-linked mammalian oligosaccharides (Kornfeld and Kornfeld, 1985), this observation indirectly suggests that these molecules contain little or no N-linked oligosaccharide chains. This conclusion is supported by the absence of alteration in electrophoretic mobility of the two proteins following N-glycanase digestion. By exclusion then, it may be expected that these proteins contain O-linked oligosaccharide chains. In Imai *et al.* (1991), the authors cite unpublished data indicating that alkaline borohydride treatment of the sulfate-labelled or fucose labelled molecules released radiolabelled oligosaccharide chains. Since this treatment is known to release O-linked oligosaccharides, this may be considered tentative evidence in support of O-linked oligosaccharide modification of these proteins.

Imai *et al.* (1991) have termed the 50 kDa molecule Sgp^{50}, for sulfated glycoprotein of 50 kDa; the 90 kDa molecule is called Sgp^{90}. They represent candidate endogenous ligands for L-selectin in the mouse. It seems likely that L-selectin primarily recognizes the glycan portions of these glycoproteins since, as summarized above, sialidase treatment prevents their recognition by the L-selectin-IgG chimera. In addition, Imai and colleagues found that this chimera can immunoprecipitate Sgp^{50} and Sgp^{90} even after these molecules have been extracted from denaturing sodium dodecyl sulphate (SDS) polyacrylamide gels. Whether this latter result can be ascribed to L-selectin recognition of denaturation-resistant protein or combined protein-carbohydrate determinants, or if it is instead solely carbohydrate dependent, is not yet defined. Structural analysis of the oligosaccharides on Sgp^{50} and Sgp^{90}, and determination of the peptide sequence(s) of these molecules, will most probably be required to fully address these issues.

6. *Expression and Regulation of L-selectin*

Study of the regulation of expression of the L-selectin molecule has revealed it to be a remarkably dynamic molecule. L-selectin is expressed both on lymphoid cells and on other circulating blood cells, including neutrophils and monocytes. On lymphocytes, L-selectin is roughly 78 kDa in size, whereas it is found as a 90–100 kDa protein on neutrophils. The difference between these sizes, and that predicted for its primary translation product (~40 kDa) can be accounted for by lineage-specific glycosylation (reviewed in Kansas *et al.*, 1991b).

Two mechanisms are known to regulate L-selectin expression. The first involves its release from the cell surface (Kishimoto *et al.*, 1989; Kishimoto *et al.*, 1990; Spertini *et al.*, 1991; Spertini *et al.*, 1991c). L-selectin release from lymphocytes is effected via activation by phorbol 12-myristate 13-acetate (PMA) or mitogens (Tedder *et al.*, 1990). Neutrophils may be induced to shed L-selectin by treatment with phorbol myristate acetate (PMA), GM-CSF, lipopolysaccharide, and calcium ionophores (Griffin *et al.*, 1990). Antigen shedding occurs within minutes of activation; shedding is inhibited by protein kinase C (PKC) inhibitors (Spertini *et al.*, 1991), implying that phosphorylation events within the cytosolic domain may be operative here (Kansas *et al.*, 1991b). It is likely that L-selectin shedding occurs as a consequence of a proteolytic event that cleaves the molecule adjacent to its transmembrane segment on the exterior of the cell. Additional details about this process are not available.

Activational stimuli can also regulate L-selectin affinity for its ligand(s). Activation of neutrophils with G-CSF, G/M-CSF, or tumor necrosis factor α results in an increase in the quantity of surface-bound polymeric polysaccharide rich in mannose-6-phosphate moieties (PPME) (Spertini *et al.*, 1991c). This binding is L-selectin dependent since it may be inhibited by EDTA, or by the anti-L-selectin antibody L1-3. Lymphocytes activated by the CD2 pathway or by T-cell receptor cross-linking also exhibit L-selectin-dependent increases in PPME binding. Since the absolute numbers of surface-expressed L-selectin molecules does not change under these conditions, it may be concluded that these stimuli yield an increase in the affinity of L-selectin for PPME. This phenomenon is similar to the increases in adhesive "activity" of the the CD11/CD18 and very late activation antigens (VLA) integrins that occurs upon leucocyte activation (Dustin and Springer, 1989; Shimizu *et al.*, 1990). These observations suggest that activational stimuli specific for different cell types may provoke transient changes in L-selectin affinity for its ligand(s), and thus may affect the tissue-specific homing patterns of the different cell types.

Recent studies suggest that L-selectin plays an important role in mediating leucocyte adhesion to extranodal endothelium. In murine models, it has been demonstrated that anti-L-selectin antibodies can block neutrophil emigration into inflamed peritoneum (Jutila *et al.*, 1989) or into inflammatory sites in the skin (Lewinsohn *et al.*, 1987). It has also been shown that neutrophil adhesion to cytokine-activated vascular endothelium may be substantially diminished by pretreating these leucocytes with anti-L-selectin antibodies (Hallmann *et al.*, 1991; Smith *et al.*, 1991). This L-selectin-dependent adhesion is manifest under conditions of flow (where CD18-dependent adhesion is minimal), suggesting an independence from the CD18-dependent adhesion pathway. This notion is supported by observations indicating that anti-L-selectin and CD18 antibodies exhibit additive blocking capabilities.

This work has been extended in *in vivo* systems that examine the participation of L-selectin in leucocyte "rolling" phenomenon (Ley *et al.*, 1991; von Adrian *et al.*, 1991). Intravital microscopic studies have shown that leukocytes "roll" along the surface of the endothelial cells lining capillaries within an inflamed tissue (Atherton and Born, 1973a; Fiebig *et al.*, 1991; Ley *et al.*, 1991). This is apparently a prerequisite for leucocyte emigration, and occurs in part because of the alteration of the vessel's flow characteristics that occurs as a consequence of vasodilatation. Under normal conditions, leucocytes normally pass through the vessel at the center, in regions where plasma is flowing most rapidly (Nobis *et al.*, 1985). Vasodilatation and plasma fluid leakage yield a more slowly flowing stream, and the leucocytes begin to occupy more peripheral positions within the stream (Chien, 1982). Marginally positioned leucocytes are subjected to rotational forces generated by the differential in fluid flow velocity between the stream center (rapid) and its margins (slow). This causes leucocytes to "roll" along the wall of the dilated vessel, and maximizes interactions between the surfaces of the leucocyte and the endothelium. Measurements of the velocities of the leucocytes indicate that they roll more slowly than would be expected if they were freely tumbling. This suggests that adhesive interactions between the tumbling cells and the vessel wall retards their progress (Atherton and Born, 1973b).

Several reports have demonstrated that L-selectin figures importantly in leucocyte rolling. Intravascular administration of anti-L-selectin antibodies or an L-selectin-IgG chimera inhibits leucocyte rolling in rat (Watson *et al.*, 1990; Ley *et al.*, 1991) and rabbit models (von Adrian *et al.*, 1991). By contrast, anti-CD18 antibodies have little effect on leucocyte rolling, although they potently inhibit the arrest of rolling leucocytes (von Adrian *et al.*, 1991). These observations suggest that leucocyte emigration is mediated by sequential processes (rolling followed by attachment) mediated by distinct classes of molecules (L-selectin, and the integrins, respectively). This model is summarized in Butcher, 1991, and will be outlined at the end of this chapter.

These observations suggest that neutrophil emigration might be diminished or even prevented altogether by blocking the L-selectin-dependent neutrophil adhesion pathway. Strong support for this hypothesis is provided by studies in which neutrophil extravasation into thioglycollate-inflammed peritoneum is inhibited by intravascular administration of an L-selectin-IgG chimera (Watson *et al.*, 1991). It might also be anticipated that similar effects could be obtained by administration of molecular mimics of the L-selectin counter-receptor expressed on inflamed endothelial cells. These molecules have not yet been defined, however. Recent studies suggest that neutrophil L-selectin displays carbohydrate ligands for E-selectin (Picker *et al.*, 1991). This observation suggests that L-selectin-dependent neutrophil adhesion to activated endothelium may in fact also be E-selectin dependent, operating via a mechanism involving the presentation of the ligand for one selectin by another. Discovery and analysis of PNHEV ligands for L-selectin, and correlation with information on corresponding molecules on inflamed vascular endothelium, will be necessary to achieve a full understanding of these issues.

7. *Oligosaccharide-dependent Cell Adhesion Mediated by E- and P-selectin*

Acute and chronic inflammatory stimuli effect the accumulation of neutrophils and monocytes into affected tissue sites (Harlan, 1985; Harlan, 1987). Accumulation of these cells first requires that they leave the blood stream. The first steps in this process involve adhesion of the cells to the endothelial cells lining capillary beds that perfuse the inflamed tissue. Since there are two cellular components to this process (myeloid lineage cells, and the endothelial cell), it is perhaps not surprising that one can also define two types of mechanisms that promote these adhesive events. One of these mechanisms involves modulation of neutrophil/monocyte-endothelial cell adhesion via processes that act directly on the blood cells. Molecules involved in this arm of the adhesion process include members of the CD11/CD18 family of integrin cell adhesion molecules (reviewed in Springer, 1990). These molecules and mechanisms are presented in detail in Chapters 3 and 5. The other arm of this process involves activational events that affect the endothelial cell. These events in turn induce the expression of two adhesion molecules (E-selectin and P-selectin) with specificity for ligands found on some circulating immune cells. These ligands contain specific oligosaccharides as essential components, whose structures and biosynthesis will be outlined in this section.

8. E-Selectin-dependent Cell Adhesion

In vitro analysis of leucocyte adherence to vascular endothelium has usually relied upon the cultured endothelial cell monolayers derived from human umbilical veins (HUVECs, Bevilacqua, 1985). Neutrophils, monocytes, and continuous cell lines derived from the myeloid lineage (i.e. HL-60 and U937) adhere tightly to HUVECs if the monolayers are first treated with the inflammatory cytokines interleukin 1β or tumor necrosis factor α, or with lipopolysaccharide (Bevilacqua *et al.*, 1985a; Bevilacqua *et al.*, 1985b; Gamble *et al.*, 1985; Pohlman *et al.*, 1986; Bevilacqua *et al.*, 1987). Cytokine-dependent adhesion peaks approximately 6 h after treatment. Increased adhesion may be prevented with protein synthesis inhibitors, suggesting that new cell surface adhesion molecules are being synthesized in response to endothelial cell activation. Two mAbs developed against such molecules potently and specifically inhibit adhesion of indicator cells (neutrophils, HL-60 cells) to activated HUVEC monolayers (Pober *et al.*, 1986; Bevilacqua *et al.*, 1987). Immunoprecipitation with these antibodies identifies a prominent cytokine-dependent endothelial cell protein of Mr 115 000, expressed precisely in temporal correspondence with cytokine-dependent increases in adhesion. This molecule was termed endothelial leucocyte adhesion molecule I, or ELAM-1. It is now known as E-selectin (Bevilacqua *et al.*, 1991).

Immunohistochemistry has been used to identify E-selectin expression by microvascular endothelium in a variety of acute and chronic inflammatory conditions. These include septic shock (Redl *et al.*, 1991), immune complex-mediated acute lung injury (Mulligan *et al.*, 1991), rheumatoid arthritis (Koch *et al.*, 1991), delayed hypersensitivity in the skin (Cotran *et al.*, 1986; Griffiths *et al.*, 1991; Groves *et al.*, 1991; Picker *et al.*, 1991; Waldorf *et al.*, 1991), and in human cardiac allografts (Briscoe *et al.*, 1991). A variety of human leucocytes express ligands for E-selectin, including neutrophils (Bevilacqua *et al.*, 1985a), monocytes (Bevilacqua *et al.*, 1985b; Carlos *et al.*, 1990), eosinophils (Weller *et al.*, 1991), and subsets of T-cells including memory T-cells (Graber *et al.*, 1990; Picker *et al.*, 1991; Shimizu *et al.*, 1991) and natural killer (NK) cells (Lobb *et al.*, 1991). Each of these cell types also may be found in inflammatory sites where E-selectin is also expressed, thus suggesting a rather broad role for E-selectin in leucocyte recruitment.

9. Molecular Cloning Identifies Sequence Similarity between E-selectin and the C-type Lectins

Two groups have independently isolated cDNAs encoding human E-selectin. Bevilacqua *et al.* (1989) used anti-E-selectin antibodies and a mammalian expression cloning approach to isolate E-selectin cDNAs. Hession *et al.* (1990) also used a mammalian cDNA library expression cloning approach, with a screen for cDNAs that conferred HL-60 adhesive properties when introduced into COS-1 cells by transfection. Sequence analysis of these cDNAs revealed that E-selectin is organized in a manner strikingly similar to L-selectin, and maintains substantial primary sequence similarity to this molecule (Fig. 6.2). Specifically, there is an amino terminal domain with sequence similarity to C-type lectins, an EGF-like domain, four CR repeat domains, a single hydrophobic and presumably membrane-spanning segment, and a short cytosolic domain.

The lectin and EGF domains of E-selectin share roughly 61% sequence identity with the L-selectin CRD and EGF segments. Substantially less sequence similarity is found when comparing the transmembrane and cytosolic portions of E- and L-selectin (less than 20%

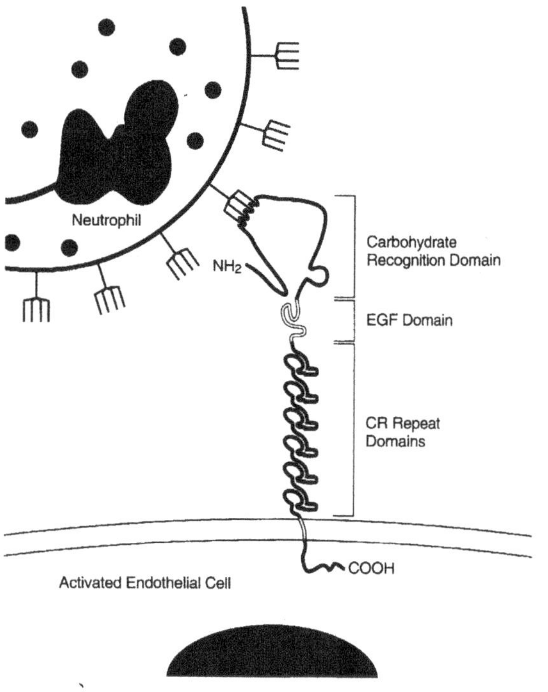

Figure 6.2 E-selectin domain structure. The lengths of each protein domain in this schematic depiction of human E-selectin as shown are drawn approximately to scale. The polypeptide domains are displayed in a folded arrangement that mirrors the disulfide bond dependent folded arrangement predicted for P-selectin by Johnston *et al.* (1989). As detailed in the text, E-selectin is expressed by activated capillary endothelial cells. Its carbohydrate recognition domain mediates adherence of circulating neutrophils (as well as monocytes, eosinophils, and some types of lymphocytes, not shown) to the endothelium via adhesive interactions with carbohydrate ligands (indicated by) expressed at the neutrophil surface.

identity). The cytosolic domain of E-selectin contains a single tyrosine residue postulated to be involved in internalization of the molecule into coated pits (Pearse, 1988). Since soluble forms of E-selectin have not been described, this mechanism may be responsible for the rapid removal of E-selectin from the endothelial cell surface although this has not yet been confirmed experimentally. E-selectin epitopes recognized by two monoclonal antibodies used to cloned E-selectin cDNAs have been mapped with recombinant E-selectin-CD7 or E-selectin-IgG1Fc chimeras expressed in COS cells (Walz *et al.*, 1990). These studies indicate that one of these antibodies (H18/7; blocks cell adhesion) maps to a position in the first 75% of the lectin domain, whereas the other (H4/18; does not block adhesion) identifies structure(s) within the three most NH2-terminal CR

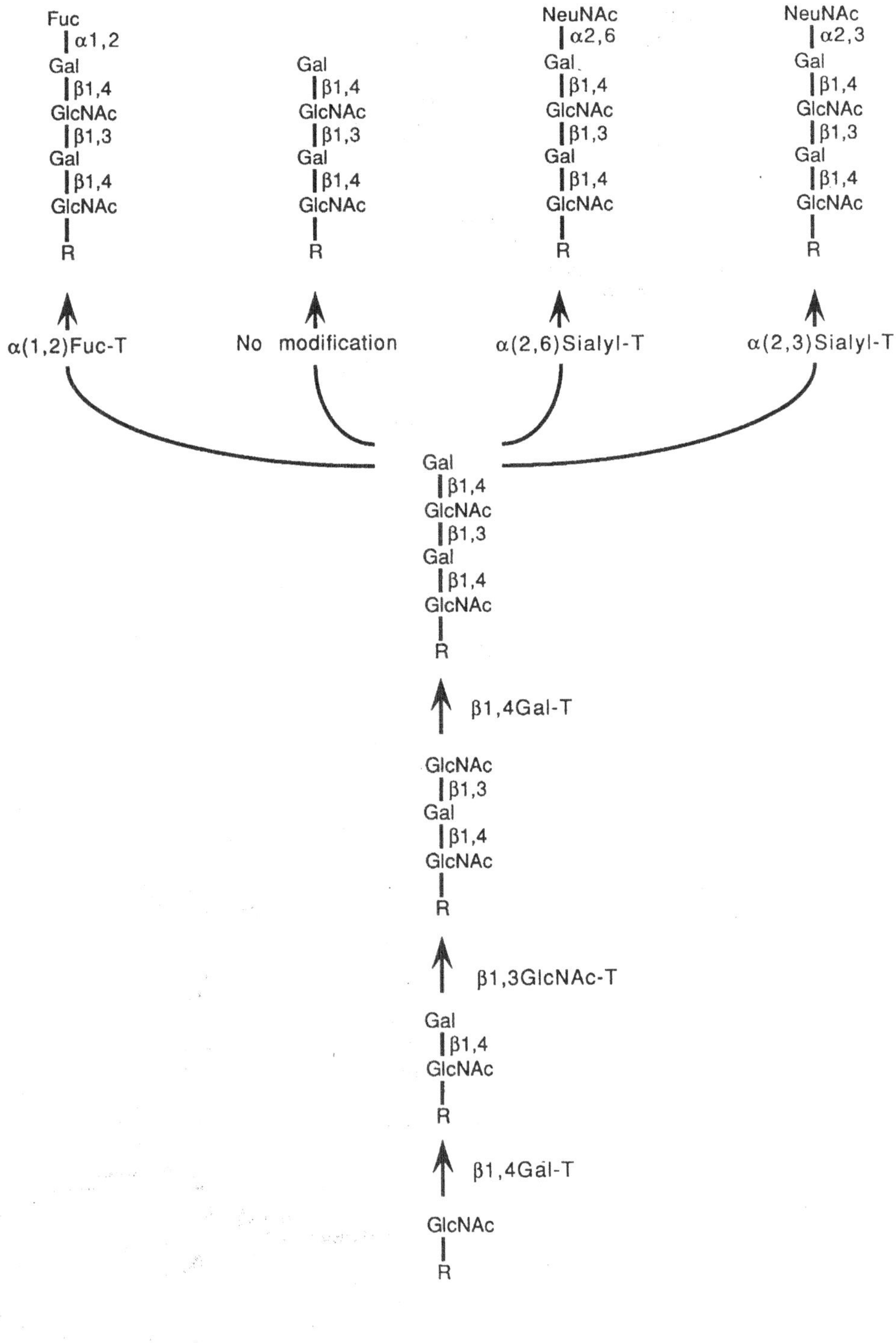

repeats. The human E-selectin gene is organized in a manner virtually identical to that of L-selectin (Collins *et al.*, 1991). Separate axons encode the leader sequence, lectin domain, CR repeats, membrane-spanning segment, while the cytosolic domain corresponds to two axons. As noted above, this gene maps to human chromosome 1 (Collins *et al.*, 1991) close to the human L-selectin gene (Tedder *et al.*, 1989).

10. *Oligosaccharide Ligands for E-selectin*

Identification of sequence similarity between E-selectin and the C-type lectins was consistent with the possibility that the E-selectin counter-receptor(s) might consist of oligosaccharide determinants. This notion was especially compelling in the context of evidence implicating carbohydrate ligands for the structurally related molecule L-selectin. Since E-selectin apparently exhibits specificity for cells of the myeloid lineage, candidate oligosaccharide counter-receptors would presumably also be restricted to these cell types, and to T-cell subsets that also adhere to E-selectin. Moreover, since the lectin domains of E- and L-selectin are so similar in their primary sequences, and since L-selectin-mediated adhesion is sialic acid-dependent, it could be imagined that the E-selectin ligand might also be sialylated.

These considerations suggested to several groups that these ligands might be found amongst a group sialylated, fucosylated glycans identified on myeloid lineage cells by Fukuda and his collaborators (Fukuda *et al.*, 1984; Fukuda *et al.*, 1985; Spooncer *et al.*, 1984), and others (Symington *et al.*, 1985). These glyco-conjugates include lipid- and protein-linked oligosaccharides whose core structures contain substantial amounts linear repeats of the lactosamine moiety [Galβ(1 → 4)Glc*N*Acβ(1 → 3)Gal]*n* (where *n*=1 to several). This polylactosaminoglycan motif may be found near the termini of numerous O-linked glycans, N-linked glycans, and lipid-linked glycans on myeloid lineage cells (Fukuda *et al.*, 1984a; Fukuda *et al.*, 1985; Spooncer *et al.*, 1984; Symington *et al.*, 1985). It is believed to be constructed by the sequential, and alternating action, of two distinct glycosyltransferases (Galβ(1 → 4)Glc*N*Ac transferase and Glc*N*Acβ(1 → 3)Gal transferase, Fig. 6.3) (van den Eijnden *et al.*, 1988). Polylactosaminoglycans, unmodified, or terminated with sialic acid in either α(2 → 3) linkage or α(2 → 6) linkage, or with fucose in α(1 → 2) linkage, are relatively common oligosaccharide motifs. They have been described on a variety of cell types, including those that do not interact with E-selectin (i.e. red cells, Fukuda *et al.*, 1984b), and by themselves represented unlikely ligand candidates.

However, these chains are found to be modified by α(1 → 3)-linked fucose residues in myeloid cell types, but not on other circulating blood cell lineages, suggesting a role for such molecules in E-selectin-dependent adhesion. This fucose modification is found on Glc*N*Ac residues of lactosamine units (Fig. 6.4). Studies of the biosynthesis of such molecules indicate that fucosylation is catalyzed by a class of glycosyltransferases called α(1 → 3)fucosyltransferases (Hansson and Zopf, 1985; Holmes *et al.*, 1985; Holmes *et al.*, 1986; Howard *et al.*, 1987; Goelz *et al.*, 1990; Kukowska-Latallo *et al.*, 1990; Mollicone *et al.*, 1990; Potvin *et al.*, 1990; Lowe, 1991; Weston *et al.*, 1992; reviewed in Lowe, 1992). Based on these studies, it is believed that fucosylation represents a terminal event in the biosynthesis of these molecules; some (but not all) α(1 → 3)fucosyltransferases can fucosylate α(2 → 3)sialylated polylactosamine moieties, whereas it has not yet been possible to demonstrate α(2 → 3)sialylation of an α(1 → 3)fucosylated glycan with the sialyltransferases that have been examined (Holmes *et al.*, 1985; Holmes *et al.*, 1986; Howard *et al.*, 1987) (Fig. 6.4).

The oligosaccharide determinants formed by these schemes include α(2 → 3)sialylated type II chains with subterminal Fucα(1 → 3)Glc*N*Ac linkages (known as the Sialyl Lewis x (SLe^x) moiety, Fig. 6.4) and analogous neutral type II fucosylated moieties (known as Lewis x (Le^x); SSEA-1, Gooi *et al.*, 1981; or CD15, Macher

Figure 6.3 Polylactosamine synthesis and chain termination. Linear polylactosamine chains are composed of alternating galactose (Gal) and *N*-acetylglucosamine (Glc*N*Ac) residues linked together by the alternating actions of a β(1 → 4) galactosyltransferase (β(1 → 4)Gal-T) and a β(1 → 3)*N*-acetylglucosaminyltransferase (β(1 → 3)GlcNAc-T) (Fukuda *et al.*, 1985; van den Eijnden, *et al.*, 1988). Linear polylactosamine chains may also be converted into branched forms by the action of β(1 → 6)*N*-acetylglucosaminyltransferases (Kapadia *et al.*, 1981) (not shown). Underlying precursor molecules (R) included glycans attached to proteins via some serine or threonine residues (Sadler, 1984), or via some asparagine residues (Kornfeld and Kornfeld, 1985), or to glycosphingolipids (Fukuda *et al.*, 1984a). The terminal lactosamine unit (Galβ(1 → 4)GlcNAcβ(1 → 3)) of a polylactosamine chain may remain unmodified after synthesis is complete, or its galactose may be further substituted with an α(1 → 2)-linked fucose residue, by the action of α(1 → 2)fucosyltransferases, to form H blood group determinants (Larsen *et al.*, 1990). This modification is found on blood cells derived from the erythroid lineage (Fukuda *et al.*, 1984b), but not on myeloid cells (Fukuda *et al.*, 1984a; Spooncer *et al.*, 1984). Alternatively, the galactose of the lactosamine moiety may be substituted with terminal sialic acid (*Neu*NAc; *N*-acetyl neuraminic acid) moieties in either α(2 → 6)- or in α(2 → 3)-linkage, via the action of α(2 → 6)- or α(2 → 3)sialyltransferases (Weinstein *et al*, 1982). In addition, the *N*-acetylglucosamine moieties within a lactosamine unit may be substituted with an α(1 → 3)-linked fucose residue by the action of α(1 → 3)fucosyltransferases, as detailed in Fig. 6.4.

et al., 1988; Fig. 6.4). Molecules have also been described that are singly fucosylated on an internal lactosamine unit of an α(2 → 3)sialylated precursor (VIM-2, Macher *et al.*, 1988; CD65, Knapp *et al.*, 1989). There is evidence that this latter structure may be synthesized by some fucosyltransferases that in turn are unable to further fucosylated the remaining unsubstituted Glc*N*Ac of the terminal lactosamine moiety (Lowe *et al.*, 1991). Alternatively, the VIM-2 determinant may function as a precursor to a difucosylated species (Fukushi *et al.*, 1985) via an ordered fucosylation scheme (Fukuda *et al.*, 1986; Howard *et al.*, 1987) (Fig. 6.4). These structures, and the multiply fucosylated species, are conveniently detected by monoclonal antibodies specific for their component terminal tri- and tetra-saccharide determinants (Fukushi *et al.*, 1984; Fukushima *et al.*, 1984).

It will be apparent from a consideration of the biosynthesis of these molecules that their expression may be controlled by regulating a(1 → 3)fucosyltransferase expression. Recent molecular cloning efforts have identified a series of four structurally distinct human α(1 → 3)fucosyltransferase genes (Goelz *et al.*, 1990; Kukowska-Latallo *et al.*, 1990; Kumar *et al.*, 1991; Lowe *et al.*, 1991; Weston *et al.*, 1992; Weston B.W. and Lowe J.B. unpublished observations). Analysis of their acceptor substrate requirements suggest that each is competent to construct a subset of the molecules shown

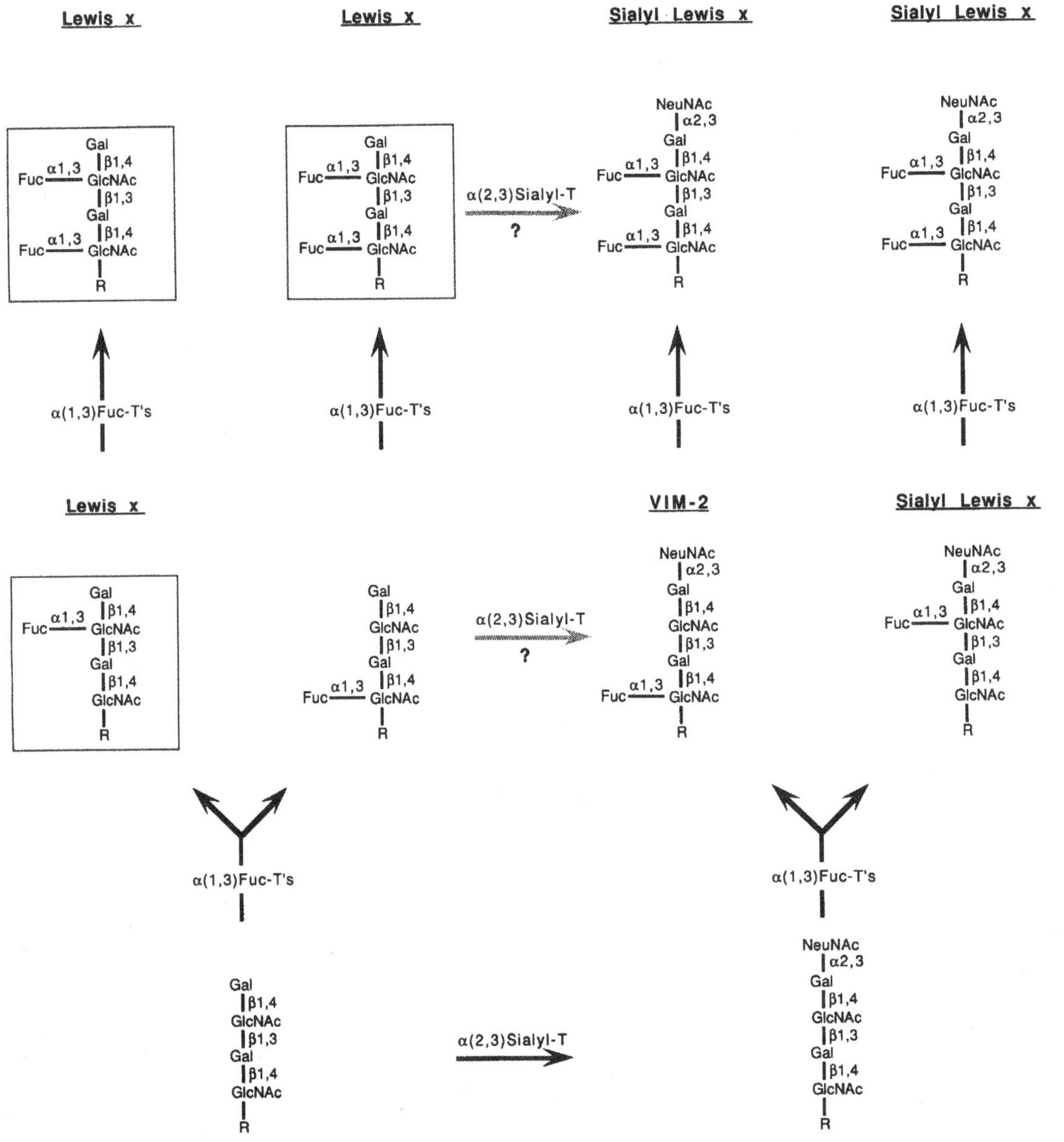

in Fig. 6.4. By contrast, little is yet known about the expression patterns of these genes, nor about mechanisms that regulate their expression. As will be noted below, it seems likely that an exploration of these issues will illuminate our understanding of mechanisms that control E-selectin ligand expression.

To test the hypothesis that one or more of these fucosylated oligosaccharides represent ligands for E-selectin, Lowe *et al.* (1990) used cloned DNA segments encoding $\alpha(1 \rightarrow 3)$fucosyltransferases to modify the cell surface glycosylation phenotype of cultured cell lines (CHO and COS-1). These cell lines do not normally adhere to E-selectin, and are deficient in $\alpha(1 \rightarrow 3)$fucosyltransferase activity. However, they do construct oligosaccharide molecules that are precursors to Le^x and SLe^x determinants. The transfected cells expressed new surface-localized oligosaccharide antigens, whose expression was then correlated with the cells' E-selectin binding properties. Each cell type was transfected with one of two different $\alpha(1 \rightarrow 3)$fucosyltransferase gene segments, yielding transfectants that expressed Le^x determinants only (via pCDNA1-α(1,3)FT; Fuc-TIV, Fig. 6.5 and Lowe *et al.*, 1991), or both Le^x and SLe^x determinants (via pCDM7-α(1,3/1,4)FT; Fuc-TIII, Fig. 6.4 and Kukowska-Latallo *et al.*, 1990). Transfectants expressing the Le^x determinant in the absence of SLe^x did not adhere to E-selectin, whereas Sle^x expression correlated with E-selectin binding. Moreover, expression of the Le^x and VIM-2 determinants, in the absence of SLe^x expression, did not confer E-selectin-dependent adhesion upon the cells (Lowe *et al.*, 1990; Lowe *et al.*, 1991).

Parallel analyses performed on two variants of the HL-60 cell line yielded results consistent with data obtained by transfection (Lowe *et al.*, 1990). One variant was competent to adhere to E-selectin whereas the other did not. The variant that did not bind expressed Le^x determinants, but no detectable SLe^x molecules. By contrast, the adhesion-competent line displayed both Le^x and SLe^x molecules. From these results, the authors concluded that one or more members of the family of $\alpha(2 \rightarrow 3)$sialylated, $\alpha(1 \rightarrow 3)$fucosylated polylactosaminoglycans, represented by the SLe^x determinant, function as endogenous ligands for E-selectin. They further concluded that regulated expression of $\alpha(1 \rightarrow 3)$fucosyltransferase(s) may control lineage-specific expression of these ligands. And, since E-selectin ligands were constructed in non-myeloid cells, and thus in the apparent absence of myeloid-specific proteins and/or lipids, they also concluded that the oligosaccharides themselves were primarily recognized by E-selectin.

Similar, though not identical, conclusions were arrived at by Goelz *et al.*, (Golez *et al.*, 1990). These investigators sought ligands for E-selectin by an approach

Figure 6.4 Synthesis of mono- and di-fucosylated oligosaccharides containing the Le^x, SLe^x and VIM-2 determinants. *N*-acetylglucosamine residues within terminal and internal lactosamine units (bottom left) may be directly fucosylated by the action of any of the known $\alpha(1 \rightarrow 3)$fucosyltransferases (Holmes *et al.*, 1985; Howard *et al.*, 1987; Goelz *et al.*, 1990; Kukowska-Latallo *et al.*, 1990; Lowe *et al.*, 1990; Lowe *et al.*, 1991; Kumar *et al.*, 1991). Monofucosylated products generated by this reaction may then be operated upon again by $\alpha(1 \rightarrow 3)$fucosyltransferases to yield difucosylated species (two vertical pathways shown at left), and also polylactosamine species containing three or more Fuc$\alpha(1 \rightarrow 3)$GlcNAc linkages (Spooncer *et al.*, 1984) (not shown). Studies with low molecular weight oligosaccharide substrates suggest that fucosylation may occur in an ordered manner that is enzyme-specific (Holmes *et al.*, 1985; Howard *et al.*, 1987), although this has yet to be studied in detail with any of the cloned $\alpha(1 \rightarrow 3)$fucosyltransferases.

Alternatively, $\alpha(2 \rightarrow 3)$sialyltransferases (arrow at bottom center) may act on a neutral polylactosamine precursor to generate an $\alpha(2 \rightarrow 3)$sialylated precursor molecule (bottom right). These sialylated molecules may then serve as substrates for the action of $\alpha(1 \rightarrow 3)$fucosyltransferases (two vertical pathways shown at right). Only a subset of known $\alpha(1 \rightarrow 3)$fucosyltransferases can efficiently fucosylate the terminal *N*-acetylglucosamine residue of this sialylated precursor, to form a monofucosylated SLe^x determinant (pathway at far right). These enzymes include, for example, Fuc-TI (Howard *et al.*, 1987), Fuc-TIII and Fuc-TV (Lowe *et al.*, 1990; Lowe *et al.*, 1991; Kumar *et al.*, 1991; Weston *et al.*, 1992), and an $\alpha(1 \rightarrow 3)$fucosyltransferase activity found in a human carcinoma cell line (Holmes *et al.*, 1985). Formation of a Fuc$\alpha(1 \rightarrow 3)$Glc*N*Ac linkages on an internal lactosamine moiety (the VIM−2 determinant) may also be completed by these α(13)fucosyltrasferases, and by at least two others that cannot efficiently construct the SLe^x determinant (Fuc-TII, Howard *et al.*, 1987; Fuc-TIV or ELFT, Goelz *et al.*. 1990; Lowe *et al.*, 1990; Potvin *et al.*, 1990, Lowe *et al.*, 1991). These monofucosylated, $\alpha(2 \rightarrow 3)$sialylated molecules can be additionally fucosylated to form difucosylated SLe^x determinants (Holmes *et al.*, 1986; Howard *et al.*, 1987). Data obtained with glycolipid acceptors indicates that some $\alpha(1 \rightarrow 3)$fucosyltransferases add fucose first to internal lactosamine units (Howard *et al.*, 1987), whereas others add preferentially to terminal units (Holmes *et al.*, 1986; Howard *et al.*, 1987). Details concerning the order of addition of the fucose residues exhibited by other $\alpha(1 \rightarrow 3)$fucosyltransferases differ are not yet available.

The mammalian $\alpha(2 \rightarrow 3)$sialyltransferase activities so far examined cannot operate on polylactosamine molecules with a terminal Fuc$\alpha(1 \rightarrow 3)$GlcNAc linkage (boxed structures) (Holmes *et al.*, 1985). It is not known if $\alpha(2 \rightarrow 3)$sialyltransferases exist that can sialylate a monofucosylated precursor to form the VIM-2 determinant. As noted in the text, there is a need to re-examine the possibility that $\alpha(2 \rightarrow 3)$sialyltransferase activities may complete these reactions (shaded arrows labelled with a question mark).

involving generation of mAbs that were specific for myeloid cell surface molecules and that also blocked adhesion of these cells to E-selectin. One such antibody, termed C2E5, was identified and then used to screen a mammalian cDNA expression library prepared from HL-60 cells. This strategy yielded a cDNA that determined C2E5 reactivity when expressed in COS cells; soluble E-selectin also bound to COS cells transfected with this cDNA. Chinese hamster cells stably transfected with this cDNA also displayed C2E5 reactivity, and bound to both soluble and solid phase E-selectin. Sequence analysis of the cDNA indicated that it maintained primary sequence similarity to the human Lewis blood group $\alpha(1 \rightarrow 3/1 \rightarrow 4)$fucosyltransferase cDNA (Kukowska-Latallo *et al.*, 1990). Extracts prepared from cells transfected with this cDNA contained substantial amounts of $\alpha(1 \rightarrow 3)$fucosyltransferase activity. This enzyme was termed ELFT, for ELAM-1 ligand fucosyltransferase. Cell surface oligosaccharides expressed by ELFT-transfected cells were not reported.

Kumar and colleagues have independently isolated a DNA segment with sequence identity to ELFT (clone D2.1, Kumar *et al.*, 1991). These two sequences are also identical to the gene termed Fuc-TIV, also cloned by Lowe *et al.* (1990 and 1991) pCDNA1-α(1,3)FT). Despite sequence identity amongst the three isolates of this gene, adhesion experiments performed by the three groups have yielded different results. In contrast to Goelz *et al.* (1990), Kukowska-Latallo *et al.* (1990), Lowe *et al.* (1991) and Kumar *et al.* (1991) were unable to demonstrate expression of E-selectin ligands when COS cells (Lowe *et al.*, 1990) or CHO cells (Kumar *et al.*, 1991; Lowe *et al.*, 1991) were transfected with this gene. Furthermore, this gene encodes an enzyme that does not efficiently construct the SLe^x determinant when tested in *in vitro* assays, and not at all in transfected cells (Kumar *et al.*, 1991; Lowe *et al.*, 1991). Several possibilities have been suggested to account for this apparent discrepancy. Kumar *et al.* (1991) note that endogenous CHO fucosyltransferase genes (like Fuc-TI; Howard *et al.*, 1987) capable of constructing SLe^x-based E-selectin ligands may be "activated" by DNA transfection approaches, and suggest that this may have occurred during amplification of the transfected ELFT cDNA in the experiments reported by Goelz *et al.* (1990). Alternatively, differences in methods used to test E-selectin-dependent adhesion may have yielded the disparate results. In particular, Goelz *et al.* (1990) used assays dependent upon both soluble and solid phase E-selectin whereas Lowe *et al.* (1990 and 1991) and Kumar *et al.* (1991) used HUVEC-based binding assay methods exclusively. In this context, it is interesting to note that the Le^x determinant itself may mediate E-selectin-dependent cell adhesion under circumstances of high-density E-selectin expression (Larkin *et al.*, 1992), likely to occur in solution-based assays but not necessarily with HUVEC-dependent methods. In any event, a final resolution of these issues will require analysis of the cell surface glyco-conjugates expressed by the ELFT transfectants, and of the fucosyltransferase(s) expressed by these cells, and quantitative determinations of E-selectin density on activated capillary endothelium *in vivo*.

Mechanisms used by myeloid lineage cells to create cell surface-localized SLe^x determinants remain incompletely understood. As noted above, and outlined in Fig. 6.4, the current understanding of this pathway would suggest that fucosylation represents the terminal event in this scheme. ELFT or Fuc-TIV would seem the logical candidate fucosyltransferase for this reaction since this enzyme is expressed in the myeloid lineage. However, this enzyme cannot create SLe^x determinants when presented with the precursor substrate Neu*N*Ac$\alpha(2 \rightarrow 3)$Gal$\beta(1 \rightarrow 4)$ Glc*N*Ac *in vitro* (Kumar *et al.*, Lowe *et al.*, 1991; Weston *et al.*, 1992), nor can it construct these molecules when expressed in transfected non-myeloid cell types (Kumar *et al.*, 1991; Lowe *et al.*, 1991; Weston *et al.*, 1992). It should also be noted, however, that extracts of granulocytes and other myeloid lineage cell types also cannot efficiently complete this type of reaction when tested *in vitro* (Mollicone *et al.*, 1990). At least two possibilities come to mind to explain these apparent paradoxes. One involves a mechanism whereby Fuc-TIV/ELFT requires one or more myeloid-specific molecules that function to allow it to efficiently fucosylate $\alpha(2 \rightarrow 3)$sialylated type II chains, in a manner that operates within intact cells but not *in vitro* when testing cell extracts. Such a molecule might represent a *trans*-acting factor that modulates acceptor substrate specificity of Fuc-TIV (analogous to α-lactalbumin modulation of $\beta(1 \rightarrow 4)$galactosyltransferase substrate specificity; reviewed in Beyer *et al.*, 1982), for example, or a molecule that preferentially "presents" the oligosaccharide acceptor substrate to the enzyme (i.e. a myeloid-specific glycan or protein, for example, analogous to the modulation of pituitary αGalNAc-transferase catalytic efficiency by the peptide that presents its glycan acceptor; Smith and Baenziger, 1992). Alternatively, an $\alpha(2 \rightarrow 3)$sialyltransferase may exist in myeloid cells with the as yet unprecedented ability to sialylate Gal$\beta(1 \rightarrow 4)$ [Fuc$\alpha(1 \rightarrow 3)$]Glc*N*Ac-terminated glycans (pathway indicated by the shaded line at the top of Fig. 6.4). Further biochemical and genetic studies will be necessary to illuminate these questions.

Three other groups have also shown that SLe^x containing oligosaccharides function as ligands for E-selectin. Phillips *et al.* (1990) first used a pair of glycosylation mutants of Chinese hamster ovary (CHO) cells that express the SLe^x and Le^x determinants in a HUVEC-based assay for E-selectin-dependent adhesion. The CHO mutant LEC11 expresses an $\alpha(1 \rightarrow 3)$fucosyltransferase termed Fuc-TI (Howard *et al.*, 1987), and displays both the SLe^x determinant and its neutral isomer Le^x. By contrast, LEC12 cells express Fuc-TII (Howard *et al.*, 1987), and display only Le^x determinants. The LEC11 mutant cells adhered to activated HUVECs in an E-selectin-dependent manner, whereas neither LEC12 cells nor wild-type CHO cells bound. E-selectin-dependent adhesion of LEC11 cells and HL-60 cells was prevented when the cells were pre-treated with anti-SLe^x antibodies, and diminished when they were subjected to sialidase treatment. By contrast, pre-treatment with anti-Le^x antibodies did not block binding of either cell type.

Purified glycolipids were then tested for their ability to inhibit E-selectin-dependent cell adhesion. Each glycolipid contained a linear lactosamine-type core, which was then substituted with fucose and/or sialic acid to form a difucosylated sialyl Le^x species, a difucosylated Le^x molecule, and an $\alpha(2 \rightarrow 3)$sialylated but nonfucosylated molecule. For testing of E-selectin interaction, each glycolipid was incorporated into liposomes, and the liposomes used were assayed for their ability to disrupt binding of HL-60 cells or neutrophils to IL-1β-activated HUVEC monolayers. Only liposomes containing the difucosylated SLe^x glycolipids exhibited significant inhibition. More recent work by this group has shown that a monofucosylated SLe^x molecule can block binding with an efficiency comparable to that shown by the difucosylated SLe^x species (Polley *et al.*, 1991).

Another group used a recombinant E-selectin-IgG chimera as a reagent to probe for ligands that bind to E-selectin (Walz *et al.*, 1990). A survey of several cell types using this chimera disclosed an imperfect correlation between their ability to bind this recombinant reagent and surface expression of the Le^x epitope. In each case, however, sialidase treatment of the cells effectively eliminated binding of the chimera, indicating that a sialylated form of the Le^x moiety was responsible for binding. Additional analyses demonstrated a perfect correlation between SLe^x expression and E-selectin binding. Binding could be inhibited by pre-treatment with anti-SLe^x antibody, but not with anti-Le^x or anti-VIM-2 antibodies. Furthermore, E-selectin adhesion was diminished by SLe^x-containing molecules isolated from amniotic fluid or on α_1-acid glycoprotein. Furthermore, fucosylation of bovine α_1-acid glycoprotein using an affinity-purified amniotic fluid $\alpha(1 \rightarrow 3)$fucosyltransferase yielded molecules with enhanced E-selectin binding activity, relative to a non-fucosylated preparation. These results together prompted the authors to conclude that L-selectin recognizes the SLe^x tetrasaccharide.

In a complementary approach, Tiemeyer and colleagues first isolated glycolipid species from human myeloid cells, and tested these for their ability to mediate E-selectin-dependent cell adhesion (Tiemeyer *et al.*, 1991). From a number of glycolipid fractions analyzed, one purified acidic fraction that supported adhesion was structurally characterized by methylation analysis and fast atom bombardment mass spectroscopy. The structure of the molecule in this fraction was proposed to be a glycolipid consisting of 4 *N*-acetyllactosamine units tethered to the membrane via a ceramide group, and terminated by $\alpha(2 \rightarrow 3)$linked sialic acid, with a single fucose in $\alpha(1 \rightarrow 3)$linkage on the Glc*N*Ac residue of the subterminal *N*-acetyllactosamine unit. This structure represents the VIM-2 determinant characterized by Macher *et al.* (1988; Fig. 6.4). This result apparently conflicted with those reported by Walz *et al.* (1990) and Lowe *et al.* (1991), who each concluded that this structure is neither necessary nor sufficient for E-selectin-dependent cell adhesion. More recent quantitative studies suggest that the VIM-2 glycolipid can in fact support E-selectin-dependent adhesion, although the affinity of E-selectin for this determinant is roughly one to two orders of magnitude less than that shown for SLe^x containing glycolipids (Tyrrel *et al.*, 1991).

Three groups have recently shown that the sialyl Lewis a (SLe^a) moiety can also support E-selectin-dependent adhesion. The SLe^a determinant is believed to be constructed from an $\alpha(2 \rightarrow 3)$sialylated type I precursor by the action of $\alpha(1 \rightarrow 3/1 \rightarrow 4)$fucosyltransferase(s) (Fig. 6.5; Hansson and Zopf, 1985). In studies reported by Berg *et al.* (1991), a panel of neoglycoproteins was screened for structures that reacted with the mAb HECA-452. This antibody reacts with a determinant on the surface of a variety of cell types that bind to E-selectin. These include skin homing memory T-cells that do not react with SLe^x (Picker *et al.*, 1991; Shimizu *et al.*, 1991). These investigators sought oligosaccharide determinants distinct from SLe^x, but which would nonetheless support both E-selectin-dependent adhesion and HECA-452 binding. The screen identified a HECA-452-reactive neoglycoprotein molecule that turned out to contain the SLe^a epitope. Quantitative binding affinity determinations indicated that the SLe^a neoglycoprotein exhibits an affinity for E-selectin that is slightly higher that that maintained by an analogous SLe^x neoglycoprotein. Computer-generated molecular models reported by this group suggest that the fucose and sialic acid residues of the two isomers are each displayed in virtually identical positions on one side of each structure.

Tyrrel *et al.* (1991) also demonstrated that the sialyl Lewis a moiety is able to support E-selectin-dependent adhesion, with an affinity approximately comparable to that shown by SLe^x-containing ligands. Substitution of a glucose for the *N*-acetylglucosamine at the reducing end of the SLe^x tetrasaccharide apparently enhances affinity. Reduction of this glucose with sodium borohydride minimally alters affinity, suggesting that a ring-closed conformation of the reducing sugar is not necessary for effective interaction with E-selectin. The authors note that the *N*-glycolyl and *N*-acetyl forms of sialic acid yield SLe^x tetrasaccharide molecules with essentially equivalent affinities, as do derivatives in which carbons 8 and 9 of the sialic acid have been removed by mild periodate oxidation. By contrast, isomeric molecules containing sialic acid in $(2 \rightarrow 6)$linkage are apparently not recognized by E-selectin. Molecular modelling studies suggest that the fucose, sialic acid, and galactose residues of these isomeric tetrasaccharides occupy relatively similar positions

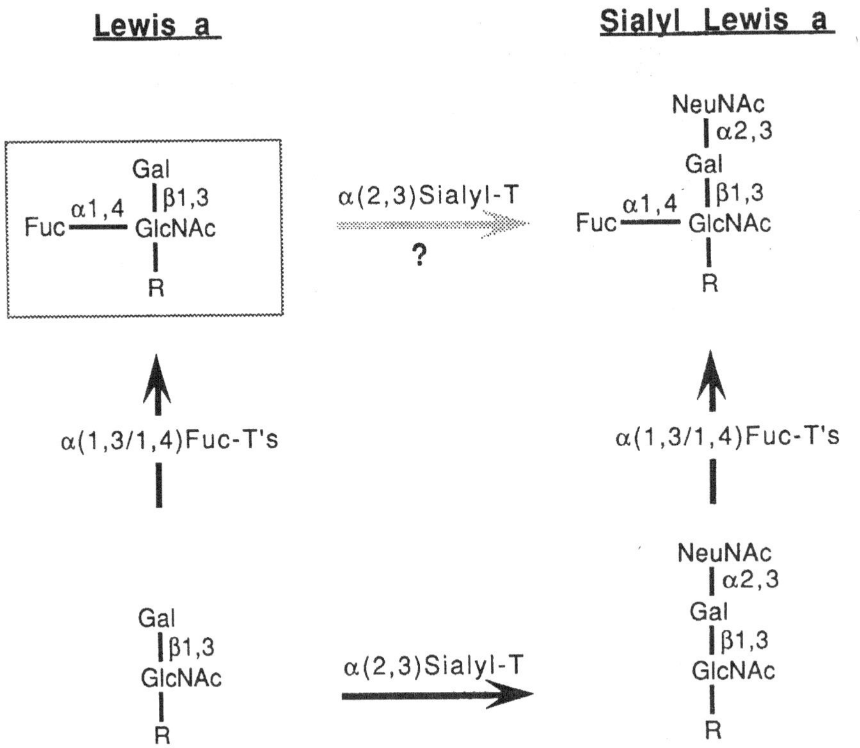

Figure 6.5 Synthesis of the Le^a and SLe^a determinants. Hansson and Zopf (1985) have shown that the SLe^a molecule is formed via fucosylation of an $\alpha(2 \rightarrow 3)$sialylated type I precursor, and not necessarily by sialylation of the Lewis a determinant. Fucosylation occurs via action of Fuc-TIII (Kukowska-Latallo *et al.*, 1990; Lowe *et al.*, 1990; Weston *et al.*, 1992), which exhibits both $\alpha(1 \rightarrow 4)$fucosyltransferase activity (shown here) and $\alpha(1 \rightarrow 3)$fucosyltransferase (detailed in Fig. 6.4). This enzyme is thought to be the product of the human Lewis blood group locus (Kukowska-Latallo *et al.*, 1990; Weston *et al.*, 1992).

in space, thus accounting for their virtual functional identity.

Takada and colleagues have also shown that the SLe^a determinant can mediate E-selectin-dependent cell adhesion (Takada *et al.*, 1991). A human colorectal carcinoma cell line Colo 201 was found to exhibit E-selectin-dependent cell adhesion which could not be diminished by pre-treatment with anti-SLe^x antibodies. By contrast, pre-treatment with SLe^a antibodies effectively diminished E-selectin-dependent adhesion. Flow cytometry analysis demonstrated that the surfaces of these cells display large amounts of SLe^a determinants, and smaller amounts of SLe^x molecules. Liposomes prepared with SLe^a glycolipids, or with SLe^x glycolipids, were each able to prevent SLe^a-mediated Colo 201 cells to E-selectin. Moreover, these reagents were also able to block SLe^x-dependent adhesion of HL-60 cells to E-selectin. These observations indicate that E-selectin recognizes both isomers. The cross-blocking properties of these two tetrasaccharides further suggest that they interact with the same site(s) on E-selectin, and maintain similar tertiary conformations.

In aggregate, the data summarized in this section provide direct evidence that at least two distinct oligosaccharides can function as ligands for E-selectin. It may also be indirectly inferred that at least one additional ligand exists; skin homing memory T-cells that adhere to E-selectin do not express either of these two determinants. Nonetheless, these cells do react with the HECA-452 antibody (as do the SLe^x and SLe^a determinants), suggesting that this as yet undefined E-selectin ligand at least maintains a conformational similarity to the two known tetrasaccharide ligands.

On myeloid lineage cells, the SLe^x determinant most certainly functions as the glycan portion of E-selectin ligands on myeloid cells, since it is abundantly represented on the surfaces of neutrophils and monocytes. By contrast, SLe^a determinants are not found on these cells, and thus play no physiological role in E-selectin-dependent myeloid cell adhesion. However, SLe^a molecules are frequently detected on carcinomas derived from pancreatic tissue, and on malignancies derived from epithelia lining other organs of the digestive tract (Kim and Itzkowitz, 1988). Such cancers may also express SLe^x determinants, as may malignancies arising from the lung and ovary, for example. This set of observations has suggested the possibility that these oligosaccharides may function *in vitro* to promote metastasis via E-selectin-dependent adhesion to vascular endothelia (Lowe *et al.*, 1990; Walz *et al.*, 1990; Berg *et al.*, 1991; Takada *et al.*, 1991). And, since P-selectin may also recognize these determinants (discussed below), such carbohydrate ligands may also operate to promote tumor spread via adhesion to P-selectin expressed by endothelial cells and by activated platelets. The actual contribution of this mechanism to tumor metastasis *in vivo* is under active investigation.

Some epithelial malignancies are known to also release mucins that display SLe^x and SLe^a moieties (Kannagi *et al.*, 1986). These molecules can be found circulating within the vascular compartment, and could in principle act to diminish normal selectin-dependent leucocyte-endothelium adhesion. Whether this actually occurs *in vivo* remains a subject for additional study.

Important unanswered questions revolve around refining our understanding of the complete structures of the oligosaccharides that display the tetrasaccharide ligands, and on further defining the proteins and lipids that in turn present these glycan structures on the exterior of the cell. Biochemical analyses of myeloid cells indicate that surface-localized SLe^x determinants are found on a diverse group of glycolipids (Symington *et al.*, 1985) and glycoproteins (Fukuda *et al.*, 1984a) (including members of the CD11/CD18 group of cell adhesion molecules, Asada *et al.*, 1991). Such molecules may figure importantly in regulating ligand affinity and or functional competence, for example. In this context, it has been demonstrated recently that substantial numbers of SLe^x determinants on granulocytes are displayed by L-selectin (Picker *et al.*, 1991). Morphological analyses indicate that the geographical location of L-selectin (on the microvillous processes at the neutrophil surface) may facilitate presentation of SLe^x moieties to E-selectin, especially under conditions of flow-dependent shear stress where carbohydrate-selectin interactions are thought to be most important (Lawrence *et al.*, 1990).

The proteins and lipids that display E-selectin carbohydrate ligand(s) may also function to transmit activational signals to the leucocyte interior following selectin engagement. For instance, it has been demonstrated that the adhesive "activity" of the neutrophil CD11b/CD18 complex may be transiently up-regulated when E-selectin interacts with the neutrophil surface (Lo *et al.*, 1991). The authors also presented evidence that human granulocytes exhibit a chemotactic response to soluble E-selectin, and thereby suggested that endothelial-expressed E-selectin may function as a "tethered chemo-attractant" for neutrophils. These observations suggested a model of neutrophil adhesion comprised of sequential adhesive events, initiated by SLe^x E-selectin-dependent adhesion, followed by up-regulation of CD11b/CD18 adhesive activity, then firm attachment mediated via CD11b/CD18-ICAM-1 adhesion, and finally transmigration to the extravascular space. Similar models, some further embellished, have been summarized by others (Butcher, 1991); each invokes an initial and essential role for E-selectin-dependent adhesion in leucocyte recruitment, and are discussed further at the end of this chapter.

These models, if correct, imply that pharmaceutical manipulation of the carbohydrate-E-selectin adhesive interaction might represent a powerful means to diminish undesired leucocyte-dependent inflammation. Intravascular delivery of the carbohydrate ligands represent one possibility. However, it has been difficult, or impossible, to chemically synthesize oligosaccharide molecules of the size and complexity of the SLe^x tetrasaccharide (let alone higher order structures) in quantities sufficient for animal testing (Kameyama *et al.*, 1991; Nicolaou, 1991). Cloned glycosyltransferase sequences (reviewed in Lowe, 1992), when coupled with enzyme-assisted synthetic schemes (Toone *et al.*, 1989; Wong, 1989; Dumas *et al.*, 1991; Ichikawa *et al.*, 1991; Ball *et al.*, 1992), have been used recently to circumvent these difficulties. It may be expected that these approaches may soon yield both natural and structurally modified (Srivastava *et al.*, 1990; Dumas *et al.*, 1991) oligosaccharides in amounts that may be tested *in vivo* for therapeutic effectiveness, and for toxicity. The availability of meaningful amounts of these molecules should also facilitate solution of their structures by nuclear magnetic resonance procedures (Ball *et al.*, 1992) and other methods. This in turn may promote the identification or construction of molecular mimics that might also exhibit anti-inflammatory properties, and that may also prove to exhibit desirable pharmacological properties.

Alternatively, it may be possible to discover or design molecules that specifically inhibit myeloid-specific glycosyltransferases involved in the construction of the SLe^x determinant, that prevent the synthesis or intracellular transport of substrates necessary for the catalytic activity of these enzymes, or that diminish the expression of the genes that encode them. These approaches must be viewed as longer-term possibilities, however, since our understanding of these areas is limited at present, and it might be expected that such approaches would represent chronic therapeutic maneuvers with a substantial potential for suppression of phagocyte-dependent immune function.

11. Oligosaccharide Ligands for P-selectin

P-selectin was discovered as a component of the α-granules in platelets. The platelet-derived molecule is a 140 kDa glycoprotein. Activation of the platelet exocytosis pathway by thrombin releases the protein from its granular location, ultimately placing the molecule within the platelet membrane (Stenberg *et al.*, 1985; Berman *et al.*, 1986). P-selectin was originally designated granule membrane protein (GMP-140; Stenberg *et al.*, 1985) and platelet activation-dependent granule to external membrane protein (PADGEM, Stenberg *et al.*, 1985). It's cluster designation is CD62 (Knapp *et al.*, 1989).

In addition to platelets and megakaryocytes (Beckstead *et al.*, 1986; McEver *et al.*, 1987), P-selectin is found in a human erythroleukemia cell line (HEL, Johnston *et al.*, 1989), and in the Weibel-Palade bodies of vascular endothelial cells (Bonfanti *et al.*, 1989; McEver, 1989). Endothelial cells may be induced to display surface-localized P-selectin by stimulation with thrombin, histamine, phorbol ester, and the calcium ionophore A23187 (Hattori *et al.*, 1989), via mobilization of preformed protein from the Weibel-Palade bodies. Expression reaches its peak at roughly 10 min following stimulation; expression returns to baseline levels within 1 h, via processes that most likely involve endocytosis (Hattori *et al.*, 1989).

Oxygen radicals also induce endothelium to express P-selectin (Patel *et al.*, 1991). Expression occurs after only brief exposure to micromolar concentrations of oxidant compounds. In contrast to histamine-dependent mechanisms, oxygen radicals cause P-selectin to be expressed for several hours. Defective endocytotic reinternalization may be responsible for this phenomenon (Patel *et al.*, 1991). These investigators postulate that oxygen radical-dependent P-selectin expression represents a potential mechanism for neutrophil-dependent amplification of P-selectin-dependent leucocyte recruitment (Johnston *et al.*, 1989).

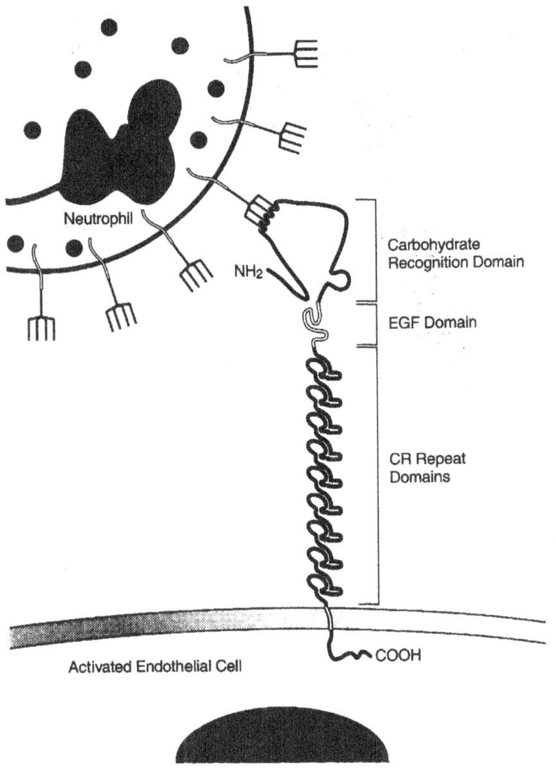

Figure 6.6 The domain structure of P-selectin. Peptide domains are drawn approximately to scale, and are folded according to predicted disulfide bond positions assigned by Johnston *et al.* (1989). As detailed in the text, P-selectin is acutely expressed by capillary endothelial cells after treatment with histamine and phorbol esters. As detailed in the text, the P-selectin carbohydrate recognition domain mediates adherence of circulating myeloid lineage cells via binding to cell surface carbohydrate ligands displayed by proteins (indicated by ⌐E).

12. *P-selectin Maintains Structural Similarity to E- and L-selectin*

Cloned cDNAs encoding human P-selectin have been isolated, using sequence information derived from platelet-derived P-selectin (Johnston *et al.*, 1989). The molecule maintains an amino terminal domain with primary sequence similarity to several C-type lectins, an adjacent domain similar to the EGF motif, nine CR segments, a single membrane-spanning segment, and a short COOH-terminal cytosolic tail (Fig. 6.6). This topology and domain structure of P-selectin is virtually identical to those exhibited by L-selectin and E-selectin. These three proteins also maintain significant primary sequence similarity between their respective amino terminal CRDs, and also between their EGF domains. The nine CR repeats in P-selectin make this molecule the largest of the selectin family members defined to date.

Careful sequence analysis of cloned P-selectin cDNAs, polymerase chain reaction (PCR) experiments, and genomic structure analyses indicate that alternative splicing mechanisms generate two additional forms of this protein (Johnston *et al.*, 1989; Johnston *et al.*, 1990). One alternately spliced transcript "skips" sequence corresponding to exon 14. The "skipped" sequence is 120 base pairs long, and corresponds to 40 amino acids that encompass the membrane spanning segment. This transcript can be found in megakaryocytes, platelets, and endothelial cells, and apparently yields a soluble form of P-selectin. The other, relatively minor transcript is found in endothelial cells, but not in platelets. This transcript is missing sequences corresponding to exon 11, and is predicted to yield a P-selectin variant that lacks the seventh CR repeat. The functional significance of these observations are not yet defined, although it can be imagined that the circulating, soluble form of P-selectin could function to dampen neutrophil adhesion during inflammation by binding to and thus masking its counter-receptor on these cells (see below).

13. *P-selectin is a Cell Adhesion Molecule*

Activated platelets are known to bind myeloid cells, including neutrophils, monocytes, and Hl-60 and U937 myeloid cell lines (Jungi *et al.*, 1986; Silverstein and Nachman, 1987). To test P-selectin's role here, Larsen *et al.* (1989) used a rosetting assay where thrombin-activated platelets served as visible markers for displaying solid phase P-selectin to myeloid cell types. These authors showed that activated platelets rosetted around HL-60 cells, whereas non-activated platelets did not. Polyclonal anti-P-selectin antisera blocked rosette formation, as did inclusion of purified P-selectin or ethylenediamine tetracetic acid (EDTA). By contrast, control antibodies directed against other platelet proteins (GPIIb/IIIa; thrombospondin; GPIV), or the cognate peptides themselves, did not inhibit rosette formation. Other types of myeloid cells also displayed these properties, whereas non-phagocytic cells (red cells, lymphocytes, Jurkatt cells, Daudi cells) did not interact with activated platelets. Moreover, the authors created fluorescently-labelled P-selectin-bearing phospholipid vesicles, and tested these for their ability to adhere to various cell types. They found that the P-selectin vesicles interacted with the same spectrum of cell types rosetted by activated platelets. Non-substituted vesicles did not bind. Vesicles reconstituted with P-selectin exhibited specific and saturable binding to U937 cells, and could be specifically inhibited by anti-P-selectin antibodies.

Hamburger and colleagues (Hamburger and McEver, 1990) have also shown that P-selectin can mediate adherence of thrombin-activated platelets to myeloid cells. This group has extended these observations in studies where P-selectin-dependent adhesive properties of neutrophils and HL-60 cells were examined (Geng *et al.*, 1990). COS cells transfected with a P-selectin cDNA bound HL-60 cells whereas control transfectants did not. Adhesion was inhibitable by a subset of the anti-P-selectin antibodies tested. Microtiter plates coated with P-selectin also effectively bound neutrophils and HL-60 cells; binding could be blocked with anti-P-selectin antibodies, and with purified P-selectin, but not with corresponding control reagents. Formalin-fixed HL-60 cells and neutrophils bound with an efficiency equivalent to that shown by unfixed cells. This observation indicates that metabolic activity is not necessary for P-selectin-dependent adhesion. Moreover, the formalin-resistant nature of the P-selectin counter-receptor on myeloid cell surfaces is consistent with subsequent observations (discussed below) indicating that it is carbohydrate in nature.

P-selectin-dependent adhesion of neutrophils to endothelial cell monolayers was also demonstrated (Geng *et al.*, 1990). In these studies, P-selectin expression was triggered by treating the HUVECs with histamine or PMA. Binding was inhibited by pre-treating with anti-P-selectin antibodies, and the time course of binding exactly paralleled the time course of P-selectin expression. Activation-dependent binding, and also P-selectin expression, were not prevented by pre-treating with RNA or protein synthesis inhibitors, indicating that preformed P-selectin is operative in this context.

An examination of P-selectin's divalent cation requirements indicated that there are at least two such metal binding sites (Geng *et al.*, 1990). Maximal adhesion occurs at calcium concentrations approximating 1 mM. In these studies, P-selectin-dependent adhesion was also supported by magnesium ions, but only when micromolar concentrations of calcium are also present. These initial observations suggested the existence of one high affinity site that binds only calcium, and a second site with a relatively lower affinity for either magnesium or calcium. Subsequent observations have refined these notions (Geng *et al.*, 1991). Equilibrium dialysis measurements with purified P-selectin identified two high affinity calcium binding sites, and that these sites do not bind magnesium ions tightly. Fluorescence spectroscopy analyses indicate that calcium binding is associated with a change in P-selectin conformation. Furthermore, a P-selectin antibody (G3) that blocks P-selectin-neutrophil adhesion was shown to recognize P-selectin only when assayed in the presence of appropriate levels of calcium. Since this antibody had been shown to interact with an epitope corresponding to a 16 residue synthetic peptide prepared from a position in the P-selectin CRD (amino acids 19–34), the authors concluded that calcium effects a conformational change in the CRD that in turn allows recognition of its leucocyte ligand(s).

A third group has also reported P-selectin-dependent adhesion of monocytes and neutrophils (Gamble *et al.*, 1990; Skinner *et al.*, 1991). Purified soluble P-selectin bound to HL-60 cells and neutrophils via a single class of high affinity binding sites. Solution phase binding studies, and platelet rosetting, were shown to be inhibitable with EDTA, and with antibody reagents directed against P-selectin. This group has also reported that soluble P-selectin could inhibit CD18-dependent adhesion of TNFα-activated neutrophils to endothelium. These latter observations remain to be reconciled with the current understanding of CD18-dependent adhesion (reviewed in Springer, 1990).

14. *Evidence for Oligosaccharide Ligands for P-selectin*

The CRD in P-selectin, and the overall structural similarity of this molecule to the other two selectins, have prompted several groups to test the possibility that P-selectin mediates cell adhesion via carbohydrate counter-receptor(s). The initial studies of Skinner *et al.* (1989) indicated that heparin binds with high affinity to P-selectin. This binding was shown to be cation-independent, however, and suggesting that heparin may interact with P-selectin via mechanisms distinct from the ones that are responsible for the calcium-dependent adhesive properties of this molecule. More recently, this group has reported that the same series of sulfated glycans previously investigated (heparin, fucoidan, and high molecular weight dextran) can also block P-selectin interactions with myeloid cells (Skinner *et al.*, 1991). Other sulfated glycans were less effective at blocking (lambda- and kappa-càrrageenan, and low molecular dextran sulfate), and others were not at all effective (chondroitin 4-sulfate and chondroitin 6-sulfate). These experiments did not address divalent cation requirements for inhibition, and thus the relationship of these observations to the cation-dependent adhesive properties of P-selectin remains to be defined.

In another line of investigation, Larsen and colleagues have reported inhibition of P-selectin-dependent cell adhesion with a trisaccharide molecule expressed by myeloid-lineage cells (Larsen *et al.*, 1990). Initially, this group tested a panel of myeloid-specific mAbs for those that would inhibit rosetting of activated platelets to myeloid cells. Three inhibiting antibodies were identified; each recognized the CD15 determinant (Fig. 6.4). Purified oligosaccharides containing this determinant (lacto-N-fucopentaose III; Fig. 6.7), or related control oligosaccharides (lacto-N-fucopentaose I and II; Fig. 6.7), were then examined for their ability to block P-selectin-dependent adhesion in assays that measured HL-60 binding to P-selectin-expressing COS cells, or in platelet rosetting assays. Lacto-N-fucopentaose III was found to inhibit half maximally at a concentration of 50 μg/ml, whereas lacto-N-fucopentaose I did not inhibit at concentrations up to 300 μg/ml, and only minimal inhibition was seen with lacto-N-fucopentaose II at these concentrations. Since rather high concentrations of the oligosaccharide were necessary to achieve inhibition, the authors speculated that the CDl5 determinant represents just a portion of the endogenous ligand.

A requirement for sialic acid in P-selectin-dependent cell adhesion has been shown by Corral *et al.* (1990). Using a platelet rosetting assay, this group demonstrated that P-selectin-dependent adhesion of HL-60 cells, and of neutrophils, can be substantially diminished if these cells are first treated with one of three different broad-

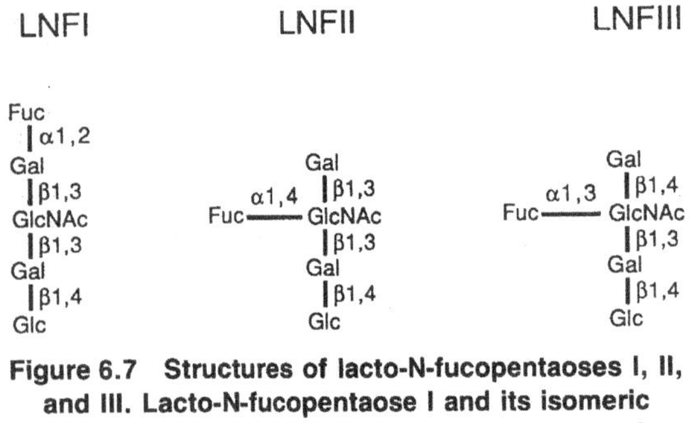

Figure 6.7 Structures of lacto-N-fucopentaoses I, II, and III. Lacto-N-fucopentaose I and its isomeric molecules lacto-N-fucopentaoses II and III, are the soluble, low molecular weight oligosaccharides tested for their ability to inhibit P-selectin-dependent cell adhesion (Larsen *et al.*, 1990).

spectrum sialidases (*Clostridium perfringens, Vibrio cholera*, or *Arthrobacter ureafaciens*). Myeloid cells are known to display glycans containing sialic acid in $\alpha(2 \rightarrow 3)$ and $\alpha(2 \rightarrow 6)$ linkage (Fukuda *et al.*, 1984a; Fukuda *et al.*, 1985; Spooncer *et al.*, 1984), suggesting that one or both types of sialic acid determinants are essential components of the P-selectin ligand(s). To address this further, cells were tested for binding after treatment with Newcastle disease virus (NDV) neuraminidase. This enzyme hydrolyzes *Neu*NAc$\alpha(2 \rightarrow 3)$ or *Neu*NAca$(2 \rightarrow 8)$ linkages, but leaves *Neu*NAc$\alpha(2 \rightarrow 6)$ linkages intact (Paulson *et al.*, 1982). Treatment with NDV neuraminidase did not diminish binding, thus indirectly implicating *Neu*NAc$\alpha(2 \rightarrow 6)$ linkages in P-selectin-dependent cell adhesion. Furthermore, it was shown that NDV neuraminidase effectively eliminated cell surface *Neu*NAc$\alpha(2 \rightarrow 3)$ linkages on cells that remained competent to bind P-selectin. This suggested that *Neu*NAc$\alpha(2 \rightarrow 3)$ determinants are not necessary for P-selectin-dependent adhesion.

A contrasting conclusion was reported by Moore and colleagues, using an approach that tests binding of purified P-selectin to myeloid cells (Moore *et al.*, 1991). In their assay, purified, soluble P-selectin exhibits specific and saturable binding to myeloid cells, with half maximal binding at P-selectin concentrations ranging between 0.9 nM and 2.2 nM. Maximal binding occurs at concentrations between 3 nM and 6 nM. The number of binding sites per cells was not calculable, since the purified P-selectin exists as a mixture of monomeric and oligomeric species.

P-selectin binding is diminished by pre-treatment with proteases, indicating that the P-selectin ligand assayed with this approach contains a protein component (Moore *et al.*, 1991). Binding was reduced to 28% of control binding if cells were first treated with a broad spectrum sialidase (*Vibrio cholera*), and a significant reduction in binding was also observed after pre-treatment with NDV sialidase (52% of control). These results implicate *Neu*NAc$\alpha(2 \rightarrow 3)$ linkages in P-selectin binding to myeloid cells, in contradistinction to those reported by Corral *et al.* (1990). This discrepancy has not yet been formally resolved in the literature, although it should be noted that NDV sialidase is a component of the viral particle itself, and thus in principle could yield effects independent of the sialidase activity itself, especially in the context of different assay systems.

Moore and colleagues also attempted to reproduce the results of Larsen *et al.* (1990), using an anti-CD15 antibody, and a multivalent Le^x-albumin conjugate, as reagents to inhibit P-selectin binding to neutrophils. The anti-CD15 antibody 82H5 used in Moore's report did not inhibit, even when tested at concentrations up to 400 μg/ml; by contrast, Larsen and colleagues reported significant inhibition (35% of control binding) with a different anti-CD15 antibody (80H5), at a concentration of 20 μg/ml. Moore *et al.* (1991) also demonstrated that a multivalent Le^x-albumin conjugate did not diminish P-selectin-dependent binding, in contrast to Larsen *et al.* (1990), who reported inhibition with a soluble Le^x reactive oligosaccharide. It seems likely that the apparently different outcomes reported by the two groups may be ascribed to assay differences, and to differences in the types of reagents used.

Polley *et al.* (1991) used a series of complementary approaches to address the nature of P-selectin ligands, using the thrombin-activated platelet rosetting assay. Their antibody blocking studies showed that anti-SLe^x antibodies could effectively inhibit P-selectin-dependent adhesion, whereas anti-Le^x antibodies were only partially inhibitory. These observations were extended with the use of glycosylation mutants of CHO cells; a SLe^x-positive mutant (LEC11) bound to activated platelets, but the Le^x-positive, SLe^x-negative mutant (LEC12) did not. Furthermore, sialidase treatment of the LEC11 cells diminished binding.

Liposomes reconstituted with SLe^x glycolipids effectively blocked P-selectin-dependent adhesion (Polley *et al.*, 1991). Likewise, a soluble hexasaccharide containing the SLe^x moiety also blocked, half maximally at 2 μg/ml. By contrast, a pentasaccharide containing the Le^x determinant exhibited half-maximal inhibition at 54 μg/ml. These results strongly suggest that the SLe^x tetrasaccharide represents an important component of the endogenous neutrophil and myeloid ligand(s) for P-selectin. Since SLe^x containing liposomes yield similar results when tested in E-selectin-dependent adhesion assays (Phillips *et al.*, 1990; Polley *et al.*, 1991), these observations suggest that the oligosaccharide portions of the ligands for these two selectins will maintain a substantial amount of structural similarity.

Similar observations, and analogous if not identical conclusions have been reported by Zhou *et al.* (1991). These investigators first sought to correlate P-selectin-dependent adhesion with cell surface glycosylation phenotype, using a panel of cultured cells differing in these properties. They found an excellent, if not perfect, correlation between P-selectin-dependent binding SLe^x expression. Binding could be diminished by pre-treating with SLe^x antibodies, but not with anti-Le^x antibodies. Likewise, Le^x-positive, SLe^x negative cells did not demonstrate P-selectin-dependent adhesion. As noted previously (Moore *et al.*, 1991), protease pre-treatment removed endogenous P-selectin ligand(s) from adherent-competent cells.

A quantitative analysis of binding site number and affinity was also reported for two different P-selectin-adherent cell lines. In these studies, HL-60 cells were shown to display a relatively low number of relatively high affinity P-selectin binding sites. By contrast, a SLe^x-positive CHO cell line expressing a cloned fucosyltransferase cDNA (CHO-FT, Lowe *et al.*, 1990) was found to express large numbers of relatively low affinity binding sites. Considered together with other observations outlined in this section, this work suggests that P-selectin recognizes protease-sensitive glycan(s) containing the SLe^x determinant. The work of Zhou *et al.* (1991) also suggests that the myeloid lineage contributes molecules that promote relatively high affinity interactions with P-selectin, that are apparently missing from non-myeloid cell type like the SLe^x-positive transfected CHO line. The structure and composition of such molecules remains to be defined.

Finally, sulfatides (3-sulfated galactosylceramides) have also been reported to be ligands for P-selectin (Aruffo *et al.*, 1991). In this study, P-selectin ligands were sought amongst HL-60 membrane glycolipids, by using a recombinant P-selectin-IgG chimera to stain glycolipids fractionated by thin layer chromatography. Sulfatides were the only group of molecules that stained with the chimera; gangliosides (presumably including those that terminate in the SLe^x moiety, Symington *et al.*, 1985) were not stained by this P-selectin probe. Purified sulfatides were found to inhibit P-selectin-dependent adhesion of U937 myeloid cells, whereas control glycolipids (lysosulfatide and galactosylceramide) did not. Finally, growth of HL-60 cells in the presence of selenate, a sulfation inhibitor, diminished both cell surface sulfatide expression and P-selectin-dependent adhesion. When considered with the extensive observations implicating a protease-sensitive, SLe^x-containing ligand for P-selectin binding, these results suggest that sulfatides may represent a second class of P-selectin ligands.

15. *Selectin-dependent Leucocyte Rolling, and a Sequential Model for Leucocyte Adhesion and Transmigration*

As noted in the sections above, under physiological flow conditions, leucocyte adhesion to activated vascular endothelium is first characterized by leucocyte rolling, and is then followed by arrest, spreading, and transmigration. A remarkable report by Lawrence and Springer (1991) presents a series of experiments demonstrating that P-selectin can also participate in leucocyte "rolling". These experiments were done using a chamber in which leucocytes are made to pass over an artificial planar lipid bilayer under conditions that mirror shear stresses found in post-capillary venules. In this system, video microscopy is used to determine the velocities of neutrophils as they flow past the experimental surface, and thus allow a discrimination between adhesion-dependent rolling and adhesion-independent tumbling. The contributions of P-selectin and ICAM-1 to leucocyte rolling were tested in this system using planar bilayers reconstituted with P-selectin alone, with ICAM-1 alone, or with both adhesion molecules. It was observed that neutrophils "rolled" on P-selectin-reconstituted bilayers, at velocities consistent with the presence of an adhesive interaction between the leucocytes' surfaces and the planar bilayer. By contrast, velocities observed when leucocytes were made to flow over ICAM-1-substituted bilayers were substantially higher, and were similar to those expected for particles that tumble freely without adhering to the adjacent surface. An interesting co-operative interaction between P-selectin and ICAM-1 was also demonstrated when leucocytes were made to flow over bilayers containing both adhesion molecules.

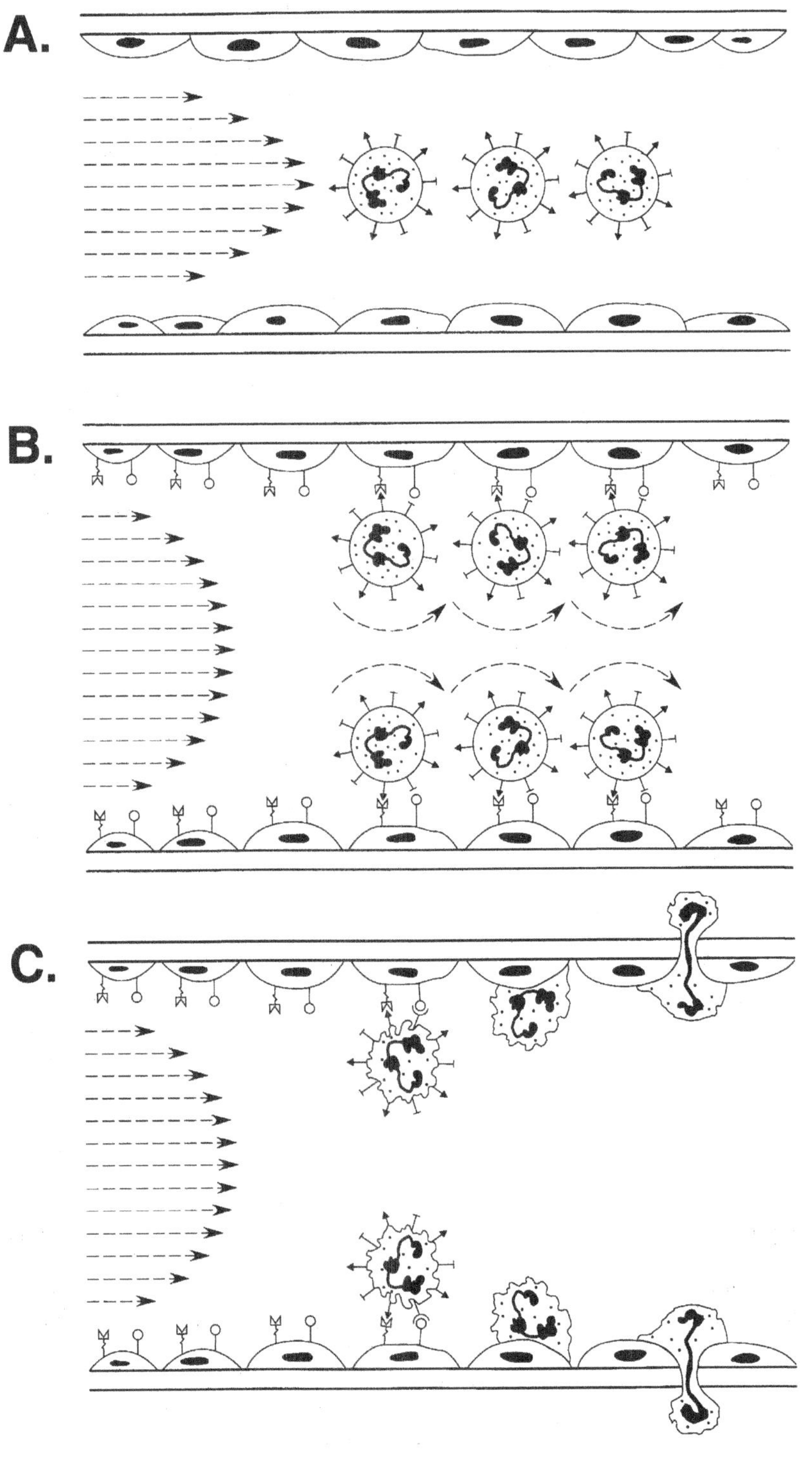

When treated with N-formyl-methionyl-leucyl-phenylalanine (FMLP), a compound that can up-regulate the avidity of the neutrophil integrin counter-receptors for ICAM-1, neutrophils rolling on this surface were seen to slow, stop, and then spread. These experiments imply that P-selectin and ICAM-1 can function in a sequentially co-operative manner, whereby P-selectin functions to bring passing leucocytes to the endothelial cell surface, at which time leucocyte integrins may be "activated" and arrest the rolling cell via adhesion to endothelial cell ICAM-1

These observations have been incorporated along with related ones into several models that invoke sequential adhesive events mediated by different classes of adhesion molecules, and modulated by a number of accessory molecules (Butcher, 1991; Lawrence and Springer, 1991; Lo *et al.*, 1991; Smith *et al.*, 1991; von Adrian *et al.*, 1991; and Fig. 6.8). In this model, inflammatory molecules act on the microvasculature to create capillary dilatation and fluid leakage. Plasma concentration ensues, and contributes to slowed flow effected by the increased diameter of the vessel. Under these conditions, a leucocyte within the vessel lumen is caused to move away from its usual central location to travel along a position near the vessel wall. It may then interact with P- and/or E-selectins expressed by activated endothelial cells. Selectin-dependent adhesion at the leucocyte-endothelial cell juncture, combined with flow-based force applied to the free, lumenal side of the leucocyte, cause the leucocyte to roll along the capillary wall. The cell does not stop, though, since the adhesive interactions between the selectins and their ligands at the leucocyte surface are inherently low affinity ones, and because the integrin receptors at the leucocyte surface are in an unactivated conformation incapable of forming an effective adhesive interaction with endothelial ICAM-1. These events have placed the leucocyte near the endothelial cell surface, however, where it may now be exposed to locally dense concentrations of a variety of molecules capable of acting to up-regulate its integrins. These "activating" substances include soluble molecules (interleukin-8, complement component C5a, N-formylated bacterial peptides, leukotriene B4), as well as membrane-associated molecules (platelet activating factor; PAF). Integrin up-regulation may also be accomplished by engagement of neutrophil ligands by E-selectin (Lo *et al.*, 1991), or by PAF-dependent events potentiated when P-selectin engages its neutrophil ligands (Lorant *et al.*, 1991). Leucocyte rolling ceases as the leucocyte now binds tightly to endothelial cell surface via integrin-ICAM-1 adhesion. The leucocyte then spreads, and proceeds to transmigrate through the endothelial cell barrier. As the experimental verification of this model proceeds and additional details are uncovered, it seems likely that many interesting opportunities will arise for therapeutic manipulation of the leucocyte recruitment process.

16. *Acknowledgements*

Appreciation is extended to Dr Ten Feizi for communicating unpublished data. This work was supported by the Howard Hughes Medical Institute. JBL is an Associate Investigator of the Howard Hughes Medical Institute.

17. *References*

Aruffo, A., Kolanus, W., Walz, G. Fredman, and Seed, B. (1991). CD62/P-selectin recognition of myeloid and tumor cell sulfatides. Cell 67, 35–44.

Asada, M., Furukawa, K., Kantor, C., Gahmberg, C. G. and Kobata, A. (1991). Structural study of the sugar chains of human leucocyte cell adhesion molecules CD11/CD18. Biochemistry 30, 1561–1571.

Ashwell, G. and Harford, J. (1982). Carbohydrate-specific receptors of the liver. Ann. Rev. Biochem. 51, 531–554.

Atherton, A. and Born, G. V. (1973). Proceedings: effects of neuraminidase and N-acetyl neuraminic acid on the adhesion of circulating granulocytes and platelets in venules. J. Physiol. (London) 234, 66P–67P.

Atherton, A. and Born, G.V. (1973). Relationship between the velocity of rolling granulocytes and that of the blood flow in venules. J. Physiol. 233, 157–165.

Baenziger, J.U., Kumar, S., Brodbeck, R.M., Smith, P.L. and Beranek, M.C. (1992). Circulatory half-life but not interaction with the lutropin/chorionic gonadotropin receptor is modulated by sulfation of bovine lutropin oligosaccharides. Proc. Natl Acad. Sci. USA 89, 334–338.

Figure 6.8 The sequential, multi-molecule model for selectin-initiated leucocyte adhesion (see text for details). (A) In normal capillaries, leucocytes travel predominantly in the central, rapidly flowing part of the stream, and interact minimally with the vessel's endothelial lining. (B) Inflammatory mediators can act upon the vascular wall in venular capillaries to dilate the capillary, and to promote capillary leakage. Flow slows due to dilatation, and to concentration of plasma constituents. The altered flow profile allows leucocytes to travel at more marginal positions, where selectin molecules expressed at the endothelial cell surface (Ý) bind to their co-receptors (↓) at the leucocyte surface. Adhesion at the leucocyte-endothelial cell interface, combined with lateral forces impinging upon the other side of the leucocyte, cause it to "roll" along the endothelial surface. Firm attachment does not occur since the leucocyte integrins are still in a low affinity (⊥), unactivated conformation and do not bind tightly to their counter-receptor (ICAM-1, ♀). (C) Soluble and membrane-associated stimuli derived from endothelial cells and the sub-endothelial environment activate the leucocyte and up-regulate its integrin receptors (⅄). The leucocyte adheres tightly to ICAM-1 and stops rolling. Spreading ensues, followed by transmigration through the capillary wall.

Ball, G.E., O'Neill, R., Schultz, J.E., Weston, B.W., Nagy, J.O., Brown, E.G., Hobbs, C. and Bednarski, M.D. (1992). Synthesis and structural analysis using 2-D NMR of sialyl Lewis x (SLe^x) and Lewis x (Le^x) oligosaccharides: ligands related to ELAM-1 binding. J. Am. Chem. Soc. (In press).

Beckstead, J.H., Stenberg, P.E., McEver, R.P., Shuman, M.A. and Bainton, D.F. (1986). Immunohistochemical localization of membrane and alpha-granule proteins in human megakaryocytes: application to plastic-embedded bone marrow biopsy specimens. Blood 67, 285–293.

Berg, E.L., Robinson, M.K., Mansson, O., Butcher, E.C. and Magnani, J.L. (1991). A carbohydrate domain common to both sialyl Le a and sialyl Le x is recognized by the endothelial cell leucocyte adhesion molecule ELAM-1. J. Biol. Chem. 266, 14 869–872.

Berman, C.L., Yeo, E.L., Wencel-Drake, J.D., Furie, B.C., Ginsberg, M.H. and Furie, B. (1986). A platelet alpha granule membrane protein that is associated with the plasma membrane after activation. Characterization and subcellular localization of platelet activation-dependent granule-external membrane protein. J. Clin. Invest. 78, 130–137.

Bevilacqua, M., Butcher, E., Furie, B., Furie, B., Gallatin, M., Gimbrone, M., Harlan, J., Kishimoto, K., Lasky, L., McEver, R., *et al.* (1991). Selectins: a family of adhesion receptors. Cell 67, 233.

Bevilacqua, M.P., Pober, J.S., Wheeler, M.E., Cotran, R.S. and Gimbrone, M.A. Jr. (1985). Interleukin 1 acts on cultured human vascular endothelium to increase the adhesion of polymorphonuclear leucocytes, monocytes, and related leucocyte cell lines. J. Clin. Invest. 76, 2003–2011.

Bevilacqua, M.P., Pober, J.S., Wheeler, M.E., Cotran, R.S. and Gimbrone, M.A. (1985). Interleukin-1 activation of vascular endothelium. Effects on procoagulant activity and leucocyte adhesion. Am. J. Pathol. 121, 394–403.

Bevilacqua, M.P., Pober, J.S., Mendrick, D.L., Cotran, R.S. and Gimbrone, M.A. (1987). Identification of an inducible endothelial-leucocyte adhesion molecule. Proc. Natl Acad. Sci. USA. 84, 9238–9242.

Bevilacqua, M.P., Stengelin, S., Gimbrone, M.A. and Seed, B. (1989). Endothelial leuocyte adhesion molecule 1: an inducible receptor for neutrophils related to complement regulatory proteins and lectins. Science 243, 1160–1165.

Beyer, T.A., Sadler, J.E., Rearick, J.I., Paulson, J.C. and Hill, R.L. (1982). Glycosyltransferases and their use in assessing oligosaccharide structure and structure-function relationships. Adv. Enzymol. 52, 23–175.

Bowen, B.R., Nguyen T. and Lasky L.A. (1989). Characterization of a human homologue of the murine peripheral lymph node homing receptor. J. Cell Biol. 109, 421–427.

Bowen, B.R., Fennie, C. and Lasky, L.A. (1990). The Mel 14 antibody binds to the lectin domain of the murine peripheral lymph node homing receptor. J. Cell. Biol. 110, 147–153.

Bonfanti, R., Furie, B.C., Furie, B. and Wagner, D.D. (1989). PADGEM is a component of Weibel-Palade bodies in endothelial cells. Blood 73, 1109–1112.

Briscoe, D.M., Schoen, F.J., Rice, G.E., Bevilacqua, M.P., Ganz, P. and Pober, J.S. (1991). Induced expression of endothelial-leucocyte adhesion molecules in human cardiac allografts. Transplantation 51, 537–539.

Butcher, E.C. (1990). Cellular and molecular mechanisms that direct leukocyte traffic. Am. J. Pathol. 136, 3–12, 1990.

Butcher, E.C. (1991). Leukocyte-endothelial cell recognition: three (or more) steps to specificity and diversity. Cell 67, 1033–1036.

Camerini, D., James, S.P., Stamenkovic, I. and Seed, B. (1989). Leu-8/TQ1 is the human equivalent of the Mel-14 lymph node homing receptor. Nature 342, 78–82.

Caras, I.W., Davitz, M.A., Rhee, L., Weddell, G., Martin, D.W. and Nussenzweig, V. (1987). Cloning of decay-accelerating factor suggests novel use of splicing to generate two proteins. Nature 325, 545–549.

Carlos, T.M., Dobrina, A., Ross, R. and Harlan, J.M. (1990). Multiple receptors on human monocytes are involved in adhesion to cultured human endothelial cells. J. Leukocyte Biol. 48, 451–456.

Cheresh, D.A., Reisfeld, R.A. and Varki, A.P. (1984). O-acetylation of disialoganglioside GD3 by human melanoma cells creates a unique antigenic determinant. Science 225, 844–846.

Chien, S. (1982). Rheology in the microcirculation in normal and low flow states. Adv. Schock Res. 8, 71–80.

Collins, T., Williams, A., Johnston, G.I., Kim, J., Eddy, R., Shows, T., Gimbrone, M.A. and Bevilacqua, M.P. (1991). Structure and chromosomal location of the gene for endothelial-leukocyte adhesion molecule 1. J. Biol. Chem. 266, 2466–2473.

Cooke, R.M., Wilkinson, A.J., Baron, M., Pastore, A., Tappin, M.J., Campbell, I.D., Gregory, H. Sheard, B. (1987). The solution structure of human epidermal growth factor. Nature 327, 339–341.

Corral, L., Singer, M.S., Macher, B.A. and Rosen, S.D. (1990). Requirement for sialic acid on neutrophils in a GMP-140 (PADGEM) mediated adhesive interaction with activated platelets. Biochem. Biophys. Res. Comm. 172, 1349–1356.

Cotran, R.S., Gimbrone, M.A. Jr, Bevilacqua, M.P., Mendrick, D.L. and Pober, J.S. (1986). Induction and detection of a human endothelial activation antigen *in vivo*. J. Exp. Med. 164, 661–666.

Davis, C.G., Goldstein, J.L., Sudhof, T.C., Anderson, R.G., Russell, D.W. and Brown, M.S. (1987). Acid-dependent ligand dissociation and recycling of LDL receptor mediated by growth factor homology region. Nature 326, 760–765.

Dowbenko, D.J., Diep, A., Taylor, B.A., Lusis, A.J. and Lasky, L.A. (1991). Characterization of the murine homing receptor gene reveals correspondence between protein domains and coding exons. Genomics 9, 270–277.

Drickamer K. (1988). Two distinct classes of carbohydrate-recognition domains in animal lectins. J. Biol. Chem. 263, 9557–9560.

Dumas, D.P., Ichikawa, Y., Wong, C.H., Lowe, J.B. and Nair, R.P. (1991). Enzymatic synthesis of sialyl SLe^x and derivatives based on a recombinant fucosyltransferase. Bioorg. Med. Chem. Lett. 1, 425–428.

Dustin, M.L. and Springer, T.A. (1989). T-cell receptor cross-linking transiently stimulates adhesiveness through LFA-1. Nature 341, 619–624.

Enfield, D.L., Ericsson, L.H., Fujikawa, K., Walsh, K.A., Neurath, H. and Titani, K. (1980). Amino acid sequence of the light chain of bovine factor X1 (Stuart factor). Biochemistry 19, 659–667.

Fiebig, E., Ley, K. and Arfors, K.E. (1991). Rapid leukocyte accumulation by "spontaneous" rolling and adhesion in the exteriorized rabbit mesentery. Int. J. Microcirc. Clin. Exp. 10, 127–144.

Fiete, D., Srivastava, V., Hindsgaul, O. and Baenziger, J.U. (1991). A hepatic reticuloendothelial cell receptor specific for SO4-4Gal*N*Ac beta 1,4Glc*N*Ac beta 1,2Man alpha that mediates rapid clearance of lutropin. Cell 67, 1103–1110.

Fukuda, M.N., Spooncer, E., Oates, J.E., Dell, A. and Klock, J.C. (1984). Structure of sialylated fucosyl lactosaminoglycan isolated from human granulocytes. J. Biol. Chem. 259, 10 925–10 935.

Fukuda, M., Dell, A., Oates, J.E. and Fukuda, M.N. (1984). Structure of branched lactosaminoglycan, the carbohydrate moiety of band 3 isolated from adult human erythrocytes. J. Biol. Chem. 259, 8260–8273.

Fukuda, M.N., Dell, A., Oates, J.E., Wu, P., Klock, J.C. and Fukuda, M. (1985). Structures of glycosphingolipids isolated from human granulocytes. The presence of a series of linear poly-N-acetyllactosaminylceramide and its significance in glycolipids of whole blood cells. J. Biol. Chem. 1985; 260, 1067–1082.

Fukuda, M.N., Dell, A., Tiller, P.R., Varki, A., Klock, J.C. and Fukuda, M. (1986). Structure of a novel sialylated fucosyl lacto-N-norhexaosylceramide isolated from chronic myelogenous leukemia cells. J. Biol. Chem. 261, 2376–2383.

Fukushi, Y., Hakomori, S., Nudelman, E. and Cochran, N. (1984). Novel fucolipids accumulating in human adenocarcinoma II. Selective isolation of hybridoma antibodies that differentially recognize mono-, di-, and trifucosylated type 2 chain. J. Biol. Chem. 259, 4681–4685.

Fukushi, Y., Kannagi, R., Hakomori, S.-I., Shepard, T., Kulander, B.G. and Singer, J.W. (1985). Location and distribution of difucoganglioside (VI3NeuAcV3III3Fuc2nLc6) in normal and tumor tissues defined by its monoclonal antibody FH6. Cancer Res. 45, 3711–3717.

Fukushima, K., Hirota, M., Terasaki. P.I., Wakisaka, A., Togashi, H., Chia D., Suyama, N., Fukushi, Y., Nudelman, E. and Hakomori, S.-I. (1984). Characterization of sialosylated Lewis X as a new tumor-associated antigen. Cancer Res. 1984; 44, 5279–5285.

Gallatin, W.M., Weissman, I.L. and Butcher, E.C. (1983). A cell-surface molecule involved in organ-specific homing of lymphocytes. Nature 304, 30–34.

Gamble, J.R., Harlan, J.M., Klebanoff, S.J. and Vada M.A. (1985). Stimulation of the adherence of neutrophils to umbilical vein endothelium by human recombinant tumor necrosis factor. Proc. Natl Acad. Sci. USA. 82, 8667–8671.

Gamble, J.R., Skinner, M.P., Berndt, M.C. and Vadas, M.A. (1990). Prevention of activated neutrophil adhesion to endothelium by soluble adhesion protein GMP140. Science 249, 414–417.

Geng, J.-G., Bevilacqua, M.P., Moore, K.L., McIntyre, T.M., Prescott, S.M., Kim, J.M., Bliss, G.A., Zimmerman, G.A. and McEver, R.P. (1990). Rapid neutrophil adhesion to activated endothelium mediated by GMP-140. Nature 343, 757–760.

Geng, J.-G., Moore, K.L., Johnson, A.E. and McEver, R.P. (1991). Neutrophil recognition requires a Ca^{2+}-induced conformational change in the lectin domain of GMP-140. J. Biol. Chem. 266, 22 313–22 318.

Goelz, S.E., Hession, C., Goff, D., Griffiths, B., Tizard, R., Newman, B., Chi-Rosso, G. and Lobb, R. (1990). ELFT: A gene that directs the expression of an ELAM-1 ligand. Cell 63, 1349–1356.

Gooi, H.C., Feizi, T., Kapadia, A., Knowles, B.B., Solter, D. and Evans, M.J. (1981). Stage-specific embryonic antigen involves α1 to 3 fucosylated type 2 blood group chains. Nature 292, 156–158.

Graber, N., Gopal, T.V., Wilson, D., Beall, L.D., Polte, T. and Newman, W. (1990). T cells bind to cytokine-activated endothelial cells via a novel, inducible sialoglycoprotein and endothelial leukocyte adhesion molecule-1. J. Immunol. 145, 819–830.

Griffin, J.D., Spertini, O., Ernst, T.J., Belvin, M.P., Levine, H.B., Kanakura, Y. and Tedder, T.F. (1990). Granulocyte-macrophage colony-stimulating factor and other cytokines regulate surface expression of the leukocyte adhesion molecule-1 on human neutrophils, monocytes, and their precursors. J. Immunol. 145, 576–584.

Griffiths, C.E., Barker, J.N., Kunkel, S. and Nickoloff, B.J. (1991). Modulation of leucocyte adhesion molecules, a T-cell chemotaxin (IL-8) and a regulatory cytokine (TNF-alpha) in allergic contact dermatitis (rhus dermatitis). Br. J. Dermatol. 124, 519–526.

Groves, R.W., Allen, M.H., Barker, J.N., Haskard, D.O. and MacDonald, D.M. (1991). Endothelial leukocyte adhesion molecule-1 (E-selectin) expression in cutaneous inflammation. Br. J. Dermatol. 124, 117–123.

Hallmann, R., Jutila, M.A., Smith, C.W., Anderson, D.C., Kishimoto, T.K. and Butcher, E.C. (1991). The peripheral lymph node homing receptor, L-selectin, is involved in CD18-independent adhesion of human neutrophils to endothelium. Biochem. Biophys. Res. Commun. 174, 236–243.

Hamburger, S.A. and McEver, R.P. (1990). GMP-140 mediates adhesion of stimulated platelets to neutrophils. Blood 75, 550–554.

Hansson, G.C. and Zopf, D. (1985). Biosynthesis of the cancer-associated sialyl-Le^a antigen. J. Biol. Chem. 260, 9388–9392.

Harlan, J.M. (1985). Leukocyte-endothelial interactions. Blood 65, 513–525.

Harlan, J.M. (1987). Consequences of leukocyte-vessel wall interactions in inflammatory and immune reactions. Sem. Thrombosis Haemostasis 13, 434–444.

Hattori, R., Hamilton, K.K., Fugate, R.D., McEver, R.D. and Sims, P.J. (1989). Stimulated secretion of endothelial von Willebrand factor is accompanied by rapid redistribution to the cell surface of the intracellular granule membrane protein GMP-140 J. Biol. Chem. 264, 7788–7771.

Hession, C., Osborn, L., Goff, D., Chi-Rosso, G., Vassallo, C., Pasek, M., Pittack, C., Tizard, R., Goelz, S., McCarthy, K., Hopple, S. and Lobb, R. (1990). Endothelial leukocyte adhesion molecule 1: Direct expression cloning and functional interactions. Proc. Natl Acad. Sci. USA. 87, 1673–1672.

Holmes, E.H., Ostrander, G.K. and Hakomori, S. (1985). Enzymatic basis for the accumulation of glycolipids with X and dimeric X determinants in human lung cancer cells (NCI-H69). J. Biol. Chem. 260, 7619–7627.

Holmes, E.H., Ostrander, G.K. and Hakomori, S. (1986). Biosynthesis of the sialyl-Lex determinant carried by type 2 chain glycosphingolipids (IV3NeuAcIII3FucnLc4, VI3NeuAcV3FucnLc6, and VI3NeuAcIII3V3Fuc2nLc6) in human lung carcinoma PC9 cells. J. Biol. Chem. 261, 3737–3743.

Hourcade, D., Holers, V.M. and Atkinson, J.P. (1989). In "Advances in Immunology", Vol 45 (ed J.I. Dixon), pp. 381–416. Academic Press, New York.

Howard, D.R., Fukuda, M., Fukuda, M.N. and Stanley, P. (1987). The GDP-fucose: *N*-acetylglucosaminide 3-α-L-fucosyltransferases of LEC11 and LEC12 Chinese hamster ovary mutants exhibit novel specificities for glycolipid substrates. J. Biol. Chem. 262, 16 830–16 837.

Ichikawa, Y., Shen, G.-J. and Wong, C.-H. (1991). Enzyme-catalyzed synthesis of sialyl oligosaccharide with in situ regeneration of CMP-sialic acid. J. Am. Chem. Soc. 113, 4698–4700.

Ichinose, A., Hendrickson, L.E., Fujikawa, K. and Davie, E.W. (1986). Amino acid sequence of the a subunit of human factor XIII. Biochemistry 25, 6900–6906.

Imai, Y., Sinfer, M.S., Fennie, C., Lasky, L.A. and Rosen, S.D. (1991). Identification of a carbohydrate-based endothelial ligand for a lymphocyte homing receptor. J. Cell. Biol. 113, 1213–1221.

Johnston, G.I., Cook, R.G. and McEver, R.P. (1989). Cloning of GMP-140, a granule membrane protein of platelets and endothelium: sequence similarity to proteins involved in cell adhesion and inflammation. Cell 56, 1033–1044.

Johnston, G.I., Kurosky, A. and McEver, R.P. (1989). Structural and biosynthetic studies of the granule membrane protein, GMP140, from human platelets and endothelial cells. J. Biol. Chem. 264, 1816–1823.

Johnston, G.I., Bliss, G.A., Newman, P.J. and McEver, R.P. (1990). Structure of the human gene encoding granule membrane protein-140, a member of the selectin family of adhesion receptors for leukocytes. J. Biol. Chem. 265, 21 381–21 385.

Jungi, T.W., Spycher, M.O. Nydegger, U.E. and Barandun, S. (1986). Platelet-leukocyte interaction: selective binding of thrombin-stimulated platelets to human monocytes, polymorphonuclear leukocytes, and related cell lines. Blood 67, 629–636.

Jutila, M.A., Rott, L., Berg, E.L. and Butcher, E.C. (1989). Function and regulation of the neutrophil MEL-14 antigen *in vivo*: comparison with LFA-1 and Mac-1. J. Immunol. 143, 3318–3324.

Kameyama, A., Ishida, H., Kiso, M. and Hasegawa, A. (1991). Chemical synthesis of sialyl Lewis X. Carbohydr. Res. 209, C1–C4.

Kannagi, R., Fukushi, Y., Tachikawa, T., Noda, A., Shin, S., Shigeta, K., Hiraiwa, N., Fukuda, Y., Inamoto, T., Hakomori, S.-I. and Imura, H. (1986). Quantitative and qualitative characterization of human cancer-associated serum glycoprotein antigens expressing fucosyl or sialyl-fucosyl type 2 chain polylactosamine. Cancer. Res. 46, 2619–2626.

Kansas, G.S., Spertini, O. and Tedder, T.F. (1991). In "Cellular and Molecular Mechanisms of Inflammation", (eds C.G. Cochrane and M.A. Gimbone), pp. 31–60, Academic Press, San Diego.

Kansas, G.S., Spertini, O., Stoolman, L.M. and Tedder, T.F. (1991). Molecular mapping of functional domains of the leukocyte receptor for endothelium, LAM-1. J. Cell. Biol. 114, 351–358.

Kapadia, A., Feizi, T. and Evans, M.J. (1981). Changes in the expression and polarization of blood group I and i antigens in post-implantation embryos and teratocarcinomas of mouse associated with cell differentiation. Exp. Cell. Res. 131, 185–195.

Kidd, S., Kelley, M.R. and Young, M.W. (1986). Sequence of the notch locus of *Drosophila melanogaster*: relationship of the encoded protein to mammalian clotting and growth factors. Mol. Cell Biol. 6, 3094–3108.

Kim, Y.S. and Itzkowitz, S. (1988). Carbohydrate antigen expression in the adenoma-carcinoma sequence. Prog. Clin. Biol. Res. 279, 241–250.

Kishimoto, T.K., Jutila, M.A., Berg, E.L. and Butcher E.C. (1989). Neutrophil Mac-1 and MEL-14 adhesion proteins inversely regulated by chemotactic factors. Science 245, 1238–1241.

Kishimoto, T.K., Jutila, M.A. and Butcher, E.C. (1990). Identification of a human peripheral lymph node homing receptor: a rapidly down-regulated adhesion molecule. Proc. Natl Acad. Sci. USA. 87, 224–228.

Klickstein, L.B., Wong, W.W., Smith, J.A., Weis, J.H., Wilson, J.G. and Fearon, D.T. (1987). Human C3b/C4b receptor (CR1). Demonstration of long homologous repeating domains that are composed of the short consensus repeats characteristics of C3/C4 binding proteins. J. Exp. Med. 165, 1095–1112.

Knapp, W., Dorken, B., Rieber, P., Schmidt, R.E., Stein, H. and von dem Borne, A.E.Kr. (1989). CD Antigens. Blood 74, 1148–1150.

Koch, A.E., Burrows, J.C., Haines, G.K., Carlos, T.M., Harlan, J.M. and Leibovich, S.J. (1991). Immunolocalization of endothelial and leukocyte adhesion molecules in human rheumatoid and osteoarthritic synovial tissues. Lab. Invest. 64, 313–320.

Kornfeld, R. and Kornfeld S. (1985). Assembly of asparagine-linked oligosaccharides. Annu. Rev. Biochem. 54, 631–664.

Kornfeld, S. (1987). Trafficking of lysosomal enzymes. FASEB J. 1, 462–468.

Kukowska-Latallo, J.F., Larsen, R.D., Nair, R.P. and Lowe, J.B. (1990). A cloned human cDNA determines expression of a mouse stage-specific embryonic antigen and the Lewis blood group α(1,3/1,4)fucosyltransferase. Genes Dev. 4, 1288–1303.

Kumar, R., Potvin, B., Muller, W.A. and Stanley, P. (1991). Cloning of a human alpha(1,3)-fucosyltransferase gene that encodes ELFT but does not confer ELAM-1 recognition on Chinese hamster ovary cell transfectants. J. Biol. Chem. 266, 21 777–21 783.

Larkin, M., Ahern, T.J., Stoll, M.S., Shaffer, M., Sako, D., O'Brien, J., Yuen, C.-T., Lawson, A.M., Childs, R.A., Barone, K.M., Langer-Safer, P.R., Hasegawa, A., Kiso, M., Larsen, G.R. and Feizi. T. (1992). Spectrum of sialylated and non-sialylated fuco-oligosaccharides bound by the endothelial leukocyte adhesion molecules E-selectin. J. Biol. Chem. (In Press).

Larsen, E., Celi, A., Gilbert, G.E., Furie, B.C., Erban, J.K., Bonfanti, R., Wagner, D.D. and Furie, B. (1989). PADGEM protein: a receptor that mediates the, interaction of activated platelets with neutrophils and monocytes. Cell. 59, 305–312.

Larsen, E., Palabrica, T., Sajer, S., Gilbert, G.E., Wagner, D.D., Furie, B.C. and Furie, B. (1990). PADGEM-dependent adhesion of platelets to monocytes and neutrophils is mediated by a lineage-specific carbohydrate, LNF III (CD15). Cell 63, 467–474.

Larsen, R.D., Ernst, L.K., Nair, R.P. and Lowe, J.B. (1990). Molecular cloning, sequence, and expression of human GDP-L-fucose:β-D-galactoside 2-α-L-fucosyltransferase cDNA that can form the H blood group antigen. Proc. Natl Acad. Sci. USA. 87, 6674–6678.

Lasky, L.A., Singer, M.S., Yednock, T.A., Dowbenko, D., Fennie, C., Rodriguez, H., Nguyen, T., Stachel, S. and Rosen, S.D. (1989). Cloning of a lymphocyte homing receptor reveals a lectin domain. Cell 56, 1045–1055.

Lawrence, M.B. and Springer, T.A. (1991). Leukocytes roll on a selectin at physiological flow rates: distinction from and prerequisite for adhesion through integrins. Cell 65, 859–873.

Lawrence, M.B., Smith, C.W., Eskin, S.G. and McIntire, L.V. (1990). Effect of venous shear stress on CD18-mediated neutrophil adhesion to cultured endothelium. Blood 75, 227–237.

Leonard, W.J., Depper, J.M., Crabtree, G.R., Rudikoff, S., Pumphrey, J., Robb, R.J., Kronke, M., Svetlik, P.B., Peffer, N.J., Waldman. T.A., *et al.* (1984). Molecular cloning and expression of cDNAs for the human interleukin-2 receptor. Nature 311, 626–631.

Lewinsohn, D.M., Bargatze, R.F. and Butcher, E.C. (1987). Leukocyte-endothelial cell recognition: evidence of a common molecular mechanism shared by neutrophils, lymphocytes, and other leukocytes. J. Immunol. 138, 4313–4321.

Ley, K., Gaehtgens, P., Fennie, C., Singer, M.S., Lasky, L.A. and Rosen, S.D. (1991). Lectin-like cell adhesion molecule 1 mediates leucocyte rolling in mesenteric venules *in vivo*. Blood 12, 2553–2555.

Ley, K., Cerrito, M. and Arfors, K.-E. (1991). Sulfated polysaccharides inhibit leukocyte rolling in rabbit mesenteric venules. Am. J. Physiol. 260, H1667–H1673.

Liener, I.E., Sharon, N. and Goldstein, I.J. (1986). "The Lectins: Properties, Functions, and Applications in Biology and Medicine". Academic Press, Orlando, Florida.

Lo, S.K., Lee, S., Ramos, R.A., Lobb, R., Rosa, M., Chi-Rosso, G. and Wright, S.D. (1991). Endothelial-leukocyte adhesion molecule 1 stimulates the adhesive activity of leukocyte integrin CR3 (CD11b/CD18, Mac-1 amb2) on human neutrophils. J. Exp. Med. 173, 1493–1500.

Lobb, R.R., Chi-Rosso, G., Leone, D.R., Rosa, M.D., Bixler, S., Newman, B.M., Luhowskyj, S., Benjamin, C.D., Dougas, I.G., Goelz, S.E., Hession, C. and Chow, E.P. (1991). Expression and functional characterization of a soluble form of endothelial-leukocyte adhesion molecule 1. J. Immunol. 147, 124–129.

Lorant, D.E., Patel, K.D., McIntyre, T.M., McEver, R.P., Prescott, S.M. and Zimmerman, G.A. (1991). Coexpression of GMP-140 and PAF by endothelium stimulated by histamine or thrombin. A juxtacrine system for adhesion and activation of neutrophils. J. Cell Biol. 115, 223–234.

Lowe, J.B., Kukowska-Latallo, J.F., Nair, R.P., Larsen, R.D., Marks, R.M., Macher, B.A., Kelly, R.J. and Ernst, L.K. (1991). Molecular cloning of a human fucosyltransferase gene that determines expression of the Lewis X and VIM-2 epitopes but not ELAM-1-dependent cell adhesion. J. Biol. Chem. 266, 17 467–17 477.

Lowe, J.B. "Molecular cloning, expression, and uses of mammalian glycosyltransferases. Seminars in Cell Biology". W.B. Saunders Company UK. (In Press).

Lowe, J.B., Stoolman, L.M., Nair, R.P., Larsen, R.D., Berhend, T.L. and Marks, R.M. (1990). ELAM-1-dependent cell adhesion to vascular endothelium determined by a transfected human fucosyltransferase cDNA. Cell 63, 475–484.

Macher, B.A., Buehler, J., Scudder, P., Knapp, W. and Feizi, T. (1988). A novel carbohydrate, differentiation antigen on fucogangliosides of human myeloid cells recognized by monoclonal antibody VIM-2. J. Biol. Chem. 263, 10 186–10 191.

McEver, R.P., Marshall-Carlson, L. and Beckstead, J.H. (1987). The platelet α-granule membrane protein GMP140 is also synthesized by human vascular endothelial cells and is present in blood vessels of diverse tissues. Blood 70, 355a.

McEver, R.P. (1989). GMP-140, a platelet α-granule membrane protein, is also synthesized by vascular endothelial cells and is localized in Weibel-Palade bodies. J. Clin. Invest. 84, 92–99.

Mian, A.J. and Percival, E. (1973). Carbohydrates of the brown seaweed Himanthalia lorea and Bifurcaria blifucata. II. Structural studies of the "fucans". Carbohyd. Res. 26, 147–161.

Mollicone, R., Gibaud, A., Francois, A., Ratcliffe. M. and Oriol, R. (1990). Acceptor specificity and tissue distribution of three human alpha-3-fucosyltransferases. Eur. J. Biochem. 191, 169–176.

Moore, K.L., Varki, A. and McEver, R.P. (1991). GMP-140 binds to a glycoprotein receptor on human neutrophils: evidence for a lectin-like interaction. J. Cell Biol. 112, 491–499.

Muchmore, E.A., Milewski, M., Varki, A. and Diaz, S. (1989). Biosynthesis of N-glycolyneuraminic acid. The primary site of hydroxylation of N-acetylneuraminic acid is the cytosolic sugar nucleotide pool. J. Biol. Chem. 264, 20 216–20 223.

Mulligan, M.S., Varani, J., Dome, M.K., Lane, C.L., Smith, C.W., Anderson, D.C. and Ward, P.A. (1991). Role of endothelial-leukocyte adhesion molecule 1 (ELAM-1) in neutrophil-mediated lung injury in rats. J. Clin. Invest. 88, 1396–1406.

Nicolaou, K.C., Hummel, C.W., Bockovich, N.J. and Wong, C.H. (1991). Stere-controlled synthesis of sialyl Lex, the oligosaccharide binding ligand to ELAM-1 (Sialyl-*N*-acetylneuramin). J. Chem. Soc. 13, 870–872.

Nobis, U., Pries, A.R., Cokelet, G.R. and Gaehtgens, P. (1985). Radial distribution of white cells during blood flow in small tubes. Microvasc. Res. 29, 25–304.

Ord, D.C., Ernst, T.J., Zhou, L.-J., Rambaldi, A., Spertini, O., Griffin, J. and Tedder, T.F. (1990). Structure of the gene encoding the human leukocyte adhesion molecule-1 (TQ1, Leu-8) of lymphocytes and neutrophils. J. Biol. Chem. 265, 7760–7767.

Patel, K.D., Zimmerman, G.A., Prescott, S.M., McEver, R.P. and McIntyre, T.M. (1991). Oxygen radicals induce human endothelial cells to express GMP-140 and bind neutrophils. J. Cell Biol. 112, 749–759.

Paulson, J.C., Weinstein, J., Dorland, L., van Halbeek, H. and Vliegenthart, J.F. (1982). Newcastle disease virus contains a linkage-specific glycoprotein sialidase. Application to the localization of sialic acid residues in N-linked oligosaccharides of alpha 1-acid glycoprotein. J. Biol. Chem. 257, 12 734–12 738.

Pearse, B.M. (1988). Receptors compete for adaptors found in plasma membrane coated pits. EMBO J. 7, 3331–3336.

Phillips, M.L., Nudelman, E., Gaeta, F.C., Perez, M., Singhal, A.K., Hakomori, S. and Paulson, J.C. (1990). ELAM-1 mediates cell adhesion by recognition of a carbohydrate ligand, sialyl-Le^x. Science 250, 1130–1132.

Picker, L.J., Warnock, R.A., Burns, A.R., Doerschuk, C.M., Berg, E.L. and Butcher, E.C. (1991). The neutrophil LECAM-1 presents carbohydrate ligands to the vascular selectins ELAM-1 and GMP-140. Cell 66, 921–933.

Picker, L.J., Kishimoto, T.K., Smith, C.W., Warnock, R.A. and Butcher, E.C. (1991). ELAM-l is an adhesion molecule for skin-homing T cells. Nature 349, 796–799.

Pober, J.S., Bevilacqua M.P., Mendrick, D.L., Lapierre, L.A., Fiers, W. and Gimbrone, M.A. (1986). Two distinct monokines, interleukin 1 and tumor necrosis factor, each independently induce biosynthesis and transient expression of the same antigen on the surface of cultured human vascular endothelial cells. J. Immunol 136, 1680–1687.

Pohlman, T.H., Stanness, K.A., Beatty, P.G., Ochs, H.D. and Harlan, J.M. (1986). An endothelial cell surface factor(s) induced *in vitro* by lipopolysaccharide, interleukin 1, and tumor necrosis factor-alpha increases neutrophil adherence by a CDw18-dependent mechanism. J. Immunol. 136, 4548–4553.

Polley, M.J., Phillips, M.L., Wayner, E., Nudelman, E., Singhal, A.K., Hakomori, S.-I. and Paulson J.C. (1991). CD62 and endothelial cell-leukocyte adhesion molecule 1 (ELAM-1) recognize the same carbohydrate ligand, sialyl-Lewis X. Proc. Natl Acad. Sci. USA. 88, 6224–6228.

Potvin, B., Kumar, R., Howard, D.R. and Stanley, P. (1990). Transfection of a human α-(1,3)fucosyltransferase gene into Chinese hamster ovary cells. Complications arise from activation of endogenous α-(1,3)fucosyltransferases. J. Biol. Chem. 265, 1615–1622.

Redl, H., Dinges, H.P., Buurman, W.A., van der Linden, C.J., Pober, J.S., Cotran, R.S. and Schlag, G. (1991). Expression of endothelial leukocyte adhesion molecule-1 in septic but not traumatic/hypovolemic shock in the baboon. Am. J. Pathol. 139, 461–466.

Rosen, S.D., Singer, M.S., Yednock, T.A. and Stoolman, L.M. (1985). Involvement of sialic acid on endothelial cells in organ-specific lymphocyte recirculation. Science 228, 1005–1007.

Rosen, S.D., Ch, S.-I., True, D.D., Singer, M.S. and Yednock, T.A. (1989). Intravenously injected sialidase inactivates attachment sites for lymphocytes on high endothelial venules. J. Immunol. 142, 1895–1902.

Sadler, J.E. (1984). In "Biology of Carbohydrates" (eds V. Ginsburg and P.W. Robbins), pp. 200–287. John Wiley and Sons, New York.

San Blas, G. and Cunningham, W.L. (1974). Structure of cell wall and exocellular mannans from the yeast *Hansenula holstii.* I. Mannans produced in phosphate-containing medium. Biochim. Biophys. Acta 354, 233–246.

Schauer, R. (1985). Sialic acids and their roles as biological masks. Trends Biochem. Sci. 10, 357–360.

Shimizu, Y., Van Seventer, G.A., Horgan, K.J. and Shaw, S. (1990). Regulated expression and binding of three VLA(Beta1) integrin receptors on T cells. Nature 345, 250–255.

Shimizu, Y., Shaw, S., Graber, N., Gopal., T.V., Horgan, K.J, Van Seventer, G.A. and Newman W. (1991). Activation-independent binding of human memory T cells to adhesion molecule ELAM-1. Nature 349, 799–802.

Siegelman, M.H., Bond, M.W., Gallatin, W.M., St. John, T., Smith, H.T., Fried, V.A. and Weissman I.L. (1986). Cell surface molecules associated with lymphocyte homing is a ubiquitinated branched-chain glycoprotein. Science 231, 823–829.

Siegelman, M.H., van de Rijn, M. and Weissman, I.L. (1989). Mouse lymph node homing receptor cDNA clone encodes a glycoprotein revealing tandem interaction domains. Science 243, 1165–1172.

Siegelman, M.H. and Weissman, I.L. (1989). Human homologue of mouse lymph node homing receptor: Evolutionary conservation at tandem cell interaction domain. Proc. Natl Acad. Sci. USA. 86, 5562–5566.

Siegelman, M.H., Cheng, I.C., Weissman, I.L. and Wakel, E.K. (1990). The mouse lymph node homing receptor is identical with the lymphocyte cell surface marker Ly-22: role of the EGF domain in endothelial binding. Cell 61, 611–622.

Silverstein R.L. and Nachman R.L. (1987). Thrombospondin binds to monocytes-macrophages and mediates platelet-monocyte adhesion. J. Clin. Invest. 79, 867–874.

Skinner, M.P., Fournier, D.J., Andrews, R.K., Gorman, J.J., Chesterman, C.N. and Berndt, M.C. (1989). Characterization of human platelet GMP-140 as a heparin-binding protein. Biochem. Biophys. Res. Comm. 164, 1373–1379.

Skinner, M.P., Lucas, C.M., Burns, G.F., Chesterrnan, C.N. and Berndt, M.C. (1991) GMP-140 binding to neutrophils is inhibited by sulfated glycans. J. Biol. Chem. 266, 5371–5374.

Smith, C.W., Kishimoto, T.K., Abbass, O., Hughes, B., Rothlein, R., McIntyre, L.V., Butcher, E. and Anderson, D.C. (1991). Chemotactic factors regulate lectin adhesion molecule 1 (LECAM-1)-dependent neutrophil adhesion to cytokine-stimulated endothelial cells *in vitro*. J. Clin. Invest. 87, 609–618.

Smith, P.L. and Baenziger, J.U. (1992). Molecular basis of recognition by the glycoprotein hormone-specific N-acetylgalactosamine-transferase, Proc. Natl Acad. Sci. USA 89, 329–333.

Spertini, O., Freedman, A.S., Belvin, M.P., Penta, A.C., Griffin, J.D. and Tedder, T.F. (1991c). Regulation of leukocyte adhesion molecule-1 (TQ1, Leu-8) expression and shedding by normal and malignant cells. Leukemia 5, 300–308.

Spertini, O., Kansas, G.S., Reimann, K.A., Mackay, C.R. and Tedder, T.F. (1991a) Function and evolutionary conservation of distinct epitopes on the leukocyte adhesion molecule-1 (TQ-1, Leu-8) that regulate leukocyte migration. J. Immunol. 147, 942–949.

Spertini, O., Kansas, G.S., Munro, J.M., Griffin, J.D. and Tedder, T.F. (1991b). Regulation of leukocyte migration by activation of the leukocyte adhesion molecule-1 (LAM-1) selectin. Nature 349, 691–694.

Spooncer, E., Fukuda, M., Klock, J.C., Oates, J.E. and Dell, A. (1984). Isolation and characterization of polyfucosylated lactosaminoglycan from human granulocytes. J. Biol. Chem. 259, 4792–4801.

Springer, T.A. (1990). Adhesion receptors of the immune system. Nature 346, 425–434.

Srivastava, G., Alton, G. and Hindsgaul, O. (1990). Combined chemical-enzymic synthesis of deoxygenated oligosaccharide analogs: transfer of deoxygenated D-GlcpNAc residues from their UDP-GlcpNAc derivatives using N-acetylglucosaminyltransferase I. Carbohydr. Res. 207, 259–276.

St. John, T., Gallatin, W.M., Siegelman, M.S., Smith, H.T., Fried, V.A. and Weissman I.L. (1986). Expression cloning of a lymphocyte homing receptor cDNA: ubiquitin is the reactive species. Science 231, 845–850.

Stamper, H.B. Jr. and Woodruff, J.J. (1977). An *in vitro* model of lymphocyte homing. (I) Characterization of the interaction between thoracic duct lymphocytes and specialized high-endothelial venules of lymph nodes. J. Immunol. 119, 772–780.

Stamper, H.B. Jr. and Woodruff, J.J. (1976). Lymphocyte homing into lymph nodes: *in vitro* demonstration of the selective affinity of recirculating lymphocytes for high endothelial venules. J. Exp. Med. 144, 828–833.

Stevens, S.K., Weissman, I.L. and Butcher, E.C. (1982). Differences in the migration of B and T lymphocytes: organ-selective localization *in vivo* and the role of lymphocyte-endothelial cell recognition. J. Immunol. 128, 844–851.

Stenberg, P.E., McEver, R.P., Shuman, M.A., Jacques, Y.V. and Bainton, D.F. (1985). A platelet alpha-granule membrane protein (GMP-140) is expressed on the plasma membrane after activation. J. Cell Biol. 101, 880–886.

Stoolman, L.M. and Rosen, S.D. (1983). Possible role for cell-surface carbohydrate-binding molecules in lymphocyte recirculation. J. Cell Biol. 96, 722–729.

Stoolman, L.M. and Ebling, H. (1989). Adhesion molecules of cultured hematopoietic malignancies. A calcium-dependent lectin is the principle mediator of binding to the high endothelial venule of lymph nodes. J. Clin. Invest. 84, 1196–1205.

Stoolman, L.M., Tenforde, T.S. and Rosen, S.D. (1984). Phosphomannosyl receptors may participate in the adhesive interaction between lymphocytes and high endothelial venules. J. Cell Biol. 99, 1535–1540.

Stoolman, L.M., Yednock, T.A. and Rosen, S.D. (1987). Homing receptors on human and rodent lymphocytes – evidence for a conserved carbohydrate-binding specificity. Blood 70, 1842–1850.

Streeter, P.R., Rouse, B.T.N. and Butcher E.C. (1988). Immunohistologic and functional characterization of a vascular addressin involved in lymphocyte homing into peripheral lymphnodes. J. Cell. Biol. 107, 1853–1862.

Symington, F.W., Hedges, D.L. and Hakomori, S-I. (1985). Glycolipid antigens of human polymorphonuclear neutrophils and the inducible HL-60 myeloid leukemia line. Immunol. 134, 2498–2506.

Takada, A., Ohmori, K., Takahashi, N., Tsuyuoka, K., Yago, A., Zenita, K., Hasegawa, A. and Kannagi, R. (1991). Adhesion of human cancer cells to vascular endothelium mediated by a carbohydrate antigen, sialyl Lewis a. Biochem. Biophys. Res. Commun. 179, 713–719.

Tedder, T.F., Isaacs, C.M., Ernst, T.J., Demetri, G.D., Adler, D.A. and Disteche C.M. (1989). Isolation and chromosomal localization of cDNAs encoding a novel human lymphocyte cell surface molecule, LAM-1. J. Exp. Med. 170, 123–133.

Tedder, T.F., Penta, A.C., Levine, H.B. and Freedman, A.S. (1990). Expression of the human leukocyte adhesion molecule, LAM-1. Identity with the TQ1 and Leu-8 differential antigens. J. Immunol. 144, 532–540.

Tiemeyer, M., Sweidler, S.J., Ishihara, M., Moreland, M., Schweingruber, H., Hirtzer P. and Brandley, B.K. (1991). Carbohydrate ligands for endothelial-leukocyte adhesion molecule 1. Proc. Natl Acad. Sci. USA. 88, 1138–1142.

Toone, E.J., Simon, E.S., Bednarski, M.D. and Whitesides, G.M. (1989). Enzyme-catalyzed synthesis of carbohydrates. Tetrahedron Rep. 45, 5365–5422.

True, D.D., Singer, M.S., Laskey, L.A. and Rosen, S.D. (1990). Requirement for sialic acid on the endothelial ligand of a lymphocyte homing receptor. J. Cell Biol. 111, 2757–2764.

Tyrrel, D., Pames, P., Rao, N., Foxall, C., Abbas, S., Dasgupta, F., Nashed, M., Hasegawa, A., Kiso, M., Asa, D., Kidd, J. and Brandley, B.K. (1991). Structural requirements for the carbohydrate ligand of E-selectin. Proc. Natl Acad. Sci. USA 88, 10 372–10 376.

van den Eijnden, D.H., Koenderman, A.H. and Schiphorst, W.E. (1988). Biosynthesis of blood group i-active polylactosaminoglycans. Partial purification and properties of a UDP-GlcNAc:N-acetyllactosaminide beta 1–3-N-acetylglucosaminyltransferase from Novikoff tumor cell ascites fluid. J. Biol. Chem. 263, 12 461–12 471.

von Adrian, U.H., Chambers, J.D., McEvoy, L., Bargatze, R.F., Arfors, K.E. and Butcher, E.C. (1991). A two step model of leukocyte-endothelial cell interaction in inflammation: distinct roles for LECAM-1 and the leukocyte $\beta 2$ integrins *in vivo*. Proc. Natl Acad. Sci. USA 88, 7538–7542.

Waldorf, H.A., Walsh, L.J., Schechter, N.M. and Murphy, G.F. (1991). Early cellular events in evolving cutaneous delayed hypersensitivity in humans. Am. J. Pathol. 138, 477–486.

Waltz, G., Aruffo, A., Kolanus, W., Bevilacqua, M. and Seed, B. (1990). Recognition by ELAM-1 of the sialyl-Lex determinant on myeloid and tumor cells. Science 250, 113–1135.

Watson, M.L., Kingsmore, S.F., Johnston, G.I., Siegelman, M.H., Le Beau, M.M., Lemons, R.S., Bora, N.S., Howard, T.A., Weissman, I.L., McEver, R.P. and Seldin, M.F. (1990). Genomic organization of the selectin family of leukocyte adhesion molecules on human and mouse chromosome 1. J. Exp. Med. 172, 263–272.

Watson, S.R., Imai, Y., Fennie, C., Geoffrey, J.S., Rosen, S.D. and Lasky, L.A. (1990). A homing receptor-IgG chimera as a probe for adhesive ligands of lymph node high endothelial venules. J. Cell. Biol. 110, 2221–2229.

Watson, S.R., Imai, Y., Fennie, C., Geoffrey, J., Rosen, S.D. and Lasky, L.A. (1991). The complement binding-like domains of the murine homing receptor facilitate lectin activity. J. Cell. Biol. 115, 234–243.

Watson. S.R., Fennie, C. and Lasky, L.A. (1991). Neutrophil influx into an inflammatory site inhibited by a soluble homing receptor-IgG chimaera. Nature 349, 164–167.

Weinstein, J., de Souza-e-Silva, U. and Paulson, J.C. (1982). Purification of a Gal $\beta_1 <$ 4GlcNAc $\alpha 2 \rightarrow$ 6sialyltransferase and a Gal $\beta 1 \rightarrow$ 3(4)GlcNAc $\alpha 2 \rightarrow$ 3sialyltransferase to homogeneity from rat liver. J. Biol. Chem. 257, 13 835–13 844.

Weis, J.H., Morton, C.C., Bruns, G.A., Weis, J.J., Klickstein, L.B., Wong, W.W. and Fearon, D.T. (1987). A complement receptor locus: genes encoding C3b/C4b receptor and C3d/*Epstein Barr* virus receptor map to 1q32. J. Immunol. 138(1), 312–315.

Weller, P.F., Rand, T.H., Goelz, S.E., Chi-Rosso, G. and Lobb, R.R. (1991). Human eosinophil adherence to vascular endothelium mediated by binding to vascular cell adhesion molecule 1 and endothelial leukocyte adhesion molecule 1. Proc. Natl Acad. Sci. USA 88, 7430–7433.

Weston, B.W., Nair, R.P., Larsen, R.D. and Lowe, J.B. (1992). Isolation of a novel human α(1,3)fucosyltransferase gene and molecular comparison to the human Lewis blood group α(1,3/1,4)fucosyltransferase gene. Syntenic, homologous, nonallelic genes encoding enzymes with distinct acceptor substrate specificities. J. Biol. Chem. In press.

Wong, C.H. (1989). Enzymaatic catalysts in organic synthesis. Science 244, 1145–1152.

Yednock, T.A. and Rosen, S.D. (1989). Lymphocyte homing. Adv. Immunol. 44, 313–378.

Yednock, T.A., Stoolman, L.M. and Rosen, S.D. (1987). Phosphomannosyl-derivatized beads detect a receptor involved in lymphocyte homing. J. Cell. Biol. 104, 713–723.

Yednock, T.A., Butcher, E.C., Stoolman, L.M. and Rosen, S.D. (1987). Receptors involved in lymphocyte homing: Relationship between a carbohydrate-binding receptor and the MEL-14 antigen. J. Cell. Biol. 104, 725–731.

Zhou, Q., Moore, K.L., Smith, D.F., Varki, A., McEver, R.P. and Cummings, R.D. (1991). The selectin GMP-140 binds to sialylated, fucosylated lactosaminoglycans on both myeloid and nonmyeloid cells. J. Cell. Biol. 115, 557–564.

7. The Role of Cell Adhesion in the Evolution of Inflammatory Arthritis

Nancy Oppenheimer-Marks and Peter E. Lipsky

1. Introduction

Cell adhesion plays an integral role in the development and progression of a variety of chronic inflammatory diseases. Such adhesive events involve cellular interactions between infiltrating and resident tissue cells or between both cell types and components of extracellular matrices (ECM). Cellular adhesive events mediate the extravasation of inflammatory cells from the blood stream, the maintenance of vascular integrity, the localization of cells within the tissue and the propagation of local inflammatory and immune responses. Each adhesive event involves the utilization of specific cell surface receptors that mediate the cell-cell or cell-ECM interactions. Expression and activity of these receptor–counter-receptor pairs are tightly regulated, providing a dynamic system that controls the various events at inflammatory sites.

Rheumatoid arthritis (RA) is the prototype of chronic inflammatory arthritis in which the role of a number of adhesive mechanisms has been analyzed. Although the stimulus that induces the initial influx of cells into the rheumatoid synovium remains unclear, it appears that the activity of inflammatory cells within the synovial tissue is responsible for the persistent up-regulation of cell adhesive mechanisms that sustains the chronic inflammation. This may result from the local secretion of mediators that modulate the function of the vascular endothelium, through which inflammatory cells migrate to enter the synovial tissue, as well as those that act directly on the inflammatory cells to stimulate their migration, maintenance, and activation within these extravascular sites. In addition, engagement of cell surface molecules, including adhesion receptors themselves, can provide a variety of activation signals that directly enhance adhesion dependent events by cells in the synovium. Thus, adhesive mechanisms play a critical role in maintaining persistent inflammation within the synovium. Developing a more complete understanding of the nature of the receptor–counter-receptor pairs

Adhesion Molecules
ISBN 0–12–741440–1

involved and their roles in specific aspects of disease pathogenesis, therefore, should provide clues concerning means to control chronic synovial inflammation more effectively.

2. *Histological Features of Rheumatoid Synovium*

Rheumatoid inflammation has characteristic histological features. One of the earliest changes is an increase in the number of synovial lining cells (Rothschild and Masi, 1982). Also evident during the early stages of rheumatoid synovitis are obliteration of small blood vessels by inflammatory cells or thrombi, and the perivascular accumulation of inflammatory cells (Bromley and Wooley, 1984). As the disease progresses, vascular congestion and edema are followed by the proliferation of endothelial cells, (EC) (neovascularization), increases in the number of synovial lining cells, cellular infiltration and the resultant chronic inflammatory state. The characteristic features of the synovium in RA, including an increase in the number and activity of the synovial lining cells, focal or segmental vascular changes, cellular infiltration and the development of lymphoid nodules evolve as the disease progresses. The development of granulation tissue (pannus) and its association with the destruction of articular cartilage and the erosion of subchondral bone are hallmarks of chronic RA.

Rheumatoid synovial tissue is richly infiltrated with mononuclear cells, the most abundant being the T lymphocyte. However, also quite evident are B lymphocytes, plasma cells, dendritic cells and cells of the monocyte/macrophage lineage. When the organization of these cells in synovial tissue is surveyed, T cell-rich areas are found to predominate around small blood vessels that express the phenotype of high endothelial venules (HEV) of lymphoid organs (Kobayashi and Ziff, 1973; Iguchi and Ziff, 1986). One other prominent and characteristic feature of rheumatoid synovitis is the presence of activated B cells and the local production of immunoglobulin and the auto-antibody, rheumatoid factor (Kobayashi and Ziff, 1973).

Several studies have identified the phenotype of the T cells that accumulate in rheumatoid synovial tissue. Thus, *in situ* studies have demonstrated a greater abundançe of CD4+ than CD8+ T cells in the lymphocyte aggregates around small blood vessels, whereas slightly more CD8+ T cells are present in the areas in between (Kurosaka and Ziff, 1983). Analysis of T cells isolated directly from synovial tissue reveals a greater total number of CD4+ than CD8+ T cells (Cush and Lipsky, 1988). These synovial tissue T cells appear to be activated since the density of CD4 and CD3 is reduced, in comparison to RA peripheral blood T cells, and they express HLA-DR, very late activation (VLA)-1 antigen (CD49a/CD29), and decreased amounts of L-selectin (Table 7.1) (Klareskog *et al.*, 1981; Forre *et al.*, 1982a; Forre *et al.*, 1982b; Poulter *et al.*, 1985; Hemler *et al.*, 1986; Burmester *et al.*, 1987; Goto *et al.*, 1987; Cush and Lipsky, 1988). Nearly all of the synovial tissue CD4+ T cells are of the memory subset since they are $CD29^{bright}$, $CD7^{dim}$, CD45RO+ and $CD45RA^{dim}$ (Pitzalis *et al.*, 1987; Cush and Lipsky, 1988; Nakao *et al.*, 1989). By contrast, rheumatoid synovium is nearly devoid of naive $CD4+/CD45RA^{bright}$ T cells. A matter of increasing interest is whether the T cells that accumulate in the synovium relate to a selective migration of a particular population of activated T cells into the tissue or the local activation of a subset of resting memory T cells with an enhanced migratory capacity.

Synovial fluids also contain large numbers of leucocytes, including lymphocytes and neutrophils. The remaining cells in joint fluid include monocytes, occasional macrophages and synovial lining cells (Zvaifler, 1973). Although the mechanism of neutrophil trafficking through synovial tissue remains to be determined, it is likely that chemo-attractants in the joint fluid facilitate their accumulation into this compartment. Unlike peripheral blood neutrophils, those present in synovial fluids appear to have been activated *in vivo*, since they lack granules and have a vacuolated appearance (Zvaifler, 1973). The trafficking of neutrophils through the tissue may play a critical role in the perpetuation of chronic inflammation, perhaps by the elaboration of mediators that affect the subsequent infiltration into tissue of specific T cell subsets. For example, in a rat model, in which neutrophils were selectively depleted by specific anti-neutrophil monoclonal antibody (mAb), CD4+ T cells were inhibited from infiltrating into tissue sites into which the chemotactic factor IL-8 was introduced (Kudo *et al.*, 1991). Infiltration of CD8+ T cells was not similarly inhibited, suggesting that neutrophils elaborate a factor(s) that specifically recruits CD4+ T cells into tissue.

The nature of the lymphocytes in rheumatoid synovial fluids has also been characterized. Unlike synovial tissue, there is a greater number of CD8+ and fewer CD4+ T cells in synovial fluid (Reme *et al.*, 1990; Smith and Roberts-Thomson, 1990; Fox *et al.*, 1982). The reasons for the differences in the distribution of these cells are not known. It is presumed that cells enter the synovium by

Table 7.1 Phenotype of CD4+ synovial tissue lymphocytes

CD45RA –
CD45RO +
$CD29^{bright}$
$L\text{-selectin}^{low}$
$CD7^{dim}$
HLA-DR +
VLA-1 +

migrating through post-capillary venules to the synovial tissue and then enter the synovial fluid from which they exit the joint. However, the exact pathway of tissue trafficking in the synovium has not been resolved. It is possible that the abundance of CD8+ T cells and neutrophils in the synovial fluid relates to their greater intrinsic motility permitting their more rapid movement out of the tissue. Alternatively, there may be mechanisms that specifically retain CD4+ T cells in the tissue or direct or retain CD8+ T cells and neutrophils in the fluid.

The rheumatoid synovium is a multi-compartment tissue with differences in the nature of cells that accumulate in the various locations. This implies that unique adhesive mechanisms are operative within the same inflammatory site to account for this non-uniform cellular distribution.

3. *The Vasculature in Rheumatoid Synovium*

In the early acute phase of RA, injured vasculature may provide an important route for the infiltration of inflammatory cells into the affected tissue. However, as the inflammation becomes chronic, the extravasation of cells into the tissue appears to result from receptor mediated events between specific cell surface adhesion molecules on the migrating cells and their counter-receptors on EC. Thus, synovial tissue EC express cell surface adhesion receptors that bind leucocytes. Moreover, the EC of rheumatoid synovium express increased adhesiveness for peripheral blood mononuclear cells (Jalkanen *et al.*, 1986; Oppenheimer-Marks and Ziff, 1986). During the course of RA, changes occur in many of the EC of post-capillary venules, presumably in response to cytokines and other factors in their environment, which tend to alter their phenotypic and functional characteristics, enhancing their capacity to mediate trans-endothelial migration of inflammatory cells.

In some areas of rheumatoid synovium, especially in areas near aggregates of lymphocytes, EC assume morphologic and cytochemical characteristics of HEV, which are specialized post-capillary venules of lymphoid tissue, and can be distinguished from other blood vessels by their characteristic cuboidal or columnar morphology (Freemont *et al.*, 1983; Iguchi and Ziff, 1986; Freemont, 1987). The development of HEV-like blood vessels has been examined in sites of delayed type hypersensitivity reactions in the skin of the baboon, and is found to depend on the exposure of the tissue to pro-inflammatory cytokines. Thus, concomitant stimulation with tumor necrosis factor-α (TNF-α) and interferonγ (IFNγ) induced the appearance of HEV as well as the adhesion and extravasation of inflammatory cells (Munro *et al.*, 1989).

Exposure to multiple cytokines in chronically inflamed tissue, such as the rheumatoid synovium is likely to induce HEV in a similar manner. The HEV-like blood vessels in RA synovial tissue, like those of lymphoid tissue, have been demonstrated to be sites of adhesion of lymphocytes, and thus have been suggested to be sites through which mononuclear cells infiltrate into the inflamed tissue (Gowans and Knight, 1964; Jalkanen *et al.*, 1986; Oppenheimer-Marks and Ziff, 1986). In other tissues, cell adhesion is not restricted to blood vessels that express the HEV phenotype and the development of HEV-like morphology is not necessary for EC to mediate lymphocyte migration (Fossum *et al.*, 1983). However, in some locations, such as the rheumatoid synovium, HEV morphology appears to identify EC capable of binding circulating inflammatory cells. HEV may also be important in regulating tissue access for only certain subsets of cells. In this regard, it recently has been suggested that HEV might be particularly involved in providing routes of extravasation for B cells, and not T cells, into sites of acute inflammation induced by infectious agents (Meeusen *et al.*, 1991). In rheumatoid synovium, however, HEV are found in areas containing aggregates of CD4+ T cells suggesting that in this tissue HEV may play a role in the trans-endothelial migration of particular subsets of T cells as well. Thus, although the HEV-like blood vessels may not be required for all inflammatory cell extravasation, in certain circumstances, such as the rheumatoid synovium, they may be utilized as specialized sites for the infiltration of specific populations of cells. HEV may offer a biological advantage over flat EC by presentation of the optimum architecture for the organization of cell surface adhesion receptors necessary for efficient trans-endothelial migration of cells.

4. *Mechanisms of Cell Adhesion During the Infiltration of Inflammatory Cells into Rheumatoid Synovium*

A number of specific adhesive mechanisms appear to play a role in the accumulation of inflammatory cells in the rheumatoid synovium. Although the data are not complete, specific recognition mechanisms that mediate the initial binding and trans-endothelial migration of leucocytes into sites of chronic inflammation have been identified and these processes are profoundly modulated by pro-inflammatory mediators. Central to these mechanisms are the activities of specific cell surface adhesion receptors. Adhesion receptors are responsible for the initial stabilized binding of leucocytes to the blood vessel wall and their subsequent trans-endothelial migration into perivascular tissue, whether it be during normal recirculation through lymphoid tissue or into sites of inflammation, including rheumatoid synovium. Not only does the binding of adhesion receptors to specific counter-receptors facilitate adhesion and trans-endothelial migration, but ligation of adhesion receptors is also likely to deliver intracellular signals as well. Thus, adhesion receptors, such as ICAM-1 and VCAM-1, as well as the extracellular matrix molecule fibronectin, engage receptors on inflammatory cells that transduce co-stimulatory signals (Altmann *et al.*, 1989; Davis *et al.*, 1990; Shimizu *et al.*, 1990a; van Seventer *et al.*, 1990; Damle and Aruffo, 1991; van Seventer *et al.*, 1991).

Tables 7.2–4 list the adhesion receptors that are likely to facilitate the infiltration of leucocytes into sites of inflammation. Adhesion receptors that are members of the immunoglobulin, selectin, and integrin superfamilies mediate much of the adhesive interactions that occur during the infiltration of leucocytes into sites of inflammation. With the exception of LFA-3 (CD58), all of the adhesion receptors are trans-membrane proteins and are members of supergene families of adhesion molecules. LFA-3 exists as either a trans-membrane protein or as a glycophosphatidylinositol linked molecule, both of which are involved in adhesive interactions (Springer, 1990). The expression of each receptor on EC is regulated by the actions of pro-inflammatory mediators. Each receptor binds at least one counter-receptor expressed by leucocytes, many of which are also functionally regulated by activation stimuli. Thus, reagents such as phorbol esters, which activate protein kinase C (PKC), as well as specific mAb directed to the T cell receptor/CD3 complex, induce changes in adhesion receptors, such as LFA-1 and VLA-4, such that they become competent to bind their counter-receptors.

4.1 IMMUNOGLOBULIN SUPERFAMILY

The immunoglobulin (Ig) gene superfamily has several members that serve adhesive and recognition functions (Table 7.2). Of particular importance to the extravasation of leucocytes into inflammatory sites are the activities of intercellular adhesion molecule (ICAM)-1 and vascular cell adhesion molecule (VCAM)-1. ICAM-2 and ICAM-3 are other members of the Ig family, but their roles in inflammation have not yet been established. LFA-3, another member of the Ig family, is expressed by both hematopoietic and non-hematopoietic cells, including EC. LFA-3 binds to CD2, an Ig family member that is expressed by T lymphocytes and some natural killer (NK) cells (Makgoba *et al.*, 1989; Springer, 1990). The trans-membrane form of LFA-3 supports the migration of large granular lymphocytes (Carpen *et al.*, 1991). Moreover, co-culture of T cells with EC induces CD2/LFA-3 mediated IL2 production by T cells (Hughes *et al.*, 1990). Although a role for this receptor–counter-receptor pair during the extravasation of T cells into inflammatory sites has not been clearly delineated, it is a candidate to play important adhesive and recognition roles.

Other members of the Ig family include the T cell receptor, class I and class II major histocompatability complex molecules, CD3, CD4, and CD8, all of which are involved in T cell recognition and activation processes and may also function to increase cellular adhesion. Finally, platelet-endothelial cell adhesion molecule (PECAM)-1 (CD31), is also an Ig family member found on leucocytes, platelets, EC and smooth muscle cells. CD31 is highly enriched at intercellular junctions of EC,

Table 7.2 Members of the Ig superfamily of adhesion receptors that mediate leucocyte interactions with EC

Receptor	*Counter-receptor*	*Distribution*	*Regulation*
ICAM-1 (CD54)	LFA-1(CD11a/CD18) Mac-1(CD11b/CD18) Leukosialin(CD43)	Endothelial cells, synovial fibroblasts, synovial macrophages, epithelial cells, germinal center dendritic cells, activated lymphocytes	Increased by IL-1, TNF-α, IFNγ
ICAM-2 (CD102)	LFA-1(CD11a/CD18)	Endothelial cells, hematopoietic cell lines	Constitutive
VCAM-1 (CD106)	VLA-4(CD49d/CD29)	Endothelial cells, synovial lining cells, germinal center dendritic cells, synovial macrophages, bone marrow fibroblasts	Induced by IL-1, TNFα, IL-4
LFA-3 (CD58)	CD2	Endothelial cells, leucocytes, epithelial cells	?
PECAM-1 (CD31)	?	Endothelial cells, platelets, leucocytes, smooth muscle cells	?
MAdCAM-1	$\alpha4\beta7$	Mucosal endothelial cells	Induced by TNFα, IL-1β

plays a critical role in adhesion between EC and the formation of confluent monolayers (Albelda, *et al.*, 1990; Muller *et al.*, 1993). In transfection studies, CD31 has been shown to mediate an adhesive function that can result in cellular aggregation (Albelda *et al.*, 1991). Thus, CD31 serves an adhesive role and may be important in the maintenance of vascular integrity by mediating endothelial cell-cell adhesion, as well as in promoting adhesive interactions of leucocytes. Mucosal addressin cell adhesion molecule-1 (MAdCAM-1) is a recently identified adhesion receptor expressed by EC of mucosal tissue. MAdCAM-1 is another member of the Ig family of adhesion receptors. The first two domains of MAdCAM-1 share strong homology with ICAM-1 and VCAM-1; the first is similar to the first domain of both ICAM-1 and VCAM-1, and the second domain shares homology with the fifth domain of VCAM-1 (Briskin *et al.*, 1993). The third domain of MAdCAM-1 is homologous to IgA. MAdCAM-1 also contains a serine/threonine rich mucin-like region which may serve to present carbohydrate ligands to cells. There are at least two counterreceptors for MAdCAM-1, the integrin, $\alpha4\beta7$ which probably binds domains 1 inasmuch as it also binds VCAM-1 and fibronectin, and L-selectin, which based upon its lectin-like activity, presumably binds the mucin-like region (see below) (Berg *et al.*, 1993; Berlin *et al.*, 1993).

ICAM-1 (CD54) is a heavily glycosylated member of the Ig superfamily (Staunton *et al.*, 1988). In addition to EC, ICAM-1 is expressed by a variety of cell types including, synovial fibroblasts, synovial macrophage-like type A cells, dermal fibroblasts, epithelial cells, tissue macrophages, germinal center dendritic cells, and activated lymphocytes (Rothlein *et al.*, 1986; Hale *et al.*, 1989; Koch *et al.*, 1991). ICAM-1 is constitutively expressed by many of these cell types, including EC, and increases following the exposure of the EC to pro-inflammatory cytokines, such as IL-1, TNF-α and IFNγ (Dustin and Springer, 1988; Renkonen, 1989). ICAM-1 possesses 5 Ig domains. It is most closely related to ICAM-2, which contains 2 Ig domains that are homologous to the 2 most amino-terminal Ig domains of ICAM-1 (Staunton *et al.*, 1989). These regions of ICAM-1 and -2 serve as the binding sites for their counter-receptors. The ICAMs are counter-receptors for members of the β_2 leucocyte integrin superfamily, which are involved in leucocyte adhesive interactions. Thus, both ICAM-1 and ICAM-2 bind leucocyte function-associated antigen-1 (LFA-1, CD11a/CD18) (Marlin and Springer, 1987; Staunton *et al.*, 1989). ICAM-1 also binds Mac-1 (CD11b/CD18) as well (Diamond *et al.*, 1990). Recently, ICAM-1 was demonstrated to bind leukosialin (CD43) expressed by T cell hybridomas as well (Rosenstein *et al.*, 1991). The role of CD43 in mediating the binding of peripheral blood T and B cells, neutrophils, monocytes and platelets to EC has not yet been determined, however. The expression of ICAM-1, but not ICAM-2 is sensitive to the actions of pro-inflammatory cytokines. Thus, IL-1, TNF-α and IFNγ up-regulate the expression of ICAM-1 at both the transcriptional and translational levels (Dustin and Springer, 1988; Renkonen, 1989; Voraberger *et al.*, 1991). ICAM-1 expression by EC is also increased by stimulation with phorbol esters that activate PKC, suggesting an involvement of a PKC mediated pathway during the induction of its expression (Voraberger *et al.*, 1991).

ICAM-1 mediates many types of cell-cell adhesive interactions. Thus, homotypic T cell adhesion, homotypic B cell adhesion, T cell-accessory cell, and T cell-synovial fibroblast interactions involve the activities of ICAM-1 and its counter-receptor, LFA-1 (Makgoba *et al.*, 1989; Springer, 1990). Particularly noteworthy, is that ICAM-1 on EC supports the adhesion of lymphocytes and neutrophils by binding to LFA-1 and Mac-1, respectively (Dustin and Springer, 1988; Smith *et al.*, 1988; Luscinskas *et al.*, 1989; Marlin and Springer, 1989; Oppenheimer-Marks, *et al.*, 1991). An additional role for ICAM-1 was recently noted when it was reported that mAb to ICAM-1 blocked the trans-endothelial migration of both T and B lymphocytes and neutrophils *in vitro* (Smith *et al.*, 1988; Luscinskas *et al.*, 1991; Oppenheimer-Marks *et al.*, 1991; Procter, J., Oppenheimer-Marks, N., and Lipsky, P.E., unpublished observations). Inhibition of trans-endothelial migration similarly was caused by a mAb against CD11a, the alpha chain of LFA-1, against CD11b, the alpha chain of macrophage-1 (Mac-1), or against CD18, the beta chain of LFA-1 and Mac-1 (Smith *et al.*, 1988; Luscinskas *et al.*, 1991; Oppenheimer-Marks *et al.*, 1991). In the case of the trans-endothelial migration of lymphocytes, a more careful dissection indicated that ICAM-1 and LFA-1 mediated interactions played a critical role in trans-endothelial migration of T cells after initial binding to EC, regardless of the cytokine exposure of the EC or the specific adhesion receptor–counter-receptor pair mediating T cell-EC binding (Oppenheimer-Marks *et al.*, 1991). The role of interactions with ICAM-1 in mediating lymphocyte migration has been documented by the findings that large granular lymphocytes and activated B cells specifically migrate on tissue culture plastic precoated with purified ICAM-1 (Dustin, 1990; Carpen *et al.*, 1991). Thus, ICAM-1 expressed by EC functions not only as a mediator of cell-cell binding, but also as a receptor that facilitates trans-endothelial migration. Above and beyond its role as an adhesion molecule, ligation of T cell or B cell LFA-1 by engagement of ICAM-1 delivers intracellular signals that facilitate functional activation of the lymphoid cell. Thus, for example, binding of LFA-1 by ICAM-1 provides a co-stimulatory signal to T cells that facilitates activation after ligation of the T cell antigen receptor (Altmann *et al.*, 1989; van Seventer *et al.*, 1990; van Seventer *et al.*, 1991). These findings imply that ICAM-1 is not merely a structural molecule with which cells interact, but it is a molecule that regulates the functional activity of lymphocytes as well. Although this has not been studied as extensively for neutropils as it has for lymphocytes, it is likely that the ligation of Mac-1 by ICAM-1 also transduces activation signals in neutrophils. Thus, adhesion of neutrophils to EC induces neutrophil degranulation, and this is dependent on the activity of Mac-1. Prevention of neutrophil-EC adhesion by mAb to Mac-1 inhibits neutrophil degranulation implying that this cellular response depends on the ligation of Mac-1 by ICAM-1 (Schleiffenbaum *et al.*, 1989).

VCAM-1 (CD106) is another member of the Ig supergene family, that has alternatively been called INCAM-110 (inducible cell adhesion molecule with a molecular weight of 110 000 Da) (Osborn *et al.*, 1989; Rice *et al.*, 1990). Unlike ICAM-1, which has 5 Ig homology domains, VCAM-1 was initially reported to have 6 such domains. However, recently evidence indicates that the major form of VCAM-1 on EC has 7 Ig domains, with the extra domain inserted between domains 3 and 4 (Hession *et al.*, 1991). The inserted domain is involved in ligand recognition (Osborn *et al.*, 1992). VCAM-1 is a counter-receptor for the beta1 integrin VLA-4 (CD49d/CD29) (Elices *et al.*, 1990). VLA-4 also is a receptor for the ECM molecule, fibronectin (Wayner *et al.*, 1989; Guan and Hynes, 1990). Using a variety of anti-VLA-4 mAb, it has been demonstrated that VCAM-1 and fibronectin, which are structurally unrelated, bind to different epitopes on the VLA-4 molecule (Campanero *et al.*, 1990; Elices *et al.*, 1990; Ferguson *et al.*, 1991). Unstimulated EC express very little, or no, cell surface VCAM-1. Upon exposure to the cytokines TNF-α, IL-1, or IL-4, however, VCAM-1 becomes prominently expressed by the EC (Osborn *et al.*, 1989; Carlos *et al.*, 1990; Masinovsky *et al.*, 1990; Rice *et al.*, 1990). This process requires protein synthesis, and thus does not reflect redistribution of preformed molecules (Masinovsky *et al.*, 1990).

VCAM-1 is expressed by synovial lining cells, germinal center dendritic cells, bone marrow derived fibroblasts, and macrophages (Koch *et al.*, 1991; Koopman *et al.*, 1991; Morales-Ducret *et al.*, 1991; Ryan *et al.*, 1991). However, there have been conflicting reports as to whether synovial EC express VCAM-1 (Koch *et al.*, 1991; Morales-Ducret *et al.*, 1991). Moreover, the extent to which VCAM-1 is involved in the infiltration of cells into inflammatory sites remains unclear. Activation of EC *in vitro* with the pro-inflammatory cytokines, IL-1 or TNF-α, induces VCAM-1 expression. Much of the binding of T cells to these activated EC is inhibited by mAb against VCAM-1 (Osborn *et al.*, 1990; Rice 1990; Oppenheimer-Marks *et al.*, 1991). The activity of VCAM-1 in supporting T cell binding has also been demonstrated by examining the adhesive behavior of cells isolated from patients with leucocyte adhesion deficiency (LAD) (Haskard *et al.*, 1989; Schwartz *et al.*, 1990; Kavanaugh *et al.*, 1991; Vennegoor *et al.*, 1991). Unlike normal cells, hematopoietic cells from these individuals do not express members of the beta2, Leu-CAM family of integrin adhesion receptors, including LFA-1 and Mac-1 (Springer, 1985). LAD T cells, but not neutrophils are, however, competent to bind to unstimulated and cytokine activated EC. Part of this binding is mediated by VCAM-1 on the EC and VLA-4 on the T cells. That LAD neutrophils are poorly adhesive for EC may be explained in part by their dependence on Leu-CAMs for stabilized adhesion to EC. Thus, normal peripheral blood neutrophils do not use VLA-4/VCAM-1 as an alternative adhesive pathway since they do not express VLA-4, nor do they bind to VCAM-1 (Osborn *et al.*, 1989; Hemler, 1990).

VCAM-1 and VLA-4 interactions appear to play a role during the extravasation of lymphocytes. Several *in vivo* studies have suggested that VLA-4 mediates the circulation of lymphocytes into sites of inflammation, since mAb to VLA-4 blocked the appearance of radiolabelled lymphocytes in these sites (Issekutz, 1991; Issekutz and Issekutz, 1991). These studies, however, did not determine whether cell infiltration was blocked because of inhibition of binding rather than trans-endothelial migration of the lymphocytes. *In vitro* studies have dissected this issue in detail and suggested that VLA-4/VCAM-1 interactions are not involved in the trans-endothelial migration of T cells that are bound to EC, even though a role in T cell binding to cytokine exposed EC can be demonstrated (Oppenheimer-Marks *et al.*, 1991) Thus, for T cells, VCAM-1 appears to play a major role in the initial binding to cytokine activated EC. It is therefore likely that these interactions are important in the accumulation of T cells at sites of inflammation but not directly in mediating migration into the extravascular compartment.

Like ICAM-1, VCAM-1 binding to its counter-receptor delivers intracellular signals to T cells. Using substrates coated with purified VCAM-1 or a soluble VCAM-1 fusion protein, a role for VCAM-1 in co-stimulating the activation of CD4+ T cells has been demonstrated (Damle and Aruffo, 1991; van Seventer *et al.*, 1991). This effect is dependent on the engagement of VLA-4, since specific mAb inhibited the response. That VLA-4 on lymphocytes mediates this extracellular signal is further implied by studies showing that fibronectin, which also binds VLA-4 as well as VLA-5, delivers co-stimulatory signals to T lymphocytes by engaging each beta1 integrin (Davis *et al.*, 1990; Shimizu *et al.*, 1990a). Thus, VLA-4 can transduce co-stimulatory signals after engaging either VCAM-1 or fibronectin.

The increasing body of data indicates that several members of the Ig superfamily of adhesion molecules, including ICAM-1 and VCAM-1, are important during recognition functions of the immune system, by not only providing adhesive substrates on which the cells adhere and migrate, but also because they engage counter-

Table 7.3 Members of the selectin superfamily of adhesion receptors that mediate leucocyte interactions with EC

Receptor	*Counter-receptor*	*Distribution*	*Regulation*
E-selectin	Sialyl Lewis X, L-selectin, CLA	Activated endothelial cells	Induced by IL-1, TNF-α
L-selectin	E-selectin, P-selectin, Peripheral node addressin, charged oligosaccharides	Resting leucocytes	Decreased by activation stimuli
P-selectin	Sialyl Lewis X, L-selectin	Activated platelets, activated endothelial cells	Induced by thrombin, histamine

receptors that deliver activation signals that up-regulate effector cell function. Not all known adhesion receptors expressed by EC serve such multiple functions, however. Thus, the binding of E-selectin, also a cytokine inducible leucocyte adhesion receptor expressed by EC, does not appear to co-stimulate T cell activation (van Seventer *et al.*, 1991).

Members of the Ig superfamily play fundamental roles in T cell adhesive interactions. For processes involving lymphocyte extravasation, both VCAM-1 and ICAM-1 are essential adhesion molecules. VCAM-1 mediated interactions are important for stabilized binding of lymphocytes to cytokine activated EC. ICAM-1 also mediates T cell-EC adhesion in certain circumstances, but also appears to promote the trans-endothelial migration of T cells already bound to EC. The increasing body of data also indicates, however, that adhesion receptors other than those previously identified members of the integrin and Ig superfamilies are also important in the process of lymphocyte extravasation.

4.2 SELECTIN SUPERFAMILY

There are three members of the selectin superfamily of adhesion receptors, each of which is important for the extravasation of leucocytes into inflammatory sites (Table 7.3). Although the name of each member has been simplified to relate them to one another, they are often referred to by their previous names. Thus, E-selectin (CD62E) was formerly called endothelial-leucocyte adhesion molecule (ELAM)-1; L-selectin (CD62L) was formerly referred to by a number of different names, including leucocyte-endothelial cell adhesion molecule (LECAM)-1, leucocyte adhesion molecule (LAM)-1, MEL-14 antigen, and Leu-8; and P-selectin was formerly referred to as granule membrane protein (GMP)-140 or platelet activation dependent granule-external membrane (PADGEM) protein. All members of the selectin family have at their amino termini a calcium dependent lectin domain, at which most of the binding activity of the molecule occurs (Bevilacqua *et al.*, 1989; Johnston *et al.*, 1989; Lasky *et al.*, 1989). Next to the lectin domain is a domain homologous to epidermal growth factor (EGF), and then variable numbers of tandem repeats homologous to complement regulatory proteins (CRP). For all selectins the EGF domains are necessary for recognition of ligand (Kansas *et al.*, 1991; 1994; Pigott *et al.*, 1991). Thus, studies examining the structure-function relationships of these molecules using various anti-selectin mAb or selectin chimeras, have indicated that the EGF domains either contribute to the formation of one or more binding domains that directly interact with ligands, or they are required for the maintenance of the functional conformation of the lectin domain. The CRP domain of L-selectin has also been suggested to serve a similar function (Kansas *et al.* 1991).

The counter-receptors for each member of the selectin superfamily contain charged oligosaccharides, with which the lectin domains interact. Thus, both E-selectin and P-selectin bind α(2-3)sialylated, α(1-3)fucosylated lactosaminoglycans, including the sialyl Lewis X (SLe^x) sequence, Neu5Acα2-3Galβ1-4(Fucα1-3)GlcNAcβ-R (Lowe *et al.*, 1990; Phillips *et al.*, 1990; Walz *et al.*, 1990; Polley *et al.*, 1991; Zhou *et al.*, 1991). CD15 (Lewis X) has also been identified as a counter-receptor for both E-selectin and P-selectin, but it seems to bind to these receptors with an affinity that is much lower than the sialylated form (Larsen *et al.*, 1990; Polley *et al.*, 1991). Particularly noteworthy is that L-selectin expressed by neutrophils contains the SLe^x sequence, and serves as one of the counter-receptors for E-selectin and P-selectin (Picker *et al.*, 1991b). On T cells, L-selectin does not contain the SLe^x sequence, and does not bind to E-selectin. Rather, another cell surface molecule, cutaneous lymphocyte antigen (CLA), that is expressed by a subset of memory T cells, serves as a counter-receptor for E-selectin (Picker *et al.*, 1991a). However, L-selectin that is expressed by lymphocytes does appear to mediate binding to cytokine activated EC, although its counter-receptor remains unidentified (Spertini *et al.*, 1991). L-selectin on neutrophils, monocytes and lymphocytes also binds to peripheral lymph node vascular addressin (PNAd), presumably by the lectin domain, since binding is abolished by prior treatment of cells with mannose-6-phosphate, fucose-1-sulfate, and related fucose-rich polysaccharides and mannose-rich phosphmannans (Stoolman *et al.*, 1984; Yednock *et al.*, 1987). Whether L-selectin binds the SLe^x sequence on PNAd has not yet been determined. Treatment of PNAd, however, with neuraminidase abolishes lymphocyte binding (True *et al.*, 1990).

The expression of E-selectin is restricted to EC, and it is found at sites of chronic inflammation, including rheumatoid synovium (Koch *et al.*, 1991; Morales-Ducret *et al.*, 1991). Normal, unstimulated EC do not express E-selectin, but it is induced by cytokines, including IL-1 and TNF (Pober *et al.*, 1986; Bevilacqua *et al.*, 1987). This has been demonstrated to occur both *in vitro* and *in vivo*, and like ICAM-1 and VCAM-1, its expression requires protein synthesis (Pober *et al.*, 1986). E-selectin was originally thought to be an adhesion receptor strictly for neutrophils (Bevilacqua *et al.*, 1987). However, it since has been determined that E-selectin also mediates the binding of a subset of memory T cells to cytokine activated EC (Picker *et al.*, 1991b; Shimizu *et al.*, 1991).

L-selectin is expressed by resting neutrophils, lymphocytes and monocytes (Lewinsohn *et al.*, 1987). Initially, L-selectin was identified as a homing receptor mediating lymphocyte trafficking into peripheral lymph node (PLN) by binding to PNAd (Berg *et al.*, 1991; Gallatin *et al.*, 1983). Studies have also shown that L-selectin binds to a number of other charged oligosaccharides expressed on EC surface receptors, including CD34 and GLYCAM-1 as well as MAdCAM-1 (Jutila, 1994; Laskey *et al.*, 1992; Berg *et al.*, 1993). As noted above, L-selectin that is expressed by neutrophils also serves as a counter-receptor for E-selectin and P-selectin, and mediates entry of neutrophils into inflammatory sites (Picker *et al.*, 1991b).

P-selectin is a membrane glycoprotein found in secretory granules of human platelets and Weibel-Palade bodies of EC (Larsen *et al.*, 1989; McEver *et al.*, 1989). When these cells are stimulated by PKC activating phorbol esters or agonists, such as thrombin or histamine, P-selectin is rapidly redistributed to the plasma membrane where it mediates the binding of myeloid cells (Geng *et al.*, 1990; Lorant *et al.*, 1991). Although much of what is known about the activity of P-selectin in mediating adhesive interactions of EC come from studies of neutrophils, recent studies have suggested that P-selectin may also mediate monocyte adhesion to synovial tissue EC (Stoolman *et al.*, 1992; Damle *et al.*, 1992; Kuzendorf *et al.*, 1993). All members of the selectin family are involved in the initial adhesion of neutrophils to the EC surface. As a result of these initial interactions additional adhesive mechanisms are induced to stabilize the binding of neutrophils to EC (Butcher, 1991).

4.3 INTEGRIN SUPERFAMILY

The term integrin was coined to denote molecules that "integrate" information from the extracellular mileau with the intracellular environment. Integrins are widely distributed being expressed on all nucleated cells. Thus, integrins mediate many cell-cell and cell-ECM adhesive interactions by binding to specific cell surface counter-receptors and/or to components of ECM (Table 7.4). Integrins have been implicated in many physiological processes including embryological development, hemostasis, thrombosis, wound healing, immune and non-immune defense mechanisms, and oncogenic transformation (Hynes, 1987; Albelda and Buck, 1990). All members of the integrin family are transmembrane glycoproteins composed of two non-covalently linked heterodimers, the α and β subunits. Integrins were initially divided into three subfamilies, β_1 (also known as very late activation antigens, VLA), β_2 (also known as Leu-CAMs), and β_3 (also known as cyto-adhesins), each having a common β subunit that is capable of associating with more than one α subunit within a specific group. However, it has since been shown that certain α subunits can associate with more than one β subunit, and that there are at least eight different identified β subunits. Moreover, several integrins have more than one counter-receptor.

The β subunits are similar, having 40–48% amino acid sequence homology, with conservation of specific structural features over a wide range of species, including mammals, birds, amphibians, insects and fungi (Marcantonio and Hynes, 1988; Hemler, 1990). Typically, a β subunit possesses one trans-membrane domain and a short cytoplasmic domain found at the carboxy terminus of the molecule (Hogg, 1989). The extracellular domains of β subunits are stabilized by intrachain disulfide bonds. The α subunits exhibit more sequence heterogeneity than do the beta subunits. Nevertheless, they each contain a short carboxy terminus domain, a single trans-membrane domain and a large globular extracellular domain, which contains divalent cation binding regions (Hogg, 1989). In the α chains of the β_2 subfamily, and the α_2 subunit of the β_1 subfamily, there is an additional amino acid sequence inserted between the last divalent cation binding site and the amino terminus. This "inserted/interactive or I" domain is homologous to repeats within von Willebrand factor, chicken cartilage matrix factor, and a region of complement factor B (Hogg, 1989; Diamond *et al.*, 1993; Randi and Hogg, 1994).

The binding of ligand by the integrins is a divalent cation dependent process, which requires the presence of both subunits (Hogg, 1989). Many of these ligands express specific amino acid sequences which are recognized by the integrins. For example, a number of ECM molecules, such as fibronectin, fibrinogen, thrombospondin, vitronectin, laminin, and type I collagen, express the arginine-glycine-aspartic acid (RGD) sequence (Hynes, 1987; Albelda and Buck, 1990). Not all integrins will, however, bind to the RGD sequence. For example, the VLA-4 ($\alpha_4\beta_1$ or CD49d/CD29) integrin binds to the variable region (connecting segment-1, CS-1) of fibronectin, which lacks an RGD sequence (Wayner et al., 1989; Guan and Hynes, 1990). The ability of integrins to mediate interactions with ligands requires the activity of the cytoskeleton. Interactions with actin-containing microfilaments have been identified and they

Table 7.4 Members of the integrin superfamily of adhesion receptors that mediate adhesive interactions of leucotyes

Sub-family	*Receptor*	*Counter-receptor*	*Distribution*	*Regulation*
Beta (CD29)	VLA-1 (CD49a/CD29)	Laminin, Collagen	Activated T Cells, fibroblasts, liver sinusoids, mesangial cells, some cell lines	Induced by mitogen, antigen
	VLA-2 (CD49b/CD29)	Collagen, laminin	Activated T cells, platelets, endothelial cells	Increased by mitogen, antigen
	VLA-4 (CD49d/CD29)	VCAM-1, fibronectin	Lymphocytes, monocytes, thymus cells, hematopoietic cell lines	Functional activation by antigen
	VLA-5 (CD49e/CD29)	Fibronectin	Lymphocytes, monocytes, thymus cells, endothelial cells, fibroblasts	Functional activation by antigen
	VLA-6 (CD49f/CD29)	Laminin	Lymphocytes, platelets	Functional activation by antigen
Beta2 (CD18)	LFA-1 (CD11a/CD18)	ICAM-1 ICAM-2 ICAM-3	All immune cells	Functional activation and increased expression by various mitogenic stimuli
	Mac-1 (CD11b/CD29)	ICAM-1, Fibrinogen, Endotoxin, Factor X, iC3b	Monocytes, granulocytes, macrophages, large granular lymphocytes, some B cells	Increased expression by chemotactic factors
	p150/95 (CD11c/CD18)	Fibrinogen, iC3b	Monocytes, granulocytes, macrophages, large granular lymphocytes, some B cells, hairy cell leukemia cells	Increased by TNF-α
Beta7	$\alpha 4\beta 7$	MAdCAM-1, VCAM-1, fibronectin	Band T cells, macrophages	Increased by activation stimuli
	$\alpha E\beta 7$	?	Intraepithelial lymphocytes	Induced by TGF-$\beta 1$

occur indirectly by association of the cytoplasmic domains of the receptors with actin binding proteins, such as talin and vinculin (Burn *et al.*, 1988; Burridge *et al.*, 1988).

In addition to binding cells to ligands, integrins deliver signals to cells. For example, LFA-1, a member of the β_2 subfamily is an accessory molecule for T-cell activation during cytotoxic T-cell binding to targets and during antigen presentation to T cells. Both VLA-4 and VLA-5 also serve as co-stimulatory molecules for T-cell activation. After interaction with VCAM-1 (VLA-4) and/or fibronectin (VLA-4, VLA-5), these integrins transduce T cell activation signals that facilitate T-cell activation when the TCR/CD3 complex is engaged that appear to involve tyrosine phosphorylation (Davis *et al.*, 1990; Shimizu *et al.*, 1990a; Damle and Aruffo, 1991; van Seventer *et al.*, 1991; Sanchez-Mateos *et al.*, 1993; Freedman *et al.*, 1993). However, the nature of these signals has not yet been determined.

The activity of many members of the integrin superfamily of adhesion receptors is essential for the infiltration of leucocytes into inflammatory sites. There are six members of the β_1 family, all of which bind components of ECM, including fibronectin, collagen and laminin (Hemler, 1990). VLA-4 also mediates cell-cell adhesion by binding to VCAM-1, expressed by EC. All but VLA-3, which has multiple ligand specificities are expressed by lymphocytes and monocytes (Hemler, 1990). VLA-3 expression is limited to a few epithelial cell types, but is absent or weakly expressed by most tissue and immune cells (Choy *et al.*, 1990; Beaulieu, 1992; Korhonen *et al.*, 1990). Memory CD4+ T cells express several fold more VLA-4, -5, and -6 than do naive T cells (Shimizu *et al.*, 1990b). VLA-1 ($\alpha_1\beta_1$) and VLA-2 ($\alpha_2\beta_1$) are expressed by T cells 2–4 weeks following their activation (Hemler and Jacobson, 1987). VLA-1 and VLA-2, which bind primarily to laminin and collagen, respectively, are reciprocally regulated in that the increased expression of

Table 7.5 Adhesion receptors utilized during neutrophil interactions with EC

	Neutrophils	Endothelial cells
EARLY	L-selectin, sialyated CD15	P-selectin (CD62P)
	?	PECAM-1 (CD31)
INTERMEDIATE	L-selectin, sialyated CD15	E-selectin (CD62E)
LATE	LFA-1 (CD11a/CD18)	ICAM-1 (CD54), ICAM-2 (CD102)
	Mac-1 (CD11b/CD18)	ICAM-1 (CD54), fibrinogen
	p150/95 (CD11c/CD18)	Fibrinogen

one is usually associated with the decreased expression of the other (Hemler and Jacobson, 1987). As noted earlier, T cells isolated form RA synovial tissue and fluid, express VLA-1, indicating that they have undergone multiple rounds of proliferation (Hemler *et al.*, 1986).

The activities of VLA-4, -5 and -6 require an activation step for them to bind their ligands. Their functional activation occurs without changes in the level of expression of the VLA molecules (Shimizu *et al.*, 1990b). This may be a common feature of integrins, since it also has been established for LFA-1 and Mac-1 of the β_2 subfamily as well. Thus, stimulation of T cells with PKC activating phorbol esters or by engaging the CD3 molecular complex induces the ability of the T cells to utilize β_1 and β_2 integrins to bind ECM molecules and cellular counter-receptors (Dustin and Springer, 1989; Shimizu *et al.*, 1990b). Also noteworthy are the findings that ligation of CD31 on a subset of T cells has been shown to lead to the functional activation of VLA-4 (Tanaka *et al.*, 1992). In addition to beta 1, the alpha 4 chain has also been shown to associate with beta 7. Moreover, beta 7 has been shown to also associate with another alpha chain, αE, which is expressed by intraepithelial lymphocytes (Cepek *et al.*, 1993). When T cells recirculate through mucosal lymphoid tissue, the adhesion receptor $\alpha 4\beta 7$ is the predominant adhesion receptor mediating interactions with the endothelium by binding its counter-receptor, MAdCAM-1 (Berlin *et al.*, 1993). MAdCAM-1 expression by endothelial cells appears to be regulated by the proinflammatory cytokines TNFα or IL-1β (Sikorski *et al.*, 1993). $\alpha 4\beta 7$ has the same binding specificity of VLA-4, recognizing VCAM-1 and fibronectin as well as MAdCAM-1 (Ruegg *et al.*, 1992). $\alpha 4\beta 7$ presumably binds the domain of MAdCAM-1 that shares homology with VCAM-1. The expression of members of the β_2 subfamily of integrin adhesion receptors is restricted to hematopoietic cells (Springer, 1990). Three leucocyte adhesion receptors, LFA-1 (CD11a/CD18), Mac-1 (CD11b/CD18) and p150/95 comprise this subfamily. In addition to mediating leucocyte adhesive interactions with EC, LFA-1 is involved in both homotypic and heterotypic interactions of cells. Like VLA, LFA-1 is constitutively expressed by lymphocytes, and it requires activation prior to binding its counter-receptors. ICAM-1 and ICAM-2 do not contain the RGD sequence, and thus recognition of these molecules by LFA-1 occurs through a separate mechanism. In contrast to LFA-1, Mac-1 (also known as CD11b/CD18, CR3), which also binds ICAM-1, has a much larger repertoire of ligands with which it interacts. Thus, Mac-1 binds iC3b, fibrinogen, coagulation factor X, and endotoxin. Both iC3b and fibrinogen contain RGD sequences. Thus, in addition to mediating cell-cell adhesion, Mac-1 also is involved in complement binding and phagocytosis. Although neutrophils express all three members of the β_2 subfamily, Mac-1 is the major one used to interact with EC. Unlike LFA-1, preformed Mac-1 is rapidly expressed on neutrophil surfaces upon their activation by PKC activating phorbol esters or chemotactic stimuli, and plays an essential role in their trans-endothelial migration (Vedder and Harlan, 1988; Dustin and Springer, 1989). However, a quantitative increase in Mac-1 expression is not necessary for it to mediate neutrophil adhesion (Vedder and Harlan, 1988; Schleiffenbaum *et al.*, 1989). Very little is known about p150/95. It is expressed by monocytes, macrophages, granulocytes, large granular lymphocytes, some B cells and some T-cell lines. Like Mac-1, p150/95 is rapidly mobilized from intracellular stores to the neutrophil surface in response to activation stimuli. Among the ligands of p150/95 are iC3b and fibrinogen (Kuypers and Roos, 1989; Albelda and Buck, 1990). Of note, activation stimuli determine the β_2 integrin that mediates the adhesion of neutrophils to substrates. Thus, PKC activating phorbol esters up-regulate Mac-1 dependent binding to fibrinogen, whereas TNF-α delivers a different signal to neutrophils and induces p150/95 mediated binding to fibrinogen (Loike *et al.*, 1991). The nature of these opposing signals has not been delineated.

4.4 REGULATION OF TRANS-ENDOTHELIAL MIGRATION OF INFLAMMATORY CELLS

In general, leucocyte extravasation involves the concerted activities of a variety of adhesion receptors. The migration of leucocytes into perivascular tissue is a multistep adhesive process that involves the induction or activation of cell surface adhesion receptors to mediate stable leucocyte binding to EC, which is followed by the migration of the bound cells through the EC layer. Central to this process is that extravasating cells possess an intrinsic migratory capacity.

The precise mechanisms involved in the establishment of stabilized lymphocyte binding to EC have not been delineated. There is a redundancy in the number of receptor–counter-receptor pairs that are capable of supporting adhesive interactions between lymphocyes and EC (Fig. 7.1). L-selectin, which does not require an activation step, has been reported to mediate lymphocyte binding to TNF-α activated EC in addition to binding PNAd (Kansas *et al.*, 1991). Thus, the ligation of L-selectin may initiate intracellular signalling events that lead to stabilized T cell-EC binding. VLA-4 and LFA-1 both require an activation step to become competent to bind their counter-receptors. This does not result from increased expression of these molecules by lymphocytes, but appears to relate to conformational changes in their structure. Both VLA-4 and LFA-1 mediate lymphocyte adhesion to EC, although all of the binding cannot be accounted for by the activities of these two molecules. VLA-4 recognition of VCAM-1 is most important in binding T cells to cytokine activated EC. On the other hand, LFA-1 functions during adhesion of resting or activated T cells to unstimulated EC by binding to both ICAM-1 and ICAM-2. Both, E-selectin and P-selectin, which are expressed by activated EC also appear to mediate lymphocyte binding (Picker *et al.*, 1990a; McEver, 1992; Damle *et al.*, 1992; Kunzendorf *et al.*, 1993). A subset of memory T cells adheres to E-selectin, and for many of these T cells this is mediated by CLA (Picker *et al.*, 1990b). Whether CLA requires an activation step to bind to E-selectin has not yet been reported. The nature of the lymphocyte counter-receptor that binds P-selectin has not been reported, but it most likely is related to SLe^x.

In contrast to the process of T cell-EC binding, there is a limited number of adhesion receptors that have been identified that mediate the trans-endothelial migration of lymphocytes (Fig. 7.1). To date, the receptor pair, LFA-1/ICAM-1, and the homing receptor CD44, have been implicated in the trans-endothelial migration of lymphocytes (Oppenheimer-Marks *et al.*, 1990; Oppenheimer-Marks, *et al.*, 1991). MAb blocking

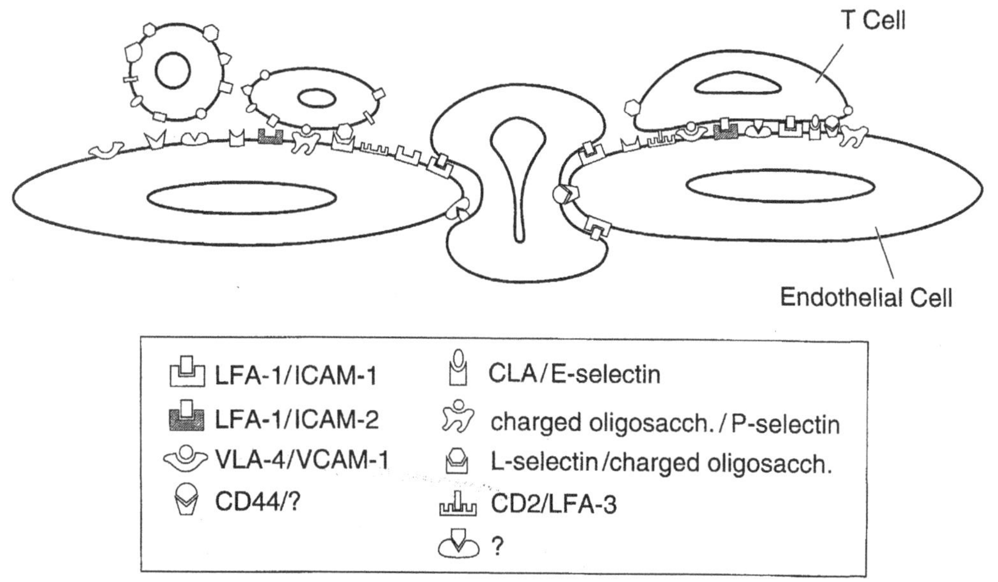

Figure 7.1 T lymphocytes express a repertoire of adhesion receptors that mediate adhesive interactions with endothelial cells at sites of inflammation Binding of T cells to EC occurs by the interaction of a variety of adhesion receptors with their specific counter-receptors. At sites of inflammation, initial adhesive interactions may result from the binding of L-selectin to charged oligosaccharide group on cytokine activated EC. Similarly, early adhesive interactions may also be mediated by charged oligosaccharides on T cells interacting with P-selectin on activated EC. Thus, interactions between LFA-1/ICAM-1, LFA-1/ICAM-2, VLA-4/VCAM-1, CD2/LFA-3, CLA/E-selectin, CD44 and its unidentified counter-receptor, as well as between other as yet unidentified receptor pairs, could then account for the stabilized adhesion of T cells to EC at sites of inflammation. The trans-endothelial migration of T cells is mediated by a variety of adhesion receptors, including LFA-1/ICAM-1, CD44 and its unidentified counter-receptor, as well as other as yet unidentified receptor pairs. Once the T cells have migrated into the perivascular tissue, additional adhesive interactions occur between the T cells and components of extracellular matrices as well as with other cells present within the tissue.

experiments conducted *in vitro* have demonstrated that the trans-endothelial migration of T cells bound to EC is inhibited by anti-LFA-1 or anti-ICAM-1 mAb, but not by mAb against VLA-4, VCAM-1, LFA-3 or E-selectin (Oppenheimer-Marks *et al.*, 1991). Thus, although a number of adhesion receptors mediate intital adhesive events between T cells and EC, they do not necessarily serve a similar function during trans-endothelial migration.

Under conditions whereby both T cells and EC have been exposed to activation stimuli, part of the trans-endothelial migration of the activated lymphocytes is mediated by CD44. CD44 is not related to the super-families described above. Rather, CD44 is more closely related to proteoglycan link protein, and it has a number of different ligand specificities, including the ECM components, hyaluronic acid and collagen, as well as a mucosal tissue specific vascular addressin (Haynes *et al.*, 1989).

It is likely that both lymphocyte-EC binding and trans-endothelial migration are redundant processes, in that there are many adhesion receptors that have overlapping functions in the facilitation of the entrance of lymphocytes into perivascular tissue. Whether this redundancy is necessary for the continuous infiltration of lymphocytes into tissue, or for the selective recruitment of specific subpopulations of T cells has not yet been determined. It is likely that the extravasation of lymphocytes into perivascular tissue is facilitated by the elaboration of mediators that possess chemotactic activity. Chemoattractants, such as RANTES specifically induce the chemotaxis of a subpopulation of memory T cells (see below) (Schall et al., 1990). Signals induced by these mediators may serve several purposes, including the activation of lymphocyte adhesion receptors, to facilitate their binding and trans-endothelial migration, and the directed movement into the tissue of specific populations of lymphocytes.

At sites of inflammation, where E-selectin and P-selectin are expressed by EC, neutrophil rolling has been shown to be mediated by either of these molecules binding to L-selectin (Table 7.5) (Butcher, 1991; Picker *et al* 1991a; Lawrence *et al.*, 1991). This type of adhesive interaction is not stable and by itself does not lead to the trans-endothelial migration of neutrophils. Rather, stable interactions occur as a result of the formation of additional adhesions between integrins on neutrophils and ICAM-1 on EC. This stabilized binding and trans-endothelial migration of neutrophils occurs as a response of neutrophils to the activities of inflammatory mediators. Thus, leukotrienes, complement components, and chemokines such as IL-8, activate neutrophils and induce changes in the adhesiveness of the cells (Kishimoto *et al.*, 1989; Huber *et al.*, 1991). In this series of events, L-selectin expression is down-regulated and concomitantly, there is an increased expression and activation of Mac-1, which by binding to ICAM-1 strengthens neutrophil interactions with EC. These mediators most likely also activate LFA-1, on neutrophils since both LFA-1 and Mac-1 have been suggested to mediate the trans-endothelial migration of neutrophils (Smith *et al.*, 1988; Smith *et al.*, 1989).

4.5 ADHESION MOLECULE EXPRESSION IN RHEUMATOID SYNOVITIS

The role of specific adhesion receptors during the extravasation of inflammatory cells into the rheumatoid synovium has not been completely delineated. The EC of rheumatoid synovium express adhesion receptors that bind lymphocytes. This is most likely a direct result of the activitites of pro-inflammatory mediators, such as IL-1 and TNF, that are elaborated within the tissue, and which, among their other activities, induce the expression of adhesion receptors by EC. Thus, several reports have indicated that the EC of rheumatoid synovium express ICAM-1, VCAM-1, E-selectin, and P-selectin (Hale et al., 1989; Koch *et al.*, 1991; Morales-Ducret *et al.*, 1991; Stoolman *et al.*, 1992). The most intensely and broadly expressed EC adhesion receptor is ICAM-1, which is present on venules, capillaries and arterioles. ICAM-2 is also expressed by RA synovial tissue EC (Szekanecz *et al.*, 1994; El-Gabalawy *et al.*, 1994). Whether ICAM-3 is expressed is a subject of debate (Szekanecz *et al.*, 1994; El-Gabalawy *et al.*, 1994). E-selectin is expressed on nearly all vessels, while VCAM-1 is expressed by EC to a lesser degree (Koch *et al.*, 1991; Veale *et al.*, 1993). Until recently, P-selectin activity was restricted to binding neutrophils. However, it has recently been reported that P-selectin also binds lymphocytes (McEver, 1992; Damle *et al.*, 1992; Kunzendorf, 1993). Particularly noteworthy is that P-selectin expressed by EC in synovial tissue binds peripheral blood monocytes, and thus may be important in facilitating the entry of this cell population into the tissue (Stoolman *et al.*, 1992). VCAM-1 expressed by synovial tissue EC has been demonstrated to mediate the binding of synovial T cells. Although a role for ICAM-1 in mediating the entrance of lymphocytes into synovial tissue has not been demonstrated, it is likely that it serves an important role in this process. The sequence of events or specific receptors utilized for entry of cells into rheumatoid synovium also has not been clearly delineated. It is likely that following their activation, VLA-4 and LFA-1 on lymphocytes serve important functions by binding to their counter-receptors, and mediating T-cell binding and trans-endothelial migration, respectively.

A subset of circulating T cells from RA patients is spontaneously competent to bind ICAM-1. This has been demonstrated by measuring interactions of RA or normal peripheral blood T cells with substrates that have been pre-coated with purified ICAM-1 (Oppenheimer-Marks, N. and Lipsky, P.E. unpublished observations). These results imply that this subset of circulating T cells in patients with RA expresses LFA-1 in an activated form and, therefore, is directly competent to bind ICAM-1. Expression of an activated form of LFA-1 may be another feature related to the overall increased activation state of circulating T cells in RA patients, and may account for their increased accumulation within rheumatoid synovium. However, the specific subset of circulating cells spontaneously expressing an activated form of LFA-1 has not been identified. Subsets of synovial tissue and fluid T cells also express activated VLA-4 as evidenced by their ability to bind to fibronectin-coated substrates (Laffon *et al.*, 1991). The nature of this subset also has not been identified. However, it is clear that subsets of circulating T cells as well as those found at inflammatory sites in RA express activated integrins that can facilitate their entry into synovial tissue.

In chronic RA, neutrophils traffick out of the tissue and accumulate in the synovial fluid compartment. The reasons for this are not certain at this time, but may be related to the lack of expression of integrins by neutrophils. As a result, adhesive mechanisms, which would mediate neutrophil binding to ECM components are absent, and neutrophils would, therefore, be subjected to modest adhesive restraints within the tissue. In contrast, strong chemotactic signals, by for example IL-8 and the activated complement component, C5a, emanating from the fluid compartment might stimulate the migration of neutrophils out of the tissue. Once in the fluid, neutrophils assume their characteristic agranular, vacuolated appearance and remain in this compartment.

4.6 CHEMOKINES

Soluble mediators emerging from the tissue are involved in the signalling events controlling the adhesiveness of leucocytes. Thus, aside from cytokines influencing adhesion receptor expression by EC, cytokines that act as chemoattractants also play a role. Two such related members of the intercrine family of cytokines, that possess chemotactic activity for leucocytes, are interleukin-8 (IL-8) (also referred to as neutrophil activating factor, NAF and monocyte derived neutrophil chemotactic factor, MDNCF, as well as others) and RANTES (Oppenheim *et al.*, 1991). "RANTES" is an acronym from several of the characteristics of the RANTES gene and its product including *R*egulated upon *A*ctivation, *N*ormal *T* *E*xpressed, and *S*ecreted (Schall *et al.*, 1988). IL-8 is a member of the alpha intercrine family, is present in synovial fluid, and is produced by a variety of cell types, including synovial tissue macrophages, EC, fibroblasts, keratinocytes, and T cells. IL-8 is abundant in fluids from RA and non-RA arthritides (Brennan *et al.*, 1990). Its concentration in these fluids has been reported to range from 1–9 ng/ml. RANTES, on the other hand, appears to be produced by antigen or mitogen activated T lymphocytes and rheumatoid synovial tissue fibroblasts, and both MIP-1α and MIP-1β are produced by many cell types including B and T lymphocytes (Lipes *et al.*, 1988; Schall *et al.*, 1990; Rathanaswami *et al.*, 1993). All of these chemokines are chemotactic for lymphocytes as well as for other leucocytes. In addition to the direct effects IL-8 has on lymphocytes *in vitro*, recent *in vivo* recirculation studies have suggested that IL-8 may actually indirectly recruit lymphocytes into extravascular sites. Thus, the selective depletion of neutrophils, *in vivo*, eliminates the IL-8 induced accumulation of CD4+, but not CD8+ lymphocytes, suggesting that neutrophil trafficking through tissue is necessary for the subsequent trans-endothelial migration of specific populations of lymphocytes (Kudo *et al.*, 1991).

It is uncertain whether the net effect of IL-8 is pro- or anti-inflammatory. On the one hand, treatment of neutrophils with IL-8 inhibited their subsequent binding to EC *in vitro* (Gimbrone *et al.*, 1989). Moreover, intravenous administration of IL-8 inhibited the trans-endothelial migration of neutrophils into sites of acute inflammation, and inhibited neutrophil responses to local subcutaneous injection of IL-8, FMLP, C5a and TNF-α as well (Hechtman *et al.*, 1991). In contrast, *in vitro* studies have demonstrated that IL-8 produced by cytokine activated EC is secreted directionally into the collagen gels upon which the EC were cultured (Huber *et al.*, 1991). The EC derived IL-8 has chemotactic activity for neutrophils, and also activated the neutrophils to decrease the expression of L-selectin and up-regulate the β_2 integrins (Huber *et al.*, 1991). Thus, in various models IL-8 may exert pro or anti-inflammatory effects. The net effect *in vivo* remains to be determined, and may differ in various kinds of inflammation. IL-8 is found at inflammatory sites, including synovial fluid from rheumatoid joints (Brennan *et al.*, 1990). Moreover, subcutaneous injection of IL-8 has been shown to induce neutrophil accumulation *in vivo* (Kudo *et al.* 1991). Thus, the compartment in which IL-8 is located may determine its ability to exert chemotactic activity. IL-8 in the blood may inhibit neutrophil chemotaxis, possibly by saturation of IL-8 receptors, whereas IL-8 in tissue could exert a pro-inflammatory effect and promote neutrophil chemotaxis.

RANTES has been demonstrated to be produced at sites of inflammation and to be chemotactic for monocytes and a subpopulation of CD4(+) memory T cells (Schall *et al.*, 1990). RANTES induces the chemotaxis of lymphocytes at 200 fold lower concentrations than the optimal concentration for monocytes (Schall *et al.*, 1990). Thus, RANTES may selectively recruit the migration into tissue of specific subpopulations of lymphocytes. Both MIP-1α and MIP-1β chemoattract specific populations of lymphocytes, although the nature of the subpopulations is a matter of debate. For example, MIP-1α was recently shown to chemoattract either CD4(+) T cells (Taub *et al.*, 1993) or CD8(+) T cells (Tanaka *et al.*,

1993) specifically. These differences may be related to the concentrations of chemokine used and the assay situation, as a recent report by Schall *et al.* demonstrated the ability of MIP-1α and MIP-1β to induce the chemotaxis of different lymphocyte populations in a concentration dependent fashion (1993).

The data support the conclusion that chemokines play an important role in facilitating the extravasation of cells into inflammatory sites, and may also have a profound effect on recruiting specific populations of cells to infiltrate into these sites.

4.7 THE MIGRATION OF SPECIFIC POPULATIONS OF T CELLS INTO RHEUMATOID SYNOVIUM

There is an enrichment in RA synovial tissue of CD4+ memory T cells. There are a number of possibilities that could explain this finding. Suggestions have been made that synovial tissue EC express unique vascular addressins that recruit only those T-cell subsets that express the appropriate counter-receptors, in an analagous manner that EC of peripheral and mucosal lymphoid tissue express addressins that bind T cells bearing appropriate counter-receptors (Butcher *et al.*, 1990). Synovial fluid T cells can bind to HEV of mucosal lymphoid tissue (Kadioglu and Sheldon, 1991). Thus, synovial fluid T cells clearly express counter-receptors for mucosal HEV suggesting that synovial and mucosal EC may express similar addressins. The nature of the vascular addressin of synovial tissue EC has, however, not been identified. The infiltration into tissue of CD4+ T cells may also result from the action of specific chemotactic factors such as RANTES, that selectively stimulate the migration of memory T cells. It has also been suggested that specific antigen, within the synovial tissue, may influence the nature of the T cell that accumulates in the tissue. In support of this are findings that EC are capable of presenting antigen to T cells, and could, therefore, recruit into tissue-only T cells that are responsive to a particular antigen (Geppert and Lipsky, 1989). Moreover, *in vitro* analyses have indicated that HLA-DR positive EC support T-cell binding more effectively than HLA-DR negative T cells, implying that specific recognition by T cells may play a role in their migration (Masuyama *et al.*, 1986). Since EC in the rheumatoid synovium are known to express HLA-DR, albeit modestly, these molecules could contribute to selective recruitment of T cells into the tissue.

Nearly all of the CD4+ T cells within RA synovial tissue are memory cells. Recent *in vivo* and *in vitro* studies indicate that it is unlikely that these T cells migrate into the tissue as CD45RA+, naive T cells and then differentiate within the tissue into CD45RA−/CD45RO+ memory T cells. Rather the data suggest that memory, and not naive, CD4+ T cells exhibit a greater intrinsic transendothelial migratory capacity, and therefore preferentially migrate into extra-lymphoid tissue sites (Pietschmann *et al.*, 1992; Oppenheimer-Marks *et al.*, 1994). Moreover, memory CD4+ T cells are more able to bind to EC than naive T cells (Damle and Doyle, 1990). Finally, cellular kinetic studies have demonstrated that CD4+/CD45RO+ T cells preferentially migrate into delayed type hypersensitivity lesions *in vivo* in humans (Pitzalis *et al.*, 1991). Thus, mechanisms other than those utilized during normal lymphocyte trafficking are likely to facilitate the infiltration of memory cells into these inflammatory sites. Memory T cells are responsive to the chemotactic effects of RANTES which is elaborated in inflammatory sites, and which would selectively promote their extravasation. Activated memory T cells also elaborate factors which could auto-regulate their extravasation. Thus, once activated in the tissue, memory T cells could increase the permeability properties of EC, and also induce the expression of adhesion receptors by EC (Damle and Doyle, 1990; Damle *et al.*, 1991).

The relationship of the phenotype of T cells to their migratory capacity has been addressed by *in vitro* studies. Phenotypic analyses of T cells that are recovered following their trans-endothelial migration into collagen gels, indicate that the predominant migrating CD4+ T cell is also CD45RO+ (Table 7.6). Not all of the CD45RO+ T cells are migratory, however, since many do not migrate. This suggests that only a subset of memory T cells in the peripheral circulation exhibits a trans-endothelial migratory capacity. Flow cytometric analyses have been conducted to delineate further the identity of the CD4+ memory T cells in the migrated population. These T cells are also CD29^{bright}/L-selectin^{low}/CD7^{dim}, and they brightly express a number of adhesion receptors, including LFA-1, VLA-4, VLA-5, CD44 and CD58. This phenotype parallels that of CD4+ T cells present in rheumatoid synovium (Cush and Lipsky, 1991). Particularly noteworthy is that IL-1 or IFNγ activation of the EC did not affect the phenotype

Table 7.6 Phenotype of CD4+ T cells that exhibit a trans-endothelial migratory capacity

Phenotype
CD45RA−
CD45RO+
CD29^{bright}
L-selectin^{low}
CD7^{dim}
CD26^{bright}
CD31^{dim}
CD44^{bright}
CD58^{bright}
VLA-2(+)
VLA-4^{bright}
VLA-5^{bright}
LFA-1^{bright}

of the migrating CD4+ T cells, even though the absolute number of migrating T cells increased (Pietschmann *et al.*, 1992). Thus, the regulatory effect on trans-endothelial migration by these cytokines may be primarily to expand the degree to which a particular memory T-cell subset that possesses an intrinsic migratory capacity infiltrates into a particular tissue, rather than inducing the migration of different T cell subsets.

Of additional importance are results from studies investigating the characteristics of RA synovial tissue and fluid T cells that exhibit a trans-endothelial migratory capacity. Thus, peripheral blood lymphocytes isolated from normal donors or RA patients migrate equally well. On the other hand, RA synovial tissue lymphocytes demonstrate a 2–3 fold increase in their trans-endothelial migratory capacity (Cush *et al.*, 1991). Greater than 90% of the CD4+ synovial fluid T cells in the migrated population are $CD29^{bright}$, whereas less than 1% are CD45RA+. The migrated T cells are also positive for expression of HLA-DR (45% positive), which is reflective of the large number of DR+ T cells in the synovial tissue and fluid (20–85% positive) (Forre *et al.*, 1982a; Forre *et al.*, 1982b; Burmester *et al.*, 1987). There also is a large number of HLA-DR positive T cells in the peripheral circulation of RA patients (10–50%) (Cush and Lipsky, 1988). The infiltration into synovial tissue of these T cells may partially account for the increased number of DR+ T cells within the tissue.

5. *Therapeutic Strategies Targeting Adhesion Receptors for the Control of RA*

New strategies which involve anti-adhesion receptor therapy to control chronic RA have been developed. One rationale behind these involves the prevention of leucocyte infiltration into synovial tissue by blocking specific cell surface adhesion receptors that mediate interactions between leucocytes and EC. Moreover, anti-adhesion receptor therapy might also block the local activation of cells that results from the ligation of specific adhesion receptors. Members of the β_2 (CD18) integrins (LFA-1 and Mac-1) play a pivotol role in leucocyte adhesion to EC and subsequent trans-endothelial migration into perivascular connective tissue. Therefore, studies have been conducted to examine the role of these molecules during the development of chronic synovitis when soluble antigen is injected into knee joints of immunized rabbits in a model of experimental arthritis (Jasin *et al.*, 1992). Intra-articular injection of antigen induces an acute inflammatory response, which resolves, followed by a chronic persistent arthritis in the injected joint. The chronic phase, which has certain similarities with RA, is a widely used example of chronic arthritis with an immune pathogenesis. The chronic synovitis in this model is characterized by the presence of dense cellular infiltrates composed of macrophages, lymphocytes and plasma cells with a tendency to form lymphoid follicles, yielding a histological picture similar to RA (Jasin, 1988).

Administration of anti-CD18 mAb during the acute phase of this model prevented the development of the chronic phase. The mechanism of action is related to prevention of the entry of neutrophils during the acute Arthus reaction. It is likely that blocking neutrophil trafficking through the tissue prevents the subsequent development of the lymphocyte mediated chronic arthritis to develop. The striking decrease in the number and proportion of neutrophils in the inflammatory synovial fluid underscores the importance of β_2 integrins in the processes of adhesion and trans-endothelial migration of these inflammatory cells. Although local production of antibodies and accumulation of immune complexes in cartilage occurred during the chronic phase in the anti-CD18 treated animals, they were insufficient to sustain chronic synovitis in the absence of antecedent PMN-mediated acute inflammation.

Other studies have utilized mAb against ICAM-1, in adjuvant arthritis in rats (Iigo *et al.*, 1991). Adjuvant arthritis is prevented when mAb to ICAM-1 is given at the time of immunization with adjuvant, presumably by inhibiting the immunization process. Administration of mAb to ICAM-1 to animals in whom adjuvant arthritis was induced by adaptive transfer of primed cells also diminished arthritis suggesting that entry of inflammatory cells into target tissue was also blocked. Also particularly noteworthy is the observation that administration of mAb against ICAM-1 to RA patients with active disease had a marked modifying effect on disease activity (Kavanaugh *et al.*, 1994). Thus, among its effects, the anti-ICAM-1 mAb altered lymphocyte trafficking into tissue as evidenced by a marked lymphocytosis in the patients and suppression of the patient's delayed type hypersensitivity response to recall antigen as well. These results support the conclusion that interference with β_2 integrin mediated adhesion might have therapeutic potential in inflammatory arthritis.

6. *Conclusions*

The activities of a variety of adhesion receptors play central roles in the pathogenesis of RA. Adhesion molecules exert pleiotropic effects during inflammatory conditions, including the mediation of inflammatory cell interactions with EC and ECM molecules, trans-endothelial migration, as well as the delivery of co-stimulatory signals to these cells. Approaches to modify the expression or utilization of adhesion molecules promises to provide new therapeutic intervention to control rheumatoid inflammation.

7. References

Albelda, S.M. and Buck, C.A. (1990). Integrins and other cell adhesion molecules. FASEB J. 4, 2868–2880.

Albelda, S.M., Oliver, P.D., Romer, L.H. and Buck, C.A. (1990). EndoCAM: a novel endothelial cell-cell adhesion molecule. J. Cell Biol. 110, 1227–1237.

Albelda, S.M., Muller, W.A., Buck, C.A. and Newman, P.J. (1991). Molecular and cellular properties of PECAM-1 (endoCAM/CD31): a novel vascular cell-cell adhesion molecule. J. Cell Biol. 114, 1059–1068.

Altmann, D.M., Hogg, N., Trowsdale, J. and Wilkinson, D. (1989). Co-transfection of ICAM-1 and HLA-DR reconstitutes human antigen-presenting cell function in mouse L cells. Nature 338, 512–514.

Baumhueter, S., Singer, M.S., Henzel, W., Hemmerich, S., Renz, M., Rosen, S.D. and Lasky, L.A. Binding of L-selectin to the vascular sialomucin CD34. Science 262, 436–438.

Beaulieu, J.-F. (1992). Differential expression of the VLA family of integrins along the crypt-villus axis in the human small intestine. J. Cell Sci. 102, 427–436.

Berg, E.L., Robinson, M.K., Warnock, R.A. and Butcher, E.C. (1991). The human peripheral lymph node vascular addressin is a ligand for LECAM-1, the peripheral lymph node homing receptor. J. Cell Biol. 114, 343–349.

Berg, E.L., McEvoy, L.M., Berlin, C., Bargatze, R.F. and Butcher, E.C. (1993). L-selectin mediated lymphocyte rolling on MAdCAM-1. Nature 366, 695–698.

Berlin, C., Berg, E.L., Briskin, M.J., Andrew, D.P., Kilshaw, P.J., Holzmann, B., Weissman, I.L., Hamann, A. and Butcher, E.C. (1993). $\alpha4\beta7$ integrin mediates lymphocyte binding to the mucosal vascular addressin MAdCAM-1. Cell 74, 185–195.

Bevilacqua, M.P., Pober, J.S., Mendrick, D.L., Cotran, R.S. and Gimbrone, Jr., M.A. (1987). Identification of an inducible endothelial-leucyocyte adhesion molecule. Proc. Natl Acad. Sci. USA 84, 9238–9242.

Bevilacqua, M.P., Stengelin, S., Gimbrone, Jr., M.A. and Seed, B. (1989). Endothelial leucocyte adhesion molecule-1: an inducible receptor for neutrophils related to complement regulatory proteins and lectins. Science 243, 1160–1165.

Brennan, F.M., Zachariae, C.O.C., Chantry, D., Larsen, C.G., Turner, M., Maini, R.N., Matsushima, and Feldman, M. (1990). Detection of interleukin 8 biological activity in synovial fluids from patients with rheumatoid arthritis and production of interleukin 8 mRNA by isolated synovial cells. Eur. J. Immunol. 20, 2141–2144.

Briskin, M.J., McEvoy, L.M. and Butcher E.C. (1993). MAdCAM-1 has homology to immunoglobulin and mucin-like adhesion receptors and to IgA. Nature 363, 461–464.

Bromley, M. and Woolley, D.E. (1984). Hisotpathology of the rheumatoid lesion: identification of cell types at sites of cartilage erosion. Arthritis Rheum. 27, 857–863.

Burmester, G.R., Jahn, B., Rohwer, P., Zacher, J., Winchester, R.J. and Kalden, J.R. (1987). Differential expression of Ia antigens by rheumatoid synovial lining cells. J. Clin. Invest. 80, 595–604.

Burn, P., Kupfer, A. and Singer, S.J. (1988). Dynamic membrane-cytoskeletal interactions: specific association of integrin and talin arises *in vivo* after phorbol ester treatmant of peripheral blood lymphocytes. Proc. Natl Acad. Sci. USA 85, 497–501.

Burridge, K., Fath, K., Kelly, T., Nuckolls, G. and Turner, C. (1988). Focal adhesion: trans-membrane junctions between the extracellular matrix and the cytoskeleton. Ann. Rev. Cell Biol. 4, 487–525.

Butcher, E.C. (1990). Cellular and molecular mechanisms that direct leucocyte traffic. Am. J. Pathol. 136, 3–11.

Butcher, E.C. (1991). Leukocyte-endothelial cell recognition: three (or more) steps for specificity and diversity. Cell 67, 1033–1036.

Campanero, M.R., Pulido, R., Ursa, M.A., Rodriguez-Moya, M., de Landazuri, M.O. and Sanchez-Madrid, F. (1990). An alternative leucocyte homotypic adhesion mechanism, LFA-1/ICAM-1-independent, triggered through the human VLA-4 integrin. J. Cell Biol. 110, 2157–2165.

Carlos, T.M., Schwartz, B.R., Kovach, N.L., Yee, E., Rosso, M., Osborn, L., Chi-Rosso, G., Newman, B., Lobb, R. and Harlan, J.M. (1990). Vascular cell adhesion molecule-1 mediates lymphocyte adherence to cytokine-activated cultured human endothelial cells. Blood 76, 965–970.

Carpen, O., Dustin, M.L., Springer, T.A., Swafford, J.A., Beckett, L.A. and Caulfield, J.P. (1991). Motility and ultrastructure of large granular lymphocytes on lipid bilayers reconstituted with adhesion receptors LFA-1, ICAM-1, and two isoforms of LFA-3. J. Cell Biol. 115, 861–871.

Cepek, K.L., Prker, C.M., Madara, J.L. and Brenner, M.B. (1993). Integrin $\alpha^E\beta7$ mediates adhesion of T lymphocytes to epithelial cells. J. Immunol. 150, 3459–3470.

Cush, J.J. and Lipsky, P.E. (1988). Phenotypic analysis of synovial tissue and peripheral blood lymphocytes isolated from patients with rheumatoid arthritis. Arthritis Rheum. 31, 1230–1238.

Cush, J.J. and Lipsky, P.E. (1991). Cellular basis for rheumatoid inflammation. Clin. Orthop. 265, 9–22.

Cush, J.J., Pietschmann, P., Oppenheimer-Marks, N. and Lipsky, P.E. (1992). The intrinsic migratory capacity of memory T cells contributes to their accumulation in rheumatoid synovium. Arthritis Rheum. 35, 1434–1444.

Damle, N.K. and Doyle, L.V. (1990). Ability of human T lymphocytes to adhere to vascular endothelial cells and to augment endothelial permeability to macromolecules is linked to their state of pot-thymic maturation. J. Immunol. 144, 1233–1240.

Damle, N.K. and Aruffo, A. (1991). Vascular cell adhesion molecule 1 induces T-cell antigen receptor-dependent activation of CD4+ T lymphocytes. Proc. Natl Acad. Sci. USA 88, 6403–6407.

Damle, N.K., Eberhardt, C. and Van der Vieren, M. (1991). Direct interaction with primed CD4+ CD45RO+ memory T lymphocytes induces expression of endothelial leucocyte adhesion molecule-1 and vascular cell adhesion molecule-1 on the surface of vascular endothelial cells. Eur. J. Immunol. 21, 2915–2923.

Damle, N.K., Klussman, K., Dietsch, M.T., Mohagheghpour, N. and Aruffo, A. (1992). GMP-140 (P-selectin, CD62) binds to chronically stimulated but not resting CD4+ T lymphocytes and regulates their production of proinflammatory cytokines. Eur. J. Immunol. 22, 1789–1793.

Davis, L.S., Oppenheimer-Marks, N., Bednarczyk, J.L., McIntyre, B.W. and Lipsky, P.E. (1990). Fibronectin promotes proliferation of naive and memory T cells by signalling through both the VLA-4 and VLA-5 integrin molecules. J. Immunol. 145, 785–793.

Diamond, M.S., Staunton, D.E., deFougerolles, A.R., Stacker, S.A., Garcia-Aguilar, J., Hibbs, M.L. and Springer, T.A. (1990). ICAM-1 (CD54): A counter-receptor for Mac-1 (CD11b/CD18). J. Cell Biol. 111, 3129–3139.

Diamond, M.S., Garcia-Aguilar, J., Bickford, J.K., Corbi, A.L. and Springer, T.A. (1993). The I domain is a major recognition site on the leukocyte integrin Mac-1 (CD11b/CD18) for four distinct adhesion ligands. J. Cell Biol. 120, 1031–1043.

Dustin, M.L. (1990). Two-way signalling through the LFA-1 lymphocyte adhesion receptor. Bioessays 12, 421–427.

Dustin, M.L. and Springer, T.A. (1988). Lymphocyte function-associated antigen-1 (LFA-1) interaction with intercellular adhesion molecule-1 (ICAM-1) is one of at least three mechanisms for lymphocyte adhesion to cultured endothelial cells. J. Cell Biol. 107, 321–331.

Dustin, M.L. and Springer, T.A. (1989). T-cell receptor cross-linking transiently stimulates adhesiveness through LFA-1. Nature 341, 619–623.

El-Gabalawy, H., Gallatin, M., Vazeux, R., Peterman, G. and Wilkins, J. (1994). Expression of ICAM-R (ICAM-3), a novel counter-receptor for LFA-1 in rheumatoid and non-rheumatoid synovium. Arthritis Rheum. 37, 846–854.

Elices, M.J., Osborn, L., Takada, Y., Crouse, C., Luhowskyj, S., Hemler, M.E. and Lobb, R.R. (1990). VCAM-1 on activated endothelium interacts with the leucocyte integrin VLA-4 at a site distinct from the VLA-4/fibronectin binding site. Cell 60, 577–584.

Ferguson, T.A., Mizutani, H. and Kupper, T.S. (1991). Two integrin-binding peptides abrogate T cell-mediated immune responses *in vivo*. Proc. Natl Acad. Sci. USA 88, 8072–8076.

Forre, O., Dobloug, J.H. and Natvig, J.B. (1982a). Augmented numbers of HLA-DR-positive T lymphocytes in the synovial fluid and synovial tissue of patients with rheumatoid arthritis and juvenile rheumatoid arthritis. Scand. J. Immunol. 15, 227–231.

Forre, O., Thoen, J., Lea, T., Dobloug, J.H., Mellbye, O.J., Natvig, J.B., Pahle, J. and Solheim, B.G. (1982b). *In situ* characterization of mononuclear cells in rheumatoid tissues, using monoclonal antibodies. No reduction of T8-positive cells or augmentation in T4-positive cells. Scand. J. Immunol. 16, 315–319.

Fossum, S., Smith, M.E. and Ford, W.L. (1983). The recirculation of T and B Lymphocytes in the athymic, nude rat. Scand. J. Immunol. 17, 551–557.

Fox, R.I., Fong, S., Sabharwal, N., Carstens, S.A., Kung, P.C. and Vaughan, J.H. (1982). Synovial fluid lymphocytes differ from peripheral blood lymphocytes in patients with rheumatoid arthritis. J. Immunol. 128, 351–354.

Freedman, A.S., Rhynhart, K., Nojima, Y., Svahn, J., Elisco, L., Benjamin, C.D., Morimoto, C. and Vivier, E. (1993). Stimulation of protein tyrosine phosphorylation in human B cells after ligation of the $\beta 1$ integrin VLA-4. J. Immunol. 150, 1645–1652.

Freemont, A.J. (1987). Molecules controlling lymphocyte-endothelial interactions in lymph nodes are produced in vessels in inflamed synovium. Ann. Rheum. Dis. 4, 924–928.

Freemont, A.J., Jones, C.J.P., Bromley, M. and Andrews, P. (1983). Changes in vascular endothelium related to lymphocyte collections in diseased synovia. Arthritis Rheum. 26, 1427–1433.

Gallatin, W.M., Weissman, I.L. and Butcher, E.C. (1983). A cell-surface molecule involved in organ-specific homing of lymphocytes. Nature 304, 30–34.

Geng, J-G., Bevilacqua, M.P., Moore, K.L., McIntyre, T.M., Prescott, S.M., Kim, J.M., Bliss, G.A., Zimmerman, G.A. and McEver, R.P. (1990). Rapid neutrophil adhesion to activated endothelium mediated by GMP-140. Nature, 343, 757–760.

Geppert, T. D. and Lipsky, P.E. (1989). Antigen presentation by cells that are not of bone marrow origin. Reg. Immunol. 2, 60–71.

Gimbrone. Jr., M.A., Obin, M.S., Brock, A.F., Luis, E.A., Hass, P.E., Hebert, C.A., Yip, Y.K., Leung, D.W., Lowe, D.G., Kohr, W.J., Darbonne, W.C., Bechtol, K.B. and Baker, J.B. (1989). Endothelial interleukin-8: a novel inhibitor of leucocyte-endothelial interactions. Science 246, 1601–1604.

Goto, M., Miyamoto, T., Nishioka, K. and Uchida, S. (1987). T cytotoxic and helper cells are markedly increased, and T suppressor and inducer cells are markedly decreased, in rheumatoid synovial fluids. Arthritis Rheum. 30, 737–743.

Gowans, J.L. and Knight, E.J. (1964). The route of recirculation of lymphocytes in the rat. Proc. R. Soc. Lond. [Biol.]. 159, 257–282.

Guan, J-L. and Hynes, R.O. (1990). Lymphoid cells recognize an alternatively spliced segment of fibronectin via the integrin receptor $\alpha_4\beta_1$. Cell 60, 53–61.

Hale, L.P., Martin, M.E., McCollum, D.E., Nunley, J.A., Springer, T.A., Singer, K.H. and Haynes, B.F. (1989). Immunohistological analysis of the distribution of cell adhesion molecules within the inflammatory synovial microenvironment. Arthritis Rheum. 32, 22–30.

Haskard, D.O., Strobel, S., Thornhill, M., Pitzalis, C. and Levinsky, R.J. (1989). Mechanisms of lymphocyte adhesion to endothelial cells: studies using a LFA-1 deficient cell line. Immunol. 66, 111–116.

Haynes, B.F., Telen, M.J., Hale, L.P. and Denning, S.M. (1989). CD44- A molecule involved in leucocyte adherence and T cell activation. Immunol. Today 10, 423–428.

Hechtman, D.H., Cybulsky, M.I., Fuchs, H.J., Baker, J.B. and Gimbrone, Jr., M.A. (1991). Intravascular IL-8. Inhibitor of polymorphonuclear leucocyte accumulation at sites of acute inflammation. J. Immunol. 147, 883–892.

Hemler, M.E. (1990). VLA proteins in the integrin family. Structures, functions and their role on leucocytes. Annu. Rev. Immunol. 8, 365–400.

Hemler, M.E. and Jacobson, J.G. (1987). Cell matrix adhesion-related proteins VLA-1 and VLA-2: regulation of expression on T cells. J. Immunol. 138, 2941–2948.

Hemler, M.E., Glass, D., Coblyn, J.S. and Jacobson, J.G. (1986). Very late activation antigens on rheumatoid synovial fluid T lymphocytes. Association with stages of T-cell activation. J. Clin. Invest. 78, 696–702.

Hession, C., Tizard, R., Vassallo, C., Schiffer, S.B., Goff, D., Moy, P., Chi-Rosso, G., Luhowskyj, S., Lobb, R. and Osborn, L. (1991). Cloning of an alternate form of vascular cell adhesion molecule-1 (VCAM1). J. Biol. Chem. 266, 6682–6685.

Hogg, N. (1989). The leucocyte integrins. Immunol. Today 10, 111–114.

Huber, A.R., Kunkel, S.L., Todd, III, R.F. and Weiss, S.J. (1991). Regulation of trans-endothelial neutrophil migration by endogenous interleukin-8. Science 254, 99–102.

Hughes, C.C.W., Savage, C.O.S. and Pober, J.S. (1990). Endothelial cells augment T cell interleukin 2 production by a contact-dependent mechanism involving CD2/LFA-3 interaction. J. Exp. Med. 171, 1453–1467.

Hynes, R.O. (1987). Integrins: a family of cell surface receptors. Cell 48, 549–554.

Iguchi, T. and Ziff, M. (1986). Electron microscopic study of rheumatoid synovial vasculature. J. Clin. Invest. 77, 355–361.

Iigo, Y., Takashi, T., Tamatani, T., Miyasaka, M., Higashida, T., Yagita, H., Okumura, K. and Tsukada, W. (1991). ICAM-1 dependent pathway is critically involved in the pathogenesis of adjuvant arthritis in rats. J. Immunol. 147, 4167–4171.

Issekutz, T.B. (1991). Inhibition of *in vivo* lymphocyte migration to inflammation and homing to lymphoid tissues by the TA-2 monoclonal antibody. A likely role for VLA-4 *in vivo*. J. Immunol. 147, 4178–4184.

Issekutz, T.B. and Issekutz, A.C. (1991). T lymphocyte migration to arthritic joints and dermal inflammation in the rat: Differing migration patterns and the involvement of VLA-4. Clin. Immunol. Immunopathol. 61, 436–447.

Jalkanen, S., Steere, A.C., Fox, R.I. and Butcher, E.C. (1986). A distinct endothelial cell recognition system that controls lymphocyte traffic into inflamed synovium. 233, 556–558.

Jasin, H.E. (1988). Chronic arthritis in rabbits. Meth. Enzymol. 162, 379–385.

Jasin, H.E., Lightfoot, E., Davis, L.S., Rothlein, R., Faanes, R.B. and Lipsky, P.E. (1992). Amelioration of antigen-induced arthritis in rabbits treated with monoclonal antibodies to leukocyte adhesion molecules. Arthritis Rheum. 35, 541–549.

Johnston, G.I., Cook, R.G. and McEver, R.P. (1989). Cloning of GMP-140, a granule membrane protein of platelets and endothelium: sequence similarity to proteins involved in cell adhesion and inflammation. Cell 56, 1033–1044.

Kadioglu, A. and Sheldon, P. (1992). Adhesion of rheumatoid peripheral blood and synovial fluid mononuclear cells to high endothelial venules of gut mucosa. Ann. Rheum. Dis. 51, 126–127.

Kansas, G.S., Spertini, O., Stoolman, L.M. and Tedder, T.F. (1991). Molecular mapping of functional domains of the leucocyte receptor for endothelium, LAM-1. J. Cell Biol. 114, 351–358.

Kansas, G.S., Saunders, K.B., Ley, K., Zakrzewicz, A., Gibson, R.M., Furie, B.C., Furie, B. and Tedder, T.F. (1994). A role for the epidermal growth factor-like domain of P-selectin in ligand recognition and cell adhesion. J. Cell Biol. 124, 609–618.

Kavanaugh, A.F., Lightfoot, E., Lipsky, P.E. and Oppenheimer-Marks, N. (1991). Role of CD11/CD18 in adhesion and transendothelial migration of T cells. Analysis utilizing CD18-deficient T cell clones. J. Immunol. 146, 4149–4156.

Kavanaugh, A.F., Davis, L.S., Nichols, L.A., Norris, S.H., Rothlein, R., Scharschmidt, L.A. and Lipsky, P.E. (1994). Treatment of refractory rheumatoid arthritis with a monoclonal antibody to intercellular adhesion molecule-1 (ICAM-1). Arthritis Rheum. 37, 992–999.

Kishimoto, T.K., Jutila, M.A., Berg, E.L. and Butcher, E.C. (1989). Neutrophil Mac-1 and MEL-14 adhesion proteins inversely regulated by chemotactic factors. Science 245, 1238–1241.

Klareskog, L., Forsum, U., Tjernlund, U.M., Kabelitz, D. and Wigren, A. (1981). Appearance of anti-HLA-DR-reactive cells in normal and rheumatoid synovial tissue. Scand. J. Immunol. 14, 183–192.

Kobayashi, I. and Ziff, M. (1973). Electron microscopic studies of lymphoid cells in the rheumatoid synovial membrane. Arthritis Rheum. 16, 471–486.

Koch, A.E., Burrows, J.C., Haines, G.K., Carlos, T.M., Harlan, J.M. and Leibovich, S.J. (1991). Immunolocalization of endothelial and leucocyte adhesion molecules in human rheumatoid and osteoarthritic synovial tissue. Lab. Invest. 64, 313–320.

Koopman, G., Parmentier, H.K., Schuurman, H-J., Newman, W., Meijer, C.J.L.M. and Pals, S.T. (1991). Adhesion of human B cells to follicular dendritic cells involves both the lymphocyte function-associated antigen 1/intercellular adhesion molecule a and very late antigen 4/vascular cell adhesion molecule 1 pathways. J. Exp. Med. 173, 1297–1304.

Korhonen, M., Ylanne, J., Laitinen, L. and Virtanen, I. (1990). The α1-α6 subunits of integrins are characteristically expressed in distinct segments of developing and adult human nephron. J. Cell Biol. 111, 1245–1254.

Kudo, C., Araki, A., Matsushima, K. and Sendo, F. (1991). Inhibition of IL-8 induced W3/25+ (CD4+) T lymphocyte recruitment into subcutaneous tissues of rats by selective depletion of *in vivo* neutrophils with a monoclonal antibody. J. Immunol. 147, 2196–2201.

Kunzendorf, U., Notter, M., Hock, H., Distler, A., Diamanstein, T. and Walz, G. (1993). T cells bind to the endothelial adhesion molecule GMP-140 (P-selectin). Transplantation 56, 1213–1217.

Kurosaka, M. and Ziff, M. (1983). Immunoelectron microscopic study of the distribution of T cell subsets in rheumatoid synovium. J. Exp. Med. 158, 1191–1210.

Kuypers, T.W. and Roos, D. (1989). Leukocyte membrane adhesion proteins LFA-1, CR3, and p150,95: a review of functional and regulatory aspects. Res. Immunol. 140, 461–486.

Laffon, A., Garcia-Vicuna, R., Humbria, A., Postigo, A.A., Corbi, A.L., de Landazuri, M.O. and Sanchez-Madrid, F. (1991). Up-regulated expression and function of VLA-4 fibronectin receptors on human activated T cells in rheumatoid arthritis. J. Clin. Invest. 88, 546–552.

Larsen, E., Celi, A., Gilbert, G.E., Furie, B.C., Erban, J.K., Bonfanti, R., Wagner, D.D. and Furie, B. (1989). PADGEM protein: a receptor that mediates the interaction of activated platelets with neutrophils and monocytes. Cell 59, 305–312.

Larsen, E., Palabrica, T., Sajer, S., Gilbert, G.E., Wagner, D.D., Furie, B.C. and Furie, B. (1990). PADGEM-dependent adhesion of platelets to monocytes and neutrophils is mediated by a lineage-specific carbohydrate, LNF III (CD15). Cell 63, 467–474.

Lasky, L.A., Singer, M.S., Yednock, T.A., Dowbenko, D., Fennie, C., Rodriguez, H., Nguyen, T., Stachel, S. and Rosen, S.D. Cloning of a lymphocyte homing receptor reveals a lectin domain. Cell 56, 1045–1055.

Lasky, L.A., Singer, M.S., Dowbenko, D., Imai, Y., Henzel, W.J., Grimley, C., Fennie, C., Gillett, N., Watson, S.R. and Rosen, S.D. (1992). An endothelial ligand for L-selectin is a novel mucin-like molecule. Cell 69, 927–938.

Lawrence M.B. and Springer, T.A. (1991). Leukocytes roll on a selectin at physiologic flow rates: Distinction from and prerequisite for adhesion through integrins. Cell 65, 859–873.

Lewinsohn, D.M., Bargatze, R.F. and Butcher, E.C. (1987). Leukocyte-endothelial cell recognition: Evidence of a common molecular mechanism shared by neutrophils, lymphocytes and other leucocytes. J. Immunol. 138, 4313–4321.

Lipes, M.A., Napolitano, M., Jeang, K.-T., Chang, N.T. and Leonard, W.J. (1988). Identification, cloning, and characterization of an immune activation gene. Proc. Natl. Acad. Sci. U.S.A. 85, 9704–9708.

Loike, J.D., Sodeik, B., Cao, L., Leucona, S., Weitz, J.I., Detmers, P.A., Wright, S.D. and Silverstein, S.C. (1991). CD11c/CD18 on neutrophils recognizes a domain at the N terminus of the Aα chain of fibrinogen. Proc. Natl Acad. Sci. USA 88, 1044–1048.

Lorant, D.E., Patel, K.D., McIntyre, T.M., McEver, R.P., Prescott, S.M. and Zimmerman, G.A. (1991). Coexpression of GMP-140 and PAF by endothelium stimulated by histamine or thrombin: A juxtacrine system for adhesion and activation of neutrophils. J. Cell Biol. 115, 223–234.

Lowe, J.B., Stoolman, L.M., Nair, R.P., Larsen, R.D., Berhend, T.L. and Marks, R.M. (1990). ELAM-1-dependent cell adhesion to vascular endothelium determined by a transfected human fucosyltransferase cDNA. Cell 63, 475–484.

Luscinskas, F.W., Cybulsky, M.I., Kiely, J-M., Peckins, C.S., Davis, V.M. and Gimbrone, Jr., M.A. (1991). Cytokine activated human endothelial monolayers support enhanced neutrophil transmigration via a mechanism involving both endothelial-leucocyte adhesion molecule-1 and intercellular adhesion molecule-1. J. Immunol. 146, 1617–1625.

Luscinskas, F.W., Brock, A.F., Arnaout, M.A. and Gimbrone, Jr., M.A. (1989). Endothelial-leukocyte adhesion molecule-1-dependent and leukocyte (CD11/CD18)-dependent mechanisms contribute to polymorphonuclear leukocyte adhesion to cytokine-activated human vascular endothelium. J. Immunol. 142, 2257–2263.

Makgoba, M.W., Sanders, M.E. and Shaw, S. (1989). The CD2-LFA-3 and LFA-1-ICAM pathways: relevance to T-cell recognition. Immunol. Today 10, 417–422.

Marcantonio, E.E and Hynes, R.O. (1988). Antibodies to the conserved cytoplasmic domain of the integrin β_1 subunit react with proteins in vertebrates, invertebrates, and fungi. J. Cell Biol. 106, 1765–1772.

Marlin, S.D. and Springer, T.A. (1987). Purified intercellular adhesion molecule-1 (ICAM-1) is a ligand for lymphocyte function-associated antigen-1 (LFA-1). Cell 51, 813–819.

Masinovsky, B., Urdal, D. and Gallatin, W.M. (1990). IL-4 acts synergistically with IL-1β to promote lymphocyte adhesion to microvascular endothelium by induction of vascular cell adhesion molecule-1. J. Immunol. 145, 2886–2895.

Masuyama, J-I., Minato, N. and Kano, S. (1986). Mechanisms of lymphocyte adhesion to human vascular endothelial cells in culture. T lymphocyte adhesion to endothelial cells through HLA-DR antigens induced by gamma interferon. J. Clin. Invest. 77, 1596–1605.

McEver, R.P. (1992). Leukocyte interactions mediated by P-selectin. In 'The Structure and Function of Molecules Involved in Leukocyte Adhesion II' (ed C.W. Smith, T.K. Kishimoto, P.E. Lipsky, R. Rothlein, and A.S. Rosenthal), Springer-Verlag pp. 135–151.

McEver, R.P., Beckstead, J.H., Moore, K.L., Marshall-Carlson, L. and Bainton, D.F. (1989). GMP-140, a platelet α-granule membrane protein, is also synthesized by vascular endothelial cells and is localized in weibel-palade bodies. J. Clin. Invest. 84, 92–99.

Meeusen, E., Lee, C-S. and Brandon, M. (1991). Differential migration of T and B cells during an acute inflammatory response. Eur. J. Immunol. 21, 2269–2272.

Miller, M.D. and Krangel, M.S. (1992). Biology and biochemistry of the chemokines: A family of chemotactic and inflammatory cytokines. Crit. Rev. Immunol. 12, 17–46.

Morales-Ducret, J., Wayner, E., Elices, M.J., Alvaro-Garcia, J.M., Zvaifler, N.J. and Firestein, G.S. (1992). Alpha 4/beta 1 integrin (VLA-4) ligands in arthritis. Vascular cell adhesion molecule-1 expression in synovium and on fibroblast-like synoviocytes. J. Immunol. 149, 1424–1431.

Muller, W.A., Weigel, S.A., Deng, X. and Phillip, D.M. (1993). PECAM-1 is required for transendothelial migration of leukocytes. J. Exp. Med. 178, 449–460.

Munro, J.M., Pober, J.S. and Cotran, R.S. (1989). Tumor necrosis factor and interferon-γ induce distinct patterns of endothelial activation and associated leukocyte accumulation in skin of *Papio anubis*. Am. J. Path. 135, 121–133.

Nakao, H., Eguchi, K., Kawakami, A., Migita, K., Otsubo, T., Ueki, Y., Shimomura, C., Tezuka, H., Mastsunaga, M., Maeda, K. and Nagataki, S. (1989). Increment of Ta1 positive cells in peripheral blood from patients with rheumatoid arthritis. J. Rheum. 16, 907–910.

Oppenheim, J.J., Zachariae, C.O.C., Mukaido, N. and Matsushima, K. (1991). Properties of the novel pro-inflammatory supergene "intercrine" cytokine family. Ann. Rev. Immunol. 9, 617–648.

Oppenheimer-Marks, N. and Lipsky, P.E. (1994). Trans-endothelial migration of T cells in chronic inflammation. The Immunologist 2, 58–64.

Oppenheimer-Marks, N. and Ziff, M. (1986). Binding of normal human mononuclear cells to blood vessels in rheumatoid arthritis synovial membrane. Arthritis Rheum. 29, 789–792.

Oppenheimer-Marks, N., Davis, L.S. and Lipsky, P.E. (1990). Human T lymphocyte adhesion to endothelial cells and trans-endothelial migration. Alteration of receptor use relates to the activation status of both the T cell and the endothelial cell. J. Immunol. 145, 140–148.

Oppenheimer-Marks, N., Davis, L.S., Bogue, D.T., Ramberg, J. and Lipsky, P.E. (1991). Differential utilization of ICAM-1 and VCAM-1 during the adhesion and transendothelial migration of human T lymphocytes. J. Immunol. 147, 2913–2921.

Osborn, L., Hession, C., Tizard, R., Vassallo, C., Luhowskyj, S., Chi-Rosso, G. and Lobb, R. (1989). Direct expression cloning of vascular cell adhesion molecule 1, a cytokine-induced endothelial protein that binds to lymphocytes. Cell 59, 1203–1211.

Osborn, L., Vassallo, C. and Benjamin, C.D. (1992). Activated endothelium binds lymphocytes through a novel binding site in the alternatively spliced domain of vascular cell adhesion molecule-1. J. Exp. Med. 176, 99–107.

Phillips, M.L., Nudelman, E., Gaeta, F.C.A., Perez, M., Singhal, A.K., Hakomori, S-I. and Paulson, J.C. (1990). ELAM–1 mediates cell adhesion by recognition of a carbohydrate ligand, SLe^x. Science 250, 1130–1132.

Picker, L.J., Kishimoto, T.K., Smith, W.C., Warnock, R.A. and Butcher, E.C. (1991a). ELAM-1 is an adhesion molecule for skin homing T cells. Nature 349, 796–799.

Picker, L.J., Warnock, R.A., Burns, A.R., Doerschuk, C.M., Berg, E.L. and Butcher, E.C. (1991b). The neutrophil selectin LECAM-1 presents carbohydrate ligands to the vascular selectins ELAM-1 and GMP-140. Cell 66, 921–933.

Pietschmann, P., Cush, J.J., Lipsky, P.E. and Oppenheimer-Marks, N. (1992). Identification of subsets of human T cells capable of enhanced transendothelial migration. J. Immunol. 149, 1170–1178.

Pigott, R., Needham, L.A., Edwards, R.M., Walker, C. and Power, C. (1991). Structural and functional studies of the endothelial activation antigen endothelial leucocyte adhesion molecule-1 using a panel of monoclonal antibodies. J. Immunol. 147, 130–135.

Pitzalis, C., Kingsley, G., Murphy, J. and Panayi, G. (1987). Abnormal distribution of the helper-inducer and suppressor-inducer T-lymphocyte subsets in the rheumatoid joint. Clin. Immunol. Immunopathol. 45, 252–258.

Pitzalis, C., Kingsley, G.H., Covelli, M., Meliconi, R., Markey, A. and Panayi, G.S. (1991). Selective migration of the human helper-inducer memory T cell subset: confirmation by *in vivo* cellular kinetic studies. Eur. J. Immunol. 21, 369–376.

Pober, J.S., Bevilacqua, M.P., Mendrick, D.L., Lapierre, L.A., Fiers, W. and Gimbrone, Jr., M.A. (1986). Two distinct monokines, interleukin 1 and tumor necrosis factor, each independently induce biosynthesis and transient expression of the same antigen on the surface of cultured human vascular endothelial cells. J. Immunol. 136, 1680–1687.

Polley, M.J., Phillips, M.L., Wayner, E., Nudelman, E., Singhal, A.K., Hakomori, S-I. and Paulson, J.C. (1991). CD62 and endothelial cell-leukocyte adhesion molecule 1 (ELAM-1) recognize the same carbohydrate ligand sialyl-Lewis X. Proc. Natl Acad. Sci. USA 88, 6224–6228.

Poulter, L.W., Duke, O., Panayi, G.S., Hobbs, S., Raftery, M.J. and Janossy, G. (1985). Activated T lymphocytes of the synovial membrane in rheumatoid arthritis and other arthropathies. Scand. J. Immunol. 22, 683–690.

Randi, A.M. and Hogg, N. (1994). I domain of B_2 integrin lymphocyte function asociated antigen-1 contains a binding site for ligand intercellular adhesion molecule-1. J. Biol. Chem. 269, 12395–12398.

Rathanaswami, P., Hachicha, M., Sadick, M., Schall, T.J. and McColl, S.R. (1993). Expression of the cytokine RANTES in human rheumatoid synovial fibroblasts. J. Biol. Chem. 268, 5834–5839.

Reme, T., Portier, M., Frayssinoux, F., Combe, B., Miossec, P., Favier, F. and Sany, J. (1990). T cell receptor expression and activation of synovial lymphocyte subsets in patients with rheumatoid arthritis. Phenotyping of multiple synovial sites. Arthritis Rheum. 33, 485–492.

Renkonen, R. (1989). Regulation of intercellular adhesion molecule-1 expression on endothelial cells with correlation to lymphocyte-endothelial binding. Scand. J. Immunol. 29, 717–721.

Rice, G.E., Munro, J.M. and Bevilacqua, M.P. (1990). Inducible cell adhesion molecule 110 (INCAM-110) is an endothelial receptor for lymphocytes. A CD11/CD18-independent adhesion mechanism. J. Exp. Med. 171, 1369–1374.

Rosenstein, Y., Park, J.K., Hahn, W.C., Rosen, F.S., Bierer, B.E. and Burakoff, S.J. (1991). CD43, a molecule defective in Wiskott-Aldrich syndrome, binds ICAM-1. Nature 354, 233–235.

Rothschild, B.M. and Masi, A.T. (1982). Pathogenesis of rheumatoid arthritis: A vascular hypothesis. Semin. Arthritis Rheum. 12, 11–31.

Rothlein, R., Dustin, M.L., Marlin, S.D. and Springer, T.A. (1986). A human intercellular adhesion molecule (ICAM-1) distinct from LFA-1. J. Immunol. 137, 1270–1274.

Ruegg, C., Portigo, A.A., Sikorski, E.E., Butcher, E.C., Pytela, R. and Erle, D. (1992). Role of integrin $\alpha4\beta7/\beta P$ in lymphocyte adherence to fibronectin and VCAM-1 and in homotypic cell clustering. J. Cell Biol. 117, 179–186.

Ryan, D.H., Nuccie, B.L., Abboud, C.N. and Winslow, J.M. (1991). Vascular cell adhesion molecule-1 and the integrin VLA-4 mediate adhesion of human B cell precursors to cultured bone marrow adherent cells. J. Clin. Invest. 88, 995–1004.

Salmi, M., Kalimo, K. and Jalkanen, S. (1993). Induction and function of vascular adhesion protein-1 at sites of inflammation. J. Exp. Med. 178, 2255–2261.

Sanchez-Mateos, P., Campanero, M.R., Balboa, M.A. and Sanchez-Madrid, F. (1993). Co-clustering of $\beta 1$ integrins, cytoskeletal proteins, and tyrosinephosphorylated substrates during integrin-mediated leukocyte aggregation. J. Immunol. 151, 3817–3828.

Schall, T.J., Jongstra, J., Dyer, B.J., Jorgensen, J., Clayberger, C., Davis, M.M. and Krensky, A.M. (1988). A human T cell-specific molecule is a member of a new gene family. J. Immunol. 141, 1018–1025.

Schall, T.J., Bacon, K., Toy, K.J. and Goeddel, D.V. (1990). Selective attraction of monocytes and T lymphocytes of the memory phenotype by cytokine RATES. Nature 347, 669–671.

Schall, T.J., Bacon, K., Camp, R.D.R., Kaspari, J.W. and Goeddel, D.V. (1993). Human macrophage inflammatory protein-1α (MIP-1α) and MIP-1β chemokines attract distinct populations of lymphocytes. J. Exp. Med. 177, 1821–1825.

Schlieffenbaum, B., Moser, R., Patarroyo, M. and Fehr, J. (1989). The cell surface glycoprotein Mac-1 (CD11b/CD18) mediates neutrophil adhesion and modulates degranulation independently of its quantitative cell surface expression. J. Immunol. 142, 3537–3545.

Schwartz, B.R., Wayner, E.A., Carlos, T.M., Ochs, H.D. and Harlan, J.M. (1990). Identification of surface proteins mediating adherence to CD11/CD18-deficient lymphoblastoid cells to cultured human endothelium. J. Clin. Invest. 85, 2019–2022.

Shimizu, Y., van Seventer, G.A., Horgan, K.J. and Shaw, S. (1990a). Co-stimulation of proliferative responses of resting CD4+ T cells by the interaction of VLA-4 and VLA-5 with fibronectin or VLA-6 with laminin. J. Immunol. 145, 59–67.

Shimizu, Y., van Seventer, G.A., Horgan, K.J. and Shaw, S. (1990b). Regulated expression and binding of three VLA (β_1) integrin receptors on T cells. Nature 345, 250–253.

Shimizu, Y., Shaw, S., Graber, N., Gopal, T.V., Horgan, K.J., van Seventer, G.A. and Newman, W. (1991). Activation-independent binding of human memory T cells to adhesion molecule ELAM-1. Nature 349, 799–802.

Sikorski, E.E., Hallmann, R., Berg, E.L. and Butcher, E.C. (1993). The Peyer's patch high endothelial receptor for lymphocytes, the mucosal vascular addressin, is induced on a murine endothelial cell line by tumor necrosis factor-α and IL-1. J. Immunol. 151, 5239–5250.

Smith, C.W., Rothlein, R., Hughes, B.J., Mariscalco, M.M., Rudloff, H.E., Schmalsteig, F.C. and Anderson, D.C. (1988). Recognition of an endothelial determinant for CD18-dependent human neutrophil adherence and transendothelial migration. J. Clin. Invest. 82, 1746–1756.

Smith, W.C., Marlin, S.D., Rothlein, R., Toman, C. and Anderson, D.C. (1989). Cooperative interactions of LFA-1 and Mac-1 with intercellular adhesion molecule-1 in facilitating adherence and trans-endothelial migration of human neutrophils *in vitro*. J. Clin. Invest. 83, 2008–2017.

Smith, M.D. and Roberts-Thompson, P.J. (1990). Lymphocyte surface marker expression in rheumatic diseases: Evidence for prior activation of lymphocytes *in vivo*. Ann. Rheum. Dis. 49, 81–87.

Spertini, O., Luscinskas, F.W., Kansas, G.S., Munro, J.M., Griffin, J.D., Gimbrone, Jr., M.A. and Tedder, T.F. (1991). Leukocyte adhesion molecule 1 (LAM 1, L selectin) interacts with an inducible endothelial cell ligand to support leukocyte adhesion. J. Immunol. 147, 2565–2573.

Springer, T.A. (1985). The LFA-1, Mac-1 glycoprotein family and its deficiency in an inherited disease. Fed. Proc. 44, 2660–2663.

Springer, T.A. (1990). Adhesion receptors of the immune system. Nature 346, 425–434.

Staunton, D.E., Marlin, S.D., Stratowa, C., Dustin, M.L. and Springer, T.A. (1988). Primary structure of ICAM-1 demonstrates interaction between members of the immunoglobulin and integrin supergene families. Cell 52, 925–933.

Staunton, D.E., Dustin, M.L. and Springer, T.A. (1989). Functional cloning of ICAM-2, a cell adhesion ligand for LFA-1 homologous to ICAM-1. Nature 339, 61–64.

Stoolman, L.M., Tenforde, T.S. and Rosen, S.D. (1984). Phosphomannosyl receptors may participate in the adhesive interaction between lymphocytes and high endothelial venules. J. Cell Biol. 99, 1535–1540.

Stoolman, L.M., Grober, J., Bowen, B., Reddy, P., Shih, J., Thompson, C., Fox, D.A. and Ebling, H. (1992). In "The Structure and Function of Molecules Involved in Leukocyte Adhesion II" (eds C.W. Smith, T.K. Kishimoto, P.E. Lipsky, R. Rothlein, and A.S. Rosenthal), Springer-Verlag, New York (In press).

Stoolman, L.M., Grober, J., Bowen, B., Reddy, P., Shih, J., Thompson, C., Fox, D.A. and Ebling, H. (1992). Selectin-dependent monocyte adhesion in frozen sections of rheumatoid synovitis. In 'The Structure and Function of Molecules Involved in Leukocyte Adhesion II' (ed C.W. Smith. T.K. Kishimoto, P.E. Lipsky, R. Rothlein, and A.S. Rosenthal), Springer-Verlag. pp. 200–213.

Szekanecz, Z., Haines, G.K., Lin, T.R., Harlow, L.A., Goerdt, S., Rayan, G. and Koch, A.E. (1994). Differential distribution of intercellular adhesion molecules (ICAM-1, ICAM-2, and ICAM-3) and the MS-1 antigen in normal and diseased human synovia. Arthritis Rheum. 37, 221–231.

Tanaka, Y., Albelda, S.M., Horgan, K.J., van Seventer, G.A., Shimizu, Y., Newman, W., Hallam, J., Newman, P.J., Buck, C.A. and Shaw, S. (1992). CD31 expressed on distinctive T cell subsets is a preferential amplifier of β1 integrin-mediated adhesion. J. Exp. Med. 176, 245–253.

Tanaka, Y., Adams, D.H., Hubscher, S., Hirano, H., Siebenlist, U. and Shaw, S. (1993). T cell adhesion induced by proteoglycan-immobilized cytokine MIP-1β. Nature 361, 79–82.

Taub, D.D., Conlon, K., Lloyd, A.R., Oppenheim, J.J. and Kelvin, D.J. (1993). Preferential migration of activated CD4+ and CD8+ T cells in response to MIP-1α and MIP-1β. Science 260, 355–358.

True, D.D., Singer, M.S., Lasky, L.A. and Rosen, S.D. (1990). Requirement for sialic acid on the endothelial ligand of a lymphocyte homing receptor. J. Cell Biol. 111, 2757–2764.

van Seventer, G.A., Shimizu, Y., Horgan, K.J. and Shaw, S. (1990). The LFA-1 ligand ICAM-1 provides an important costimulatory signal for T cell receptor-mediated activation of resting T cells. J. Immunol. 144, 4579–4586.

van Seventer, G.A, Newman, W., Shimizu, Y., Nutman, T.B., Tanaka, Y., Horgan, K.J., Gopal, T.V., Ennis, E., O'Sullivan, D., Grey, H. and Shaw, S. (1991). Analysis of T cell stimulation by superantigen plus major histocompatability complex class II molecules or by CD3 monoclonal antibody: co-stimulation by purified adhesion ligands VCAM-1, ICAM-1, but not ELAM-1. J. Exp. Med. 174, 901–913.

Veale, D., Yanni, G., Rogers, S., Barnes, L., Bresnihan, B. and Fitzgerald, O. (1993). Reduced synovial membrane macrophage numbers, ELAM-1 expression, and lining layer hyperplasia in psoriatic arthritis as compared with rheumatoid arthritis. Arthritis Rheum. 36, 893–899.

Vedder, N.B. and Harlan, J.M. (1988). Increased surface expression of CD11b/CD18 (Mac-1) is not required for stimulated neutrophil adherence to cultured endothelium. J. Clin. Invest. 81, 676–682.

Vennegoor, C.J.G.M., van de Wiel-van Kemenade, E., Huijbens, R.J.F., Sanchez-Madrid, F., Melief, C.J.M. and Figdor, C.G. (1992). Role of LFA-1 and VLA-4 in the adhesion of cloned normal and LFA-1 (CD11/CD18)-deficient T cells to cultured endothelial cells. Indication of a new adhesion pathway. J. Immunol. 148, 1093–1101.

Voraberger, G., Schafer, R. and Stratowa, C. (1991). Cloning of the human gene for intercellular adhesion molecule 1 and analysis of its 5'-regulatory region. Induction by cytokines and phorbol ester. J. Immunol. 147, 2777–2786.

Walz, G., Aruffo, A., Kolanus, W., Bevilacqua, M. and Seed, B. (1990). Recognition by ELAM-1 of the sialyl-LeX determinant on myeloid and tumor cells. Science 250, 113–1135.

Wayner, E.A., Garcia-Pardo, A., Humphries, M.J., McDonald, J.A. and Carter, W.G. (1989). Identification and characterization of the T lymphocyte adhesion receptor for an alternative cell attachment domain (CS-1) in plasma fibronectin. J. Cell Biol. 109, 1321–1330.

Yednock, T.A., Stoolman, L.M. and Rosen, S.D. (1987). Phosphomannosyl-derivatized beads detect a receptor involved in lymphocyte homing. J. Cell Biol. 104, 713–723.

Zhou, Q., Moore, K.L., Smith, D.F., Varki, A., McEver, R.P. and Cummings, R.D. (1991). The selectin GMP-140 binds to sialylated fucosylated lactosaminoglycans on both myeloid and non-myeloid cells. J. Cell Biol. 115, 557–564.

Zvaifler, N. (1973). The immunopathology of joint inflammation in rheumatoid arthritis. Adv. Immunol. 16, 265–336.

8. Pathogenesis of Ischemia/Reperfusion: Role of Neutrophil-endothelial Cell Adhesion

Ronald J. Korthuis and D. Neil Granger

1. Introduction

Polymorphonuclear leucocytes are armed with an impressive arsenal of bactericidal agents that allow these cells to play a vital role in host defense against invading pathogens. These same agents can produce extensive cellular damage in host tissues when leucocytes are activated during inflammatory conditions such as ischemia/reperfusion (I/R). Neutrophil-mediated destruction of host tissue involves a complex series of pathologic events that are initiated at the level of the microcirculation. Indeed, there is a growing body of evidence which indicates that neutrophil adherence to vascular endothelium is a prerequisite to the recruitment

Adhesion Molecules
ISBN 0-12-741440-1

and accumulation of these phagocytic cells in post-ischemic tissues where they act to exacerbate the cellular dysfunction induced by ischemia. As a result, pharmacological and immunological inhibition of neutrophil/endothelial cell adhesive processes has become the central focus of research directed at attenuating granulocyte-mediated tissue injury in I/R. The objective of this chapter is to provide a comprehensive and integrative review of the available literature on the role of neutrophils in the pathogenesis of post-ischemic cellular dysfunction with a particular emphasis on the factors that regulate leucocyte/endothelial interactions during ischemia and reperfusion.

2. *Neutrophil Recruitment in Post-ischemic Tissues*

2.1 QUANTITATION OF NEUTROPHIL INFLUX IN POST-ISCHEMIC TISSUES

Several approaches have been used to quantitate leucocyte recruitment in post-ischemic tissues. One strategy involves monitoring the appearance of radio-labelled leucocytes (Mullane *et al.*, 1984; Engler *et al.*, 1986b; Chatelain *et al.*, 1987; Go *et al.*, 1988; Simpson *et al.*, 1988; Dreyer *et al.*, 1991) or granulocyte specific enzymes, such as myeloperoxidase (Allan *et al.*, 1985; Grisham *et al.*, 1986; Granger, 1988; Carden *et al.*, 1989; Smith *et al.*, 1989; Carden *et al.*, 1990; Rubin *et al.*, 1990; Zimmerman and Granger, 1990; Zimmerman *et al.*, 1990a and 1990b; Barone *et al.*, 1991; Horgan *et al.*, 1991; Korthuis *et al.*, 1991; Schoenberg *et al.*, 1991; Kurtel *et al.*, 1992), in post-ischemic tissues. Histologic assessment of neutrophil influx in tissue biopsies is another widely used technique to examine granulocyte sequestration in I/R (Engler *et al.*, 1984, 1986a, 1986b; Arranda *et al.*, 1988; Jaeshke *et al.*, 1990; Barone *et al.*, 1991; del Zoppo, 1991; Kloner *et al.*, 1991; Oliver *et al.*, 1991; Olof *et al.*, 1991). Although these techniques have proven useful for quantitating neutrophil sequestration in post-ischemic tissues, they do not readily allow separation of the contribution of neutrophil adherence and emigration to total granulocyte accumulation. However, this limitation has been overcome by application of *in vivo* microscopic techniques to examine leucocyte interactions with the endothelium during ischemia and reperfusion (Yamakawa *et al.*, 1987; Granger *et al.*, 1989; Grisham *et al.*, 1989; Suzuki *et al.*, 1989; Kubes *et al.*, 1990c; Lehr *et al.*, 1991a; Lehr *et al.*, 1991b; Nolte *et al.*, 1991a; Nolte *et al.*, 1991b; Perry and Granger, 1991; Suzuki *et al.*, 1991a; Suzuki *et al.*, 1991b; Villringer *et al.*, 1991; Kurose *et al.*, 1994). This approach is particularly advantageous since: (1) leucocyte influx can be continuously monitored and (2) leucocyte adhesion and emigration can be quantified. A disadvantage of the intravital microscopic approach is that the resolution of the system does not allow for determination of the specific population of leucocyte that is adhering or emigrating. However, this problem can be overcome by use of histological techniques to determine which type(s) of leucocyte infiltrated the trans-illuminated tissue (Oliver *et al.*, 1991).

Data derived using the aforementioned approaches indicate that while ischemia is associated with enhanced granulocyte sequestration, reperfusion greatly exacerbates neutrophil infiltration (Grisham *et al.*, 1986; Granger *et al.*, 1989; Smith *et al.*, 1989; Zimmerman *et al.*, 1989) (Fig. 8.1). Moreover, the magnitude of granulocyte recruitment varies with the severity of the ischemic insult (Fig. 8.2) and is confined specifically to the portions of the tissue subjected to ischemia (Go *et al.*, 1988; Dreyer *et al.*, 1991). These studies also indicate that the rate of neutrophil sequestration is greatest during the first hour

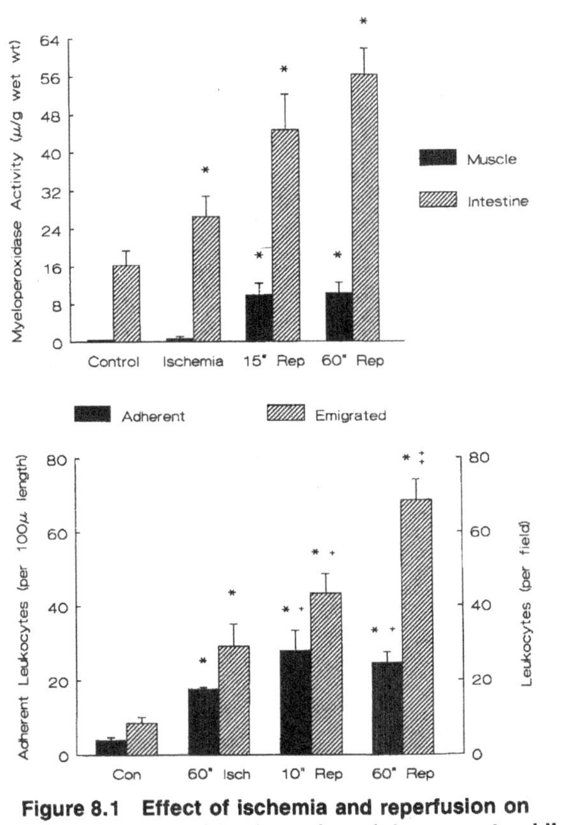

Figure 8.1 Effect of ischemia and reperfusion on myeloperoxidase activity (a marker of tissue neutrophil sequestration) in skeletal muscle and small intestine (upper panel). Data from Grisham *et al.*, 1986; Smith *et al.*, 1989; and Zimmerman *et al.*, 1989. Tissue neutrophil content increases in post-ischemic tissues as a consequence of increases in the number of both adherent and emigrated leukocytes, as exemplified by this data obtained in the feline mesentery (lower panel). Data from Granger *et al.*, 1989.

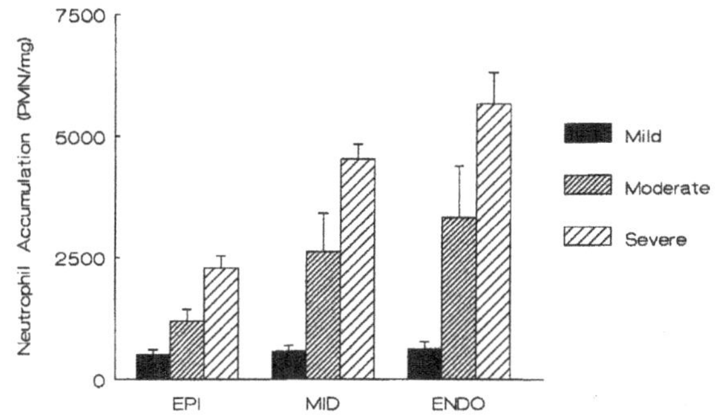

Figure 8.2 Neutrophil accumulation in the epi- (EPI), mid- (MID), and endocardium (ENDO) of hearts subjected to mild (12 min), moderate (40 min), and severe (90 min) ischemia. Data from Go *et al.*, 1988.

of reperfusion and declines thereafter (Fig. 8.3), although the cumulative influx continues for 18 to 48 h, depending on the tissue (Dreyer *et al.*, 1991; Rubin *et al.*, 1992).

2.2 PHYSIOLOGICAL REGULATION OF NEUTROPHIL/ENDOTHELIAL CELL INTERACTIONS IN I/R

The egress of neutrophils from the vascular to extravascular compartment requires that the granulocytes first make intimate contact with endothelial cells. Thus, increased neutrophil adhesiveness for microvascular endothelium is a critical first step in the sequence of events leading to granulocyte-dependent tissue injury. *In vitro* studies suggest that radial dispersive forces play an important role in establishing this adhesive interaction by promoting close apposition between the circulating granulocyte and the endothelium. That is, as blood vessel diameter increases in moving from the capillaries to post-capillary venules, the more flexible erythrocytes begin to pass the leucocyte which, in turn, deflects the white cells towards the vessel walls and margination occurs (Schmid-Schonbein *et al.*, 1980).

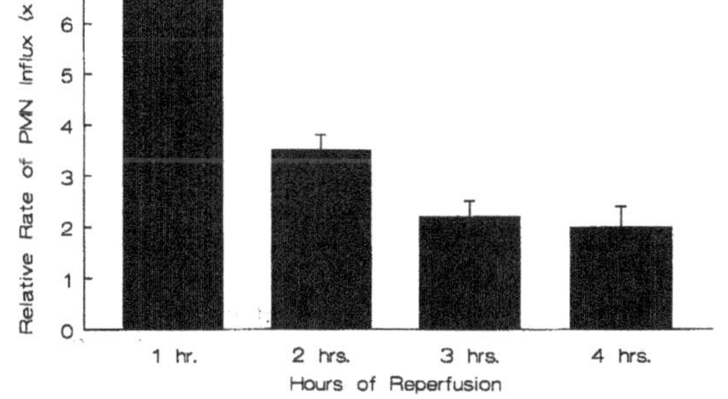

Figure 8.3 Comparisons of the rates of neutrophil (PMN) influx during the first four hours of reperfusion in the heart (bottom panel). Data from Dreyer *et al.*, 1991.

Granulocyte adhesion begins as a rolling movement along post-capillary venular endothelium in which the leucocyte moves slower than erthrocytes in the same stream of blood (Atherton and Born, 1973). As inflammation progresses, the number of rolling leucocytes increases and leucocyte rolling velocity decreases. As a result of these changes, the number of rolling leucocytes that become stationary (firmly adherent) increases and a proportion of adherent leucocytes migrate out of the vessels into the tissues.

Although leucocyte rolling is commonly observed in venules, the adherence of leucocytes to venular endothelium is rare in the absence of a pro-inflammatory stimulus. It is now apparent that neutrophil adhesion in post-capillary venules is a well-controlled physiological phenomenon that is influenced by both physical and chemical forces acting on cell (neutrophil and endothelium) surfaces. Physiological factors that may exert a significant influence on neutrophil sequestration during I/R include hydrodynamic dispersal forces (e.g. wall shear stress), physical impaction (leucocyte capillary plugging), electrostatic charge interactions between the neutrophil and endothelial cell surfaces, liberation of chemotactic stimuli, generation of reactive oxygen metabolites and release of granule-associated proteins from activated neutrophils, and expression of adhesion molecules on the surface of neutrophils and/or endothelial cells.

2.2.1 Hydrodynamic Dispersal Forces in I/R

Whether leucocytes adhere to venular endothelium largely depends on the balance between adhesion forces generated by the leucocyte and endothelium and the hydrodynamic dispersal forces (e.g. blood flow velocity and shear rate) that tend to sweep leucocytes away from the vascular wall (Atherton and Born, 1973; House and Lipowsky, 1987; Perry and Granger, 1991). Thus, in order for a leucocyte to remain stationary within the vasculature, it must bind to endothelium with sufficient strength to withstand the shear rates and associated forces acting to sweep it away. Thus, one might predict that the reduction in blood flow and shear rate associated with ischemia would tend to promote leucocyte adhesion to endothelium. Intravital microscopic examination of the influence of shear forces on leucocyte adherence to endothelium support this prediction. Perry and Granger (1991) have demonstrated that as shear stress in mesenteric venules (25–35 μm diameter) is reduced from its normal value of 18 dyn/cm^2 to 6 dyn/cm^2, adherence increased progressively, and below 6 dyn/cm^2, adherence increased dramatically (Fig. 8.4). The shear dependence of leucocyte adherence is abolished by immunoneutralization of the neutrophil surface CD11/CD18 glycoprotein complex (Fig. 8.4). This data indicates that adhesive forces are far greater than hydrodynamic

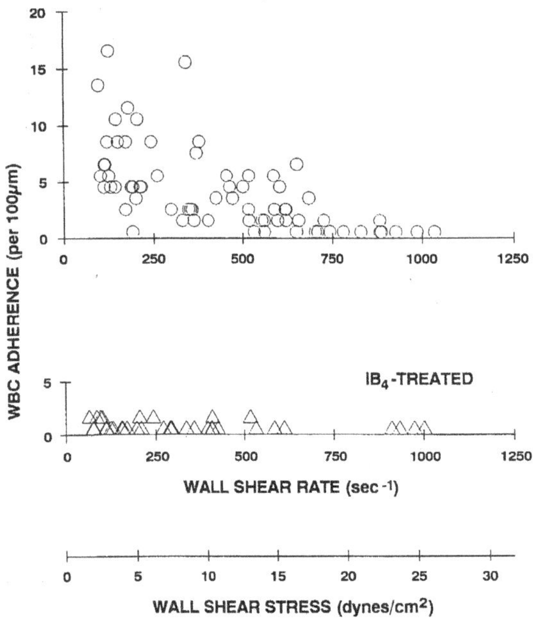

Figure 8.4 Effects of wall shear rate and wall shear stress on leucocyte (WBC) adherence to post-capillary venular endothelium in the presence and absence of a CD18-specific monoclonal antibody (IB_4). Data from Perry and Granger, 1991.

dispersal forces at shear stresses encountered in ischemic tissues *in vivo*. Shear-dependent leucocyte adhesion has also been examined *in vitro*. However, these studies indicate that there is little adherence of leucocytes to cultured endothelium at shear stresses above 2 dyn/cm^2 (Worthen *et al.*, 1987; Lawrence *et al.*, 1990) and occurs by CD18-independent mechanisms except at shear stresses below 0.5 dyn/cm^2 (Worthen *et al.*, 1987). Although the lower shear stress values observed *in vitro* may reflect a lower expression of adhesion receptors on cultured endothelium compared with vessels *in vivo*, this remains to be investigated.

The mechanisms underlying leucocyte adhesion at low shear rates *in vivo* are not clear. One explanation for this behavior is that there is a low level of CD11/CD18 expression in normal circulating leucocytes. When tissue blood flow is normal, the pro-adhesive force generated by these adhesion molecules is not sufficient to overcome the anti-adhesive force generated by shear. However, as shear rate is reduced during ischemia, the balance between pro-adhesive and hydrodynamic dispersal forces changes such that leucocytes adhere. Thus, while a low level expression of CD18 may result in minimal adhesion at normal shear stresses, a strong adhesive interaction may develop between leucocytes and endothelium at low shear rates because the contact time is increased, allowing greater opportunity for adhesive bonds to be formed between the leucocyte and the endothelial cells (Perry and Granger, 1991). An alternative explanation for the increased adherence at low shear rates is that there may be minimal or no expression of CD11/CD18 on circulating leucocytes, but as venular blood flow is reduced there is a concomitant reduction in the washout of pro-inflammatory agents normally produced by endothelial and/or parenchymal cells. This accumulation of chemotactic stimuli would lead to expression of CD11/CD18 on rolling leucocytes and ultimately to adherence. The lower the shear rate, the greater the adherence. Whichever mechanism accounts for this phenomenon, it is clear that the reductions in hydrodynamic dispersal forces may contribute to neutrophil adherence during the ischemic period.

An interesting and consistent observation is that leucocyte rolling and adhesion is much less common in arterioles than in venules (Atherton and Born, 1973; Granger *et al.*, 1989; Perry and Granger, 1991; Ley and Gaehtgens, 1991). The traditional explanation for this phenomenon has been that since shear rates tend to be lower in venules than in arterioles, the propensity for leucocyte/endothelial cell adhesive forces to overcome hydrodynamic dispersal forces may be much greater in post-capillary than pre-capillary vessels (Atherton and Born, 1973). This explanation no longer appears tenable in view of recent studies which demonstrate that the magnitude of leucocyte rolling and adhesion is dramatically different between arterioles and venules exposed to the same shear rate (Perry and Granger, 1991, Ley and Gaehtgens, 1991) (Fig. 8.5). The low level of adhesion in arterioles may reflect a lower density of a ligand for CD11/CD18 on arteriolar versus venular endothelium. It is also possible that arterioles may normally produce an anti-adhesive factor that is not produced in venules. Radial dispersive forces may also favor the establishment

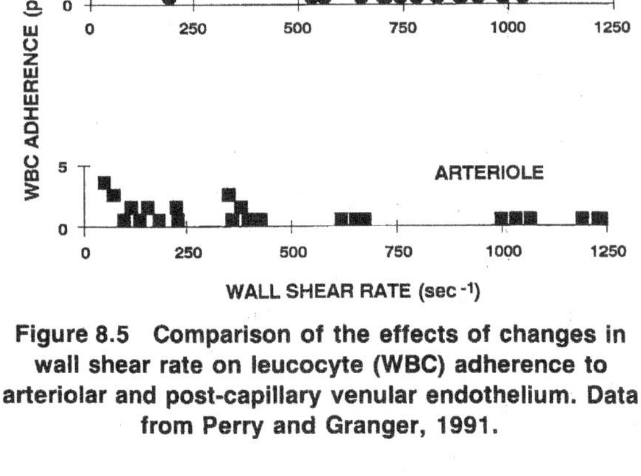

Figure 8.5 Comparison of the effects of changes in wall shear rate on leucocyte (WBC) adherence to arteriolar and post-capillary venular endothelium. Data from Perry and Granger, 1991.

of margination in venules (Schmid-Schonbein *et al.*, 1980). That is, the hydrodynamic interaction of leucocytes with red blood cells as they exit capillaries into the larger post-capillary venules results in leucocytes being preferentially displaced towards the vessel wall. This phenomenon would enhance the likelihood of leucocyte rolling and adhesion in post-capillary venules but not in arterioles.

2.2.2 Post-ischemic Leucocyte-capillary Plugging

Neutrophils are large, stiff, viscoelastic cells which have to undergo substantial deformation during their transit through capillaries (Dahlgren *et al.*, 1983; Engler *et al.*, 1983; Engler *et al.*, 1986a; Engler *et al.*, 1986b; Worthen, *et al.*, 1989). When activated during I/R, neutrophils become stiffer, making it harder for these cells to traverse capillaries. Ischemia-induced tissue hypoxia and acidosis may also increase neutrophil stiffness. These biophysical changes in the neutrophil, coupled with the reduced blood flow and shear rate, act to impair their transit through capillaries in ischemic regions of the tissue. In addition, I/R may induce a reduction in capillary diameter as a consequence of: (1) extravascular compression secondary to edema formation; and (2) endothelial cell or parenchymal swelling. By these mechanisms, leucocytes may become entrapped in capillaries (leucocyte-capillary plugging) during ischemia (Dahlgren *et al.*, 1983; Engler *et al.*, 1983; Engler *et al.*, 1986a; Engler *et al.*, 1986b; Worthen *et al.*, 1989; Carden *et al.*, 1990). If the hydrodynamic dispersal forces associated with reperfusion are insufficient to dislodge these entrapped leucocytes, the distribution of blood flow in the capillary network remains impaired despite the restoration of perfusion.

2.2.3 Electrostatic Charge Interactions Between the Neutrophil and Endothelial Cell Surfaces

Neutrophils and endothelial cells normally exhibit a negative surface charge that tends to promote cell separation through electrostatic repulsion (Harlan, 1985). However, factors (e.g. chemotactic peptides) that decrease the negative surface charge on neutrophils generally tend to promote their adhesiveness to endothelium and biologically inert surfaces (Gallin, 1980). Similarly, it might be expected that I/R-induced disruption of the endothelial cell glycocalyx would also tend to promote neutrophil adhesion by diminishing electrostatic repulsion. Cationic proteins (e.g. lactoferrin) released from activated neutrophils also tend to promote neutrophil adhesion to endothelial cell monolayers and post-capillary venules (Oseas *et al.*, 1981; Boxer *et al.*, 1982). Since similar concentrations of albumin or ferritin, both of which exhibit a net negative charge, had no effect on leucocyte adhesion, the effect of lactoferrin to promote adherence has been attributed to neutralization of negative surface charges on endothelium (Boxer *et al.*, 1982). If this were the case, then one would expect myeloperoxidase and elastase to exert similar pro-adhesive actions since these enzymes are even more positively charged than lactoferrin. It has not been determined whether any macromolecule with a similar positive charge to lactoferrin (e.g. cationic ferritin) also exerts a pro-adhesive effect *in vivo*.

2.2.4 Chemical Mediators of Ischemia/Reperfusion-induced Neutrophil Infiltration

The recognition that reperfusion of ischemic tissues is associated with an infiltration of granulocytes has led to several studies designed to identify the factor(s) that mediate this response. Of the mediators that have been implicated in reperfusion-induced neutrophil accumulation, platelet activating factor (PAF), leukotriene B_4 (LTB_4), complement fragments, and oxygen radicals have received the greatest attention.

2.2.4.1 Platelet Activating Factor

The phospholipid PAF is a potent inflammatory mediator which is formed by a variety of cell types, including endothelial cells, neutrophils, macrophages, and platelets (Lefer, 1989). PAF and its precursor, lyso-PAF are normal constituents of cell membranes. When released from the cell membrane, PAF can exert a number of biological effects, including platelet aggregation and degranulation, smooth muscle contraction, neutrophil activation and chemotaxis, and increased vascular permeability (Bjork *et al.*, 1983; Lefer, 1989). The effects of PAF on vascular tone are variable, with vasodilation observed at low concentrations and constriction observed in some tissues at high concentrations (Lefer, 1989).

PAF is known to cause marked hypotension, mesenteric ischemia, interstitial edema, hemoconcentration, and leucocyte adherence to microvascular endothelium (Braquet *et al.*, 1987). The PAF-induced hemoconcentration is considered to result from an increased fluid and protein filtration across a leaky microvasculature. Adherent leucocytes appear to mediate the PAF-induced vascular leakage since pretreatment with monoclonal antibodies (mAbs) directed against the common β-subunit (CD18) of CD11/CD18 largely prevent the leucocyte adherence and emigration in mesenteric venules and increased vascular protein leakage induced by local intra-arterial infusion of PAF (Kubes *et al.*, 1990a). The ability of exogenously administered PAF to mimic the altered leucocyte kinetics and microvascular dysfunction normally associated with ischemia/reperfusion provides indirect evidence implicating the phospholipid in I/R-induced neutrophil infiltration (Kubes *et al.*, 1990b).

Reports that tissue and venous plasma PAF levels are elevated following reperfusion of ischemic tissues provide additional support for PAF as a mediator of I/R-induced neutrophil infiltration. Montrucchio and coworkers (1989) have demonstrated that reperfusion of the ischemic rabbit heart is associated with significant release of PAF (1.2 to 37 ng/ml) into the coronary effluent, with undetectable levels noted in the perfusate during both the control and ischemic periods. A similar outpouring of PAF into venous blood was noted in the superior mesenteric vein of dogs that were subjected to 2 h occlusion of the superior mesenteric artery (Mozes *et al.*, 1989). Ten fold and 4-fold increments in venous PAF levels were measured at 5 and 60 min after reperfusion. Increased levels of PAF have also been observed in the post-ischemic intestinal mucosa, i.e. 3.5-fold and 6.5-fold increases in mucosal PAF were noted at 10 and 60 min of reperfusion, following 1 h of partial superior mesenteric artery occlusion (Kubes *et al.*, 1990c).

The major line of evidence that supports a role for PAF in I/R-induced neutrophil infiltration is the observation that PAF receptor antagonists are effective in attenuating the reperfusion-induced inflammatory response. Kubes *et al.* (1990c) used two structurally unrelated PAF receptor antagonists (WEB 2086 and BN 52021) to assess the contribution of the phospholipid to I/R-induced leucocyte adherence and emigration in mesenteric venules (Fig. 8.6). In untreated animals, 1 h of ischemia (blood flow reduced to 20% of baseline) was associated with significant levels of leucocyte adherence and emigration, with reperfusion greatly enhancing these responses. Pretreatment with either PAF antagonist resulted in a significant reduction in the rates of leucocyte adherence (35–40%) and emigration (54–91%), indicating PAF plays a major role in mediating the leucocyte-endothelial cell adhesive interactions and neutrophil infiltration elicited by ischemia/reperfusion. It is interesting that the PAF receptor antagonists, (particularly WEB 2086), caused a more dramatic reduction in leucocyte emigration than adherence, suggesting that PAF directly promotes emigration. Although an explanation for this observation is not readily available, it may reflect a propensity of the phospholipid to stimulate neutrophil degranulation and the release of proteases which could facilitate the migration of phagocytic cells across restrictive barriers such as the basement membrane (Zimmerman and Granger, 1990).

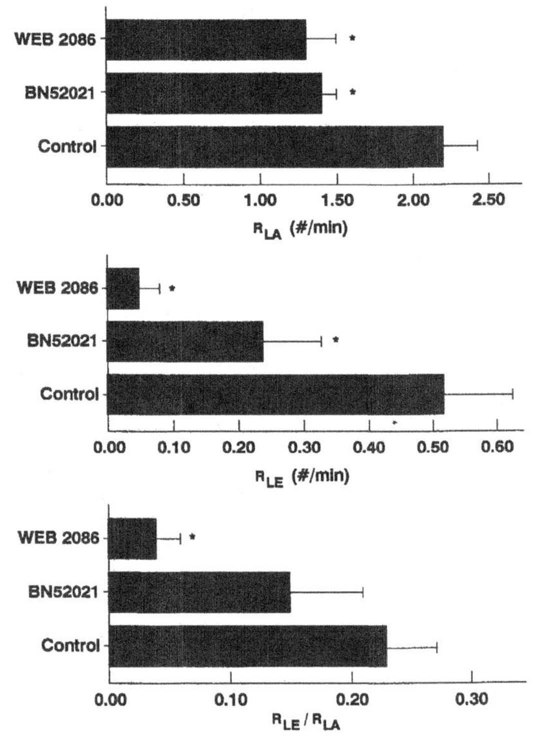

Figure 8.6 Effects of platelet activating factor receptor antagonists (WEB 2086 and BN 52021) on reperfusion-induced leucocyte adherence and emigration in mesenteric venules. R_{LA} represents the rate of leucocyte adherence and R_{LE} represents the rate of leucocyte emigration. Data from Kubes *et al.*, 1990c.

2.2.4.2 *Arachidonic Acid Metabolites*

Arachidonic acid, a normal constituent of cell membrane phospholipids, is the primary substrate for production of eicosanoids. Metabolism of arachidonate via the lipoxygenase pathway leads to the formation of leukotrienes (e.g. LTB4) while thromboxanes (e.g. TxA_2) are synthesized by the cyclo-oxygenase pathway. Both LTB4 and TxA_2 have been implicated as mediators of the neutrophil recruitment associated with reperfusion of ischemic tissues. A role for LTB4 is based on three lines of evidence: (1) exogenously administered LTB4 is a potent inflammatory mediator which can cause most of the microvascular alterations induced by I/R; (2) LTB4 levels are elevated in venous blood and tissues samples obtained from organs subjected to I/R; and (3) receptor antagonism of LTB4 or inhibition of its synthesis reduces I/R-induced neutrophil infiltration.

LTB4 is known to activate neutrophils and to promote chemotaxis (Lefer, 1989). The eicosanoid promotes leucocyte adherence and emigration in post-capillary venules (Dahlen *et al.*, 1981; Kubes *et al.*, 1991a), and it increases microvascular protein efflux in the presence of a transvascular chemotactic gradient (Dahlen *et al.*, 1981). Elevated LTB4 levels are detected in venous blood draining post-ischemic limbs (Klausner *et al.*, 1989a) and kidneys (Klausner *et al.*, 1989b), and in the ischemic and post-ischemic intestinal mucosa (Mangino *et al.*, 1989; Zimmerman *et al.*, 1990b). The rise in renal effluent LTB4 levels after ischemia is significantly attenuated in animals rendered neutropenic with anti-neutrophil serum (Klausner *et al.*, 1989b), suggesting that neutrophils sequestered in post-ischemic kidneys are an important source of the eicosanoid.

When plasma samples derived from venous blood of post-ischemic limbs are injected into skin abrasion blister chambers, neutrophil accumulation is noted (Goldman *et al.*, 1990). Pre-treatment with a lipoxygenase inhibitor prevents the I/R-induced rise in LTB4 and also prevents the diapedesis elicited by post-ischemic plasma. A similar degree of inhibition of neutrophil accumulation is observed in plasma containing either an LTB4 antagonist or a mAb directed against CD18 (mAb 15.7) or ICAM-1 (Rl.1) (Welbourn *et al.*, 1990; Goldman *et al.*, 1991a). The rise in mucosal myeloperoxidase activity observed in post-ischemic intestinal mucosa is also prevented by pre-treatment with an LTB4 antagonist (SC41930) (Zimmerman *et al.*, 1990b) or with mAb IB4, a CD18-specific antibody (Kurtel *et al.*, 1992) (Fig. 8.7). Overall, these observations indicate that I/R leads to the formation and release of LTB4, which in turn, promotes neutrophil accumulation via up-regulation/activation of the leucocyte adhesion glycoprotein CD11/CD18.

There is also evidence in the literature which indicates that LTB4 mediates I/R-induced neutrophil accumulation by promoting leucocyte adherence and emigration in post-capillary venules. Using a dorsal skinfold chamber model for interval microscopic assessment of leucocyte–endothelial cell adhesion in the micro-circulation of muscle, Lehr and coworkers (Lehr *et al.*, 1991) demonstrated that 4 h of ischemia and 30 min of reperfusion results in an 8–fold increase in tissue LTB4, an increased microvascular permeability, and ≥30–fold increase in the number of adherent leucocytes in post-capillary venules. They also observed that pre-treatment with the 5-lipoxygenase inhibitor MK-886 prevented the rise in tissue LTB4 and significantly attenuated I/R-induced alterations in leucocyte adherence and microvascular permeability (Fig. 8.8). These observations support the view that lipoxygenase products play an important role in mediating the tissue leukosequestration associated with ischemia/reperfusion.

Another arachidonic acid metabolite which has been implicated in I/R-induced neutrophil accumulation is thromboxane. Klausner and coworkers (Klausner *et al.*, 1989a, 1989c) have demonstrated that reperfusion after limb ischemia leads to sequestration of neutrophils in the lungs, increased pulmonary microvascular permeability, and elevated plasma levels of LTB4 and TxB2. Inhibition of thromboxane synthesis or thromboxane receptor antagonists attenuate neutrophil limb I/R-induced leukosequestration in the lung (Klausner *et al.*, 1989c). However, there is evidence which indicates that thromboxane contributes to the I/R-induced leukosequestration by stimulating LTB4 synthesis, which directly mediates CD11/CD18 activation and neutrophil-endothelial cell adhesion (Goldman *et al.*, 1991b). Thus, the thromboxanes that are co-released with LTB4 in post-ischemic tissues appear to act synergistically to promote neutrophil infiltration.

2.2.4.3 *Complement*

There is a limited amount of evidence in the literature which indicates that complement activation occurs in tissues exposed to ischemia-reperfusion. Activation of the serum complement system results in proteolytic generation of the anaphylatoxins C3a and C5a. C5a is a potent chemo-attractant for neutrophils, which promotes leucocyte adherence and increases macromolecule permeability in post-capillary venules (Bjork *et al.*, 1985). The C5a-induced increase in microvascular permeability is attenuated by neutropenia, which indicates that neutrophils mediate at least part of the permeability changes caused by C5a (Bjork *et al.*, 1985).

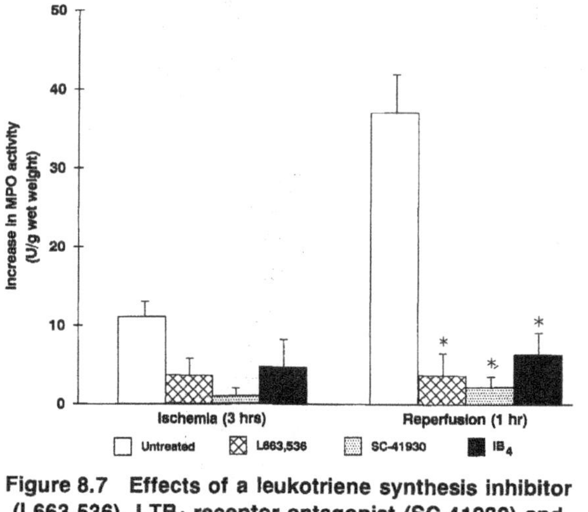

Figure 8.7 Effects of a leukotriene synthesis inhibitor (L663,536), LTB_4 receptor antagonist (SC 41930) and CD18-specific antibody (IB_4) on ischemia-reperfusion induced increases in myeloperoxidase activity in feline intestinal mucosa. Data derived from Zimmerman *et al.*, 1990b; Kurtel *et al.*, 1992.

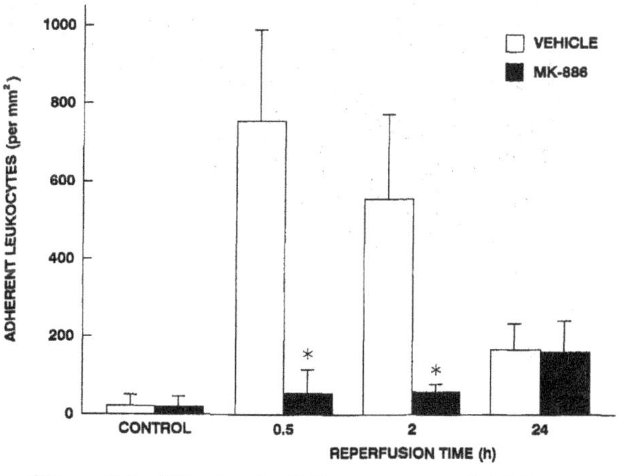

Figure 8.8 Effects of a 5-lipooxygenase inhibitor (MK-886) on ischemia/reperfusion induced leucocyte adherence in postcapillary venules of the dorsal skinfold. Data derived from Lehr *et al.*, 1991.

A role for complement in reperfusion injury is supported by reports describing an elevated level of C1q binding molecules in the circulation of patients after acute myocardial infarction (Rossen *et al.*, 1988). Changes in plasma complement and localization of C1q in ischemic myocardium has been demonstrated in a canine model of myocardial ischemia (Rossen *et al.*, 1985). C1q localization was significantly correlated with the location of neutrophil accumulation. Reperfusion of ischemic skeletal muscle is associated with an 18% reduction in complement protein factor B (alternative pathway) and no change in C4 (classical pathway) levels in systemic plasma (Rubin *et al.*, 1990), suggesting that I/R leads to activation of the alternative but not the classical complement pathway.

The pulmonary leukosequestration elicited by reperfusion of ischemic limbs does not appear to be mediated by complement (Klausner *et al.*, 1989a). C3 levels in plasma are unaffected by limb ischemia-reperfusion. Furthermore, decomplementation with cobra venom factor, while reducing plasma C3 levels to 10% of baseline, does not alter limb I/R-induced pulmonary neutrophil accumulation. Finally, bilateral hind limb ischemia led to a comparable level of lung leukosequestration in both genetically sufficient or deficient C5 mice.

2.2.4.4 Reactive Oxygen Metabolites

Both neutrophils and endothelial cells are capable of producing reactive oxygen metabolites (ROM) such as superoxide and hydrogen peroxide. Neutrophils produce ROM via the membrane associated enzyme NADPH oxidase, while xanthine oxidase is considered to be a major source of ROM produced by vascular endothelial cells (Weiss, 1986; Weiss, 1989; Reilly *et al.*, 1991). Activated neutrophils also secrete myeloperoxidase (MPO), an enzyme which catalyzes the formation of hypochlorous acid (HOCl) from H_2O_2 and chloride ions. The hydrogen peroxide which fuels this reaction is derived from the spontaneous dismutation of neutrophil-derived superoxide. HOCl is a potent oxidizing and chlorinating agent which can react with primary amines to yield N-chloramines. There is a large body of evidence which indicates that ROM are produced by activated endothelial cells and neutrophils in tissues exposed to I/R (Weiss, 1986; Granger, 1988). Furthermore, there is a growing body of evidence which suggests that the ROM generated in post-ischemic tissues may play an important role in promoting the neutrophil infiltration associated with I/R.

There are several reports based on *in vitro* experiments which indicate that ROM can interact with specific components of extracellular fluid to generate substances that are chemotactic for neutrophils. It has been demonstrated that chemotactic activity is generated in plasma when it is treated with a superoxide-generating system, an effect which is completely inhibited by superoxide dismutase but not by catalase (Petrone *et al.*, 1980). Furthermore, intradermal injection of superoxide-treated human plasma or hypoxanthine-xanthine oxidase (HX-XO) causes heavy infiltration of neutrophils at the injection site. These observations, coupled to the ability of superoxide dismutase (SOD) to inhibit neutrophil migration into sites of inflammation (Petrone *et al.*, 1980), has led to the proposal that superoxide reacts with certain substances in extracellular fluid to produce a potent chemo-attractant. This superoxide-dependent chemo-attractant consists of an as yet undefined lipid component (possibly arachidonate) that is bound to serum albumin (Petrone *et al.*, 1980). However, a weakness of this hypothesis is that the existence of a superoxide-dependent chemo-attractant cannot be consistently demonstrated using human plasma and attempts to demonstrate this phenomenon in certain species (e.g. cat) have failed (Zimmerman *et al.*, 1990a).

Other studies indicate that hydrogen peroxide (H_2O_2) may play a more important role than superoxide in the formation of chemo-attractants in extracellular fluid. Normal human serum exposed to H_2O_2 concentrations as low as 70 μM (activated neutrophils can generate H_2O_2 levels as high as 300 μM) result in the generation of chemotactic activity for human neutrophils (Shingu and Nobunaga, 1984), which is inhibited by the addition of either catalase or antibodies directed against C5 (Fig. 8.9). These observations, coupled to the finding that exposure of purified C5 to H_2O_2 generates chemotactic activity, has led to the proposal that H_2O_2 generates C5a-like chemotactic factor through hydrolysis of C5 (Vogt *et al.*, 1987). The ability of H_2O_2 to activate complement (Shingu and Nobunaga, 1984) and generate a chemotactic agent in plasma is enhanced by the presence of catalytically active iron (e.g. Fe EDTA, hemoglobin) (Zimmerman *et al.*, 1990a), suggesting that

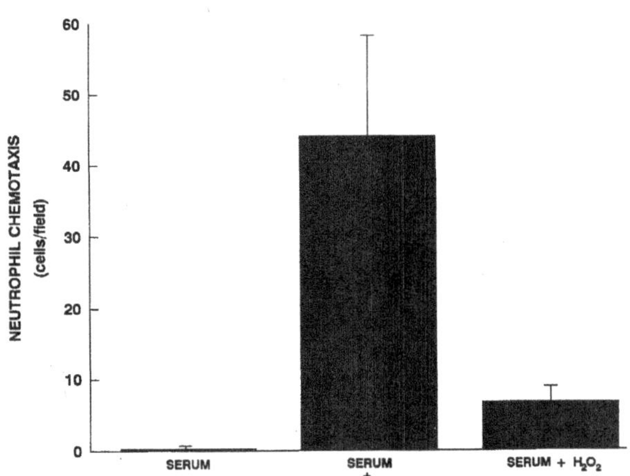

Figure 8.9 Role of complement C5 in hydrogen peroxide-induced increase in the chemotactic potential of human plasma. Data from Shingu and Nobunaga, 1984.

the process of complement activation is not brought about by H_2O_2 *per se*, but by ROM derived from H_2O_2. The ability of heme-iron to enhance H_2O_2-mediated formation of chemotactic agents in plasma may be important in conditions (ischemia/reperfusion) which are associated with hemolysis and the liberation of hemoglobin in extracellular fluid.

Reactive oxygen metabolites have also been shown to promote neutrophil adherence in post-capillary venules. Exposure of post-capillary venules in hamster cheek pouch to superoxide and hydrogen peroxide, generated from hypoxanthine-xanthine oxidase (HX-XO), results in a 50% reduction in leucocyte rolling velocity and causes a concomitant increase in the number of adherent leucocytes (Del Maestro *et al.*, 1982). These HX-XO induced alterations in leucocyte rolling velocity and leucocyte adhesion were completely prevented by superoxide dismutase. However, neither catalase nor L-methionine were able to attenuate the changes induced by HX-XO, suggesting that superoxide may be an important modulator of leucocyte-endothelial cell adhesive interactions during acute and/or chronic inflammation.

Although the negative findings with catalase in the hamster cheek pouch model would suggest that H_2O_2 does not exert a significant influence on leucocyte adherence, evidence derived from both *in vitro* and *in vivo* models of leucocyte adherence suggest otherwise. Hydrogen peroxide and monochloramine (NH_2Cl), at concentrations produced by activated neutrophils, promote leucocyte adherence to microvascular endothelium in post-capillary venules (Suzuki *et al.*, 1991b). Hypochlorous acid (HOCl), another potent oxidant produced by neutrophils, does not enhance leucocyte adhesion. Incubation of isolated cat neutrophils with either NH_2Cl or H_2O_2 results in up-regulation of CD11/CD18, as assessed by flow cytometry. Although the leucocyte adhesion induced by both H_2O_2 and NH_2Cl are associated with a reduction in venular wall shear rate, corresponding decrements in shear rate induced by partial occlusion of the mesenteric artery do not lead to similar levels of leucocyte adherence. The leucocyte adherence induced by either H_2O_2 or NH_2Cl were largely prevented by the CD18-specific mAb IB4 (Fig. 8.10), which indicates that both oxidants promote leucocyte adherence via up-regulation of CD11/CD18. Since PAF receptor antagonists also attenuate H_2O_2-induced leucocyte adherence, it is likely that PAF mediates this response (Suzuki *et al.*, 1991b). The latter contention is consistent with *in vitro* experiments which demonstrate increased PAF production and enhanced neutrophil adherence in endothelial cell monolayers (Lewis *et al.*, 1988; Patel *et al.*, 1991) or isolated perfused canine arteries exposed to H_2O_2 (Gasic *et al.*, 1991). Thus, it appears that H_2O_2 can promote leucocyte adhesion to vascular endothelium at physiologically relevant concentrations and that the adhesion results in PAF-mediated up-regulation and/or activation of CD11/CD18.

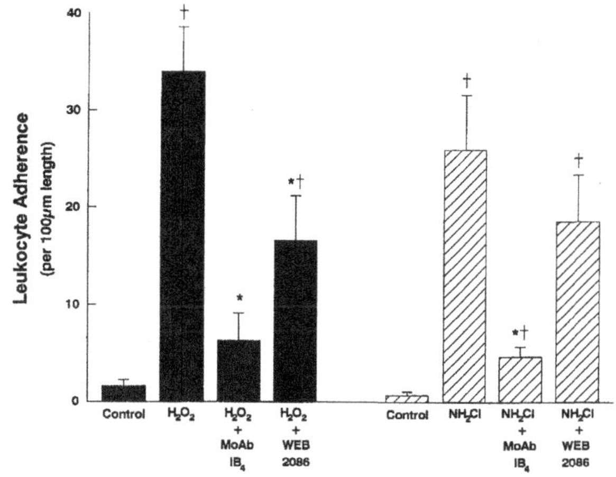

Figure 8.10 Effects of a CD18-specific antibody (MoAb IB_4) and a PAF receptor antagonist on H_2O_2- and NH_2Cl-induced leucocyte adherence in mesenteric venules. Data derived from Suzuki *et al.*, 1991b.

It has also been proposed that xanthine oxidase-derived oxygen radicals contribute significantly to the neutrophil recruitment elicited by I/R. This hypothesis is based on reports which demonstrate that both inhibitors of xanthine oxidase and oxygen radical scavengers significantly attenuate the granulocyte accumulation in intestinal mucosa (Grisham *et al.*, 1986; Zimmerman *et al.*, 1990a) and skeletal muscle (Carden *et al.*, 1989) exposed to ischemia and reperfusion. For example, pre-treatment with either SOD or allopurinol, at concentrations which were previously shown to attenuate I/R-induced microvascular injury (Granger, 1988; Carden *et al.*, 1989) largely abolishes the rise in tissue myeloperoxidase (MPO) activity normally observed in the post-ischemic tissue (for example of intestine data, see Fig. 8.11). Similar observations have been reported using catalase, desferrioxamine, dimethylsulfoxide (DMSO), and dimethylthiourea (DMTU) (Zimmerman *et al.*, 1990a). While it was initially held that the superoxide, hydrogen peroxide, and hydroxyl radicals (termed via the iron-catalyzed Haber-Weiss reaction) produced by xanthine oxidase are directly responsible for the microvascular dysfunction induced by ischemia-reperfusion (Granger, 1989), the aforementioned studies indicate that xanthine oxidase-derived oxidants may play a more important role in initiating the formation of substances that attract, activate, and promote the adherence of granulocytes to microvascular endothelium. This revised version of the xanthine oxidase hypothesis is summarized in Fig. 8.12.

Additional support for the view that xanthine oxidase-derived oxidants mediate reperfusion-induced neutrophil infiltration is provided by intravital microscopy studies of mesenteric venules which demonstrate that allopurinol and SOD attenuate the increases in leucocyte adherence and emigration caused by ischemia and reperfusion

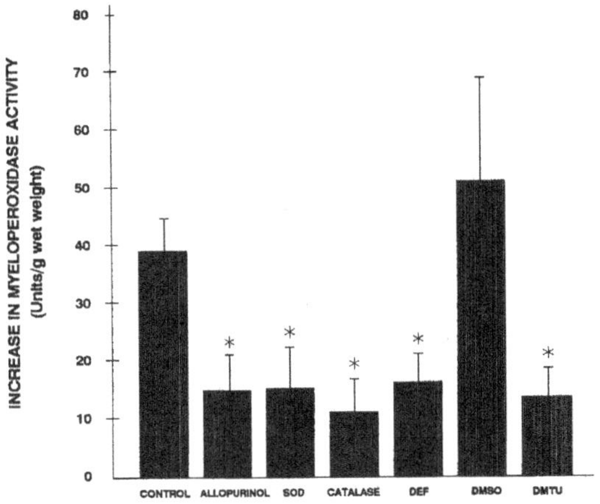

Figure 8.11 Effects of allopurinol, superoxide dismutase (SOD), catalase, desferioxamine (Def), dimethylsulfoxide (DMSO), and dimethylthiourea (DMTU) on ischemia/reperfusion induced increase in mucosal myeloperoxidase activity in feline intestine. Data from Zimmerman *et al.*, 1990a.

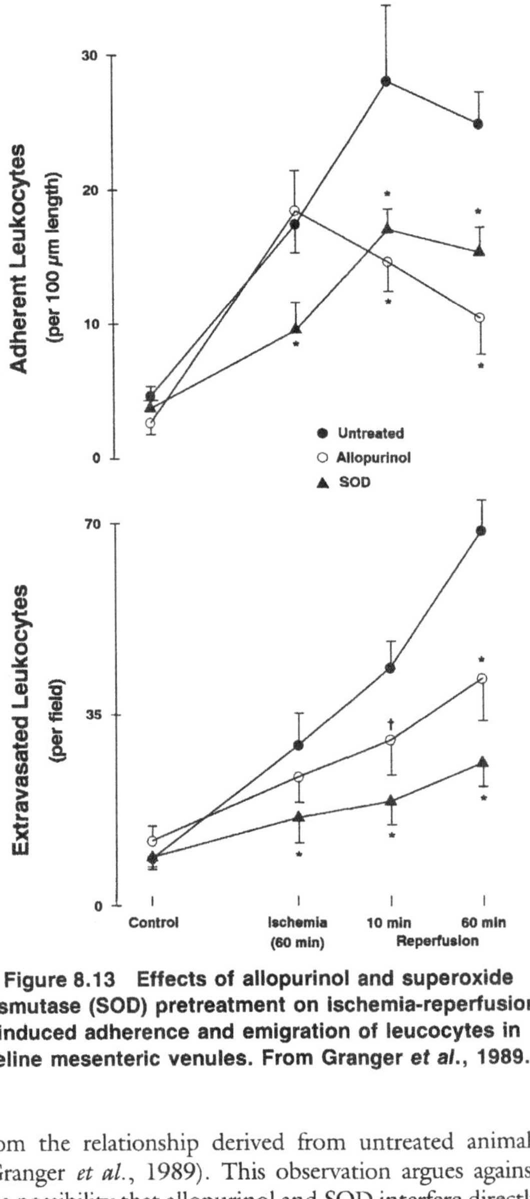

Figure 8.13 Effects of allopurinol and superoxide dismutase (SOD) pretreatment on ischemia-reperfusion induced adherence and emigration of leucocytes in feline mesenteric venules. From Granger *et al.*, 1989.

(Granger *et al.*, 1989). Allopurinol pretreatment did not alter the responses to ischemia *per se*, however it largely prevented the further increment in leucocyte adherence and emigration associated with reperfusion (Fig. 8.13). However, pretreatment with CuZn–SOD attenuated the leucocyte responses elicited by both ischemia and reperfusion. The reductions in reperfusion-induced leucocyte emigration in SOD and allopurinol treated animals suggests that these treatments interfere either with the ability of leucocytes to adhere to endothelium or with their ability to move across the endothelial cell barrier, or both. However, it was noted that the linear relationships between leucocyte emigration and adherence in allopurinol and SOD treated animals were not different from the relationship derived from untreated animals (Granger *et al.*, 1989). This observation argues against the possibility that allopurinol and SOD interfere directly with the leucocyte's ability to emigrate from the vasculature.

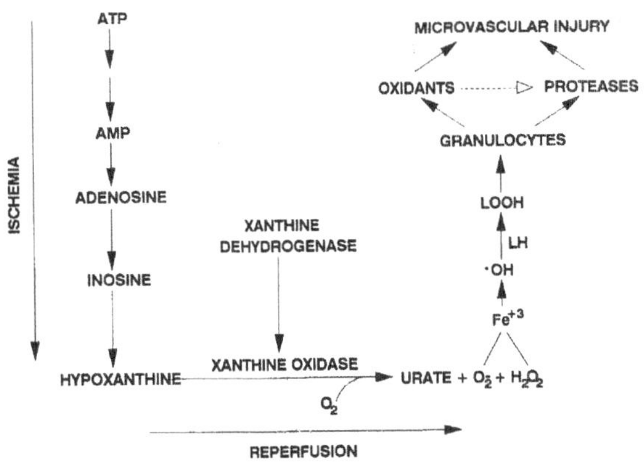

Figure 8.12 Mechanism proposed to explain the involvement of xanthine oxidase, oxygen radicals, and granulocytes in the microvascular dysfunction associated with reperfusion of ischemic tissues. Adapted from Granger, 1988.

In the aforementioned studies on I/R-induced, superoxide-mediated leucocyte adherence, CuZn–SOD and allopurinol were administered prior to the induction of ischemia (Granger *et al.*, 1989). It has also been demonstrated that CuZn–SOD, Mn–SOD, oxypurinol, and catalase, but not desferrioxamine, leads to a rapid and significant reduction of I/R-induced leucocyte adherence in mesenteric venules when either of the agents is administered at one hour after reperfusion (Suzuki *et al.*, 1989; 1991a). These observations suggest that xanthine oxidase-derived oxidants may be responsible for both the initiation of I/R-induced leucocyte adherence and for the sustained leucocyte adhesion observed long after reperfusion. Furthermore, these observations indicate that at least part of the protective effects previously observed with oxypurinol, SOD and catalase in models of I/R injury may be attributable to the anti-adhesive actions of these drugs, while the beneficial effects of desferrioxamine cannot be ascribed to an influence on leucocyte adhesion.

The anti-adhesive effects of superoxide dismutase has also been demonstrated using cultured microvascular endothelium exposed to anoxia (simulating ischemia) and reoxygenation (simulating reperfusion) (Suzuki *et al.*, 1989). After 30 min of anoxia, followed by 60 min of reoxygenation, the endothelial cell monolayers were exposed to either feline neutrophils alone or neutrophils in the presence of either hSOD (4 and 8 mg/kg) peroxide-inactivated hSOD, or the CD18-specific mAb IB4. The results obtained in the cell culture model of ischemia-reperfusion were similar to those obtained in *in vivo* experiments, i.e. hSOD decreased neutrophil adhesion by 19% (low dose) and 27% (high dose), while inactive hSOD was ineffective. MAb IB4 had a more profound effect (38% reduction) on anoxia/reoxygenation-induced neutrophil adhesion. Neutrophil adherence to a biologically inactive surface (plastic) was unaffected by hSOD yet IB4 virtually eliminated adherence. These observations indicate that the presence of endothelial cells are required for O_2-mediated neutrophil adhesion.

The mechanism(s) involved in superoxide-mediated leucocyte adherence remain undefined, however, there are several possibilities that warrant consideration. Superoxide dismutase may interfere with either superoxide-induced: (1) formation of inflammatory mediators; (2) expression of leucocyte and/or endothelial cell adhesion molecules; or (3) inactivation of an endothelial cell-derived anti-adhesive substance. The possibility that nitric oxide is an endogenous molecule that accounts for the anti-adhesive actions of SOD has recently been addressed (Kubes *et al.*, 1991b). Nitric oxide (NO), which is known to be inactivated by superoxide, is normally produced by vascular endothelium. If NO normally attenuates leucocyte adhesion to vascular endothelium, then conditions associated with enhanced formation of superoxide (e.g. ischemia/reperfusion) should lead to increased leucocyte adherence. This hypothesis was addressed by assessing the influence of inhibitors of nitric oxide production on leucocyte adherence in normal post-capillary venules. N^G-nitro-L-arginine methyl ester (L-NAME) and N^G-monomethyl-L-arginine (L-NMMA), analogues of L-arginine that inhibit NO production, were superfused onto the surface of cat mesentery. Both inhibitors increased leucocyte adherence more than 15-fold (Fig. 8.14). Leucocyte emigration was also increased and venular wall shear rate was reduced by nearly half. The CD18-specific antibody mAb IB4 abolished the leucocyte adherence induced by both L-NAME and L-NMMA. Incubation of isolated feline neutrophils with L-NMMA, but not L-NAME,

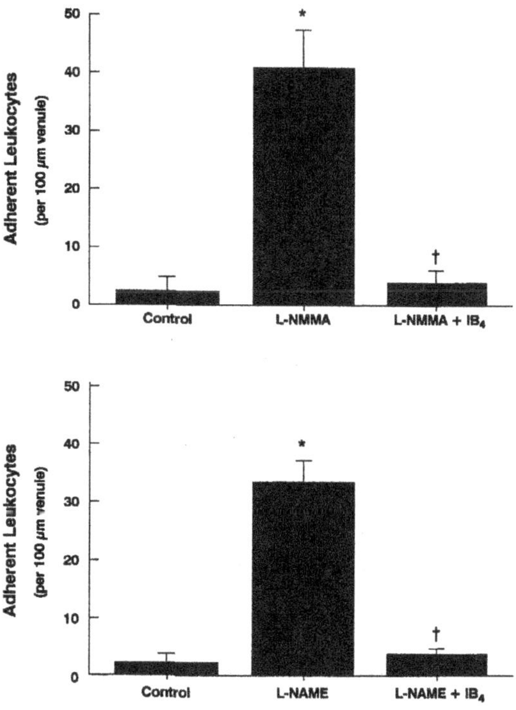

Figure 8.14 Effects of inhibitors of nitric oxide synthase (L-NMMA and L-NAME) on leucocyte adherence in post-capillary venules both in the presence and absence of a mAb directed against CD11/CD18 (IB_4). From Kubes *et al.*, 1991b.

lead to up-regulation of CD11/CD18 as assessed by flow cytometry. Only a minor component of the increased leucocyte adherence induced by NO inhibitors could be attributed to the reductions in venular shear rate. The L-NAME-induced leucocyte adherence was inhibited by L-arginine but not by D-arginine. Overall, these results indicate that NO may be an important endogenous inhibitor of leucocyte adherence in post-capillary venules and that inactivation of NO by superoxide may contribute to the increased leucocyte adherence observed in conditions associated with enhanced production of superoxide by endothelial cells and/or neutrophils.

2.2.4.5 *Neutrophil-derived Proteins*

Pro-inflammatory agents such as fMLP, LTB_4, and PAF are known to differentially stimulate the exocytosis of primary (azurophilic) and secondary (specific) granules of neutrophils (Harlan, 1985). The release of neutrophilic proteases may serve to amplify the inflammatory response by inducing the formation of chemotactic stimuli. For example, Perez *et al.* (1987) have shown that cathepsin H can generate complement C5-derived proteins. In addition, the cationic nature of some of the enzymes (MPO and elastase) and other proteins (lactoferrin) released from these granules may promote the adhesive interactions between leucocytes and vascular endothelium during ischemia/reperfusion by attenuating the electrostatic repulsion between these cells, as discussed above. Whether granule-associated cationic proteins such as lactoferrin and MPO play a significant role in modulating leucocyte adhesion during acute inflammatory conditions such as ischemia/reperfusion remains unclear.

Elastase release from activated neutrophils may also play an important role in the emigration of leucocytes across post-capillary venules. Elastase has been shown *in vitro* to degrade several components of vascular basement membrane (collagen, fibronectin, laminin) (Weiss *et al.*, 1986b). Thus, one might expect that limited release of elastase by neutrophils during diapedesis could effectively "digest" a route of passage from the vascular compartment into the interstitium. This hypothesis is not supported by the work of Huber and Weiss (1989), which demonstrates that the migration of activated neutrophils across monolayers of cultured endothelial cells with a basement membrane is not altered when neutrophils are pretreated with a serine protease inhibitor. Similar conclusions were drawn from intravital microscopic studies of LTB_4-stimulated leucocyte emigration in hamster cheek pouch. Rosengren and Arfors (1991) observed that elastase inhibitors (Eglin C and L658,758) did not alter the LTB_4-induced leucocyte adherence and emigration, and vascular leakage of FITC-dextran. In contrast to the results obtained with LTB_4, ischemia/reperfusion induced leucocyte emigration across cat mesenteric venules is decreased by 90% and leucocyte adherence by 50% in response to pretreatment with a specific elastase inhibitor (L658,758) (Zimmerman and Granger, 1990), which does not alter the isoelectric point of the enzyme. This disproportionate effect of elastase inhibition on leucocyte emigration indicates that the enzyme may play an important role in regulating the rate of trans-endothelial migration following reperfusion of ischemic tissues. The discrepancy between the LTB_4 and ischemia/reperfusion emigration responses to elastase inhibition may be explained by the relative magnitude of neutrophil degranulation induced by the inflammatory mediator released during ischemia/reperfusion vs. LTB_4. However, a potential criticism of the ischemia/reperfusion data is that the elastase inhibitor may act by interfering with the production of an elastase-dependent chemoattractant during ischemia-reperfusion, rather than by inhibition of elastase-mediated digestion of vessel wall components.

2.2.5 Molecular Determinants of Ischemia/Reperfusion-induced Leucocyte Adherence

Over the past several years, it has become apparent that leucocyte adhesion to endothelium involves several different adhesion molecules expressed both by endothelium and granulocytes. Indeed, the molecular events associated with the establishment of this adhesive interaction appears to involve at least three families of adhesion molecules including the leucocyte β_2 integrins (CD11a/CD18 (LFA-1), CD11b/CD18 (Mac-1, MO1, or CR3), and CD11c/CD18 (p150,95)), a member of the immunoglobulin gene superfamily (intercellular adhesion molecule 1 (ICAM-1 or CD54)), and the lectin-like cell adhesion molecule (LEC-CAM or selectin) family (leucocyte adhesion molecule-1 (LAM-1 or L-selectin), endothelial-leucocyte adhesion molecule 1 (ELAM-1 or E-selectin), and granule membrane protein-140 (GMP-140 or P-selectin)) (Bevilacqua *et al.*, 1989; Kishimoto *et al.*, 1989; Springer *et al.*, 1989; Kishimoto, 1991; McEver, 1991; Smith, 1990).

Evidence derived from a large number of *in vitro* studies suggest that localization of neutrophils to sites of inflammation involves a complex, highly co-ordinated, and dynamic sequence of adhesive steps in which each of the known adhesive molecules play distinct roles (Bevilacqua *et al.*, 1989; Kishimoto *et al.*, 1989; Springer *et al.*, 1989; Smith, 1990; Kishimoto, 1991; McEver, 1991). The initial adhesive interaction between neutrophils and the endothelium involves up-regulation of adhesion molecules (ELAM-1, and GMP-140) on the surface of endothelium adjacent to sites of inflammation. Neutrophilic LAM-1 appears to mediate the initial adhesive interaction (leucocyte rolling and reversible sticking) between non-activated neutrophils and the stimulated endothelium by interacting with GMP-140 and/or ELAM-1. Only these transiently bound neutrophils are exposed to low concentrations of chemo-attractants for a sufficient time to allow for activation. Subsequent steps involve strengthening of this adhesive interaction and trans-endothelial neutrophil migration resulting from engagement of CD11/CD18- and ICAM-1-dependent mechanisms. According to this model (Kishimoto *et al.*, 1989; Smith, 1990; Kishimoto, 1991), the initial adhesion mediated by LAM-1 is down-regulated by release of this leucocyte associated selectin, thereby allowing the neutrophil to proceed from binding to the endothelium to extravasation. Thus, this model suggests that the selectins (LAM-1, GMP-140, and ELAM-1) may play a crucial role in modulating neutrophil margination while the leucocyte integrins, CD11a/CD18 and CD11b/CD18, and the endothelial cell adhesive structure, ICAM-1, are critical to extravasation.

It is not yet clear whether such a model applies to inflammatory conditions such as I/R *in vivo*. However, there have been a number of recent attempts to determine the relative contributions of both leucocyte and endothelial cell adhesion glycoproteins to the leucocyte adherence elicited by reperfusion (reoxygenation) of ischemic (hypoxic) tissues (Granger *et al.*, 1991; Yoshida *et al.*, 1992). In cat mesenteric venules exposed to 1 h of ischemia followed by 1 h of reperfusion, pre-treatment with a CD18-specific antibody (mAb IB_4) attenuated the reperfusion-induced increments in both leucocyte adherence and emigration by ~95%, while CD11a- and

CD11b-specific antibodies reduced leucocyte adherence by 60–65% and emigration by 45–50% (Granger *et al.*, 1991). An ICAM-1 specific antibody reduced reperfusion-induced adherence by 40% and emigration by 50% (Granger *et al.*, 1991). These results indicate that the leucocyte adherence and emigration observed in post-capillary venules exposed to ischemia/reperfusion are mediated by both CD11/CD18 on neutrophils and ICAM-1 on endothelial cells.

The contribution of different adhesive glycoproteins to ischemia/reperfusion induced leucocyte adhesion has also been addressed in an *in vitro* model that utilizes monolayers of cultured human umbilical vein endothelial cells (HUVEC) and human neutrophils (Yoshida *et al.*, 1992). In this model, the HUVEC monolayers are exposed to sublethal episodes of anoxia and subsequently reoxygenated. Radio-labelled neutrophils are added immediately upon reoxygenation and neutrophil-endothelial cell interactions can be quantitated. Under basal (normoxia) conditions, neutrophil adherence to endothelial cell monolayers is inhibited by mAbs directed to CD11a and CD18, but not by mAb directed to CD11b. Inhibition is also observed following administration of mAb directed against ICAM-1, but not by mAbs against ELAM-1. Anoxia/reoxygenation leads to a significant increase in neutrophil adherence to the endothelial cell monolayers. The hyperadhesivity of neutrophils on anoxia/reoxygenation conditioned endothelial cells is inhibited by mAbs against either CD11a, CD11b, or CD18 (Fig. 8.15). The inhibitory effects of CD11a- and CD11b-specific antibodies were additive when used in combination and this additive effect was equivalent to that of CD18, which completely prevented the hyper-adhesivity. The anoxia/reoxygenation-induced increase in neutrophil adherence to endothelial cells was inhibited by mAb directed to ICAM-1, but not by mAb directed to ELAM-1. Enzyme immunoassay studies indicated that ICAM-1 but not ELAM-1 was basally expressed on endothelial cells and that anoxia/reoxygenation did not increase the expression of either of these endothelial cell determinants. Supernatants recovered from endothelial cells exposed to anoxia/reoxygenation increased the expression of CD11b and CD18 on the surface of neutrophils as assessed by flow cytometry. Taken together these findings indicate that endothelial cells exposed to anoxia/reoxygenation release a soluble factor(s) which results in the expression and/or activation of CD11b/CD18 on neutrophils and the increased neutrophil adherence to endothelial cells is a result of interactions between CD11a/CD18 and CD11b/CD18 adherence glycoproteins on neutrophils and ICAM-1 on endothelial cells.

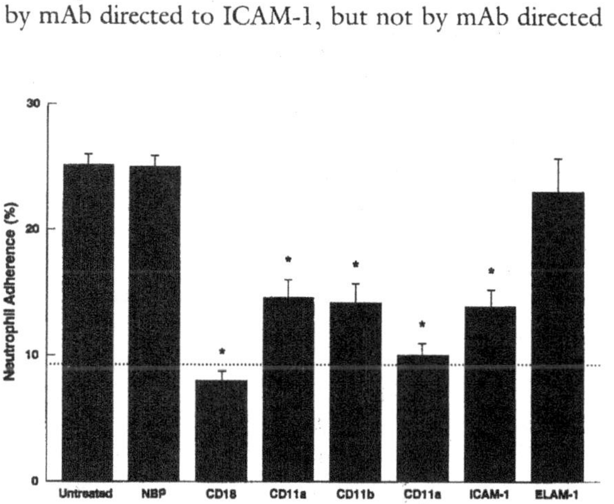

Figure 8.15 Role of neutrophil and endothelial cell adhesion glycoproteins in the hyperadhesivity of neutrophils to endothelial cell monolayers exposed to anoxia-reoxygenation. NBP refers to non-binding antibody. Data from Yoshida *et al.*, 1992.

3. *Role of Neutrophil/Endothelial Cell Adhesion in Post-ischemic Tissue Injury*

3.1 ISCHEMIA-REPERFUSION (I/R)

Ischemia and reperfusion play an important role in the pathogenesis of a wide variety of clinical conditions including myocardial infarction, stroke, mesenteric, renal, and peripheral vascular disease, compartment syndrome, shock and resuscitation, hemorrhagic "stress" gastritis, cardiac and pulmonary complications associated with cardiopulmonary bypass, frostbite, and transplantation. The main goal of traditional therapy for such ischemic disorders has been to reestablish perfusion to the affected region thereby alleviating tissue hypoxia. However, it is now apparent that reperfusion may result in a cascade of pathological events that produce the same end-result as prolonged hypoxia, namely, cellular injury and necrosis.

Although it is clear that reperfusion can exacerbate tissue injury after ischemia, the mechanisms involved are not certain. However, the observation that neutrophils accumulate in post-ischemic tissues in proportion to the severity of the ischemic insult (references cited above), when coupled with the recognition that inflammatory granulocytes can damage host tissues, is often cited as *prima facie* evidence supporting a role for these phagocytic cells in the pathogenesis of reperfusion injury. More compelling evidence for this concept is provided by the following observations: (1) neutrophil depletion greatly attenuates post-ischemic cellular dysfunction in the heart (Romson *et al.*, 1983; Lucchesi *et al.*, 1987; Kofsky *et al.*, 1991; Mullane, 1991), brain (Shiga *et al.*, 1991), small intestine (Hernandez *et al.*, 1987), stomach (Smith *et al.*, 1988), liver (Jaeschke *et al.*, 1990), kidneys (Hellberg *et al.*, 1988; Olaf *et al.*, 1990), lungs (Adkins and Taylor, 1991), and skeletal muscle (Carden *et al.*, 1991);

and (2) the extent of tissue damage and cellular dysfunction is reduced by agents that prevent neutrophil activation or accumulation (Romson *et al.*, 1982; Mullane *et al.*, 1984; Mitsos *et al.*, 1986; Korthuis *et al.*, 1987; Simpson *et al.*, 1987a and b; Grisham *et al.*, 1988; Korthuis *et al.*, 1991; Lehr *et al.*, 1991a; Lehr *et al.*, 1991b; Mullane, 1991; Nolte *et al.*, 1991a; Nolte *et al.*, 1991b). The aforementioned studies suggest that elucidation of the mechanisms involved in the regulation of neutrophil sequestration in post-ischemic tissues may play a critical role in the genesis of I/R injury. This postulate suggests that pharmacological or immunological inhibition of neutrophil adhesion to the endothelium may represent an important avenue for adjunctive therapies that enhance the benefits of restoring the blood supply to ischemic tissues.

Whereas leucocyte/endothelial cell adhesion reactions have been extensively studied in several *in vitro* models of acute inflammation (Springer *et al.*, 1989; Tonnenson, 1989; Kishimoto *et al.*, 1991; McEver, 1991; Smith *et al.*, 1991), comparatively less attention has been devoted to elucidating the role of the molecular determinants of neutrophil/endothelial adhesion in I/R-induced tissue dysfunction. Perhaps the strongest support for this concept is available with regard to the leucocyte integrin, CD11/CD18. For example, a variety of chemotactic agents (C5a, LTB_4, and PAF), cytokines (IL-1, TNF), and oxidants (superoxide, hydrogen peroxide) have been implicated in the pathogenesis of I/R and are known to induce neutrophil adherence by CD11/CD18-dependent mechanisms (Rossen *et al.*, 1985 and 1988; Springer *et al.*, 1989; Suzuki *et al.*, 1989, 1991a, 1991b; Kubes *et al.*, 1990c; Zimmerman *et al.*, 1990a and 1990b; Entman *et al.*, 1991). In addition, antibodies recognizing functional epitopes on the common β-chain (CD18) of the CD11/CD18 glycoprotein adherence complex have been shown to attenuate neutrophil sequestration in a variety of tissues subjected to I/R (Fig 8.16), neutrophil adherence and emigration in post-ischemic mesentery (Suzuki *et al.*, 1989 and 1991b; Oliver *et al.*, 1991) (Fig. 8.16), reperfusion-induced myocardial necrosis (Seewaldt-Becker *et al.*, 1989; Gomoll *et al.*, 1991; Ma *et al.*, 1991) (Fig. 8.17), alterations in coronary endothelium-dependent vascular reactivity associated with I/R (Ma *et al.*, 1991), post-ischemic skeletal muscle contractile dysfunction (Weselcouch *et al.*, 1991), neurological deficits associated with spinal cord ischemia (Clark *et al.*, 1991a), post-ischemic intestinal mucosal injury (Schoenberg *et al.*, 1991), and microvascular barrier disruption in ischemic lung (Adkins and Taylor, 1990; Horgan *et al.*, 1990 and 1991), intestine (Hernandez *et al.*, 1987), mesentery (Oliver, *et al.*, 1991), skin (Vedder *et al.*, 1990a), and skeletal muscle (Carden *et al.*, 1990) (Fig. 8.18).

The aforementioned studies clearly establish that the common β-subunit of CD11/CD18 plays an important role in the pathogenesis of ischemia/reperfusion. However, the contribution of the different α-subunits (CD11a, CD11b, and CD11c) is less clear. Recent studies indicate that exposing neutrophils to lymph draining reperfused myocardium leads to a significant increase in expression of CD11b/CD18 relative to pre-ischemic lymph (Dreyer *et al.*, 1989) (Fig. 8.19). In addition, immunoneutralization of the CD11a or CD11b subunits reduces infarct size in experimental models of myocardial ischemia/reperfusion in which the ischemic period exceeds 30–45 min (Simpson *et al.*, 1988 and 1990; Seewaldt-Becker *et al.*, 1989) (Fig. 8.17). However, immunoneutralization of CD11b does not reduce myocardial dysfunction associated with shorter periods of ischemia that induce myocardial stunning (Schott *et al.*, 1989).

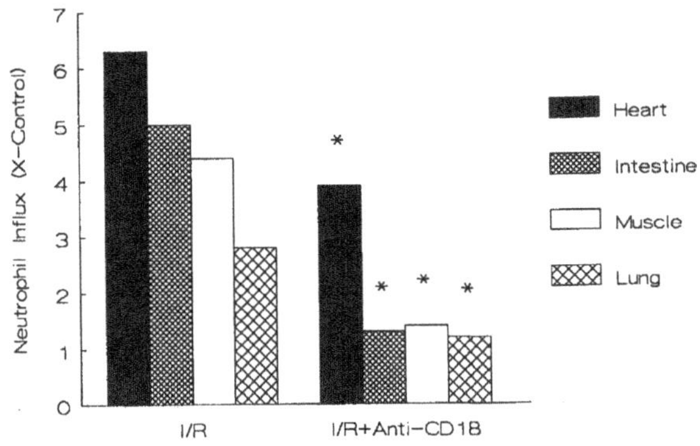

Figure 8.16 Ischemia/reperfusion (I/R) induced neutrophil influx is attenuated by administration of anti-CD18 antibodies in heart, small intestine, skeletal muscle, and lung. Data from Carden *et al.*, 1990; Dreyer *et al.*, 1991; Schoenberg *et al.*, 1991; and Horgan *et al.*, 1991.

Relatively little information is available regarding the role of adhesion molecules expressed by endothelial cells in the pathogenesis of post-ischemic cellular dysfunction. However, the site-specificity of neutrophil adhesion implies that it results from alterations in the endothelium. Indeed, immunoneutralization of ICAM-1 attenuates I/R-induced neutrophil adherence in the mesentery (Granger *et al.*, 1991), reduces post-ischemic pulmonary neutrophil sequestration and edema (Horgan *et al.*, 1991) (Fig. 8.20), decreases infarct size in reperfused rabbit hearts (Seewaldt-Becker *et al.*, 1989) (Fig. 8.17), and diminishes the neurological deficit associated with reversible spinal cord ischemia (Clark *et al.*, 1991a). Neutrophil adherence to cardiac myocytes has been shown to be ICAM-1 dependent (Smith *et al.*, 1991). Goldman and coworkers (1991a) have shown that anti-ICAM antibodies abolish neutrophil migration into dermabrasion chambers induced by plasma obtained from reperfused skeletal muscle. Up-regulation of ICAM-1 does not appear to be required for this process since treatment with protein synthesis inhibitors prior to

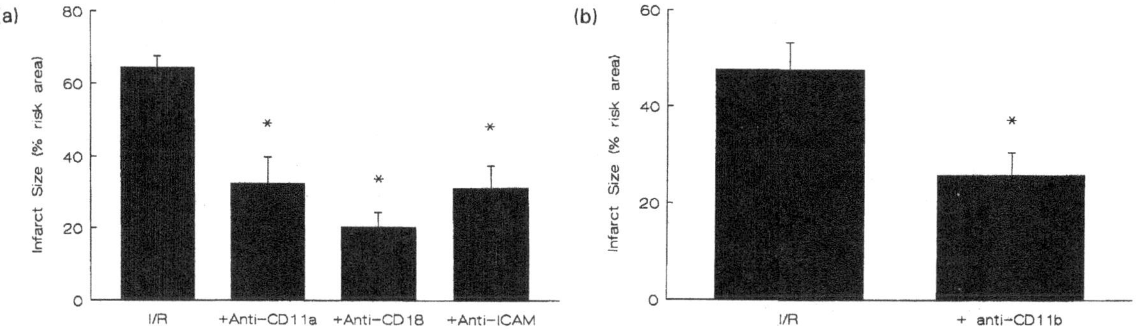

Figure 8.17 Reduction in post-ischemic myocardial infarct size by administration of CD11a-, CD18-, and ICAM-1-specific antibodies (a). Data from Seewaldt-Becker *et al.*, 1989. Similar results were obtained with anti-CD11b antibodies (b). Data from Simpson *et al.*, 1988. I/R refers to ischemia/reperfusion.

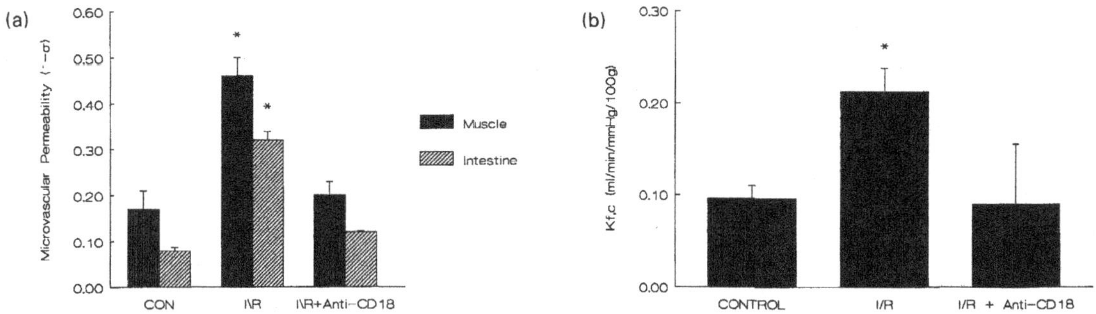

Figure 8.18 Immunoneutralization of CD18 attenuates ischemia/reperfusion (I/R) induced increases in microvascular permeability (a) in skeletal muscle and small intestine. Data from Hernandez *et al.*, 1987 and Carden *et al.*, 1990. (b) depicts changes in the pulmonary capillary filtration coefficient (Kfc, a measure of microvascular permeability) induced by I/R in the presence and absence of CD18 specific antibodies. Data from Adkins and Taylor, 1990.

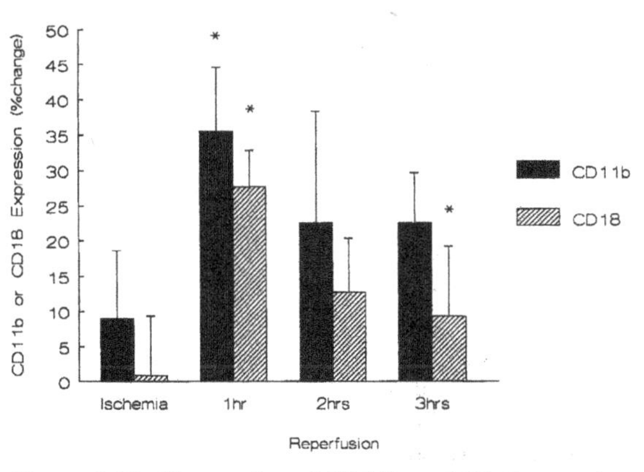

Figure 8.19 Expression of CD11b and CD18 on naive neutrophils exposed to cardiac lymph obtained during ischemia and after 1, 2, or 3 h of reperfusion. Data from Dreyer *et al.*, 1989.

ischemia did not attenuate neutrophil diapedesis induced by post-ischemic plasma. Although the role of ICAM-1 in neutrophil-mediated post-ischemic microvascular barrier disruption remains uncertain, recent evidence indicates that anti-ICAM antibodies inhibit neutrophil adherence to endothelial cell monolayers exposed to 30 min of anoxia and reoxygenation by ~50% (Yoshida *et al.*, 1992).

No information is available regarding the role of ELAM-1 in I/R. Indeed, the participation of ELAM-1 in the pathogenesis of post-ischemic cellular dysfunction is often excluded because the molecule is not constitutively expressed and the time course for ELAM-1 expression frequently exceeds the duration of ischemia/reperfusion studied (Bevilacqua *et al.*, 1989). However, it is possible that ELAM-1 plays an important role in more delayed adherence reactions induced by I/R. Furthermore, since tissues such as skin and skeletal muscle are very resistant to the short periods of ischemia that are so devastating to the heart and brain, the duration of ischemia required to produce injury in these tissues (⩾4 h) may allow for changes that ultimately promote ELAM-1 expression during reperfusion. However, these issues have not yet been experimentally addressed.

In some tissues (e.g. the heart, brain, liver, kidney, and skeletal muscle), reperfusion is associated with the development of the no-reflow phenomenon (Ames *et al.*, 1968; Strock and Majno, 1969; Dahlgren *et al.*, 1983; Engler *et al.*, 1983; Engler *et al.*, 1986a; Engler *et al.*,

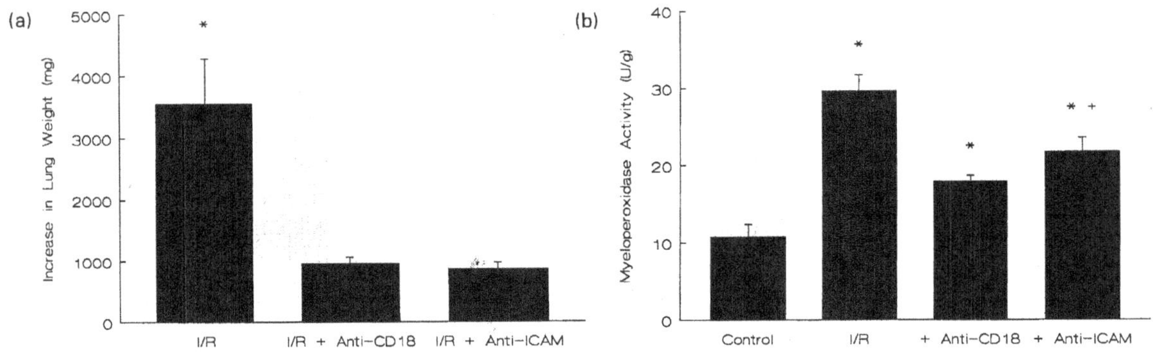

Figure 8.20 ICAM-1 and CD18-specific antibodies attenuate neutrophil sequestration (a) and edema (b) in lungs subjected to ischemia/reperfusion (I/R). Data from Horgan *et al.*, 1991.

1986b; Carden *et al.*, 1989; del Zoppo *et al.*, 1991). That is, when blood flow to these organs is reinstituted after ischemia, a large proportion of capillaries fail to reperfuse. Although several potential mechanisms for the development of no-reflow have been postulated (Strock and Majno, 1968; Carden *et al.*, 1990), it appears likely that the compressive effect of edema on the vasculature contributes to the problem in tissues that cannot readily expand as fluid accumulates in the interstitial spaces. For example, expansion of the brain is limited by the cranial vault while the kidney is limited by its capsular investment. Many skeletal muscles are surrounded by a tight fascial sheath which prevents tissue swelling. As a consequence, edema formation in these tissues is associated with a marked increase in interstitial fluid pressure which acts to compress the vasculature. This coupled with the fact that ischemia reduces neutrophil deformability and induces endothelial cell and neutrophil swelling tends to promote granulocyte trapping within the microvasculature (Dahlgren, *et al.*, 1983; Worthen, *et al.*, 1989). This notion is supported by the strong correlation between the percent of capillaries exhibiting no-reflow and the percentage of capillaries that contain granulocytes in heart, brain, kidney, pancreas, and skeletal muscle biopsies obtained from animals subjected to I/R or hemorrhagic shock and resuscitation (Engler, *et al.*, 1983; Engler, *et al.*, 1986a; Engler, *et al.*, 1986b; Barrosa-Aranda, *et al.*, 1988; del Zoppo *et al.*, 1991). Furthermore, the development of no-reflow associated with skeletal muscle I/R or hemorrhagic shock is virtually eliminated by prior neutrophil depletion (Barrosa Aranda *et al.*, 1988; Carden *et al.*, 1990). Neutrophil/endothelial cell adhesion reactions appear to be a prerequisite for the development of no-reflow since administration of an anti-CD18 antibody moAb IB_4 prevents the development of this phenomenon in post-ischemic skeletal muscle (Carden *et al.*, 1990) (Fig. 8.21).

The accumulation of neutrophils in post-ischemic tissues is well-documented and it is assumed that these extravasated leucocytes are responsible for the production of parenchymal cell injury. Recent studies indicate that alterations in the endothelial cell cytoskeleton play an important role in neutrophil accumulation in inflamed tissues. For example, Lewis and Granger (1988) have elegantly demonstrated that microfilament-rich endothelial cell lamellopodia extend luminally and envelope leucocytes as they emigrate. In addition, stabilization of F-actin in endothelial cells with phalloidin prevents the alterations in cytoskeletal architecture and leucocyte emigration that occurs in response to a variety of pro-inflammatory agents both *in vitro* and *in vivo* (Doukas *et al.*, 1987; Paterson *et al.*, 1989). Recent studies indicate that post-ischemic microvascular dysfunction and neutrophil sequestration can be attenuated by stabilization of microfilamentous cytoskeletal elements in the endothelium (Fig. 8.22) (Korthuis *et al.*, 1991). Intravital microscopic studies indicate that phalloidin does not alter the increases in leucocyte adherence to venular endothelium induced by PAF, LTB_4, and FMLP (Asako *et al.*, 1992). However, this agent completely prevented leucocyte emigration and increased vascular albumin clearance induced by these pro-inflammatory stimuli. Taken together, these studies

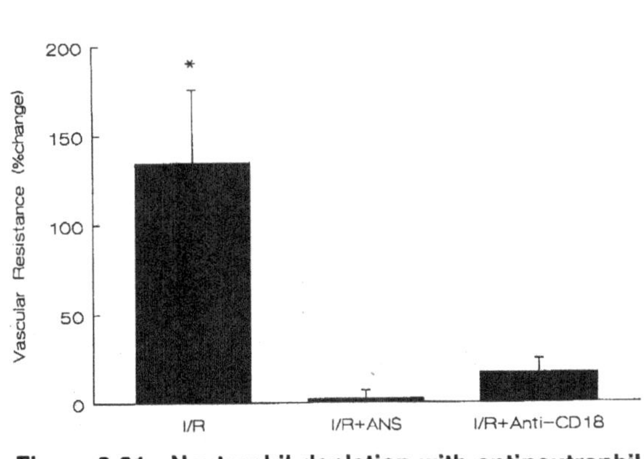

Figure 8.21 Neutrophil depletion with antineutrophil serum (ANS) or administration of CD18-specific antibodies abolishes the rise in vascular resistance (an index of the no-reflow phenomenon) induced by ischemia/reperfusion (I/R) in skeletal muscle. Data from Carden *et al.*, 1991.

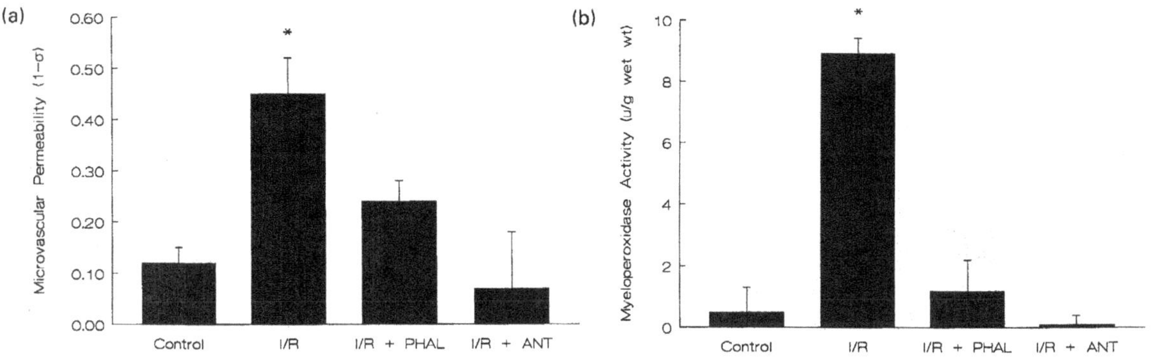

Figure 8.22 Stabilization of F-actin in the endothelial cell cytoskeleton with phalloidin (PHAL) or antamanide (ANT) attenuates ischemia/reperfusion (I/R) induced microvascular dysfunction (a) and neutrophil sequestration (b) in canine skeletal muscle. Data from Korthuis *et al.*, 1991.

suggest that prevention of leucocyte emigration by use of compounds which stabilize the architecture of the endothelial cell cytoskeleton may represent a useful new therapeutic approach to ameliorate I/R injury.

If the mass of tissue subjected to I/R is sufficiently large, neutrophil-dependent injury can occur in organ sites distant from the tissue subjected to the ischemic insult. For example, the remote consequences of I/R affecting large portions of the lower limbs are neutrophil sequestration in the lungs and the development of non-cardiogenic pulmonary edema (Welbourn *et al.*, 1991a). This remote injury appears to be neutrophil-dependent since granulocyte depletion effectively eliminates the formation of pulmonary edema induced by lower torso ischemia (Klausner *et al.*, 1988). Immunoneutralization of CD18 also eliminates the disruption of the pulmonary microvascular barrier indicating that neutrophil adhesion to microvascular endothelium plays a very important role in this distant site injury (Welbourn *et al.*, 1991b) (Fig. 8.23). Although the mechanisms responsible for the selective accumulation of neutrophils in the lungs are not clear, it seems likely that proinflammatory mediators produced in the post-ischemic limb muscles may up-regulate the expression of adhesive structures on neutrophils and induce their sequestration in the first microvascular bed they encounter after activation. In addition, it is possible that significant plasma concentrations of pro-inflammatory mediators (e.g. thromboxane, cytokines) can be achieved if the mass of tissue subjected to I/R is large as in lower torso ischemia. As a consequence, these agents may directly activate the endothelium in organs distant to the ischemic tissue to become adhesive for neutrophils. Such a scenario may explain the coincident injury to and sequestration of neutrophils in the lungs, liver, kidneys and myocardium that can occur after reperfusion of the ischemic small intestine (Schmeling *et al.*, 1989; Horton *et al.*, 1991; Turnag *et al.*, 1991).

3.2 CIRCULATORY SHOCK

Circulatory shock may be defined as any state in which perfusion is inadequate to meet the metabolic requirements of the tissues. The etiologies of circulatory shock are many and include cardiogenic, hypovolemic, and septic shock. Despite diverse initiating events, neutrophils

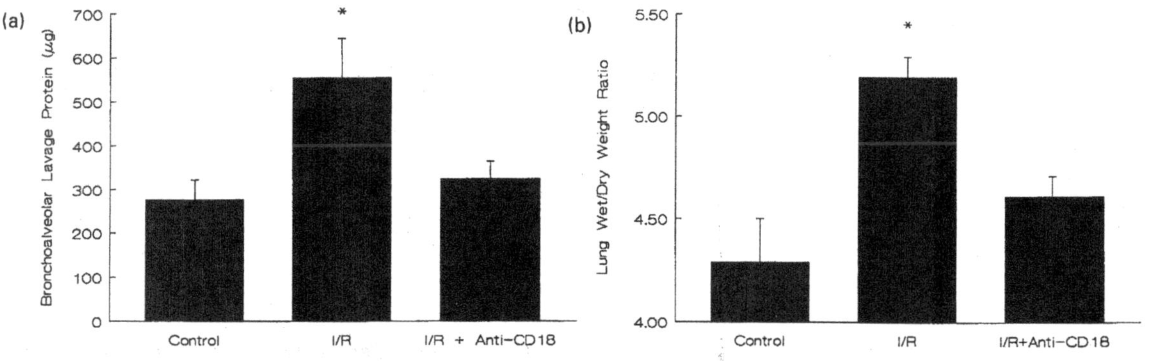

Figure 8.23 Changes in bronchoalveolar lavage protein (a) and pulmonary edema (b) induced by hindlimb ischemia/reperfusion (I/R) in the presence and absence of CD18-specific antibody treatment. Data from Welbourn *et al.*, 1991.

have been implicated as common mediators of the cellular dysfunction that occurs in various experimental shock models. For example, acute myocardial infarction is the predominant cause of cardiogenic shock. The current routine use of thrombolytic therapy in the setting of acute myocardial infarction sets the stage for neutrophil-induced reperfusion injury in previously ischemic myocardium, as discussed above. Resuscitation after cardiac arrest or hypovolemic shock is widely regarded as an event that results in "whole body ischemia/reperfusion" (Vedder *et al.*, 1989a). As such, it has been postulated that the period of generalized ischemia followed by reperfusion may play a critical role in the genesis of multiple organ injury associated with such states. Thus, it should not be surprising that neutrophils have been implicated as important mediators of the cellular dysfunction associated with circulatory collapse. Plasma obtained after hemorrhage causes neutrophil aggregation and tissue infiltration (Hammerschmidt, 1984). More compelling support for the notion that neutrophils play an active role in hemorrhagic shock is provided by studies in which neutrophil depletion is produced prior to ischemia. Barrosa-Arranda *et al.* (1987) have demonstrated that survival rates averaged 36% and 0% in rats subjected to hemorrhagic shock protocols in which systemic arterial pressure was reduced to 40 mm Hg for 3 h or 30 mm Hg for 7 h followed by resuscitation with the shed blood, respectively. However, production of neutropenia by administration of antineutrophil serum prior to hemorrhage increased survival rates to 100% in both groups. This finding is consistent with the observations of Smith *et al.* (1987) who demonstrated that neutropenia effectively eliminates gastric mucosal bleeding induced by hemorrhagic shock and resuscitation.

While the aforementioned studies clearly demonstrate that neutrophils are important mediators of the cellular dysfunction associated with hypovolemic shock, the mechanisms whereby activated granulocytes produce this injury is uncertain. However, recent studies indicate that granulocyte adherence to vascular endothelium may be a prerequisite to the production of neutrophil-mediated cellular dysfunction in individuals subjected to hypovolemic shock and resuscitation. Vender and associates (1988) observed that immunoneutralization of the common β chain of CD11/CD18 increased the survival rate of rabbits subjected to 1 h of hypovolemic hypotension (blood pressure was reduced to 45 mm Hg) followed by resuscitation from 29% to 100% when assessed five days later (Fig. 8.24). Gross and histological examination of the lungs, liver, and gastrointestinal mucosa revealed extensive accumulation of neutrophils, endothelial cell disruption, edema, and hemorrhagic necrosis. With the exception of the lung, this evidence of tissue destruction was prevented by administration of CD18-specific antibodies prior to hemorrhage. Recent studies from this same group indicate that administration

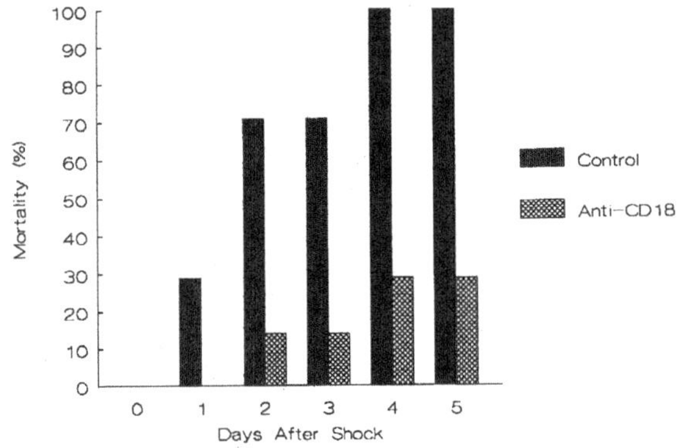

Figure 8.24 Effect of treatment with a CD18-specific antibody on mortality rates associated with hemorrhagic shock. Data from Vedder *et al.*, 1988.

of anti-CD18 antibodies at the time of resuscitation improved survival rates, reduced visceral organ dysfunction, and reduced fluid requirements in rabbits and monkeys subjected to hemorrhagic shock (Vedder *et al.*, 1989a and b; Mileski *et al.*, 1990). The observation that administration of the anti-CD18 antibody at resuscitation is nearly as effective as pre-treatment prior to the hypotensive episode (Vedder *et al.*, 1989a and 1989b; Mileski *et al.*, 1990) is important since administration of the CD18-specific antibody at the time of resuscitation is a more clinically relevant treatment protocol. In addition, this observation indicates that a major proportion of cellular injury induced by hemorrhagic shock and resuscitation occurs after the hypotensive episode.

It is important to note that the pulmonary injury induced by hemorrhagic shock was not attenuated by immunoneutralization of CD11/CD18 (Vedder *et al.*, 1989; Vedder *et al.*, 1990). This result suggests that the development of acute respiratory distress syndrome secondary to hypovolemia may occur by neutrophil-independent mechanisms or that hemorrhage-induced neutrophil adherence and emigration is not dependent on CD11/CD18. In any event, the failure to protect the lungs from damage arising secondary to hemorrhagic shock and resuscitation by CD18 antibody treatment is in direct contrast to the observation that immunoneutralization of CD18 ameliorates the pulmonary injury induced by lower torso ischemia/reperfusion (Welbourn *et al.*, 1991). This dichotomy suggests that fundamental differences exist with regard to the regulation of neutrophil adherence reactions in these conditions.

Another important conclusion from the aforementioned studies is that increased neutrophil adhesiveness may play a key role in the multiple organ failure that is a frequent complication of shock states (Vedder *et al.*, 1989). In addition, intestinal mucosal injury induced by hemorrhagic shock may provide a route for translocation of bacteria and endotoxin from the intestinal lumen to the portal and lymphatic circulations (Deitch *et al.*,

1990). This has led to the concept that the gastrointestinal complications associated with shock states may act as "the engine driving multiple organ failure syndrome" (Carrico *et al.*, 1986). The marked attenuation in gastrointestinal mucosal injury associated with inhibition of neutrophil adhesiveness may limit the development of septic complications arising secondary to hemorrhage or trauma by reducing the likelihood for bacterial translocation. However, while inhibition of granulocyte adherence may prevent the intestinal injury that leads to bacterial translocation, this intervention may also impair the bactericidal activity of neutrophils. Thus, inhibition of granulocyte adhesion may represent a double-edged sword in this context. However, recent studies indicate that transient inhibition of CD18-dependent neutrophil adherence does not increase the susceptibility to sepsis (see below).

Septic shock is a particularly devastating form of shock that afflicts over 200 000 people per year and is associated with a mortality rate of approximately 50% (Dal Nogare, 1991). The fact that neutrophil activation is a key component of host defense against invading pathogens strongly suggests that granulocytes contribute to the pathogenesis of this shock state. Much of the experimental evidence in favor of the view that neutrophils participate in the pathophysiology of sepsis is derived from studies in which known mediators of the septic state are administered. For example, administration of endotoxin tumor necrosis factor (TNF) produces tissue injury coincident with prominent neutrophil infiltration in affected organs (Doerschuk *et al.*, 1990; Linas *et al.*, 1991; Matsuda *et al.*, 1991; Munro *et al.*, 1991; Dal Nogare, 1991). Moreover, neutrophil depletion attenuates tissue injury induced by endotoxin in some, but not all, models of endotoxemia (Heflin and Brigham, 1981; Winn *et al.*, 1987; Matsuda *et al.*, 1991). Perhaps the most compelling support for a role for endotoxin and TNF in sepsis is provide by the observation that antibodies to these mediators prevent the development of shock following administration of lethal doses of *E. coli* (Ziegler *et al.*, 1982; Baumgartner *et al.*, 1985; Calandra *et al.*, 1988; Sheppard *et al.*, 1989; Dal Nogare, 1991).

Both TNF and endotoxin induce the up-regulation of CD11/CD18 on neutrophils and ELAM-1 on endothelial cells. Similar observations have been made in septic animal models suggesting that neutrophil adherence may play an important role in the pathogenesis of this shock state (Tonnenson, 1989; Morisaki *et al.*, 1991; Munro *et al.*, 1991; Nathan and Sporan, 1991; Redl *et al.*, 1991; Smith *et al.*, 1991; Springer *et al.*, 1991; Walsh *et al.*, 1991a and 1991b). This notion is supported by the observations that neutrophils obtained from septic patients demonstrate increased adhesiveness *in vitro* while naive neutrophils exposed to plasma from septic patients makes them more adherent (Venezio *et al.*, 1982). Increased neutrophil adhesiveness in septic patients would induce neutrophil margination and extravasation, thereby accounting for the neutropenia associated with this shock state, a notion supported by recent studies which demonstrate that CD18 antibodies attenuate neutropenia associated with gram negative sepsis (Walsh *et al.*, 1991). Although these observations suggest that inhibition of neutrophil adherence may represent a promising new therapeutic approach to the treatment of septic shock, the clinical utility of this treatment regimen has been questioned because of the potential risk for uncontrolled bacterial invasion (Vedder *et al.*, 1989). Indeed, neutrophil depletion has been shown to result in higher mortality rates in one model of septic shock, an effect presumably due to inhibition of neutrophil-dependent bacterial cytolysis (Crowley *et al.*,1987). In addition, CD18 antibodies increase the incidence and severity of subcutaneous abscesses in rabbits inoculated with *S. aureus* (Sharar *et al.*, 1991). However, Mileski and coworkers (1991) have demonstrated that *transient* immunoneutralization of the CD18-dependent neutrophil adherence does not increase the susceptibility to sepsis (Fig. 8.25). This conclusion was based on the observation that the survival rate for rabbits in which abdominal sepsis was produced by appendiceal ligation averaged 50% whether animals received no treatment or were administered an anti-CD18 antibody. Treatment with the antibiotic cefazolin increased survival rates to 80% while concomitant administration of the antibiotic with CD18-specific antibodies was associated with a 90% survival rate. More recently, Walsh and coworkers (1991) have demonstrated that immunoneutralization of CD18 attenuates alveolar capillary membrane injury in a porcine model of gram-negative sepsis. Thus, transient inhibition of neutrophil adherence may represent a useful adjunct to conventional antibiotic therapy in sepsis.

3.3 TRANSPLANTATION

Leucocyte/endothelial cell interactions are also thought

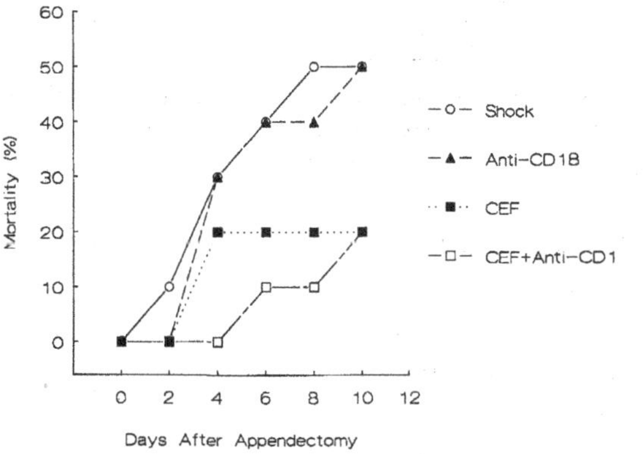

Figure 8.25 Transient immunoneutralization of CD18 does not increase the susceptibility to septic shock induced by appendiceal ligation. CEF refers to the antibiotic cefazolin. Data from Mileski *et al.*, 1991.

to play a very important role in the pathogenesis of allograft rejection. Endothelial cells of the transplanted organ form the major interface or boundary between the allograft and the recipient's immune defences and as such represent the principal target in hyperacute and acute rejection. The pivotal role of microvascular endothelium in the pathogenesis of acute rejection is further exemplified by the fact that its large surface area presents a considerable target for activated neutrophils.

The endothelial cells of the donor organ constitutively express a variety of alloantigens that are recognized by preformed antibodies in the recipient (Sedmak and Orosz, 1991). Antibody binding to these antigenic sites leads to complement activation with subsequent adherence of neutrophils to endothelial cells. Complement and neutrophil-dependent endothelial cell lysis exposes the basement membrane which in turn, is associated with platelet margination, activation of the clotting cascade, and widespread formation of fibrin thrombi all of which contribute to the pathogenesis of hyperacute rejection.

Although the endothelium is the principal target in acute rejection, the aforementioned mechanism suggests that endothelial cells play only a passive role in this process. However, the recent observation that endothelial expression of adhesive molecules for inflammatory cells is induced in donor organs suggests that the endothelium may be a more active participant in the pathogenesis of acute rejection. For example, ICAM-1 is constitutively expressed in cardiac, renal, hepatic, and corneal allografts and is up-regulated during cellular rejection, an effect that correlates with inflammatory cell infiltration (Adams *et al.*, 1989; Adams *et al.*, 1990; Briscoe *et al.*, 1991; Elner *et al.*, 1991; Hubscher and Adams, 1991; Sedmak and Orosz, 1991; Takei *et al.*, 1991). The most compelling support for a role for ICAM-1 in allograft rejection is derived from studies in which it was demonstrated that mABs against this adhesion molecule prolong the survival of cardiac and renal allografts (Flavin *et al.*, 1991; Wee *et al.*, 1991) (Fig. 8.26). CD11a does not appear to act as a ligand for ICAM-1 in acute rejection since mAbs directed against this alpha subunit do not improve renal function during acute rejection after kidney transplantation (Le Mauff *et al.*, 1991). Nevertheless, the success of the anti-ICAM antibody treatment studies suggest that use of mAbs for targeted disruption of the adhesive interaction between the recipient's inflammatory cells and the endothelium in the donor organ may represent an important and more selective strategy for prolonging allograft survival than conventional immunosuppressive therapy. In addition, since the molecular determinants of neutrophil/endothelial cell adhesion are up-regulated in acute rejection, quantitation of adhesion molecule expression may represent an important diagnostic tool for prediction of impending allograft rejection.

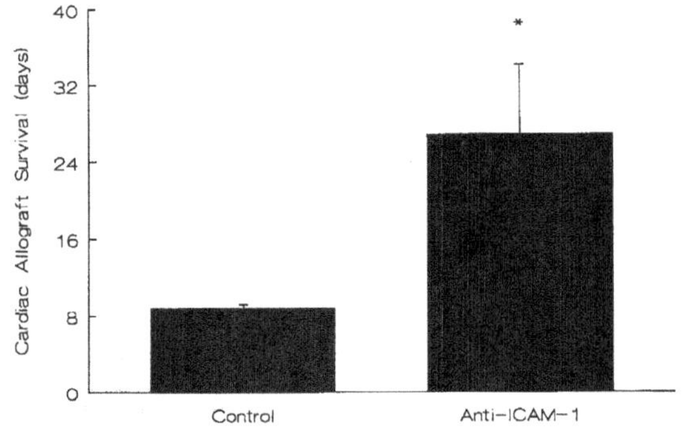

Figure 8.26 ICAM-1 specific antibodies prolong cardiac allograft survival. Data from Cosimi *et al.*, 1991.

3.4 CARDIOPULMONARY BYPASS

One of the most serious potential complications of cardiopulmonary bypass is the development of acute respiratory distress. This manifestation of the "post perfusion syndrome" is characterized by increased pulmonary capillary permeability, a marked accumulation of interstitial fluid in the lungs, complement activation, and pulmonary leucocyte retention (Chenowith *et al.*, 1981; Kirklin, 1989). In addition to inducing life-threatening pulmonary dysfunction, the post-operative course of patients who undergo cardiopulmonary bypass is often complicated by cardiac, renal, hepatic, and/or neural damage that is associated with marked leucocyte accumulation and inflammation. These observations have given rise to the concept that cardiopulmonary bypass induces a "whole body inflammatory reaction" (Kirklin, 1989).

Although it has been suggested that complement activation is the initial step in the pathogenesis of multi-organ dysfunction induced by cardiopulmonary bypass (Chenowith *et al.*, 1981; Kirklin, 1989), events associated with pulmonary and myocardial ischemia and reperfusion may also contribute to the injury. As a consequence, leucocytes become activated and release their cytotoxic arsenal resulting in microvascular barrier disruption, enhanced transcapillary fluid filtration, and edema (Kirklin, 1989; Dreyer *et al.*, 1991; Faymonville *et al.*, 1991). Bando and coworkers (1990) have recently provided strong support for this postulate in the lung. Using an experimental preparation that mimics the extracorporeal circuitry employed in cardiopulmonary bypass, these investigators demonstrated that bypass induced complement activation and a concurrent increase in plasma conjugated diene levels, a marker of oxidant production. These changes were associated with significant pulmonary dysfunction, intravascular leucocyte aggregation, perivascular hemorrhage, and focal alveolar injury. With the exception of complement activation, these effects were eliminated by leucocyte depletion prior to bypass. Although the role of leucocyte/endothelial cell adhesion reactions in post-bypass pulmonary dysfunction has not yet been examined, the fact that complement activation modulates the surface expression and function of adhesion molecules expressed by both neutrophils and endothelial cells strongly supports this concept (Dreyer *et al.*, 1991). Examination of molecular mechanisms involved in the regulation of the adhesive interaction between neutrophils and the endothelium may provide the rationale for pursuit of new therapeutic approaches to ameliorate post-bypass tissue injury.

3.5 FROSTBITE INJURY

The pathogenesis of tissue injury from frostbite is complicated and appears to result as a consequence of events occurring during the initial freezing and the vascular and inflammatory changes that occur during rewarming (Heggers *et al.*, 1987; Britt *et al.*, 1991). Frostbite is associated with increased vascular permeability, progressive vasoconstriction, and cellular disruption induced by ice crystal formation. During rewarming, blood flow is initially restored, interstitial edema worsens, and leucocyte activation becomes apparent. As the frostbitten tissue continues to rewarm, thromboxane is generated resulting in vasoconstriction. The ensuing progressive ischemia is exacerbated by leucocyte and platelet aggregation with subsequent formation of microvascular thrombi. Recent studies indicate that administration of anti-CD18 antibodies reduce frostbite injury in rabbit ears indicating the leucocyte adherence reactions play an important role in the pathogenesis of this problem (Carlos and Harlan, 1990).

4. *Mechanisms of Neutrophil-Mediated Tissue Injury in I/R*

Phagocytic leucocytes secrete a number of cytotoxic agents which can produce extensive cellular injury in host tissues. For example, neutrophils contain a membrane bound NADPH oxidase which is activated when neutrophils are exposed to chemo-attractants, immune stimuli, or bacteria. This enzyme oxidizes cytoplasmic NADPH to $NADP^+$ and generates two molecules of superoxide which can subsequently dismutate to form hydrogen peroxide. While it has been suggested that neutrophil-derived superoxide and hydrogen peroxide can interact with low molecular weight iron to yield hydroxyl radicals, recent data indicates that granulocytes produce very little (if any) hydroxyl radicals *in vivo* (Winterbourne 1986). This appears to be related to the fact that myeloperoxidase (a granular enzyme released by activated granulocytes, see below) can consume virtually all of the hydrogen peroxide, thereby limiting the interaction of this oxidant with superoxide and iron.

Activated neutrophils also secrete myeloperoxidase (MPO), an enzyme which catalyzes the formation of hypochlorous acid (HOCl) from hydrogen peroxide and chloride ions. As noted above, the hydrogen peroxide which fuels this reaction is derived from the spontaneous disputation of superoxide. HOCl is a potent oxidizing and chlorinating agent that reacts rapidly with primary amines to yield cytotoxic N-chloramines, which contain the two oxidizing equivalents of hydrogen peroxide and HOCl. The cellular toxicity associated with HOCl and N-chloramines appears to depend on their relative lipophilicities and probably occurs as a result of sulfhydryl oxidation, hemeprotein and cytochrome inactivation, chlorination of purine bases on DNA, and degradation of amino acids and proteins.

Neutrophil-derived HOCl and hydrogen peroxide can also initiate hydrolytic reactions catalyzed by neutrophilic proteases. For example, HOCl can alter the balance between tissue proteases and antiproteases either by oxidative inactivation of several plasma protease inhibitors (e.g. alpha-1-proteinase inhibitor) or by activation of latent proteases (e.g. collagenase or gelatinase) or both. Several reports indicate that neutrophil-derived proteases may play an important role in the cellular dysfunction induced by I/R (See Inauen *et al.*, 1990 for references). Thus, it is apparent that the neutrophil utilizes the NADPH oxidase system and granular constituents in a coordinated fashion to facilitate the destructive potential of the granulocyte.

The cumulative effect of the cascade of cytotoxic reactions associated with neutrophil activation is damage to the entire array of biomolecules found in tissues including structural, contractile, and transport proteins, enzymes, receptors, membrane lipids, glycosaminoglycans, and nucleic acids (Halliwell and Gutteridge, 1989). The molecular pathology that results as a consequence of oxidant formation includes DNA "nicking" and other abnormalities, peroxidation of membrane lipid components, and crosslinking and degradation of proteins. These oxidant-induced alterations in molecular structure and organization can decimate cellular function. For example, peroxidation of linolenic acid in the acyl chains of cardiolipin, which is essential for the activity of cytochrome oxidase, may lead to a loss of function of this critical cellular enzyme (Soussi *et al.*, 1990). Peroxidation of membrane lipids can also produce devastating effects on membrane fluidity and cell compartmentalization, leading to altered mobility of membrane receptors, impaired second messenger function, and leakage of intracellular compounds (e.g. lactate dehydrogenase and creatine kinase) into the plasma. In addition, lipid peroxidation often precedes cell lysis in post-ischemic tissues and is associated with increased rates of protein degradation. By attacking the glycosaminoglycans and proteins in the substrata supporting intestinal epithelial cells, reactive oxygen metabolites may contribute to the exfoliation of the intestinal mucosa noted in severe I/R. Oxidant-mediated damage to structural, transport, and contractile proteins and inactivation of receptors and essential enzymes in post-ischemic myocardium and skeletal muscle may explain the contractile dysfunction noted in these tissues during reperfusion.

Although it is clear that oxidants play an important role in the genesis of post-ischemic cellular dysfunction, the contribution of neutrophil-derived reactive oxygen metabolites versus those generated from other sources such as xanthine oxidase is unclear. Schoenberg *et al.* (1991) have recently demonstrated that immunoneutralization of CD18 prevents neutrophil infiltration but does not attenuate the increase in conjugated dienes in post-ischemic small intestine. This result suggests that reperfusion-induced lipid peroxidation is not dependent on reactive oxygen metabolites generated from the infiltrating neutrophils. Since allopurinol, a competitive antagonist of xanthine oxidase, prevents the increase on conjugated dienes in the same experimental model, it appears that this enzyme is the source of oxidants responsible for post-ischemic lipid peroxidation, at least in the small intestine (Schoenberg *et al.*, 1985).

The establishment of adhesive interactions between neutrophils and endothelium not only allows these phagocytic cells to marginate and extravasate at sites of inflammation but also appears to facilitate the destructive potential of the granulocyte. For example, neutrophil-endothelial cell adherence reactions may establish a "protected microenvironment" which shields granulocyte-derived mediators from inactivation by plasma inhibitors (Weiss, 1989). Establishing an intimate area of contact between granulocytes and target cells also allows the cytotoxic arsenal of these phagocytic cells to be focused at a specific cellular site within the neutrophil's local environment. The concept that granulocyte/target cell adhesion promotes focused mediator release takes on an even greater significance with regard to facilitation of the neutrophil's cytotoxic potential since these phagocytic cells release far greater quantities of oxidants when adherent to cells (Nathan *et al.*, 1987; Nathan *et al.*, 1989; Entman *et al.*, 1990; Shappell *et al.*, 1990).

5. *Summary*

It is now clear that neutrophils play a critical role in the pathogenesis of I/R. Although the pathogenetic mechanisms involved in neutrophil-dependent, reperfusion-induced cellular dysfunction is complex and not well-understood, a growing body of evidence indicates reperfusion is associated with the formation and release of pro-inflammatory agents which subsequently attract and activate neutrophils. The activated granulocytes adhere to vascular endothelium, extravasate, and release cytotoxic oxidants and hydrolytic enzymes that are capable of damaging the entire array of biomolecules found in tissues. In addition, neutrophils may contribute to the development of the no-reflow phenomenon in post-ischemic heart and skeletal muscle by becoming entrapped in capillaries and/or by adhering to venular endothelium, which reduces vascular caliber and increases the resistance to blood flow.

While the association of inflammation with I/R-induced cellular injury has long been appreciated, only in recent years has the essential interaction between neutrophils and the endothelium captured the attention of a large number of investigators. This renewed interest regarding the role neutrophil/endothelial cell interactions has led to the recognition that a number of physiological factors may contribute to I/R-induced neutrophil sequestration, including alterations in hydrodynamic dispersal forces, physical impaction (plugging), reduced electrostatic repulsion between the neutrophil and endothelial cell surfaces, liberation of chemotactic stimuli such as LTB_4 and PAF, generation of reactive oxygen metabolites and release of granule-associated proteins from activated neutrophils, and expression of adhesion molecules on the surface of neutrophils and/or endothelial cells.

There is a wealth of experimental evidence which indicates that neutrophil/endothelial cell adhesive interactions may play an essential role in the genesis of tissue injury in a number of clinically relevant ischemic disorders. This data suggests that the pharmacological or immunological inhibition of neutrophil adherence to the endothelium may represent an important new adjunctive therapy that may enhance the benefits of restoring the blood supply to ischemic tissues.

6. *Acknowledgement*

Dr. Korthuis is the recipient of an Established Investigatorship from the American Heart Association. This work was supported by grants from the National Heart, Lung, and Blood Institute (HL-36069 and HL26441).

7. *References*

Adams, D.H., Hubscher, S.G., Shaw, J., Rothlein, R. and Neuberger, J.M. (1989). Intercellular adhesion molecule 1 on liver allografts during rejection. Lancet ii, 1122–1125.

Adams, D.H., Wang, L.F., Burnett, D., Stockley, R.A. and Neuberger, J.M. (1990). Neutrophil activation – an important cause of tissue damage during liver allograft rejection? Transplantation 50, 86–91.

Adkins, W.K. and Taylor, A.E. (1990). Role of xanthine oxidase and neutrophils in ischemia-reperfusion injury in rabbit lung. J. Appl. Physiol. 69, 2012–2018.

Allan, G., Bhattacherjee, P., Brook, C.D., Read, N.G. and Parke. A.J. (1985). Myeloperoxidase activity as a quantitative marker of polymorphonuclear leucocyte accumulation into an experimental myocardial infarct – the effect of ibuprofen on infarct size and polymorphonuclear leucocyte accumulation. J. Cardiovasc. Pharmacol. 7, 1154–1160.

Ames, A., Wright, L.W., Kowada, M., Thurston, J.M. and Majno, G.M. (1968). Cerebral ischemia: II. The no-reflow phenomenon. Am. J. Pathol. 52, 437–453.

Asako, H., Korthuis, R.J., Wolf, R. and Granger, D.N. (1992). Phalloidin prevents leucocyte emigration induced by pro-inflammatory stimuli in rat mesentary. Am. J. Physiol. 263, H1637–H1642.

Atherton, A. and Born, G.V.R. (1973). Relationship between the velocity of rolling granulocytes and that of blood flow in venules. J. Physiol. 233, 157–165.

Bando, K., Pillai, R., Cameron, D.E., Brawn, J.D., Winkelstein, J.A., Hutchins, G.M., Reitz, B.A. and Baumgartner, W.A. (1990). Leukocyte depletion ameliorates free radical-mediated lung injury after cardiopulmonary bypass. J. Thorac. Cardiovasc. Surg. 99, 873–877.

Barone, F.C., Hillegass, L.M., Price, W.J., White, R.F., Lee, E.V., Feuerstein, G.Z., Sarau, H.M., Clark, R.K. and Griswold, D.E. (1991). Polymorphonuclear leucocyte infiltration into cerebral focal ischemic tissue: myeloperoxidase activity assay and histologic verification. J. Neurosci. Res. 29, 336–345.

Barrosa-Arranda, J., Schmid-Schonbein, G.W., Zweifach, B.W. and Engler, R.L. (1988). Granulocytes and the no-reflow phenomenon in irreversible hemorrhagic shock. Circ. Res. 63, 437–447.

Baumgartner, J., McCutchan, J.A., Van Melle, G., Vogt, M., Luethy, R., Glauser, M.P., Ziegler, E.P., Klauber, M.R., Muehlen, E., Chiolero, R. and Geroulanos, S. (1985). Prevention of gram-negative shock and death in surgical patients by antibody to endotoxin core glycolipid. Lancet II. 59–63.

Bevilacqua, M.P., Strengelin, S., Gimbrone, M.A. and Seed, B. (1989). Endothelial leucocyte adhesion molecule 1: an inducible receptor for neutrophils related to complement regulatory proteins and lectins. Science 243, 1160–1165.

Bjork, J., Hugli, T.E. and Smedegard, G. (1985). Microvascular effects of anaphylatoxin C3a and C5a. J. Immunol. 134, 1115–1119.

Bjork, J., Lindbom, L., Gerdin, B., Smedegard, G., Arfors, K.E. and Benveniste, J. (1983). PAF-acether increases microvascular permeability and affects endothelium-granulocyte interactions in microvascular beds. Acta. Physiol. Scand. 119, 305–308.

Boxer, L.A., Bjorksten, B., Bjork, J., Yang, H., Allen, J.M. and Baehner, R.L. (1982). Neutropenia induced by systemic infusion of lactoferrin. J. Lab. Clin. Med. 99, 866–872.

Braquet, P., Touqui, L., Shen, T.Y. and Vargaftig, B.B. (1987). Perspectives in platelet activating factor research. Pharmacol. Rev. 39, 97–145.

Briscoe, D.M., Schoen, F.J., Rice, G.E., Bevilacqua, M.P. and Pober, J.S. (1991). Induced expression of endothelial-leucocyte adhesion molecules in human cardiac allografts. Transplantation 51, 537–547.

Britt, L.D., Dascombe, W.H. and Rodriguez, A. (1991). New horizons in management of hypothermia and frostbite injury. Surg. Clin. N. Amer. 71, 345–370.

Calandra, T., Glauser, M.P., Schellekens, J., Verhoef, J. and the Swiss-Dutch J5 Immunoglobulin Study Group. (1988). Treatment of gram-negative septic shock with human IgG antibody to *Escherichia coli* J5. A prospective, double-blind, randomized trial. J. Infect. Dis. 158, 312–318.

Carden, D.L. and Korthuis, R.J. (1989). Mechanisms of post-ischemic vascular dysfunction in skeletal muscle: implications for therapeutic intervention. Microcirc. Endo. Lymph. 5, 277–298.

Carden, D.L., Smith, J.K. and Korthuis, R.J. (1990). Neutrophil-mediated microvascular dysfunction in post-ischemic canine skeletal muscle: role of granulocyte adherence. Circ. Res. 66, 1436–1444.

Carlos, T.M. and Harlan, J.M. (1990). Membrane proteins involved in phagocyte adherence to endothelium. Immunol. Rev. 114, 5–28.

Carrico, C.J., Meakins, J.L., Marshall, J.C., Fry, D. and Maier, R.V. (1986). Multiple organ failure syndrome. Arch. Surg. 121, 196–208.

Chatelain, P., Latour, J.-G., Tran, D., de Lorgeril, M., Dupras, G. and Bourassa, M. (1987). Neutrophil accumulation in experimental myocardial infarcts: relation with the extent of injury and effect of reperfusion. Circulation 75, 1083–1090.

Chenowith, D.E., Cooper, S.W., Hugli, T.E., Stewart, R.W., Blackstone, E.H. and Kirklin, J.W. (1981). Complement activation during cardiopulmonary bypass: evidence for generation of C3a and C5a anaphylatoxins. N. Engl. J. Med. 304, 497–503.

Clark, W.M., Madden, K.P., Rothlein, R. and Zivin, J.A. (1991a). Reduction of central nervous system ischemic injury in rabbits using leucocyte adhesion antibody treatment. Stroke 22, 877–883.

Clark, W.M., Madden, K.P., Rothlein, R. and Zivin, J.A. (1991b). Reduction of central nervous system ischemic injury by monoclonal antibody to intercellular adhesion molecule. J. Neurosurg. 75, 623–627.

Crowley, J.P., Dennis, R.C., Pivacek, L., Metzger, J., Carvalho, A. and Valeri, C.R. (1987). Effects of granulocytopenia on the hemodynamic responses of dogs during *E. coli* bacteremia. Circ. Shock 22, 91–104.

Currie, M.S., Simel, D.L., Christenson, R.H., Holmes, C., Crawford, J., Cohen, H.J. and Rao, K.M.K. (1991). Anti-inflammatory effects of pentoxifylline in claudication. Am. J. Med. Sci. 301, 85–90.

Dahlen, S.E., Bjork, J., Hedqvist, P., Arfors, K.E., Hammarstrom, S., Lindgren, J.A. and Samuelsson, B. (1981). Leukotrienes promote plasma leakage and leukocyte adhesion in post-capillary venules, *in vivo* effects with relevance to the acute inflammatory response. Proc. Natl Acad. Sci. USA. 78, 3887–3891.

Dahlgren, M.D., Peterson, M.A., Engler, R.L. and Schmid-Schonbein, G.W. (1983). In "Whlte Cell Mechanlcs, Basic Science and Clinical Aspects", pp. 271–283, Alan R. Liss, Inc., New York.

Dal Nogare, A.R. (1991). Southwestern Internal Medicine Conference: Septic shock. Am. J. Med. Sci. 302, 50–65.

Deitch, E.A., Bridges, W., Ma, L., Berg, R., Specian, R.D. and Granger, D.N. (1990). Hemorrhagic shock-induced bacterial translocation: the role of neutrophils and hydroxyl radicals. J. Trauma 30, 942–952.

Del Maestro, R.F., Planker, M. and Arfors, K.E. (1982). Evidence for the participation of superoxide anion radical in altering the adhesive interaction between granulocytes and endothelium, *in vivo*. Int. J. Microcirc., Clin. Exp. 1, 105–119.

del Zoppo, G.J., Schmid-Schonbein, G.W., Mori, E., Copeland, B.R. and Chang, C.-M. (1991). Polymorphonuclear leucocytes occlude capillaries following middle cerebral artery occlusion and reperfusion in baboons. Stroke 22, 1276–1283.

Doerschuk, C.M., Winn, R.K., Coxson, H.O. and Harlan, J.M. (1990). CD18-dependent and -independent mechanisms of neutrophil emigration in the pulmonary and systemic microcirculation of rabbits. J. Immunol. 144, 2327–2333.

Doukas, J., Shepro, D. and Hechtman, H.B. (1987). Vasoactive amines directly modify endothelial cells to affect polymorphonuclear leucocyte diapedesis *in vitro*. Blood 69, 1563–1569.

Dreyer, W.J., Smith, C.W. and Entman, M.L. (1991). Neutrophil activation during cardiopulmonary bypass. J. Thorac. Cardiovasc. Surg. 102, 318–320.

Dreyer, W.J., Smith, C.W., Michael, L.H., Rossen, R.D., Hughes, B.J., Entman, M.L. and Anderson, D.C. (1989). Canine neutrophil activation by cardiac lymph obtained during reperfusion of ischemic myocardium. Circ. Res. 65, 1751–1762.

Dreyer, W.J., Michael, L.H., West, M.S., Smith, C.W., Rothlein, R.D., Rossen, R.D., Anderson, D.C. and Entman, M.L. (1991). Neutrophil accumulation in ischemic myocardium. Insights into time course, distribution, and mechanism of localization during early reperfusion. Circulation 84, 400–411.

Elner, V.M., Elner, S.G., Pavilack, M.A., Todd, R.F., III, Yue, B.Y.J.T. and Huber, A.R. (1991). Intercellular adhesion molecule-1 in human corneal endothelium. Modulation and function. Am. J. Pathol. 138, 525–536.

Engler, R.L., Schmid-Schonbein, G.W. and Pavelec, R.S. (1983). Leukocyte capillary plugging in myocardial ischemia and reperfusion in the dog. Am. J. Pathol. 111, 98–111.

Engler, R.L., Dahlgren, M.D., Morris, D.D., Peterson, M.A. and Schmid-Schonbein, G.W. (1986a). Role of leucocytes in response to acute myocardial ischemia and reflow in dogs. Am. J. Physiol. 251, H314–H323.

Engler, R.L., Dahlgren, M.D., Peterson, M.A., Dobbs, A. and Schmid-Schonbein, G.W. (1986b). Accumulation of polymorphonuclear leucocytes during 3-h experimental myocardial ischemia. Am. J. Physiol. 251, H93–H100.

Entman, M.L., Youker, K., Shappell, S.B., Siegel, C., Rothlein, R., Dreyer, W.J., Schmalstieg, F.C. and Smith, C.W. (1990). Neutrophil adherence to isolated adult canine myocytes: evidence for a CD18-dependent mechanism. J. Clin. Invest. 85, 1497–1506.

Entman, M.L., Michael, L., Rossen, R.D., Dreyer, W.J., Anderson, D.C., Taylor, A.R. and Wayne Smith, C. (1991). Inflammation in the course of early myocardial ischemia. FASEB J. 5, 2529–2537.

Faymonville, M.E., Pincemail, J., Duchateau, J., Paulus, J.M., Adam, A., Deby-Dupont, G., Deby, C., Albert, A., Larbuisson, R., Limet, R. and Lamy, M. (1991). Myeloperoxidase and elastase as markers of leucocyte activation during cardiopulmonary bypass in humans. J. Thorax. Cardiovasc. Surg. 102, 309–317.

Flavin, T., Ivens, K., Rothlein, R., Faanes, R., Clayberger, C., Billingham, M. and Starnes, V.A. (1991). Monoclonal antibodies against intercellular adhesion molecule 1 prolong cardiac allograft survival in cynomolgus monkeys. Transplant. Proc. 23, 533–534.

Gallin, J.I. (1980). Degranulating stimuli decrease the negative surface charge and increase the adhesiveness of human neutrophils. J. Clin. Invest. 65, 298–306.

Gasic, A.C., McGuire, G., Krater, S., Farhood, A.I., Goldstein, M.A., Smith, C.W., Entman, M.L. and Taylor, A.A. (1991). Hydrogen peroxide pretreatment of perfused canine vessels induces ICAM-1 and CD18 dependent neutrophil adherence. Circulation 84, 2154–2166.

Go, L.O., Murray, C.E., Richard, V.J., Weischedel, G.R., Jennings, R.B. and Reimer, K.A. (1988). Myocardial neutrophil accumulation during reversible or irreversible ischemic injury. Am. J. Physiol. 255, H1188–H1198.

Goldman, G., Welbourn, R., Paterson, I.S., Klausner, J.M., Kobzik, L., Valeri, C.R., Shepro, D. and Hechtman, H.B. (1990). Ischemia-induced neutrophil activation and diapedesis is lipoxygenase dependent. Surgery 107, 428–433.

Goldman, G., Welbourn, R., Klausner, J.M., Valeri, C.R., Shepro, D. and Hechtman, H.B. (1991a). Thromboxane mediates diapedesis after ischemia by activation of neutrophil adhesion receptors interacting with basally expressed intercellular adhesion molecule-1. Circ. Res. 68, 1013–1019.

Goldman, G., Welbourn, R., Valeri, C.R., Shepro, D. and Hechtman, H.B. (1991b). Thomboxane A_2 induces leukotriene B_4 synthesis that in turn mediates neutrophil diapedesis via CD18 activation. Microvasc. Res. 41, 367–375.

Gomoll, A.W., Lekich, R.F. and Grove. R.I. (1991). Efficacy of a monoclonal antibody (mAb 60.3) in reducing myocardial injury resulting from ischemia/reperfusion in the ferret. J. Cardiovasc. Pharmacol. 17, 873–878.

Granger, D.N. (1988). Role of xanthine oxidase and granulocytes in ischemia-reperfusion injury. Am. J. Physiol. 255, H1269–1275.

Granger, D.N., Benoit, J.N., Suzuki, M. and Grisham, M.B. (1989). Leukocyte adherence to venular endothelium during ischemia-reperfusion. Am. J. Physiol. 257, G683–G688.

Granger, D.N., Russell, J.M., Arfors, K.E., Rothlein, R. and Anderson, D.C. (1991). Role of CD18 and ICAM-1 in ischemia/reperfusion-induced leukocyte adherence and emigration in mesenteric venules. FASEB J. 5, A1753.

Grisham, M.B., Hernandez, L.A. and Granger, D.N. (1986). Xanthine oxidase and neutrophil infiltration in intestinal ischemia. Am. J. Physiol. 251 G567–G573.

Grisham, M.B., Hernandez, L.A. and Granger, D.N. (1989). Adenosine inhibits ischemia/reperfusion-induced leukocyte adherence and extravasation. Am. J. Physiol. 257, H1334–H1339.

Halliwell, B. and Gutteridge, J.M.C. (1989). Free radicals in Biology and Medicine. Clarendon Press, Oxford, England.

Hammerschmidt, D.E., Redl, H. and Schlag, G. (1984). Quantifying pulmonary leukostasis in hypovolemic/traumatic shock. Acta Chir. Scand. 31, 113–117.

Harlan, J.M. (1985). Leukocyte-endothelial cell interactions. Blood 65, 513–525.

Heflin, A.C. and Brigham, K.L. (1981). Prevention by granulocyte depletion of increased vascular permeability of sheep lung following endotoxemia. J. Clin. Invest. 68, 1253–1260.

Heggers, J.P., Robson, M.C., Manavalen, K., Weingarten, M.D., Carethers, J.M., Boertman, J.A., Smith, D.J. and Sachs, R.J. (1987). Experimental and clinical observations on frostbite. Ann. Emerg. Med. 16, 1056–1062.

Hellberg, P.O.A., Kallskog, O., Wolgast, M. and Ojteg, G. (1988). Effects of neutrophil granulocytes on the inulin barrier of renal tubular epithelium after ischaemic damage. Acta Physiol. Scand. 134, 313–315.

Hernandez, L.A., Grisham, M.B., Twohig, B., Arfors, K.E., Harlan, J.M. and Granger, D.N. (1987). Role of neutrophils in ischemia/reperfusion-induced microvascular injury. Am J. Physiol. 253, H699–H703.

Horgan, M.J., Wright, S.D. and Malik, A.B. (1990). Antibody against leukocyte integrin (CD18) prevents reperfusion-induced lung vascular injury. Am. J. Physiol. 259, L315–L319.

Horgan, M.J., Ge, M., Gu, J., Rothlein, R. and Malik, A.B. (1991). Role of ICAM-1 in neutrophil-mediated lung vascular injury after occlusion and reperfusion. Am. J. Physiol. 261, H1578–H1584.

Horgon, J.W. and White, D.J. (1991). Cardiac contractile injury after intestinal ischemia-reperfusion. Am. J. Physiol. 261, H1164–H1170.

House, S.D. and Lipowsky, H.H. (1987). Leukocyte-endothelium adhesion: microhemodynamics in mesentery of the cat. Microvasc. Res. 34, 363–379.

Huber, A.R. and Weiss, S.J. (1989). Disruption of the sub-endothelial cell basement membrane during neutrophil diapedesis in an *in vitro* construct of a blood vessel wall. J. Clin. Invest. 83, 1122–1136.

Hubscher, S.G. and Adams, D.H. (1990). ICAM-1 expression in normal liver. J. Clin. Pathol. 44, 438–439.

Jaeschke, H., Farhood, A. and Smith, C.W. (1990). Neutrophils contribute to ischemia/reperfusion injury in rat liver *in vivo*. FASEB J. 4, 3355–3359.

Kirklin, J.K. (1989). In "Cardiopulmonary Bypass: Current Concepts and Controversies" (ed J.H. Tinger) pp. 131–151. W.B. Saunders Co., Philadelphia.

Kishimoto, T.K. (1991). A dynamic model for neutrophil localization to inflammatory sites. J. NIH Res. 3, 75–77.

Kishimoto, T.K., Jutila, M.A., Berg, E.L. and Butcher, E.C. (1989). Neutrophil Mac-1 and MEL-14 adhesion proteins are inversely regulated by chemotactic factors. Science 245, 1238–1241.

Klausner, J.M., Anner, H., Paterson, I.S., Shepro, D. and Hechtman, H.B. (1988). Lower torso ischemia-induced lung injury is leukocyte dependent. Ann. Surg. 209, 231–236.

Klausner, J.M., Paterson, I.S., Kobzik, L., Valeri, C.R., Shepro, D. and Hechtman, H.B. (1989a). Leukotrienes but not complement mediate limb ischemia-induced lung injury. Ann. Surg. 209, 462–470.

Klausner, J.M., Paterson, I.S., Goldman, G., Kobzik, L., Rodzen, C., Lawrence, R., Valeri, C.R., Shepro, D. and Hechtman, H.B. (1989b). Post-ischemic renal injury is mediated by neutrophils and leukotrienes. Am. J. Physiol. 256, F794-F802.

Klausner, J.M., Paterson, I.S., Golman, G., Kobzik, L., Valeri, C.R., Shepro, D. and Hechtman, H.B. (1989c). Thomboxane A_2 mediates increased pulmonary microvascular permeability following limb ischemia. Circ. Res. 64, 1178–1189.

Kloner, R.A., Giacomelli, F., Alker, K.J., Hale, S.L., Mathews, R. and Bellows, S. (1991). Influx of neutrophils into the walls of large epicardial coronary arteries in response to ischemia/reperfusion. Circulation 84, 1758–1772.

Kofsky, E.R., Julia, P.L., Buckberg, G.D., Quillen, J.E. and Acar, C. (1991). Studies of controlled reperfusion after ischemia. XXII. Reperfusate composition: Effects of leukocyte depletion of blood and blood cardioplegic reperfusates after acute coronary occlusion. J. Thorac. Cardiovasc. Surg. 101, 350–359.

Korthuis, R.J., Grisham, M.B., Zimmerman, B.J., Granger, D.N. and Taylor, A.E. (1988). Vascular injury in dogs during ischemia/reperfusion: improvement with ATP-$MgCl_2$ pretreatment. Am. J. Physiol. 254, H702–H708.

Korthuis, R.J., Carden, D.L., Kvietys, P.R., Shepro, D. and Fuseler, J. (1991). Phalloidin attenuates post-ischemic neutrophil infiltration and increased microvascular permeability. J. Appl. Physiol. 71, 1261–1269.

Kubes, P., Suzuki, M. and Granger, D.N. (1990a). Platelet activating factor-induced microvascular dysfunction: role of adherent leukocytes. Am. J. Physiol. 258, G158–G163.

Kubes, P., Suzuki, M. and Granger, D.N. (1990b). Modulation of PAF-induced leukocyte adherence and increased microvascular permeability. Am. J. Physiol. 259, G859–G864.

Kubes, P., Ibbotson, G., Russell, J., Wallace, J.L. and Granger, D.N. (1990c). Role of platelet activating factor in ischemia/reperfusion-induced leucocyte adherence. Am. J. Physiol. 259, G300–G305.

Kubes, P., Grisham, M.B., Barrowman, J.A., Gaginella, T. and Granger, D.N. (1991a). Leukocyte-induced vascular protein leakage in cat mesentery. Am. J. Physiol. 261, H1872–H1879.

Kubes, P., Suzuki, M. and Granger, D.N. (1991b). Nitric oxide: an endogenous modulator of leucocyte adhesion. Proc. Natl Acad. Sci. USA 88, 4651–4655.

Kurose, I., Anderson, D.C., Miuosaka, M., Tamatani, T., Paulson, J.C., Todd, R.F., Rusche, J.R. and Granges, D.N. (1994). Molecular determinants of reperfusion-induced leukocyle adhesion and vasaclar protein leakage, Circ. Res. 74, 336–343.

Kurtel, H., Tso, P. and Granger, D.N. (1992). Granulocyte accumulation in postischemic intestine: role of leukocyte adhesion glycoprotein CD11/CD18. Am. J. Physiol. 262: G878–G882.

Lawrence, M.B., McIntire, L.V. and Esken, S.G. (1987). Effect of flow on polymorphonuclear leukocyte endothelial cell adhesion. Blood 75, 227–237.

Lefer, A.M. (1989). Significance of lipid mediators in shock states. Circ. Shock 27, 3–12.

Lehr, H.A., Guhlmann, A., Nolte, D., Keppler, D. and Messmer, K. (1991a). Leukotrienes as mediators in ischemia-reperfusion injury in a microcirculation model in the hamster. J. Clin. Invest. 87, 2036–2041.

Lehr, H.-A., Hubner, C., Nolte, D., Kohlschutter, A. and Messmer, K. (1991b). Dietary fish oil blocks the micro-circulatory manifestations of ischemia-reperfusion injury in striated muscle in hamsters. Proc. Natl Acad. Sci. USA 88, 6726–6730.

Le Mauff, B., Hourmant, M., Rougier, J.-P., Hirn, M., Dantal, J., Baatard, R., Cantarovich, D., Jacques, Y. and Soulillou, J.P. (1991). Effect of anti-LFA1 (CD11a) monoclonal antibodies in acute rejection in human kidney transplantation. Transplantation 52, 291–296.

Lewis, M.S., Whatley, R.E., Cain, P., McIntyre, T.M., Prescott, S.M., and Zimmerman, G.A. (1988). Hydrogen peroxide stimulates the synthesis of platelet-activating factor by endothelium and induces endothelial cell-dependent neutrophil adhesion. J. Clin. Invest. 82, 2045–2055.

Lewis, R.E. and Granger, H.J. (1988). Diapedesis and the permeability of venous microvessels to protein macromolecules: the impact of leukotriene β_4. Microvasc. Res. 35, 27–47.

Ley, K. and Gaehtgens, P. (1991). Endothelial, not hemodynamic, differences are responsible for preferential leukocyte rolling in rat mesenteric venules. Circ. Res. 69: 1034–1041.

Linas, S.L., Whittenberg, D. and Repine, J.E. (1991). Role of neutrophil derived oxidants and elastase in lipopolysaccharide-mediated renal injury. Kid. International 39, 618–623.

Lucchesi, B.R., Mickelson, J.K., Homeister, J.W. and Jackson, C.V. (1987). Interaction of formed elements of blood with the coronary vasculature *in vivo*. Fed. Proc. 46, 63–72.

Ma, X.-L., Tsao, P.S. and Lefer, A.M. (1991). Antibody to CD-18 exerts endothelial and cardiac protective effects in myocardial ischemia and reperfusion. J. Clin. Invest. 88, 1237–1243.

Mangino, M.J., Anderson, C.B., Murphy, M.K., Brunt, E. and Turk, J. (1989). Mucosal arachidonate metabolism and intestinal ischemia-reperfusion injury. Am. J. Physiol. 257, G299–G307.

Matsuda, T., Rubinstein, I., Robbins, R.A., Koyama, S., Joyner, W.L. and Rennard, S.I. (1991). Role of neutrophils in endotoxin-mediated microvascular injury in hamsters. J. Appl. Physiol. 71, 307–313.

McEver, R.P. (1991). Selectins: Novel adhesion receptors that mediate leukocyte adhesion during inflammation. Thromb. Haem. 65, 223–228.

Mileski, W.J., Winn, R.K., Vedder, N.B., Pohlman, T.H., Harlan, J.M. and Rice, C.L. (1990). Inhibition of CD18-dependent neutrophil adherence reduces organ injury after hemorrhagic shock in primates. Surgery 108, 206–212.

Mileski, W.J., Winn, R.K., Harlan, J.M. and Rice, C.L. (1991). Transient inhibition of neutrophil adherence with the anti-CD18 monoclonal antibody 60.3 does not increase mortality rates in abdominal sepsis. Surgery 109, 497–501.

Mitsos, S.E., Askew, T.E., Fantone, J.C., Kunkel, S.L., Abrams, G.D., Shork, A. and Lucchesi, B.R. (1986). Protective effects of N-2-mercaptopropionyl glycine against myocardial reperfusion injury after neutrophil depletion in the dog: evidence for the role of intracellular-derived free radicals. Circulation 73, 1077–1086.

Montrucchio, G., Alloatti, G., Tetta, C., De Luca, R., Saunders, R.N., Emanuelli, G. and Camussi, G. (1989). Release of platelet-activating factor from ischemic-reperfused rabbit heart. Am. J. Physiol. 256, H1236–H1246.

Morisaki, T., Goya, T., Toh, H., Nishihara, K. and Torisu, M. (1991). The anti Mac-1 monoclonal antibody inhibits neutrophil sequestration in lung and liver in a septic murine model. Clin. Immunol. Immunopathol. 61, 365–375.

Mozes, T., Braquet, P. and Filip, J. (1989). Platelet-activating factor: an endogenous mediator of mesenteric ischemia-reperfusion induced shock. Am. J. Physiol. 257, R872–R877.

Mullane, K. (1991). Neutrophil and endothelial changes in reperfusion injury. TCM 1, 282–289.

Mullane, K.M., Read, N., Salmon, J.A. and Moncada, S. (1984). Role of leukocytes in acute myocardial infarction in anesthetized dogs: relationship to myocardial salvage by anti-inflammatory drugs. J. Pharmacol. Exp. Therap. 228, 510–522.

Munro, J.M., Pober, J.S. and Cotran, R.S. (1991). Recruitment of neutrophils in the local endotoxin response: association with *de novo* endothelial expression of endothelial leukocyte adhesion molecule-1. Lab. Invest. 64, 295–299.

Nathan, C.F. (1987). Neutrophil activation on biological surfaces: massive secretion of hydrogen peroxide in response to products of macrophages and lymphocytes. J. Clin. Invest. 80, 1550–1560.

Nathan, C.F. (1989). Respiratory burst in adherent human neutrophils: triggering by colony stimulating factors CSF-GM and CSF-G. Blood 73, 301–306.

Nathan, C. and Sporn, M. (1991). Cytokines in context. J. Cell Biol. 113, 981–986.

Nolte, D., Lehr, H.-A. and Messmer, K. (1991a). Adenosine inhibits postischemic leukocyte-endothelium interaction in post-capillary venules of the hamster. Am. J. Physiol. 261, H651–H655.

Nolte, D., Lehr, H.-A., Sack, F.-U. and Messmer, K. (1991b). Reduction of post-ischemic reperfusion injury by the vaso-active drug buflomedil. Blood Vessels 28, 8–14.

Oliver, M.G., Specian, R.D., Perry, M.A. and Granger, D.N. (1991). Morphologic assessment of leukocyte-endothelial cell interactions in mesenteric venules subjected to ischemia and reperfusion. Inflammation 15, 331–346.

Olof, P., Hellberg, A., Kallskog, O.T., Ojteg, G. and Wolgast, M. (1990). Peritubular capillary permeability and intra-vascular RBC aggregation after ischemia: effects of neutrophils. Am. J. Physiol. 258, F1018–F1025.

Oseas, R., Yang, H., Baehner, R.L. and Boxer, L.A. (1981). Lactoferrin: a promoter of polymorphonuclear leukocyte adhesiveness. Blood 57, 939–945.

Patel, K. D., Zimmerman, G.A., Prescott, S.M., McEver, R.P. and McIntyre, T.M. (1991). Oxygen radicals induce human endothelial cells to express GMP-140 and bind neutrophils. J. Cell Biol. 112, 749–759.

Paterson, I.S., Klausner, J.M., Goldman, G., Welbourn, R., Alexander, J.S., Shepro, D. and Hechtman, H.B. (1989). The endothelial cell cytoskeleton modulates extravascular polymorphonuclear leukocyte accumulations *in vivo* Microvasc. Res. 38: 49–56.

Perez, H.D., Ohtani, O., Banda, D., Ong, R., Fukuyama, K. and Goldstein, I.M. (1983). Generation of biologically active complement-(C5) derived peptides by cathepsin H. J. Immunol. 131, 397–402.

Perry, M.A. and Granger, D.N. (1991). Role of CD11/CD18 in shear rate-dependent leukocyte-endothelial cell interactions in cat mesenteric venules. J. Clin. Invest. 87, 1798–1804.

Petrone, W.F., English, D.K., Wong, K. and McCord, J.M. (1980). Free radicals and inflammation: superoxide-dependent activation of a neutrophil chemotactic factor in plasma. Proc. Natl Acad. Sci. USA 77, 1150–1163.

Redl, H., Dinges, H.P., Buurman, W.A., van der Linden, C.J., Pober, J.S., Cotran, R.S. and Schlag. G. (1991). Expression of endothelial leukocyte adhesion molecule-1 in septic but not traumatic/hypovolemic shock in the baboon. Am. J. Pathol. 139, 461–466.

Reilly, P.M., Schiller, H.J. and Bulkley, G.B. (1991). Pharmacological approach to tissue injury by free radicals and other reactive oxygen metabolites. Am. J. Surgery 161, 488–503.

Romson, J.L., Hook, B.G., Rigot, V.H., Schork. M.A., Swanson, D.P. and Lucchesi, B.R. (1982). The effect of ibuprofen on accumulation of ^{111}indium labelled platelets and leukocytes in experimental myocardial infarction. Circulation 66, 1002–1011.

Romson, J.L., Hook, B.G., Kunkel, S.L., Abrams, G.D., Schork, M.A. and Lucchesi, B.R. (1983). Reduction of the extent of ischemic myocardial injury by neutrophil depletion in the dog. Circulation 67, 1016–1023.

Rosengren, S. and Arfors, K.E. (1990). Neutrophil-mediated vascular leakage is not suppressed by leukocyte elastase inhibitors. Am. J. Physiol. 259, H1288–H1294.

Rossen, R.D., Swain, J.L., Michael, L.H., Weakly, S., Giannini, E. and Entman, M.L. (1985). Selective accumulation of the first component of complement and leukocytes in ischemic cannine muscle: a possible initiator of an extra myocardial mechanism of ischemic injury. Circ. Res. 57, 119–130.

Rossen, R.D., Micheal, L.H., Kagiyama, A., Savage, H.E., Hanson, G., Reisbery, J.N., Moake, J.N., Kim, S.H., Weakly, S., Giannini, E. and Entman, M.L. (1988). Mechanism of complement activation following coronary artery occlusion: evidence that myocardial ischemia caused release of constituents of myocardial subcellular origin which complex with the first component of complement. Circ. Res. 62, 572–584.

Rubin, B.B, Smith, A., Liauw, S., Isenman, D., Romaschin, A.D. and Walker, P.M. (1990). Complement activation and white cell sequestration in postischemic skeletal muscle. Am. J. Physiol. 259, H525–H531.

Schmeling, D.J., Caty, M.G., Oldham, K.T., Guice, K.S. and Hinshaw, D.B. (1989). Evidence for neutrophil-related acute lung injury after intestinal ischemia-reperfusion. Surgery 106, 195–202.

Schmid-Schonbein, G.W., Usami, S., Skalak, R. and Chien, S. (1980). The interaction of leukocytes and erythrocytes in capillary and post-capillary vessels. Microvasc. Res. 19, 45–70.

Schoenberg, M.H., Fredholm, B.B., Haglund, U., Younes, M., Sellin, D. and Schildberg, F.W. (1985). Studies on oxygen radical mechanism involved in the small intestinal reperusion damage. Acta Physiol. Scand. 124, 581–589.

Schoenberg, M.H., Poch, B., Younes, M., Schwartz, A., Baczako, K., Lundberg, C., Haglund, U. and Beger, H.G. (1991). Involvement of neutrophils in postischaemic damage to the small intestine. Gut 32, 905–912.

Schott, R.J., Nao, B.S., McClanahan, T.B., Simpson, P.J., Stirling, M.C., Todd, R.F. III, and Gallagher, K.P. (1989). (F(ab′)2 fragments of anti-Mo1 (904) monoclonal antibodies do not prevent myocardial stunning. Circ. Res. 65, 1112.

Sedmak, D.D. and Orosz, C.G. (1991). The role of vascular endothelial cells in transplantation. Arch. Pathol. Lab. Med. 115, 260–265.

Seewaldt-Becker, E., Rothlein, R. and Dammgen, J.W. (1989). In "Leukocyte Adhesion Molecules" (eds. T. Springer, D.C. Anderson, A. Rosenthal, and R. Rothlein), pp. 138–148, Springer-Verlag, New York.

Shappell, S.B., Toman, C., Anderson, D.C., Taylor, A.A., Entman, M.L. and Smith, C.W. (1990). Mac-1 (CD11b/CD18) mediates adherence-dependent hydrogen peroxide production by human and canine neutrophils. J. Immunol. 144, 2702–2711.

Sharer, S.R., Winn, R.K., Murray, C.E., Harlan, J.M. and Rice, C.L. (1991). A CD18 monoclonal antibody increases the severity of subcutaneous abscess formation after high-dose *Staphylococcus aureus* injection in rabbits. Surgery 110, 213–220.

Sheppard, B.C., Fraker, D.L. and Norton, J.A. (1989). Prevention and treatment of endotoxin and sepsis lethality with recombinant human tumor necrosis factor. Surgery 106, 156–162.

Shiga, Y., Onodera, H., Kogure, K., Yamasaki, Y., Yashima, Y., Syozuhara, H. and Sendo, F. (1991). Neutrophil as a mediator of ischemic edema formation in the brain. Neurosci. Let. 125, 110–112.

Shingu, M. and Nobunaga, M. (1984). Chemotactic activity generated in human serum from the fifth component of complement by hydrogen peroxide. Am. J. Pathol. 117, 201–206.

Simpson, P.J., Mickelson, J., Fantone, J.C., Gallagher, K.P. and Lucchesi, B.R. (1987a). Iloprost inhibits neutrophil function *in vitro* and *in vivo* and limits experimental infarct size in canine heart. Circ. Res. 60, 666–673.

Simpson, P.J., Mitsos, S.E., Ventura, A., Gallagher, K.P., Fantone, J.C., Abrams, G.D., Shork, M.A. and Lucchesi, B.R. (1987b). Prostacyclin protects ischemic reperfused myocardium in the dog by inhibition of neutrophil activation. Am. Heart J. 113, 129–137.

Simpson, P.J., Todd III, R.F., Fantone, J.C., Mickelson, J.K., Griffen, J.D. and Griffen, B.R. (1988). Reduction of experimental canine myocardial reperfusion injury by a monoclonal antibody (anti-Mo1, anti-CD11b) that inhibits leukocyte adhesion. J. Clin. Invest. 81, 624–629.

Simpson, P.J., Todd III, R.F., Mickelson, J.K., Fantone, J.C., Gallagher, K.P., Lee, K.A., Tamura, Y., Cronin, M. and Lucchesi, B.R. (1990). Sustained limitation of myocardial reperfusion injury by a monoclonal antibody that alters leukocyte function. Circulation 81, 226–237.

Smith, C.W. (1990). Molecular determinants of neutrophil-endothelial cell adherence reactions. Am. J. Respir. Cell. Molec. Biol. 2. 487–499.

Smith, S.M., Rutili, L.H., Perry, M.A., Grisham, M.B., Arfors, K.-E., Granger, D.N. and Kvietys, P.R. (1987). Role of neutrorphils in hemorrhagic shock-induced gastric mucosal injury in the rat. Gastroenterology 93, 466–471.

Smith, J.K., Grisham, M.B., Granger, D.N. and Korthuis, R.J. (1989). Free radical defense mechanisms and neutrophil infiltration in post-ischemic skeletal muscle. Am. J. Physiol. 256, H789–H793.

Smith, C.W., Entman, M.L., Lane, C.L., Beaudet, A.L., Ty, T.I., Youker, K., Hawkins, H.K. and Anderson, D.C. (1991). Adherence of neutrophils to canine cardiac myocytes *in vitro* is dependent on intercellular adhesion molecule-1 J. Clin. Invest. 88, 1216–1223.

Soussi, B., Idstrom, J.-P., Schersten. T. and Bylund-Fellenius, A.-C. (1990). Cytochrome c oxidase and cardiolipin alterations in response to skeletal muscle ischemia and reperfusion. Acta Physoiol. Scand. 138, 107–114.

Springer, T., Anderson, D.C., Rosenthal, A. and Rothlein, R. (1989). In "Leukocyte Adhesion Molecules". Springer-Verlag, New York.

Strock, P.E. and Majno, G.M. (1969). Vascular responses to experimental tourniquet ischemia. Surg. Gynecol. Obstet. 129, 309–318.

Suzuki, M., Inauen, W., Kvietys, P.R., Grisham, M.B., Meininger, C.J., Schelling, Granger, H.J., and Granger, D.N. (1989). Superoxide mediates reperfusion-induced leukocyte-endothelial cell interactions. Am. J. Physiol. 257, H1740–H1745.

Suzuki, M., Grisham, M.B. and Granger, D.N. (1991a). Leukocyte-endothelial cell interactions: role of xanthine oxidase-derived oxidants. J. Leukocyte Biol. 50, 488–494.

Suzuki, M., Asako, H., Kubes, P., Jennings, S., Grisham, M. and Granger, D.N. (1991b). Neutrophil-derived oxidants promote leukocyte adherence in post-capillary venules. Microvasc. Res. 42, 125–138.

Takei, Y., Marzi, I., Gao, W., Gores, G.J., Lemasters, J.J. and Thurman, R.G. (1991). Leukocyte adhesion and cell death following orthotopic liver transplantation in the rat. Transplantation 51, 959–965.

Tonnenson, M.G. (1989). Neutrophil-endothelial cell interactions: Mechanisms of neutrophil adherence to vascular endothelium. J. Invest. Dermatol. 93, 53S–58S.

Turnage, R.H., Bagnasco, J., Berger, J., Guice, K.S., Oldham, K.T. and Hinshaw, D.B. (1991). Hepatocellular oxidant stress following intestinal ischemia-reperfusion injury. J. Surg. Res. 51, 467–471.

Vedder, N.B., Winn, R.K., Rice, C.L. and Harlan, J.M. (1989). In "Perspectives in Shock Research: Metabolism, Immunology, Mediators, and Models". pp. 181–191. Alan R. Liss, Inc, New York.

Vedder, N.B., Winn, R.K., Rice, C.L., Chi, E.Y., Arfors, K.-E. and Harlan, J.M. (1988). A monoclonal antibody to the adherence-promoting leukocyte glycoprotein, CD18, reduces organ injury and improves resuscitation in rabbits. J. Clin. Invest. 81, 939–944.

Vedder, N.B., Fouty, B.W., Winn, R.K., Harlan, J.M. and Rice, C.L. (1989b). Role of neutrophils in generalized perfusion injury associated with resuscitation from shock. Surgery 106, 509–516.

Vedder, N.B., Winn, R.K., Rice, C.L., Chi, E.Y., Arfors, K.-E. and Harlan, J.M. (1990a). Inhibition of leucocyte adherence by anti-CD18 monoclonal antibody attenuates reperfusion injury in the rabbit ear. Proc. Natl Acad. Sci. USA 87, 2543–2646.

Venezio, F.R., Westenfelder, G.O. and Phair, J.P. (1982). The adherence of polymorphonuclear leucocytes in patients with sepsis. J. Infect. Dis. 145, 351–357.

Villringer, A., Dimagi, U., Them, A., Schurer, L., Krombach, F. and Einhaupl, K.M. (1991). Imaging of leucocytes with the rat brain cortex *in vivo*. Microvasc. Res. 42, 305–315.

Vogt, W., von Zabern, I., Hesse, D., Nolte, R. and Haller, Y. (1987). Generation of an activated form of human C5 (C5b-like C5) by oxygen radicals. Immunol. Letters 14, 209–215.

Walsh, C.J., Carey, P.D., Cook, D.J., Bechard, D.E., Fowler, A.A. and Sugerman, H.J. (1991). Anti-CD18 antibody attenuates neutropenia and alveolar-capillary membrane injury during gram-negative sepsis. Surgery 110, 205–212.

Walsh, C.J., Leeper-Woodford, S.K., Carey, P.D., Cook, D.J., Bechard, D.E., Fowler, A.A. and Sugerman, H.J. (1991). CD18 adhesion receptors, tumor necrosis factor, and neutropenia during septic lung injury. J. Surg. Res. 50, 323–329.

Wee, S.-L., Cosimi, A.B., Preffer, F.I., Faanes, R., Conti, D. and Colvin, R.B. (1991). Functional consequences of anti-ICAM-1 (CD54) in cynomolgus monkeys with renal allografts. Transplant. Proc. 23, 279–280.

Weiss, S. (1986). Oxygen, ischemia, and inflammation. Acta Physiol. Scand. 126 (Suppl. 548), 9–38.

Weiss, S.J. (1989). Tissue destruction by neutrophils. N. Engl. J. Med. 320, 365–376.

Weiss, S.J., Curnute, J.T. and Regiani, S. (1986). Neutrophil-medated solubilization of the subendothelial matrix, oxidative and non-oxidative mechanisms of proteolysis used by normal and chronic granulomatous disease phagocytes. J. Immunol. 136, 636–641.

Welbourn, R., Goldman, G., Kobzik, L., Paterson, I., Valeri, C.R., Shepro, D. and Hechtman, H.B. (1990). Neutrophil adherence receptors (CD18) in ischemia. Dissociation between quantitative cell surface expression and diapedesis mediated by leukotriene B_4 J. Immunol. 145, 1906–1911.

Welbourn, C.R.B., Goldman, G., Paterson, I.S., Shepro, D. and Hechtman, H.B. (1991a). Pathophysiology of ischaemia reperfusion injury, central role of the neutrophil. Br. J. Surg. 78, 651–655.

Welbourn, R., Goldman, G., Hill, J., Lindsay, T., Shepro, D. and Hechtman, H. B. (1992). Role of neutrophil adherence receptors (CD) in lung permeability following lower torso ishemia. Circ. Res. 71, 82–86.

Weselcouch, E.O., Grove, R.I., Demusz C.D. and Baird, A.J. (1991). Effect of *in vivo* inhibition of neutrophil adherence on skeletal muscle function during ischemia in ferrets. Am. J. Physiol. 261, H1178–H1183.

Winn, R., Maunder, R., Chi, E. and Harlan. J. (1987). Neutrophil depletion does not prevent lung edema after endotoxin infusion in goats. J. Appl. Physiol. 62, 116–121.

Winterbourne, C.C. (1986). Myeloperoxidase as an effective inhibitor of hydroxyl radical production. Implications for the oxidative reactions of neutrophils. J. Clin. Invest. 78, 545–550.

Worthen, G.S., Smedley, L.A., Tonnenson, M.G., Ellis, D., Voelkel, N.F., Reeves, J.J. and Henson. P.M. (1987). Effects of shear stress on adhesive interaction between neutrophils and cultured endothelial cells. J. Appl. Physiol. 63, 2031–3041.

Worthen, G.S., Schwab, B., Elson, E.L. and Downey, G.P. (1989). Cellular mechanics of stimulated neutrophils: stiffening of cells induces retention in pores *in vitro* and lung capillaries *in vivo*. Science 245, 183–186.

Yamakawa, T., Yamaguchi, S., Niimi, H., Sugiyama. I. (1987). White blood cell plugging and blood flow maldistribution in the capillary network of cat cerebral cortex in acute hemorrhagic hypotension: an intravital microscopic study. Circ. Shock 22, 323–332.

Yoshida, N., Granger, D.N., Anderson, D.C., Rothlein, R., Lane, C. and Kvietys, P.R. (1992). Anoxia/reoxygenation induced neutrophil adherence to cultured endothelial cells. Am. J. Physiol. 262, H1891–H1898.

Ziegler, E.J., McCutchan, J.A., Fierer, J., Glauser, M.P., Sadoff, J.C., Douglas, H. and Braude, A.I. (1982). Treatment of gram-negative bacteremia and shock with human antiserum to a mutant *Escherichia coli*. N. Engl. J. Med. 307, 1225–1230.

Zimmerman, B.J. and Granger, D.N. (1990). Reperfusion-induced leukocyte infiltration: role of elastase. Am. J. Physiol. 259, H390–H394.

Zimmerman, B.J., Guillory, D.J., Gaginella, T., Grisham, M.B. and Granger, D.N. (1990a). Role of leukotriene B_4 in granulocyte infiltration into the postischemic feline intestine. Gastroenterology 99, 1358–1363.

Zimmerman, B.J., Grisham, M.B. and Granger, D.N. (1990b). Role of oxidants in ischemia/reperfusion-induced granulocyte infiltration. Am. J. Physiol. 258, G185–G190.

9. Lung Inflammation

Craig D. Wegner

1. Introduction

With the exception of some acute lung injuries and rare cases of pulmonary vasculitis, the major foci of inflammation in lung diseases is in or just adjacent to the airspaces. Asthma and cystic fibrosis are characterized by inflammation within and surrounding the epithelium lining the larger and more central airways. In respiratory viral infections, peribronchiolar infiltrates impair gas exchange to a greater degree, and thus are more troublesome, than the cytopathetic effects of the virus. Finally, pneumonia, acute lung injuries and pulmonary oxygen toxicity exhibit an often life threatening alveolitis. As depicted in Fig. 9.1, adhesion molecules control the retention, locomotion and activation of leucocytes at several (at least four) phases during their transition from a quiescent state circulating in the blood to their effector action at inflammatory foci.

In this chapter, the adhesion molecules, processes and concepts possibly involved at each of these phases based on studies performed *in vitro*, on systemic microcirculatory beds, and in organs other than the lungs will be briefly reviewed. Processes unique to the lung microvasculature and tissue will then be expounded on.

Adhesion Molecules
ISBN 0-12-741440-1

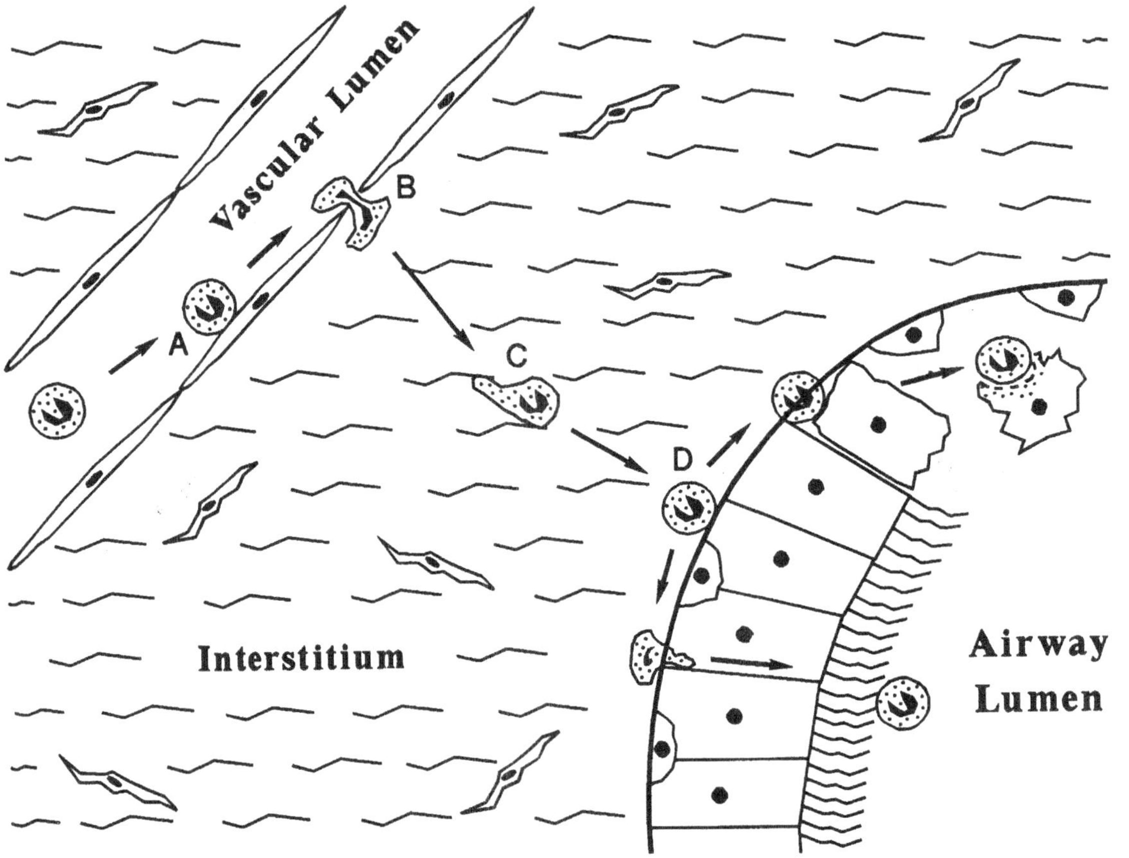

Figure 9.1 Four of the phases in the migration of leucocytes from the circulation to the airway lumen: (A) the initial adhesion, "rolling", of leucocytes to the endothelium (margination); (B) trans-endothelial migration; (C) migration through the interstitium; and (D) retention and/or activation at, migration through, or leucocyte-mediated desquamation of, the airway epithelium. Each of these phases utilizes a different set of adhesion receptors and ligands (see text and Fig. 9.2).

With this as background, the potential role of specific adhesion molecules in representative inflammatory lung diseases will be examined with the aid of results from animal models and human tissue (biopsy) samples. Finally and in conclusion, the implications of our present understanding of cell adhesion pathways in lung inflammation, approaches to the identification of inhibitors of these pathways, and the plausible effects of such inhibitors on host defence are considered.

2. *Microvascular Adhesion: Leucocyte Margination and Diapedesis*

2.1 Systemic Microvasculature (Post-Capillary Venules): Overview[1]

By studying the adhesion of neutrophils to confluent monolayers of human umbilical vein endothelial cells (HUVECs), planar membranes containing specific purified adhesion molecule(s), or monolayers of COS cells transfected with specific adhesion molecules, followed by confirming intravital microscopy of various systemic microvascular post-capillary venules, the molecules and steps involved in neutrophil margination have been elucidated (Fig. 9.2A). At sites adjacent to foci of tissue inflammation, the expression of members of two families of adhesion molecules are up-regulated on the endothelium, presumably by cytokine stimulation, to direct leucocyte margination and eventual diapedesis at these sites. Members of the selectin family, P-selectin and E-selectin, so named because of their carbohydrate or "lectin" binding domain, cause a rapid but weak adhesion resulting in slowing or "rolling" of neutrophils along the endothelial lined vessel walls (Bevilacqua *et al.*, 1989; Lasky *et al.*, 1989; Siegelman *et al.*, 1989; Geng *et al.*, 1990; Lawrence and Springer, 1991; von Andrian *et al.*, 1993). The counter-receptors for these molecules on neutrophils consist of a variety of sialylated glycoproteins

[1] See Chapter 3 for more details.

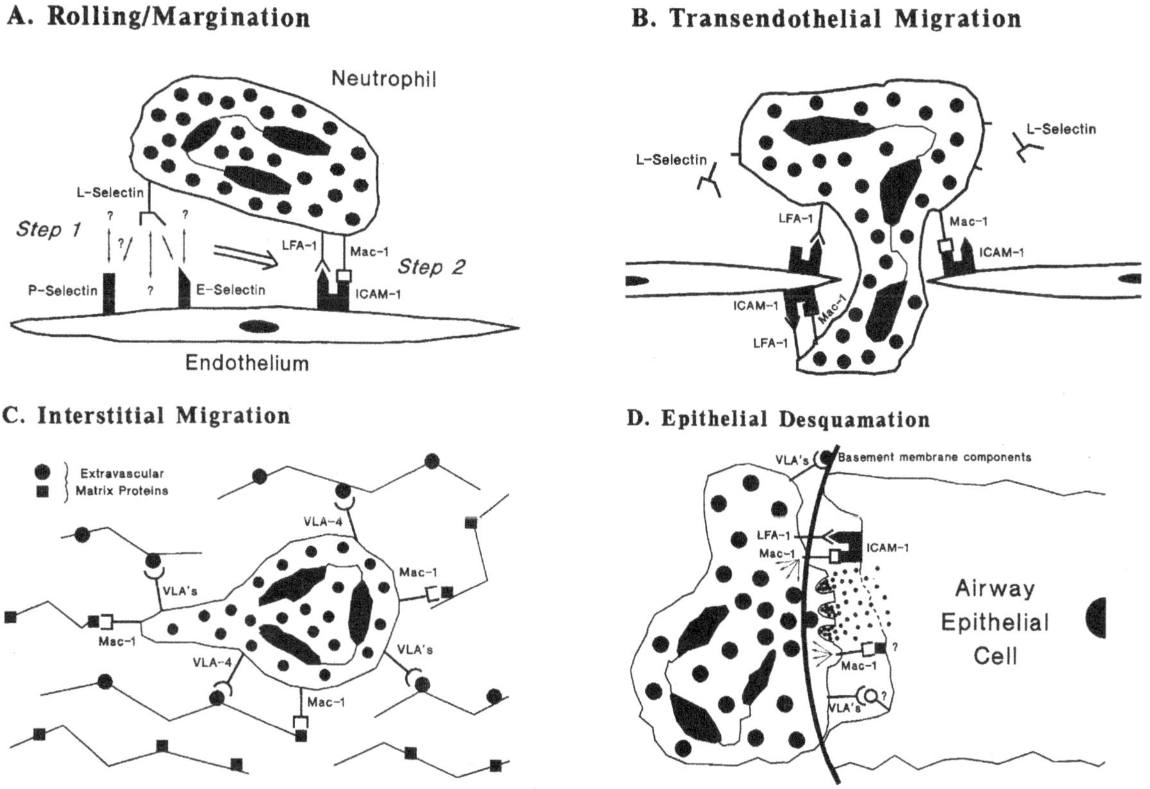

Figure 9.2 Adhesion molecules and steps believed to regulate granulocyte (A) margination and (B) diapedesis in systemic microvascular beds. Possible adhesion molecules that may regulate granulocyte (C) migration through and priming in the interstitium and (D) retention, activation and tissue mediated injury proximal to the airway epithelium.

including L-selectin (Lowe *et al.*, 1990; Imai *et al.*, 1991; Kishimoto *et al.*, 1991; Ley *et al.*, 1991; Spertini *et al.*, 1991; Erbe *et al.*, 1993).

Firm adhesion and stopping, "margination", is then conferred by the slower in onset but stronger adhesion mediated by a member of the immunoglobulin supergene family, intercellular adhesion molecule-1 (ICAM-1), acting as a ligand for the neutrophil β_2 integrins CD11a/CD18 (also termed LFA-1) and CD11b/CD18 (also termed Mac-1, Mo-1, or CR3) (von Andrian *et al.*, 1993). Binding of these integrin receptors to their ligands requires the conversion of these receptors to an activation-specific phenotype (change in conformation?) (Hibbs *et al.*, 1991; Diamond and Springer, 1993b). It is presumed that the initial slowing/rolling mediated by the selectins facilitates the conversion of a subpopulation of the neutrophil integrins to this activation-specific (increased binding affinity) phenotype by stimulation of the neutrophil via: (i) E-selectin mediated adhesion (Lo *et al.*, 1991); (ii) endothelial bound PAF (Kuijpers *et al.*, 1992; Casale *et al.*, 1993); (iii) endothelial secreted IL-8 (Kuijpers *et al.*, 1992); and/or (iv) diffusion of chemoattractants from the surrounding inflamed tissue (Kishimoto *et al.*, 1989). Neutrophil shedding of L-selectin (Kishimoto *et al.*, 1989) and translocation of CD11b/CD18 (Mac-1) from intracellular pools across the surface (from head to tail) (Hughes *et al.*, 1992) then control trans-endothelial migration (Fig. 9.2B).

Although less well studied, these same molecules, steps and processes are probably also utilized by other leucocyte subtypes (Rosen and Gordon, 1987; Imai *et al.*, 1991; Picker *et al.*, 1991b; Shimizu *et al.*, 1991; Spertini *et al.*, 1991; Wegner and Wallace, 1992c; Kansas *et al.*, 1993). The only notable addition is that, unlike the neutrophil, most other leucocyte subtypes (most notably eosinophils and lymphocytes) also express the β_1 integrin VLA-4 (CD49d/CD29) which binds to the immunoglobulin supergene family member vascular cell adhesion molecule-1 (VCAM-1) expressed on inflamed endothelium to, teaming with Mac-1/ICAM-1 and LFA-1/ICAM-1 (CD18/ICAM-1), mediated firm adhesion and possibly diapedesis (Masinovsky *et al.*, 1990; Bochner *et al.*, 1991; Kyan-Aung *et al.*, 1991; Shimizu *et al.*, 1991; Thornhill *et al.*, 1991; Walsh *et al.*, 1991; Schleimer *et al.*, 1992).

The complete dependence of neutrophils on the β_2 integrins for diapedesis *in vivo* is exemplified in the CD18 (LFA-1) deficient leucocyte adhesion deficiency (LAD) type 1 patients (Anderson *et al.*, 1985; Anderson and Springer, 1987). In severe phenotypes of these patients, neutrophils and monocytes do not accumulate at sites of infection despite a moderate, chronic, neutrophilic leucocytosis that becomes striking during episodes of acute infection. The importance of another pathway, possibly VLA-4/VCAM-1, for the diapedesis of other leucocyte subtypes is also illustrated by these patients since eosinophils and lymphocytes do emigrate sites of infection. Similarly, the importance of the selectin molecules in leucocyte migration is illustrated by patients unable to assemble sialylated glycoproteins, LAD type 2 (Etzioni *et al.*, 1992).

2.2 PULMONARY MICROVASCULATURE

2.2.1 Unique Capillary System: Adhesion Independent Margination

Unlike the systemic circulation where margination of leucocytes occurs almost exclusively in the post-capillary venules (Schmid-Schonbein *et al.*, 1980; Perry and Granger, 1991), the major site of the margination in the lungs is the capillary bed (Doerschuk *et al.*, 1987; Lein *et al.*, 1987). This maybe partially due to the much lower pressures and more pulsatile flow in the pulmonary circulation. However, the principal cause seems to be the smaller average diameter of the pulmonary capillaries and the extensive number of capillary segments (~60) that must be transversed from arteriole to venule (Hogg, 1987). The mean capillary diameter has been reported to be 2.0 μm in zone II conditions and 5.8 μm in zone III (Glazier *et al.*, 1964). The mean diameter of circulating neutrophils, monocytes and lymphocytes have been estimated to be 7.0 μm, 6.4 μm and 5.4 μm, respectively (Schmid-Schonbein *et al.*, 1985; Doerschuk *et al.*, 1990a). Thus, leucocytes must deform, in some cases substantially, to transit through the pulmonary capillaries resulting in a concentration of neutrophils with respect to red cells within the pulmonary circulation to produce an adhesion independent "marginated pool". Stimulation of circulating neutrophils with a variety of activating agents increases their retention with the lung initially (0–2 min) due to an adhesion independent decrease in their deformability (Schmid-Schonbein *et al.*, 1985; Worthen *et al.*, 1989) followed by an increased CD18-dependent adhesiveness (Doerschuk, 1992).

Thus, it is interesting to speculate that the diameter which produced slowing of leucocytes in the capillaries of the pulmonary circulation may make a selectin-mediated step unnecessary for their infiltration into the alveoli. In support of this theory, inhaled allergen-induced inflammation in primates has been reported to enhance ICAM-1 and E-selectin expression on blood vessels supplying the main stem bronchus, but only ICAM-1 expression in the peripheral lung units (Wegner *et al.*, 1992b).

2.2.2 CD18-Independent Migration: Alveolar Macrophage Regulated?

While the unconditional CD18 dependence of neutrophil emigration in systemic microvascular beds is universally supported by *in vitro* studies with HUVEC monolayers (Smith *et al.*, 1989; Furies *et al.*, 1991), *in vivo* studies in animal models (Harlan *et al.*, 1992), and severe phenotypes of LAD type 1 (CD18 deficiency) patients, neutrophil infiltration into the alveoli via a CD18 independent pathway has been identified in animals (Doerschuk *et al.*, 1990b) and was noted in the post-mortem histological examination of a patient with severe LAD type 1 (Harlan *et al.*, 1992). Two sets of elegant experiments in rabbits suggest that the presence of a significant population of alveolar macrophages in normal lungs may explain the lung specific CD18-independent pathway.

In the first set, the effect of an anti-CD18 monoclonal antibody (mAb) on the neutrophil accumulation in the alveoli and abdominal wall induced by various inflammatory stimuli was evaluated (Doerschuk *et al.*, 1990b). Neutrophil accumulation into the lungs induced by the intrabronchial instillation of *Escherichia coli* endotoxin or PMA was CD18-dependent while that induced by *Streptococcus pneumoniae* or hydrochloric acid was not. Neutrophil infiltration in the abdomen was CD18-dependent despite the stimuli. The CD18 independence of the *S. pneumoniae* induced lung neutrophil influx was conclusively demonstrated by showing that the abdominal influx induced by *S. pneumoniae* or the lung influx induced in the opposite lung by *E. coli* in dual challenged animals was completely CD18-dependent.

In the second set of experiments, a *S. pneumoniae* induced CD18-independent pathway was created in the peritoneum by the prior recruitment or injection of "primed" macrophages (Mileski *et al.*, 1990). The *E. coli* induced neutrophil infiltration remained CD18-dependent. Likewise, removal of these primed macrophages by washing the peritoneum returned the CD18 dependence of the *S. pneumoniae* induced influx. The macrophage-generated product(s) elicited by *S. pneumoniae* and the adhesion molecules involved in this CD18-independent pathway remain to be identified.

More recently, the neutrophil influx and associated late phase broncho-obstruction at 6–8 h post allergen inhalation in primates has been shown to be ICAM-1 (Gundel *et al.*, 1991b) and CD18 independent (R.H. Gundel, personal communication). Whether this involves the same macrophage generate product(s) and adhesion molecules, or E-selectin and hemorrhage (Wegner *et al.*, 1993b), also remains to be determined.

2.2.3 Arterioles Contribute Too!

As mentioned above, rolling, margination and subsequent diapedesis of leucocytes in the systemic microvasculature is confined to the post-capillary venules.

Observations using intravital microscopy under normal and retrograde flow conditions at various rates have demonstrated that adherence is always much greater in the venules than the arterioles (Perry and Granger, 1991). Thus, a site specific expression of endothelial adhesion molecules in the systemic microvascular beds is indicated.

Two preliminary but convincing observations suggest that this may not be the case in lung microvascular beds. First, using intravital microscopy through pleural skin windows to examine microvascular beds of the pulmonary circulation, rolling of leucocytes along vessel walls has been observed to commonly occur in the arterioles in addition to the venules (W.W. Wagner and C.M. Doerschuk, personal communication). In addition, we have noted VCAM-1 (but not ICAM-1 or E-selectin) expression on arterioles of the bronchial circulation in the central airways following repeat allergen inhalations in primates (unpublished data).

3. *Interstitial migration and priming*

After diapedesis, chemo-attractant gradients presumably direct leucocyte migration through the extravascular tissue to the foci of inflammation. This movement probably occurs through the interaction of β_1 integrins (very late activation antigen, VLAs) with tissue matrix components and cells (Fig. 9.2C) (Hemler, 1990). In addition, CD11b/CD18 (Mac-1) has been demonstrated to bind to fibrinogen (Wright *et al.*, 1988) and other yet unidentified matrix proteins (Wegner *et al.*, 1989; Wright *et al.*, 1989; Thompson and Matsushima, 1992; Diamond *et al.*, 1993a). In addition to locomotion, adhesion via the integrin Mac-1 or VLA-4 receptors have been shown to enhance neutrophil, basophil and eosinophil oxidative burst, mediator release and/or survival (Shappell *et al.*, 1990; Watanabe *et al.*, 1992; Anwar *et al.*, 1993). Thus, transit through the interstitium may contribute an important preparatory priming of leucocytes for activation at the foci of inflammation.

4. *Epithelium: The Site of Action*

The epithelium lining the airspaces of the lungs plays a vital role as the first line defence (protective barrier) against airborne particles, allergens, foreign proteins, chemicals, microbes and viruses. Congruently, the epithelium is the primary foci of most inflammatory responses and immune reactions in the lungs. Thus, knowledge of the expression and function(s) of the adhesion molecules that are found on the surface of these cells is essential to the understanding of host defence as well as the pathogenesis of many pulmonary diseases.

4.1 AIRWAYS

4.1.1 ICAM-1 Expression

Of the endothelial adhesion molecules mentioned above, only ICAM-1 is found on the surface of lung epithelial cells. Constitutive ICAM-1 expression is small on airway epithelial cells cultured to confluent monolayers *in vitro* but can be impressively enhanced by cytokine stimulation (Fig. 9.3) (Wegner *et al.*, 1990; Tosi *et al.*, 1992a; Churchill *et al.*, 1993). IFNγ is the strongest cytokine enhancer of epithelial ICAM-1 expression with IL-1, tumor necrosis factor α (TNFα), and IL-4 (but not LPS) also effective. The time course of enhanced ICAM-1 expression [onset at $\geqslant$2 h and peak at 12 to 16 h (Wegner *et al.*, 1991b)] is similar to that reported for HUVECs *in vitro* (Smith *et al.*, 1988) and human skin keratinocytes *in vivo* (Vejlsgaard *et al.*, 1989) and indicative of the required *de novo* protein synthesis. In primary tracheal epithelial cells as well as in true airway epithelial cell lines, ICAM-1 expression is not directly inhibited by steroids at non-toxic concentrations (Wegner *et al.*, 1993b). However, *in vivo* steroid treatment probable does minimize ICAM-1 expression by attenuating the generation of the cytokines that upregulated it.

The probable functions of airway epithelial ICAM-1 are presented in the next section.

4.1.2 Leucocyte Retention and Activation

Using immunohistochemical staining, the *in vivo* enhancement of airway epithelial ICAM-1 expression has been impressively demonstrated following repeated

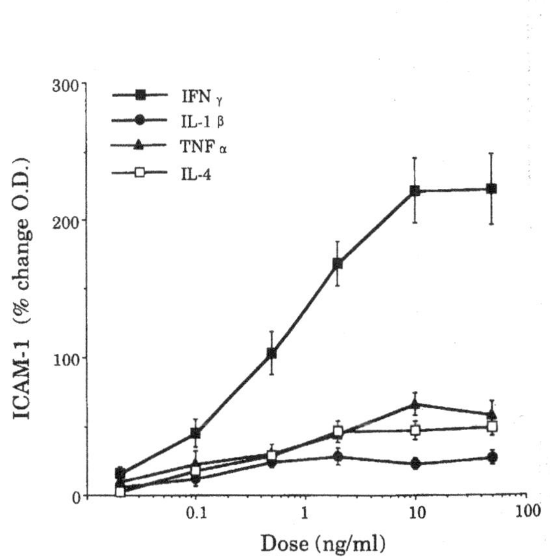

Figure 9.3 Dose response effects of 16 h of stimulation with inflammatory cytokines on ICAM-1 expression on confluent monolayers of the rhesus monkey bronchus epithelial cell line 4MBr-5. ICAM-1 is plotted as percentage change above basal expression (n $\geqslant$ 5). From (Churchill *et al.*, 1993) with permission.

allergen inhalation (i.e. associated with chronic inflammation) in primates (Wegner *et al.*, 1990; Wegner *et al.*, 1992b). ICAM-1 expression was most strikingly upregulated on the basilateral surface of the tracheal epithelium (compare A and D in Color Plate I. Interestingly, staining for CD11a, a receptor for ICAM-1 found on all leucocytes and thus simply a marker of leucocyte influx, revealed a marked leucocyte infiltrate just below the epithelial basement membrane and within the basilateral tracheal epithelium (i.e. just adjacent to the most pronounced ICAM-1 staining) (Color Plate I(B)). In similarly stained sections from an animal 20 min after a single allergen inhalation (i.e. an acute inflammation) a leucocyte emigration was also notable (Color Plate I(E)) but not accumulated (localized) adjacent to the epithelium which only weakly, if at all, expressed ICAM-1 (Color Plate I(D)). Staining with non-specific mouse serum (Color Plate I(C) and Color Plate I(F)) was used to demonstrate the specificity of the enhanced ICAM-1 staining.

These results strongly suggest a role for ICAM-1 in the retention of the infiltrating leucocytes (mostly eosinophils in this case) at the foci of inflammation, the airway epithelium. In support of this, human neutrophil adhesion to human airway epithelial cells *in vitro* has been shown to be enhanced and largely ICAM-1-dependent when epithelial ICAM-1 expression was induced by IL-1, TNFα or infection with para-influenza virus type-2 (Tosi *et al.*, 1992a; Tosi *et al.*, 1992b).

In addition to contributing to leucocyte retention, epithelial ICAM-1 and other yet undefined epithelial ligands, by engaging CD11a (LFA-1), CD11b (Mac-1), other leucocytes integrins (e.g. VLA-4), etc., probably also act to prime the leucocyte for enhanced mediated release and other effector functions. Adhesion via Mac-1 has been shown to enhance the respiratory burst (Shappell *et al.*, 1990) and, in separate studies, the killing of alveolar epithelial cells (Simon *et al.*, 1986) by neutrophils *in vitro*. An example of this principle is depicted in Fig. 9.2D where adhesion via Mac-1/ICAM-1, etc., provides both a stimulatory signal (spread arrows) for granulocyte degranulation as well as a protected micro-environment (pocket) between the two cells to amplify the direction and extracellular concentrations of the mediators released while limiting the effects on bystanders (i.e. targeted cell mediated cytotoxicity). The basilateral specific surface generation and expression of fibronectin by the airway epithelium (Wang *et al.*, 1990) may act similarly to prime eosinophils (Anwar *et al.*, 1993) and/or lymphocytes (Shimizu *et al.*, 1992; Masumoto and Hemler, 1993) via engagement of their VLA-4 receptors.

Finally, lymphocyte adhesion via CD11a (LFA-1), VLA-4, or CD28 to epithelial ICAM-1, fibronectin, and/or B7 (found on dendritic cells within the airway epithelium), respectively, provides a similar co-stimulatory signal to both T- and B-lymphocytes also engaging antigen via the class II major histocompatibility complex (Dustin and Springer, 1989; Singer, 1990; Koopman *et al.*, 1991; van Seventer *et al.*, 1991; Shimizu *et al.*, 1992; Wegner and Wallace, 1992c; Augustin *et al.*, 1993). When these "accessory" and co-stimulatory adhesions are blocked with mAbs *in vitro*, greater than 100-fold higher concentrations of antigen are required to initiate lymphocyte responses (Regnier-Vigouroux *et al.*, 1986; Martz, 1987) (see also Chapters 2 and 5).

4.2 ALVEOLI

Of all the organs/tissues examined to date by immunostaining, the acinus of the lungs contain by far the highest constitutive ICAM-1 expression (Wegner *et al.*, 1992a; Christensen *et al.*, 1993; Kang *et al.*, 1993). Electron microscopy and immunogold staining of normal mouse lung tissue has shown that the majority of this constitutive expression is on the luminal (airspace) surface of type I pneumocytes, with little if any on the subliminal surface, the type II pneumocytes or capillary endothelium (Kang *et al.*, 1993). Detailed morphometric evaluation of the luminal distribution of ICAM-1 on these cells demonstrated its graded escalation toward interepithelial junctions. This suggests, although yet to be proven, that in the normal lung, constitutive alveolar ICAM-1 expression may serve an anti-inflammatory haptotaxis function of directing infiltrated leucocytes out of the gas exchange units.

Alveolar epithelial ICAM-1 expression can also be enhanced *in vivo*, as shown after repeated allergen inhalations in primates (Wegner *et al.*, 1992b), 48 h of pure oxygen breathing in mice (Wegner *et al.*, 1992a), or infection with respiratory syncytial virus in mice (Raymond *et al.*, 1994). Again using electron microscopy and immunogold staining, the increase in ICAM-1 induced by hyperoxic exposure was found to be the result of induction of expression on the luminal surface of type II pneumocytes (Kang *et al.*, 1993), possibly a preliminary indicator of their eventual differentiation into type I pneumocytes (Christensen *et al.*, 1993) in response to this injurious insult. The total expression on the type I pneumocytes was unchanged while the gradient of distribution was lost. Under these conditions, alveolar ICAM-1 seems to play a pro-inflammatory role similar to that on the airway epithelium described above. Evidence for this is the finding that treatment with an anti-ICAM-1 mAb attenuated the lung injury and dysfunction induced by hyperoxia in mice by partially (~50%) blocking neutrophil infiltration as well as by apparently completely preventing the degranulation of the infiltrating neutrophils as measured by the amount of myeloperoxidase activity found in the cell free lung lavage supernatant (Wegner *et al.*, 1992a).

5. *Role of Specific Adhesion Molecules in Inflammatory Lung Disease*

5.1 ASTHMA

Asthma is characterized by episodes of reversible airway obstruction causing chest tightness, dyspnea and cough. In mild asthmatics, these episodes or "attacks" are relatively infrequent and well treated (reversed) with inhaled bronchodilators. However, in individuals with more moderate to severe disease, the intensity of an underlying, distinctive and chronic airway inflammation is associated and seemingly linked to more frequent, intense and prolonged attacks that are less reversible by bronchodilators. The reasons for this have become increasing clear in recent years. The inflammation, which consists principally of an activated or primed infiltrate of Th2-lymphocytes, eosinophils, mast cells and possibly platelets, causes an expansion of the perivascular (interstitial) spaces and release of mediators/growth factors which cause thickening of the basement membrane, epithelial damage and shedding, production of viscous mucus, and hyperplasia, priming as well as partial constriction of airway smooth muscle (Dunnill, 1960; Laitinen *et al.*, 1985; Barnes *et al.*, 1988; Beasley *et al.*, 1989; Jeffery *et al.*, 1989; Azzawi *et al.*, 1990; Brewster *et al.*, 1990; Kay, 1991). All of these outcomes support an increase in airway responsiveness which lowers the threshold for response to an environmental stimuli, thus making attacks occur more frequently (Boushey *et al.*, 1980). The presence and enhanced sensitivity (responsiveness) of the infiltrate also primes the airways for a subsequent or late-phase broncho-obstructive response on the influx of additional leucocytes 4 to 6 h after an "attack." Persistent late-phase responses may relate directly to the chronicity of asthma symptoms as well as the occurrence of night time episodes (O'Byrne *et al.*, 1987). Thus, understanding the molecules and mediators that regulate the induction and sustaining of this airway inflammation is key to the management of moderate to severe asthma.

Using mAb to specific adhesion glycoproteins and primate models of asthma-like induced and persistent airway inflammation and responses, the expression and possible roles of specific cell adhesion molecules in asthma have been illuminated. Below is a summary of these primate results followed by supportive findings in atopic humans. Finally, the VLA-4/VCAM-1, IL-4, Th2-lymphocyte hypothesis for asthma is reviewed.

5.1.1 Primate Allergen Inhalation Models

5.1.1.1 Induced Airway Hyper-responsiveness

In cynomolgus monkeys with near normal airway cell composition ($\geqslant 90\%$ alveolar macrophages), a single *Ascaris* inhalation induced an acute (peak 6 h, gone by 24 to 48 h) and variable (usually small) neutrophil infiltration, a more consistent and prolonged (6 h to 7 days) eosinophil infiltration (Wegner *et al.*, 1991b), and only rarely a small increase in airway responsiveness (Wegner *et al.*, 1991a). In contrast, three alternate day *Ascaris* inhalations produced a more consistent (usually 10–fold) increase in airway responsiveness (decrease in the concentration of inhaled methacholine required to produce a 100% increase in respiratory system resistance, provocation concentration 100, PC_{100}) associated with a massive and essentially pure eosinophil infiltration at three days after the last allergen challenge (Wegner *et al.*, 1991a). The magnitude of the induced airway hyper-responsiveness was found to be significantly correlated to the degree of airway epithelial desquamation induced (Wegner *et al.*, 1991b). In addition, a causal, as opposed to simply associative, role for eosinophils was indicated by the finding that the intratracheal instillation of purified eosinophil major basic protein (MBP) by itself produced an acute but marked (mean 10–fold) increase in airway responsiveness (Gundel *et al.*, 1991a). Thus, these results suggest that repeated allergen inhalations in these monkeys induce an infiltration and activation of eosinophils that produce, possibly through the release of MBP, airway epithelial desquamation causing, through a number of possible mechanisms, airway hyper-responsiveness.

To explore the role of specific adhesion glycoproteins in this model, mouse anti-human mAbs to ICAM-1, E-selectin or Mac-1 (CD11b), that showed functional cross-reactivity in cynomolgus monkeys, were administered intravenously, daily beginning one day prior to the three alternate day *Ascaris* inhalations. The monkeys were rested five or more weeks between each study, to allow the induced airway inflammation and hyper-responsiveness to resolve, and the mAb treatment study was compared to bracketing control studies in each animal. Allergen inhalation was found to increase ICAM-1 expression on airway and alveolar endothelium and epithelial as well as inducing E-selectin expression only on airway endothelium (Wegner *et al.*, 1990; Wegner *et al.*, 1992b). Treatment with the anti-ICAM-1 mAb R6.5 attenuated both the eosinophil infiltration and increase in airway responsiveness ($60 \pm 8\%$ and $92 \pm 34\%$, respectively) (Wegner *et al.*, 1990). In contrast, administration of the anti-E-selectin mAb CL2 did not significantly inhibit the eosinophil infiltration ($\times 10^3$/ml BAL: 507 ± 118 in control vs. 430 ± 114 in CL2 treated) or the increase in airway responsiveness (change in log. PC_{100}: -1.16 ± 0.27 in control vs. -1.24 ± 0.29 in CL2 treated) (Wegner *et al.*, 1991b). Finally, the anti-Mac-1 mAb LM2, while not reducing the eosinophil infiltration, significantly inhibited the activation of airway eosinophils [BAL supernatant eosinophil peroxidase (EPO) activity in optical density (OD) units: 618 ± 210 in control vs. 185 ± 41 in LM2 treated] as well as the increase in airway responsiveness (change in log. PC_{100}: -0.99 ± 0.16 in control vs. -0.14 ± 0.15 in LM2 treated) (Wegner *et al.*, 1993a).

Results with the anti-Mac-1 mAb indicate that the inhibition of induced airway hyper-responsiveness by treatment with the anti-ICAM-1 mAb may have been due to an attenuation of an epithelial ICAM-1 binding eosinophil Mac-1 co-stimulatory, eosinophil activation signal in addition to its attenuation of eosinophil infiltration via blocking endothelial ICAM-1. To investigate this possibility further, a placebo controlled cross-over study design was used to evaluate the effects of daily inhalation administration of the anti-ICAM-1 mAb R6.5 to selectively target epithelial ICAM-1. In animals who experienced greater than a half-log increase in airway responsiveness during either arm of the study, inhaled R6.5 treatment completely blocked the induced airway hyper-responsiveness in two animals, blocked by ~50% in two additional animals, and was ineffective in two others (Fig. 9.4). However, serum levels of immunoglobulin M (IgM) anti-R6.5 antibodies (possibly neutralizing the administered R6.5) were detected in both animals in which R6.5 was ineffective, and in one of the two in which it inhibited only 50%. Inhaled R6.5 also attenuated eosinophil activation [bronchoalveolar (BAL) supernatant eosinophil peroxidase (EPO) activity]

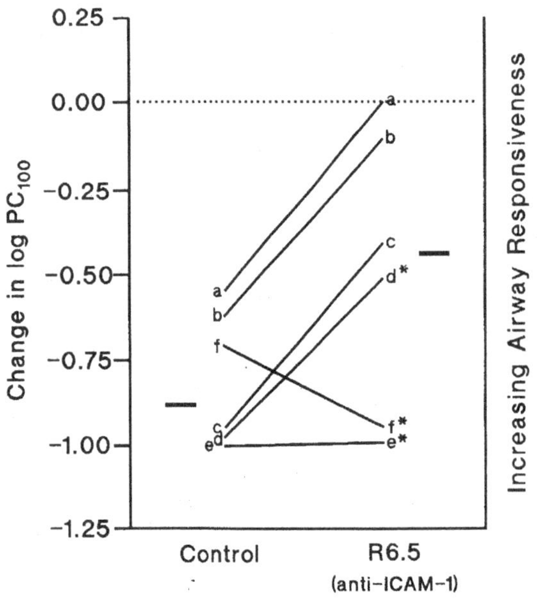

Figure 9.4 Effects of inhalation administration of the mouse anti-human ICAM-1 mAb R6.5 versus placebo (cross-over study design) on the airway hyper-responsiveness (concentration of inhaled methacholine needed to produce a 100% increase in respiratory system resistance, PC_{100}) induced by three alternate-day allergen inhalations in primates. Bars represent the median for all animals. Asterisks identify animals in whom anti-R6.5 IgM (neutralizing?) antibodies were detectable in the serum by day 6 of R6.5 administration. Modified from (Wegner *et al.*, 1993b) with permission.

in most animals consistent with the antagonism of a co-stimulatory activation of eosinophils via their adhesion to epithelial ICAM-1 (Wegner *et al.*, 1993b). Interesting eosinophil infiltration seemed also to be attenuated. Since only small amounts of R6.5 were found to reach the airway endothelium in inhaled R6.5 treated animals (as determined by immunostaining of frozen airway sections and biopsies), the attenuation of eosinophil infiltration (biopsies consistent with BAL results) may have been in part due to the reduction of eosinophil activation, and thus inflammation and desquamation, proximal to the airway epithelium.

5.1.1.2 Late-Phase Broncho-obstruction

Like in atopic asthmatics, allergen inhalation in *Ascaris* hypersensitive cynomolgus monkeys induces an immediate bronchoconstriction followed in some individuals 6 to 10 h later by a second "late-phase" broncho-obstruction. In these primates, dual (immediate and late-phase responses) compared to single (immediate only response) responders are characterized by enhanced expression of ICAM-1, but not E-selectin, on airway epithelium and endothelium, increased numbers and activation (BAL supernatant EPO activity) of airway eosinophils and a 10- or more, fold greater airway responsiveness to inhaled methacholine prior to antigen challenge (Gundel *et al.*, 1991b; Gundel *et al.*, 1992a). That is, dual responders have a baseline, persisting airway inflammation and hyper-responsiveness. After antigen inhalation, airway eosinophils acutely decrease, airway endothelial E-selectin expression is induced, and the severity of the late-phase airway obstruction significantly correlates with the intensity of an acute neutrophil influx and increase in airway myeloperoxidase (MPO) activity (BAL at 6 h post antigen).

To determine the role of ICAM-1 and E-selectin in the late-phase response, R6.5 (anti-ICAM-1) and CL2 (anti-E-selectin) were administered intravenously prior to *Ascaris* inhalation in dual responder monkeys. R6.5 had little if any effect on the neutrophil influx (BAL at 6 h post challenge) and only a slight (~15–20%) but consistent attenuation of the early and late phase airway narrowing (Gundel *et al.*, 1991b). In contrast, CL2 significantly and almost completely inhibited the neutrophil infiltration, increase in neutrophil activation (BAL supernatant MPO activity) and the associated late-phase airway obstruction (Table 9.1).

Thus, the roles of ICAM-1 and E-selectin in the late-phase response induced by a single allergen inhalation in monkeys with pre-existing airway inflammation were found to be opposite from those in the induction of airway hyper-responsiveness by multiple allergen inhalations in monkeys without baseline airway inflammation. The reasons for these different effects are not clear, but certainly merit further investigation because of their obvious importance to the understanding of apparently distinct pathways involved in the generation of asthma-like airway inflammation and dysfunction.

Table 9.1 Effects of antibodies to ICAM-1 and E-selectin on the allergen-induced late-phase response in primates

Treatment	Infiltrating neutrophils ($\times 10^4$/ml of BAL)	Late-phase BAL MPO Activity (OD units $\times 10^{-3}$)	Percentage inhibition of the late-phase broncho-obstruction
Placebo	33 ± 21	423 ± 87	—
Anti-ICAM-1	36 ± 23	388 ± 92	19 ± 15
Anti-E-selectin	6 ± 9*	101 ± 20*	72 ± 8*

* $p < 0.05$ compared to placebo control in each animal.
All values are means ±S.E.M.
Modified from (Wegner *et al.*, 1993b) with permission

The consistent, although small, attenuation of both the early- and late-phase airway narrowing by treatment with the anti-ICAM-1 mAb is again in agreement with epithelial ICAM-1 contributing to leucocyte priming and thus increased responsiveness to allergen challenge. The finding that this treatment effect was small is also compatible with the fact that eosinophils were already present and activated before anti-ICAM-1 administration as well as the fact that the amounts of chemo-attractants released by antigen inhalation were so large that the co-stimulatory signal provided by ICAM-1-mediated adhesion was comparatively small.

5.1.1.3 Persistent Airway Hyper-responsiveness

Finally, the abilities of R6.5 (anti-ICAM-1) and CL2 (anti-E-selectin) to reverse, or to inhibit the re-establishment of, a persistent airway inflammation and hyper-responsiveness were evaluated in the above described dual responder monkeys (Gundel *et al.*, 1992b; Letts *et al*,, 1993). Daily intravenous administration of R6.5 or CL2 for seven days failed to significantly reverse the persistent airway eosinophilia, eosinophil activation or hyper-responsiveness in these animals. In contrast, daily intramuscular (systemic) dexamethasone impressively lessened all of these parameters. However, after removal of the persistent airway inflammation and hyper-responsiveness by dexamethasone treatment, the daily intravenous administration of R6.5, but not CL2, was effective in inhibiting their re-establishment. That is, and consistent with the above mentioned results and conclusions, anti-ICAM-1 treatment effectively inhibits the induction of airway eosinophilia, and more importantly, airway hyper-responsiveness. However, once established, persistent airway inflammation and hyper-responsiveness is only weakly, if at all, displaced by at least a short term (≤1 week) antagonism of ICAM-1. Whether a more efficient antagonism of airway epithelial ICAM-1 (than achieved by intravenous R6.5) or longer duration of treatment might be more effective deserves further investigation.

5.1.1.4 What About Effects On Lymphocytes?

While the animal studies mentioned above have focused on the contribution of specific adhesion molecules to granulocyte infiltration and activation, the adhesive interactions that regulate lymphocyte infiltration, cytokine release, cytotoxic activity, and antibody production could certainly also play a prominent role in the responses monitored. The unique and characteristic eosinophilic airway inflammation found in asthma, as well as in the primate models presented above, may be the consequence of the cytokines IL-4 and IL-5 released from selectively recruited and activated lymphocytes of the Th2-like subtype (Kay, 1991) (also see below). Thus, many of the above described effects of anti-ICAM-1 treatment attributed to the attenuation of eosinophil infiltration and activation may rather, or in addition, be secondary to a reduction of lymphocyte infiltration and activation. Indeed, in recent studies we have found an increase in lymphocytes of the CD4 phenotype and expressing the IL-2 receptor (CD25) in the BAL obtained from primates with either induced or persistent airway eosinophilia and hyper-responsiveness (Souza *et al.*, 1993).

5.1.2 Supportive Findings in Atopic Humans

Using immunostaining of nasal biopsies from patients with perennial allergic rhinitis and from non-atopic healthy volunteers, the following observations have been reported: (i) in the normal nasal mucosa, ICAM-1 expression on the endothelium occurred more frequently than E-selectin with VCAM-1 expression minimal or absent; (ii) in perennial rhinitis, the number of endothelial cells expressing ICAM-1 and VCAM-1 were significantly elevated; (iii) the number of vessels positive for ICAM-1 correlated with the number of infiltrated (CD11a staining) leucocytes; and (iv) the number of infiltrated neutrophils correlated with the number of E-selectin expressing vessels (Montefort *et al.*, 1992). Immunostaining of bronchial biopsies from atopic patients with asthma, from atopic patients with rhinitis but not asthma, and from non-atopic healthy volunteers demonstrated the following: (i) ICAM-1 expression on the airway epithelium and endothelium was elevated in the asthmatics and correlated to both the number of infiltrated eosinophils and the degree of airway hyper-responsiveness; and (ii) epithelial ICAM-1 expression as well as endothelial ICAM-1 and E-selectin expression were elevated 6 h post allergen inhalation in the atopic non-asthmatic patients (Ando *et al.*, 1992; T. Fukuda, personal communication). Finally, BAL cell-free supernatant E-selectin levels were found to be increased 18 h following segmental antigen challenge in allergic subjects (Georas *et al.*, 1992) suggestive of the up-regulated expression and subsequent shedding of this endothelial specific adhesion molecule during allergen-induced airway inflammation. All of these findings are consistent with the results found in the primate allergen inhalation models.

5.1.3 VLA-4/VCAM-1, IL-4, Th2-Lymphocyte Hypothesis

The chronic airway inflammation found in asthma is characterised by a strikingly selective accumulation of eosinophils and lymphocytes (Dunnill, 1960; Azzawi *et al.*, 1990). This characteristic and selective infiltrate has been hypothesized to be generated by a VLA-4/VCAM-1 emigration pathway induced by the release of IL-4 and IL-5 from Th2-lymphocytes (Tepper *et al.*, 1990; Kay, 1991; Walker *et al.*, 1991). This hypothesis is based on the following observations: (i) the marked expression of VLA-4 on basophils, eosinophils and lymphocytes, but not on neutrophils and only minimally on monocytes (Bochner *et al.*, 1991; Carlos *et al.*, 1991; Walsh *et al.*, 1991; Weller *et al.*, 1991; Vachula and Vanepps, 1992); (ii) the selective up-regulation of endothelial VCAM-1 by singular or co-stimulation with IL-4 (Masinovsky *et al.*, 1990; Thornhill *et al.*, 1991; Briscoe *et al.*, 1992; Schleimer *et al.*, 1992); (iii) the principal source of IL-4 is the Th2-lymphocyte (Paul, 1991); (iv) IL-4 also enhances, and induces class switching to, IgE production by B-lymphocytes (Coffman *et al.*, 1986; Paul, 1991) and up-regulated ICAM-1 on airway epithelium (Churchill *et al.*, 1993); and (v) Th2-lymphocytes are also the major source IL-5, an eosinophil-selective stimulator of differentiation, adhesion and activation (Lopez *et al.*, 1988; Clutterbuck *et al.*, 1989; Fujisawa *et al.*, 1990; Walsh *et al.*, 1990).

Support of this hypothesis is found in two animal models. In guinea pigs, the eosinophil accumulation induced in a passive cutaneous anaphylaxis (PCA) reaction or by the intradermal injections of platelet-activating factor (PAF), leukotriene (LT) B_4 or C5a des Arg was in all cases inhibited by an anti-VLA-4 mAb (Weg *et al.*, 1993). In inhaled allergen challenged sheep, treatment with an anti-VLA-4 mAb prevented the late-phase broncho-obstruction and subsequent increase in airway responsiveness (Abraham *et al.*, 1994). However, in conditions where ICAM-1 is expressed (i.e. most, since there is a small constitutive expression of ICAM-1 on most endothelial cells *in vivo*), the VLA-4/VCAM-1 mediated adhesion to, and migration through, endothelial monolayers by eosinophils and/or lymphocytes is dominated by the CD18/ICAM-1 pathway (Ebisawa *et al.*, 1992; Vachula and Vanepps, 1992; Vankooyk *et al.*, 1993). That is, the VLA-4/VCAM-1-dependent pathway is only notable in CD18/ICAM-1 blocked or deficient (i.e. LAD type 1) conditions. In addition, in contrast to ICAM-1, VCAM-1 expression is not enhanced in the bronchial biopsies of asthmatics even during acute exacerbations requiring emergency room attention (S. Montefort and S.T. Holgate, personal communication). Furthermore, in preliminary studies a mouse anti-cynomolgus monkey VCAM-1 mAb failed to inhibit the eosinophil infiltration, eosinophil or lymphocyte activation, or increase in airway responsiveness induced by repeated allergen inhalations in the above mentioned primate model (C.D. Wegner, unpublished data). Finally, in the allergen challenge sheep model, despite efficacy on functional parameters (see above), treatment with the anti-VLA-4 mAb had no effect on the eosinophil or total leucocyte influx. Therefore, the positive effects of the anti-VLA-4 mAbs in the above mentioned animal models may be the consequence of a VLA-4/fibronectin mediated migration and/or activation of eosinophils within the interstitium rather than an attenuation of a VLA-4/VCAM-1 trans-endothelial diapedesis pathway.

5.2 ACUTE RESPIRATORY DISTRESS SYNDROME (ARDS)

The acute (or adult) respiratory distress syndrome (ARDS) is a "common clinical catastrophe" following acute lung injury (Petty, 1990). A multiplicity of causes can lead to ARDS including bypass surgery, trauma, head injuries, sepsis, emboli, pneumonia, smoke or chemical inhalation, acid aspiration, and premature birth. The most common cause, gram negative sepsis, is estimated to be responsible for 300 000–500 000 cases each year in the USA alone, with mortality rates $\geqslant 50\%$ (Bone, 1989). While the lungs are only one of many organs failing in established ARDS, the lungs are usually the first organ to fail (Welbourn and Young, 1992) and acute lung injury remains the most common initial precipitating factor (Petty, 1990). Thus, understanding the pathways involved in the pathogenesis of acute lung injury are key to future interventions into this devastating disease.

5.2.1 Rat Complement Activation Models

The major source of insight into the roles of various adhesion molecules in the pathogenesis of acute lung injury has come from rat models of cobra venom factor (CVF), IgG immune complex, and IgA immune complex induced complement activation. A summary of the findings in these studies is shown in Table 9.2.

The rapid in onset (minutes), neutrophil-mediated injury to the lung vasculature (increased permeability, hemorrhage and neutrophil emigration) induced by the

Table 9.2 Contribution of various adhesion molecules, cytokines and leucocytes to the lung vascular injury induced by cobra venom factor, IgG immune complexes or IgA immune complexes in rats

	Factor Inducing Lung Injury		
	CVF	*IgG IC*	*IgA IC*
Peak injury time	*20 min*	*4 h*	*4 h*
Role of alveolar macrophages	–	+ +	+ + +
Prevention by:			
anti-CD18	+ + +	+ + +	+ + +
anti-CD11a	+	+ +	+
anti-CD11b	+ +	+	+ +
anti-ICAM-1	+ + +	+ + +	+ +
anti-E-selectin	ND	+ + +	–
anti-P-selectin	+ +	ND	ND
anti-VLA-4	–	+ +	+ + +
anti-IL-1	–	+ + +	–
anti-TNFα	–	+ + +	–
Neutrophil depletion	+ + +	+ + +	–

Abbreviations: CVF = cobra venom factor; IC = immune complexes; – no inhibition; + significant but minimal (<50%) inhibition; + + moderate but incomplete inhibition; + + + marked (≥70%) inhibition; ND = no data published yet.
Summary of results from (Mulligan *et al.*, 1992c; Mulligan *et al.*, 1993b; Mulligan *et al.*, 1991; Mulligan *et al.*, 1992b; Mulligan *et al.*, 1992a; Mulligan *et al.*, 1993a).

intravenous injection of CVF is apparently governed by the constitutive expression of ICAM-1 and rapid induction of P-selectin on the pulmonary endothelium as well as by the β_2 (CD11/CD18) integrins on neutrophils (Mulligan *et al.*, 1992c; Mulligan *et al.*, 1993b). In contrast, a similar injury after IgG immune complex deposition in the lungs is more delayed in onset (h), employs the generation of cytokines (IL-1 and TNFα) from alveolar macrophages and is governed by the induction of E-selectin and ICAM-1 on the pulmonary endothelium as well as by β_1 (VLA-4) and β_2 (mostly CD11a) integrin mediated leucocyte adhesions (Mulligan *et al.*, 1991; Mulligan *et al.*, 1992b; Mulligan *et al.*, 1993a). Finally, in the IgA immune complex model, the acute lung injury (increased vascular permeability and hemorrhage) also peaks at 4 h, but is independent of IL-1/TNFα generation and neutrophil-mediated vascular injury/infiltration. The ability of alveolar macrophages to complement-opsonize the IgA immune complexes via their CD11b (Mac-1, CR3) receptors to generate oxidants (e.g. superoxide), and possibly proteases, as well as macrophage emigration and/or activation via CD11a, ICAM-1 and VLA-4 dependent adhesions seem to be critical in this model (Mulligan *et al.*, 1992a; Mulligan *et al.*, 1993a).

The induction of E-selectin on the pulmonary capillary endothelium during septic, but not traumatic/hypovolemic, shock in baboons (Redl *et al.*, 1991) as well as elevated serum E-selectin levels in patients with septic shock (Newman *et al.*, 1993), further implicate an important role for this endothelial specific adhesion molecule in ARDS. In contrast, despite indisputable evidence of systemic neutrophil activation and heightened CD18 expression during sepsis (Walsh *et al.*, 1991), anti-CD18 mAbs have been disappointing in their protection of the lung sequelae (e.g. permeability) in septic animal models, even though protective of the failure by other organs (Thomas *et al.*, 1992). Thus, under conditions of pronounced vascular neutrophil activation, adhesion to pulmonary microvascular endothelium via a selectin-dependent, CD18-independent mechanism may be sufficient to mediate vascular injury and neutrophil emigration. Such a mechanism has previously been also suggested as an explanation for the E-selectin-dependent, ICAM-1-independent infiltration of neutrophils during the inhaled allergen-induced late-phase response in primates (Wegner *et al.*, 1993b).

5.2.2 Remote Neutrophil Activation

Commonly (as in bypass and other surgeries, trauma, hypovolemic shock, limb re-attachment or other re-establishments of organ perfusion) acute lung injury is initiated by neutrophils moderately activated (presumably less deformable and expressing more CD18; see Section 2.2.1) at inflammatory sites remote from the lungs. Lung injury initiated in this manner is apparently highly CD18-dependent, as increases in lung permeability induced by ischemia followed by reperfusion of the rat hind limb (Seekamp and Ward, 1993) or the rabbit intestine (Hill *et al.*, 1993) have been found to be almost completely prevented by anti-CD18 mAbs.

5.2.3 Pulmonary Oxygen Toxicity

The impaired alveolar gas exchange in ARDS necessitates the use of elevated levels of inhaled oxygen to achieve acceptable blood oxygen saturation. However, prolonged exposure to high concentrations of oxygen can precipitate an acute edematous lung injury in any remaining normal lung regions (Clark and Lambertsen, 1971; Jackson, 1990) as well as enhance the injury in already damaged lung regions (Witschi *et al.*, 1981; Tryka *et al.*, 1982; Crouse *et al.*, 1990). Initiation of hyperoxia-induced lung injury occurs through a direct increase in the intracellular production of oxygen radicals, by cells within and lining the gas exchange units (acinus), that overwhelm intracellular antioxidant defence mechanisms (Martin *et al.*, 1981; Crapo, 1986). Once these effects are initiated, the lung injury is amplified and morphological changes become pronounced exponentially associated with a marked infiltration of neutrophils after 48 to 72 h of pure oxygen breathing (Crapo *et al.*, 1983; Crapo, 1986).

As mentioned above, hyperoxic exposure for 48 h strikingly enhances acinar ICAM-1 expression in the lungs of mice (see Section 4.2.) (Wegner *et al.*, 1992a). In addition, daily treatment with an anti-ICAM-1 mAb attenuated the lung injury and dysfunction induced by 84 h of pure oxygen breathing in these animals. This protection was congruent with the partial blocking of a small neutrophil infiltration as well as the apparent complete prevention of degranulation of the emigrated neutrophils as measured by the increase in MPO activity in the cell-free lung lavage supernatant. These data once again indicate the important contribution of adhesion molecules to leucocyte activation in addition to trafficking.

5.2.4 Pulmonary Fibrosis

Finally, if the patient survives an acute lung injury (ARDS) as well as its enhancement by pulmonary oxygen toxicity, a scarring or fibrosis of lung regions is common (Witschi *et al.*, 1981; Crapo, 1986; Petty, 1990). Recent studies have demonstrated that treatment of rats with anti-CD11a or anti-CD11b mAbs completely prevented the enhanced lung collagen deposition induced by silica or bleomycin (Piguet *et al.*, 1993). Interestingly, treatment with the anti-CD11a mAb was claimed to also reverse an established pulmonary fibrosis in these studies. The relevance of these observations is enhanced by the finding that serum levels of ICAM-1 (a ligand for CD11a and CD11b) appear to be elevated in patients with idiopathetic pulmonary fibrosis (Shijubo *et al.*, 1992).

5.3 Respiratory Viral Infection/Sarcoidosis/ Transplantation: Cell Mediated Immunity

While *in vivo* pulmonary studies to date have focused on the role of adhesion molecules in granulocyte-mediated dysfunction and injury, the adhesive interactions that regulate lymphocyte infiltration, cytokine/antibody release (activation), and cytotoxic activity are clearly also prominent contributors to a variety of lung diseases. In respiratory viral infections, the impairment of gas exchange and other symptoms seem to be more related to the severity of the immune/inflammatory response than the cytopathetic effects of the virus (Graham *et al.*, 1991). In mice infected with respiratory syncytial virus (RSV), ICAM-1 expression has recently been reported to be impressively up-regulated on lung endothelial cells as well as on both bronchiolar and alveolar epithelia (Raymond *et al.*, 1994). Daily treatment of these mice with an anti-ICAM-1 mAb was found to attenuate the RSV-induced mononuclear cell influx by ~80%. Whether this attenuation was due to prevention of endothelial ICAM-1 mediated diapedesis or secondary to the prevention of ICAM-1-dependent, CD11a-mediated lymphocyte activation and cytokine release is not known and deserves further evaluations. Yet, these findings, combined with the fact that most respiratory viral infections are associated with the enhanced production of IFNγ (the most effective up-regulator of epithelial ICAM-1 expression), strongly suggest that ICAM-1 will be found to contribute prominently to the pathogenesis and morbidity of respiratory viral infections. Likewise, lymphocytes, monocytes, and T-cell mediated immunity are certain to have important roles in lung granulomatous diseases (e.g. sarcoidosis) and transplantation.

In non-pulmonary models of cell-mediated immunity/ transplantation, mAbs to CD11a, ICAM-1, VLA-4 and VCAM-1 have all yielded impressive immunomodilatory effects. Treatment with CD11a mAbs improve engraftment and survival after allogeneic bone marrow transplantation in mice (van Dijken *et al.*, 1990) as well as after human leucocyte antigen (HLA) mismatched bone marrow transplantation in children with variety of immunodeficient and neoplastic disorders (Fischer *et al.*, 1986). The severity of allogeneic graft-versus-host-disease was reduced and survival increase by anti-CD11a as well as anti-ICAM-1 mAbs in mice (Buyon *et al.*, 1990). In a cynomolgus monkey model of allogeneic kidney transplantation, prophylactic administration of an anti-ICAM-1 mAb prolonged graft survival while a therapeutic administration of this mAb completely, and possibly permanently, reversed a pre-existing rejection produced by tapering Cyclosporine to subtherapeutic levels (Cosimi *et al.*, 1990). Finally, anti-VLA-4 mAbs prevent the vasculitis (but interestingly not the lymphocyte emigration) in both rat and rabbit cardiac allografts (Paul *et al.*, 1993; Sadahiro *et al.*, 1993) and a anti-VCAM-1 mAb has been reported to attenuate murine cardiac allograft rejection (Pelletier *et al.*, 1993).

6. Conclusion

It is now well established that the cell surface adhesive glycoproteins that mediate cell-cell and cell-matrix interactions are principal regulators of nearly all aspects of immune/inflammatory responses. Many of the molecules and concepts described from studies performed *in vitro* and *in vivo* in the systemic microvasculature and non-pulmonary tissue are certainly also employed in the lungs. However, there is also convincing evidence of lung specific pathways and properties predominately related to differences in (uniqueness of) the pulmonary microvascular beds (e.g. capillaries and possibly arterioles) and the striking expression of ICAM-1 on the airspace lining epithelium, the most common foci of inflammatory responses in lung disease.

Results gained from animal models with the aid of mAbs against specific adhesion molecules indicate that antagonism of but one of the many key cell surface adhesive glycoproteins can have dramatic inhibitory efficacy and thus may lead to new and novel therapies for lung diseases. However, the dominant molecule will vary with the timing, intensity, conditions, components and injury associated with the underlying inflammatory/immune processes. In addition to antagonism, attenuation of adhesion mediated pathways could be achieved by: (i) suppression of constitutive cell surface expression; (ii) inhibition of up-regulated expression; (iii) decrease in constitutive binding avidity; (iv) prevention of up-regulated binding avidity; or (v) blockage of intracellular stimulatory signals generated (e.g. by the leucocyte integrins) by ligand engagement (adhesion) (Wegner and Wallace, 1992c).

Currently, adhesion molecules are being utilized as markers of disease. Serum levels of soluble, presumably shed, forms of the endothelial adhesion molecules ICAM-1, E-selectin and VCAM-1, as well as the leucocyte adhesion molecule L-selectin are being measured as indicators of inflammatory onset, severity and/or reversal with therapy (Gearing *et al.*, 1992; Georas *et al.*, 1992; Shijubo *et al.*, 1992; Newman *et al.*, 1993; Stockenhuber *et al.*, 1993). In addition, imaging of ICAM-1 induction in the heart following cardiac transplantation has proven an effective, non-invasive method for detecting early allograft rejection (Ohtani *et al.*, 1993).

Finally, as inhibitors of adhesion molecules are developed, their effects on host defence will need to be closely monitored. Obviously, inflammatory responses have their beneficial features in addition to their well documented deleterious creation of tissue injury and organ dysfunction. An example of such a beneficial effect was recently reported in ozone-induced lung injury in primates (Hyde *et al.*, 1994). Treatment with an anti-CD18 mAb markedly delayed repair of the ozone injured bronchiolar epithelium by preventing the influx of neutrophils.

7. References

Abraham, W.M., Sietczak, M.W., Ahmed, A., Cortes, A., Lauredo, I.T., Kim, J., Pepinsky, B., Benjamen, C.D., Leone, D., Lobb, R.R., and Weller, P.F. (1994). Alpha-4-Integrins mediate antigen-induced late bronchial responses and prolonged airway hyperresponsiveness in sheep. J. Clin. Invest. 93, 776–787.

Anderson, D.C. and Springer, T.A. (1987). Leukocyte adhesion deficiency: an inherited defect in the Mac-1, LFA-1 and p150,95 glycoproteins. Ann. Rev. Med. 38, 175–194.

Anderson, D.C., Schmalstieg, F.C., Finegold, M.J., Hughes, B.J., Rothlein, R., Miller, L.J., Kohl, S., Tosi, M.F., Jacobs, R.L., Waldrop, T.C., Goldman, A.S., Shearer, W.T. and Springer, T.A. (1985). The severe and moderate phenotypes of heritable Mac-1, LFA-1 deficiency: their quantitative definition and relation to leukocyte dysfunction and clinical features. J. Infect. Dis. 152, 668–689.

Ando, N., Fukuda, T., Nakajima, H. and Makino, S. (1992). Expression of intercellular adhesion molecule-1 (ICAM-1) is upregulated in the bronchial mucosa of symptomatic asthmatics. J Allergy Clin. Immunol. 89, 215 (Abstr).

Anwar, A.R.F., Moqbel, R., Walsh, G.M., Kay, A.B. and Wardlaw, A.J. (1993). Adhesion to fibronectin prolongs eosinophil survival. J Exp. Med. 177, 839–843.

Augustin, M., Dietrich, A., Niedner, R., Kapp, A. and Schopf, E. (1993). Phorbol-12-myristate-13-acetate-treated human keratinocytes express B7-like molecules that serve a costimulatory role in T-cell activation. J. Invest Dermatol, 100, 275–281.

Azzawi, M., Bradley, B., Jeffery, P.K., Frew, A.J., Wardlaw, A.J., Knowles, G., Assoufi, B., Collins, J.V., Durham, S. and Kay, A.B. (1990). Identification of activated T lymphocytes and eosinophils in bronchial biopsies in stable atopic asthma. Am. Rev. Respir. Dis. 142, 1407–1413.

Barnes, P.J., Chung, K.F. and Page, C.P. (1988). Inflammatory mediators and asthma. Pharmacol. Rev. 40, 49–84.

Beasley, R., Roche, W.R., Roberts, J.A and Holgate, S.T. (1989). Cellular events in the bronchi in mild asthma and after bronchial provocation. Am. Rev. Respir. Dis. 139, 806–817.

Bevilacqua, M.P., Stengelin, S., Gimbrone, M.A. Jr. and Seed, B. (1989). Endothelial leukocyte adhesion molecule 1: an inducible receptor for neutrophils related to complement regulatory proteins and lectins. Science 243, 1160–1165.

Bochner, B.S., Luscinskas, F.W., Gimbrone, M.A., Newman, W., Sterbinsky, S.A., Derse-Anthony, C.P., Klunk, D. and Schleimer, R.P. (1991). Adhesion of human basophils, eosinophils, and neutrophils to interleukin 1-activated human vascular endothelial cells: Contribution of endothelial cell adhesion molecules. J. Exp. Med. 173, 1553–1556.

Bone, R.C. (1989). In "Second Vienna Shock Forum", pp. 857–865, New York. Alan R. Liss.

Boushey, H.A., Holtzman, M.J., Sheller, J.R. and Nadel, J.A. (1980). Bronchial hyperreactivity. Am. Rev. Respir. Dis. 121, 389–413.

Brewster, C.E., Howarth, P.H., Djukanovic, R., Wilson, J., Holgate, S.T. and Roche, W.R. (1990). Myofibroblasts and subepithelial fibrosis in bronchial asthma. Am J. Respir. Cell Molecular Biol. 3, 507–511.

Briscoe, D.M., Cotran, R.S. and Pober, J.S. (1992). Effects of tumor necrosis factor, lipopolysaccharide, and IL-4 on the expression of vascular cell adhesion molecule-1 *in vivo*: correlation with CD3+ T cell infiltration. J. Immunol. 149, 2954–2960.

Buyon, J.P., Slade, S.G., Reibman, J., Abramson, S.B., Philips, M.R., Weissmann, G. and Winchester, R. (1990). Constitutive and induced phosphorylation of the alpha- and beta-chains of the CD11/CD18 leukocyte integrin family. J. Immunol. 144, 191–197.

Carlos, T., Kovach, N., Schwartz, B., Rosa, M., Newman, B., Wayner, E., Benjamin, C., Osborn, L., Lobb, R. and Harlan, J. (1991). Human monocytes bind to two cytokine-induced adhesive ligands on cultured endothelial cells: endothelial-leukocyte adhesion molecule-1 and vascular cell adhesion molecule-1. Blood 77, 2266–2271.

Casale, T.B., Erger, R.A. and Little, M.M. (1993). Platelet-activating factor-induced human eosinophil trans-endothelial migration: evidence for a dynamic role of the endothelium. Am. J. Respir. Cell Molecular Biol. 8, 77–82.

Christensen, P.J., Kim, S., Simon, R.H., Toews, G.B. and Paine III, R. (1993). Differentiation-related expression of ICAM-1 by rat alveolar epithelial cells. Am. J. Respir. Cell Molecular Biol. 8, 9–15.

Churchill, L., Gundel, R.H., Letts, L.G. and Wegner, C.D. (1993). Contribution of specific cell-adhesive glycoproteins to airway and alveolar inflammation and dysfunction. Am. Rev. Respir. Dis. 148, S83–S87.

Clark, J.M. and Lambertsen, C.J. (1971). Pulmonary oxygen toxicity: a review. Pharmacol. Rev. 23, 37–133.

Clutterbuck, E.J., Hirst, E.M.A. and Sanderson, C.J. (1989). Human interleukin-5 (IL-5) regulates the production of eosinophils in human bone marrow cultures: comparison and interaction with IL-1, IL-3, IL-6, and GMCSF. Blood 73, 1504–1512.

Coffman, R., Ohara, J., Bond, M.W., Carty, J., Zlotnik, A. and Paul, W.E. (1986). B cell stimulatory factor-1 enhances the IgE response of lipopolysaccharide-activated B cells. J. Immunol. 136, 4538–4541.

Cosimi, A.B., Conti, D., Delmonico, F.L., Preffer, F.I., Wee, S.-L., Rothlein, R., Faanes, R. and Colvin, R.B. (1990). *In vivo* effects of monoclonal antibody to ICAM-1 (CD54) in non-human primates with renal allographs. J. Immunol. 144, 4604–4612.

Crapo, J.D. (1986). Morphologic changes in pulmonary oxygen toxicity. Ann. Rev. Physiol. 48, 721–731.

Crapo, J.D., Freeman, B.A., Barry, B.E., Turrens, J.F. and Young, S.L. (1983). Mechanisms of hyperoxic injury to the pulmonary micro-circulation. The Physiologist 26, 170–176.

Crouse, D.T., Cassell, G.H., Waites, K.B., Foster, J.M. and Cassady, G. (1990). Hyperoxia potentiates ureaplasma urealyticum pneumonia in newborn mice. Infect. Immun. 58, 3487–3493.

Diamond, M.S. and Springer, T.A. (1993b). A subpopulation of Mac-1 (CD11b/CD18) molecules mediates neutrophil adhesion to ICAM-1 and fibrinogen. J Cell Biol. 120, 545–556.

Diamond, M.S., Garciaaguilar, J., Bickford, J.K., Corbi, A.L., and Springer, T.A. (1993a). The I-domain is a major recognition site on the leukocyte integrin Mac-1 (Ca11b/CD18) for 4 distinct adhesion ligands. J. Cell Biol. 120, 1031–1043.

Doerschuk, C.M. (1992). The role of CD18-mediated adhesion in neutrophil sequestration induced by infusion of activated plasma in rabbits. Am. J. Respir. Cell Molecular Biol. 7, 140–148.

Doerschuk, C.M., Allard, M.F., Martin, B.A., MacKenzie, A., Autor, A.P. and Hogg, J.C. (1987). The marginated pool of neutrophils in the lungs of rabbits. J. Appl. Physiol. 63, 1806–1815.

Doerschuk, C.M., Downey, G.P., Doherty, D.E., English, D., Gie, R.P., Ohgami, M., Worthen, G.S., Henson, P.M. and Hogg, J.C. (1990a). Leukocyte and platelet margination within microvasculature of rabbit lungs. J. Appl. Physiol. 68, 1956–1961.

Doerschuk, C.M., Winn, R.K., Coxson, H.O. and Harlan, J.M. (1990b). CD18-dependent and -independent mechanisms of neutrophil emigration in the pulmonary and systemic microcirculation of rabbits. J. Immunol. 144, 2327–2333.

Dunnill, M.S. (1960). The pathology of asthma with special reference to changes in the bronchial mucosa. J. Clin. Pathol. 13, 27–33.

Dustin, M.L. and Springer, T.A. (1989). T-cell receptor cross-linking transiently stimulates adhesiveness through LFA-1. Nature 341, 619–624.

Ebisawa, M., Bochner, B.S., Georas, S.N. and Schleimer, R.P. (1992). Eosinophil transendothelial migration induced by cytokines: role of endothelial and eosinophil adhesion molecules in IL-1beta-induced trans-endothelial migration. J. Immunol. 149, 4021–4028.

Erbe, D.V., Watson, S.R., Presta, L.G., Wolitzky, B.A., Foxall, C., Brandley, B.K. and Lasky, L.A. (1993). P-Selectin and E-selectin use common sites for carbohydrate ligand recognition and cell adhesion. J. Cell Biol. 120, 1227–1235.

Etzioni, A., Frydman, M., Pollack, S., Avidor, I., Phillips, M.L., Paulson, J.C. and Gershoni-Baruch, R. (1992). Brief report: recurrent severe infections caused by a novel leukocyte adhesion deficiency. N. Engl. J. Med. 25, 1789–1792.

Fischer, A., Griscelli, C., Blanche, S., LeDeist, F., Verber, F., Lopez, M., Delaage, M., Oliver, D., Mawas, C. and Janossy, G. (1986). Prevention of graft failure by an anti-LFA-1 monoclonal antibody in HLA-mismatched bone marrow transplantation. Lancet 2, 1058–1062.

Fujisawa, T., Abu-Ghazaleh, R., Kita, H., Sanderson, C.J. and Gleich, G.J. (1990). Regulatory effect of cytokines on eosinophil degranulation. J. Immunol. 144, 642–646.

Furie, M.B., Tancinco, M.C.A., and Smith, C.W. (1991). Monoclonal antibodies to leukocyte integrins CD11a/CD18 and CD11b/CD18 or intercellular adhesion molecule-1 inhibit chemo-attractant-stimulated neutrophil trans-endothelial migration *in vitro*. Blood 78, 2089–2097.

Gearing, A.J.H., Hemingway, I., Pigott, R., Hughes, J., Rees, A.J. and Cashman, S.J. (1992). Soluble forms of vascular adhesion molecules, E-selectin, ICAM-1, and VCAM-1: pathological significance. Ann. NY Acad. Sci. 667, 324–331.

Geng, J.G., Bevilacqua, M.P., Moore, K.L., McIntyre, T.M., Prescott, S.M., Kim, J.M., Bliss, G.A., Zimmerman, G.A. and McEver, R.P. (1990). Rapid neutrophil adhesion to activated endothelium mediated by GMP-140. Nature 343, 757–760.

Georas, S.N., Liu, M.C., Newman, W., Beall, L.D., Stealey, B.A. and Bochner, B.S. (1992). Altered adhesion molecule expression and endothelial cell activation accompany the recruitment of human granulocytes to the lung after segmental antigen challenge. Am. J. Respir. Cell Molecular Biol. 7, 261–269.

Glazier, J.B., Hughes, J.M.B., Maloney, J.E. and West, J.B. (1964). Measurement of capillary dimensions and blood volume in rapidly frozen lungs. J. Appl. Physiol. 19, 713–724.

Graham, B.S., Bunton, L.A., Wright, P.F. and Karzon, D.T. (1991). Role of T lymphocyte subsets in the pathogenesis of primary infection and rechallenge with respiratory syncytial virus in mice. J. Clin. Invest. 88, 1026–1033.

Gundel, R.H., Letts, L.G. and Gleich, G.J. (1991a). Human eosinophil major basic protein induces airway constriction and airway hyper-responsiveness in primates. J. Clin. Invest. 87, 1470–1473.

Gundel, R.H., Wegner, C.D., Torcellini, C.A., Clarke, C.C., Haynes, N., Rothlein, R., Smith, C.W. and Letts, L.G. (1991b). Endothelial leukocyte adhesion molecule-1 mediates antigen-induced acute airway inflammation and late-phase airway obstruction in monkeys. J. Clin. Invest. 88, 1407–1411.

Gundel, R.H., Wegner, C.D. and Letts, L.G. (1992a). Antigen-induced acute and late-phase responses in primates. Am. Rev. Respir. Dis. 146, 369–373.

Gundel, R.H., Wegner, C.D., Torcellini, C.A. and Letts, L.G. (1992b). The role of intercellular adhesion molecule-1 in chronic airway inflammation. Clin. Exp. Allergy 22, 569–575.

Harlan, J.M., Winn, R.K., Vedder, N.B., Doerschuk, C.M. and Rice, C.L. (1992). In "Adhesion: its Role in Inflammatory Disease", (eds J.M. Harlan and D.Y. Lui), pp. 117–150, Freeman and Co., New York.

Hemler, M.E. (1990). VLA proteins in the integrin family: structures, functions, and their role on leukocytes. Annu. Rev. Immunol. 8, 365–400.

Hobbs, M.A., Jakes, S., Stacker, S.A., Wallace, R.W. and Springer, T.A. (1991). The cytoplasmic domain of the integrin lymphocyte function-associated antigen 1 beta subunit: sites required for binding to intercellular adhesion molecule 1 and the phorbol ester-stimulated phosphorylation site. J. Exp. Med. 174, 1227–1238.

Hill, J., Lindsay, T., Valeri, C.R., Shepro, D. and Hechtman, H.B. (1993). A CD18 antibody prevents lung injury but not hypotension after intestinal ischemia-reperfusion. J. Appl. Physiol. 74, 659–664.

Hogg, J.C. (1987). Neutrophil kinetics and lung injury. Physiol. Rev. 67, 1249–1295.

Hughes, B.J., Hollers, J.C., Crockett Torabi, E. and Smith, C.W. (1992). Recruitment of CD11b/CD18 to the neutrophil surface and adherence-dependent cell locomotion. J. Clin. Invest. 90, 1687–1696.

Hyde, D.M., Pinkerton, K.E., Wegner, C.D., Rothlein, R. and Plopper, C.G. (1994). Neutrophils aid epithelial repair following short-term ozone exposure in rhesus monkeys. J. Respire. Critical Care Med. 149, A231 (Abstr).

Imai, Y., Singer, M.S., Fennie, C., Lasky, L.A. and Rosen, S.D. (1991). Identification of a carbohydrate-based endothelial ligand for a lymphocyte homing receptor. J. Cell Biol. 113, 1213–1221.

Jackson, R.M. (1990). Molecular, pharmacological and clinical aspects of oxygen-induced lung injury. Clin. Chest Med. 11, 73–86.

Jeffery, P.K., Wardlaw, A.J., Nelson, F.C., Collins, J.V. and Kay, A.B. (1989). Bronchial biopsies in asthma. An ultra structural, quantitative study and correlation with hyperreactivity. Am. Rev. Respir. Dis. 140, 1745–1753.

Kang, B.-H., Crapo, J.D., Wegner, C.D., Letts, L.G. and Chang, L.-Y. (1993). Intercellular adhesion molecule-1 expression on the alveolar epithelium and its modification by hyperoxia. Am. J. Respir. Cell Molecular Biol. 9, 350–355.

Kansas, G.S., Ley, K., Munro, J.M. and Tedder, T.F. (1993). Regulation of leukocyte rolling and adhesion to high endothelial venules through the cytoplasmic domain of L-selectin. J. Exp. Med. 177, 833–838.

Kay, A.B. (1991). T lymphocytes and their products in atopic allergy and asthma. Int. Arch. Allergy Appl. Immunol. 94, 189–193.

Kishimoto, T.K., Jutila, M.A., Berg, E.L. and Butcher, E.C. (1989). Neutrophil Mac-1 and MEL-14 adhesion proteins inversely regulated by chemotactic factors. Science 245, 1238–1241.

Kishimoto, T.K., Warnock, R.A., Jutila, M.A., Butcher, E.C., Lane, C., Anderson, D.C. and Smith, C.W. (1991). Antibodies against human neutrophil LeCAM-1 (LAM-1/Leu-8/DREG-56 antigen) and endothelial cell ELAM-1 inhibit a common CD18-independent adhesion pathway *in vitro*. Blood 78, 805–811.

Koopman, G., Parmentier, H.K., Schuurman, H.-J., Newman, W., Meijer, C.J.L.M. and Pals, S.T. (1991). Adhesion of human B cells to follicular dendritic cells involves both the lymphocyte function-associated antigen 1/intercellular adhesion molecule 1 and very late antigen 4/vascular cell adhesion molecule 1 pathways. J. Exp. Med. 173, 1297–1304.

Kuijpers, T.W., Hakkert, B.C., Hart, M.H.L. and Roos, D. (1992). Neutrophil migration across monolayers of cytokine-prestimulated endothelial cells: role for platelet-activating factor and IL-8. J. Cell Biol. 117, 565–572.

Kyan-Aung, U., Haskard, D.O., and Lee, T.H. (1991). Vascular cell adhesion molecule-1 and eosinophil adhesion to cultured human umbilical vein endothelial cells *in vitro*. Am. J. Respir. Cell Molecular Biol. 5, 445–450.

Laitinen, L.A., Heino, M., Laitinen, A., Kava, T. and Haahtela, T. (1985). Damage of the airway epithelium and bronchial reactivity in patients with asthma. Am. Rev. Respir. Dis. 131, 599–606.

Lasky, L.A., Singer, M.S., Yednock, T.A., Dowbenko, D., Fennie, C., Rodriguez, H., Nguyen, T., Stachel, S. and Rosen, S.D. (1989). Cloning of a lymphocyte homing receptor reveals a lectin domain. Cell 56, 1045–1055.

Lawrence, M.B., and Springer, T.A. (1991). Leukocytes roll on a selectin at physiologic flow rates: distinction from and prerequisite for adhesion through integrins. Cell 65, 859–873.

Lein, D.C., Wagner Jr., W.W., Capen, R.L., Haslett, C., Hanson, W.L., Hofmeister, S.E., Henson, P.M. and Worthen, G.S. (1987). Physiological neutrophil sequestration in the lung: visual evidence for localization in capillaries. J. Appl. Physiol. 62, 1236–1243.

Letts, L.G., Wegner, C.D. and Gundel, R.H. (1993). In "Chemotactic Cytokines 2", (ed I.L.D. Lindley), Plenum, New York. (In press).

Ley, K., Gaehtgens, P., Fennie, C., Singer, M.S., Lasky, L.A. and Rosen, S.D. (1991). Lectin-like cell adhesion molecule 1 mediates leukocyte rolling in mesenteric venules *in vivo*. Blood 77, 2553–2555.

Lo, S.K., Lee, S., Ramos, R.A., Lobb, R., Rosa, M., Chi-Rosso, G. and Wright, S.D. (1991). Endothelial-leukocyte adhesion molecule 1 stimulates the adhesive activity of leukocyte integrin CR3 (CD11b/CD18, Mac-1, alpha-m beta-2) on human neutrophils. J. Exp. Med. 173, 1493–1500.

Lopez, A.F., Sanderson, C.J., Gamble, J.R., Campbell, H.D., Young, I.G. and Vadas, M.A. (1988). Recombinant human interleukin 5 is a selective activator of human eosinophil function. J. Exp. Med. 167, 219–224.

Lowe, J.B., Stoolman, L.M., Nair, R.P., Larsen, R.D., Berhend, T.L. and Marks, R.M. (1990). ELAM-1-dependent cell adhesion to vascular endothelium determined by a transfected human fucosyltransferase cDNA. Cell 63, 475–484.

Martin, W.J., Gadek, J.E., Hunninghake, G.W. and Crystal, R.G. (1981). Oxidant injury of lung parenchymal cells. J. Clin. Invest. 68, 1277–1288.

Martz, E. (1987). LFA-1 and other accessory molecules functioning in adhesions of T and B lymphocytes. Human Immunol. 18, 3–37.

Masinovsky, B., Urdal, D., and Gallatin, W.M. (1990). IL-4 acts synergistically with IL-1 beta to promote lymphocyte adhesion to microvascular endothelium by induction of vascular cell adhesion molecule-1. J. Immunol. 145, 2886–2895.

Masumoto, A. and Hemler, M.E. (1993). Multiple activation states of VLA-4. J. Biol. Chem. 268, 228–234.

Mileski, W., Harlan, J., Rice, C. and Winn, R. (1990). Streptococcus pneumoniae-stimulated macrophages induce neutrophils to emigrate by a CD18-independent mechanism of adherence. Circ. Shock 31, 259–267.

Montefort, S., Feather, I.H., Wilson, S.J., Haskard, D.O., Lee, T.H., Holgate, S.T. and Howarth, P.H. (1992). The expression of leukocyte-endothelial adhesion molecules is increased in perennial allergic rhinitis. Am. J. Respir. Cell Molecular Biol. 7, 393–398.

Mulligan, M.S., Varani, J., Dame, M.K., Lane, C.L., Smith, C.W., Anderson, D.C. and Ward, P.A. (1991). Role of endothelial-leukocyte adhesion molecule 1 (ELAM-1) in neutrophil-mediated lung injury in rats. J. Clin. Invest. 88, 1396–1406.

Mulligan, M.S., Warren, J.S., Smith, C.W., Anderson, D.C., Yeh, C.G., Rudolph, A.R. and Ward, P.A. (1992a). Lung injury after deposition of IgA immune complexes: requirement for CD18 and L-arginine. J. Immunol. 148, 3086–3092.

Mulligan, M.S., Varani, J., Warren, J.S., Till, G.O., Smith, C.W., Anderson, D.C., Todd III, R.F. and Ward, P.A. (1992b). Role of β_2 integrins of rat neutrophils in complement and oxygen radical-mediated acute inflammatory injury. J. Immunol. 148, 1847–1857.

Mulligan, M.S., Polley, M.J., Bayer, R.J., Nunn, M.F., Paulson, J.C. and Ward, P.A. (1992c). Neutrophil-dependent acute lung injury: requirement for P-selectin (GMP-140). J. Clin. Invest. 90, 1600–1607.

Mulligan, M.S., Wilson, G.P., Todd, R.F., Smith, C.W., Anderson, D.C., Varani, J., Issekutz, T.B., Myasaka, M., Tamatani, T., Rusche, J.R., Vaporciyan, A.A. and Ward, P.A. (1993a). Role of β_1, β_2 integrins and ICAM-1 in lung injury after deposition of IgA and IgA immune complexes. J. Immunol. 150, 2407–2417.

Mulligan, M.S., Smith, C.W., Anderson, D.C., Todd III, R.F., Miyasaka, M., Tamatani, T., Issekutz, T.B. and Ward, P.A. (1993b). Role of leukocyte adhesion molecules in complement-induced lung injury. J. Immunol. 150, 2401–2406.

Newman, W., Beall, L.D., Carson, C.W., Hunder, G.G., Graben, N., Randhawa, Z.I., Gopal, T.V., Wienerkronish, J. and Matthay, M.A. (1993). Soluble E-selectin is found in supernatants of activated endothelial cells and is elevated in the serum of patients with septic shock. J. Immunol. 150, 644–654.

O'Byrne, P.M., Dolovich, J. and Hargreave, F.E. (1987). Late asthmatic responses: state of art. Am. Rev. Respir. Dis. 136, 740–751.

Ohtani, H., Strauss, H.W., Southern, J.F., Tamatani, T., Miyasaka, M. and Isobe, M. (1993). Imaging of intercellular adhesion molecule-1 induction in rejecting heart – a new scintigraphic approach to detect early allograft rejection. Transplant. Proc. 25, 867–869.

Paul, L.C., Davidoff, A., Paul, D.W., Benediktsson, H. and Issekutz, T.B. (1993). Monoclonal antibodies against LFA-1 and VLA-4 inhibit graft vasculitis in rat cardiac allografts. Transplant. Proc. 25, 813–814.

Paul, W.E. (1991). Interleukin-4: A prototypic immunoregulatory lymphokine. Blood 77, 1859–1870.

Pelletier, R., Ohye, R., Kincade, P., Ferguson, R. and Orosz, C. (1993). Monoclonal antibody to anti-VCAM-1 interferes with murine cardiac allograft rejection. Transplant. Proc. 25, 839–841.

Perry, M.A., and Granger, D.N. (1991). Role of CD11/CD18 in shear rate-dependent leukocyte-endothelial cell interactions in cat mesenteric venules. J. Clin. Invest. 87, 1798–1804.

Petty, T.L. (1990). Acute respiratory distress syndrome (ARDS). Disease-a-Month January 1990, 1–58.

Picker, L.J., Warnock, R.A., Burns, A.R., Doerschuk, C.M., Berg, E.L. and Butcher, E.C. (1991a). The neutrophil selectin LECAM-1 presents carbohydrate ligands to the vascular selectins ELAM-1 and GMP-140. Cell 66, 921–933.

Picker, L.J., Kishimoto, T.K., Smith, C.W., Warnock, R.A. and Butcher, E.C. (1991b). ELAM-1 is an adhesion molecule for skin-homing T cells. Nature 349, 796–799.

Piguet, P.F., Rosen, H., Vesin, C. and Grau, G.E. (1993). Effective treatment of the pulmonary fibrosis evicted in mice by bleomycin or silica with anti-CD11 antibodies. Am. Rev. Respir. Dis. 147, 435–441.

Raymond, E.L., McFarland, M.L., Van, G.Y., Churchill, L., Letts, L.G. and Wegner, C.D. (1994). Role of ICAM-1 in respiratory syncytial virus (RSV) infection in mice. J. Respir. Critical Care Med. 149, A48 (Abstr).

Redl, H., Dinges, H.P., Buurman, W.A., van der Linden, C.J., Pober, J.S., Cotran, R.S. and Schlag, G. (1991). Expression of endothelial leukocyte adhesion molecule-1 in septic but not traumatic/hypovolemic shock in the baboon. Am. J. Pathol. 139, 461–466.

Regnier-Vigouroux, A., Blanc, D., Pont, S., Marchetto, S. and Pierres, M. (1986). Accessory molecules and T cell activation I. Antigen receptor avidity differentially influences T cell sensitivity to inhibition by monoclonal antibodies to LFA-1 and L3T4. Eur. J. Immunol. 16, 1385–1390.

Rosen, H. and Gordon, S. (1987). Monoclonal antibody to the murine type 3 complement receptor inhibits adhesion of myelomonocytic cells *in vitro* and inflammatory cell recruitment *in vivo*. J. Exp. Med. 166, 1685–1701.

Sadahiro, M., McDonald, T.O. and Allen, M.D. (1993). Reduction in cellular and vascular rejection by blocking leukocyte adhesion molecule receptors. Am. J. Pathol. 142, 675–683.

Schleimer, R.P., Sterbinsky, S.A., Kaiser, J., Bickel, C.A., Klunk, D.A., Tomioka, K., Newman, W., Luscinskas, F.W., Gimbrone, M.A., McIntyre, B.W. and Bochner, B.S. (1992). IL-4 induces adherence of human eosinophils and basophils but not neutrophils to endothelium: Association with expression of VCAM-1. J. Immunol. 148, 1086–1092.

Schmid-Schonbein, G.W., Usami, S., Skalak, R. and Chien, S. (1980). The interaction of leukocytes and erythrocytes in capillary and postcapillary vessels. Microvas. Res. 19, 45–70.

Schmid-Schonbein, G.W., Shih, Y.Y. and Chien, S. (1985). Morphometry of human leukocytes. Blood 56, 866–875.

Seekamp, A. and Ward, P.A. (1993). Ischemia-reperfusion injury. Agents Actions Suppl. 41, 137–152.

Shappell, S.B., Toman, C., Anderson, D.C., Taylor, A.A., Entman, M.L. and Smith, C.W. (1990). Mac-1 (Ca11b/CD18) mediates adherence-dependent hydrogen peroxide production by human and canine neutrophils. J. Immunol. 144, 2702–2711.

Shijubo, N., Imai, K., Aoki, S., Hirasawa, M., Sugawara, H., Koba, H., Tsujisaki, M., Sugiyama, T., Hinoda, Y., Yachi, A. *et al.*, (1992). Circulating intercellular adhesion molecule-1 (ICAM-1) antigen in sera of patients with idiopathic pulmonary fibrosis. Clin. Exp. Immunol. 89, 58–62.

Shimizu, Y., Newman, W., Gopal, T.V., Horgan, K.J., Graber, N., Beall, L.D., van Seventer, G.A. and Shaw, S. (1991). Four molecular pathways of T cell adhesion to endothelial cells: roles of LFA-1, VCAM-1, and ELAM-1 and changes in pathway hierarchy under different activation conditions. J. Cell Biol. 113, 1203–1212.

Shimizu, Y., van Seventer, G.A., Ennis, E., Newman, W., Horgan, K.J. and Shaw, S. (1992). Crosslinking of the T cell-specific accessory molecules CD7 and CD28 modulate T cell adhesion. J. Exp. Med. 175, 577–582.

Siegelman, M.H., van de Rijn, M. and Weissman, I.L. (1989). Mouse lymph node homing receptor cDNA clone encodes a glycoprotein revealing tandem interaction domains. Science 243, 1165–1172.

Simon, R.H., DeHart, P.D. and Todd III, R.F. (1986). Neutrophil-induced injury of rat pulmonary alveolar epithelial cells. J. Clin. Invest. 78, 1375–1386.

Singer, K.H. (1990). Interactions between epithelial cells and T lymphocytes: role of adhesion molecules. J. Leuko. Biol. 48, 367–374.

Smith, C.W., Rothlein, R., Hughes, B.J., Mariscalco, M.M., Rudloff, H.E., Schmalstieg, F.C. and Anderson, D.C. (1988). Recognition of an endothelial determinant for CD18-dependent human neutrophil adherence and transendothelial migration. J. Clin. Invest. 82, 1746–1756.

Smith, C.W., Marlin, S.D., Rothlein, R., Toman, C., and Anderson, D.C. (1989). Co-operative interactions of LFA-1 and Mac-1 with intercellular adhesion molecule-1 in facilitating adherence and trans-endothelial migration of human neutrophils *in vitro*. J. Clin. Invest. 83, 2008–2017.

Souza, D.J., Gundel, R.H., Barton, R.W., Stearns, C.D., Torcellini, C.A., Miner, E.J., Letts, L.G. and Wegner, C.D. (1993). Changes in T-lymphocyte subsets (CD4/CD8) and activation (IL-2R) following chronic antigen inhalations in monkeys. Chest 103, In press (Abstr).

Spertini, O., Luscinskas, F.W., Kansas, G.S., Munro, J.M., Griffin, J.D., Gimbrone, M.A. and Tedder, T.F. (1991). Leukocyte adhesion molecule-1 (LAM-1, L-selectin) interacts with an inducible endothelial cell ligand to support leukocyte adhesion. J. Immunol. 147, 2565–2573.

Stockenhuber, F., Kramer, G., Schenn, G., Pec, M., Keil, M., and Gnant, M. (1993). Circulating ICAM-1 – novel parameter of renal graft rejection. Transplant. Proc. 25, 919–920.

Tepper, R.I., Levinson, D.A., Stanger, B.Z., Campos-Torres, J., Abbas, A.K., and Leder, P. (1990). IL-4 induces allergic-like inflammatory disease and alters T cell development in transgenic mice. Cell 62, 457–467.

Thomas, J.R., Harlan, J.M., Rice, C.L., and Winn, R.K. (1992). Role of leukocyte CD11/CD18 complex in endotoxic and septic shock in rabbits. J. Appl. Physiol. 73, 1510–1516.

Thompson, H.L. and Matsushima, K. (1992). Human polymorphonuclear leucocytes stimulated by tumor necrosis factor-alpha show increased adherence to extracellular matrix proteins which is mediated via the CD11b/CD18 complex. Clin. Exp. Immunol. 90, 280–285.

Thornhill, M.H., Wellicome, S.M., Mahiouz, D.L., Lanchbury, J.S., Kyan-Aung, U. and Haskard, D.O. (1991). Tumor necrosis factor combines with IL-4 or IFN-gamma to selectively enhance endothelial cell adhesiveness for T-cells. The contribution of vascular cell adhesion-1-dependent and -independent binding mechanisms. J. Immunol. 146, 592–598.

Tosi, M.F., Stark, J.M., Smith, C.W., Hamedani, A., Gruenert, D.C. and Infeld, M.D. (1992a). Induction of ICAM-1 expression on human airway epithelial cells by inflammatory cytokines: effects on neutrophil-epithelial cell adhesion. Am. J. Respir. Cell Molecular Biol. 7, 214–221.

Tosi, M.F., Stark, J.M., Hamedani, A., Smith, C.W. and Gruenert, D.C. (1992b). Intercellular adhesion molecule-1 (ICAM-1)-independent and ICAM-1-independent adhesive interactions between polymorphonuclear leukocytes and human airway epithelial cells infected with para-influenza virus type-2. J. Immunol. 149, 3345–3349.

Tryka, A.F., Skornik, W.A., Godleski, J.J. and Brain, J.D. (1982). Potentiation of bleomycin-induced lung injury by exposure to 70% oxygen: morphologic assessment. Am. Rev. Respir. Dis. 126, 1074–1079.

Vachula, M. and Vanepps, D.E. (1992). *In vitro* models of lymphocyte trans-endothelial migration. Invasion & Metastasis 12, 66–81.

van Dijken, P.J., Gharyur, T., Mauch, P., Down, J., Burakoff, S.J. and Ferrara, J.L. (1990). Evidence that anti-LFA-1 *in vivo* improves engraftment and survival after allogeneic bone marrow transplantation. Transplantation 49, 882–886.

van Seventer, G.A., Newman, W., Shimizu, Y., Nutman, T.B., Tanaka, Y., Horgan, K.J., Gopal, T.V., Ennis, E., O'Sullivan, D., Grey, H. and Shaw, S. (1991). Analysis of T cell stimulation by superantigen plus major histocompatibility complex class II molecules or by CD3 monoclonal antibody: Costimulation by purified adhesion ligands VCAM-1, ICAM-1, but not ELAM-1. J. Exp. Med. 174, 901–913.

Vankooyk, Y., Vandewielvankemenade, E., Weder, P., Huijbens, R.J.F. and Figdor, C.G. (1993). Lymphocyte function-associated antigen-1 dominates very late antigen-4 in binding of activated T-cells to endothelium. J. Exp. Med. 177, 185–190.

Vejlsgaard, G.L., Ralfkiaer, E., Avnstorp, C., Czajkowski, M., Marlin, S.D. and Rothlein, R. (1989). Kinetics and characterization of intercellular adhesion molecule-1 (ICAM-1) expression on keratinocytes in various inflammatory skin lesions and malignant cutaneous lymphomas. J. Am. Acad. Dermatol. 20, 782–790.

von Andrian, U.H., Berger, E.M., Ramezani, L., Chambers, J.D., Ochs, H.D., Harlan, J.M., Paulson, J.C., Etzioni, A. and Arfors, K.E. (1993). *In vivo* behavior of neutrophils from two patients with distinct inherited leukocyte adhesion deficiency syndromes. J. Clin. Invest. 91, 2893–2897.

Walker, C., Virchow Jr., J.-C., Bruijnzeel, P.L.B. and Blaser, K. (1991). T cell subsets and their soluble products regulate eosinophilia in allergic and non-allergic asthma. J. Immunol. 146, 1829–1835.

Walsh, C.J., Leeper-Woodford, S.K., Carey, P.D., Cook, D.J., Bechard, D.E., Fowler, A.A. and Sugerman, H.J. (1991). CD18 adhesion receptors, tumor necrosis factor, and neutropenia during septic lung injury. J. Surg. Res. 50, 323–329.

Walsh, G.M., Hartnell, A., Wardlaw, A.J., Kurihara, K., Sanderson, C.J. and Kay, A.B. (1990). IL-5 enhances the *in vitro* adhesion of human eosinophils, but not neutrophils, in a leucocyte integrin (CD11/18)-dependent manner. Immunology 71, 258–265.

Walsh, G.M., Mermod, J.-J., Hartnell, A., Kay, A.B. and Wardlaw, A.J. (1991). Human eosinophil, but not neutrophil, adherence to IL-1-stimulated human umbilical vascular endothelial cells is $\alpha_4\beta_1$ (very late antigen-4) dependent. J. Immunol. 146, 3419–3423.

Wang, A., Palmer, E., Cohen, D.S., and Sheppard, D. (1990). Transforming growth factor-beta induces polarized secretion of fibronectin by airway epithelial cells. Am. Rev. Respir. Dis. 141, A467 (Abstr).

Watanabe, A., Tominaga, T., Tsuji, J., Yanagihara, Y. and Koda, A. (1992). Effects of anti-CD11a, anti-CD11b and anti-CD18 on histamine release from human basophils primed with IL-3. Int. Arch. Allergy Immunol. 98, 308–310.

Weg, V.B., Williams, T.J., Lobb, R.R. and Nourshargh, S. (1993). A monoclonal antibody recognizing very late activation antigen-4 inhibits eosinophil accumulation in vivo. J. Exp. Med. 177, 561–566.

Wegner, C.D. and Wallace, R.W. (1992c). In "Pharmacology of the Respiratory Tract: Clinical and Experimental", (eds. F.K. Chung and P.J. Barnes), pp. 223–252, Marcel Dekker, Inc., New York.

Wegner, C.D., Smith, C.W. and Rothlein, R. (1989). In "Leukocyte Adhesion Molecules: Structure, Function, and Regulation", (eds. T.A. Springer, D.C. Anderson, A.S. Rosenthal and R. Rothlein), pp. 208–214, Springer-Verlag, New York.

Wegner, C.D., Gundel, R.H., Reilly, P., Haynes, N., Letts, L.G. and Rothlein, R. (1990). Intercellular adhesion molecule-1 (ICAM-1) in the pathogenesis of asthma. Science 247, 456–459.

Wegner, C.D., Torcellini, C.A., Clarke, C.C., Letts, L.G. and Gundel, R.H. (1991a). Effects of single and multiple inhalations of antigen on airway responsiveness in monkeys. J. Allergy Clin. Immunol. 87, 835–841.

Wegner, C.D., Rothlein, R. and Gundel, R.H. (1991b). Adhesion molecules in the pathogenesis of asthma. Agents Actions Suppl. 34, 529–544.

Wegner, C.D., Wolyniec, W.W., LaPlante, A.M., Marschman, K., Lubbe, K., Haynes, N., Rothlein, R. and Letts, L.G. (1992a). Intercellular adhesion molecule-1 (ICAM-1) contributes to pulmonary oxygen toxicity in mice: role of leukocytes revised. Lung 170, 267–279.

Wegner, C.D., Gundel, R.H., Rothlein, R. and Letts, L.G. (1992b). Expression and probable roles of cell adhesion molecules in lung inflammation. Chest 101, 34S–39S.

Wegner, C.D., Gundel, R.H. and Letts, L.G. (1993a). Efficacy of monoclonal antibodies against adhesion molecules in animal models of asthma. Agents Actions Suppl. 43, 151–162.

Wegner, C.D., Gundel, R.H., Churchill, L. and Letts, L.G. (1993b). In "ASTHMA: Physiology, Immunopharmacology, and Treatment", Vol. 4. (eds S.T. Holgate, K.F. Austen, L.M. Lichtenstein, and A.B. Kay), pp. 227–241, Academic Press Ltd, London.

Welbourn, C.R.B. and Young, Y. (1992). Endotoxin, septic shock and acute lung injury – neutrophil, macrophages and inflammatory mediators. British J. Sur. 79, 998–1003.

Weller, P.F., Rand, T.H., Goelz, S.E., Chi-Rosso, G., and Lobb, R.R. (1991). Human esoinophil adherence to vascular endothelium mediated by binding to vascular cell adhesion molecule 1 and endothelial leukocyte adhesion molecule 1. Proc. Natl Acad. Sci. USA 88, 7430–7433.

Witschi, H.R., Haschek, W.M., Klein-Szanto, A.J. and Hakkinen, P.J. (1981). Potentiation of diffuse lung damage by oxygen: determining variables. Am. Rev. Respir. Dis. 123, 98–103.

Worthen, G.S., Schwab, B., Elson, E.L. and Downey, G.P. (1989). Mechanics of stimulated neutrophils: Cell stiffening induces retention in capillaries. Science 245, 183–186.

Wright, S.D., Weitz, J.I., Huang, A.J., Levin, S.M., Silverstein, S.C. and Loike, J.D. (1988). Complement receptor type three (CD11b/CD18) of human polymorphonuclear leukocytes recognizes fibrinogen. Proc. Natl Acad. Sci. USA 85, 7734–7738.

Wright, S.D., Lo, S.K. and Detmers, P.A. (1989). In "Leukocyte adhesion molecules: structure, function, and regulation", (eds T.A. Springer, D.C. Anderson, A.S. Rosenthal and R. Rothlein), pp. 190–207, Springer-Verlag, New York.

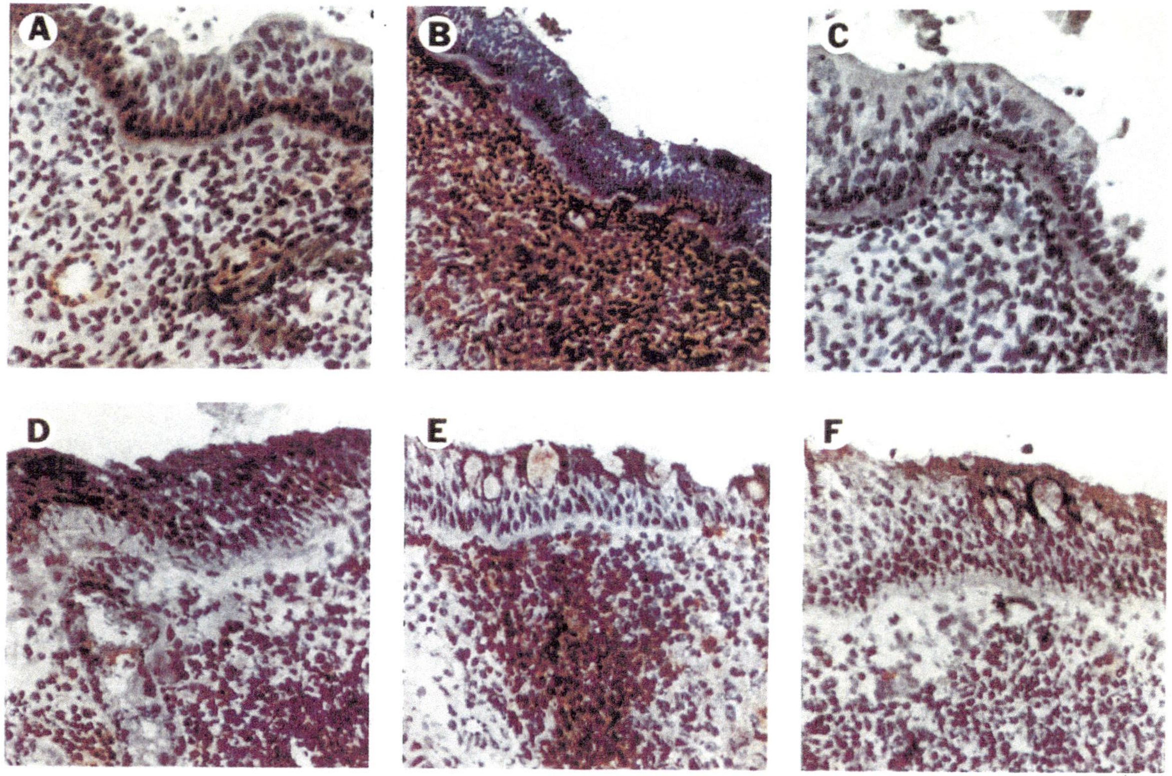

COLOR PLATE I Immunohistochemical staining, with (A and D) mouse anti-human ICAM-1 mAbs, (B and E) a mouse anti-human CD11a mAb, and (C and F) normal mouse serum, of a tracheal section taken from a cynomolgus monkey 20 min after the third of three alternate-day allergen inhalations (A–C) or from another monkey 20 min after a single allergen inhalation. From (Wegner *et al.*, 1990) with permission.

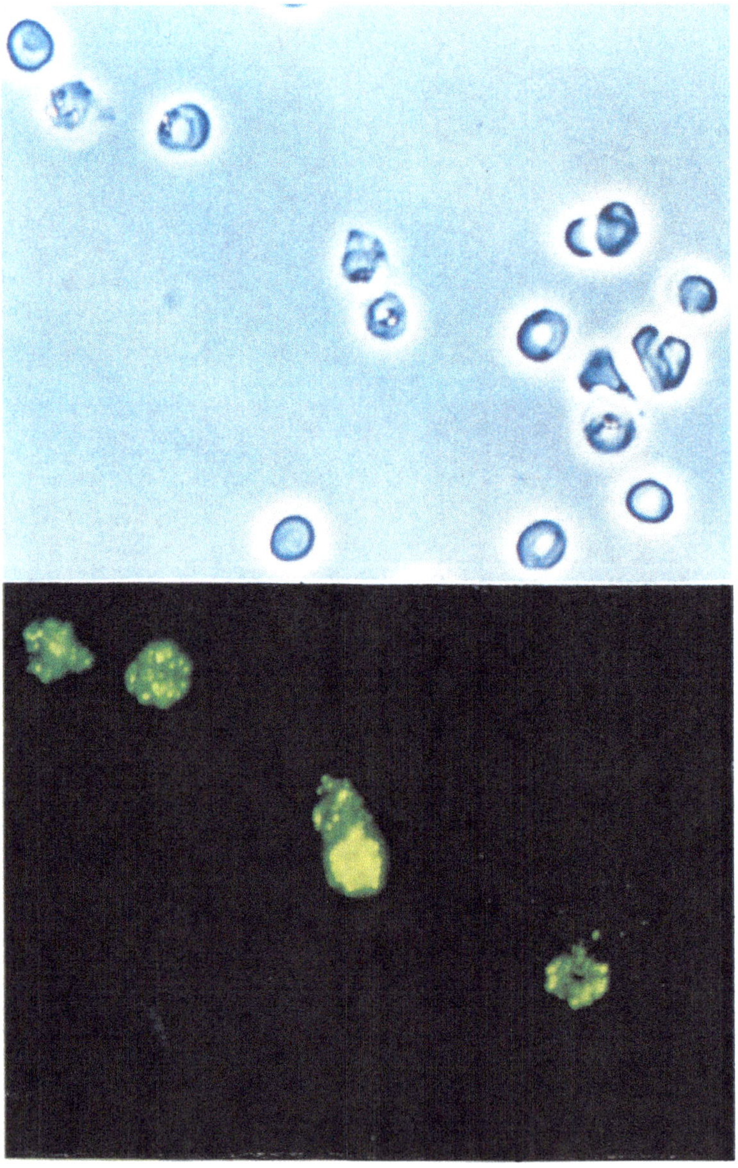

COLOR PLATE II Detection of sequestrin on the surface of malaria-infected erythrocytes but not uninfected erythrocytes by indirect immunofluorescence with anti-idiotype antibodies. The punctuate pattern of immunofluorescent staining corresponds to the distribution of "knobs" on the surface of parasitized red cells. Knobs are electron dense protusions of the surface of malaria-infected red cells which contain sequestrin, a counter-receptor for the endothelial cell adhesion molecule CD36.

10. *Leucocyte-Endothelial Adhesion Molecules* in vivo: *A Role in Physiological Functions and Pathophysiological Vasculidities*

Lawrence W. Argenbright and Thomas C. Noonan

In the broadest sense, the term vasculitis refers to a diverse group of clinical conditions which is characterized by inflammation and damage to blood vessels. Vasculitic events may produce local tissue dysfunction and may also compromise function of distal tissue by reducing blood flow and promoting ischemia.

The inflammatory events which damage the vessel wall depend on critical interactions between circulating leucocytes and the endothelial cells lining the vessel lumen. While the emigration of leucocytes from the bloodstream into tissues is crucial to a successful host response to infection, the inflammatory process itself can, in many cases, cause significant damage to host tissue. While other contributors to this volume have addressed the cellular and biochemical mechanisms underlying the activation of both leucocytes and endothelial cells with reference to the membrane surface molecules mediating their interactions, we will try to place that work in the context of tissue and organ physiology. First, we will discuss observations of leucocyte-endothelial interactions which have been made by the technique of intravital microscopy, then we will discuss quantitation of inflammatory vasculitic processes and their consequences.

1. *Physiological Interactions Between Leucocytes and the Blood Vessel Wall*

An inflammatory response is primarily a cascade of events centred around a terminal vascular bed. Appreciation of this has come largely from direct *in vivo* observation of living vascular beds. Since Leewenhoeck (1907) first ground lenses with sufficient magnification and resolution to observe capillary circulation, and Conheim's description of leucocyte behavior in "inflammation" (Cohnheim, 1882), those interested in both the normal physiology and pathology of small blood vessels have realized the advantages of real-time observation of circulatory phenomena. While a great deal of information regarding the cellular and molecular mechanisms underlying leucocyte–endothelial adhesion has been learned over the past few years, most of this information has been generated in *in vitro* settings of isolated, purified cell populations where experimental conditions can be rigorously controlled. Such studies have yielded specific, detailed information regarding such things as synthesis, expression, regulation of conformation and activity, etc.

Adhesion Molecules
ISBN 0-12-741440-1

of adhesion molecules. Much less is known, however, about cell adhesion mechanisms in the normal and pathophysiological *in vivo* situation. Intravital microscopy offers a method of observing the micro-circulation of a living animal. The technique involves the exteriorization of a relatively transparent tissue (we and many others use the mesentery, but other tissues are available; Messmer, 1983), and keeping the tissue viable by bathing it in a

(a)

(b)

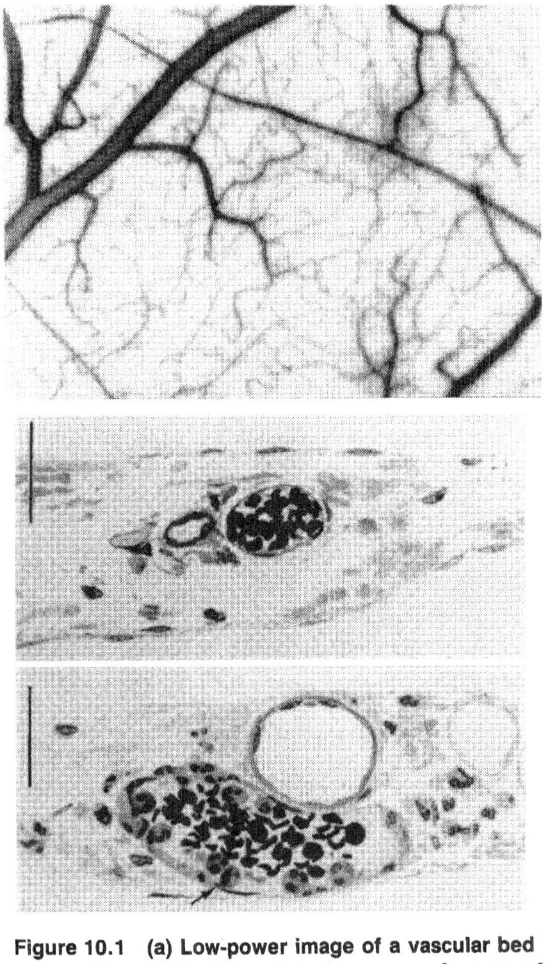

Figure 10.1 (a) Low-power image of a vascular bed as seen through the intravital microscope. A range of blood vessel types and sizes is available for observation. A terminal arteriole cuts across the top of the picture from 11 o'clock to 3 o'clock. A small vein (from 12 o'clock to 9 o'clock) receives blood from a large capillary network. Bar = 100 μm. (b) Transmission electron micrograph of a cross-section of the cat mesentery. Top panel – illustrates a normal arteriole/venule pair. The arteriolar lumen is empty while the larger venule contains erythrocytes. Bottom panel – illustrates an inflammatory response. Dilation of both vessel types is seen. Numerous neutrophils are seen adherent to the venous endothelium (arrow), as well as in the interstitium on either side of the vessel. Bar = 100 μm. (From Oliver *et al.*, 1991, with permission).

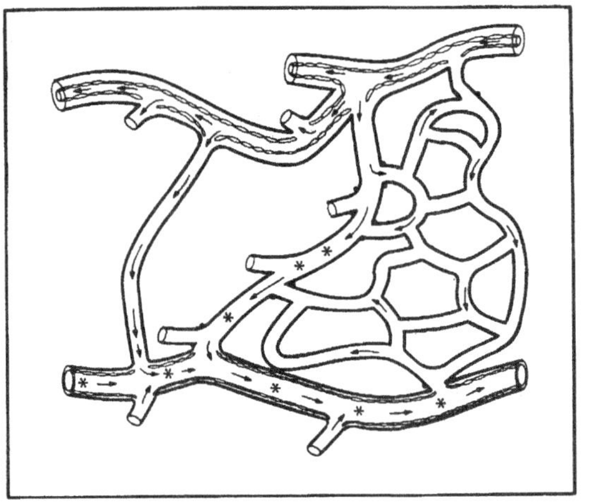

Figure 10.2 Illustration of an idealized terminal vascular bed. Blood circulates from the arterial vessel (top) to the venous vessel (bottom) by moving through a network of capillaries as controlled by the activity of precapillary sphincters (black dots). The asterisks denote post-capillary venous vessels where, under normal conditions, leucocytes can be observed rolling along the endothelial surface, and during inflammation, adhering to and emigrating across the vessel wall.

physiological salt solution where temperature and pH are carefully controlled. The tissue is held across the optical axis of a suitable microscope and observed by trans-illumination where vascular anatomy and interactions between blood cells and the vessel wall can be seen (Fig. 10.1). Still micrography, video microscopy and newer computer-based image processing and analysis allow recording and quantitation of many microvascular functions.

As one observes the ideal terminal vascular bed of a living animal, capillary beds (exchange vessels) are seen interposed between the arterial (supply) vessels, and the venous (collecting) vessels (Fig. 10.2). Leucocytes, rather than flowing freely in the blood with other cells and plasma components, interact preferentially with specific blood vessel types and segments. It is in the post-capillary venous vessels where leucocytes can clearly be seen rolling along the endothelial surface of the vessels at a velocity considerably less than that of the plasma and other cellular blood elements. The rheology of leucocytes certainly plays a role in the initiation of rolling behavior. The flow dynamics through a micro-circulatory network are affected by the low deformability and slow transit times of leucocytes through narrow vessels. As arteriolar segments diverge into narrow capillaries, the entry of a leucocyte into one branch will increase resistance (and decrease flow) in that branch (Fig. 10.3A). Arriving erythrocytes will therefore be preferentially diverted into the other branch, increasing flow there. As capillaries converge into post-capillary venules (Fig. 10.3B), the interaction between the leucocyte and erythrocytes arriving from the feeding capillaries tends to displace the

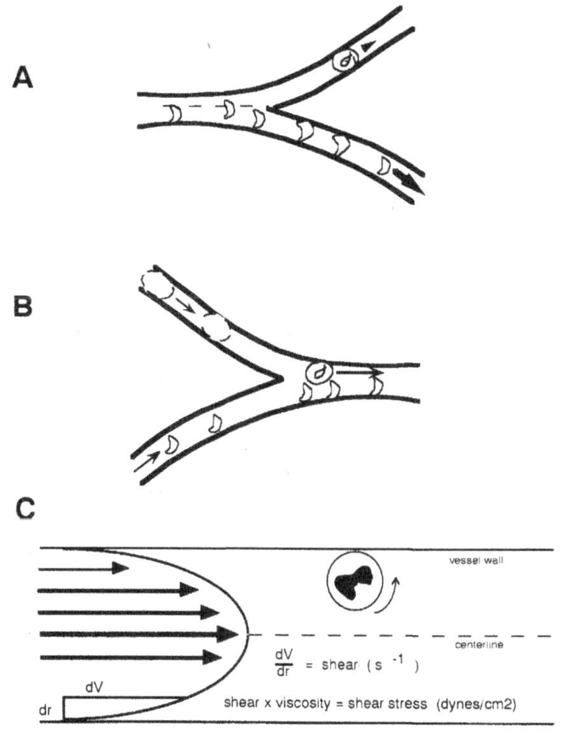

Figure 10.3 (A) Schematic drawing of a divergent capillary bifurcation showing the influence of the entry of a leucocyte into one branch on the flow rates (indicated by the relative sizes of the arrows), and the preferential entry of the subsequent arriving red blood cells (RBC's) into the other branch. (B) Schematic drawing of a convergent microvascular branch point showing the influence of RBC's on the leucocyte, causing the leucocyte to remain in a flow stream very close to the endothelium. (From Chien *et al.*, 1984). (C) Schematic representation of the parabolic flow profile in a blood vessel. Flow velocity is highest at the centerline, and decreases toward the vessel wall. The change in velocity per change in radial displacement is called shear and is highest at the wall. The product of shear and viscosity is called shear stress. Shear imposes a torque on the leucocyte near the vessel wall, promoting a rolling behavior. (From Lawrence and Springer, 1991).

leucocyte toward the vessel wall (Bagge and Branemark, 1977; Chien *et al.*, 1984). Blood is considered a Newtonian fluid as it flows through vessels, and therefore the velocity profile within the vessel is parabolic, with the maximal velocity at the center of the vessel, with a decrease in velocity, because of fluid drag, closer to the vessel wall (Fig. 10.3C). Such a velocity profile imparts a torque on the leucocytes near the wall which tends to make them roll. In addition to these hydrodynamic forces on the leucocyte from the plasma stream, interactions with erythrocytes and erythrocyte aggregates will displace the larger, less deformable leucocytes from the axial flow toward the vessel wall. Most intravital microscopic preparations require some degree of tissue handling, and these preparations also generally show rolling of leucocytes in post-capillary venules. Controversy exists as to whether in fact leucocyte rolling is a normal physiological process, or whether it is inducible by even delicate tissue handling. Observation of the circulation in the bat wing requires no surgical preparation, and rolling of leucocytes is generally not seen (Mayrovitz *et al.*, 1977), whereas it is seen in the dermal circulation of anesthetized mice, which also requires no surgical manipulation (Tangelder and Arfors, 1988).

While hydrodynamic interactions may promote contact of leucocytes entering post-capillary vessels, it is not sufficient to maintain leucocyte rolling to the extent usually seen by intravital microscopy. If these forces were sufficient to maintain leucocyte rolling, then rolling should be seen in occasional arterioles where shear stresses may be equal to those in venous vessels. This has not been observed (Schmid-Schoenbein *et al.*, 1980; Gaehtgens *et al.*, 1985). Further, leucocyte rolling still occurs in venous vessels where calculated shear stress is higher than in some arterioles where rolling is not seen (Atherton and Born, 1972; Mayrovitz *et al.*, 1980; House and Lipowski, 1987). Perry and Granger (1991) modified shear rates in the cat mesentery microvessels by graded occlusion of arterial vessels and found that in arterioles, leucocyte rolling was never seen at shear rates as low as 385 s^{-1}, while leucocyte rolling continued in venules at shear rates approaching 900 s^{-1}. Gaehtgens *et al.* (1985) has calculated the minimum velocity a free-flowing leucocyte could assume in the parabolic flow of small blood vessels, and termed this minimum velocity V_{crit}:

$$V_{crit} = V_{CL} \cdot (r_{wbc}/r) \cdot (2 - r_{wbc}/r)$$

where V_{CL} is the centerline blood velocity, r_{wbc} is leucocyte radius, and r is vessel radius. V_{crit} represents a conservative estimate of the minimum velocity a free-flowing leucocyte with radius r_{wbc} can assume. Any leucocyte that is observed rolling slower than V_{crit} can be assumed to be interacting with the blood vessel wall. In an elegant experiment, Ley and Gaehtgens (1991a) compared leucocyte rolling behavior in arterioles and post-capillary venules of the rat mesentery. An arteriole was identified, and by stroboscopic illumination, fluorescent leucocytes could be seen travelling through the vessel segment. No leucocytes travelled less than V_{crit} suggesting that all leucocytes were free-flowing and not interacting with the vessel wall (Fig. 10.4). Micro-occlusion of the vessel with a micropipet decreased flow velocity and shear rate without inducing any leucocyte rolling. In a post-capillary venule, a population of the leucocytes passing through the vessel were travelling below V_{crit}, and continued to do so after micro-occlusion. These observations suggest that leucocyte rolling is not determined sole by hydrodynamic factors, but may reflect

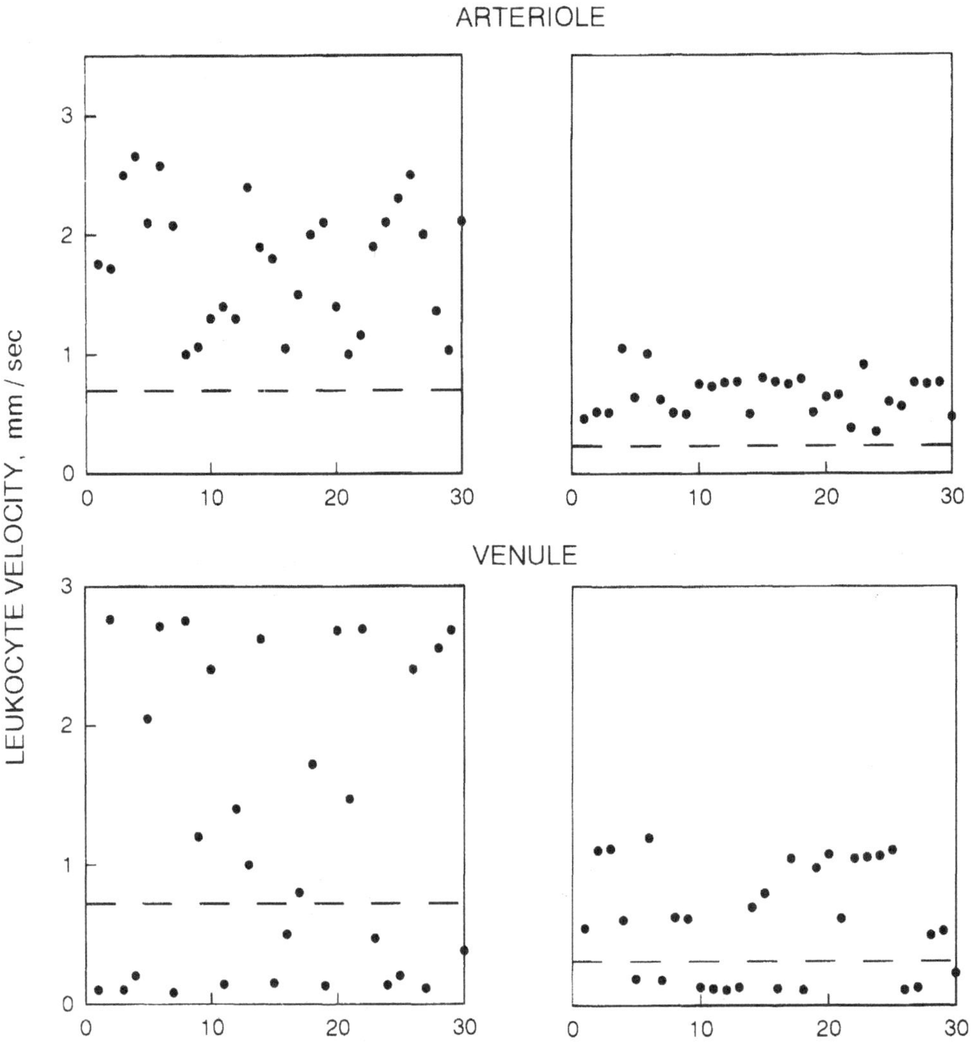

Figure 10.4 Velocities of successive leucocytes observed in a mesenteric arteriole (top panels) and venule (bottom panels); critical velocity, v_{crit} indicated by broken line. Arteriole: diameter, 28.9 μm; mean blood flow velocity, *v*, 1.38 before (left) and 0.52 mm/sec after (right) micro-occlusion (wall shear rate *w*, 382 and 144 sec^{-1}, respectively); no rolling leucocytes. Venule, diameter, 29.6 μm; *v*, 1.40 and 0.67 mm/sec; *w*, 378 and 181 sec^{-1}; 13 and 11 rolling leucocytes, respectively. (From Ley and Gaehtgens, 1991a, with permission).

fundamental differences in arteriolar and venous endothelium, and that rolling implies a transient interaction between leucocytes and endothelial cells.

1.1 SELECTIN-MEDIATED ADHESION

Leucocyte rolling has been shown to be calcium dependent (Thompson *et al.*, 1967), and inhibited *in vivo* by the synthetic sulfated polysaccharide, dextran sulfate, as well as the naturally occurring sulfated polyasccharides, heparin, chondroitin sulfate, keratan sulfate, and heparan sulfate (Ley *et al.*, 1991b; 1993; Tangelder and Arfors, 1991). Intravenous administration of neuraminidase but not sialic acid also blocks leucocyte rolling (Atherton and Born, 1973). These observations suggest that leucocyte rolling is mediated by specific cell-surface adhesion molecules with an affinity for charged carbohydrates. There is now evidence from several laboratories that the selectin family of adhesion molecules plays a role in leucocyte rolling *in vivo*. L-selectin was first discovered by Gallatin *et al.* (1983) and described as a tissue-specific lymphocyte homing receptor, in that the MEL-14 mAb (raised against mouse L-selectin) inhibited binding of mouse lymphocytes to endothelium of frozen, thin sections of lymphoid tissue. L-selectin is also expressed on unstimulated neutrophils, monocytes and eosinophils (Lasky, 1991). Lewinsohn *et al.* (1987) showed that *in vivo* administration of anti-L-selectin mAb inhibits neutrophil localization to cutaneous inflammatory sites, and Jutila and colleagues inhibited neutrophil migration into the inflamed peritoneum of mice with anti-L-selectin (Jutila *et al.*, 1989).

Several groups have now provided convincing evidence that L-selectin mediates leucocyte rolling in post-capillary venules. Ley *et al.* (1991c) reported that a polyclonal rabbit antiserum reactive with L-selectin significantly reduced leucocyte rolling (by about 80%) as evaluated by intravital microscopy. In addition, soluble L-selectin, in the form of an L-selectin-IgG chimeric molecule, also blocked leucocyte rolling, whereas a CD4-IgG chimeric control did not. von Andrian *et al.* (1991) also showed that the DREG-200 mAb, which cross-reacts with rabbit L-selectin, reduces the number of rolling leucocytes as well as increasing the rolling velocity of those cells which maintain rolling contact with the vessel wall. While the specific endothelial ligand for L-selectin has not been identified, Imai *et al.* (1991), using soluble L-selectin, have immunoprecipitated a heavily glycosylated 50 kDa protein from lymphoid tissue. Binding of this protein to L-selectin is blocked by neuraminadase treatment. Independently, Butcher and colleagues (Streeter *et al.*, 1988) have developed a mAb, MECA-79 which stains lymph node endothelium, and also immunoreacts with the 50 kDa glycoprotein isolated by Rosen's group (Imai *et al.*, 1991). It has recently been shown that L-selectin and ELAM-1 [an inducible endothelial antigen (Pober *et al.*, 1986a; Bevilacqua *et al.*, 1987)] operate in the same adhesion pathway *in vitro*, possibly as receptor counter-structures (Kishimoto *et al.*, 1991). Since there is no evidence that there is constitutive ELAM-1 expression *in vivo*, the role, if any, ELAM-1 plays in mediating the normal rolling of leucocytes *in vivo* remains questionable.

P-selectin (GMP-140, PADGEM, CD62) was originally identified on activated platelets (Hsu-Lin *et al.*, 1984; Steinberg *et al.*, 1985), and is found in Weibel-Palade bodies of endothelial cells (Hattori *et al.*, 1989; McEver *et al.*, 1989). This selectin is rapidly expressed on the surface of these cells following exposure to histamine, thrombin, or phorbol esters (Lorant *et al.*, 1991), and is expressed predominantly on venous endothelium in a variety of tissues (McEver *et al.*, 1989). Human neutrophils have been shown to roll on purified P-selectin *in vitro* (Lawrence and Springer, 1991), but to date, no data is available regarding the involvement of P-selectin in leucocyte rolling *in vivo*.

2. *Inflammation-induced Leucocyte Adhesion*

The development of an inflammatory response depends on a change in the way circulating leucocytes and the vascular endothelium interact. Before the initiation of inflammation, it seems that hemodynamic factors present in the micro-circulation, and specific cell adhesion molecules (L-selectin and unidentified endothelial ligand) combine to promote contact (rolling) between leucocytes and the endothelium of post-capillary venules *in vivo*.

One of the earliest responses of a vascular network to an inflammatory stimulus is an alteration in the interaction between this population of rolling leucocytes and the endothelium. A wide range of inflammatory mediators has been shown to bring about changes on the leucocyte and/or endothelial membranes which increase adhesive forces between these two cell types. If these adhesive forces become sufficient to overcome shear forces exerted on the leucocyte by the bloodstream, leucocytes will become firmly attached to the endothelium. This leucocyte–endothelial adhesion can result in several possible outcomes in an *in vivo* situation: (1) if

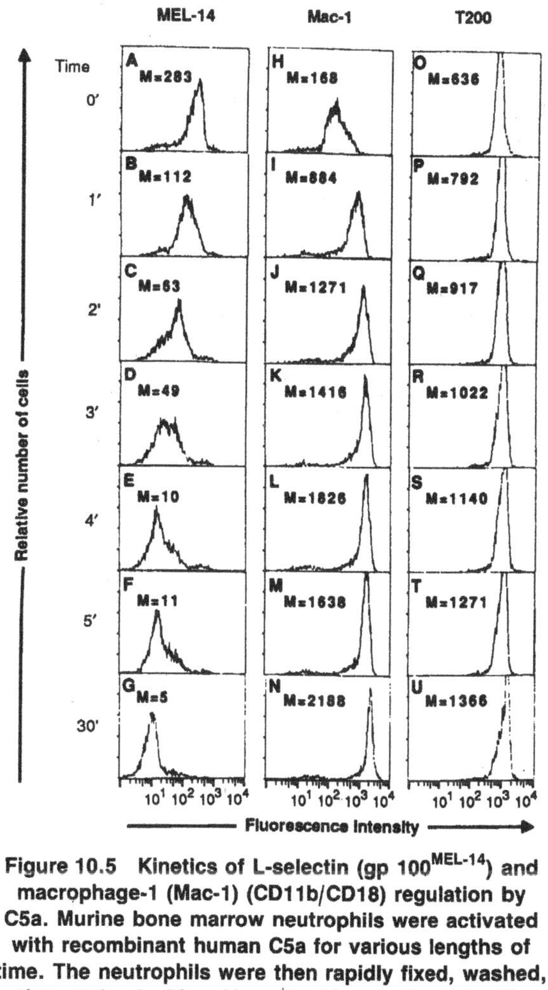

Figure 10.5 Kinetics of L-selectin (gp 100^{MEL-14}) and macrophage-1 (Mac-1) (CD11b/CD18) regulation by C5a. Murine bone marrow neutrophils were activated with recombinant human C5a for various lengths of time. The neutrophils were then rapidly fixed, washed, then stained with mAb against L-selectin (A to G), Mac-1 (H to N), or T200, a non-adhesion membrane protein, then analyzed by flow cytometry. Down-regulation of L-selectin was rapid; 60% reduction 1 min after exposure to C5a (compare A and B). Expression of Mac-1 was concurrently up-regulated. (From Kishimoto *et al.*, 1989).

the nature of the inflammatory event is either mild or brief enough, these adherent leucocytes may eventually detach from the vessel wall and be swept away by the blood. This brief interaction may have no obvious effect on the vessel wall, or the leucocyte may mediate a short-lived increase in vascular permeability to plasma proteins. This loss of plasma protein, and consequent water loss to satisfy oncotic pressure will produce edema. The resolution of edema then being largely determined by the clearance of extravascular protein by lymphatics; (2) leucocytes may both adhere to the endothelium and also aggregate with each other to such an extent as to significantly alter flow dynamics of the micro-circulation, rendering distal tissue ischemic. This leucocyte aggregation induced reduction of blood flow may underlie the wide range of clinical expressions of vasculitis. The circulation of each organ system and its ability to cope with ischemia will determine specific pathologic entities; (3) once adherent to the endothelium, leucocytes may be called through the vessel wall by inflammatory mediators or bacterial products to emigrate to sites of focal infection, etc; (4) if adherent leucocytes are appropriately stimulated, they may move into or through the vessel wall considerably damaging (beyond repair) the vessel wall itself. The focus of a large part of the remainder of this chapter will be to discuss several experimental models in which the vessel wall itself becomes the target of leucocyte destruction, and the role played by certain leucocyte-endothelial cell adhesion molecules in the mediation of these vasculitides.

2.1 INTEGRIN/ICAM-1 ADHESION

In experimental conditions, firm attachment of leucocytes to the endothelium occurs only in those venous vessels where rolling of leucocytes was observed. It is as if the rolling of leucocytes along the vessel wall serves a "surveillance" function. Leucocytes are on the lookout for an in inflammatory condition which may require their accumulation (participation?). Earlier chapters have established that leucocytes, when activated, bind to endothelial cells via the CD18 family of integrins. Integrin activity can be increased very rapidly (sec) (Fig. 10.5). We, and others, have shown by intravital microscopy, that within 1 min of exposing the rabbit mesenteric micro-circulation to the complement split-product C5a, (which activates leucocytes, thus up-regulating integrin activity) these cells adhere to and accumulate on the endothelium of post-capillary venules (Arfors *et al.*, 1987; Argenbright *et al.*, 1991) (Fig. 10.6). This adhesion can be dramatically blocked by pre-treatment of the animal with mAb reactive with either

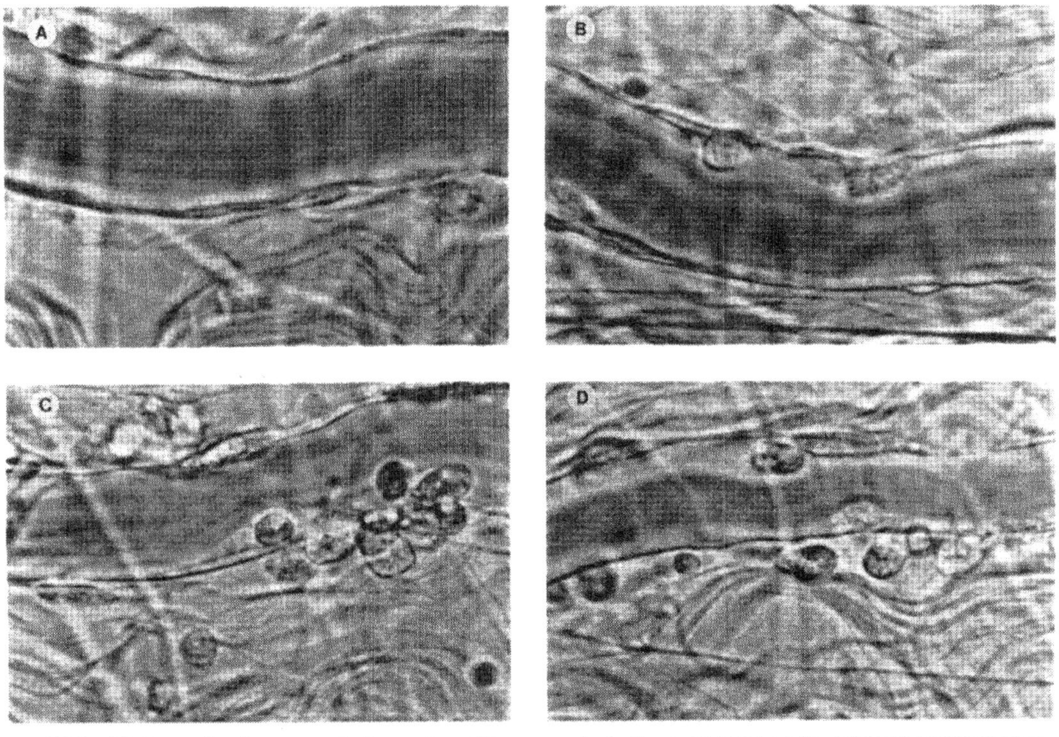

Figure 10.6 Photograph of a mesenteric post-capillary venule in the rabbit taken by intravital microscopy. (A) normal vessel segment; no adherent leucocytes. (B) 2 min after application of the complement activation product C5a; several neutrophils have become firmly adherent to the endothelial surface. (C) 10 min after C5a; several neutrophils have emigrated across the vessel wall. (D) 30 min after C5a; extravasated neutrophils have moved along the abluminal surface of the vessel into the interstitium. (From Wegner, 1992).

CD18, the β-subunit common to the integrins, or to ICAM-1. While we have inhibited the adhesion of activated leucocytes to post-capillary venules with mAb against either CD18 or ICAM-1, the endothelial ligand for CD18 binding, neither antibody alters the normal "surveillance", that is, the rolling of inactivated leucocytes through these vessels. This suggests that the change from rolling to sticking is more than a quantitative change, but that a qualitative change in leucocyte–endothelial interaction takes place following activation of leucocytes. That is, the adhesion glycoproteins mediating firm adhesion *in vivo* are not mediating the weaker attachment which allows rolling to occur.

Of particular importance to this discussion is the demonstration that L-selectin, which is expressed on unstimulated neutrophils, is quickly shed following activation of the cell by the same agents which up-regulate integrin activity such as C5a, N-formyl-methionyl-leucyl-phenylalanine (fMLP), leukotrienes B_4 (LTB_4), IL-1, and tumor necrosis factor (TNF), among others (Fig. 10.7). There is a simultaneous shedding of L-selectin, and up-regulation of integrin activity on the surface of neutrophils which suggests functionally distinct roles for these molecules. These observations support the concept that firm adhesion of leucocytes to endothelium *in vivo* is a two-step process (von Andrian *et al.*, 1991). That is, the "surveillance" function of normal rolling leucocytes along the vascular endothelium depends on L-selectin activity on the unstimulated leucocyte membrane. Rolling behavior gives way to firm attachment following leucocyte activation as integrin activity is quickly up-regulated, and L-selectin is shed from the leucocyte surface (Fig.10.8). The initial rolling event allows the neutrophil just enough contact with the endothelium to be affected by tissue or endothelial-derived inflammatory (chemotactic) mediators. These mediators are then the trigger for the transition from L-selectin-mediated rolling to CD18-mediated firm adhesion. That the endothelium may play a role in the activation of the leucocyte seems an attractive hypothesis since it would be a means by which leucocyte adhesion and trans-endothelial migration would be limited to discreet sites of inflammation. Huber *et al.* (1991) recently reported that IL-8 secreted by stimulated endothelial cells induced rapid shedding of L-selectin, up-regulation of CD18-containing surface molecules, and attachment and transmigration of neutrophils, suggesting that endogenous endothelial IL-8 may regulate transvenular migration of leucocytes. Also, platelet activating factor (PAF) and P-selectin are induced within minutes on thrombin- or histamine-activated endothelium, and are associated with increased leucocyte adhesion, therefore, it has been suggested that P-selectin acts as a tether to stop the circulating leucocyte, allowing PAF to trigger the transition to CD18-dependent transmigration.

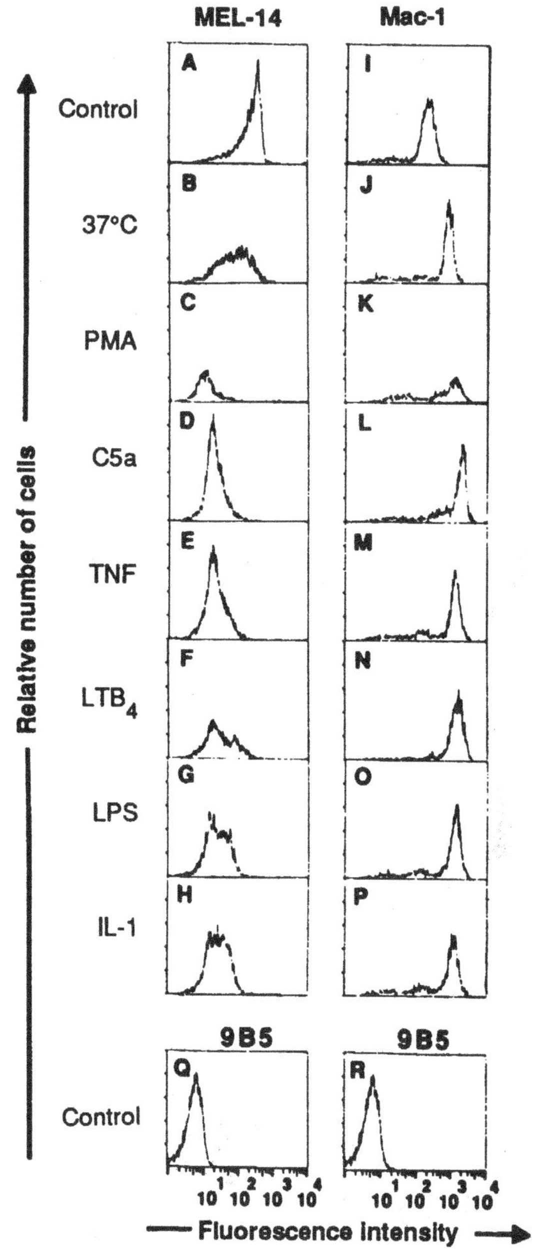

Figure 10.7 Chemotactic factors are potent regulators of L-selectin [metabolic equivalent level-14 (MEL-14)] expression. Murine bone marrow leucocytes (80% neutrophils) were untreated (A and I), or incubated *in vitro* for 1 h at 37°C in the absence of stimulus (B and J), or in the presence of various chemotactic mediators. The cells were then stained with mAb against L-selectin (MEL-14), against Mac-1, or with an irrelevant antibody, before analysis by quantitative flow cytometry. All chemotactic factors caused down regulation of L-selectin and simultaneous increases in Mac-1 activity. (From Kishimoto *et al.*, 1989)

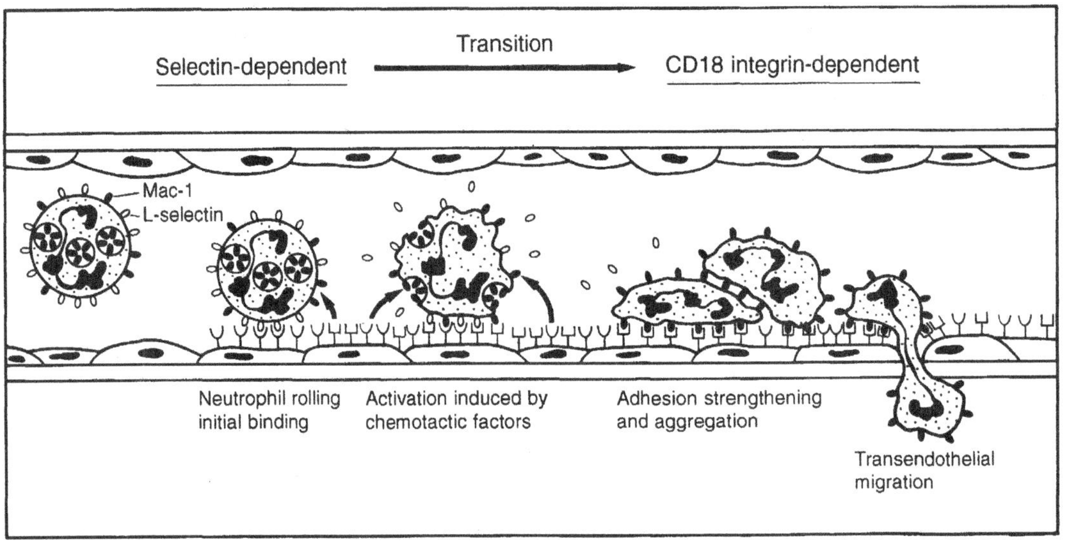

Figure 10.8 A model for neutrophil interaction with inflamed endothelium. Neutrophil localization and activation are multi-step processes, which must be tightly regulated to minimize injury to healthy tissue. Endothelial cells adjacent to a site of inflammation are stimulated by cytokines such as TNF and interleukin-1 to synthesize the adhesion molecules endothelial-leucocyte adhesion molecule-1 (ELAM-1) and intercellular adhesion molecule-1 (ICAM-1)(the two Y-shaped symbols), which guide localization of leucocytes. The adhesion molecule LECAM-1 (L-selectin)(open symbols) present on the neutrophil surface mediate the initial interaction between the inactivated neutrophil and the endothelium – observed as neutrophil rolling and reversible sticking. (Both ELAM-1 and ICAM-1 appear to have multiple ligands. This illustration is not intended to show all possible molecular pairings.) Only these transiently bound neutrophils are exposed to low concentrations of chemoattractants (large arrows) for a sufficient time to be programmed for activation. The activation triggers a rapid transition in neutrophil adhesiveness: there is a transient increase in LECAM-1-mediated adhesion, followed by a rapid shedding of the cell-surface LECAM-1 and, concomitantly, activation of Mac-1-mediated adhesiveness (closed symbols). This is accompanied by a rapid increase in cell-surface expression of Mac-1 from intracellular stores. Thus, activation rapidly shuts down one adhesion pathway (LECAM-1) and turns on another (Mac-1). Engagement of Mac-1 results in adhesion strengthening, neutrophil aggregation, and transendothelial migration, allowing neutrophils to enter the inflamed tissue. (From Kishimoto *et al.*, 1991)

2.2 LEUCOCYTE-MEDIATED VASCULAR DAMAGE

While our discussions so far have concerned the mechanisms of the normal rolling "surveillance" function of leucocytes, and the qualitative changes occurring between circulating leucocytes and the endothelium, we have not addressed the damage to the vessel wall itself (vasculitis) which may be of significant clinical importance. The Arthus reaction has for many years been considered a model of clinically relevant vasculidities, where the deposition of immune complexes in the vessel wall initiates destruction of the vessel itself (Humphrey, 1955a, 1955b; Cochrane *et al.*, 1959). We have recently produced Arthus lesions in rabbit skin by the sequential intradermal injection of anti-bovine IgG (antibody) and intravenous injection of bovine IgG (antigen). Four to six hours later, the injected skin sites were grossly hemorrhagic, thus we have quantitated vascular damage by the accumulation of radio-labelled autologous erythrocytes. We have investigated the role of selected adhesion molecules in both the Arthus reaction and the Shwartzman response in rabbit skin. The Shwartzman response is produced by the intradermal injection of a suitable "preparative" agent, then 24 h later, intravenous challenge with a provocative agent. Four to six h after challenge, hemorrhagic skin lesions appear which are grossly and histologically indistinguishable from Arthus lesions (Stetson, 1951). It is our feeling that agents which successfully prepare the skin for the Shwartzman response do so by up-regulating ICAM-1 on vascular endothelium (Argenbright and Barton, 1992). The intravenous challenge can be any agent which up-regulates integrin activity on circulating leucocytes. We routinely use the intradermal injection of endotoxin as preparative agent. Literature references also include the use of IL-1, TNF (alone or in combination), and bacteria (Movat *et al.*, 1980; Movat *et al.*, 1987; Beck *et al.*, 1986), which are also known to stimulate ICAM-1 expression on cultured endothelium (Dustin *et al.*, 1986; Pober *et al.*, 1986b; Pohlman *et al.*, 1986). Fig. 10.9 shows that the hemorrhage of the Shwartzman response is dependent on the amount of preparative agent (endotoxin) given, and that the Arthus response varies with the amount of intradermal antibody. Results of pre-treatment of rabbits with selected mAb are illustrated

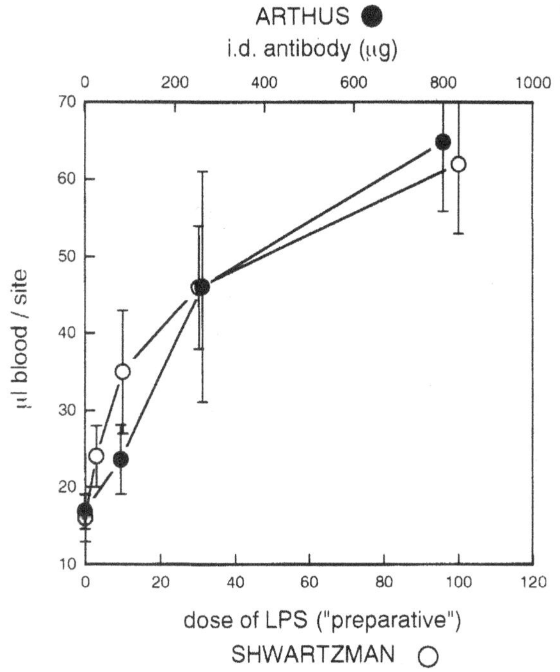

Figure 10.9 Accumulation of RBCs (hemorrhage) in rabbit skin as an index of vascular damage. ARTHUS reaction; intradermal injection of anti-bovine IgG into an animal passively sensitized by intravenous injection of bovine IgG. SHWARTZMAN reaction; intradermal injection of LPS (endotoxin, *S. typhosa*) into rabbit skin, followed 24 h later by systemic activation of complement (intravenous, i.v. zymosan). Both procedures produce an acute necrotizing vasculitis of which hemorrhage is a cardinal feature. Hemorrhage is quantitated by radio-labelled autologous RBC's.

in Fig. 10.10. The hemorrhage of both Arthus and Shwartzman responses is inhibited ⩾90%, by anti-CD18, clearly indicating a leucocyte dependence in both reactions. Of great interest is the observation that anti-ICAM-1 blocked the Shwartzman response by 85%, while there was not a statistically significant inhibition in the Arthus response. These data support the concept that "preparation" for the Shwartzman response equates with ICAM-1 up-regulation, which we have confirmed immunohistochemically. In addition to ICAM-1/CD18 dependent leucocyte–endothelial adhesion, an intravascular leucocyte–leucocyte aggregation appears to be required. Histology of a 1 h old Schwartzman lesion (Fig. 10.11) shows dramatic intravascular leucocyte plugging. Anderson and colleagues (1986) have shown that neutrophil homotypic aggregation *in vitro* is Mac-1 (CD11b)-dependent. Our finding that anti-CD11b was effective in blocking the Shwartzman response suggests the requirement for intravascular leucocyte aggregation in the development of Shwartzman lesions, as well as a leucocyte–endothelial adhesion mediated by CD18/ICAM-1 interaction.

A significant difference between these two models of leucocyte-mediated vascular damage is that anti-ICAM-1 very effectively blocked the Shwartzman response, but showed no significant inhibition of the Arthus response. If the preparation injection in the Shwartzman response induces ICAM-1 expression, then it follows that this reaction should be blocked by anti-ICAM-1 mAb. In the Arthus response, immune-complexes are deposited in the wall of a normal blood vessel, and hemorrhagic events take place before significant endothelial ICAM-1, or ELAM-1 could be induced. Therefore, if leucocytes are binding via integrins, but not to ICAM-1 or ELAM-1, to what are they initially binding? Cochrane, (1963) has shown immune-complexes, in experimental Arthus lesions, appearing in the vessel wall as well as extending intraluminally, where antigen, antibody, and host complement were found to co-localize by immunohistochemistry (Fig. 10.12). When an Arthus reaction was produced in the hamster cheek-pouch and observed by intravital microscopy, Bjork and Smedegard (1984) showed that immune-complexes were only deposited in the wall of post-capillary venules, the same vessels where leucocytes were adherent (Fig. 10.13). We feel it is likely then that leucocytes are binding to opsonized immune-complexes and/or activated complement products on the endothelial surface. In an *in vitro* system, Marks *et al.* (1989) showed that complement fixation on the endothelial surface is a rapid and potent stimulus for neutrophil adhesion. Using a range of specific mAb's, they showed that the attachment of leucocytes to

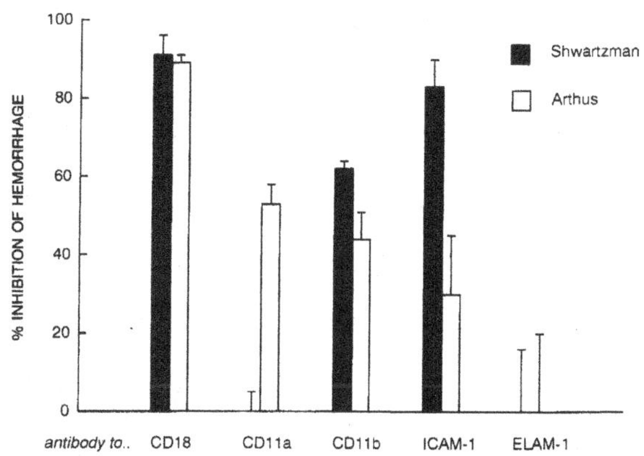

Figure 10.10 Effect of selected anti-adhesion mAbs on the hemorrhagic component of the Shwartzman and Arthus reactions. Groups of 4–6 rabbits each were pretreated with mAb to CD18 (R15.7), CD11a (R7.1), CD11b (LM2), ICAM-1 (R6.5), and ELAM-1 (CL26). Greater than 80% inhibition of hemorrhage was seen, in both models, with antiCD18, clearly indicating a critical role of the leucocyte. Binding of leucocytes to ICAM-1 appears to be more important in the Shwartzman response (anti-ICAM-1, ⩾80% inhibition), whereas some inhibition was suggested in the Arthus response, this inhibition was not statistically significant.

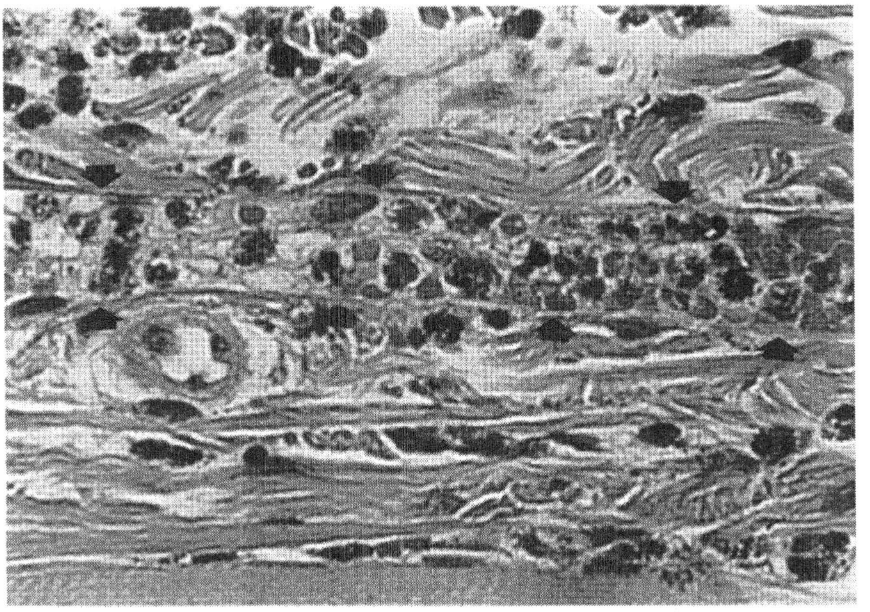

Figure 10.11 Histology of a 1 h old Shwartzman lesion in rabbit skin. A venous vessel is completely occluded by leucocyte aggregates and fibrin deposition. Arrows indicate vessel outline. Notice that an arteriole (seen in cross-section) is patent, with no intravascular plugging present.

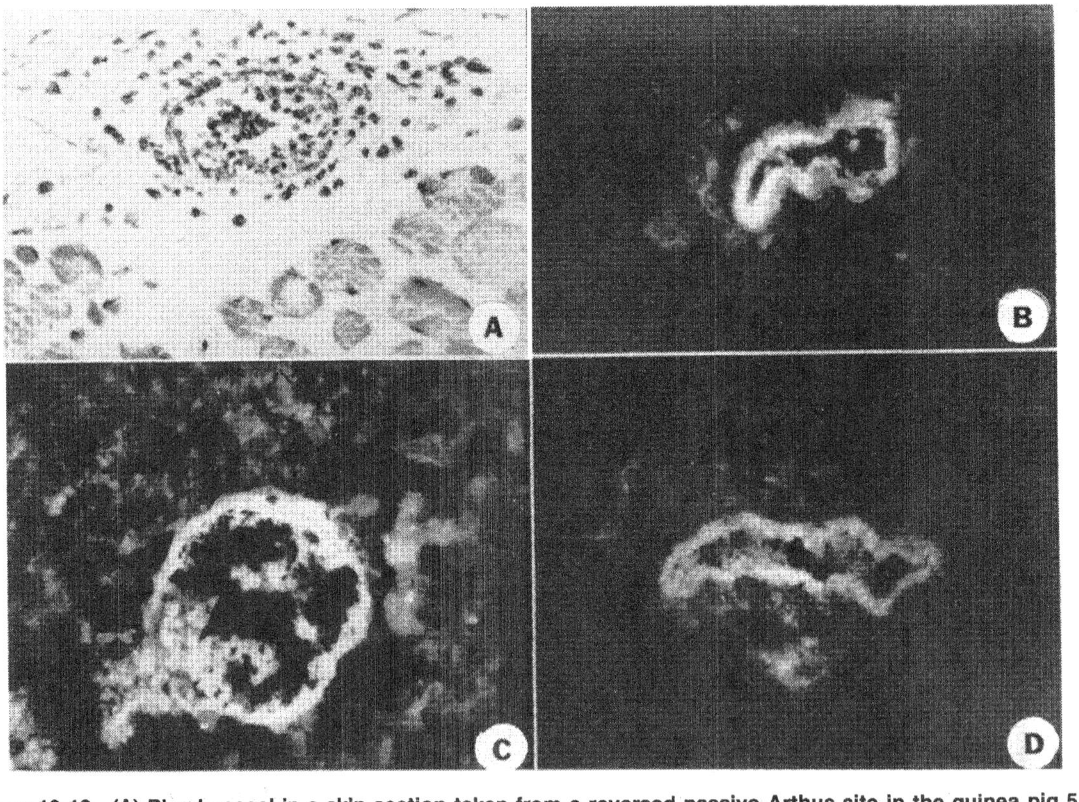

Figure 10.12 (A) Blood vessel in a skin section taken from a reversed passive Arthus site in the guinea pig 5 h after injection. 20 μg anti-human gamma globulin N injected intradermally, and 1.0 mg human gamma globulin intravenously to provoke the reaction. Note typical inflammation and necrosis marked by the influx and agglomeration of polymorphs. X200. (B) A similar vessel to that seen in A, but treated with fluorescent anti-human gamma globulin to show the presence of the antigen in the vessel wall. X250. (C) Similar to B, but showing deposition of antigen within the lumen of the vessel, extending out from the walls. X250. (D) Another vessel similar to those in B and C, but treated with fluorescent anti-guinea pig (host) complement (C'3c). Hence antigen, gamma globulin of the antibody donor species, and host complement were localized. (From Cochrane, 1963).

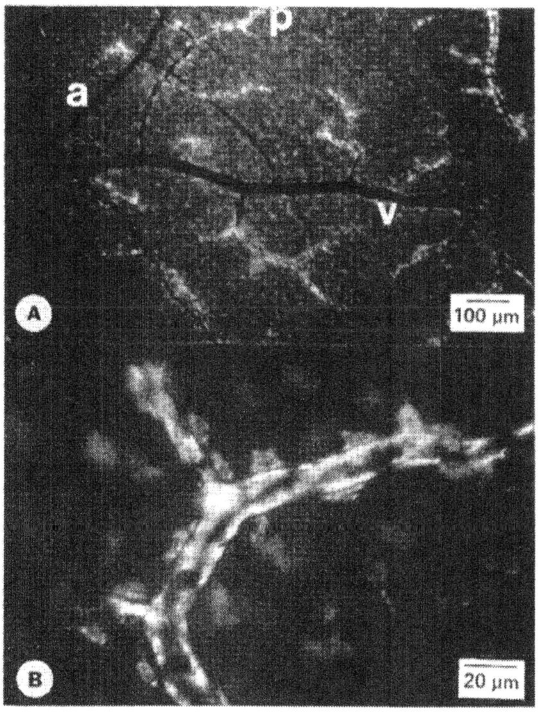

Figure 10.13 (A) Fluorescent micrograph of a hamster cheek-pouch preparation from an ovalbumin-immunized animal after superfusion with fluorescein isothiocyanate (FITC)-ovalbumin. Note the deposition of FITC-OA at the wall of post-capillary and small venules (p). Note also the absence of FITC-OA at arterioles (a) and large venules (v). (B) A post-capillary venule from a preparation treated as in A, showing FITC-OA deposition at its wall. (From Bjork and Smedegard, 1984)

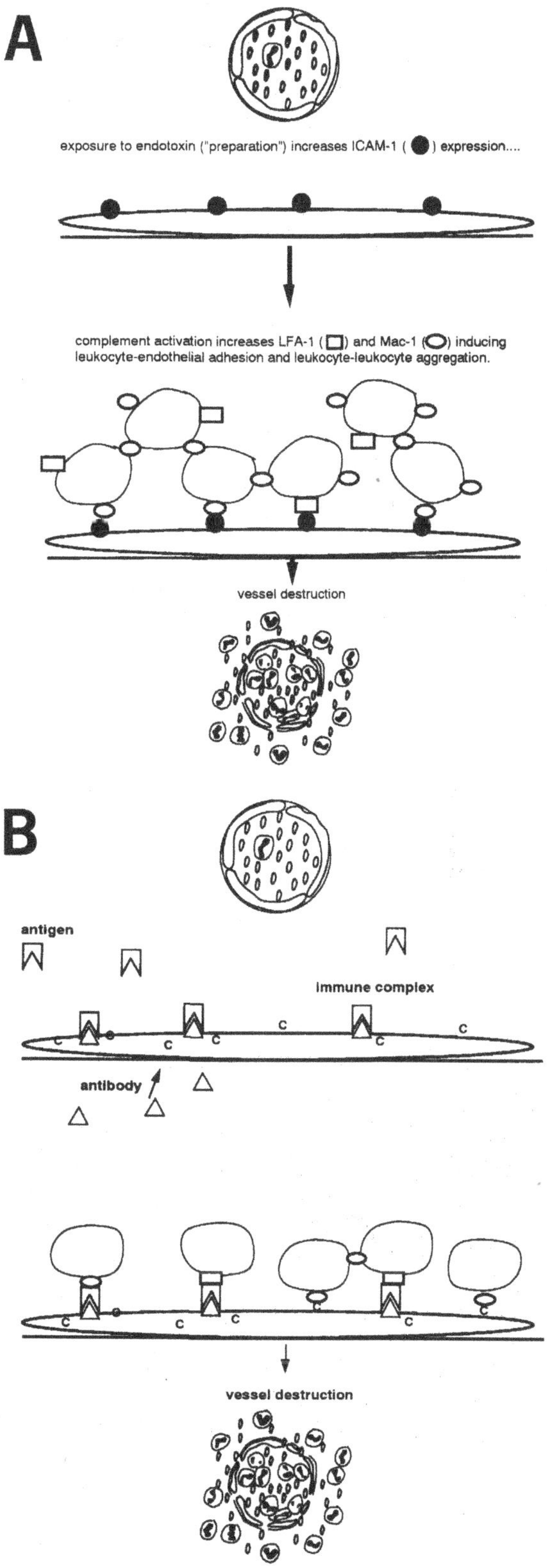

Figure 10.14 (A) Schematic drawing illustrating a proposed mechanism of the development of Shwartzman lesions. Critical for the "preparative" injection appears to be the ability to induce ICAM-1 expression on the vascular endothelium. Complement activation (or any other agent which activates neutrophils) initiates binding of leucocytes via Mac-1 (O) and/or LFA-1 (•) to the endothelium, as well as leucocyte-leucocyte aggregation mediated by Mac-1. This type of inflammatory vascular damage is quite susceptible to intervention with anti-ICAM-1 therapy. (B) Schematic drawing illustrating a proposed mechanism for the development of the Arthus response. Intradermal injection of antibody followed by intravenous injection of antigen causes immune-complexes to be deposited in the vessel wall, extending into the vessel lumen. Immune-complexes stimulate local activation of complement with subsequent complement split-products (C) bound to the endothelial surface. Leucocytes become activated by these complement products and increase Mac-1 and LFA-1 activity on the cell membrane. Leucocytes then bind directly to immune-complexes or to complement products on the endothelium, via Mac-1 (complement receptor type 3, CR3). This inflammatory event is not nearly as dependent on endothelial ICAM-1 expression as is the Shwartzman response, thus anti-ICAM-1 mAb shows only slight inhibitory activity.

complement-treated endothelium was mediated by CD11b, initially identified as the complement type 3 receptor, CR3. Even when immune-complexes are present, significant neutrophil accumulation in the Arthus response is blocked in complement-depleted animals (Ward and Cochrane, 1965). These findings are consistent with our observation that vascular damage in the Arthus response is dependent on CD11b and CD18, but not ICAM-1 or ELAM-1.

Taken together, these results suggest that leucocyte-mediated vasculitis may be produced by at least two different mechanisms (Fig. 10.14). When immune-complex deposition is an initiating event, leucocytes may bind to immune-complexes (opsonized), or to complement products on the endothelium and not to endothelial adhesion molecules. In inflammatory situations where prior exposure of the vascular bed to various cytokines has induced ICAM-1 expression, any leucocyte activation will lead to an integrin-ICAM-1 mediated adhesion. While the pathway from adhesion to vascular destruction is still not clear, and may be common to both types of vasculitis, one type (Shwartzman-like) can be dramatically inhibited by anti-ICAM-1, whereas the other (Arthus-like) can not.

3. *References*

Anderson, D., Miller, L., Schmalstieg, R., Rothlein, R. and Springer, T. (1986). Contributions of the Mac-1 glycoprotein family to adherence-dependent granulocyte functions: structure-function assessments employing sub-unit-specific monoclonal antibodies. J. Immunol. 137, 15–27.

Arfors, K., Lundberg, C., Lindblom, L., Lundberg, K., Beatty, P. and Harlan, J. (1987). A monoclonal antibody to the membrane glycoprotein complex CD18 inhibits polymorphonuclear leukocyte accumulation and plasma leakage *in vivo*. Blood 69(1), 338–340.

Argenbright, L. and Barton, R. (1992). Interactions of leukocyte integrins with intercellular adhesion molecule 1 in the production of inflammatory vascular injury *in vivo*: the Shwartzman reaction revisited. J. Clin. Invest. 89, 259–272.

Argenbright, L., Letts, G. and Rothlein, R. (1991). Monoclonal antibodies to the leukocyte membrane CD18 glycoprotein complex and to intercellular adhesion molecule-1 inhibit leukocyte-endothelial adhesion in rabbits. J. Leuk. Biol. 49, 253–257.

Atherton, A. and Born, G. (1972). Quantitative investigations of the adhesiveness of circulating polymorphonuclear leukocytes to blood vessels. J. Physiol. (London), 222, 447–474.

Atherton, A. and Born. G. (1973). Relationship between the velocity of rolling granulocytes and that of the blood flow in venules. J. Physiol. (London), 233, 157–165.

Bagge, U. and Branemark. P. (1977). White blood cell rheology: an intravital study in man. Adv. Microcirc. 7, 1–17.

Beck, G., Habicth, G., Benach, J. and Miller, F. (1986). Interleukin-1 a common endogenous mediator of inflammation and the local Shwartzman reaction. J. Immunol. 136, 3025–3031.

Bevilacqua, M., Prober, J., Mendrick, D., Cotran, R. and Gimbrone, M. (1987). Identification of an inducible endothelial-leukocyte adhesion molecule, E-LAM 1. Proc. Natl Acad. Sci. USA 84, 9238–9242.

Bjork, J and Smedegard, G. (1984). The microvasculature of the hamster cheek-pouch as a model for studying acute immune-complex-induced inflammatory reactions. Archs. Allergy Appl. Immun. 74, 178–185.

Chien, S., Schmid-Schoenbein, G., Sung, K., Schmalzer, E. and Skalak, R. (1984). In "White Cell Mechanics: Basic Science and Clinical Aspects", (eds. H. Meiselman, M. Lichtman and P. LaCelle), pp. 19–51. Alan R. Liss, New York.

Cochrane, C. (1963). Studies on the localization of circulating antigen-antibody complexes and other macromolecules in vessels. I. Structural studies. J. Exp. Med. 118, 489–509.

Cochrane, C., Weigle, W. and Dixon, F. (1959). The role of polymorphonuclear leukocytes in the initiation and cessation of the Arthus vasculitis. J. Exp. Med. 110, 481–488.

Cohnheim, J. (1882). In "Lectures in General Pathology", 2nd edition (translated from 2nd German edition), Vol. 1. The New Sydenham Society. London.

Dustin, M., Rothlein, R., Bhan, C., Dinarello, C. and Springer, T. (1986). Induction by IL-1 and interferon-gamma: tissue distribution, biochemistry, and function of a natural adhesion molecule (ICAM-1). J. Immunol. 137, 245–254.

Gaehtgens, P., Ley, K., Pries, A. and Muller, R. (1985). Mutual interactions between leukocytes and microvascular blood flow. Prog. Appl. Microcirc. 7, 15–28.

Gallatin, W., Weissman, I. and Butcher, E. (1983). A cell-surface molecule involved in organ-specific homing of lymphocytes. Nature 304, 30–32.

Hattori, R., Hamilton, K., Fugate, R., McEver, R. and Sims, P. (1989). Stimulated secretion of endothelial von Willebrand factor is accompanied lay rapid redistribution to the cell surface of the intracellular granule membrane protein (GMP-140). J. Biol. Chem. 264, 7768–7776.

House, S. and Lipowski, H. (1987). Leukocyte-endothelium adhesion: microhemodynamics in mesentery of the cat. Microvasc. Res. 34, 363–379.

Hsu-Lin, S., Berman, C., Furie, B.C., August, D. and Furie, B. (1981). A platelet membrane protein expressed during platelet activation and secretion. Studies using a monoclonal antibody specific for thrombin-activated platelets. J. Biol. Chem. 259, 9121–9128.

Huber, A., Kunkel, S., Todd, R. and Weiss, S. (1991). Regulation of trans-endothelial migation by endogenous IL-8. Science 25, 99–102.

Humphrey, J. (1955a). The mechanism of Arthus reactions. I. The role of polymorphonuclear leukocytes and other factors in reversed passive Arthus reactions in rabbits. Brit. J. Exp. Path. 36, 268–282.

Humphrey, J. (1955b). The mechanism of Arthus reactions. II. The role of polymorphonuclear leukocytes and platelets in reversed passive reactions in the guinea pig. Brit. J. Exp. Path. 36, 283–297.

Imai, Y., Singer, M., Fannie, C., Lask, L. and Rosen, S. (1991). Identification of a carbohydrate-based endothelial ligand for a lymphocyte homing receptor. J. Cell Biol. 113, 1213–1220.

Jutila, M., Roth, L., Berg, E. and Butcher, E. (1989). Function and regulation of the neutrophil Mel-14 antigen *in vivo*: comparison with LFA-1 and Mac-1. J. Immunol. 10, 3318–3324.

Kishimoto, T., Jutila, M., Berg, E. and Butcher, E. (1989). Neutrophil Mac-1 and MEL-14 adhesion proteins inversely regulated by chemotactic factors. Science. 245, 1238–1241.

Kishimoto,.T., Warnock, R., Jutila, M., Butcher, E., Lane, C., Anderson, D. and Smith, V. (1991). Antibodies against human neutrophil LECAM-1 (LAM-1/Leu-8/DREG-56 antigen) and endothelial cell ELAM-1 inhibit a common CD18-independent adhesion pathway *in vitro*. Blood 78(3), 805–807.

Lasky, L. (1991). Lectin cell adhesion molecules (LEC-CAMs): a new family of cell adhesion proteins involved with inflammation. J. Cell. Biochem. 45, 139–146.

Lawrence, M. and Springer, T. (1991). Leukocytes roll on a selectin at physiologic flow rates: distinction from and prerequisite for adhesion through integrins. Cell 65, 1–20.

Leeuwenhoek, A. van. (1907). Den waaragtigenomloop des bloeds, als mededat de arterien en venae gecontinueerde bloed-vaten zijn, klaar voor de oogen gestelt. Verhandelt in een brief, geschreven aan de koninglijke. Societeit tot London. 65th Missive, Sept. 7, 1688. Opuscula select a Neerlandicorum de Arte medica. Fasc. 1, 92.

Lewinsohn, D., Bargatze, R. and Butcher, E. (1987). Leukocyte-endothelial cell recognition: evidence of a common molecular mechanism shared by neutrophils, lymphocytes, and other leukocytes. J. Immunol. 138, 4313–4318.

Ley, K. and Gaehtgens, P. (1991a). Endothelial, not hemodynamic, differences are responsible for preferential leukocyte rolling in rat mesenteric venules. Circ. Res. 69, 1034–1041.

Ley, K., Cerrito, M. and Arfors, K-E. (1991b). Sulfated polysaccharides inhibit leukocyte rolling in rabbit mesentery venules. Heart Circ. Res. 29, H1667–H1673.

Ley, K., Gaehtgens, P., Fennie, C., Singer, M., Lasky, L. and Rosen, S. (1991c). Lectin-like cell adhesion molecule 1 mediates leukocyte rolling in mesenteric venules *in vivo*. Blood 77, 2553–2555.

Ley, K., Linneman, G., Meinen, M., Stoolman, L. and Gaehtgens, P. (1993). Fucoidan, but not yeast polyphosphomannan PPME inhibits leukocyte rolling in venules of the rat mesentery. Blood 81(1), 177–185.

Lorant, D., Patel, K., McIntyre, T., McEver, R., Prescott, S. and Zimmerman, G. (1991). Coexpression of GMP-140 and PAF by endothelium stimulated by histamine or thrombin: a juxtacrine system for adhesion and activation of neutrophils. J. Cell Biol. 115(1), 223–234.

Marks, R., Todd, R. and Ward, P. (1989). Rapid induction of neutrophil-endothelial adhesion by endothelial complement fixation. Nature. 339, 314–317.

Mayrovitz, H., Wiedeman, M. and Tuma, R. (1977). Factors influencing leukocyte adherence in microvessels. Thromb. Haemostas. 38, 823–830.

Mayrovitz, H., Tuma, R. and Wiedeman, M. (1980). Leukocyte adherence in arterioles following extravascular tissue trauma. Microvasc. Res. 20, 264–274.

McEver, R., Beckstead, J., Moore, K., Marshall, C. and Bainton, D. (1989). GMP-140, a platelet alpha-granule membrane protein, is also synthesized by vascular endothelial cells and is localized in Weibel-Palade bodies. J. Clin. Invest. 84, 92–97.

Messmer, K., Hammersen, F. and Zweifach, B. (eds). (1983). "Perspectives in methodology for studies of the microcirculation". Karger, Basel.

Movat, H., Jaynes, B., Wasi, S., Movat, K. and Kopaniak, M. (1980). In "Bacterial Endotoxins and Host Response", (ed M.K. Agarwal), pp. 179–201, Elsevier/North-Holland, Amsterdam.

Movat, H., Burrows, C., Cybulsky, M. and Dinarello, C. (1987). Acute inflammation and a Shwartzman reaction induced by interleukin-1 and tumor necrosis factor. Am. J. Pathol. 129, 463–476.

Oliver, M., Specian, R., Perry, M. and Granger, D. (1991). Morphologic assessment of leukocyte-endothelial cell interactions in mesenteric venules subjected to ischemia and reperfusion inflammation, 15(5), 331–346.

Perry, M. and Granger, N. (1991). Role of CD11/CD18 in shear rate-dependent leukocyte-endothelial cell interactions in cat mesenteric venules. J. Clin. Res. 87, 1798–1804.

Pober, J., Bevilacqua, M., Mendrick, D., Lapierre, L., Fiers, W. and Gimbrone, M. (1986a). Two distinct monokines, interleukin 1 and tumor necrosis factor, each independently induce biosynthesis and transient expression of the same antigen on the surface of cultured human vascular endothelial cells. J. Immunol. 136, 1680–1687.

Pober, J., Gimbrone, M., Lapierre, L., Mendrick, D., Fiers, W., Rothlein, R. and Springer, T. (1986b). Overlapping patterns of activation of human endothelial cells by interleukin-1, tumor necrosis factor and immune interferon. J. Immunol. 137, 1893–1896.

Pohlman, T., Stanness, K., Beatty, P., Ochs, H. and Harlan, J. (1986). An endothelial cell surface factor induced *in vitro* by lipopolysaccharide, interleukin-1, and tumor necrosis factor increases neutrophil adherence by a CDw18-dependent mechanism. J. Immunol. 136, 4548–4553.

Schmid-Schoenbein, G., Usami, S., Skalak, R and Chien, S. (1980). Cell distribution in capillary networks. Microvasc. Res. 19, 18–44.

Stenberg, P., McEver, R., Shuman, M., Jacques, Y. and Bainton, D. (1985). A platelet alpha granule membrane protein (GMP-140) is expressed on the plasma membrane after activation. J. Cell Biol. 101, 880–885.

Stetson, C. (1951). Similarities on the mechanisms determining the Arthus and Shwartzman phenomena. J. Exp. Med. 94, 347–357.

Streeter, P., Rouse, B. and Butcher, E. (1988). Immunohistologic and functional characterization of a vascular addressin involved in lymphocyte homing into peripheral lymph nodes. J. Cell Biol. 107, 1853–1858.

Tangelander, G. and Arfors, K-E. (1988). Monoclonal antibody Mel-14 cannot abolish leukocyte rolling in mouse venules. Int. J. Microcirc: Clin. Exp. 7 Suppl: S68 (Abstr).

Tangelder, J. and Arfors, K-E. (1991). Inhibition of leukocyte rolling in venules by protamine and sulfated polysaccharides. Blood 77, 1565–1571.

Thompson, P., Papadimitriuo, J. and Walters, M. (1967). Suppression of leukocyte sticking and emigration by chelation of calcium. J. Pathol. Bacteriol. 94, 389–396.

von Andrian, U., Chambers, J., McEvoy, L., Bargatze, R., Arfors, K-E. and Butcher, E. (1991). Two-step model of leukocyte-endothelial cell interaction in inflammation: distinct roles for LECAM-1 and the leukocyte β_2 integrins in vivo. Proc. Natl Acad. Sci. USA 88, 7538–7542.

Ward, C. and Cochrane, C. (1965). Bound complement and immunologic injury of blood vessels. J. Exp. Med. 121, 215–237.

Wegner, C. (1992). In "Update: Pulmonary Diseases and Disorders", (ed. A. Fishman). pp. 125–135. McGraw-Hill, Inc. New York.

11. *Molecular Interaction of Viruses with Host-cell Receptors*

Thomas L. Lentz

1. *Introduction*

The initial stage in the infection of cells by viruses is attachment of the virus to the surface of the host cell (Dales, 1973; Lonberg-Holm and Philipson, 1974; Dimmock, 1982; Tardieu *et al.*, 1982; Kuhn, 1985; Paulson, 1985; Marsh and Helenius, 1989; Lentz, 1990b; Haywood, 1994). Attachment is mediated by the binding of a protein on the virus surface, referred to as the viral attachment protein (VAP), to a molecule on the cell surface acting as a virus receptor. In enveloped viruses, the VAPs are the glycoproteins forming the coat of the virus. The glycoproteins can often be visualized by electron microscopy as spikes protruding from the membrane of the virion. In non-enveloped viruses, the capsid proteins serve as the VAP. In some viruses, a capsid protein takes the form of a fiber or spike which may be involved in initial attachment to cells (Philipson *et al.*, 1968).

The virus receptor is potentially any normal constituent comprising or associated with the cell membrane of the host cell. These components include phospholipids, glycolipids, and integral membrane glycoproteins such as enzymes, ion channels, transporters, adhesion molecules, major histocompatibility complex (MHC) antigens, immunoglobulins, and receptors for substances such as neurotransmitters, growth factors and hormones. Thus, viruses have opportunistically usurped and used as receptors normal surface components serving useful

Adhesion Molecules
ISBN 0-12-741440-1

physiological functions. The attachment step of the viral infectious cycle, though, may represent an Achilles heel at which therapeutic agents can be targeted (Lentz, 1988). These agents would ideally prevent infection by inhibiting attachment of virus to the host cell receptor. In order to rationally develop or design agents that prevent attachment, it is necessary to identify the binding domains of both the VAP and the host cell receptor and determine how they interact at a molecular level. Increasing numbers of host cell receptors for viruses as well as the specific regions of the receptor and VAP that mediate binding have been identified in recent years. In addition, many agents that prevent infection by targeting the recognition event between virus and host cell have been identified. This chapter summarizes current information on the binding domains of host cell virus receptors and the VAP. Examples are presented where this knowledge serves as a basis for the rational development of agents that block the interaction of viruses with host cell receptors.

2. *Virus Entry into Cells*

Binding of virus particles to host cell receptors brings them into immediate physical contact with the cell surface and sets the stage for crossing the membrane barrier. Attachment of virus to host cell receptors may be followed by a conformational alteration in the virus which is essential for internalization (Flynn *et al.*, 1990; Fricks and Hogle, 1990; Moss and Racaniello, 1991). Viruses then enter the cell either by direct fusion with the plasma membrane or by adsorptive- or receptor-mediated, endocytosis (Fig. 11.1) (see Paulson, 1985; Marsh and Helenius, 1989 for review of entry of enveloped and non-enveloped viruses into cells). In adsorptive endocytosis, the receptor-virus complex is internalized in clathrin-coated pits which pinch off from the surface to form coated vesicles. Coated vesicles are then delivered to, and fuse with, endosomes. Endosomes comprise a membrane compartment where internalized ligands and receptors are sorted and targeted to lysosomes, the Golgi apparatus, or back to the cell surface (see Goldstein *et al.*, 1985; Mellman *et al.*, 1986). Endosomes have an acidic interior generated by an adenosine triphosphate (ATP) driven proton pump. The acidic interior of the endosome triggers a conformational change in the viral fusion protein. Exposed hydrophobic

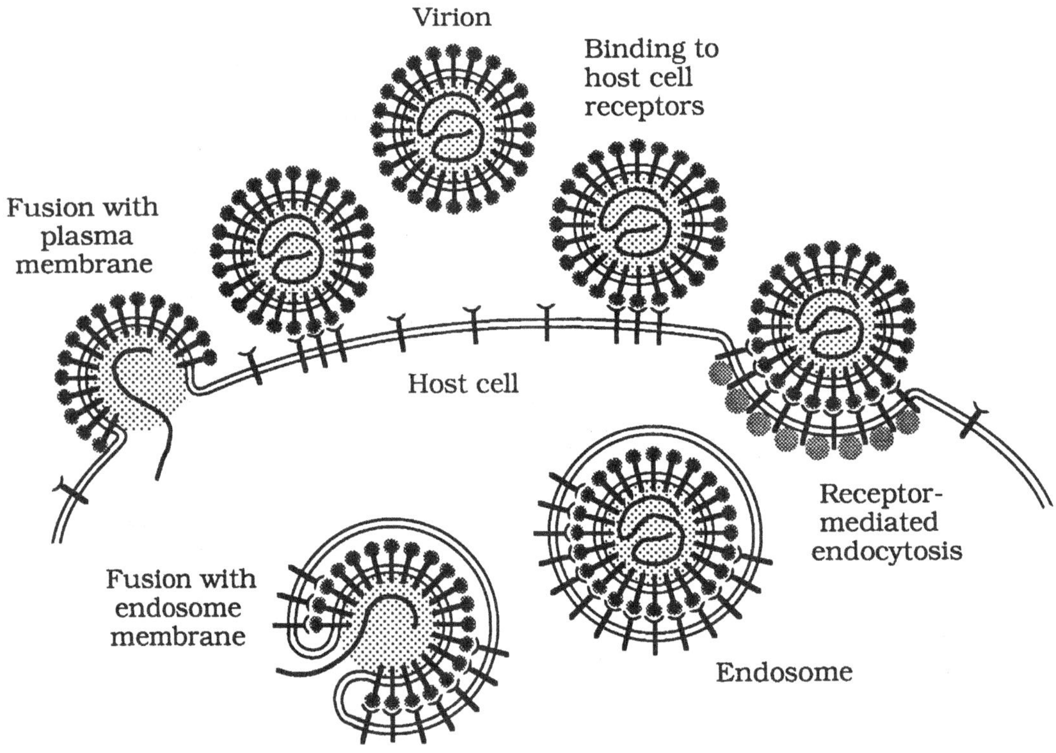

Figure 11.1 Mechanisms of entry of enveloped viruses into cells. The VAPs are spike glycoproteins embedded in the membrane of the virus. The VAPs bind to host cell surface molecules which act as virus receptors. Entry can occur by fusion of the membrane of the virus with the cell membrane releasing the genome into the cytoplasm. Alternatively, the virus is internalized by receptor-mediated-endocytosis in which the virus is internalized in clathrin-coated pits. The coated pit forms a coated vesicle which loses its coat and fuses with an endosome. The acidic interior of the endosome triggers the fusion of the viral membrane with the endosome membrane releasing the genetic material into the cytoplasm. From Lentz (1990b).

sequences induce fusion of the viral membrane or viral particle with the endosome membrane allowing the viral genome to escape into the cytoplasm. Some viruses require proteolytic cleavage of a fusion protein or hemagglutination protein by a host protease before fusion occurs. This cleavage exposes a new hydrophobic N-terminal region. Direct fusion of virions with the plasma membrane similarly releases the genome into the cytoplasm. Some viruses may preferentially utilize one of the two pathways, but many viruses have been shown to enter by both mechanisms. With either process, the viral genome enters the cell without lysing the cell membrane, which would be fatal to the cell. Replication of the genome may take place in the cytoplasm, or transport to the nucleus may be necessary. Viral proteins are synthesized in the cytoplasm and re-assembled with the replicated genome to produce infectious virions which are released into the extracellular space.

Viruses in some cases may be able to adhere to and enter cells through mechanisms that do not involve attachment to specific receptors. As noted below, viruses may adhere to cell surfaces through non-specific interactions. For example, non-saturable binding of foot-and-mouth disease virus to BHK cells has been described (Spier *et al.*, 1983). Solutes and small particles are internalized by cells through fluid-phase or constitutive endocytosis (Steinman *et al.*, 1983) and it might be expected that virus particles adhering non-specifically to the cell surface might be internalized through this mechanism.

3. *Host-cell Receptors for Viruses*

A virus receptor is defined here as a component of the cell surface that contains a domain which by virtue of a structural and conformational complementarity with a domain on the VAP forms a stable complex with the VAP. The initial recognition event may be followed by conformational changes in the VAP or receptor resulting in an induced fit and multipoint interaction. The virus-receptor interface is formed by complementary surfaces and clefts where the VAP and receptor form electrostatic, hydrogen, and van der Waals bonds. Association of the VAP with the receptor is followed by a biological response which may include internalization and replication of the virus, cytopathic changes in the cell, activation of second messenger systems, inhibition of host protein synthesis, or down-regulation of receptors.

3.1 VIRUS RECEPTORS

Virus receptors and their binding domains are listed in Table 11.1. The examples in Table 11.1 are limited to those where the specific identity of the receptor molecule is known or postulated. For many other viruses not listed, the general nature of the receptor, e.g. lipid, protein, has been reported. The experimental evidence is stronger for some of the receptors than for others and relatively few have been identified with certainty. In some cases, the receptors have been hypothesized based, for example, on sequence similarities between portions of the VAP and normal ligands of receptors. Thus, some of the receptors listed here should be considered hypothetical, but hopefully this will stimulate additional research. Future studies will undoubtedly lead to revisions and additions to the list of virus receptors. Whereas five years ago, very few virus receptors were known, receptors are now being identified at an increasing rate.

Techniques involving the isolation of cDNA clones encoding virus receptors promise to lead to the identification of many additional virus receptors, especially because it is not necessary to know the identity of the receptor beforehand. These techniques were used to identify the rhinovirus and poliovirus receptors. In one approach used to identify the rhinovirus receptor, mAbs (monoclonal antibodies) were raised against HeLa cells and antibodies were identified that protected cell monolayers against infection by human rhinovirus 14 (Greve *et al.*, 1989). L cells were transfected with high molecular weight genomic DNA from HeLa cells and cells over-expressing the rhinovirus-binding phenotype were selected. The mABs were used to isolate the receptor protein from HeLa cells and from the transformed cells by affinity chromatography. The purified receptor protein was subjected to proteolytic digestion and the resulting fragments sequenced. The sequences were searched against protein sequence data banks and found to align with the sequence of intercellular adhesion molecule-1 (ICAM-1). In the case of poliovirus, mouse L cells were transfected with HeLa cell DNA, and transformants obtained which expressed the poliovirus receptor and were sensitive to infection by poliovirus (Mendelsohn *et al.*, 1989). Genomic libraries of transformant DNA were probed with a human repetitive probe to rescue the human receptor gene from the genome of the mouse cell transformants. The genomic probe obtained was then used to isolate cDNA clones encoding the poliovirus receptor from HeLa libraries. Nucleotide and predicted amino acid sequence analysis of the receptor cDNA clones revealed the receptor to be a member of the immunoglobulin family of cell surface molecules.

Carbohydrates, lipids, and proteins have been found to act as virus receptors (Table 11.1). Many viruses bind to the sialyloligosaccharides of glycoproteins and glycolipids (gangliosides). These include encephalomyocarditis virus, murine megalovirus, orthomyxoviruses (influenza viruses), paramyxoviruses (Sendai virus, Newcastle disease virus), polyoma virus, rabies virus, reovirus, and rotaviruses. Vesicular stomatitis virus binds to phosphatidylserine and phosphatidylinositol. Most other viruses attach to integral membrane proteins present on the cell surface.

Table 11.1 Host cell receptors for viruses

Virus	*Host-cell receptor*	*Viral attachment protein*
DNA viruses		
Papovaviridae		
Polyomavirus		
Polyomavirus	Sialyloligosaccharides (NeuAcα2,3Galβ1,3GalNAc) (Fried *et al.*, 1981)	
Simian virus (SV) 40	Class I major histocompatibility complex (MHC) molecule (Attwood and Norkin, 1989)	
Adenoviridae		
Mastadenovirus		
Human adenovirus	Class I HLA MHC molecule (Chatterjee and Maizel, 1984)	Residues 1-141 of E3gp (Chatterjee and Maizel, 1984)
Herpesviridae		
Simplexvirus		
Herpes simplex virus type 1	Heparan sulfate moieties of proteoglycans (WuDunn and Spear, 1989) Fibroblast growth factor receptor (Kaner *et al.*, 1990)	Glycoprotein C (Herold *et al.*, 1991)
Cytomegalovirus		
Human cytomegalovirus	Class I HLA MHC molecule via β_2-microglobulin (Grundy *et al.*, 1987) Heparan sulfate (Compton *et al.*, 1993) CD13 (aminopeptidase N) (Soderberg *et al.*, 1993)	Residues 204-297 of H301 α-domain gene product (Beck and Barrell, 1988)
Muromegalovirus		
Murine megalovirus	Sialic acid (*N*-acetylneuraminic acid, NeuAc) and *N*-acetylglucosamine (GlcNAc) (Ravindranath and Graves, 1990) Class I MHC molecule (Wykes *et al.*, 1993)	
Lymphocryptovirus		
Epstein-Barr virus	C3d receptor CR2 (CD21) of B lymphocyte (Fingeroth *et al.*, 1984) 2 N-terminal short consensus repeats (SCR) of CR2 (Lowell *et al.*, 1989; Carel *et al.*, 1990; Molina *et al.*, 1991)	Residues 21–29 and 372–378 of gp350 (Nemerow *et al.*, 1987) Residues 21–30, 219–222, and 372–460 of gp350 (Tanner *et al.*, 1987) Residues 21–29 of gp350 (Nemerow *et al.*, 1989)
Poxviridae		
Orthopoxvirus		
Vaccinia virus	Epidermal growth factor receptor (Eppstein *et al.*, 1985)	Residues 71–80 of VGF protein (Epstein *et al.*, 1985) 50 C-terminal residues of the 32 KDa protein (Maa *et al.*, 1990)
Hepadnaviridae		
Hepatitis B virus	Hepatocyte receptor for polymerized serum albumin via albumin (Machida *et al.*, 1984) Hepatocyte receptor for polymeric IgA (Neurath *et al.*, 1986) Sialoglycoprotein (Komai *et al.*, 1988) Endonexin II (Hertogs *et al.*, 1993)	PreS portion of Env protein (Machida *et al.*, 1984) Residues 21–47 of preS1 portion of L env protein (Neurath *et al.*, 1986) PreS1 sequence of large S protein (hepatocyte receptor) and preS2 sequence of middle and large S proteins (polymerized serum albumin) (Pontisso *et al.*, 1989a) Residues 120–145 of preS2 (polymerized serum albumin (Dash *et al.*, 1991)

(continued)

Table 11.1 (*continued*)

Virus	*Host-cell receptor*	*Viral attachment protein*
DNA viruses		
Parvoviridae		
Parvovirus		
Canine parvovirus		Residues of VP2 capsid protein lining a canyon on the virus surface (Tsao *et al.*, 1991)
Human parvovirus B19	Erythrocyte P antigen (Brown *et al.*, 1993)	
RNA viruses		
Picornaviridae		
Enterovirus		
Poliovirus	Member of immunoglobulin superfamily (Mendelsohn *et al.*, 1989) N-terminal immunoglobulin like domain (V domain, domain 1) (Koike *et al.*, 1991b; Selinka *et al.*, 1991) Gln-55 and other residues in predicted C′-C″-D edge of domain 1 (Morrison *et al.*, 1994) Gly-73, Ser-74, Gln-82, Leu-99-Glu-102, and Gln-130-Ser-132 of domain 1 (Aoki *et al.*, 1994)	Residues 95–105 of VP1 capsid protein (Murray *et al.*, 1988)
Coxsackievirus A9		ArgGlyAsp sequence in C-terminal region of VP1 (Chang *et al.*, 1989)
Echovirus 1	Integrin VLA-2 (Bergelson *et al.*, 1992)	
Rhinovirus		
Human rhinovirus	Intercellular adhesion molecule-1 (ICAM-1) (Greve *et al.*, 1989; Staunton *et al.*, 1989; Tomassini *et al.*, 1989) 82 N-terminal residues of ICAM-1 (Lineberger *et al.*, 1990) First N-terminal immunoglobulin-like domain of ICAM-1 (Staunton *et al.*, 1990)	Residues of VP1 and VP3 major capsid proteins lining a canyon on the virus surface (Rossmann *et al.*, 1985) Lys-103, Pro-155, His-220, Ser-223 of VP1 protein (Colonno *et al.*, 1988)
Human rhinovirus-87	Sialic acid (Uncapher *et al.*, 1991)	
Human rhinovirus-2	Low density lipoprotein receptor (Hoffer *et al.*, 1994)	
Cardiovirus		
Encephalomyocarditis virus	Sialoglycoproteins (Burness and Pardoe, 1981) Sialic acid (Tavakkol and Burness, 1990)	
Mengo virus		Residues of VP 1 and VP 3 major capsid proteins lining a pit on the virus surface (Luo *et al.*, 1987)
Aphthovirus		
Foot-and-mouth disease virus	Integrins (adhesion proteins) (Fox *et al.*, 1989)	Residues 145–147 (ArgGlyAsp) and 203–213 of VP 1 protein (Fox *et al.*, 1989) Residues 133–158 and C-terminal region of VP 1 protein (Acharya *et al.*, 1989)
Reoviridae		
Reovirus		
Reovirus 3	β Adrenergic receptor (Co *et al.*, 1985) Sialoglycoproteins (Paul and Lee, 1987) α-Anomeric form of sialic acid (Paul *et al.*, 1989) Sialic acid (Williams *et al.*, 1991)	C-terminal portion of hemaglutinin (Bassel-Duby *et al.*, 1985) Tyr-326, Ser-327, Ser-329, Ser-325 in hemagglutinin type 3 (Williams *et al.*, 1991)

(*continued*)

Table 11.1 *(continued)*

Virus	*Host-cell receptor*	*Viral attachment protein*
RNA viruses		
Rotavirus		
Simian and bovine rotaviruses	Sialic acid (Fukudome *et al.*, 1989; Bass *et al.*, 1991)	
SA11 rotavirus	Acetylated sialic acid (Willoughby and Yolken, 1990)	
Group A porcine rotavirus	Sialic acid of monosialoganglioside (Rolsma *et al.*, 1994)	
Togaviridae		
Alphavirus		
Semliki Forest virus	Class I HLA and H-2 MHC molecules (Helenius *et al.*, 1978)	
Lactate dehydrogenase-elevating virus	Class II Ia MHC molecule of macrophage (Inada and Mims, 1984)	
Sindbis virus	Laminin receptor (Wang *et al.*, 1992)	
Orthomyxoviridae		
Influenzavirus		
Influenza virus	Sialyloligosaccharides (Paulson *et al.*, 1979; Paulson, 1985) Influenza A: NeuAcα2,6Galβ1,4GlcNAc NeuAcα2,3Galβ1,3GlcNAc NeuAcα2,3Galβ1,4GlcNAc (Carroll *et al.*, 1981; Paulson *et al.*, 1979; Rogers and Paulson, 1983) CD43 (sialophorin) of human neutrophils (Rothwell and Wright, 1994) Influenza C: 9-*O*-AcNeuAc (Rogers, 1986)	Residue 226 of hemagglutinin (HA) (Rogers *et al.*, 1983) Tyr-98, Trp-153, His-183, Glu-190, Leu-194 and neighboring pocket residues of HA (Wiley and Skehel, 1987) Tyr-98, Gly-134, Gly-135, Ser-136, Asn-137, Ala-138, Trp-153, Thr-155, His-183, Ser-186, Glu-190, Leu-194, Tyr-195, Leu-226, Ser-228 of HA (Weis *et al.*, 1988)
Paramyxoviridae		
Paramyxovirus		
Sendai virus	Sialyloligosaccharides (NeuAcα2,3Galβ1,3GalNAc) Paulson *et al.*, 1979) NeuAcα2,8NeuAcα2,3Galβ1,3GalNAc (Holmgren *et al.*, 1980)	
Newcastle disease virus	Sialyloligosaccharides (NeuAcα2,3Galβ1,3GalNAc) (Paulson *et al.*, 1979)	Residue 193 of hemagglutinin-neuraminidase glycoprotein (Iorio *et al.*, 1991)
Morbillivirus		
Measles virus	CD46 (membrane cofactor protein) (Dorig *et al.*, 1993; Naniche *et al.*, 1993; Manchester *et al.*, 1994)	
Coronoviridae		
Coronavirus		
Mouse hepatitis virus	Member of carcinoembryonic antigen family (Williams *et al.*, 1991)	
Transmissible gastroenteritis virus	CD13 (aminopeptidase N) (Delmas *et al.*, 1992)	
Human coronavirus 229E	CD13 (aminopeptidase N) (Yeager *et al.*, 1992)	
Rhabdoviridae		
Vesiculovirus		
Vesicular stomatitis virus	Phosphatidylserine (Schlegel *et al.*, 1983; Mastromarino *et al.*, 1987) Phosphatidylinositol (Mastromarino *et al.*, 1987) GM3 ganglioside (Mastromarino *et al.*, 1987)	

(continued)

Table 11.1 *(continued)*

Virus	*Host-cell receptor*	*Viral attachment protein*
RNA viruses		
Lyssavirus		
Rabies virus	Acetylcholine receptor (Lentz *et al.*, 1982)	Residues 151–238 of G protein (Lentz *et al.*, 1984)
	Residues 173–204 of acetylcholine receptor α-subunit (Lentz, 1990a)	Residues 175–203 of G protein (Lentz *et al.*, 1987)
	Sialylated gangliosides (Superti *et al.*, 1986)	Residues 192–203 of G protein (Bracci *et al.*, 1988)
Retroviridae		
Oncovirinae		
Human T-cell leukemia virus (HTLV-1)	Class I HLA MHC molecule (Clarke *et al.*, 1983)	Residues 246–253 of envelope gyp (Kohtz *et al.*, 1988)
	Interleukin 2 receptor (Kohtz *et al.*, 1988; Lando *et al.*, 1983)	
Murine leukemia virus	Lymphoma cell surface IgM (McGrath *et al.*, 1987)	
Ecotropic murine leukemia virus	622 amino acid, hydrophobic protein (Albritton *et al.*, 1989), possibly an amino acid transporter (Kim *et al.*, 1991; Wang *et al.*, 1991)	245 residue N-terminal domain of gp70 (Heard and Danos, 1991)
Radiation leukemia virus	T cell receptor-L3T4 molecule complex (O'Neill *et al.*, 1987)	
Avian leukosis virus		Residues 100–271 of gp85 (Dorner *et al.*, 1985)
Gibbon ape leukemia virus (GALV)	679 amino acid protein (O'Hara *et al.*, 1990), possibly a phosphate transporter (Johann *et al.*, 1992)	
Feline leukemia virus	Same receptor as for GALV (Takeuchi *et al.*, 1992)	
Amphotropic murine retroviruses	Same receptor as for GALV (Miller *et al.*, 1994)	
Rous sarcoma virus-A	Low density lipoprotein receptor (Bates *et al.*, 1993)	
Lentivirinae		
Lentivirus		
Visna virus	Ovine class II MHC molecule (Dalziel *et al.*, 1991)	Residues 193–200 of gp 120 (HTLV-III$_b$ strain) (Pert *et al.*, 1986)
Human immunodeficiency virus (HIV-1)	CD4 molecule of T lymphocyte (Dalgleish *et al.*, 1984; Klatzmann *et al.*, 1984b; McDougal *et al.*, 1986a)	Residues 426–437 of gp 120 (HTLV-III$_b$ strain) (Lasky *et al.*, 1987)
	Residues 37–53 of CD4 (Jameson *et al.*, 1988)	Regions of gp 120 including residues 363, 419, 473, and 39 C-terminal residues (HTLV-III$_b$ strain) (Kowalski *et al.*, 1987)
	Residues 37–83 of CD4 Landau *et al.*, 1988	Residues 393–417 of gp120 (HTLV-III$_b$ strain) (Dowbenko *et al.*, 1988)
	Residues 31, 44, 48, 52, 55, 57 of CD4 (Mizukami *et al.*, 1988)	Residues 468–511 of gp 120 (BRU strain) (Linsley *et al.*, 1988)
	Residues 42–49 of CD4 (Peterson and Seed, 1988)	Trp 432 of gp 120 (BRU strain) (Cordonnier *et al.*, 1989a)
	Residues 41–55 of CD4 (Arthos *et al.*, 1989)	Residues 82–95, 386–389, 424–432, and 487–499 of gp120 (BRU strain) (Cordonnier *et al.*, 1989b)
	Residues 48, 50, 51, 121–123 of CD4 (Clayton *et al.*, 1989)	Residues 368, 370, and 427 of gp120 (HXBc2 strain) (Thali *et al.*, 1991)
	CD4 molecule interacting with class II HLA-DR MHC molecule (Mann *et al.*, 1988)	
	Acetylcholine receptor (Neri *et al.*, 1990; Bracci *et al.*, 1992)	
	Galactosyl ceramide (Bhat *et al.*, 1991)	
HIV-2	CD4 molecule (Guyader *et al.*, 1987)	
	Residues 44 and 48 of CD4 (Chakrabarti *et al.*, 1990)	
Simian immunodeficiency virus (SIV)	CD4 molecule (Kannagi *et al.*, 1985; Kornfeld *et al.*, 1987; Hoxie *et al.*, 1988)	

Many of these proteins are members of the immunoglobulin superfamily (White and Littman, 1989). These include the CD4 receptor which serves as the receptor for human immunodeficiency virus-1, -2 (HIV-1, HIV-2) and simian immunodeficiency virus (SIV), IgM suggested to be a receptor for murine leukemia virus, ICAM-1 which is the rhinovirus receptor, the receptor for poliovirus, a member of the carcino-embryonic antigen family which is the receptor for mouse hepatitis virus, and MHC molecules which have been suggested to act as receptors for several viruses including human adenovirus, cytomegalovirus, lactate dehydrogenase-elevating virus, Semliki Forest virus, simian virus 40, and visna virus. Other proposed virus receptors function normally as receptors for hormones or neurotransmitters. These include epidermal growth factor receptor for vaccinia virus, β adrenergic receptor for reovirus 3, acetylcholine (ACh) receptor for rabies virus, IL-2 receptor for human T-cell leukemia virus-1 (HTLV-1), and fibroblast growth factor receptor for herpes simplex virus. Foot-and-mouth disease virus utilizes members of the integrin family of adhesion receptors. Finally, the receptor for ecotropic murine leukemia virus (Albritton *et al.*, 1989) may be an amino acid transporter (Kim *et al.*, 1991; Wang *et al.*, 1991), indicating transmembrane transporters or channels can be utilized as receptors (Vile and Weiss, 1991).

The most frequently used receptors for viruses appear to be sialyloligosaccharides of glycoproteins and glycolipids, and members of the immunoglobulin superfamily. The latter includes immunoglobulins, MHC molecules, some CD molecules, ICAMs, T cell receptors, lymphocyte function-related antigens, and some growth factor receptors (Springer, 1990). Together, the two groups comprise a major portion of cell surface molecules.

3.2 Modes of Virus Attachment

Binding of viruses to cells is considerably more complicated than the interaction of relatively small ligands such as neurotransmitters or hormones with their receptors. Techniques to analyze the latter such as Scatchard analysis may not always be applicable to virus binding data. It has been noted that virus binding is affected by the number of VAPs (12–60 for picornaviruses to around 1000 for some enveloped viruses), the number of receptors on the host cell (generally in the range of 10^3–10^6 per cell), affinity of the VAP for the receptor, and non-specific binding (Wickham *et al.*, 1990). Binding is also influenced by the multiplicity of input and by factors such as pH, temperature, and ions. Modes of virus attachment to the cell surface are shown in Fig. 11.2.

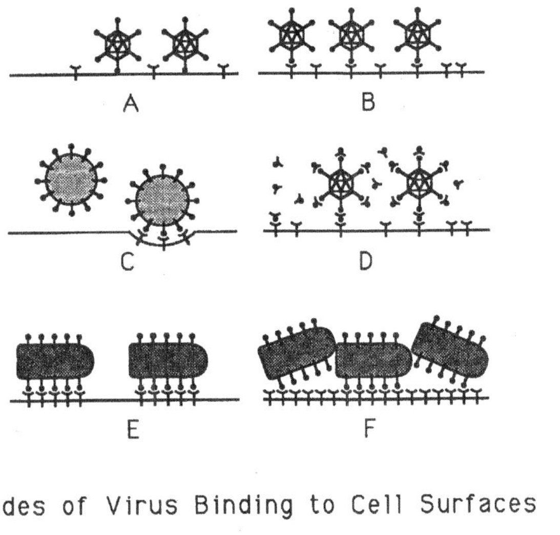

Figure 11.2 Mechanisms by which viruses may attach to cell surfaces. A, Non-specific binding not mediated by specific receptors. B, Monovalent binding of the VAP with host cell receptors. C, Binding to receptors restricted to a localized domain on the cell surface. D, Binding mediated by an intermediate molecule. E, Multivalent binding between VAPs and receptors. F, Multivalent binding with saturation of space on cell surface without saturation of receptors. Binding modes B, D, and E based on Wickham *et al.* (1990).

Three modes of virus binding were proposed by Wickham *et al.* (1990) and each mode analyzed mathematically. Mode 1 consists of monovalent binding of the viral attachment protein (VAP) to a receptor. In mode 2, binding is multivalent with multiple interactions between copies of the VAP and receptors. In mode 2, the receptors can become saturated. In mode 3, binding is multivalent as in mode 2 but the space on the cell surface becomes saturated with virus before the receptors become saturated.

It has been noted that for some viruses saturation of binding occurs when only 5–10% of the cell surface is covered by virions attached to the surface (Lipkind and Urbakh, 1988). It was suggested that this was due to the existence of patches or domains of receptors on the cell surface. Another explanation is that attached virions recruit and aggregate cell surface receptors. However, it is likely that some receptors are localized to specific domains of the cell surface in polarized cells such as epithelial cells and neurons. In these cells, some cell surface molecules such as receptors, enzymes, and ion channels are localized in specific domains such as apical or basolateral surfaces or axons and dendrites. Receptors may also be restricted to even more localized regions such as coated pits, microvilli, or synaptic regions.

Multivalent interactions of the VAP with host cell receptors potentially can bind the virus particle to the cell surface with extremely high affinity. The affinity of intact Semliki Forest virus is two to three orders of magnitude greater than that of soluble, spike glycoprotein complexes (Fries and Helenius, 1979). Soluble gp120 of HIV-1 binds to the CD4 molecule with relatively high affinity ($K_D = 4 \times 10^{-9}$M) (Lasky *et al.*, 1987). However, virus adsorption and penetration requires multiple gp120-CD4 interactions (Layne *et al.*, 1990). The multiple interactions with receptors, besides increasing affinity of binding, may also aggregate receptors and induce endocytosis.

It appears that some viruses may utilize different receptors on different cell types. For example, some cell lines derived from human malignant gliomas which do not bear the CD4 molecule, the receptor for HIV-1 on T lymphocytes, could be infected by HIV-1 (Cheng-Mayer *et al.*, 1987; Chiodi *et al.*, 1987; Clapham *et al.*, 1989; Weber *et al.*, 1989). Infection was not blocked by anti-CD4 antibodies or soluble CD4 (Clapham *et al.*, 1989). It was concluded that virus gains entry into these cells by a non-specific mechanism or that a second, low affinity receptor for HIV-1 exists (Weber *et al.*, 1989). Separate regions of the VAP may be able to interact with different receptors. For example, a determinant on HIV-1 gp120 responsible for mononuclear phagocyte tropism is distinct from the CD4 binding domain (O'Brien *et al.*, 1990; Shioda *et al.*, 1991). In view of the large size of most viral envelope proteins, it should not be surprising that different binding domains may be present. It is also possible that some viruses can utilize different receptors on the same cell type. Reovirus has been shown to bind to multiple plasma membrane proteins on mouse L fibroblasts (Choi *et al.*, 1990). If reovirus utilizes sialic acid as a receptor as has been proposed (Paul *et al.*, 1989), it may be able to bind to a variety of sialic acid-bearing glycoproteins.

3.3 CRITERIA FOR IDENTIFICATION OF VIRUS RECEPTORS

Identification of virus receptors has proven extremely difficult for several reasons. Virus particles can adhere to many substances, including inert materials, and it is often difficult to distinguish between non-specific binding and specific interaction with a receptor. Viruses may adhere to cell surfaces through non-specific electrostatic or physico–chemical interactions (Tardieu *et al.*, 1982; Spier *et al.*, 1983). Specific virus receptors may be present in low numbers on the cell surface or may be labile making identification difficult. In addition, it appears that some viruses can utilize more than one type of receptor. Finally, viruses may bind to intermediate molecules which in turn can bind to cell surface components.

Because of the difficulties encountered in the identification of virus receptors, a number of criteria should be satisfied before the receptor can be considered to be conclusively identified. These criteria and types of experimental evidence required for identification of receptors are listed in Table 11.2. The most conclusively identified virus receptors include the CD4 molecule (HIV-1 receptor), CR2 (CD21) (C3d portion of complement component 3 and Epstein-Barr virus (EBV) receptor), immunoglobulin superfamily member (poliovirus receptor), ICAM-1 (rhinovirus receptor), and sialyloligosaccharides (influenza virus receptor).

Table 11.2 Criteria for the identification of host cell receptors for viruses

Binding to receptor
- Virus binds to cells expressing receptor but not to cells lacking receptor
- Binding is followed by a biological effect including internalization and replication of virus and cellular cytopathic changes
- Transfection of non-permissive cells with receptor confers susceptibility to infection
- Virus binds to receptor in a saturable manner
- Virus-receptor complexes can be immunoprecipitated

Competition
- Binding to receptor is competed by cold homologous virus particles but not by unrelated viruses
- Antibodies against the binding domain of the receptor block attachment
- Virus competes with antibodies against the binding domain of the receptor for binding to the receptor
- Anti-idiotypic antibodies against antibodies to the binding domain of the VAP bind to the receptor
- Ligands of the binding domain of the receptor block virus attachment
- Soluble receptor blocks virus attachment

Physico-chemical basis
- A structural or conformational domain on the VAP binds to a complementary site on the receptor
- Deletion of the binding domain on the receptor blocks binding
- Site-directed mutagenesis of the binding domain on the receptor alters virus binding characteristics
- Mutation of the binding domain on the VAP alters binding

The evidence for CD4 as the HIV-1 receptor is reviewed against the criteria outlined in Table 11.2. The CD4 molecule is a glycoprotein of 60 000 molecular weight present on the surface of T lymphocytes and some other cells (Sattentau and Weiss, 1988; Springer, 1990). The CD4 molecule has four extracellular immunoglobulin-like domains, a transmembrane segment, and a cytoplasmic tail. It is believed to function as an adhesion molecule by binding to class II MHC molecules present on B lymphocytes and antigen-presenting cells. This binding facilitates T cell activation. First, HIV infects cells that express the CD4 molecule. These cells include helper/inducer T lymphocytes (Klatzmann *et al.*, 1984a) and cells of the monocyte/macrophage lineage (Gartner *et al.*, 1986), including those in the brain (Koenig *et al.*, 1986; Maddon *et al.*, 1986; Nicholson *et al.*, 1986; Wiley *et al.*, 1986; Vazeux *et al.*, 1987), and skin (Tschachler *et al.*, 1987). Binding of virus to CD4+ cells is followed by internalization, apparently by direct fusion with the plasma membrane (Maddon *et al.*, 1988), cytopathic changes including cell fusion and syncytia formation, and replication with release of virus particles. Exposure of cells to HIV-1 causes down regulation of CD4 (Dalgleish *et al.*, 1984; Maddon *et al.*, 1988).

Human cells that lack the CD4 molecule and are refractory to infection by HIV acquire the ability to bind virus and become permissive for infection following transfection with a cDNA clone encoding the CD4 protein (Maddon *et al.*, 1986). Direct binding of CD4 to HIV-1 virions (Moore *et al.*, 1991) and of labelled HIV envelope glycoprotein (gp120) to CD4+ cells (Lundin *et al.*, 1987) has been demonstrated. Immunoprecipitation of infected cells with HIV antibodies precipitated a complex of the gp120 and the CD4 molecule (McDougall *et al.*, 1986a).

Competition studies provide further evidence for CD4 as the HIV-1 receptor. Monoclonal antibodies directed against the CD4 molecule block binding, syncytia formation, or infection of CD4+ cells (Dalgleish *et al.*, 1984; Klatzmann *et al.*, 1984b; McDougall *et al.*, 1986a). Conversely, HIV blocks binding of a mAb against the CD4 molecule (McDougall *et al.*, 1986b). Soluble forms of the CD4 molecule inhibit HIV-1 infection of cultured cells (Smith *et al.*, 1987; Bergen *et al.*, 1988; Deen *et al.*, 1988; Fisher *et al.*, 1988; Hussy *et al.*, 1988; Traunecker *et al.*, 1988; Arthos *et al.*, 1989; Byrn *et al.*, 1989; Chao *et al.*, 1989; Clapham *et al.*, 1989). Soluble CD4 binding to HIV-1 virions results in dissociation of gp120 from its complex with the transmembrane glycoprotein gp41 (Moore *et al.*, 1990). It was suggested that a similar dissociation takes place when HIV-1 binds to CD4 on the cell surface and that this exposes the fusigenic domain on gp41.

The interaction between gp120 and CD4 is being characterized at a molecular level. Studies using truncated derivatives of CD4 have shown that a major recognition site for HIV lies within the first 106 amino acids of CD4 (Berger *et al.*, 1988; Deen *et al.*, 1988; Maddon *et al.*, 1988; Traunecker *et al.*, 1988; Arthos *et al.*, 1989). This region, the V1 domain, has significant sequence homology with immunoglobulin light chain variable regions. Using mutational analysis, several studies have further localized the major HIV binding determinant to a region encompassing residues 31 to 57 (see Table 11.1). This region is homologous to the second complementarity-determining region (CDR2) of immunoglobulin light chains. However, some residues outside this region contribute to binding (Ashkenazi *et al.*. 1990; Brodsky *et al.*, 1990). The major determinant for CD4 binding lies near the C terminal region of gp120 (Table 11.1). The crystal structure of the two N-terminal immunoglobulin-like domains (V1 and V2) of the CD4 molecule have been determined (Ryu *et al.*, 1990; Wang *et al.*, 1990). Most of the residues known to interact with gp120 lie in a ridge which is considered to interact with a complementary groove on the gp120 surface. The groove is estimated to comprise a region 25Å long and 12Å wide (Wang *et al.*, 1990). The tertiary structure of CD4 maintained by disulfide bonds is required for gp120 binding (Richardson *et al.*, 1988; Ibegbu *et al.*, 1989).

These studies strongly indicate the CD4 molecule is a part of the receptor site for HIV. Other evidence, however, indicates the situation is more complicated or that the virus can utilize alternate receptors. In cells exposed to HIV, binding of antibodies to MHC class II antigen HLA-DR is blocked and antibodies to HLA-DR block HIV binding (Mann *et al.*, 1988). These findings suggest that HLA-DR is involved in the receptor site for HIV. As discussed above, HIV can infect glioma and rhabdomyosarcoma cell lines some of which do not appear to express surface CD4 molecules. Infection of these cells is not inhibited by soluble forms of the CD4 molecule or anti-CD4 antibodies. These findings indicate the CD4 molecule is not required for HIV infection of these cells and that the virus can gain entry through an alternate receptor or a non-specific mechanism. It has been proposed that galactosyl ceramide can serve as an alternate receptor for HIV-1 in the nervous system (Bhat *et al.*, 1991).

4. *Viral Attachment Protein (VAP)*

In order to construct a molecular model of the virus-receptor interface, the binding domain on the VAP must also be identified. Several approaches are used to identify binding sites on VAPs. Comparison of the amino acid sequence of the VAP with sequenced vertebrate proteins has in some cases revealed structural similarities with natural ligands of receptors. These regions of the VAP may mediate binding of the virion to the host cell receptor. Determination of the three-dimensional crystal structure of the virus has revealed exposed projections or depressions that could function as attachment sites. The epitope of antibodies that inhibit virus attachment can be mapped on the coat protein. Finally, the effects of site-directed mutagenesis of the putative binding site on attachment and infectivity can yield direct evidence on structure-function relationships providing the conformation of the VAP is not altered. Binding domains of VAPs identified by these approaches are listed in Table 11.1. While the identity of the coat protein involved in binding is known for many other viruses, the list here is restricted to examples where a specific region of the VAP has been identified. For most viruses, it is likely that a specific amino acid sequence and conformation of the VAP interacts specifically with a complementary binding site on the receptor. In addition, oligosaccharides which may be present on the VAP can affect receptor binding properties (Aytay and Schulze, 1991).

5. *Binding to Receptors Through an Intermediate Molecule*

While many viruses interact directly with a receptor, some viruses bind to a host cell molecule indirectly via an intermediate molecule attached to the virus. The best known example of this type of interaction is enhancement of viral infection by antibodies. Antibody enhancement has been demonstrated for several viruses (Halstead, 1988) including, for example, dengue virus (Halstead and O'Rourke, 1977), West Nile virus (Peiris *et al.*, 1981), and most recently HIV (Homsy *et al.*, 1989). The crystallizable fraction of immunoglobulin molecule (Fc) portion of anti-viral antibodies coating the virus interacts with cell surface Fc receptors, crosslinking virions to the cell surface and inducing uptake of the virus-antibody complex. Fc receptor-mediated entry of HIV-1 has been described (Takeda *et al.*, 1988). In addition, antibody enhancement of HIV-1 independent of the Fc receptor has been described (Lenert *et al.*, 1990). In this case, it was suggested that antibodies enhance infection by binding to CD4 thereby forming a trimolecular complex of HIV, antibody to gp120, and CD4. In addition, complement-dependent enhancement of West Nile virus infectivity has been described in which the C3b component of complement binds complexes of virus and IgM to type 3 complement receptors of macrophages (Cardosa *et al.*, 1986). Similarly, complement was shown to enhance infectivity of HIV-1 (Robinson *et al.*, 1988; Soelder *et al.*, 1989; Gras and Dormont, 1991). Although entry could be mediated by complement receptors, it was also noted that complement by increasing virus affinity for CR-2 bearing cells might bring the gp120 into close proximity to CD4 molecules on the cell surface (Gras and Dormont, 1991). Polymerized serum albumin can act as an intermediate receptor for hepatitis B virus by binding to polymeric albumin receptors on hepatocytes (Machida *et al.*, 1984). Hepatitis B virus can also bind directly to other receptors in the membranes of hepatocytes. The binding sites for albumin and the hepatocyte receptor are located on different regions of the viral envelope proteins (Pontisso *et al.*, 1989a). β_2-microglobulin may serve as an intermediate receptor for cytomegalovirus (Grundy *et al.*, 1987). β_2-microglobulin coating the virus is considered to displace β_2-microglobulin from the class I HLA heavy chain-β_2-microglobulin complex and bind the virus to the HLA molecule. Fibroblast growth factor associated with herpes simplex type 1 mediates binding to fibroblast growth factor receptor and penetration of vascular cells (Baird *et al.*, 1990). Viruses attached to host cell receptors may be able to act as a bridge for entry by other viruses. The sialic acid residues on the envelopes of vesicular stomatitis virus and Semliki Forest virus can bind to the hemagglutinin of influenza virus present on the surface of infected cells (Fuller *et al.*, 1985).

6. *Receptors and Viral Tropisms*

The nature, number, and distribution of host cell receptors is a major determinant of the host range and tissue tropisms of viruses (Holland, 1961; Lonberg-Holm and Philipson, 1974; Crowell and Landau, 1979; Paulson, 1985; Mims, 1986; Lentz, 1990b). For example, viruses infecting lymphocytes bind to molecules expressed on the surface of these cells. Examples of virus-lymphocyte receptor interactions are EBV and C3d receptor CR2 of B lymphocytes (Fingeroth *et al.*, 1984) and HIV-1 and the CD4 molecule of T lymphocytes (Dalgleish *et al.*, 1984; Klatzmann *et al.*, 1984b; McDougall *et al.*, 1986b). It is clear, however, that factors besides cellular receptors contribute to tissue tropisms. As mentioned above, HIV can infect cells lacking the CD4 molecule. In addition, HIV-1 does not infect mouse cells transfected with a CD4 cDNA, although it does bind to the surface of mouse cells bearing CD4, indicating other factors in addition to CD4 are required for infection (Maddon *et al.*, 1986), see also Levy (1988). Many factors acting on the stages of the infectious cycle prior to and subsequent to attachment are required for viral replication and these also affect cell tropism. These factors include dose and site of exposure to virus, extracellular barriers affecting access of virus to cells, whether the receptor is occupied by its natural ligand, host defense mechanisms, entry mechanisms, and factors influencing each step of replication and assembly following entry. The VAP also plays a major role in tropisms and mutations in the VAP can affect receptor binding properties and host range (Aytay and Schulze, 1991). The affinity of the VAP for the host cell receptor plays a role in determining cell tropisms. It has been shown that diverse isolates of immunodeficiency viruses (HIV-1, HIV-2, SIV) which exhibit different host cell ranges and cytopathicities have a wide range of affinities for CD4 (Ivey-Hoyle *et al.*, 1991).

For some viruses, the host range or the cell types infected is narrow. In these cases, the receptor determinant that binds virus is restricted to one type of molecule that occurs on a few specific cell types. HIV-1 infects helper/inducer T lymphocytes and cells of the monocyte-macrophage lineage which express the CD4 molecule. EBV infects B lymphocytes expressing the C3d receptor CR2 (CD21). In these cases, the virus receptor is the major determinant of tropism. Other receptors for viruses with restricted tropisms are widely distributed among cell types or species making it more difficult to understand how they alone can confer specificity. For example, poliovirus replicates in the gut and motor neurons of primates. The poliovirus receptor has been identified as a previously unidentified member of the immunoglobulin superfamily (Mendelsohn *et al.*, 1989). However, messenger RNA encoding the receptor is present in cells to which the virus does not attach and in which it does not replicate. It is possible that post-translational modifications are necessary for expression of the poliovirus receptor sites or that additional proteins interacting with the receptor are necessary for virus binding activity (Mendelsohn *et al.*, 1989). Nonetheless, the poliovirus receptor appears to be the major determinant of poliovirus host cell range because transgenic mice containing the human poliovirus receptor gene are susceptible to poliovirus infection (Ren *et al.*, 1990; Koike *et al.*, 1991a).

Other viruses can infect a broader range of hosts or cell types. In these cases, the receptor determinant is shared by many different molecules. Influenza viruses bind the sialic acid (*N*-acetylneuraminic acid) of oligosaccharides on cell surface glycoproteins and glycolipids (Paulson, 1985). Sialic acid binds to a pocket of conserved residues on the top of the virus hemagglutinin molecule (Weiss *et al.*, 1988). Sialyloligosaccharide structures of proteins and lipids are widely distributed among different tissues and animal species. They also show diversity in structure and linkages which confers some specificity. Different strains of influenza viruses as well as other viruses binding sialic acids are able to recognize specific sialyloligosaccharide structures (Rogers and Paulson, 1983). Amino acid substitutions in the conserved pocket of the hemagglutinin result in changes in specificity for different sialosides (Rogers, 1983). A substitution of Leu to Gln at position 226 results in a change in preferential binding specificity from NeuAcα2,6Gal linkage to NeuAcα2,3Gal. Similarly, a single amino acid change in the gp120 of HIV causes a restriction in viral tropism (Cordonnier *et al.*, 1989a).

Viruses often do not show the same tropisms *in vitro* as *in vivo* (Mims, 1989). This presents difficulties for the identification of virus receptors because cultured cells are frequently used for these studies. Poliovirus, for example, was reported to replicate in tissues lacking neurons (Enders *et al.*, 1949). Rabies virus infects muscle fibers, neurons, and salivary glands *in vivo* (Murphy, 1977). *In vitro*, however, it infects a wide range of cell types (Wiktor and Clark, 1975), including cells lacking ACh receptors (Reagan and Wunner, 1985), which have been proposed as possible rabies virus receptors (Lentz *et al.*, 1982). Some of the differences in infectivity of cells *in vitro* and *in vivo* may be due to the absence *in vitro* of some of the factors besides specific receptors that influence susceptibility *in vivo*. In addition, as discussed above, viruses may enter cultured cells through non-specific mechanisms. Viruses adapted to cultured cells may change their receptor specificities through mutation of their VAPs. For example, it has been shown that human H1 influenza viruses change their sialyloligosaccharide specificities with continued laboratory passage (Rogers and D'Souza, 1989).

7. *Conservation of Binding Domains*

A major advantage of intervention at the attachment stage of the viral infectious cycle is that the binding domain on the VAP is likely to be conserved among strains of virus that maintain the same tropism. Mutations of binding domains may often be lethal. Conservation of binding regions is particularly important for possible therapies in the case of viruses such as HIV-1 and influenza viruses that evade immune clearance by frequent mutation. It has been noted that if all pathogenic strains of HIV-1 display the same cell tropism (CD4+ cells), then the region of gp120 interacting with CD4 is likely to be highly conserved (McDougal *et al.*, 1986a). A segment of the HIV-1 gp120 considered to interact with the CD4 molecule contains regions that are highly conserved among various HIV-1 isolates as well as in HIV-2 and SIV-1 which also bind to CD4 (Lasky *et al.*, 1987). Similarly, the receptor-binding residues of the hemagglutinin glycoprotein of influenza A virus are conserved among different serotypes (Nobusawa *et al.*, 1991). In the case of poliovirus, though, mutants resistant to neutralization with soluble poliovirus receptor have been isolated (Kaplan *et al.*, 1990b). A single mutation (Trp 432) in the region of gp120 of HIV-1 considered to interact with CD4 blocks binding to CD4 (Cordonnier *et al.*, 1989a). Mutations of other regions of coat proteins not directly involved in binding may also affect tropisms and attachment to receptors by altering the conformation of the protein. It is also likely that the binding domains on the host cell receptors are conserved if these domains are binding sites of normal ligands. It has been reported that the region on the CD4 molecule that binds gp120 appears to be conserved (Sattentau *et al.*, 1986).

The receptor-binding sites of human rhinovirus 14 (Rossmann *et al.*, 1985), Mengo virus (Luo *et al.*, 1987), and on the hemagglutinin glycoprotein of influenza virus (Weis *et al.*, 1988) lie within a canyon, pit, or pocket on the virus surface. A similar depression is present on the surface of poliovirus (Hogle *et al.*, 1985). Determination of the three-dimensional structure of the influenza hemagglutinin complexed with sialyllactose reveals the sialic acid filling the pocket (Weis *et al.*, 1988). Sialic acid occupies the entire pocket, with one face of the pyranose ring facing the bottom of the depression and the other face exposed to solution. Ring substituents of sialic acid interact with amino acids lining the pocket through hydrogen bonds, van der Waals contacts, and non-polar interactions. The amino acids lining these depressions are highly conserved whereas the surrounding surface residues are variable and could represent the binding sites of neutralizing antibodies. According to the canyon hypothesis, these viruses can accept mutations in the hypervariable antibody binding sites thereby escaping the immune response while at the same time maintaining a constant receptor binding site which is physically inaccessible to the antigen-binding region of antibodies (Rossmann and Palmenberg, 1988; Rossmann, 1989) It remains to be determined whether antibodies that prevent attachment are directed to the variable regions around the canyon. Antigenic sites, identified on the basis of amino acid substitutions in different strains of influenza virus, do occur in proximity to the receptor-binding pocket of the hemagglutinin (Wiley *et al.*, 1981). Experimental studies on neutralization of influenza virus have shown that neutralizing monomeric IgG and IgA do not inhibit binding, and therefore act on a stage subsequent to attachment and penetration, while larger polymeric IgA and IgM do reduce virus attachment to cells (Taylor and Dimmock, 1985a, 1985b). Eisenlohr *et al.* (1987) reported that attachment of influenza virus to antigen-presenting cells is inhibited by anti-hemagglutinin IgG. In human rhinovirus 14, neutralizing immunogenic sites correspond to hypervariable regions and are located on protrusions from the viral surface (Sherry *et al.*, 1986).

The canyon or pit found in other picornaviruses is absent in foot-and-mouth disease virus. The receptor-binding region of this virus is a loop projecting from the surface of the virus (Acharya *et al.*, 1989). This region, which contains the highly-conserved arginine-glycine-aspartic acid (RGD) sequence considered an important binding determinant, is situated within an immunodominant hypervariable region (Fox *et al.*, 1989). Thus, it may not be necessary for the binding site to be inaccessible to avoid neutralization by antibodies if the site is smaller than than an antibody "footprint" (Colman *et al.*, 1983; Harrison, 1989).

8. Molecular Mimicry of Ligands by the VAP

The ability of the virus to bind to a host cell receptor may be conferred by a domain on the VAP that resembles in amino acid sequence or conformation the normal ligand of the cellular receptor. Many regions of viruses and other pathogens share amino acid similarities with normal host proteins, a phenomenon termed molecular mimicry (Damian, 1987). It was found that of over 600 mAb against 11 different viruses, 21 (or 3.5%) reacted against normal tissues (Srinivasappa *et al.*, 1986). Molecular mimicry can cause autoimmune disease as a result of cross-reaction of antibodies against the virus with similar epitopes on the host-cell proteins (Oldstone, 1987). Mimicry of the binding domain of a ligand by the VAP may confer the ability to bind to the host cell receptor for the ligand.

Table 11.3 Structural similarities between viral attachment proteins and ligands of receptors. Identities are shown in bold type. Regions of greatest similarity are shown. In some instances, the similarities extend beyond the regions shown

Adenovirus glycoprotein (102–116)	DITMYMSKQYKLWPP
Immunoglobulin Mκ chain, human (83–97) (Chatterjee and Maizel, 1984)	DIATYYCQQYNNWPP
Epstein-Barr virus gp350 (372–378), (21–29)	TPSGCEN EDPG--FFNVE
Complement C3d fragment (1006–1012), (1221–1231) (Nemerow *et al.*, 1987)	TPSGCGE EDPGKQLYNVE
Epstein-Barr virus BCRFI protein (41–67)	DFKGYLGCQALSEMIQFYLEEVMPQAE
Cytokine synthesis inhibitory factor (interleukin-10), mouse (46–72) (Moore *et al.*, 1990)	DFKGYLGCQALSEMIQFYLVEVMPQAE
Foot-and-mouth disease virus VP1 (145–147)	RGD
Fibronectin, fibrinogen, type 1 collagen, and other proteins binding to integrins (Fox *et al.*, 1989)	RGD
Hepatitis B virus preS1 env protein (21–47)	PLGFFPDHQLDPAFGANSNN-PDWDFNP
Immunoglobulin A α1 constant region (28–55) (Neurath and Strick, 1990)	VQGFFPQQPLSVTWSESGEGVTARDFPP
Human cytomegalovirus H301 gene product (261–286)	DGTFHQ--GCYV-AIFCNQNYTCRVTH
Class I MHC molecule α-I domain, human (239–264) (Beck and Barrell, 1988)	DGTFQKWAAVVV-PSG-EQRYTCHVQH
Human immunodeficiency virus-1 gp120 (193–200)	ASTTTNYT
Vasoactive intestinal polypeptide (4–11) (Ruff *et al.*, 1987)	AVFTDNYT
Human immunodeficiency virus-1 gp120 (61–88)	CASDAKAYDTEVHKVWATHACVPTDPNP
Immunoglobulin γ2 heavy chain, human (84–111) (Maddon *et al.*, 1986)	CNVDHKPSNTKVDKTVERKCCVECPPCP
Human immunodeficiency virus-1 gp120 (245–273)	VQCTHGIRPVVSTQLLLNGSLAEEEVVIR
Neuroleukin (410–438) (Lee *et al.*, 1987)	VQTQHPIRKGLHHKILLANFLAQTEALMR
Human immunodeficiency virus-1 gp120 (164–174)	FNISTSIRGKV
α-Cobratoxin (30–40) (Neri *et al.*, 1990)	CDAFCSIRGKR
Human T-cell leukemia virus-1 gp (246–253)	SVPSSSST
Interleukin 2, human (20–27) (Kohtz *et al.*, 1988)	SAPTSSST
Rabies virus glycoprotein (189–199)	CDIFTNSRGKR
Naja naja naja (India) toxin b (30–40) (Lentz *et al.*, 1984)	CDGFCSSRGKR
Vaccinia virus VGF protein (69–80)	CRCSHGYTGIRC
Epidermal growth factor, human (31–42) (Eppstein *et al.*, 1985)	CNCVVGYIGERC

Examples of similarities between portions of VAPs and physiological ligands of receptors are shown in Table 11.3. In some cases, these regions of the VAP may mediate binding to a host cell receptor. For example, portions of the gp350 of EBV show structural similarities with complement fragment C3d (Nemerow *et al.*, 1987). These regions of gp350 could mediate binding of EBV to the C3d receptor, CR2, of B lymphocytes. Similarly, regions of the HIV gp120 that are structurally similar to immunoglobulins may bind to the CD4 molecule of T lymphocytes (Maddon *et al.*, 1986). A segment of the rabies virus glycoprotein bears a structural similarity to snake venom curaremimetic neurotoxins which bind with high affinity to the nicotinic ACh receptor (Lentz *et al.*, 1984). Some of the similarities between VAPs and ligands may not be related to receptor attachment, but may be responsible for biological effects as described below.

The mechanisms that led to the structural similarities between portions of VAPs and normal ligands for host cell receptors and the ability to bind cell surface molecules are not clear. It would be expected that within a large group of protein sequences, some similarities are simply the result of chance. The molecules most frequently used as receptors by viruses, glycoproteins and adhesion molecules of the immunoglobulin superfamily, are widespread throughout cell types and the animal kingdom. Thus, viruses have had ample opportunity to evolve mechanisms for interacting with these molecules. Some sequence similarities may be the result of divergence from a common ancestor. In this case, the virus may have acquired genetic material (DNA or messenger RNA) coding the binding domain of the normal ligand from the host during the course of infection. The VAP could also acquire the binding domain of another virus due to genetic recombination between two viruses infecting the same cell.

Viruses could also develop binding domains similar to that of a ligand and capable of binding to a host cell receptor through evolutionary convergence. Mutations that enhance binding might have survival value. In the case of both convergent and divergent evolution, the acquisition of ability to bind to a new host cell receptor might confer a selective advantage by providing another mechanism for binding and entry into cells, the ability to bind to a new, more hospitable cell or host, or suppression of defense mechanisms.

9. *Biological Effects of VAP–Receptor Interaction*

Similarities between the VAP and normal ligands besides mediating entry of virus into the cell, may also be responsible for certain biological effects of viruses that are unrelated to the infectious cycle. Some of the examples of virus-receptor interactions in Table 11.1 may actually represent interactions of the VAP with cell surface molecules that result in biological effects but which may not be followed by internalization of virus. In some viral infections, the glycoprotein is synthesized in excess and secreted or shed extracellularly. The coat proteins may also dissociate from virus particles. These soluble glycoproteins may be responsible for some of the observed biological effects. Biological effects exerted by viruses during the course of infection include toxic effects on cells, activation or inactivation of the immune response, activation of second messenger systems, suppression or induction of protein synthesis and secretion, and mitogenesis. Vaccinia virus encodes a polypeptide (VGF) that is structurally similar to epidermal growth factor and transforming growth factor α (Brown *et al.*, 1985; Reisner, 1985; King *et al.*, 1986). VGF has been shown to stimulate tyrosine kinase activity of epidermal growth factor receptors (King *et al.*, 1986) and to stimulate mitogenesis (Twardzck *et al.*, 1985). The receptor-mediated leukemogenesis hypothesis proposes that murine leukemia viruses induce proliferation of lymphocytes by binding to antigen-specific receptors complementary to virus envelope gene products (McGrath *et al.*, 1987). Similarly, gp55 of Friend spleen focus-forming virus binds directly to the erythropoietin receptor experimentally expressed in non-erythroid cells and triggers their proliferation. It was suggested that the first step in leukemogenesis induced by Friend virus is mimicry of erythropoietin by gp55 and stimulation of proliferation of erythroid cells (Li *et al.*, 1990).

Some similarities of regions of the VAP and host cell proteins may aid the virus in evading host immune or defense responses. For example, the vaccinia virus 35 kDa major secretory protein bears a structural similarity to complement control proteins, and it was shown that the viral protein binds to the fourth component of complement and inhibits the complement cascade (Kotwal *et al.*, 1990). Epstein-Barr virus protein BCRF1 exhibits a sequence similarity to cytokine synthesis inhibitory factor (IL-10) and blocks the production of IFNγ, thereby interfering with host defences (Hsu *et al.*, 1990).

Synthetic peptides of a region of rabies virus glycoprotein that is structurally similar to snake venom curare mimetic neurotoxins (Lentz *et al.*, 1984) act as competitive antagonists of the nicotinic ACh receptor and inhibit carbachol-induced ^{22}Na flux into BC3H-1 cells (Donnelly-Roberts and Lentz, 1989). Thus, soluble glycoprotein or proteolytic fragments of the glycoprotein could be responsible for some of the bizarre behavioral manifestations of this disease.

The gp120 of HIV as a result of interaction with the CD4 protein causes activation of some T cell functions and suppression of others. Binding of HIV to CD4 induces phosphorylation of CD4 possibly mediated by protein kinase C (Fields *et al.*, 1988). HIV activates lymphocytes with increases of intracellular IP_3 and calcium, expression of IL-2 receptors, and induction of a motile response (Kornfeld *et al.*, 1988). In addition,

HIV-1 binding to CD4 on mononuclear phagocytes also induces production of the cytokines IL-1 and tumor necrosis factor-α (Merrill *et al.*, 1989). Gp120 also mediates immune suppression by different mechanisms. Soluble gp120 inhibits the binding of CD4 with class II MHC molecules of B lymphocytes (Clayton *et al.*, 1989) and inhibits the CD4-dependent response to antigen (Diamond *et al.*, 1988). The MHC class II molecule- and gp120-binding sites are separate so that it should be possible to design CD4 analogs that inhibit HIV infectivity but do not affect the immune response (Lamarre *et al.*, 1989; Fleury *et al.*, 1991). Gp120 also suppresses T-cell activation and expression of cytolytic actions by binding to CD4 and renders activated CD4 positive lymphoblasts susceptible to immune cytolysis (Weinhold *et al.*, 1989). The receptor can be affected by antibodies against the VAP. It has been reported that antibodies to gp120 form a trimolecular complex, with gp120 bound to CD4, and suppress the activation of T cells (Mittler and Hoffmann, 1989). In addition, HIV has been proposed to interfere with the neurotrophic factors neuroleukin (Lee *et al.*, 1987) and VIP (Brenneman *et al.*, 1988) by virtue of regions of the gp120 that bear an amino acid similarity to these factors. HIV-1-infected monocytic cells were reported to release small, heat stable, protease-resistant molecules that are toxic to neurons in culture (Giulian *et al.*, 1990). These toxins may act by activating the *N*-methyl-D-aspartate (NMDA) receptor. Thus, some of the neurological effects and dementia observed in patients with acquired immune deficiency syndrome (AIDS) could be the result of inhibition of normal neurotrophic factors or of adverse effects on neurons, glial cells, or blood vessels in the central nervous system.

Finally, viruses may exert biological effects that enhance the infectivity of another virus. Human herpes virus 6 infection of a T-cell line greatly increases expression of CD4, rendering the cells susceptible to infection by HIV-1 (Lusso *et al.*, 1991). Certain combinations of ovine-caprine lentiviruses resulted in enhanced binding and replication of the second virus (Jolly and Narayan, 1989).

Proteins that are expressed in greater amounts on the cell surface in response to viral infection could themselves be used as targets for antiviral agents. As noted above, HIV-1 infection leads to cellular expression of IL-2 receptors in monocytes. A toxin composed of the toxic portion of diphtheria toxin and the receptor-binding domain of IL-2 killed cells expressing IL-2 receptors (Finberg *et al.*, 1991).

10. *Inhibition of the Virus-Receptor Interaction*

The binding of the VAP to host cell receptors represents a target for agents that block this interaction. Two general classes of agents should theoretically prevent viral entry into cells by blocking attachment of virus particles to host cell receptors (Fig. 11.3). The first group of agents structurally or conformationally resembles the binding domain on the VAP and should competitively inhibit binding of the VAP to the cellular receptor. Substances mimicking the binding domain of the VAP are: (1) antibodies directed against the binding site of the receptor; (2) anti-idiotypic antibodies against antibodies to the binding domain of the VAP; (3) natural ligands of the binding site of the receptor, and; (4) designed ligand-mimics. The latter include synthetic peptides of the VAP or designed drugs that mimic the VAP. All four of these types of agents have been shown to inhibit virus binding or infectivity.

Regions of the VAP not involved in receptor binding, but bearing a similarity to host cell proteins, may also be potential targets of antiviral agents. For example, the genome of human cytomegalovirus encodes polypeptides similar to mammalian G protein-coupled receptors (Chee *et al.*, 1990). If these proteins retain the ability to bind ligands, e.g. neurotransmitters or hormones, it may be possible to target pharmacological agents to these proteins and disrupt the virus.

The second group of agents that should prevent attachment of virus to host cell receptors are agents that resemble the binding domain on the receptor to which the VAP binds (Fig. 11.3). This group of agents includes: (1) antibodies directed against the binding domain of the ligand; (2) anti-idiotypic antibodies against antibodies to the binding site on the receptor; (3) the receptor itself, and; (4) designed receptor-mimics [also referred to as molecular decoys (Gershoni and Aronheim, 1988)]. Again, there are many examples of the potential effectiveness of all four types of agents, some of which are described below as examples. An additional mechanism by which agents can inhibit attachment is by inducing conformational changes in the binding domain of the VAP, thereby preventing binding to the receptor. It has been shown that agents called WIN compounds inhibit rhinovirus binding by causing conformational changes in the canyon floor (Pevear *et al.*, 1989).

As noted earlier, a virion is a multivalent ligand capable of binding to the cell with extremely high affinity. It follows, therefore, that agents which are used to inhibit attachment of pathogens to target cells are more likely to be effective if these agents are designed to be polyvalent (Matrosovich, 1989). It has been shown that synthetic polymers of sialic acid were three orders of magnitude more effective than monosialosides as inhibitors of influenza attachment (Matrosovich *et al.*, 1990).

The utilization of intermediate receptors by viruses increases the range of susceptible cell types and complicates the strategy of preventing infection by blocking attachment. However, it has been suggested that inhibiting the interaction of HIV with CD4, e.g. with soluble CD4, could be complemented with blocking the

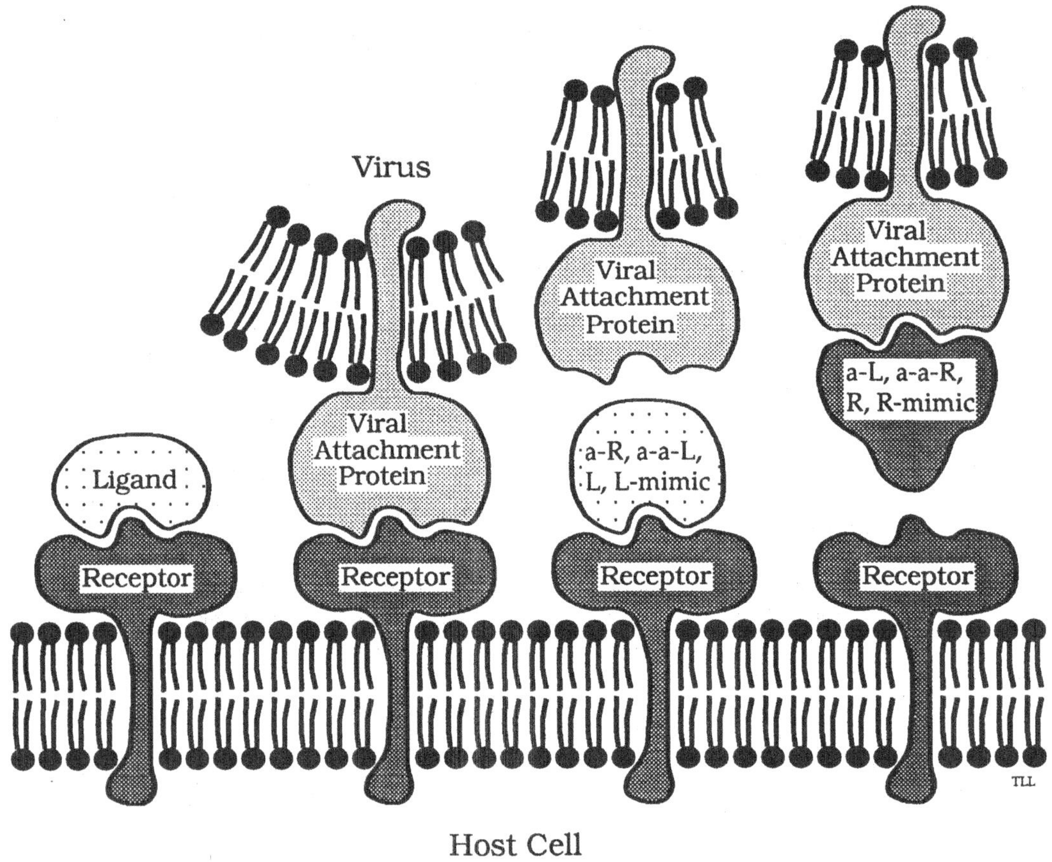

Figure 11.3 **Agents with the capability of inhibiting viral infection of cells by inhibiting the attachment of the VAP to host cell receptors. The VAP attaches to cell surface molecules that are often the site of binding of natural ligands such as hormones, neurotransmitters, immunoglobulins, or other substances. The virus may have the ability to bind to a receptor by virtue of mimicry of the binding site of a normal ligand. Substances that resemble in structure or conformation the binding domain of the ligand will bind to the receptor and compete binding of the virus. These agents include antibodies to the binding domain of the receptor (a-R), anti-idiotypic antibodies against antibodies to the binding domain of the ligand (a-a-L), the normal ligands themselves (L), and ligand-mimics (L-mimic) which may be drugs or synthetic peptides designed to resemble the binding domain of the ligand. A second group of agents resembles the binding domain of the receptor. These agents will bind to the VAP and prevent binding to the cellular receptor. These substances include antibodies directed against the binding domain of the VAP (a-L), anti-idiotypic antibodies against antibodies to the binding site of the receptor (a-a-R), soluble forms of the receptor (R), and designed or engineered substances mimicking the receptor binding domain (R-mimic). Modified from Lentz (1990b).**

relevant Fc receptors to prevent uptake of the virus-immune complexes (Homsy *et al.*, 1989).

10.1 AGENTS MIMICKING THE VAP (LIGAND-MIMICS)

10.1.1 Anti-Receptor Antibodies

Antibodies directed to the binding site of the receptor will resemble in structure the binding domain of the ligand (VAP). As described above, antibodies against the CD4 molecule inhibit HIV binding and infectivity. Some of the antibodies that block HIV infection are directed against the gp120 binding site on CD4 (Landau *et al.*, 1988; Mizukami *et al.*, 1988). Monoclonal anti-receptor antibodies inhibit poliovirus (Minor *et al.*, 1984) and rhinovirus (Colonno *et al.*, 1986) attachment. Antibodies to the binding domain of ICAM-1 block attachment of rhinovirus (Lineberger *et al.*, 1990). Such antibodies may be useful in preventing or treating viral diseases. For example, antibody fragments, expressed in bacteria, of an antibody that blocks virus attachment to receptors have been shown to retain the ability to bind to the receptor and protect cells from rhinovirus (Condra *et al.*, 1990). Davis *et al.* (1992), on the other hand, have argued that this approach is not likely to be successful because of the improbability of producing antibodies that closely mimic the binding of a ligand to its receptor.

10.1.2 Anti-idiotypic Antibodies

Anti-idiotypic antibodies contain an “internal image” of

external ligands. Thus, anti-idiotypic antibodies directed against antibodies reacting with the binding domain of the VAP will resemble the binding domain of the ligand and bind to the receptor. It has been noted that production of anti-idiotypic antibodies does not require characterization and purification of the receptor and that the antibodies can be used to identify the receptor (Krah and Choppin, 1988). Binding of reovirus type 3 to its receptor is blocked by a monoclonal anti-idiotypic antibody directed against an idiotypic determinant on a mAb against reovirus hemagglutinin (Kauffman *et al.*, 1983). As described below, this antibody was used to model a peptide that resembled the ligand. An anti-idiotypic antibody against a neutralizing antibody against rabies virus glycoprotein reacted with the nicotinic ACh receptor (Hanham *et al.*, 1993).

10.1.3 Natural Ligands

Naturally-occurring ligands of the receptor potentially can inhibit binding of the virus. Ligands mimicked by the virus which have been reported to inhibit infection of cultured cells are α-bungarotoxin in the case of rabies virus (Lentz *et al.*, 1982; Tsiang *et al.*, 1986) and epidermal growth factor in the case of vaccinia virus (Eppstein *et al.*, 1985). Lectins specific for *N*-acetylglucosamine inhibit infection of cells by murine cytomegalovirus (Ravindranath and Graves, 1990). Platelet factor 4, a protein with specific affinity for glycosaminoglycans including heparan sulfate, inhibits attachment of herpes simplex virus to cells (WuDunn and Spear, 1989). Natural ligands may have limited usefulness as therapeutic agents because they exert toxic or biological effects through their interaction with the receptor.

10.1.4 Designed Ligand-Mimics

Agents designed to resemble the ligand or VAP may represent useful anti-viral agents. Synthetic peptides of the binding domains of the VAP inhibit binding of vaccinia virus (Eppstein *et al.*, 1985), HIV (Pert *et al.*, 1986), hepatitis B virus (Neurath *et al.*, 1988), foot-and-mouth disease virus (Fox *et al.*, 1989), and EBV (Nemerow *et al.*, 1989) to their receptors. Fibroblast growth factor residues 103 to 120 inhibited herpes simplex virus type 1 uptake by cultured cells (Kaner *et al.*, 1990). An HIV-1 gp120 octapeptide (peptide T) inhibited HIV infection of human T cells *in vitro* (Pert *et al.*, 1986). In an initial trial, peptide T was reported to result in improvement in AIDS patients (Bridge *et al.*, 1989). Peptide T, however, does not correspond to the region of gp120 several other laboratories have identified as being the major binding determinant (Table 11.1), so it is uncertain whether its effects are due to inhibiting attachment. A potential difficulty with synthetic peptides as competitors of virus attachment is that a single linear sequence may not be effective if the binding site consists of several discontinuous segments brought into proximity in the native, folded structure of the protein.

It has been demonstrated that biologically active peptides can be modelled based on the structure of the hypervariable regions of antibodies (Bruck *et al.*, 1986; Williams *et al.*, 1989, 1991). It was found that there is an amino acid similarity between residues 317–332 of the reovirus hemagglutinin and a combined determinant comprised of the heavy and light chain second complementarity-determining regions of the anti-idiotypic antibody described above. A synthetic peptide of this region of the antibody inhibited the interaction of both the antibody and reovirus type 3 with the reovirus receptor, indicating the peptide interacts with the receptor. The peptide also exerted biological effects similar to those exerted by reovirus type 3.

Synthetic peptides have also been used to inhibit the interaction of viral fusion proteins with the plasma membrane. Oligopeptides resembling the N-terminal regions, generated by proteolytic cleavage, of paramyxovirus F_1 polypeptide and myxovirus HA_2 polypeptide, inhibited cell fusion and infectivity (Richardson *et al.*, 1980). These peptides exerted their effects on cells and not virus suggesting they might inhibit membrane-fusing activity by competing with the fusion proteins for a membrane "receptor" site (Richardson and Choppin, 1983).

10.2 AGENTS MIMICKING THE RECEPTOR (RECEPTOR-MIMICS)

10.2.1 Anti-VAP Antibodies

Antibodies against the binding site of the ligand will resemble in structure the binding domain of the receptor. Antibodies against the region of the viral coat protein thought to be involved in binding to the receptor have been shown to block attachment of virus. There are many other examples of antibodies directed against components of the viral capsid or envelope that inhibit adsorption of virus to cells, but the precise locations of their epitopes have not been determined. Antibodies to the binding domain of the VAP can be raised by immunization with synthetic peptides. Immunization of chimpanzees with a synthetic peptide comprising a portion of the preS2 region of the envelope protein of hepatitis B virus protected animals from hepatitis B virus (Itoh *et al.*, 1986). The preS2 sequence serves as the receptor for polymerized serum albumin (Pontisso *et al.*, 1989a). Antibodies directed to the preS1 sequence (21–47) of hepatitis B virus also inhibit binding (Pontisso *et al.*, 1989b; Neurath *et al.*, 1990; Petit *et al.*, 1991). The preS1 sequence binds to the hepatocyte receptor for hepatitis B virus. Immunization of animals with synthetic peptides corresponding to the binding region of VP1 polypeptide (Acharya *et al.*, 1989; Fox *et al.*, 1989) of foot-and-mouth disease virus resulted in production of neutralizing antibody (Bittle *et al.*, 1982). Immunization of cattle with the peptide protected them against inoculation with infectious virus (DiMarchi *et al.*, 1986). The antibodies to this region blocked attachment of virus to cells (Fox *et al.*, 1989). Antibodies against a putative binding domain of HIV gp120 blocked the interaction of gp120 with the CD4 molecule (Lasky *et al.*, 1987; Dowbenko *et al.*, 1988; Sun *et al.*, 1989). Antibodies against synthetic peptides of the proposed capsid binding sites of human rhinovirus bound to the capsid proteins and had neutralizing activity (McCray and Werner, 1987). These studies demonstrate the potential usefulness of synthetic peptides of the binding domains of the VAP as immunogenic agents in the development of safe and specific subunit vaccines. On the other hand, it has been noted that in some cases the antibodies raised against VAP synthetic peptides have very low neutralizing titers (Palmenberg, 1987). Most likely, this is because the binding domain is often formed by discontinuous regions on the primary sequence brought into proximity in the folded state of the protein. In addition, it has been shown that antibodies to conserved conformational epitopes neutralize a wider range of divergent virus isolates than antibodies directed to linear determinants (Steimer *et al.*, 1991).

Antibodies that inhibit virus attachment do not have to be directed precisely against the receptor-binding region of the VAP. In the case of rhinovirus (Rossmann *et al.*, 1985) and influenza virus (Weis *et al.*, 1988), the variable antigenic sites surround the recessed receptor-binding site. Thus, antibodies binding to the antigenic sites can block binding to the receptor by steric hindrance. Similarly, β inhibitors of influenza virus in bovine and mouse serum are lectins that bind to a high mannose carbohydrate at the tip of the hemagglutinin spike and block access to the receptor-binding site (Anders *et al.*, 1990).

10.2.2 Anti-idiotypic Antibodies

Anti-idiotypic antibodies against antibodies to the binding site of the receptor will contain an internal image of the receptor and potentially bind to the ligand. A monoclonal anti-idiotypic antibody generated against a mAb specific for the CD4 molecule mimicked the CD4 molecule and bound to the HIV glycoprotein (Chanh *et al.*, 1987). The antibody also partially neutralized HIV infection of T cells *in vitro*. This approach in which anti-idiotypic antibodies are elicited by immunization with antibodies directed against the binding site of the receptor has the advantage that it is not necessary to expose the host to virus.

10.2.3 Receptors

The virus receptors themselves in soluble form should bind to the virus and prevent attachment to host-cell receptors. It has been shown that soluble, secreted, forms of the CD4 molecule neutralize infectivity of HIV *in vitro* (Smith *et al.*, 1987; Deen *et al.*, 1988; Fisher *et al.*, 1988; Hussy *et al.*, 1988; Traunecker *et al.*, 1988). The whole receptor is not essential as fragments of the CD4 molecule containing the *N*-terminal immunoglobulin-like domains of CD4 were also effective (Berger *et al.*, 1988; Traunecker *et al.*, 1988; Arthos *et al.*, 1989; Chao *et al.*, 1989). Besides acting as a competitive inhibitor, another effect of CD4 may be to strip the gp120 from the viral envelope (Moore *et al.*, 1990). Recombinant soluble CD4 is now undergoing clinical trials in patients with AIDS or AIDS-related complex (Kahn *et al.*, 1990; Schooley *et al.*, 1990). Soluble recombinant CR2 (CD21) inhibits Epstein-Barr virus infection of B lymphocytes (Nemerow *et al.*, 1990). A soluble form of ICAM-1 inhibits rhinovirus infection of cultured cells (Marlin *et al.*, 1990). Soluble receptor neutralizes poliovirus infectivity of cultured cells (Kaplan *et al.*, 1990a). Alpha and γ inhibitors of influenza virus present in normal sera are sialylated glycoproteins which compete binding of the hemagglutinin with host cell receptors (see Anders *et al.*, 1990). The γ inhibitors are α_2-macroglobulins (Hanaoka *et al.*, 1989). Resistance to serum inhibitors by influenza variants is associated with mutations in the hemagglutinin receptor-binding site, supporting the notion that serum inhibitors act as receptor analogs (Ryan-Poirier and Kawaoka, 1991).

10.2.4 Designed Receptor-Mimics

Agents that contain the binding site of the receptor have been shown to have anti-viral activity. Sialylated gangliosides have been proposed to comprise part of a rabies receptor complex (Superti *et al.*, 1986), and it was shown that gangliosides were effective in inhibiting rabies virus infection of cultured cells (Conti *et al.*, 1988). Heparin inhibits attachment of herpes simplex virus to cells (Vaheri, 1964), presumably by blocking the interaction of virus with heparan sulfate on the cell surface (WuDunn and Spear, 1989). In the case of influenza virus, the ability of glycoproteins to inhibit hemagglutination was related to the sialic acid content of the glycoproteins (Springer *et al.*, 1969). A number of monovalent sialosides were tested for their ability to inhibit influenza virus adsorption to erythrocytes (Pritchett *et al.*, 1987). The most effective sialic acid derivatives were NeuAcαCH$_3$ (α-methyl-*N*-acetylneuraminic acid) and NeuAcαC7H7. Sialyloligosaccharide sequences were also tested. The most effective contained the NeuAcα2,6GlcNAc and NeuAcα2,Gal linkages, in accordance with the receptor specificity of the influenza strain used. However, the sialosides were effective at millimolar concentrations which may limit their usefulness as therapeutic agents.

Synthetic peptides of CD4 residues 25–58 of CD4 (Jameson *et al.*, 1988) and a benzylated peptide of residues 81–92 (Lifson *et al.*, 1988) specifically inhibited HIV-1 induced cell fusion and infection *in vitro*. The latter agent also had a virostatic effect on cultures already infected with HIV-1 (Nara *et al.*, 1989). Derivatives of phenylalanine dipeptides were effective in inhibiting gp120 binding to CD4 and inhibited infection by HIV-1 (Finberg *et al.*, 1990). Relatively high concentrations of these compounds were required. Mutation of Phe 43 of

CD4 interferes with gp120 binding (Arthos *et al.*, 1989). Thus, the effect of these agents could be due to mimicry of an important determinant of the binding site of the receptor.

Another application of receptor-mimics is to utilize them to target toxic agents to infected cells. The first 178 amino acids of CD4 linked to the active regions of *Pseudomonas* exotoxin A (Chaudhary *et al.*, 1988) and soluble CD4 conjugated to the active subunit of the toxin ricin (Till *et al.*, 1988) killed HIV-infected cells. In cocultures of HIV-infected and uninfected cells, the *Pseudomonas* exotoxin hybrid protein reduced the rate of spread of infection (Berger *et al.*, 1989). Because CD4 interacts with MHC class II molecules, the CD4-toxin hybrids could interfere with functions mediated by these molecules. However, in both cases the hybrids did not affect cells expressing MHC class II antigens.

Based on these examples, receptor-mimics may represent useful anti-viral agents (Lentz, 1988; Rossmann, 1988). These agents by binding to the conserved binding domains of the VAP should be effective against different strains of virus. Infection of cells and cytopathic effects may be prevented or suppressed. Receptor-mimics could suppress or delay infection sufficiently long for the normal defensive mechanisms to clear the virus. The virus-receptor complex could be degraded or cleared from the body by normal mechanisms. These inhibitory agents can be produced in large quantities by recombinant DNA techniques or by chemical synthesis. It is also possible that they can be designed or engineered to bind with even higher affinity than the native receptor and to avoid adverse effects.

There are also several potential difficulties in the use of receptor-mimics as antiviral agents. The receptors or receptor fragments may be rapidly degraded necessitating frequent administration. In order to overcome this difficulty with the CD4 molecule, Capon *et al.* (1989) produced hybrid CD4-immunoglobulin molecules (immuno-adhesins) composed of the CD4 binding domain for gp120 and IgG1 heavy-chain constant region. The hybrid molecules bound gp120 and blocked infection of T cells and monocytes by HIV-1. Significantly, the immuno-adhesins, unlike soluble CD4 alone, showed the long half-life characteristics of antibodies *in vivo*. Traunecker *et al.* (1989) produced a CD4-IgM chimera which was highly effective in the syncytium inhibition assay and which retained effector functions including the binding of Fc receptors and the first component of the complement cascade.

Other possible adverse effects are that these substances could elicit immune responses including an anaphylactic response, hypersensitivity reaction, or production of auto-antibodies resulting in autoimmune disease. They could bind the natural ligand of the receptor interfering with a normal physiological process. Soluble receptors could also lead to the selection of mutants that are resistant to neutralization by the receptor analog or that can utilize other receptors. As discussed above, it could be expected that the receptor binding domain of the VAP is likely to be highly conserved. However, it has been shown that infectious poliovirus mutants resistant to neutralization with soluble poliovirus receptor can be readily isolated (Kaplan *et al.*, 1990b). Finally, soluble CD4 which blocks HIV-1 infection, significantly enhances infection of a human T-cell line by SIV (Allan *et al.*, 1990). It was suggested that binding of CD4 to gp120 exposed fusigenic sites thereby enhancing viral fusion.

11. Conclusion

The viral infectious cycle begins with attachment of the virus particle to the host cell membrane after which the virion is internalized by fusion with the membrane or by receptor-mediated endocytosis. Attachment is mediated by the specific interaction of a viral surface protein, the VAP, with a complementary site on a normal component of the plasma membrane, either protein, carbohydrate, or lipid, which acts as a virus receptor. Information is accumulating rapidly not only on the nature of the VAPs and of viral receptors but also on the specific domains involved in binding. This knowledge is allowing the development of molecular models for the virus-host cell interaction. Such models will permit the rational design of agents that might be effective in preventing viral infection by blocking the interaction of the VAP with the host-cell receptor. One group of agents, termed ligand-mimics, chemically or structurally resemble the binding domain of the ligand, in this case the VAP, and competitively inhibit binding of virus to the receptor. Ligand-mimics are antibodies directed against the binding site of the receptor, anti-idiotypic antibodies against anti-ligand antibodies, natural ligands of receptors, and designed or engineered agents such as synthetic peptides of the VAP. The second group of agents, termed receptor-mimics, will resemble the binding domain of the receptor and prevent binding of virus by attaching to the binding domain of the VAP. Receptor-mimics are antibodies directed against the binding domain of the ligand, anti-idiotypic antibodies against anti-receptor antibodies, soluble receptors, or agents designed to mimic the receptor. All of these classes of agents have been shown to inhibit binding of viruses to host cell receptors, demonstrating the potential usefulness of targeting the attachment step in the prevention of virus infection. However, there are also disadvantages to each approach which may limit their usefulness. It is possible that combinations of these approaches with each other or with other antiviral agents acting on stages subsequent to attachment could prove highly effective.

12. Acknowledgement

The author's research is supported by grants NS 21896

from the National Institute of Health and BNS 88-05780 from the National Science Foundation.

13. *References*

Acharya, R., Fry, E., Stuart, D., Fox, G., Rowlands, D. and Brown, F. (1989). The three-dimensional structure of foot-and-mouth disease virus at 2.9 Å resolution. Nature 337, 709–716.

Allan, J.S., Strauss, J. and Buck, D.W. (1990). Enhancement of SIV infection with soluble receptor molecules. Science 247, 1084–1088.

Albritton, L.M., Tseng, L., Scadden, D. and Cunningham, J.M. (1989). A putative murine ecotropic retrovirus receptor gene encodes a multiple membrane-spanning protein and confers susceptibility to virus infection. Cell 57, 659–666.

Anders, E.M., Hartley, C.A. and Jackson, D.C. (1990). Bovine and mouse serum β inhibitors of influenza A viruses are mannose-binding lectins. Proc. Natl Acad. Sci. USA 87, 4485–4489.

Aoki, J., Koike, S., Ise, I., Sato-Yoshida, Y. and Nomoto, A. (1994). Amino acid residues on human poliovirus receptor involved in interaction with poliovirus. J. Biol. Chem. 269, 8431–8438.

Arthos, J., Deen, K.C., Chaikin, M.A., Fornwald, J.A., Sathe, G., Sattentau, Q.J., Clapham, P.R., Weiss, R.A., McDougal, J.S., Pietropaolo, C., Axel, R., Truneh, A., Maddon, P.J. and Sweet, R.W. (1989). Identification of the residues in human CD4 critical for the binding of HIV. Cell 57, 469–481.

Ashkenazi, A., Presta, L.G., Marsters, S.A., Camerato, T.R., Rosenthal, K.A., Fendly, B.M. and Capon, D.J. (1990). Mapping the CD4 binding site for human immunodeficiency virus by alanine-scanning mutagenesis. Proc. Natl Acad. Sci. USA 87, 7150–7154.

Attwood, W.J. and Norkin, L.C. (1989). Class I major histocompatibility proteins as cell surface receptors for simian virus 40. J. Virol. 63, 4474–4477.

Aytay, S. and Schulze, I.T. (1991). Single amino acid substitutions in the hemagglutinin can alter the host range and receptor binding properties of H1 strains of influenza A virus. J. Virol. 65, 3022–3028.

Baird, A., Florkiewicz, R.Z., Maher, P.A., Kaner, R.J. and Hajjar, D.P. (1990). Mediation of virion penetration into vascular cells by association of basic fibroblast growth factor with herpes simplex virus type 1. Nature 348, 344–346.

Bass, D.M., Mackow, E.R. and Greenberg, H.B. (1991). Identification and partial characterization of a rhesus rotavirus binding glycoprotein on murine enterocytes. Virology 183, 602–610.

Bassel-Duby, R., Jayasuriya, A., Chatterjee, D., Sonenberg, N., Maizel Jr., J.V. and Fields, B.N. (1985). Sequence of a reovirus haemagglutinin predicts a coiled-coil structure. Nature 315, 421–423.

Bates, P., Young, J.A.T. and Varmus, H.E. (1993). A receptor for subgroup A Rous sarcoma virus is related to the low density lipoprotein receptor. Cell 74,1043–1051.

Beck, S. and Barrell, B.G. (1988). Human cytomegalovirus encodes a glycoprotein homologous to MHC class-I antigens. Nature 331, 269–272.

Bergelson, J.M., Shepley, M.P., Chan, B.M.C., Hemler, M.E. and Finberg, R.W. (1992). Identification of the integrin VLA-2 as a receptor for echovirus 1. Science 255, 1718–1720.

Berger, E.A., Fuerst, T.R. and Moss, B. (1988). A soluble recombinant polypeptide comprising the amino-terminal half of the extracellular region of the CD4 molecule contains an active binding site for human immunodeficiency virus. Proc. Natl Acad. Sci. USA 85, 2357–2361.

Berger, E.A., Clouse, K.A., Chaudhary, V.K., Chakrabarti, S., FitzGerald, D.J., Pastan, I. and Moss, B. (1989). CD4-*Pseudomonas* exotoxin hybrid protein blocks the spread of human immunodeficiency virus infection *in vitro* and is active against cells expressing the envelope glycoproteins from diverse primate immunodeficiency retroviruses. Proc. Natl Acad. Sci. USA 86, 9539–9543.

Bhat, S., Spitalnik, S.L., Gonzalez-Scarano, F. and Silberberg, D.H. (1991). Galactosyl ceramide or a derivative is an essential component of the neural receptor for human immunodeficiency virus type 1 envelope glycoprotein gp120. Proc. Natl Acad. Sci. USA 88, 7131–7134.

Bittle, J.L., Houghten, R.A., Alexander, H., Shinnick, T.M., Sutcliffe, J.G., Lerner, R.A., Rowlands, D.J. and Brown, F. (1982). Protection against foot-and-mouth disease by immunization with a chemically synthesized peptide predicted from the viral nucleotide sequence. Nature 298, 30–33.

Bracci, L., Antoni, G., Cusi, M.G., Lozzi, L., Niccolai, N., Petrini, S., Rustici, M., Santucci, A., Soldani, P., Valensin, P.E. and Neri, P. (1988). Antipeptide monoclonal antibodies inhibit the binding of rabies virus glycoprotein and alpha-bungarotoxin to the nicotinic acetylcholine receptor. Mol. Immunol. 25, 881–888.

Bracci, L., Lozzi, L., Rustici, M. and Neri, P. (1992). Binding of HIV-1 gp120 to the nicotinic receptor. FEBS Lett. 311,115–118.

Brenneman, D.E., Westbrook, G.L., Fitzgerald, S.P., Ennist, D.L., Elkins, K.L., Ruff, M.R. and Pert, C.D. (1988). Neuronal cell killing by the envelope protein of HIV and its prevention by vasoactive intestinal peptide. Nature 335, 639–642.

Bridge, T.P., Heseltine, P.N.R., Parker, E.S., Eaton, E., Ingraham, L.J., Gill, M., Ruff, M. and Pert, C.B. (1989). Improvement in AIDS patients on peptide T. Lancet ii, 226–227.

Brodsky, M.H., Warton, M., Myers, R.M. and Littman, D.R. (1990). Analysis of the site in CD4 that binds to the HIV envelope glycoprotein. J. Immunol. 144, 3078–3086.

Brown, J.P., Twardzik, D.R., Marquardt, H. and Todaro, G.J. (1985). Vaccinia virus encodes a polypeptide homologous to epidermal growth factor and transforming growth factor. Nature 313, 491–492.

Brown, K.E., Anderson, S.M. and Young, N.S. (1993). Erythrocyte P antigen: cellular receptor for B19 parvovirus. Science 262,114–117.

Bruck, C., Co, M.S., Slaoui, M., Gaulton, G.N., Smith, T., Fields, B.N., Mullins, J.I. and Greene, M.I. (1986). Nucleic acid sequence of an internal image-bearing monoclonal anti-idiotype and its comparison to the sequence of the external antigen. Proc. Natl Acad. Sci. USA 83, 6578–6582.

Burness, A.T.H. and Pardoe, I.U. (1981). Effect of enzymes on the attachment of influenza and encephalomyocarditis viruses to erythrocytes. J. Gen. Virol. 55, 275–288.

Byrn, R.A., Sekigawa, I., Chamow, S.M., Johnson, J.S., Gregory, T.J., Capon, D.J. and Groopman, J.E. (1989). Characterization of in vitro inhibition of human immunodeficiency virus by purified recombinant CD4. J. Virol. 63, 4370–4375.

Capon, D.J., Charnow, S.M., Mordenti, J., Marsters, S.A., Gregory, T., Mitsuya, H., Byrn, R.A., Lucas, C., Wurm, F.M., Groopman, J.E., Broder, S. and Smith, D.H. (1989). Designing CD4 immunoadhesins for AIDS therapy. Nature 337, 525–531.

Cardosa, M.J., Gordon, S., Hirsch, S., Springer, T.A. and Porterfield, J.S. (1986). Interaction of West Nile virus with primary murine macrophages: role of cell activation and receptors for antibody and complement. J. Virol. 57, 952–959.

Carel, J.-C., Myones, B.L., Frazier, B. and Holers, V.M. (1990). Structural requirements for C3d,g/Epstein-Barr virus receptor (CR2/CD21) ligand binding, internalization, and viral infection. J. Biol. Chem. 265, 12293–12299.

Carroll, S.M., Higa, H.H. and Paulson, J.C. (1981). Different cell-surface receptor determinants of antigenically similar influenza virus hemagglutinins. J. Biol. Chem. 256, 8357–8363.

Chakrabarti, S., Mizukami, T., Franchini, G. and Moss, B. (1990). Synthesis, oligomerization, and biological activity of the human immunodeficiency virus type 2 envelope glycoprotein expressed by a recombinant vaccinia virus. Virology 178, 134–142.

Chang, K.H., Auvinen, P., Hyypiä, T. and Stanway, G. (1989). The nucleotide sequence of coxsackievirus A9; Implications for receptor binding and enterovirus classification. J. Gen. Virol. 70, 3269–3280.

Chanh, T.C., Dreesman, G.R. and Kennedy, R.C. (1987). Monoclonal anti-idiotypic antibody mimics the CD4 receptor and binds human immunodeficiency virus. Proc. Natl Acad. Sci. USA 84, 3891–3895.

Chao, B.H., Costopoulos, D.S., Curie, T., Bertonis, J.M., Chisholm, P., Williams, C., Schooley, R.T., Rosa, J.J., Fisher, R.A. and Maraganore, J.M. (1989). A 113-amino acid fragment of CD4 produced in *Escherichia coli* blocks human immunodeficiency virus-induced cell fusion. J. Biol. Chem. 264, 5812–5817.

Chatterjee, D. and Maizel Jr., J.V. (1984). Homology of adenoviral E3 glycoprotein with HLA-DR heavy chain. Proc. Natl Acad. Sci. USA 81, 6039–6043.

Chaudhary, V.K., Mizukami, T., Fuerst, T.R., Fitzgerald, D.J., Moss, B., Pastan, I. and Berger, E.A. (1988). Selective killing of HIV-infected cells by recombinant human CD4-*Pseudomonas* exotoxin hybrid protein. Nature 335, 369–372.

Chee, M.S., Satchwell, S.C., Preddie, E., Weston, K.M. and Barrell, B.G. (1990). Human cytomegalovirus encodes three G protein-coupled receptor homologues. Nature 344, 774–777.

Cheng-Mayer, C., Rutka, J.T., Rosenblum, M.L., McHugh, T., Stites, D.P. and Levy, J.A. (1987). Human immunodeficiency virus can productively infect cultured human glial cells. Proc. Natl Acad. Sci. USA 84, 3526–3530.

Chiodi, F., Fuerstenberg, S., Gidlund, M., Åsjö, B. and Fenyö, E.M. (1987). Infection of brain-derived cells with the human immunodeficiency virus. J. Virol. 61, 1244–1247.

Choi, A.H.C., Paul, R.W. and Lee, P.W.K. (1990). Reovirus binds to multiple plasma membrane proteins of mouse L fibroblasts. Virology 178, 316–320.

Clapham, P.R., Weber, J.N., Whitby, D., McIntosh, K., Dalgleish, A.G., Maddon, P.J., Deen, K.C., Sweet, R.W. and Weiss, R.A. (1989). Soluble CD4 blocks the infectivity of diverse strains of HIV and SIV for T cells and monocytes but not for brain and muscle cells. Nature 337, 368–370.

Clarke, M.F., Gelmann, E.P. and Reitz Jr., M.S. (1983). Homology of human T-cell leukaemia virus envelope gene with class I HLA gene. Nature 305, 60–62.

Clayton, L.K., Sieh, M., Pious, D.A. and Reinherz, E.L. (1989). Identification of human CD4 residues affecting class II MHC versus HIV-1 gp120 binding. Nature 339, 548–551.

Co, M.S., Gaulton, G.N., Tominaga, A., Homcy, C.J., Fields, B.N. and Greene, M.I. (1985). Structural similarities between the mammalian β-adrenergic and reovirus type 3 receptors. Proc. Natl Acad. Sci. USA 82, 5315–5318.

Colman, P.M., Varghese, J.N. and Laver, W.G. (1983). Structure of the catalytic and antigenic sites in influenza virus neuraminidase. Nature 303, 41–44.

Colonno, R.J., Callahan, P.L. and Long, W.J. (1986). Isolation of a monoclonal antibody that blocks attachment of the major group of human rhinoviruses. J. Virol. 57, 7–12.

Colonno, R.J., Condra, J.H., Mizutani, S., Callahan, P.L., Davies, M.-E. and Murcko, M.A. (1988). Evidence for the direct involvement of the rhinovirus canyon in receptor binding. Proc. Natl Acad. Sci. USA 85, 5449–5453.

Compton, T., Nowlin, D.M. and Cooper, N.R. (1993). Initiation of human cytomegalovirus infection requires initial interaction with cell surface heparan sulfate. Virology 193, 834–841.

Condra, J.H., Sardana, V.V., Tomassini, J.E., Schlabach, A.J., Davies, M.-E., Lineberger, D.W., Graham, D.J., Gotlib, L. and Colonno, R.J. (1990). Bacterial expression of antibody fragments that block human rhinovirus infection of cultured cells. J. Biol. Chem. 265, 2292–2295.

Conti, C., Hauttecoeur, B., Morelec, M.J., Bizzini, B., Orsi, N. and Tsiang, H. (1988). Inhibition of rabies virus infection by a soluble membrane fraction from the rat central nervous system. Arch. Virol. 98, 73–86.

Cordonnier, A., Montagnier, L. and Emerman, M. (1989a). Single amino-acid changes in HIV envelope affect viral tropism and receptor binding. Nature 340, 571–574.

Cordonnier, A., Riviere, Y., Montagnier, L. and Emerman, M. (1989b). Effects of mutations in hyperconserved regions of the extracellular glycoprotein of human immunodeficiency virus type 1 on receptor binding. J. Virol. 63, 4464–4468.

Crowell, R.L. and Landau, B.L. (1979). In "Receptors and Human Diseases" (eds. A.G. Bearn and P.W. Choppin), pp. 1–33. Macy Foundation, New York.

Dales, S. (1973). Early events in cell-animal virus interactions. Bact. Rev. 37, 103–135.

Dalgleish, A.G., Beverley, P.C.L., Clapham, P.R., Crawford D.H., Greaves, M.F. and Weiss, R.A. (1984). The CD4 (T4) antigen is an essential component of the receptor for the AIDS retrovirus. Nature 312, 763–767.

Dalziel, R.G., Hopkins, J., Watt, N.J., Dutia, B.M., Clarke, H.A. and McConnell, I. (1991). Identification of a putative cellular receptor for the lentivirus visna virus. J. Gen. Virol. 72, 1905–1911.

Damian, R.T. (1987). Molecular mimicry revisited. Parasitol. Today 3, 263–266.

Dash, S., Rao, K.V.S., Joshi, B., Nayak, N.C. and Panda, S.K. (1991). Significance of natural polymerized albumin and its receptor in hepatitis B infection of hepatocytes. Hepatology 13, 134–142.

Davis, S.J., Schockmel, G.A., Somoza, C., Buck, D.W., Healey, D.G., Rieber, E.P., Reiter, C. and Williams, A.F. (1992). Antibody and HIV-1 gp120 recognition of CD4 undermines the concept of mimicry between antibodies and receptors. Nature 358, 76–79.

Deen, K.C., McDougal, J.S., Inacker, R., Folena-Wasserman, G., Arthos, J., Rosenberg, J., Maddon, P.J., Axel, R. and Sweet, R.W. (1988). A soluble form of CD4 (T4) protein inhibits AIDS virus infection. Nature 331, 82–84.

Delmas, B., Gelfi, J., L'Haridon, R., Vogel, L.K., Sjöström, H., Norén, O. and Laude, H. (1992). Aminopeptidase N is a major receptor for the enteropathic coronavirus TGEV. Nature 357, 417–420.

Diamond, D.C., Sleckman, B.P., Gregory, T., Lasky, L.A., Greenstein, J.L. and Burakoff, S.J. (1988). Inhibition of CD4+ T cell function by the HIV envelope protein, gp120. J. Immunol. 141, 3715–3717.

DiMarchi, R., Brooke, G., Gale, C., Cracknell, V., Doel, T. and Mowat, N. (1986). Protection of cattle against foot-and-mouth disease by a synthetic peptide. Science 232, 639–641.

Dimmock, N.J. (1982). Initial stages in infection with animal viruses. J. Gen. Virol. 59, 1–22.

Donnelly-Roberts, D.L. and Lentz, T.L. (1989). Synthetic peptides of neurotoxins and rabies virus glycoprotein behave as antagonists in a functional assay for the acetylcholine receptor. Peptide Res. 2, 221–226.

Dörig, R.E., Marcil, A., Chopra, A. and Richardson, C.D. (1993). The human CD46 molecule is a receptor for measles virus (Edmonston strain). Cell 75, 295–305.

Dorner, A.J. and Coffin, J.M. (1986). Determinants for receptor interaction and cell killing on the avian retrovirus glycoprotein gp85. Cell 45, 365–374.

Dowbenko, D., Nakamura, G., Fennie, C., Shimasaki, C., Riddle, L., Harris, R., Gregory, T. and Lasky, L. (1988). Epitope mapping of the human immunodeficiency virus type 1 gp120 with monoclonal antibodies. J. Virol. 62, 4703–4711.

Eisenlohr, L.C., Gerhard, W. and Hackett, C.J. (1987). Role of receptor-binding activity of the viral hemagglutinin molecule in the presentation of influenza virus antigens to helper T cells. J. Virol. 61, 1375–1383.

Enders, J.F., Weller, T.H. and Robbins, F.C. (1949). Cultivation of the Lansing strain of poliomyelitis virus in cultures of various human embryonic tissues. Science 109, 85–87.

Eppstein, D.A., Marsh, Y.V., Schreiber, A.B., Newman, S.R., Todaro, G.J. and Nestor, Jr. J.J. (1985). Epidermal growth factor receptor occupancy inhibits vaccinia virus infection. Nature 318, 663–665.

Fields, A.P., Bednarik, D.P., Hess, A. and May, W.S. (1988). Human immunodeficiency virus induces phosphorylation of its cell surface receptor. Nature 333, 278–280.

Finberg, R.W., Diamond, D.C., Mitchell, D.B., Rosenstein, Y., Soman, G., Norman, T.C., Schreiber, S.L. and Burakoff, S.J. (1990). Prevention of HIV-1 infection and preservation of CD4 function by the binding of CPFs to gp120. Science 249, 287–291.

Finberg, R.W., Wahl, S.M., Allen, J.B., Soman, G., Strom, T.B., Murphy, J.R. and Nichols, J.C. (1991). Selective elimination of HIV-1-infected cells with an interleukin-2 receptor-specific cytotoxin. Science 252, 1703–1705.

Fingeroth, J.D., Weis, J.J., Tedder, T.F., Strominger, J.L., Biro, P.A. and Fearon, D.T. (1984). Epstein-Barr virus receptor of human B lymphocytes is the C3d receptor CR2. Proc. Natl Acad. Sci. USA 81, 4510–4514.

Fisher, R.A., Bertonis, J.M., Meier, W., Johnson, V.A., Costopoulos, D.S., Liu, T., Tizard, R., Walker, B.D., Hirsch, M.S., Schooley, R.T. and Flavell, R.A. (1988). HIV infection is blocked *in vitro* by recombinant soluble CD4. Nature 331, 76–78.

Fleury, S., Lamarre, D., Meloche, S., Ryu, S.-E., Cantin, C., Hendrickson, W.A. and Sekaly, R.-P. (1991). Mutational analysis of the interaction between CD4 and class II MHC: class II antigens contact CD4 on a surface opposite the gp120-binding site. Cell 66, 1037–1049.

Flynn, D.C., Meyer, W.J., Mackenzie Jr., J.M. and Johnston, R.E. (1990). A conformational change in Sindbis virus glycoproteins E1 and E2 is detected at the plasma membrane as a consequence of early virus-cell interaction. J. Virol. 64, 3643–3653.

Fox, G., Parry, N.R., Barnett, P.V., McGinn, B., Rowlands, D.J. and Brown, F. (1989). The cell attachment site on foot-and-mouth disease virus includes the amino acid sequence RGD (Arginine-Glycine-Aspartic Acid). J. Gen. Virol. 70, 625–637.

Fricks, C.E. and Hogle, J.M. (1990). Cell-induced conformational change in poliovirus: externalization of the amino terminus of VP1 is responsible for liposome binding. J. Virol. 64, 1934–1945.

Fried, H., Cahan, L.D. and Paulson, J.C. (1981). Polyoma virus recognizes specific sialyloligosaccharide receptors on host cells. Virology 109, 188–192.

Fries, E. and Helenius, A. (1979). Binding of Semliki Forest virus and its spike glycoproteins to cells. Eur. J. Biochem. 97, 213–220.

Fukudome, K., Yoshie, O. and Konno, T. (1989). Comparison of human, simian, and bovine rotaviruses for requirement of sialic acid in hemagglutination and cell adsorption. Virology 172, 196–205.

Fuller, S.D., von Bonsdorff, C.-H. and Simons, K. (1985). Cell surface influenza hemagglutinin can mediate infection by other animal viruses. EMBO J. 4, 2475–2485.

Gartner, S., Markovits, P., Markovitz, D.M., Kaplan, M.H., Gallo, R.C. and Popovic, M. (1986). The role of mononuclear phagocytes in HTLV-III/LAV infection. Science 233, 215–219.

Gershoni, J.M. and Aronheim, A. (1988). Molecular decoys: Ligand-binding recombinant proteins protect mice from curarimimetic neurotoxins. Proc. Natl Acad. Sci. USA 85, 4087–4089.

Giulian, D., Vaca, K. and Noonan, C.A. (1990). Secretion of neurotoxins by mononuclear phagocytes infected with HIV-1. Science 250, 1593–1596.

Goldstein, J.L., Brown, M.S., Anderson, R.G.W., Russell, D.W. and Schneider, W.J. (1985). Receptor-mediated endocytosis: Concepts emerging from the LDL receptor system. Ann. Rev. Cell Biol. 1, 1–39.

Gras, G.S. and Dormont, D. (1991). Antibody-dependent and antibody-independent complement-mediated enhancement of a human immunodeficiency virus type 1 infection in a human, Epstein-Barr virus-transformed B-lymphocytic cell line. J. Virol. 65, 541–545.

Greve, J.M., Davis, G., Meyer, A.M., Forte, C.P., Yost, S.C.,

Marlor, C.W., Kamarck, M.E. and McClelland, A. (1989). The major human rhinovirus receptor is ICAM-1. Cell 56, 839–847.

Grundy, J.E., McKeating, J.A., Ward, P.J., Sanderson, A.R. and Griffiths, P.D. (1987). β_2 microglobulin enhances the infectivity of cytomegalovirus and when bound to the virus enables class I HLA molecules to be used as a virus receptor. J. Gen. Virol. 68, 793–803.

Guyader, M., Emerman, M., Sonigo, P., Clavel, F., Montagnier, L. and Alizon, M. (1987). Genome organization and transactivation of the human immunodeficiency virus type 2. Nature 326, 662–669.

Halstead, S.B. (1988). Pathogenesis of dengue: challenges to molecular biology. Science 239, 476–481.

Halstead, S.B. and O'Rourke, E.J. (1977). Antibody-enhanced dengue virus infection in primate leukocytes. Nature 265, 739–741.

Hanaoka, K., Pritchett, T.J., Takasaki, S., Kochibe, N., Sabesan, S., Paulson, J.C. and Kobata, A. (1989). 4-*O*-acetyl-*N*-acetylneuraminic acid in the *N*-linked carbohydrate structures of equine and guinea pig α_2-macroglobulins, potent inhibitors of influenza virus infection. J. Biol. Chem. 264, 9842–9849.

Hanham, C.A., Zhao, F. and Tignor, G.H. (1993). Evidence from the anti-idiotypic network that the acetylcholine receptor is a rabies virus receptor. J. Virol. 67, 530–542.

Harrison, S.C. (1989). Finding the receptors. Nature 338, 205–206.

Haywood, A.M. (1994). Virus receptors: binding, adhesion strengthening, and changes in viral structure. J. Virol. 68, 1–5.

Heard, J.M. and Danos, O. (1991). An amino-terminal fragment of the Friend murine leukemia virus envelope glycoprotein binds the ecotropic receptor. J. Virol. 65, 4026–4032.

Helenius, A., Morein, B., Fries, E., Simons, K., Robinson, P., Schirrmacher, V., Terhorst, C. and Strominger, J.L. (1978). Human (HLA-A and HLA-B) and murine (H-2K and H-2D) histocompatibility antigens are cell surface receptors for Semliki Forest virus. Proc. Natl Acad. Sci. USA 75, 3846–3850.

Herold, B.C., WuDunn, D., Soltys, N. and Spear, P.G. (1991). Glycoprotein C of herpes simplex virus type 1 plays a principal role in the adsorption of virus to cells and in infectivity. J. Virol. 65, 1090–1098.

Hertogs, K., Leenders, W.P.J., Depla, E., DeBruin, W.C.C., Meheus, L., Raymackers, J., Moshage, H. and Yap, S.H. (1993). Endonexin II, present on human liver plasma membranes, is a specific binding protein of small hepatitis B virus (HBV) envelope protein. Virology 197, 549–557.

Hofer, F., Gruenberger, M., Kowalski, H., Machat, H., Huettinger, M., Kuechler, E. and Blaas, D. (1994). Members of the low density lipoprotein receptor family mediate cell entry of a minor-group common cold virus. Proc. Natl. Acad. Sci. USA 91, 1839–1842.

Hogle, J.M., Chow, M. and Filman, D.J. (1985). Three-dimensional structure of poliovirus at 2.9Å resolution. Science 229, 1358–1365.

Holland, J.J. (1961). Receptor affinities as major determinants of enterovirus tissue tropisms in humans. Virology 15, 312–326.

Holmgren, J., Svennerholm, L., Elwing, H., Fredman, P. and Strannegård, Ö. (1980). Sendai virus receptor: proposed recognition structure based on binding to plastic-adsorbed gangliosides. Proc. Natl Acad. Sci. USA 77, 1947–1950.

Homsy, J., Meyer, M., Tateno, M., Clarkson, S. and Levy, J.A. (1989). The Fc and not CD4 receptor mediates antibody enhancement of HIV infection in human cells. Science 244, 1357–1360.

Hoxie, J.A., Haggarty, B.S., Bonser, S.E., Rackowski, J.L., Shan, H. and Kanki, P.J. (1988). Biological characterization of a simian immunodeficiency virus-like retrovirus (HTLV-IV): evidence for CD4-associated molecules required for infection. J. Virol. 62, 2557–2568.

Hsu, D.-H., Malefyt, R.d.W., Fiorentino, D.F., Dang, M.-N., Vieira, P., deVries, J., Spits, H., Mosmann, T.R. and Moore, K.W. (1990). Expression of interleukin-10 activity by Epstein-Barr virus protein BCRF1. Science 250, 830–832.

Hussy, R.E., Richardson, N.E., Kowalski, M., Brown, N.R., Chang, H.-C., Siliciano, R.F., Dorfman, T., Walker, B., Sodroski, J. and Reinherz, E.L. (1988). A soluble CD4 protein selectively inhibits HIV replication and syncytium formation. Nature 331, 78–81.

Ibegbu, C.C., Kennedy, M.S., Maddon, P.J., Deen, K.C., Hicks, D., Sweet, R.W. and McDougal, J.S. (1989). Structural features of CD4 required for binding to HIV. J. Immunol. 142, 2250–2256.

Inada, T. and Mims, A. (1984). Mouse Ia antigens are receptors for lactate dehydrogenase virus. Nature 309, 59–61.

Iorio, R.M., Syddall, R.J., Sheehan, J.P., Bratt, M.A., Glickman, R.L. and Riel, A.M. (1991). Neutralization map of the hemagglutinin-neuraminidase glycoprotein of Newcastle disease virus: domains recognized by monoclonal antibodies that prevent receptor recognition. J. Virol. 65, 4999–5006.

Itoh, Y., Takai, E., Ohnuma, H., Kitajima, K., Tsuda, F., Machida, A., Mishiro, S., Nakamura, T., Miyakawa, Y. and Miyumi, M. (1986). A synthetic peptide vaccine involving the product of the pre-S(2) region of hepatitis B virus DNA: protective efficacy in chimpanzees. Proc. Natl Acad. Sci. USA 83, 9174–9178.

Ivey-Hoyle, M., Culp, J.S., Chaikin, M.A., Hellmig, B.D., Matthews, T.J., Sweet, R.W. and Rosenberg, M. (1991). Envelope glycoproteins from biologically diverse isolates of immunodeficiency viruses have widely different affinities for CD4. Proc. Natl Acad. Sci. USA 88, 512–516.

Jameson, B.A., Rao, P.E., Kong, L.I., Hahn, B.H., Shaw, G.M., Hood, L.E. and Kent, S.B.H. (1988). Location and chemical synthesis of a binding site for HIV-1 on the CD4 protein. Science 240, 1335–1339.

Johann, S.V., Gibbons, J.J. and O'Hara, B. (1992). GLVR1, a receptor for gibbon ape leukemia virus, is homologous to a phosphate permease of *Neurospora crassa* and is expressed at high levels in the brain and thymus. J. Virol. 66, 1635–1640.

Jolly, P.E. and Narayan, O. (1989). Evidence for interference, coinfections, and intertypic virus enhancement of infection by ovine-caprine lentiviruses. J. Virol. 63, 4682–4688.

Kahn, J.O., Allan, J.D., Hodges, T.L., Kaplan, L.D., Arri, C.J., Fitch, H.F., Izu, A.E., Mordenti, J., Sherwin, S.A., Groopman, J.E. and Volberding, P.A. (1990). The safety and pharmacokinetics of recombinant soluble CD4 (rCD4) in subjects with the acquired immunodeficiency syndrome (AIDS) and AIDS-related complex. A phase 1 study. Ann. Intern. Med. 112, 254–261.

Kaner, R.J., Baird, A., Mansukhani, A., Basilico, C., Summers, B.D., Florkiewicz, R.Z. and Hajjar, D.P. (1990). Fibroblast growth factor receptor is a portal of cellular entry for herpes simplex virus type 1. Science 248, 1410–1423.

Kannagi, M., Yetz, J.M. and Letvin, N.L. (1985). *In vitro* growth characteristics of simian T-lymphotropic virus type III. Proc. Natl Acad. Sci. USA 82, 7053–7057.

Kaplan, G., Freistadt, M.S. and Racaniello, V.R. (1990a). Neutralization of poliovirus by cell receptors expressed in insect cells. J. Virol. 64, 4697–4702.

Kaplan, G., Peters, D. and Racaniello, V.R. (1990b). Poliovirus mutants resistant to neutralization with soluble cell receptors. Science 250, 1596–1599.

Kauffman, R.S., Noseworthy, J.H., Nepom, J.T., Finberg, R., Fields, B.N. and Greene, M.I. (1983). Cell receptors for the mammalian reovirus II. Monoclonal anti-idiotypic antibody blocks viral binding to cells. J. Immunol. 131, 2539–2541.

Kim, J.W., Closs, E.I., Albritton, L.M. and Cunningham, J.M. (1991). Transport of cationic amino acids by the mouse ecotropic retrovirus receptor. Nature 352, 725–728.

King, C.S., Cooper, J.A., Moss, B. and Twardzck, D.R. (1986). Vaccinia virus growth factor stimulates tyrosine protein kinase activity of A431 cell epidermal growth factor receptors. Mol. Cell Biol. 6, 332–336.

Klatzmann, D., Barre-Sinoussi, F., Nugeyre, M.T., Dauguet, C., Vilmer, E., Griselli, C., Brun-Vezinet, F., Rouzioux, C., Gluckman, J.C., Chermann, J.-C. and Montagnier, L. (1984a). Selective tropism of lymphadenopathy associated virus (LAV) for helper-inducer T lymphocytes. Science 225, 59–63.

Klatzmann, D., Champagne, E., Chamaret, S., Gruest, J., Guetard, D., Hercend, T., Gluckman, J.-C. and Montagnier, L. (1984b). T-lymphocyte T4 molecule behaves as the receptor for human retrovirus LAV. Nature 312, 767–768.

Koenig, S., Gendelman, H.E., Orenstein, J.M., Dal Canto, M.C., Pezeshkpour, G.H., Yungbluth, M., Janotta, F., Aksamit, A., Martin, M.A. and Fauci, A.S. (1986). Detection of AIDS virus in macrophages in brain tissue from AIDS patients with encephalopathy. Science 233, 1089–1093.

Kohn, A. (1985). Membrane effects of cytopathogenic viruses. Prog. Med. Virol. 31, 109–167.

Kohtz, D.S., Altman, A., Kohtz, J.D. and Puszkin, S. (1988). Immunological and structural homology between human T-cell leukemia virus type I envelope glycoprotein and a region of human interleukin-2 implicated in binding the β receptor. J. Virol. 62, 659–662.

Koike, S., Taya, C., Kurata, T., Abe, S., Ise, I., Yonekawa, H. and Nomoto, A. (1991a). Transgenic mice susceptible to poliovirus. Proc. Natl Acad. Sci. USA 88, 951–955.

Koike, S., Ise, I. and Nomoto, A. (1991b). Functional domains of the poliovirus receptor. Proc. Natl Acad. Sci. USA 88, 4104–4108.

Komai, K., Kaplan, M. and Peeples, M.E. (1988). The Vero cell receptor for the hepatitis B virus small S protein is a sialoglycoprotein. Virology 163, 629–634.

Kornfeld, H., Riedel, N., Viglianti, G.A., Hirsch, V. and Mullins, J.I. (1987). Cloning of HTLV-4 and its relation to simian and human immunodeficiency viruses. Nature 326, 610–613.

Kornfeld, H., Cruikshank, W.W., Pyle, S.W., Berman, J.S. and Center, D.M. (1988). Lymphocyte activation by HIV-1 envelope glycoprotein. Nature 335, 445–448.

Kotwal, G.J., Isaacs, S.N., McKenzie, R., Frank, M.M. and Moss, B. (1990). Inhibition of the complement cascade by the major secretory protein of vaccinia virus. Science 250, 827–830.

Kowalski, M., Potz, J., Basiripour, L., Dorfman, T., Goh, W.C., Terwilliger, E., Dayton, A., Rosen, C., Haseltine, W. and Sodroski, J. (1987). Functional regions of the envelope glycoprotein of human immunodeficiency virus type 1. Science 237, 1351–1355.

Krah, D.L. and Choppin, P.W. (1988). Mice immunized with measles virus develop antibodies to a cell surface receptor for binding virus. J. Virol. 62, 1565–1572.

Lamarre, D., Ashkenazi, A., Fleury, S., Smith, D.H., Sekaly, R.-P. and Capon, D.J. (1989). The MHC-binding and gp120-binding functions of CD4 are separable. Science 245, 743–746.

Landau, N.R., Warton, M. and Littman, D.R. (1988). The envelope glycoprotein of the human immunodeficiency virus binds to the immunoglobulin-like domain of CD4. Nature 334, 159–162.

Lando, Z., Sarin, P., Megson, M., Greene, W.C., Waldman, T.A., Gallo, R.C. and Broder, S. (1983). Association of human T-cell leukaemia/lymphoma virus with the Tac antigen marker for the human T-cell growth factor receptor. Nature 305, 733–736.

Lasky, L.A., Nakamura, G., Smith, D.H., Fennie, C., Shimasaki, C., Patzer, E., Berman, P., Gregory, T. and Capon, D.J. (1987). Delineation of a region of the human immunodeficiency virus type 1 gp120 glycoprotein critical for interaction with the CD4 receptor. Cell 50, 975–985.

Layne, S.P., Merges, M.J., Dembo, M., Spouge, J.L. and Nara, P.L. (1990). HIV requires multiple gp120 molecules for CD4-mediated infection. Nature 346, 277–279.

Lee, M.R., Ho, D.D. and Gurney, M.E. (1987). Functional interaction and partial homology between human immunodeficiency virus and neuroleukin. Science 237, 1047–1051.

Lenert, P., Kroon, D., Spiegelberg, H., Golub, E.S. and Zanetti, M. (1990). Human CD4 binds immunoglobulins. Science 248, 1639–1643.

Lentz, T.L. (1988). Binding of viral attachment protein to host-cell receptor: the Achilles heel of infectious viruses. Trends Pharmacol. Sci. 9, 247–252.

Lentz, T.L. (1990a). Rabies virus binding to an acetylcholine receptor α-subunit peptide. J. Mol. Recognit. 3, 82–88.

Lentz, T.L. (1990b). The recognition event between virus and host cell receptor: a target for antiviral agents. J. Gen. Virol. 71, 751–766.

Lentz, T.L., Burrage, T.G., Smith, A.L., Crick, J. and Tignor, G.H. (1982). Is the acetylcholine receptor a rabies virus receptor? Science 215, 182–184.

Lentz, T.L., Wilson, P.T., Hawrot, E. and Speicher, D.W. (1984). Amino acid sequence similarity between rabies virus glycoprotein and snake venom curaremimetic neurotoxins. Science 226, 847–848.

Lentz, T.L., Hawrot, E. and Wilson, P.T. (1987). Synthetic peptides corresponding to sequences of snake venom neurotoxins and rabies virus glycoprotein bind to the nicotinic acetylcholine receptor. Proteins: Struct. Funct. Genet. 2, 298–307.

Levy, J.A. (1988). Mysteries of HIV: challenges for therapy and prevention. Nature 333, 519–522.

Li, J.-P., D'Andrea, A.D., Lodish, H.F. and Baltimore, D. (1990). Activation of cell growth by binding of Friend spleen focus-forming virus gp55 glycoprotein to the erythropoietin receptor. Nature 343, 762–764.

Lifson, J.D., Hwang, K.M., Nara, P.L., Fraser, B., Padgett, M., Dunlop, N.M. and Eiden, L.E. (1988). Synthetic CD4 peptide derivatives that inhibit HIV infection and cytopathicity. Science 241, 712–716.

Lineberger, D.W., Graham, D.J., Tomassini, J.E. and Colonno, R.J. (1990). Antibodies that block rhinovirus attachment map to domain 1 of the major group receptor. J. Virol. 64, 2582–2587.

Linsley, P.S., Ledbetter, J.A., Kinney-Thomas, E. and Hu, S.-L. (1988). Effects of anti-gp120 monoclonal antibodies on CD4 receptor binding by the env protein of human immunodeficiency virus type 1. J. Virol. 62, 3695–3702.

Lipkind, M. and Urbakh, V. (1988). Dependence of virus adsorption to the cell surface on the input multiplicity of infection. Zbl. Bakt. Hyg. A 269, 501–505.

Lonberg-Holm, K. and Philipson, L. (1974). "Early interaction between animal viruses and cells", 148pp. S. Karger, Basel.

Lowell, C.A., Klickstein, L.B., Carter, R.H., Mitchell, J.A., Fearon, D.T. and Ahearn, J.M. (1989). Mapping of the Epstein-Barr virus and C3dg binding sites to a common domain on complement receptor type 2. J. Exp. Med. 170, 1931–1946.

Lundin, K., Nygren, A., Arthur, L.O., Robey, W.G., Morein, B., Ramstedt, U., Gidlund, M. and Wigzell, H. (1987). A specific assay measuring binding of ^{125}I-Gp 120 from HIV to T4+/CD4+ cells. J. Immunol. Meth. 97, 93–100.

Luo, M., Vriend, G., Kamer, G., Minor, I., Arnold, E., Rossmann, M.G., Boege, U., Scraba, D.G., Duke, G.M. and Palmenberg, A.C. (1987). The atomic structure of Mengo virus at 3.0 Å resolution. Science 235, 182–191.

Lusso, P., DeMaria, A., Malnati, M., Lori, F., DeRocco, S.E., Baseler, M. and Gallo, R.C. (1991). Induction of CD4 and susceptibility to HIV-1 infection in human CD8+ T lymphocyte by human herpesvirus 6. Nature 349, 533–535.

Maa, J.-S., Rodriguez, J.F. and Esteban, M. (1990). Structural and functional characterization of a cell surface binding protein of vaccinia virus. J. Biol. Chem. 265, 1569–1577.

Machida, A., Kishimoto, S., Ohnuma, H., Baba, K., Ito, Y., Miyamoto, H., Funatsu, G., Oda, K., Usuda, S., Togami, S., Nakamura, T., Miyakawa, Y. and Mayumi, M. (1984). A polypeptide containing 55 amino acid residues coded by the pre-S region of hepatitis B virus deoxyribonucleic acid bears the receptor for polymerized human as well as chimpanzee albumins. Gastroenterology 86, 910–918.

Maddon, P.J., Dalgleish, A.G., McDougal, J.S., Clapham, P.R., Weiss, R.A. and Axel, R. (1986). The T4 gene encodes the AIDS virus receptor and is expressed in the immune system and the brain. Cell 47, 333–348.

Maddon, P.J., McDougal, J.S., Clapham, P.R., Dalgleish, A.G., Jamal, S., Weiss, R.A. and Axel, R. (1988). HIV infection does not require endocytosis of its receptor, CD4. Cell 54, 865–874.

Manchester, M., Liszewski, M.K., Atkinson, J.P. and Oldstone, M.B.A. (1994). Multiple isoforms of CD46 (membrane cofactor protein) serve as receptors for measles virus. Proc. Natl. Acad. Sci. USA 91, 2161–2165.

Mann, D.L., Read-Connole, E., Arthur, L.O., Robey, W.G., Wernet, P., Schneider, E.M., Blattner, W.A. and Popovic, M. (1988). HLA-DR is involved in the HIV-1 binding site on cells expressing MHC class II antigens. J. Immunol. 141, 1131–1136.

Marlin, S.D., Staunton, D.E., Springer, T.A., Stratowa, C., Sommergruber, W. and Merluzzi, V.J. (1990). A soluble form of intercellular adhesion molecule-1 inhibits rhinovirus infection. Nature 344, 70–72.

Marsh, M. and Helenius, A. (1989). Virus entry into animal cells. Adv. Virus Res. 36, 107–151.

Mastromarino, P., Conti, C., Goldoni, P., Houttecoeur, B. and Orsi, N. (1987). Characterization of membrane components of the erythrocyte involved in vesicular stomatitis attachment and fusion at acidic pH. J. Gen. Virol. 68, 2359–2369.

Matrosovich, M.N. (1989). Towards the development of antimicrobial drugs acting by inhibition of pathogen attachment to host cells: a need for polyvalency. FEBS Lett. 252, 1–4.

Matrosovich, M.N., Mochalova, L.V., Marinina, V.P., Byramova, N.E. and Bovin, N.V. (1990). Synthetic polymeric sialoside inhibitors of influenza virus receptor-binding activity. FEBS Lett. 272, 209–212.

McCray, J. and Werner, G. (1987). Different rhinovirus serotypes neutralized by antipeptide antibodies. Nature 329, 736–738.

McDougal, J.S., Kennedy, M.S., Sligh, J.M., Cort, S.P., Mawle, A. and Nicholson, J.K.A. (1986a). Binding of HTLV-III/LAV to T4+ T cells by a complex of the 110K viral protein and the T4 molecule. Science 231, 382–385.

McDougal, J.S., Nicholson, J.K.A., Cross, G.D., Cort, S.P., Kennedy, M.S. and Mawle, A.C. (1986b). Binding of the human retrovirus HTLV-III/LAV/ARV/HIV to the CD4 (T4) molecule: conformation dependence, epitope mapping, antibody inhibition, and potential for idiotypic mimicry. J. Immunol. 137, 2937–2944.

McGrath, M.S., Tamura, G. and Weissman, I.L. (1987). Receptor mediated leukemogenesis: Murine leukemia virus interacts with BCL_1 lymphoma cell surface IBM. J. Mol. Cell. Immunol. 3, 227–242.

Mellman, I., Fuchs, R. and Helenius, A. (1986). Acidification of the endocytic and exocytic pathways. Ann. Rev. Biochem. 55, 663–700.

Mendelsohn, C.L., Wimmer, E. and Racaniello, V.R. (1989). Cellular receptor for poliovirus: Molecular cloning, nucleotide sequence, and expression of a new member of the immunoglobulin superfamily. Cell 56, 855–865.

Merrill, J.E., Koyanagi, Y. and Chen, I.S.Y. (1989). Interleukin-1 and tumor necrosis factor α can be induced from mononuclear phagocytes by human immunodeficiency virus type 1 binding to the CD4 receptor. J. Virol. 63, 4404–4408.

Miller, D.G., Edwards, R.H. and Miller, A.D. (1994). Cloning of the cellular receptor for amphotropic murine retroviruses reveals homology to that for gibbon ape leukemia virus. Proc. Natl. Acad. Sci. USA 91, 78–82.

Mims, C.A. (1986). Virus receptors and cell tropisms. J. Infect. 12, 199–203.

Mims, C.A. (1989). The pathogenetic basis of viral tropism. Am. J. Pathol. 135, 447–455.

Minor, P.D., Pipkin, P.A., Hockley, D., Schild, G.C. and Almond, J.W. (1984). Monoclonal antibodies which block cellular receptors of poliovirus. Virus Res. 1, 203–213.

Mittler, R.S. and Hoffmann, M.K. (1989). Synergism between HIV gp120 and gp120-specific antibody in blocking human T-cell activation. Science 245, 1380–1382.

Mizukami, T., Fuerst, T.R., Berger, E.A. and Moss, B. (1988). Binding region for human immunodeficiency virus (HIV) and epitopes for HIV-binding monoclonal antibodies of the CD4 molecule defined by site-directed mutagenesis. Proc. Natl Acad. Sci. USA 85, 9273–9277.

Molina, H., Brenner, C., Jacobi, S., Gorka, J., Carel, J.-C., Kinoshita, T. and Holers, V.M. (1991). Analysis of Epstein-Barr virus-binding sites on complement receptor 2 (CR2/CD21) using human-mouse chimeras and peptides. J. Biol. Chem. 266, 12 173–12 179.

Moore, J.P., McKeating, J.A., Weiss, R.A. and Sattentau, Q.J. (1990). Dissociation of gp120 from HIV-1 virions induced by soluble CD4. Science 250, 1139–1142.

Moore, J.P., McKeating, J.A., Norton, W.A. and Sattentau, Q.J. (1991). Direct measurement of soluble CD4 binding to human immunodeficiency virus type 1 virions: gp120 dissociation and its implications for virus-cell binding and fusion reactions and their neutralization by soluble CD4. J. Virol. 65, 1133–1140.

Moore, K.W., Vieira, P., Fiorentino, D.F., Trounstine, M.L., Khan, T.A. and Mosmann, T.R. (1990). Homology of cytokine synthesis inhibitory factor (IL-10) to the Epstein-Barr virus gene BCRFI. Science 248, 1230–1234.

Morrison, M.E., He, Y.-J., Wien, M.W., Hogle, J.M. and Racaniello, V.R. (1994). Homolog-scanning mutagenesis reveals poliovirus receptor residues important for virus binding and replication. J. Virol. 68, 2578–2588.

Moss, E.G. and Racaniello, V.R. (1991). Host range determinants located on the interior of the poliovirus capsid. EMBO J. 10, 1067–1074.

Murphy, F.A. (1977). Rabies pathogenesis. Brief review. Arch. Virol. 54, 279–297.

Murray, M.G., Bradley, J., Yang, X.-F., Wimmer, E., Moss, E.G. and Racaniello, V.R. (1988). Poliovirus host range is determined by a short amino acid sequence in neutralization antigenic site I. Science 241, 213–215.

Naniche, D., Varior-Krishnan, G., Cervoni, F., Wild, T.F., Rossi, B., Rabourdin-Combe, C. and Gerlier, D. (1993). Human membrane cofactor protein (CD46) acts as a cellular receptor for measles virus. J. Virol. 67, 6025–6032.

Nara, P.L., Hwang, K.M., Rausch, D.M., Lifson, J.D. and Eiden, L.E. (1989). CD4 antigen-based antireceptor peptides inhibit infectivity of human immunodeficiency virus *in vitro* at multiple stages of the viral life cycle. Proc. Natl Acad. Sci. USA 86, 7139–7143.

Nemerow, G.R., Mold, C., Schwend, V.K., Tollefson, V. and Cooper, N.R. (1987). Identification of gp350 as the viral glycoprotein mediating attachment of Epstein-Barr virus (EBV) to the EBV/C3d receptor of B cells: sequence homology of gp350 and C3 complement fragment C3d. J. Virol. 61, 1416–1420.

Nemerow, G.R., Houghten, R.A., Moore, M.D. and Cooper, N.R. (1989). Identification of an epitope in the major envelope protein of Epstein-Barr virus that mediates viral binding to the B lymphocyte EBV receptor (CR2). Cell 56, 369–377.

Nemerow, G.R., Mullen III, J.J., Dickson, P.W. and Cooper, N.R. (1990). Soluble recombinant CR2 (CD21) inhibits Epstein-Barr virus infection. J. Virol. 64, 1348–1352.

Neri, P., Bracci, L., Rustici, M. and Santucci, A. (1990). Sequence homology between HIV gp120, rabies virus glycoprotein, and snake venom neurotoxins. Is the nicotininc acetylcholine receptor an HIV receptor? Arch. Virol. 114, 265–269.

Neurath, A.R. and Strick, N. (1990). Antigenic mimicry of an immunoglobulin A epitope by a hepatitis B virus cell attachment site. Virology 178, 631–634.

Neurath, A.R., Kent, S.B.H., Strick, N. and Parker, K. (1986). Identification and chemical synthesis of a host cell receptor binding site on hepatitis B virus. Cell 46, 429–436.

Neurath, A.R., Kent, S.B.H., Strick, N. and Parker, K. (1988). Delineation of contiguous determinants essential for biological functions of the pre-S sequence of the hepatitis B virus envelope protein: its antigenicity, immunogenicity and cell-receptor recognition. Ann. Inst. Pasteur/Virol. 139, 13–38.

Neurath, A.R., Strick, N., Sproul, P., Ralph, H.E. and Valinsky, J. (1990). Detection of receptors for hepatitis B virus on cells of extrahepatic origin. Virology 176, 448–457.

Nicholson, J.K.A., Cross, G.D., Callaway, C.S. and McDougal, J.S. (1986). *In vitro* infection of human monocytes with human T lymphotropic virus type III/lymphadenopathy-associated virus (HTLV-III/LAV). 1986 J. Immunol., 323–329.

Nobusawa, E., Aoyama, T., Kato, H., Suzuki, Y., Tateno, Y. and Nakajima, K. (1991). Comparison of complete amino acid sequences and receptor-binding properties among 13 serotypes of hemagglutinins of influenza A viruses. Virology 182, 475–485.

O'Brien, W.A., Koyanagi, Y., Namazie, A., Zhao, J.-Q., Diagne, A., Idler, K., Zack, J.A. and Chen, I.S.Y. (1990). HIV-1 tropism for mononuclear phagocytes can be determined by regions of gp120 outside the CD4-binding domain. Nature 348, 69–73.

O'Hara, B., Johann, S.V., Klinger, H.P., Blair, D.G., Rubinson, H., Dunn, K.J., Sass, P., Vitek, S.M. and Robins, T. (1990). Characterization of a human gene conferring sensitivity to infection by gibbon ape leukemia virus. Cell Growth Different. 1, 119–127.

O'Neill, H.C., McGrath, M.S., Allison, J.P. and Weissmann, I.L. (1987). A subset of T cell receptors associated with L3T4 molecules mediates C6VL leukemia cell binding of its cognate retrovirus. Cell 49, 143–151.

Oldstone, M.B.A. (1987). Molecular mimicry and autoimmune disease. Cell 50, 819–820.

Palmenberg, A. (1987). A vaccine for the common cold? Nature 329, 668–669.

Paul, R.W. and Lee, P.W. (1987). Glycophorin is the reovirus receptor on human erythrocytes. Virology 159, 94–101.

Paul, R.W., Choi, A.H.C. and Lee, P.W.K. (1989). The α-anomeric form of sialic acid is the minimal receptor determinant recognized by reovirus. Virology 172, 382–385.

Paulson, J.C. (1985). In "The Receptors" (ed. P.M. Conn), pp. 131–219. Academic Press, Orlando.

Paulson, J.C., Sadler, J.E. and Hill, R.L. (1979). Restoration of specific myxovirus receptors to asialoerythrocytes by incorporation of sialic acid with pure sialyltransferases. J. Biol. Chem. 254, 2120–2124.

Peiris, J.S.M., Gordon, S., Unkeless, J.C. and Porterfield, J.S. (1981). Monoclonal anti-Fc receptor IgG blocks antibody enhancement of viral replication in macrophages. Nature 289, 189–191.

Pert, C.B., Hill, J.M., Ruff, M.R., Berman, R.M., Robey, W.G., Arthur, L.O., Ruscetti, F.W. and Farrar, W.L. (1986). Octapeptides deduced from the neuropeptide receptor-like pattern of antigen T4 in brain potently inhibit human immunodeficiency virus receptor binding and T-cell infectivity. Proc. Natl Acad. Sci. USA 83, 9254–9258.

Peterson, A. and Seed, B. (1988). Genetic analysis of monoclonal antibody and HIV binding sites on the human lymphocyte antigen CD4. Cell 54, 65–72.

Petit, M.-A., Dubanchet, S., Capel, F., Voet, P., Dauguet, C. and Hauser, P. (1991). HepG2 cell binding activities of different hepatitis B virus isolates: inhibitory effect of anti-HBs and anti-preS1(21–47). Virology 180, 483–491.

Pevear, D.C., Fancher, M.J., Felock, P.J., Rossmann, M.G., Miller, M.S., Diana, G., Treasurywala, A.M., McKinley, M.A. and Dutko, F.J. (1989). Conformational change in the floor of the human rhinovirus canyon blocks adsorption to HeLa cell receptors. J. Virol. 63, 2002–2007.

Philipson L., Lonberg-Holm, K. and Pettersson, U. (1968). Virus-receptor interaction in an adenovirus system. J. Virol. 2, 1064–1075.

Pontisso, P., Petit, M.-A., Bankowski, M.J. and Peeples, M.E. (1989a). Human liver plasma membranes contain receptors for the hepatitis B virus pre-S1 region and, via polymerized human serum albumin, for the pre-S2 region. J. Virol. 63, 1981–1988.

Pontisso, P., Ruvoletto, M.G., Gerlich, W.H., Heermann, K.-H., Bardini, R. and Alberti, A. (1989b). Identification of an attachment site for human liver plasma membranes on hepatitis B virus particles. Virology 173, 522–530.

Pritchett, T.J., Brossmer, R., Rose, U. and Paulson, J.C. (1987). Recognition of monovalent sialosides by influenza virus H3 hemagglutinin. Virology 160, 502–506.

Ravindranath, R.M.H. and Graves, M.C. (1990). Attenuated murine cytomegalovirus binds to N-acetylglucosamine, and shift to virulence may involve recognition of sialic acids. J. Virol. 64, 5430–5440.

Reagan, K.J. and Wunner, W.H. (1985). Rabies virus interaction with various cell lines is independent of the acetylcholine receptor. Arch. Virol. 84, 277–282.

Reisner, A.H. (1985). Similarity between the vaccinia virus 19K early protein and epidermal growth factor. Nature 313, 801–803.

Ren, R., Costantini, F., Gorgacz, E.J., Lee, J.J. and Racaniello, V.R. (1990). Transgenic mice expressing a human poliovirus receptor: a new model for poliomyelitis. Cell 63, 353–362.

Richardson, C.D. and Choppin, P.W. (1983). Oligopeptides that specifically inhibit membrane fusion by paramyxoviruses: Studies on the site of action. Virology 131, 518–532.

Richardson, C.D., Scheid, A. and Choppin, P.W. (1980). Specific inhibition of paramyxovirus and myxovirus replication by oligopeptides with amino acid sequences similar to those at the N-termini of the F1 or HA2 viral polypeptides. Virology 105, 205–222.

Richardson, N.E., Brown, N.R., Hussey, R.E., Vaid, A., Matthews, T.J., Bolognesi, D.P. and Reinherz, E.L. (1988). Binding site for human immunodeficiency virus coat protein gp120 is located in the NH_2-terminal region of T4 (CD4) and requires the intact variable-like domain. Proc. Natl Acad. Sci. USA 85, 6102–6106.

Robinson Jr., W.E., Montefiori, D.C. and Mitchell, W.M. (1988). Antibody-dependent enhancement of human immunodeficiency virus type 1 infection. Lancet i, 790–794.

Rogers, G.N. and D'Souza, B.L. (1989). Receptor binding properties of human and animal H1 influenza virus isolates. Virology 173, 317–322.

Rogers, G.N. and Paulson, J.C. (1983). Receptor determinants of human and animal influenza virus isolates: Differences in receptor specificity of the H3 hemagglutinin based on species of origin. Virology 127, 361–373.

Rogers, G.N., Paulson, J.C., Daniels, R.S., Skehel, J.J., Wilson, I.A. and Wiley, D.C. (1983). Single amino acid substitutions in influenza hemagglutinin change receptor binding specificity. Nature 304, 76–78.

Rogers, G.N., Herrler, G., Paulson, J. C., and Klenk, H.-D. (1986). Influenza C virus uses 9-*O*-acetyl-*N*-acetylneuraminic acid as a high affinity receptor determinant for attachment to cells. J. Biol. Chem. 261, 5947–5951.

Rolsma, M.D., Gelberg, H.B. and Kuhlenschmidt, M.S. (1994). Assay for evaluation of rotavirus-cell interactions: Identification of an enterocyte ganglioside fraction that mediates group A porcine rotavirus recognition. J. Virol. 68, 258–268.

Rossmann, M.G. (1988). Viral receptors and drug design. Nature 333, 392–393.

Rossmann, M.G. (1989). The canyon hypothesis. Hiding the host cell receptor attachment site on a viral surface from immune surveillance. J. Biol. Chem. 264, 14 587–14 590.

Rossmann, M.G. and Palmenberg, A.C. (1988). Conservation of the putative receptor attachment site in picornaviruses. Virology 164, 373–382.

Rossmann, M.G., Arnold, E., Erickson, J.W., Frankenberger, E.A., Griffith, J.P., Hecht, H.-J., Johnson, J.E., Kamer, G., Luo, M., Mosser, A.G., Rueckert, R.R., Sherry, B. and Vriend, G. (1985). Structure of a human common cold virus and functional relationship to other picornaviruses. Nature 317, 145–153.

Rothwell, S.W. and Wright, D.G. (1994). Characterization of influenza A virus binding sites on human neutrophils J. Immunol. 152, 2358–2367.

Ruff, M.R., Martin, B.M., Ginns, E.I., Farrar, W.L. and Pert, C.B. (1987). CD4 receptor binding peptides that block HIV infectivity cause human monocyte chemotaxis. Relationship to vasoactive intestinal polypeptide. FEBS Lett. 211, 17–22.

Ryan-Poirier, K. and Kawaoka, Y. (1991). Distinct glycoprotein inhibitors of influenza A virus in different animal sera. J. Virol. 65, 389–395.

Ryu, S.-E., Kwong, P.D., Truneh, A., Porter, T.G., Arthos, J., Rosenberg, M., Dai, X., Xuong, N.-h., Axel, R., Sweet, R.W. and Hendrickson, W.A. (1990). Crystal structure of an HIV-binding recombinant fragment of human CD4. Nature 348, 419–426.

Sattentau, Q.J. and Weiss, R.A. (1988). The CD4 antigen: physiological ligand and HIV receptor. Cell 52, 631–633.

Sattentau, Q.J., Dalgleish, A.G., Weiss, R.A. and Beverly, P.C.L. (1986). Epitopes of the CD4 antigen and HIV infection. Science 234, 1120–1123.

Schlegel, R., Tralka, T.S., Willingham, M.C. and Pastan, I. (1983). Inhibition of VSV binding and infectivity by phosphatidylserine: is phosphatidylserine a VSV-binding site? Cell 32, 639–646.

Schooley, R.T., Merigan, T.C., Gaut, P., Hirsch, M.S., Holodniy, M., Flynn, T., Liu, S., Byington, R.E., Henochowicz, S., Gubish, E., Spriggs, D., Kufe, D., Schindler, J., Dawson, A., Thomas, D., Hanson, D.G., Letwin, B., Liu, T., Gulinello, J., Kennedy, S., Fisher, R. and Ho, D.D. (1990). Recombinant soluble CD4 therapy in patients with the acquired immunodeficiency syndrome (AIDS) and AIDS-related complex. A phase I-II escalating dosage trial. Ann. Intern. Med. 112, 247–253.

Selinka, H.-C., Zibert, A. and Wimmer, E. (1991). Poliovirus can enter and infect mammalian cells by way of an intercellular adhesion molecule 1 pathway. Proc. Natl Acad. Sci. USA 88, 3598–3602.

Sherry, B., Mosser, A.G., Colonno, R.J. and Rueckert, R.R. (1986). Use of monoclonal antibodies to identify four neutralization immunogens on a common cold picornavirus human rhinovirus 14. J. Virol. 57, 246–257.

Shioda, T., Levy, J.A. and Cheng-Mayer, C. (1991). Macrophage and T cell-line tropisms of HIV-1 are determined by specific regions of the envelope gp120 gene. Nature 349, 167–169.

Smith, D.H., Byrn, R.A., Marsters, S.A., Gregory, T., Groopman, J.E. and Capon, D.J. (1987). Blocking of HIV-1 infectivity by a soluble, secreted form of the CD4 antigen. Science 238, 1704–1707.

Söderberg, C., Giugni, T.D., Zaia, J.A., Larsson, S., Wahlberg, J.M. and Möller, E. (1993). CD13 (human aminopeptidase N) mediates human cytomegalovirus infection. J. Virol. 67, 6576–6585.

Soelder, B.M., Reisinger, E.C., Koefler, D., Bitterlich, G., Wachter, H. and Dierich, M.P. (1989). Complement receptors: another port of entry for HIV. Lancet ii, 271–272.

Spier, R.E., Murdin, A. and Whittle, C.J. (1983). The attachment of the foot-and-mouth disease virus Asia I Iran 1/73 to BHK suspension cells does not require virus specific cell receptors. Arch. Virol. 77, 97–108.

Springer, G.F., Schwick, H.G. and Fletcher, M.A. (1969). The relationship of the influenza virus inhibitory activity of glycoproteins to their molecular size and sialic acid content. Proc. Natl Acad. Sci. USA 64, 634–641.

Springer, T.A. (1990). Adhesion receptors of the immune system. Nature 346, 425–434.

Srinivasappa, J., Saegusa, J., Prabhakar, B.S., Gentry, M.K., Buchmeier, M.J., Wiktor, T.J., Koprowski, H., Oldstone, M.B.A. and Notkins, A.L. (1986). Molecular mimicry: frequency of reactivity of monoclonal antiviral antibodies with normal tissues. J. Virol. 57, 397–401.

Staunton, D.E., Merluzzi, V.J., Rothlein, R., Barton, R., Marlin, S.D. and Springer, T.A. (1989). A cell adhesion molecule, ICAM-1, is the major surface receptor for rhinoviruses. Cell 56, 849–853.

Staunton, D.E., Dustin, M.L., Erickson, H.P. and Springer, T.A. (1990). The arrangement of the immunoglobulin-like domains of ICAM-1 and the binding sites for LFA-1 and rhinovirus. Cell 61, 243–254.

Steimer, K.S., Scandella, C.J., Skiles, P.V. and Haigwood, N.L. (1991). Neutralization of divergent HIV-1 isolates by conformation-dependent human antibodies to gp120. Science 254, 105–108.

Steinman, R.M., Mellman, I.S., Muller, W.A. and Cohn, Z.A. (1983). Endocytosis and recycling of plasma membrane. J. Cell Biol. 96, 1–27.

Sun, N.-C., Ho, D.D., Sun, C.R.Y., Liou, R.-S., Gordon, W., Fung, M.S.C., Li, X.-L., Ting, R.C., Lee, T.-H., Chang, N.T. and Chang, T.-W. (1989). Generation and characterization of monoclonal antibodies to the putative CD4-binding domain of human immunodeficiency virus type 1 gp120. J. Virol. 63, 3579–3585.

Superti, F., Hauttecoeur, B., Morelec, M.-J., Goldoni, P., Bizzini, B. and Tsiang, H. (1986). Involvement of gangliosides in rabies virus infection. J. Gen. Virol. 67, 47–56.

Takeda, A., Tuazon, C.U. and Ennis, F.A. (1988). Antibody-enhanced infection by HIV-1 via Fc receptor-mediated entry. Science 242, 580–583.

Takeuchi, Y., Vile, R.G., Simpson, G., O'Hara, B., Collins, M.K.L. and Weiss, R.A. (1992). Feline leukemia virus subgroup B uses the same cell surface receptor as gibbon ape leukemia virus. J. Virol. 66, 1219–1222.

Tanner, J., Weiss, J., Fearon, D., Whang, Y. and Kieff, E. (1987). Epstein-Barr virus gp350/220 binding to the B lymphocyte C3d receptor mediates adsorption, capping, and endocytosis. Cell 50, 203–213.

Tardieu, M., Epstein, R.L. and Weiner, H.L. (1982). Interaction of viruses with cell surface receptors. Int. Rev. Cytol. 80, 27–61.

Tavakkol, A. and Burness, T.H. (1990). Evidence for a direct role for sialic acid in the attachment of encephalomyocarditis virus to human erythrocytes. Biochemistry 29, 10 684–10 690.

Taylor, H.P. and Dimmock, N.J. (1985a). Mechanism of neutralization of influenza virus by secretory IgA is different from that of monomeric IgA or IgG. J. Exp. Med. 161, 198–209.

Taylor, H.P. and Dimmock, N.J. (1985b). Mechanisms of neutralization of influenza virus by IgM. J. Gen. Virol. 66, 903–907.

Thai, M., Olshevsky, U., Furman, C., Gabuzda, D., Li, J. and Sodroski, J. (1991). Effects of changes in gp120-CD4 binding affinity on human immunodeficiency virus type 1 envelope glycoprotein function and soluble CD4 sensitivity. J. Virol. 65, 5007–5012.

Till, M.A., Ghetie, V., Gregory, T., Patzer, E.J., Porter, J.P., Uhr, J.W., Capon, D.J. and Vitetta, E.S. (1988). HIV-infected cells are killed by rCD4-ricin A chain. Science 242, 1166–1168.

Tomassini, J.E., Graham, D., Dewitt, C.M., Lineberger, D.W., Rodkey, J.A. and Colonno, R.J. (1989). cDNA cloning reveals that the major group rhinovirus receptor on HeLa cells is intercellular adhesion molecule 1. Proc. Natl Acad. Sci. USA 86, 4907–4911.

Traunecker, A., Luke, W. and Karjalainen, K. (1988). Soluble CD4 molecules neutralize human immunodeficiency virus type 1. Nature 331, 84–86.

Traunecker, A., Schneider, J., Kiefer, H. and Karjalainen, K. (1989). Highly efficient neutralization of HIV with recombinant CD4-immunoglobulin molecules. Nature 339, 68–70.

Tsao, J., Chapman, M.S., Agbandje, M., Keller, W., Smith, K., Wu, H., Luo, M., Smith, T.J., Rossmann, M.G., Compans, R.W. and Parrish, C.R. (1991). The three-dimensional structure of canine parvovirus and its functional implications. Science 2541, 1456–1464.

Tschachler, E., Groh, V., Popovic, M., Mann, D.L., Konrad, K., Safai, B., Eron, L., Veronese, F.d., Wolff, K. and Stingl, G. (1987). Epidermal Langerhans cells-A target for HTLV-III/LAV infection. J. Invest. Derm. 88, 233–237.

Tsiang, H., de la Porte, S., Ambroise, D.J., Derer, M. and Koenig, J. (1986). Infection of cultured rat myotubes and neurons from the spinal cord by rabies virus. J. Neuropath. Exp. Neurol. 45, 28–42.

Twardzck, D.R., Brown, J.P., Ranchalis, J.E., Todaro, G.J. and Moss, B. (1985). Vaccinia virus-infected cells release a novel polypeptide functionally related to transforming and epidermal growth factors. Proc. Natl Acad. Sci. USA. 82, 5300–5304.

Uncapher, C.R., Dewitt, C.M. and Colonno, R.J. (1991). The major and minor group receptor families contain all but one human rhinovirus serotype. Virology 180, 814–817.

Vaheri, A. (1964). Heparin and related polyionic substances as virus inhibitors. Acta Pathol. Microbial. Scand. Suppl. 171, 1–98.

Vazeux, R., Brousse, N., Jarry, A., Henin, D., Marche, C., Vedrenne, C., Mikol, J., Wolff, M., Michon, C., Rozenbaum, W., Bureau, J.-F., Montagnier, L. and Brahic, M. (1987). AIDS subacute encephalitis. Identification of HIV-infected cells. Am. J. Pathol. 126, 403–410.

Vile, R.G. and Weiss, R.A. (1991). Virus receptors as permeases. Nature 352, 666–667.

Wang, H., Kavanaugh, M.P., North, R.A. and Kabat, D. (1991). Cell-surface receptor for ecotropic murine retroviruses is a basic amino-acid transporter. Nature 352, 729–731.

Wang, J., Yan, Y., Garrett, T.P.J., Liu, J., Rodgers, D.W., Garlick, R.L., Tarr, G.E., Husain, Y., Reinherz, E.L. and Harrison, S.C. (1990). Atomic structure of a fragment of human CD4 containing two immunoglobulin-like domains. Nature 348, 411–418.

Wang, K.-S., Kuhn, R.J., Strauss, E.G., Ou, S. and Strauss, J.H. (1992). High-affinity laminin receptor is a receptor for Sindbis virus in mammalian cells. J. Virol. 66, 4992–5001.

Weber, J., Clapham, P., McKeating, J., Stratton, M., Robey, E. and Weiss, R. (1989). Infection of brain cells by diverse human immunodeficiency virus isolates: role of CD4 as receptor. J. Gen. Virol. 70, 2653–2660.

Weinhold, K.J., Lyerly, H.K., Stanley, S.D., Austin, A.A., Matthews, T.J. and Bolognesi, D.P. (1989). HIV-1 gp120-mediated immune suppression and lymphocyte destruction in the absence of viral infection. J. Immunol. 142, 3091–3097.

Weis, W., Brown, J.H., Cusack, S., Paulson, J.C., Skehel, J.J. and Wiley, D.C. (1988). Structure of the influenza virus haemagglutinin complexed with its receptor, sialic acid. Nature 333, 426–431.

White, J.M. and Littman, D.R. (1989). Viral receptors of the immunoglobulin superfamily. Cell 56, 725–728.

Wickham, T.J., Granados, R.R., Wood, H.A., Hammer, D.A. and Shuler, M.L. (1990). General analysis of receptor-mediated viral attachment to cell surfaces. Biophys. J. 58, 1501–1516.

Wiktor, T.J. and Clark, H.F. (1975). Growth of rabies virus in cell culture. In "The Natural History of Rabies" (ed. G.M. Baer), pp. 155–179. Academic Press, New York.

Wiley, C.A., Schrier, R.D., Nelson, J.A., Lampert, P.W. and Oldstone, M.B.A. (1986). Cellular localization of human immunodeficiency virus infection within the brains of acquired immune deficiency syndrome patients. Proc. Natl Acad. Sci. USA 83, 7089–7093.

Wiley, D.C. and Skehel, J.J. (1987). The structure and function of the hemagglutinin membrane glycoprotein of influenza virus. Ann. Rev. Biochem. 56, 365–394.

Wiley, D.C., Wilson, I.A. and Skehel, J.J. (1981). Structural identification of the antibody binding sites of Hong Kong influenza haemagglutinin and their involvement in antigenic variation. Nature 289, 373–378.

Williams, R.K., Jiang, G.-S. and Holmes, K.V. (1991). Receptor for mouse hepatitis virus is a member of the carcino-embryonic antigen family of glycoproteins. Proc. Natl Acad. Sci. USA 88, 5533–5536.

Williams, W.V., Moss, D.A., Kieber-Emmons, T., Cohen, J.A., Myers, J.N., Weiner, D.B. and Greene, M.I. (1989). Development of biologically active peptides based on antibody structure. Proc. Natl Acad. Sci. USA 86, 5537–5541.

Williams, W.V., Kieber-Emmons, T., Weiner, D.B., Rubin, D.H. and Greene, M.I. (1991). Contact residues and predicted structure of the reovirus type 3-receptor interaction. J. Biol. Chem. 266, 9241–9250.

Willoughby, R.E. and Yolken, R.H. (1990). SA11 rotavirus is specifically inhibited by an acetylated sialic acid. J. Infect. Dis. 161, 116–119.

WuDunn, D. and Spear, P.G. (1989). Initial interaction of herpes simplex virus with cells is binding to heparan sulfate. J. Virol. 63, 52–58.

Wykes, M.N., Shellam, G.R., McCluskey, J., Kast, W.M., Dallas, P.B. and Price, P. (1993). Murine cytomegalovirus interacts with major histocompatibility complex class I molecules to establish cellular infection. J. Virol. 67, 4182–4189.

Yeager, C.L., Ashmun, R.A., Williams, R.K., Cardellichio, C.B., Shapiro, L.H., Look, A.T. and Holmes, K.V. (1992). Human aminopeptidase N is a receptor for human coronavirus 229E. Nature 357, 420–422.

12. *Bacterial Adhesins*

Soman N. Abraham

1. *Bacterial adhesins*

The host is constantly exposed to a wide range of potentially pathogenic bacteria, but only a small fraction of these organisms cause infection. A common property of many successful infectious agents is their adhesive ability. The adherence of bacteria to host tissue is achieved by highly specific interactions (Fig. 12.1) between special adhesive structures (called adhesins) and complementary structures (called receptors) on host cells (Kabat, 1978). The strength of this attachment is dependent, in part, on the nature and number of the adhesins expressed by the bacteria. For many adhesins, limited information is available on their structure and receptor specificity. Some adhesins have, however, been studied in greater detail. Examples of some bacterial adhesins that have been studied and their complementary receptors on host cells is provided in Table 12.1 Recent studies employing powerful tools of molecular biology, protein and carbohydrate chemistry have provided interesting insights into the morphogenesis, structure and function of some of these bacterial adhesins as well as the structure of the complementary receptors. These findings have also stimulated interest in the design of novel strategies directed at blocking bacterial adherence for the prevention of infections in man and animals.

In this review, I do not intend to describe all the bacterial adhesins and compare them to one another, as these have already been dealt with elsewhere (Klemm, 1985; Mirelman, 1986; Ceri *et al.*, 1988; Finlay and Falkow, 1989; Baddour *et al.*, 1990; Ofek and Sharon, 1990; Tennent *et al.*, 1990; Isberg, 1991; Stromberg *et al.*, 1992). Instead, I will focus on the type 1 and P fimbriae of *Escherichia coli*, and the multiple adhesins of *Streptococcus pyogenes*, since these have been investigated in considerable detail and serve as paradigms for bacterial adhesions. I will also discuss the role of bacterial adhesions in infection and some novel approaches to anti-adhesive therapy.

1.1 TYPE 1 AND P FIMBRIAE

Two of the best studied bacterial adhesins are the type 1 and P fimbriae of *E. coli*, a frequent pathogen of humans.

Adhesion Molecules
ISBN 0-12-741440-1

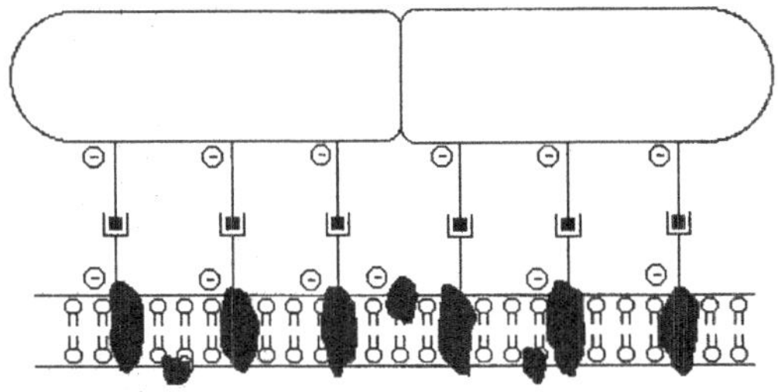

Figure 12.1 Attachment of a bacterial cell (top) via an organelle presenting a terminal adhesin (straight lines terminating in a black square) to complementary receptors (branched vertical lines) on the host cell membrane (bottom). To overcome the net negative charges (circled minus signs) on both the bacterial and host cell surfaces, the bacterial adhesin must bind avidly to its complementary receptor carbohydrate conjugated to either protein or lipid (irregular black structures) in the host cell membrane. Redrawn from Abraham *et al.*, (1983).

E. coli is the most frequent cause of urinary tract infections, bacteremia, bacterial-related travellers diarrhea and neonatal meningitis (Eisenstein, 1990). The initiation of many of these infections is thought to be by the molecular recognition between these cell-surface fimbriae and complementary receptors on the host (Beachey, 1980, 1981; Abraham and Beachey, 1985). The type 1 and P fimbriae of *E. coli* are filamentous structures which are morphologically identical; they are both long (up to 2 μm) with a diameter of 5–10 nm (Brinton, 1959, 1965; Duguid *et al.*, 1979; Duguid and Old, 1980). There can be up to 200 of these structures arranged peritrichously on the surface of bacteria (Fig. 12.2). Historically, type 1 and P fimbriae have been distinguished based on the sensitivities of their hemagglutination reactions to specific sugars (Duguid and Old, 1980). Type 1 are sensitive to D-mannose and its analogues (Duguid and Old, 1980; Firon *et al.*, 1987), whereas P fimbriae are sensitive to Galα(1–4)Gal-containing sugars (Duguid and Old, 1980; Leffler and Svanborg-Eden, 1980, 1986; Kallenius *et al.*, 1981). Unlike P fimbriae, which are expressed exclusively by *E. coli*, type 1 fimbriae are expressed by most species in the family *Enterobacteriaceae*, including *Escherichia*, *Klebsiella*, *Citrobacter*, *Serratia*, *Salmonella* and *Enterobacter* (Duguid and Old, 1980).

1.1.1 Genetic Organization

Both type 1 and P fimbriae have been cloned from different sites of the *E. coli* chromosome and extensively characterized (Hull *et al.*, 1981; Korhonen *et al.*, 1982; Normark *et al.*, 1983; Orndorff and Falkow, 1984; Clegg, Hull *et al.*, 1985; Clegg *et al.*, 1985; van Die *et al.*, 1985; Minion *et al.*, 1986; Rhen *et al.*, 1986; Maurer and Orndoff, 1987; Gerlach *et al.*, 1989). Non-fimbriated and non-adhesive *E. coli* transformed with

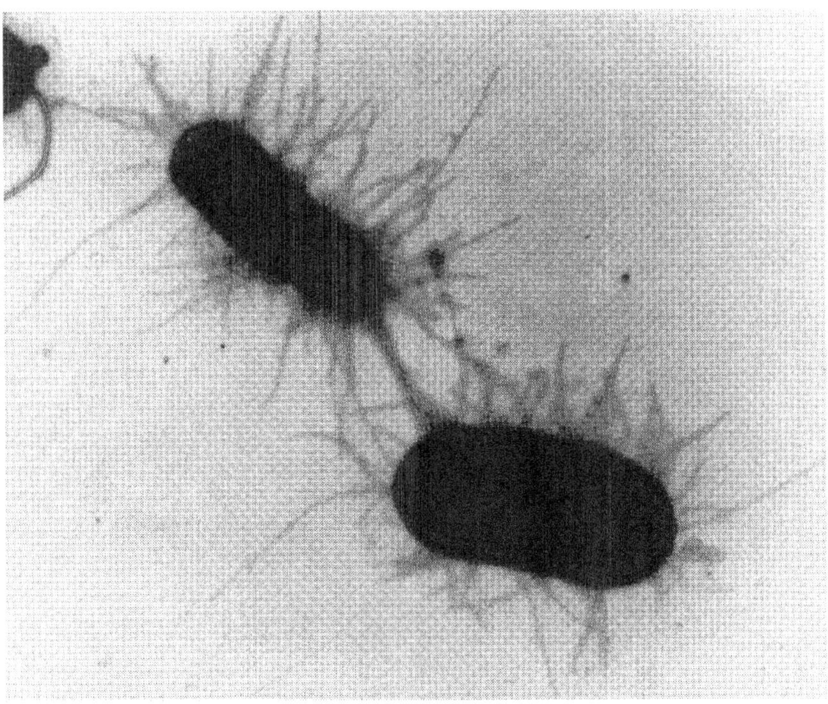

Figure 12.2 Electron micrograph of type 1 fimbriated *E. coli* showing the fairly rigid fimbriae radiating uniformly from the surface of the bacteria.

plasmids containing cloned fragments of DNA express fimbriae that were capable of adhering to surfaces containing the appropriate receptor. It is clear that several genes clustered together are involved in the biosynthesis and function of each fimbriae (Normark *et al.*, 1983; Hull *et al.*, 1984; Orndorff and Falkow, 1984; Maurer and Orndorff, 1987). The functions of each of these genes were determined largely by inactivating each gene and then examining the effects of the mutation. Eight genes make up the type 1 gene cluster, whereas the synthesis of P fimbriae involves 11 genes (Hultgren *et al.*, 1991). The genetic organization of the genes of type 1 and P fimbriae are similar; and since the characterization of the latter is more complete, I shall discuss the *pap* gene cluster in greater detail. Fig. 12.3 summarizes the genetic organization and function of the *pap* gene cluster The structural gene that encodes the major subunit, which makes up over 95% of the bulk of the P fimbriae, is encoded by PapA. Deletion of this gene results in loss of fimbrial structure. However, this did not result in loss in adhesive function which was the first indication that the major subunit was not the determinant of adhesion. Subsequent studies showed that the Galα(1–4)Gal binding property of P fimbriae depended on the expression of a 35 kDa protein, PapG, encoded by a gene located at the distal end of the *pap* operon (Lindberg *et al.*, 1984; Lund *et al.*, 1987; Hoschutzky *et al.*, 1989). It was also shown that PapG along with two other proteins, PapE and PapF, are located solely at the tip of the fimbriae (Lindberg *et al.*, 1987), but unlike PapG, PapE and PapF proteins do not appear to play a role in mediating the adhesive property of the fimbriae because deleting either of the genes appears not to affect the adhesive function (Lindberg *et al.*, 1987). Interestingly, inactivation of *papF* results in dramatic reduction in the amount of expression of P fimbriae, suggesting that it plays an important role in initiating the assembly of the fimbrial filament (Lindberg *et al.*, 1984). The products of two genes, *papC* and *papD*, play a crucial role in translocation and assembly of fimbrial subunits (Normark *et al.*, 1983, 1986). Inactivation of either gene completely abrogates expression of cell surface fimbriae as well as the ability to adhere to red blood cells. PapC is a relatively large (88 kDa) protein which is localized in the outer membrane and is thought to serve as a base for fimbrial polymerization. The *papD* gene encodes a 28.5 kDa protein located in the periplasmic space (Hultgren *et al.*, 1991). A mutation in *papD* results in rapid proteolytic degradation of the major and minor fimbrial subunits, including PapG (Lund *et al.*, 1985; Hultgren *et al.*, 1989). PapD appears to function by interacting with different subunits of the P fimbriae in the periplasm, stabilizing them in an assembly competent form (Hultgren *et al.*, 1991). The role of PapJ is unclear, but it may also be involved in fimbrial assembly. PapH encodes for a protein which is thought to be at the base of the fimbrial filament (Baga *et al.*, 1987). Its presence at the base is thought to signal termination of fimbrial growth. Deletion of this gene results in extremely long fimbriae, and over-production of this gene results in very short fimbriae (Baga *et al.*, 1987). Two genes, *papI* and *papB*, regulate over-all expression of P fimbrial proteins, because inactivation of either of these genes dramatically affects the transcription of the *pap* operon (Baga *et al.*, 1985; Normark *et al.*, 1986; Hultgren *et al.*, 1991). Fig 12.4 depicts a general model of fimbrial synthesis and assembly.

1.1.2 Structural Organization

Considerable information is now available on the structure of both type 1 and P fimbriae. Both consist of a quaternary assembly of several thousand copies of the major subunit that typically is 17 kDa for type 1 and 20 kDa for P fimbriae. X-ray crystallography of the type 1 fimbrial filament by Brinton has revealed that the type 1 fimbrial filament is comprised of subunits that are

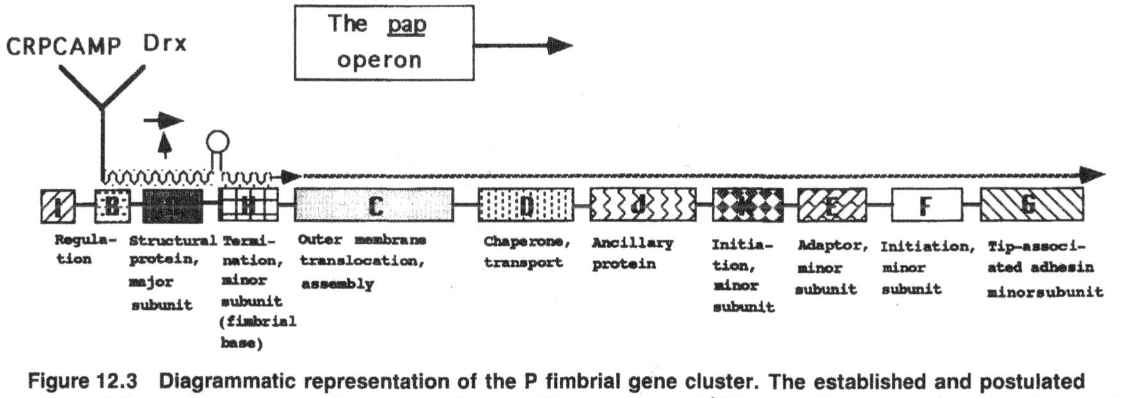

Figure 12.3 Diagrammatic representation of the P fimbrial gene cluster. The established and postulated functions of the various gene products are indicated. The messenger RNA transcripts emanating from the *papB* promoter are depicted by a wavy line followed by a dashed line. Regulatory targets of the CRP protein, protein DRX, and cAMP are also shown. A transcriptional terminator (circle on stilts) occurs between *papA* and *papH*. Redrawn from Hultgren et *al.* (1991).

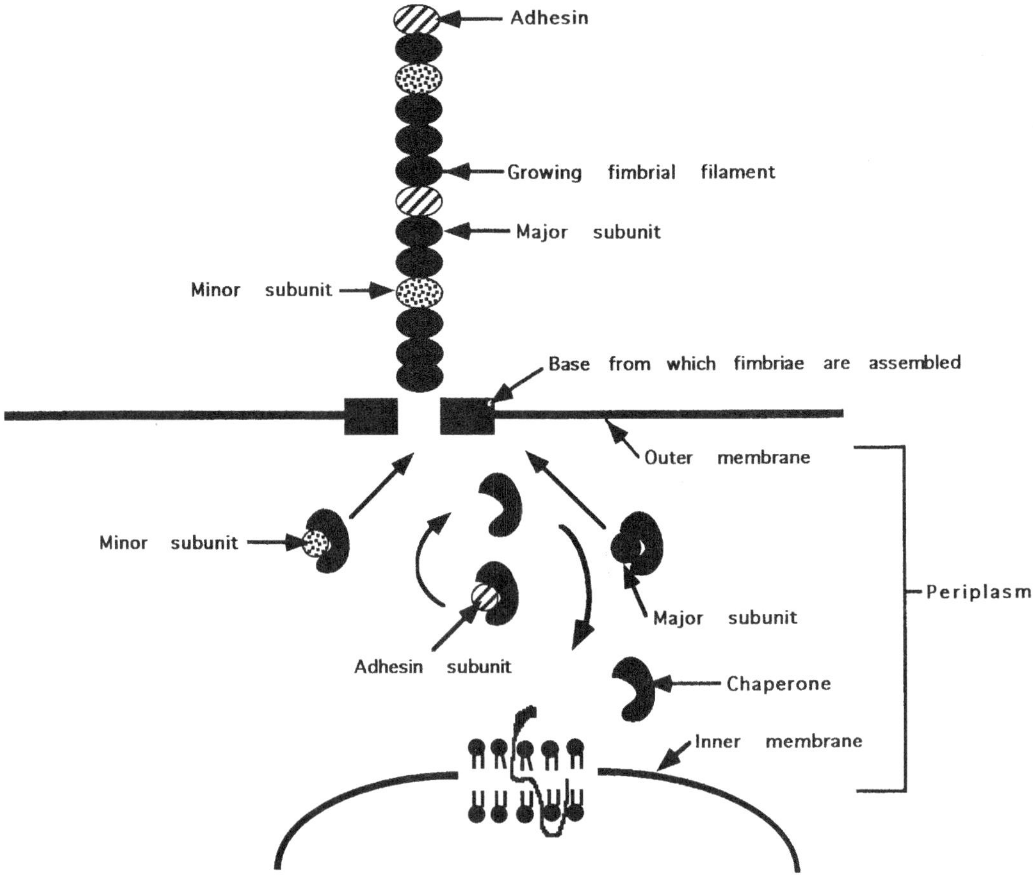

Figure 12.4 General model of fimbrial assembly. The major and minor fimbrial subunits including the adhesin are synthesized in the cytoplasm and secreted through the inner membrane. The chaperone binds to the nascent subunits as they enter the periplasm and transports them to the outer membrane for assembly. The polymerization of the fimbrial subunits occurs through a large transmembrane porin-like structure. The growth of the fimbrial filament occurs from their base and the frequency of incorporation of different subunits appears to be highly regulated. The first subunits to be incorporated are usually the tip proteins which are minor fimbrial components. The growth of the filament is terminated usually by the incorporation of a specific subunit at the base of the filament. Redrawn from Hultgren *et al.* (1991).

arranged in a simple, tight right-handed helix with a central axial hole (Brinton, 1965). He predicted that each turn of the helix consisted of 3.14 subunits and the helical pitch distance was 2.3 nm. Since these studies were undertaken long before the presence of minor subunits in the fimbriae were known, he did not account for these subunits in his prediction. Recently generated X-ray crystallographic data on P fimbriae suggest that the fimbriae is composed of 2.3 subunits per turn with a 236.8 Å axial repeat (Gong *et al.*, 1992).

Lindberg *et al.*, 1987, were the first to localize minor components of the fimbriae in the fimbrial structure. They prepared non-specific antiserum against each of the minor fimbrial components to probe the surface of the P fimbrial filament. Immunogold electron microscopy revealed that three of the minor proteins of the P fimbriae, including PapG, the adhesin moiety, were exclusively located at the distal tip of the fimbriae (Lindberg *et al.*, 1987). Recently it was discovered that the tip of the fimbriae is morphologically distinct from the rest of the fimbrial filament. The tip comprising of PapE, PapF and PapG forms a linear fiber approximately 1/3 the diameter of the fimbrial shaft comprised of PapA (Kuehn *et al.*, 1992). Unlike the rigid rod-like shaft, the tip fibrillum appears to be flexible. Mutation analysis followed by electron microscopy has revealed that the majority of the fibrillum structure is composed of PapE and that the PapF and PapG adhesin components are located at the distal tip of the fibrillum structure (Kuehn *et al.*, 1992).

In contrast to the P fimbriae, the tip of type 1 fimbriae does not appear to have a linear fibrillum structure. The receptor-binding component of type 1 fimbriae is a minor adhesin protein, FimH, which is not exclusively localized at the tips. Immune electron microscopy with FimH-specific monoclonal antibodies (mAbs) revealed that

FimH was also located along the length of the fimbrial filament (Abraham *et al.*, 1987; Abraham, Goguen *et al.*, 1988; Abraham *et al.*, 1988). The location of FimH along the fimbrial filament did not appear to have any regular periodicity (Abraham *et al.*, 1988).

Although FimH is undoubtedly associated with the type 1 fimbrial filament, whether it is intercalated into the filament or merely tightly associated with the lateral sides of the fimbriae is unknown. Some evidence, albeit of an indirect nature, suggests that FimH is an integral part of the fimbrial filament. For example, fimbriae from a hyper-adhesive mutant that contains five times more FimH than wild type appear distorted and fragmented compared to the wild type (Abraham *et al.*, 1988a). Furthermore, fimbriae from a FimH$^-$ mutant appear longer and less fragmented than wild type (Klemm and Christensen, 1987). Both of these observations imply that the association of FimH with the fimbrial filament enhances the destabilization and fragmentation of the latter, which is most consistent with a model of physical insertion of FimH rather than lateral ssociation with the filament. The notion that the presence of FimH in the fimbrial filament leads to increased fragility was tested by subjecting wild type and mutant FimH$^-$ fimbriae to several cycles of freezing and thawing (Ponniah *et al.*, 1991). Electron microscopic measurements of both fimbrial types before and after freeze-thawing revealed that wild-type fimbriae were at least five times more fragmented than mutant fimbriae (Ponniah *et al.*, 1991). Even more intriguing was the finding associated with this study that the hemagglutinating titers of wild-type fimbriae were at least 16 times greater after freeze thawing. Taken together, these data argue that fragmentation increased the total number of available receptor binding FimH molecules. This increase can be explained only if the FimH molecules exposed at the tips are functional or if they at least have a higher hemagglutination titer than FimH integrated within the fimbrial structure. The idea that FimH exposed at the tips are functionally more active has been supported by other observations that show increased potency of fragmented fimbriae in mediating other FimH-associated functions of type 1 fimbriae (Ponniah *et al.*, 1991). The concept that only the fimbrial subunits exposed at the tip are capable of receptor-binding is supported by the findings of Hanson and Brinton (1988) and by the recent observations of Buhler *et al.* (1991) that showed that terminally exposed fimbrial subunits but not those intercalated into the filament of CFA/I fimbriae of enterotoxigenic *E. coli* were capable of mediating adherence.

Valuable information regarding the molecular arrangement of fimbrial subunits and the conditions that affect subunit polymerization were derived by completely dissociating fimbriae into their subunits and then determining if and how reconstitution could best take place. Such studies have be undertaken with type 1 fimbriae of *E. coli* (Eshdat *et al.*, 1981; Abraham *et al.*, 1983), 987P fimbriae of enterotoxigenic *E. coli* (Schifferli *et al.*, 1987) and PAK fimbriae of *Pseudomonas aeruginosa* (Watts *et al.*, 1982). The quaternary structure of type 1 fimbriae is extremely stable and, with the exception of guanidine hydrochloride, is resistant to actions of all common dissociating agents. Saturated guanidine hydrochloride, however, completely dissociates type 1 fimbriae without evoking irreversible structural damage to the subunits. Eshdat *et al.* (1981), showed that type 1 fimbriae depolymerized by guanidine hydrochloride can be reconstituted by the removal of the denaturant. They found that the reconstituted fimbriae had the same thickness but were often shorter than native fimbriae as determined by electron microscopy. Abraham *et al.* (1983), monitored the denaturation and renaturation of type 1 fimbriae by using a panel of mAbs that were directed towards conformation-specific-epitopes. They demonstrated that depolymerization of fimbriae resulted in the loss of several quaternary structure-specific epitopes and in the exposure of several new epitopes on the major subunit that were previously buried in the quaternary fimbrial conformation (Abraham *et al.*, 1983). In addition, reconstituted fimbriae exhibited the same level of reactivity as did native fimbriae when probed with a panel of quaternary structure-specific mAbs (Abraham *et al.*, 1983), supporting Eshdat's findings that reconstituted fimbriae were morphologically similar to wild-type fimbriae. For example, mAbs specific for quaternary structural epitopes displayed the same periodicity and spiral pattern of binding to reconstituted fimbriae as with native fimbriae, indicating that the highly ordered subunit packing in the native fimbriae was restored in the *in vitro* reconstituted fimbriae. While no systematic study was undertaken, it was found that pH, temperature, and ionic conditions were critical factors in determining the rate of fimbrial repolymerization *in vitro*. These reconstitution studies also highlighted the requirement for divalent cations in the polymerization of type 1 fimbriae; magnesium ions greatly enhanced assembly whereas EDTA inhibited this process (Abraham *et al.*, 1983). Similarly interesting is that *in vitro* polymerization of fimbrial subunits does not require adenosine triphosphate (ATP). An intriguing question is whether reconstituted fimbriae retain any of the adhesive capability of native fimbriae.

1.1.3 Receptor Binding

In light of the discovery that the adhesins of fimbriae are not the major subunits but the minor components located at the tips of the fimbriae, interest has focused on the structure and function of these minor components. By studying the ability of various truncates of PapG, isolated from the periplasmic space of P fimbriated bacteria, to bind to Galα(1–4)Gal immobilized onto Sepaharose beads, Hultgren *et al.*, (1989), were able to show that the binding domain resided within the amino terminal region of the PapG molecule.

By employing inhibition assays involving glycosides of galabiose and *in vitro* binding assays to glycolipids immobilized on chromatograms, it was recognized that not all P fimbriae share the same binding specificity (Stromberg *et al.*, 1990). In a study involving three serologically distinct P fimbrial types, it was discovered that all three strains displayed differences in hemagglutination reactions as well as in binding patterns to immobilized glycolipids (Stromberg *et al.*, 1990; Hultgren *et al.*, 1991). Since PapG was their putative adhesin moiety of the P fimbriae, the *papG* genes of each fimbrial type were cloned and the DNA sequence determined. From this information, the primary structure of PapG from each clone was established and revealed that there were only five amino acids different between the sequences of two of the PapG proteins (Hultgren *et al.*, 1991). However, the third PapG molecule was only 50% homologous with the others, indicating considerable structural heterogeneity amongst PapG from different P fimbriated *E. coli*. After a review of the recent findings on the binding characteristics of different P fimbriated clones and their derivatives, Hultgren *et al.* (1991), concluded that the receptor binding property and the variation in binding seen among different P fimbriated *E. coli* depended solely on the PapG subunit and that some of the variability in receptor specificity could be localized to sequence variation in the amino terminal region of different PapG proteins.

Although FimH has definitively been identified as the mannose-binding moiety of type 1 fimbriae (Christensen *et al.*, 1982; Maurer and Orndorff 1985; Abraham, Goguen *et al.*, 1988; Abraham *et al.*, 1988; Krogfelt *et al.*, 1990), very little is known about the nature of the binding domain within the molecule. However, interesting insight into differences in binding specificities of type 1 fimbriae was recently obtained by evaluating the binding characteristics of clones expressing *E. coli* and *K. pneumoniae* type 1 fimbriae (B. Madison, I. Ofek and S.N. Abraham, submitted for publication). Although their respective FimH proteins were structurally and antigenically similar, several distinct differences in the binding specificities between the two type 1 fimbriated clones were noted, including the observation that type 1 fimbriae of *K. pneumoniae*, but not *E. coli*, mediated mannose-sensitive agglutination of sheep erythrocytes. Since the putative mannose-binding moiety of type 1 fimbriae is the FimH protein, an attempt was made to convert the binding characteristics of *E. coli* into that of *K. pneumoniae* and vice versa by exchanging their respective FimHs with each other. This was achieved by deleting the *fimH* gene in each *fimH* cluster and then complementing each deletion in trans with the heterologous *fimH* borne on a compatible plasmid. In this manner, two hybrid forms of type 1 fimbriae were generated; in one case the *E. coli* FimH was presented on a filament of *K. pneumoniae* structural subunits (EcFimH-KpFimA); in the other case the *K. pneumoniae* FimH was presented on a filament of *E. coli* structural subunits (KpFimH-EcFimA). When the adhesive properties of these hybrid fimbriae were evaluated, the workers found that EcFimH-KpFimA fimbriae but not KpFimH-EcFimA fimbriae mediated mannose-sensitive agglutination of sheep erythrocytes (B. Madison, I. Ofek and S.N. Abraham, submitted for publication). This finding was unexpected because it suggested that the sheep agglutinating property of *K. pneumoniae* was apparently not linked to its FimH protein but, rather, appeared to be associated with the filament composed of the *K. pneumoniae* structural subunits. However, FimH$^-$ mutant fimbriae of *K. pneumoniae* have no hemagglutinating activity, arguing that the *K. pneumoniae* fimbrial filament, in the absence of FimH, does not contain a sheep erythrocyte-binding adhesin moiety. One possible explanation for the unexpected agglutination reactions is that the filament can indirectly affect the adhesin moiety by altering the conformation or presentation of the heterologous *E. coli* FimH molecule. For example, since the structural subunits of *K. pneumoniae* and *E. coli* are different, *E. coli* FimH must undergo some conformational alteration to fit the quaternary constraints of the *K. pneumoniae* fimbriae. Regions of the heterologous FimH molecule that are possibly altered could include the mannose-binding site. This model was supported by the data showing that some of the FimH-specific mAbs had a dramatically lower reactivity to isolated EcFimH-KpFimA hybrid fimbriae compared to the wild-type EcFimH-EcFimA (S.N. Abraham, unpublished observations). Experiments are underway to determine if the D-mannose binding pocket of the hybrid EcFimH-KpFimA was converted into a receptor-binding pocket that mimics the wild-type *K. pneumoniae* fimbriae (KpFimH-KpFimA) by probing their respective combining sites with various analogues of D-mannose as described previously by Firon *et al.* (1982, 1983, 1984). The notion that the fimbrial filament can influence the binding properties of the adhesin moiety by modulating its configuration is novel and contrasts with the P fimbrial system in which PapG is the sole determinant of binding specificity.

1.2 LIPOTEICHOIC ACID AND OTHER ADHESINS OF *STREPTOCOCCUS PYOGENES*

The gram positive bacterium *S. pyogenes* is the causative agent of numerous suppurative diseases that are sometimes followed by the onset of rheumatic fever or acute glomerulonephritis. Its niche in man is the upper respiratory tract, specifically the pharynx and tonsils (Hasty *et al.*, 1992). A fine irregular fuzz or fibrillae covers the surface of the organisms and is thought to contain the adhesins responsible for promoting adherence of *S. pyogenes* to buccal, pharyngeal and endothelial cells as well as to matrix proteins such as fibronectin (Beachey and Ofek, 1976; Beachey *et al.*, 1982). The adherence mechanisms of *S. pyogenes* have not been as clearly defined as those of *E. coli*. To date, several adhesins have been described and at least six different mechanisms of adhesion have been proposed (Ellen and Gibbons, 1974; Tebi *et al.*, 1987; Tylewska *et al.*, 1988; Hasty *et al.*, 1992).

Perhaps the best characterized adhesin is Lipoteichoic acid (LTA). This is an amphipathic hydrophobic molecule which is capable of recognizing a wide range of cell types (Alkan, 1977; Alkan and Beachey, 1978; Ofek and Beachey, 1980; Courtney *et al.*, 1981, 1983; Ofek *et al.*, 1982; Baird *et al.*, 1990). An ester bond links LTA's hydrophilic polyglycerolphosphate "backbone" to its hydrophobic glycolipid "tail". The molecule is found with its tail buried in the bacterial cytoplasmic membrane of most gram positive bacteria (Ofek and Beachey, 1980; Ofek *et al.*, 1982). However, the organism constantly leaks LTA into the surrounding media (Alkan and Beachey, 1978); and as LTA crosses the cell wall, it forms complexes with various surface proteins, including M protein, which enables LTA to make an about-face, where the hydrophobic tail is now exposed and able to promote procaryotic-eucaryotic cell contact (Ofek *et al.*, 1982). The adherence mediated by LTA is dependent on its hydrophobicity which displays broad specificity for different cell types and different molecules (Simpson *et al.*, 1980; Miorner *et al.*, 1983; Nealon and Mattingly, 1984; Courtney *et al.*, 1990). Thus, LTA cannot be entirely responsible for the specific adherence characteristics of *S. pyogenes*. For example, streptococcal cutaneous strains preferentially bind to cutaneous epithelial cells (Alkan, 1977; Alkan and Beachey, 1978). Furthermore, rheumatogenic strains bind preferentially to pharyngeal cells of rheumatic heart disease patients (Selinger *et al.*, 1978), and streptococci bind to adult buccal epithelial cells but not to neonatal epithelial cells (Ofek *et al.*, 1977). The specificity of this cellular attachment indicates that additional adhesins might be present on streptococci which interact with cellular receptors. Different cell surface proteins on *S. pyogenes* have been implicated as adhesins for specific tissue (Hasty *et al.*, 1992).

Recently, Courtney *et al.* (1992), suggested that M protein promoted the adherence of *S pyogenes* to Hep-2 tissue culture cells, because M protein expressing streptococci bound to these cells 10 times more than M protein deficient streptococci. However, these authors also reported that the adherence of *S. pyogenes* to pharyngeal and buccal epithelial cells was independent of M protein (Courtney *et al.*, 1992). It is thus clear that *S. pyogenes* express several adhesins, and the relative contribution of these adhesins to the attachment process depends on the type of host cells. Hanski and Caparon (1992) recently described the molecular cloning and expression of a novel gene from *S. pyogenes* whose product, protein F, mediated high affinity binding to fibronectin, one of the putative receptors for *S. pyogenes* on mucosal cells. The cloned gene product exhibited binding to both soluble and immobilized fibronectin. Insertional mutagenesis of the cloned gene in *S. pyogenes* generated a mutation that resulted in the loss of fibronectin binding activity (Hanski and Caparon, 1992). When the inactivated gene was introduced into the *S. pyogenes* chromosome by homologous recombination with the wild-type allele, the resulting recombinant strains no longer bound fibronectin. Furthermore, these authors showed that mutants lacking protein F had a much lower capacity to adhere to respiratory epithelial cells than the wild type (Hanski and Caparon, 1992).

In order to accommodate all the findings on the multiple adherence interactions of *S. pyogenes* with various host cells, a two-step model was proposed recently (Hasty *et al.*, 1992). In this model, the first step of adhesion is relatively weak and reversible and broadly specific and is probably mediated by hydrophobic interactions of LTA. The second step involves adhesins that are specialized for receptors expressed by host cells within specific niches (Hasty *et al.*, 1992). The outcome of this second step is a firm practically irreversible adhesion (mediated either by composite multivalent interactions or by a high affinity adhesin/receptor interaction). A diagrammatic illustration of this model is depicted in Fig. 12.5.

2. *Receptor Specificity*

The presence on host cell surfaces of receptors that interact in a complementary manner with specific bacterial adhesins is an important determinant of stable bacterial adherence and colonization (Beachey *et al.*, 1982; Abraham and Beachey, 1985). If the receptors on the host cells are present at low density, are inaccessible or are not expressed at all, binding of bacteria and subsequent colonization cannot take place.

The receptivity of host cells, including epithelial cells, erythrocytes, and leucocytes for various bacterial adhesins has been associated, for the most part, with carbohydrate residues (Table 12.1). Traditional methods for revealing carbohydrate receptors for bacterial adhesins include indirect approaches based on selective chemical or enzymatic modification of target cell carbohydrates and direct assaying of inhibition of binding with natural and synthetic mono- and oligosaccharides (Duguid and Old, 1980; Andersson *et al.*, 1983; Firon *et al.*, 1983, 1984; Dahmen *et al.*, 1984; Holgersson *et al.*, 1985; Neeser *et al.*, 1986; Giampapa *et al.*, 1988; Kehlberg *et al.*, 1988; van Alpen *et al.*, 1991). In this way, many carbohydrate structures have been shown to block bacterial attachment in a highly specific manner. More recent approaches include the exquisite technique of sialidase treatment, followed by re-sialylation with specific

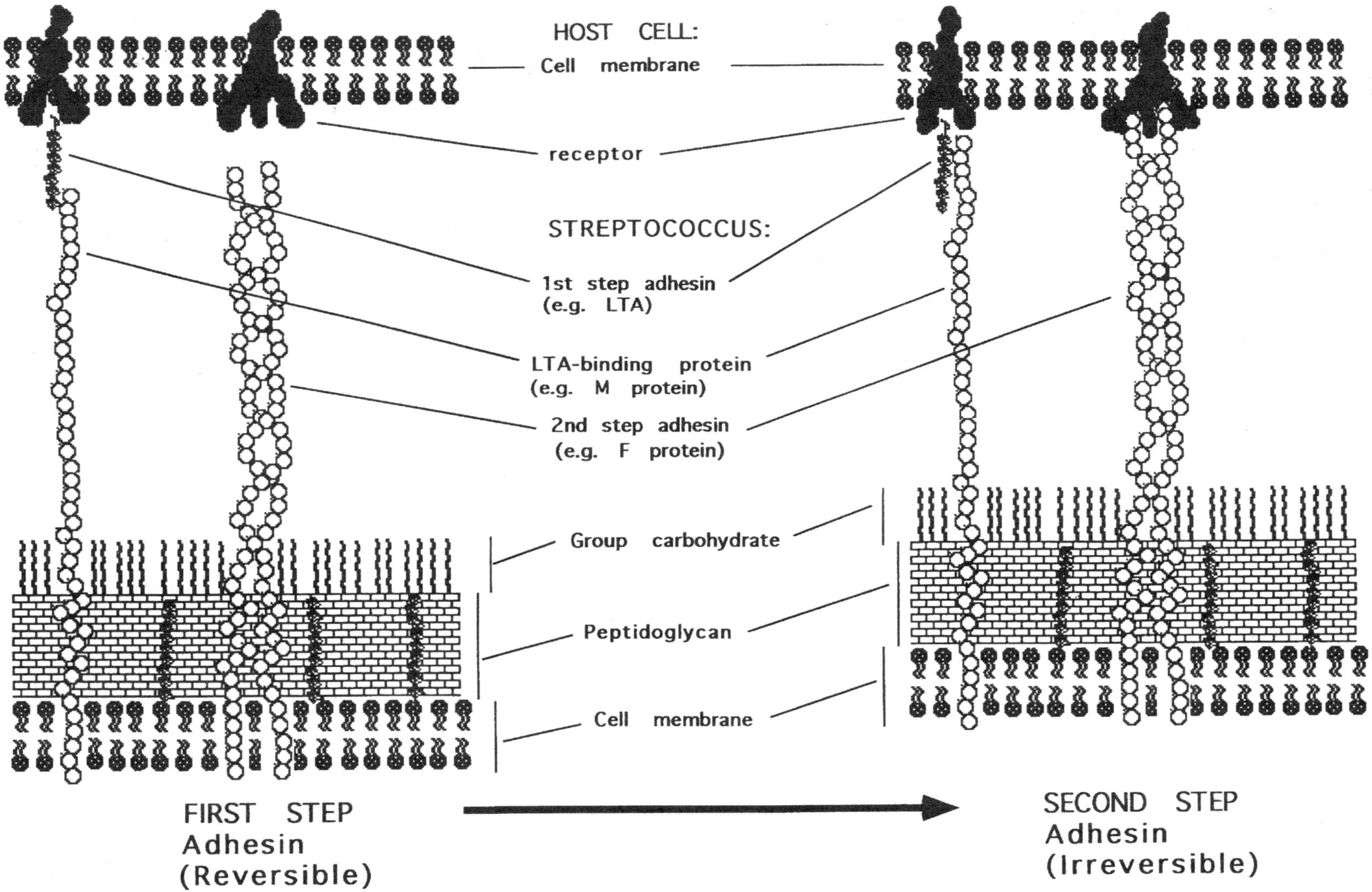

Figure 12.5 Two-step model for the interaction of *S. pyogenes* to host cells. Two sequential steps occur: The first step is relatively weak, broadly specific and reversible and probably involves LTA which is strategically anchored by M protein (or other proteins) to the bacterial cell surface with its hydrophobic region exposed. The second step involves the interaction of cell surface proteins such as protein F which can mediate high affinity binding to complementary receptors on host cells such as fibronectin. The second step of attachment is relatively strong, highly specific and irreversible. It is proposed that the completion of the first step is a requisite for the second step to occur. Redrawn from Hasty *et al.* (1992).

Table 12.1 Examples of bacterial adhesins and their receptors on mucosal cells*

Bacteria	*Adhesin*	*Receptor specificity*	*Common target sites*
Escherichia coli	P fimbriae	Galα(1–4)Gal	Urinary tract
E. coli (Enterotoxigenic)	CFA/1 fimbriae CFA/2 fimbriae	NeuAcα8 Sialic acid containing glycoconjugates	Intestinal tract Intestinal tract
E. coli *Klebsiella* sp. *Serrattia* sp. *Citrobacter* sp. *Edwardsiella* sp. *Salmonella* sp.	Type 1 fimbriae	Mannose	Urinary tract Respiratory tract Intestinal tract
Haemophilus influenzae	fimbriae	Ganglioside GM3	Respiratory tract
Pseudomonas aeruginosa	fimbriae	GalNAcβ4Gal	Respiratory tract
Neisseriae gonorrhoeae	fimbriae	Galβ4Glcβ	Genital tract
Actinomyces naeslundii	Type 2 fimbriae	Galβ3GalNAcβ, GalNAcβ3Galβ	Oral cavity
Bordetella pertussis	Filamentous haemagglutinin	GalNAcβ4Gal Sulphated glycolipids	Respiratory tract

*This table was compiled from Ref: Andersson *et al.* (1986); Badour *et al.* (1990); Beachey (1980); Brennan *et al.* 1991; Buhler *et al.* (1991); Duguid and Old (1980); Firen *et al.* (1987); Garegg and Oscarsson (1985); Kabat (1978); Karlsson (1988, 1989); Krivan *et al.* (1988); Krivan and Roberts *et al.* (1988); Leffler and Svanborg-Eden (1980); Moch *et al.* (1987); Ramphal and Ryle (1983); Stromberg and Hendley (1983); Tylewska *et al.* (1988).

transferases (Loomes *et al.*, 1984) and solid-phase methods for determining binding specificities of bacteria based on separated glycolipids on thin layer chromatograms (Karlan *et al.*, 1980; Hansson *et al.*, 1985; Karlsson, 1988; Krivan *et al.*, 1988; Stromberg *et al.*, 1990; Brennan *et al.*, 1991). The general consensus from these analyses is that the glycoconjugate receptors for bacterial adhesins are either associated with proteins or lipids and that the precise orientation of the sugar moieties in the receptor structure and the presence of neighboring submolecular structures determines the binding affinities and specificity of the receptor for its complementary adhesin.

Differences in the specificity and distribution of receptors account, in part, for tissue tropisms wherein different bacterial species attach selectively to one tissue niche in preference to another (Holgersson *et al.*, 1985). For example, *E. coli* bacteria are the most common cause of urinary tract infections and are seldom found in the upper respiratory tract. Conversely, *S. pyogenes* organisms are the most common cause of bacterial pharyngitis and are seldom found in the urinary tract.

The binding of *E. coli* to the uroepithelium has been related to the presence of mannose residues and Galα(1–4)Gal receptors respectively on uroepithelial cells (Leffler and Svanborg-Eden, 1980, 1986; Kallenius *et al.*, 1981). Similarly the binding of *S. pyogenes* to the upper respiratory tract has been associated with the presence of fibronectin on the surfaces of upper respiratory epithelial cells (Simpson and Beachey, 1982, 1983; Abraham *et al.*, 1983; Babu *et al.*, 1983; Hasty *et al.*, 1989). Interestingly fibronectin appears to serve as a receptor on oral cell epithelial cells for *S. pyogenes* but appears to hinder the adhesion. of *P. aeruginosa* and *E. coli* (Woods *et al.*, 1981, 1981; Abraham *et al.*, 1983). The mechanism by which fibronectin hinders adherence of the latter organisms remains unclear. It can be speculated, however, that, while providing receptors for the attachment of streptococci, fibronectin masks the receptors for these gram-negative bacteria. Another example of receptor mediated-tissue tropisms is illustrated by the discovery that the predominating anaerobes of the normal flora of human large intestine, including *Bacteroides*, *Clostridium*, *Fusobacterium*, and *Lactobacillus*, express binding specificity for lactosylceramides (Karlsson and Stromberg, 1987; Karlsson, 1989). Interestingly, lactosylceramide is present in the epithelium of the colon and not in other regions of the gastrointestinal tract such as the small intestine (Bjork *et al.*, 1987; Holgersson *et al.*, 1988).

The presence of specific receptors on the cells of a host appears to be genetically determined. For example, Sellwood *et al.* (1975) have shown that the intestinal epithelial cells of pigs that are highly resistant to diarrheagenic K88 fimbriated *E. coli* lack complementary receptors, whereas those from susceptible pigs are rich in K88 receptors. Cross-breeding of resistant and susceptible pigs has demonstrated that susceptibility is coded by autosomal-dominant genes (Sellwood *et al.*, 1975). Equally intriguing observations have been made in humans. The presence of specific glycolipid receptors on the membranes of human uroepithelium that recognize the P fimbriae appears to be genetically determined (Svanborg-Eden *et al.*, 1981; Lindstedt *et al.*, 1991). A certain population of people in the world are known to lack P antigens. P fimbriated *E. coli* do not adhere to mucosal cells or red blood cells of these individuals. However, persons of blood-group P_1 phenotype have a higher density of receptor glycolipid in their erythrocyte membrane than do individuals of the P_2 phenotype. It has been suggested that the P_1 phenotype may increase the host's susceptibility to upper urinary tract infections (Kallenius *et al.*, 1980, 1981). Attempts have been made to determine the distribution and relative density of receptors in human urogenital tissue. Galα(1–4)Gal and mannose-containing receptors for P and type 1 fimbriae have been detected in epithelial cells obtained from the vagina, bladder, ureter, renal pelvis, and kidney but not in the glomerulus and loop of Henle (O'Hanley *et al.*, 1985; Karr *et al.*, 1989). P fimbriae receptors were more prevalent in the collecting ducts than were type 1 fimbriae receptors and were less frequently found in transitional epithelium of the bladder and ureter. The finding of receptors for P fimbriae in the upper urinary tract may explain their frequent association with *E. coli* causing pyelonephritis.

Developmental changes also may have an important bearing on the bacterial receptivity of a particular host cell. For example, Ofek *et al.* (1977), demonstrated that oral epithelial cells of newborn infants were poor binders of streptococci. Within 3 days, however, the epithelial cells rapidly gained the capacity to bind adult levels of streptococci (Ofek *et al.*, 1977). Although age related resistance to infections by enterotoxigenic *E. coli* bearing 987P fimbriae have been demonstrated experimentally in pigs, no age related differences in receptors for the 987P fimbriae were found in the brush borders of the small intestines of neonates ($\leqslant 6$ day-old), and older ($\geqslant 3$ week-old) animals. Interestingly, Dean (Dean, 1990) isolated several receptor molecules for the 987P fimbriae from the intestinal mucin of older pigs but not from the mucin of neonates. The author hypothesized that the soluble receptors in intestinal mucin protect older pigs from 987P mediated diarrhea by preventing contact between 987P fimbriae and other receptors on the brush border (Dean, 1990). The attachment of *E. coli* to adult mucin could also facilitate bacterial agglutination and clearing from the animal.

In addition to naturally occurring receptors for bacteria, new receptors that recognize different species have been observed with cells infected with certain viruses (Sanford *et al.*, 1978, 1980; Abramson, 1988). Thus, a previously resistant host cell may be rendered susceptible to the adherence of pathogenic bacteria during certain viral infections. Evidence in support of this concept includes data from *in vitro* and *in vivo* studies that examined the effect of viruses on bacteria as well as on the host, both locally as well as systemically. A number of viruses, including influenza, parainfluenza, and respiratory syncytial virus have been observed to increase the attachment and colonization of *Haemophilus influenzae*, *Streptococcus pneumoniae* and *Staphylococcus aureus* (Fainstein *et al.*, 1980; Smith, 1987). Viral induced changes in the cells lining the mucosa of the gastrointestinal tract have been associated with an increase in the occurrence of bacterial diarrhea (Lecce *et al.*, 1982).

3. *Role of Adhesins in Infection*

It is known that bacterial adhesins can promote bacterial virulence and influence disease in the host through a variety of ways including: (i) adherence to mucosal surfaces; (ii) adherence to neutrophils; (iii) adherence to effector cells involved in inflammation; (iv) adherence to glycoconjugates present in body fluids, and; (v) adherence to foreign bodies.

3.1 ADHERENCE TO MUCOSAL SURFACES

Most infectious diseases originate on mucosal surfaces where the pathogenic bacteria encounter the mucosal barrier for the first time (Savage, 1972). The result of the initial encounter depends on the ability of the organisms to adhere to the mucosal surface; non-adherent bacteria are simply swept away in the fluids that constantly flow across the mucous membranes (Savage, 1972; Gibbons and Van Houte, 1975, 1980). Not only must successful pathogens adhere, they must be able to multiply at a rate sufficient to overcome the clearing effect brought about by the desquamation of colonized epithelial cells. Once the pathogen has successfully anchored itself, it proceeds to progressively colonize the surrounding mucosal surface. Certain bacteria (e.g. *Vibrio cholerae* and *Corynebacterium diphtheriae*) cause disease without any further invasion of tissue by elaborating toxins that penetrate the mucosal barrier and produce local and systemic tissue injury. In order to deliver these toxins in the most efficient manner, it is advantageous for the organisms to be attached closely to the epithelial cells of the mucosae. In this way, the excreted toxins evade destructive enzymes in the mucosal secretions and are delivered in concentrated form directly to toxin receptors on the epithelial cells. The tight association with mucosal surface also has a growth advantage to the bacterium. Zafriri *et al.* (1987), compared the growth of a fimbriated and a non-fimbriated clone of *E. coli* in tissue culture cells and detected a marked growth advantage for the fimbriated organisms. They concluded that the growth advantage for the bacteria was due to the assimilation of nutrients secreted by the host cell as a result of its close association.

Several organisms from the family *Enterobactericeae* cause infections which are initiated at the mucosal surface (Elsenstein, 1990). Interest in the role of their cell surface fimbriae has centred around infections in the gastro-intestinal and urinary tract (Gaastra and deGraaf, 1982; Jayappa *et al.*, 1985; Clegg and Gerlach, 1987; Reid and Sobel, 1987).Perhaps the best studied is the role of type 1 and P fimbriae in *E. coli* infections of the urinary tract. The infectivity of fimbriated and non-fimbriated phenotypes was compared in a number of animal models, and the infectivity of the former was always found to be significantly higher (Alkan *et al.*, 1986; Marre and Hacker, 1987; Ofek and Sharon, 1990). This was found for strains expressing type 1 fimbriae in *K. pneumoniae* in rats (Fader and Davis, 1980), by *E. coli* expressing type 1 fimbriae in mice (Iwaki *et al.*, 1983; Hultgren *et al.*, 1985; Orndorff and Bloch, 1990) and P fimbriae expressing *E. coli* in mice (Hagberg *et al.*, 1983; Hagberg, Hull *et al.*, 1983).

Another experimental approach to assess the role of fimbrial adhesins is to examine the effects of specific antibodies directed at the adhesins in animals that were passively or actively immunized against the corresponding fimbriae. This approach was employed in the study of type 1 and P fimbriated *E. coli* in urinary tract infections. In all cases, infection was prevented in the immunized animals challenged with strains expressing the fimbriae (Abraham *et al.*, 1985; Schmidt *et al.*, 1988). In addition, fimbriae-specific antibodies directed specifically at the receptor of type 1 fimbriae were shown to be protective when passively applied in mice (Abraham *et al.*, 1985). Taken together these findings provide the most convincing evidence for the central role of bacterial adhesins in initiating infection on mucosal surfaces.

In an interesting study, O'Hanley and colleagues compared the infectivity of two recombinant strains encoding type 1 and P fimbriae against the non-fimbriated host strain (O'Hanley *et al.*, 1985). They administered 10^6 organisms of each strain intravesicularly and examined the subsequent renal colonization in each mouse. They determined that the P fimbriated but not the type 1 or non-fimbriated strains colonized the kidneys. However, at higher concentrations of bacteria, type 1 fimbriated strains were also able to colonize the kidneys. These studies support the epidemiological association of P fimbriated organisms with upper urinary tract infection and type 1 fimbriae with lower urinary tract infection (Svanborg-Eden *et al.*, 1981).

3.2 ADHERENCE TO NEUTROPHILS

After colonization of the mucosal surface the invading pathogen may penetrate the musosal surface and infect deep tissue. This assault evokes an immune and inflammatory response from the host. The capacity of the pathogen to avoid, subvert or nullify this response largely determines its success in establishing itself in the host. Professional phagocytic cells such as neutrophils and macrophages are perhaps the first cells to respond to the invading pathogen. In this regard, it would appear that the presence of adhesins on the bacterial surface would be detrimental to its survival, as it will promote association with phagocytic cells which could result in ingestion and intracellular destruction. Indeed, the expression of type 2 fimbrial adhesions by *A. viscosus* and *A. naeslundii* have been shown to promote bacterial killing by human peripheral blood neutrophils (Sandberg *et al.*, 1986, 1988). In the absence of serum components, incubation of type 2 expressing *Actinomyces* species resulted in a 44% decrease in viability (Sandberg *et al.*, 1986, 1988). The type 2 adhesins were clearly implicated in the induction of phagocyte-mediated bactericidal activity, since the loss of viability of a mutant of *A. viscosus* lacking the type 2 but possessing type 1 fimbriae and a non-fimbriated mutant is minimal or not detectable (Sandberg *et al.*, 1986, 1988). The initiation of the phagocytic process is attributable to the interaction of type 2 fimbriae and a β-linked galactose-containing receptor present on neutrophil membranes, because the attachment and internalization of *A. viscosus* or *A. naeslundii* by neutrophils is inhibited by lactose and β-methyl-galactoside but not by α-methyl galactoside and cellobiose (Sandberg *et al.*, 1986, 1988).

The interaction of bacterial fimbrial adhesins with neutrophils does not always result in bacterial killing. Recent evidence from this and other laboratories suggests that bacteria utilize their adhesins in various intriguing yet highly effective ways to circumvent bactericidal activity of neutrophils. It is well known that type 1 fimbriae of *E. coli* promote the association of bacteria with phagocytic cells; and consequently, it was assumed that these adhesins promoted bacterial phagocytosis (Bar-Shavit *et al.*, 1977; Goetz and Silverblatt, 1987; Ofek and Sharon, 1990). However, recent studies suggest that, although type 1 fimbriae promote *E. coli* attachment to neutrophils, some of these adherent organisms remain viable (Goetz *et al.*, 1987; Keith *et al.*, 1990). By employing scanning and transmission electron microscopy, it was observed that the majority of type 1 fimbriated *E. coli* associated with neutrophils remained on the extracellular surface of the phagocytes and resisted ingestion. It was shown that even after prolonged incubation with neutrophils, viable bacteria can be eluted from neutrophil surface with the receptor analogue α-methyl mannoside (MacGregor and Abraham, unpublished observations). It would appear that, although type 1 fimbriae promote *E. coli* attachment to neutrophils, some of them adhere to receptors which fail to trigger intracellular signals for the initiation of phagocytic mechanisms. The identity of these mannose-containing receptors on neutrophil has not been definitively determined, although it was recently reported that monoclonal antibodies directed at the leucocyte adhesion molecules CD11/CD18 which belong to the integrin receptor family blocked up to 80% of the binding of type 1 fimbriated *E. coli* to neutrophils (Gbarah *et al.*, 1991).

The Galα(1–4)-Gal binding P fimbriae have been shown to confer on *E. coli* resistance to bactericidal activity of neutrophils (Svanborg-Eden *et al.*, 1984; Tewari *et al.*, submitted). The mechanism of this action was related to reduced adherence of P fimbriated bacteria compared to nonfimbriated bacteria. Studies with recombinant P fimbriated strains where a single P fimbrial subunit gene was inactivated showed that the minor subunit and adhesin PapG was responsible for the reduced interaction with neutrophil (Tewari *et al.*, submitted). The lack of adherence of PapG-containing clones is attributed to the highly negative charge on PapG which appears to augment the electrostatic repulsion between the negatively charged bacterial surface and similarly charged neutrophil membrane (Tewari *et al.*, submitted). This conclusion is based on the finding that poly-L-lysine, a positively charged cation, markedly increased the neutrophil response of P fimbriated or PapG-expressing clones of *E. coli* to levels elicited by nonfimbriated *E. coli*. Additional evidence comes from the observation that a PapG-containing clone but not a PapG deficient mutant was reversibly retained in a positively charged DE52 resin column when both strains were subjected to anion exchange chromatography (Tewari *et al.*, submitted). Furthermore, the isolated PapG protein has a pI of 5.2, as determined by isoelectric focusing, which is consistent with a protein with a net negative charge under physiological conditions. The neutrophil-repellent effect of PapG is pronounced, perhaps, because it is strategically situated at the distal tip of the radially arranged fimbriae and is the first bacterial component encountered by the neutrophils.

The M protein of *S. pyogenes* is another example where a bacterial adhesin has been employed to escape opsonization and phagocytosis. This surface protein on *S. pyogenes* confers resistance to phagocytosis by preventing opsonization by complement (Lancefield, 1962; Whitnack *et al.*, 1988). Avirulent *S. pyogenes* lacking M protein are readily opsonized by complement which has been activated by the alternate pathway, whereas M protein-bearing organisms resist opsonization because of the ability of their M protein to bind fibrinogen and its breakdown product fibrin (Peterson *et al.*, 1979; Whitnack *et al.*, 1988). Binding of fibrinogen to M protein impedes subsequent attachment of complement and, thereby, prevents opsonization. Thus it would appear that once the mucosal barrier is penetrated, certain adhesins, which normally facilitate the adherence of bacteria to mucosal surface, could switch roles and serve a diametric function of impeding bacterial adherence to phagocytic cells. Both roles are clearly crucial for promoting the survival and proliferation of the pathogen.

3.3 ADHERENCE TO EFFECTOR CELLS INVOLVED IN INFLAMMATION

The inflammatory response of the host to the bacteria is often dependent on the interaction of various bacterial cell surface components, including adhesins with inflammatory cells that elicit mediators, such as cytokines (Kenig *et al.*, 1985, 1989; Ventur *et al.*, 1990). Interleukins such as IL-1, IL-6 and tumor necrosis factor TNF-α are released by stimulated host cells and are involved in the recruitment and chemotaxis of neutrophils as well as in the proliferation and function of a variety of effector cells (Rugo *et al.*, 1992). In an interesting series of experiments, Svanborg-Eden and others have shown that certain bacterial adhesins can activate mucosal cells and organs (Linder, Jann *et al.*, 1988; Linder *et al.*, 1988; Rugo *et al.*, 1992), such as kidneys, to elicite the release of IL-6 in the urine. It is not clear what the role of IL-6 is in the urinary tract, but it is known to activate febrile responses in rabbits and to stimulate hepatocytes to produce C-reactive protein and fibrinogen, and mucosal B cells to produce immunoglobulin A (Linder, Jann *et al.*, 1988; Linder *et al.*, 1988). It may, therefore be responsible for some of the symptoms that characterize urinary tract infections. Interestingly, these authors determined that elicitation of IL-6 is specific to P fimbriae and other fimbriae such as type 1, and S fimbriae fail to evoke this response (Linder, Jann *et al.*, 1988; Linder *et al.*, 1988).

Mast cells are one of the cell types activated during the development of an inflammatory response. It has been suggested that their strategic location along the skin; blood vessels, especially the post-capillary venules; gastrointestinal tract; and upper and lower respiratory tract makes them extremely important (Rothe *et al.*, 1990). Interaction of bacterial adhesins with mast cells can result in the release of an enormous panoply of mediators with their numerous intertwined biological effects in the tissue micro-environment, as well as in the entire body (Williams *et al.*, 1986). Some of the inflammatory mediators secreted by mast cells are preformed such as histamines, heparin, and chemotactically active peptides (Wasserman, 1989; Marshall and Bienenstock, 1990; Rothe *et al.*, 1990). Histamine induces bronchoconstriction and causes vasodilation and increases in vascular permeability which allows for the influx of phagocytic cells. Newly generated mediators consist of leukotrienes, prostaglandins, thromboxane and the platelet activating factor (Wassern, 1989; Marshall and Bienenstock, 1990; Rothe *et al.*, 1990). The extent of release of mediators by effector cells is highly dependent on the stimulus. It has been shown that certain bacterial fimbriae can adhere to rat peritoneal mast cells and specifically elicite release of histamine (Scheffer *et al.*, 1985, 1986; Konig *et al.*, 1986). Fig. 12.6 shows a type 1 fimbriated *E. coli* binding to a mast cell obtained from a mouse. Certain fimbriae have also been reported to induce the release of leukotrienes (LTB_4) which are chemotactic for phagocytic cells from neutrophils (Scheffer *et al.*, 1985, 1986; Konig *et al.*, 1986).

Figure 12.6 Scanning electron micrograph of the binding of type 1 fimbriated *E. coli* to a mouse mast cell.

In addition to the physiological events associated with inflammation such as vascular permeability, bronchoconstriction, mucus production and fever, it is known that subtle alterations in receptivity of endothelial cells occur which allow for increased attachment of certain bacteria as well as for neutrophils, monocytes and lymphocytes. For example, it has recently been shown that the cytokine, TNF-α, secreted by monocytes during inflammation can promote the attachment of *S. aureus* to human cultured endothelial cells (Cheung *et al.*, 1991). One can speculate that the increased susceptibility of septic patients with *S. aureus* to the fulminant, life-threatening disease of acute infective endocarditis may be the consequence of cytokine induced attachment of *S. aureus* to the endocardium.

3.4 ADHERENCE TO GLYCOCONJUGATES IN BODY FLUIDS

Various body fluids are replete with glycoconjugates which could serve as receptors for bacterial adhesins. Such moieties may interfere with the adherence of the bacteria to host tissue and influence its infective capability. Tamm-Horsfall glycoprotein is the most abundant glycoprotein in normal human urine, which contains N-linked oligomannose units (Serafini-Cessi *et al.*, 1984; Dell'Olio *et al.*, 1988). It is known to be capable of binding specifically to type 1 fimbriae, a reaction which is completely inhibitable by D-mannose (Orskov *et al.*, 1980; Kuriyama and Silverblatt, 1986). It has been suggested that in the urinary tract, specifically in the bladder, Tamm-Horsfall glycoprotein could potentially facilitate the bacterial colonization of the mucosal surface, since it adsorbs to the mucosal walls and provides a matrix to which the bacteria can readily attach (Chick *et al.*, 1981). Alternatively the soluble Tamm-Horsfall could attach to type 1 fimbriated organisms, aggregate them and facilitate their elimination with voided urine (Orshov *et al.*, 1980). Thus, the physical state of the Tamm-Horsfall protein can potentiate or block the development of urinary tract infection. The Tamm-Horsfall has also been shown to modulate the interaction of type 1 fimbriated bacteria with phagocytic cells (Kuriyama and Silverblatt, 1986). Cells pretreated with Tamm-Horsfall appeared to interact less with neutrophils than untreated controls (Kuriyama and Silverblatt, 1986). These studies show that interaction of soluble glycoconjugates carrying suitable sugars may affect the infectious process not only at the mucosal surface but also in deep tissue.

3.5 CO-AGGREGATION WITH OTHER BACTERIA

A common but less well-recognized application of bacterial adhesins is in promoting polymicrobial communities on host surfaces. Perhaps the best studied example is in the microbiology of the human oropharynx (Gibbons and van Houte, 1979, 1980; van Houte, 1982; Mergenhagen *et al.*, 1987). There is now ample evidence that specific interactions between bacterial adhesins and complementary cell surface molecules on the same or other genera of bacteria contribute to the development of microbial communities in dental plaque. Studies of oral ecology have shown that the first colonizers of the freshly cleaned tooth surface are the oral streptococci, including *Streptococcus sanguis* (Mergenhagen *et al.*, 1987; Baddour *et al.*, 1990). The *Actinomyces* that arrive as secondary colonizers may bind to the buccal epithelium and other available sites on oral tissue (Mergenhagen *et al.*, 1987; McIntire *et al.*, 1992). In addition, however, these organisms bind to the streptococci already adherent on the tooth surface. Type 2 fimbriae of *A. viscosus* and *A. naeslundii* have been implicated in these reactions because mutants that lack these fimbriae fail to participate in lactose-sensitive co-aggregation with *S. sanguis*. McIntire and others have shown that lactose and other β-linked galactosides inhibited the interaction of various species of *Actinomyces* with *Streptococcus sanguis* (McIntire *et al.*, 1982). It has also been determined that the complementary receptor on *S. sanguis* for type 2 fimbriae is a polysaccharide composed of a repeating hexasaccharide-containing N-acetyl-galactosamine at the non-reducing end in α-linkage to L-rhamnose (Mergenhagen *et al.*, 1987). The ability of the secondary colonizers such as

Actinomyces species to bind to streptococci obviously increases the number of surface habitats available to the organisms and is clearly an important determinant of its success as a colonizer of the oral cavity.

3.6 ADHESION TO FOREIGN BODIES

The insertion of implants and devices has become a common and often life-saving procedure. Bacteria that arrive randomly either by direct contamination or hematogenous seeding attach through their adhesions to these foreign bodies; and if conditions are favorable, these organisms establish colonization (Gristina *et al.*, 1990). Commonly infected devices include urinary catheters (Sugarman, 1982); prosthetic vascular grafts (Sugarman, 1982), suture material (Sugarman and Musher, 1981); myringotomy tubes (Karlan *et al.*, 1980) and intravascular catheters (Locci *et al.*, 1981). *In vitro* studies indicate that the device's roughness, chemical composition, and hydrophobicity increase bacterial colonization (Gristina *et al.*, 1990). Although the precise mechanisms of bacterial adherence to these devices is unknown, it has been suggested that hydrophobicity of cell surface moieties plays an important role (Baddour *et al.*, 1990). Studies on the attachment of coagulase-negative staphylococci to catheters suggest that, in most cases, the attachment is mediated by an extracellular polymeric material or slime (Christensen *et al.*, 1982; Baddour *et al.*, 1990). Epidemiological and animal studies have shown that this slime plays a significant role in the pathogenesis of coagulase negative staphylococci to foreign bodies (Diaz-Mitoma *et al.*, 1987; Schadow *et al.*, 1988). Clinical evidence comes from the finding that intravascular catheters and cerebrospinal fluid shunts removed from patients are covered with coagulase-negative staphylococci enmeshed in a slimy polysaccharide deposit (Christensen *et al.*, 1982; Hogt *et al.*, 1982). Furthermore, medical treatment for cerebrospinal fluid shunt complication is more likely to occur in infections caused by slime-producing rather than non-slime-producing strains of coagulase negative staphylococci (Hogt *et al.*, 1982; Baddour *et al.*, 1990).

4. *Anti-adhesive Therapy*

Now that several adhesins have been well defined, it is possible to develop anti-adhesive strategies that can interrupt bacterial adherence and circumvent the infectious disease at an early stage. Some of the strategies have been directed at: (i) immunization (active and passive) against bacterial adhesins; (ii) application of adhesin or receptor analogues; (iii) use of autochthonous microbial flora to exclude potential pathogens, and; (iv) fortification of mucosal surfaces.

4.1 ACTIVE AND PASSIVE IMMUNIZATION AGAINST BACTERIAL ADHESINS

Perhaps the best characterized of these strategies has been the use of purified adhesins to evoke protective and long-lasting immunity. One of the earliest fimbrial vaccines to be tested was the K88 antigen isolated from *E. coli* causing diarrhea in young pigs (Rutter and Jones, 1973). In these studies, immunization was undertaken by inoculating purified fimbrial preparations into the posterior mammary glands of gilts approximately 11 weeks before parturition. The piglets of vaccinated and unvaccinated control sows were orally inoculated with K88-positive *E. coli* at birth before suckling. Although all piglets demonstrated clinical symptoms of infection, piglets from vaccinated gilts had markedly less severe symptoms, and their mortality after 72 h was significantly lower than that of the control piglets (Rutter and Jones, 1973). Only four of 31 vaccinated piglets died, whereas 20 of 29 of the non-vaccinated controls succumbed to infection. No differences were detected in the bacteriostatic and bactericidal activities in the serum and mammary secretions of vaccinated and unvaccinated gilts. Nevertheless, anti-K88 antibodies were detected in the serum, colostrum, and milk of vaccinated gilts, but not in the unvaccinated gilts. The K88-specific antibodies were found to block adhesion of the related *E. coli* to epithelial cells. The logical conclusion was that the ingestion of K88-specific antibodies in maternal colostrum and milk protected the suckling piglets by preventing adhesion and colonization (Rutter and Jones, 1973). Similar trials were conducted with purified K99 and 987P fimbriae of *E. coli*, which cause diarrhea in lambs, calves and piglets. In each case, vaccination of the mothers resulted in protection of the suckling offspring (Isaacson *et al.*, 1980; Nagy *et al.*, 1985). Maternal immunization with P fimbriae for the prevention of neonatal pyelonephritis have also been attempted (Kaack *et al.*, 1988). Protection against ascending pyelonephritis in different animals including primates using purified type 1 or P fimbriae as vaccine has been achieved, and in some of these studies this protection has been correlated with the presence of fimbriae-specific antibodies in the urine (Silverblatt *et al.*, 1982; Svanborg-Eden *et al.*, 1983; Roberts *et al.*, 1984, 1984; Abraham *et al.*, 1985; Ofek and Sharon, 1990). Protection has also been achieved against ascending urinary tract infection in mice using synthetic peptides corresponding to certain regions of the major subunit of P fimbriae (Schmidt *et al.*, 1988; Pecha *et al.*, 1989).

Passive administration of fimbriae-specific antibody to mice also protected them against ascending pyelonephritis induced by type 1 fimbriated *E. coli* instilled into their bladders (Abraham *et al.*, 1985). Unlike active immunization, which involved stimulation of antibody production in the host, passive transfer of anti-adhesive antibodies provides immediate immunity and is not dependent on immunocompetence of the host. Passive immunization with fimbriae-specific antibodies may be particularly useful when employed in high-risk populations, such as neonates and immunologically compromised patients. Although the potential value of passive immunization has been clearly demonstrated in protecting neonates from gastrointestinal infection (Isaacson *et al.*, 1980; Nagy *et al.*, 1985), its applications at other mucosal sites has certain limitations. These limitations concern delivery of antibodies to a particular site and at concentrations sufficient to be effective in the relatively short life-span of the administered antibodies.

Vaccine trials in humans employing fimbriae have also been performed (Brinton *et al.*, 1982; Siegel *et al.*, 1982). Purified gonococcal fimbriae were employed to vaccinate humans against *N. gonorrhoeae* infections. The vaccine evoked high levels of antifimbrial antibodies that enhanced the phagocytosis of the homologous infecting strains (Siegel *et al.*, 1982). In most cases, protection was acquired against only the homologous strains, with little or no protection against heterologous strains possessing antigenically cross-reactive fimbriae (Brinton *et al.*, 1982; Siegel *et al.*, 1982) In another vaccine trial in humans, Levine *et al.* (1983) observed that, although parenteral administration of type 1 fimbriae isolated from *E. coli* evoked high serum levels of antifimbrial antibodies, it failed to protect against colonization of the gastrointestinal tract.

Thus, successful use of fimbrial vaccines has been hindered by their limited protective effect. The lack of protective activity is due, in part, to a high level of antigenic heterogeneity among fimbriae exhibiting the same binding specificity. Gonococcal fimbriae represent one of the best examples of antigenic heterogeneity of adhesion molecules. Extensive serological studies by Brinton *et al.* (1982) have demonstrated the existence of hundreds of gonococcal fimbrial serotypes. The fimbriae and the outer membrane proteins (particularly OMP2) of the gonococci have a tremendous tendency to vary, and multiple antigenic shifts are common during *in vitro* culturing as well as during natural infections (Brinton *et al.*, 1982; Meyer *et al.*, 1990). Evidence for such variation in natural infections was provided by close examination of *N. gonorrheae* cultured from the urethra in male patients and from the cervix and urethra in female contacts (Duckworth *et al.*, 1983; Zak *et al.*, 1984). Although the same strain was isolated from a given group, it was noted that the molecular weights of the outer membrane proteins and the fimbriae varied considerably among different individuals in the group (Duckworth *et al.*, 1983; Zak *et al.*, 1984). Furthermore, antibodies evoked to a particular fimbrial variant showed limited cross-reactivity with other variants produced by the same strain (Zak *et al.*, 1984). No particular relationship was seen between the expression of proteins of different molecular weights and the host site of isolation (Duckworth *et al.*, 1983; Zafriri *et al.*, 1987). Such intrastrain variations in surface proteins may be related to phase variation in the adhesive properties of the strain. Such alterations may permit variants to selectively colonize different mucosal surfaces. *In vivo* variations of surface antigens may also be a manifestation of the host's immune response directed toward specific epitopes. Such immune pressures may further enhance the evolution or selection of new antigenic variants.

Despite the antigenic heterogeneity seen in the gonococcal fimbriae, peptide mapping of the fimbrial subunits has revealed considerable structural similarities between fimbrial variants (Schoolnick *et al.*, 1982). Amino acid sequence analyses of peptide fragments obtained by cyanogen bromide cleavage of fimbrial subunits have revealed that most of the variation occurs in the primary structure of the carboxy terminal region of the subunit molecule (Schoolnick *et al.*, 1982). Considerable antigenic heterogeneity has also been observed among the P and type 1 fimbriae of *E. coli*; a great number of serotypes can be predicted from past serological studies (Dodd and Eisenstein, 1982; Lund *et al.*, 1983; O'Hanley *et al.*, 1983; Parry *et al.*, 1983; Schmidt *et al.*, 1984; Adegbola and Old, 1987; Denich *et al.*, 1991). Indeed several antigenetically distinct types of P fimbriae are expressed simultaneously by the same strain (Rhen *et al.*, 1983; Lund *et al.*, 1988).

It would appear that the existence of multiple variants of the fimbriae is a major problem in the development of a "broad-spectrum" anti-adhesive vaccine. However, much of the antigenic variation described thus far has been observed with the major subunit of the fimbriae but it is now recognized that adhesive capacity of many fimbriae reside within minor proteins which may be antigenically less variable. As a result of the identification of some of these minor adhesins it might be possible to identify the binding domain within the adhesin and to prepare a vaccine either by cleaving the natural protein or by chemically synthesizing the peptide. Such a peptide vaccine should display the receptor binding domain of the adhesin to the host's immune system in a more concentrated and effective manner and thereby evoke higher titers of the functionally relevant antibodies. Abraham *et al.* (1988), demonstrated that antibodies generated against residues 1–25 of the mannose-binding adhesin FimH of *E. coli* type 1 fimbriae cross-reacted with FimH molecules on type 1 fimbriae of clinical strains of *K. pneumoniae*, *S. marcescens*, and *C. freundii*. Furthermore these authors were able to demonstrate that these FimH specific antibodies were capable of blocking the *in vitro* adherence of all these strains to mucosal cells (Abraham *et al.*, 1988). A similar cross-reactive epitope localized to the amino-terminal domains (residues 1–25) was recently identified on the subunits of type 4 fimbriae expressed by *P. aeruginosa*, *Moraxella bovis*, *N. gonorrhoeae*, *Dichelobacter nodosus* and *V. cholerae*. (Patel *et al.*, 1991).

Whether or not this antigenically conserved region is involved in adherence and whether antibody directed at this conserved epitope has broad anti-adhesive capabilities remains to be determined.

Another reason for the limited protection observed with the use of gonococcal and other fimbrial preparations in humans is that antifimbrial antibody levels in local mucosal secretions have been either non-detectable or present at very low levels, even though serum antibody levels have been high (Siegel *et al.*, 1982). Current lines of research are now directed toward raising the duration and height of mucosal immune responses by presenting the purified fimbrial antigens directly to the mucosal surfaces. Because the most common and effective antibody on mucosal surfaces is IgA, a number of laboratories proposed directly eliciting mucosal antibodies by supplying antigens directly to immunologically relevant sites on the mucosal surface such as the peyers patches (also known as gut associated lymphoid tissue). They reasoned that, by presenting antigens to T and B lymphocytes contained in the peyers patches, they could trigger local production of specific IgA antibody as well as a generalized response on other mucosal surfaces because stimulated lymphocytes were expected to seed all of the secretory tissues of the body, including lamina propria of the gut, respiratory and urogenital tract, mammary glands and lacrimal glands (Tacket *et al.*, 1992). To achieve their goal they chose to employ *Salmonella* strains as a vehicle to deliver cloned protective vaccine antigens (Attridge *et al.*, 1988; Newton *et al.*, 1989; Schmidt *et al.*, 1989) to the immune system of the host. The *Salmonella* strains were genetically attenuated by making them nutritional auxotrophs by interrupting the pathway for biosynthesis of aromatic metabolites (Tacket *et al.*, 1992). As a consequence of this auxotrophy the mutated *Salmonella* are not expected to proliferate within mammalian cells. However, they should be viable long enough to invade the peyers patches and stimulate lymphocytes with their antigens.

Thus these organisms were expected to effect only a transient colonization of the gut, but during this period they were expected to stimulate local and generalized production of IgA antibodies directed specifically at cloned antigens presented by the vaccine vehicle. There are several advantages to this approach over parenteral vaccination; for example, antibodies of this class might be more effective in the intestines than the more (proteolytically) labile IgG primarily induced by parenteral vaccination. And oral vaccination would avoid the local reactions sometimes associated with injection of conventional vaccines. Recently, the safety and immunogenicity of attenuated *Salmonella typhi* strains in adult human volunteers were undertaken (Tacket *et al.*, 1992). Very few incidences of fever or bacteremia were detected in these volunteers. Clearly, research along these lines should eventually lead to the replacement of parenteral vaccination strategy by employing live oral attenuated *Salmonella* strains as carriers for foreign antigens.

4.2 USE OF RECEPTOR ANALOGUES AND LECTINS

In addition to the traditional approaches of vaccination, another approach has been the use of receptor analogues to block adherence and colonization *in vivo* (Andersson *et al.*, 1986). Aronson *et al.* (Aronson *et al.*, 1979), attempted to see if mannose and its analogues could act *in vivo* and block the attachment and colonization of mice bladders by type one fimbriated *E coli*. To test their hypothesis, they instilled *E. coli* with either the receptor analogue, α-methyl mannoside or a control α-methyl glucoside (Aronson *et al.*, 1979). Animals receiving the receptor analogue were colonized to a far fewer extent than animals treated with amethyl glucoside. A similar experiment was shown to be effective for P fimbriated *E. coli* in the urinary tract using Galα(1–4)Gal containing globosides as receptor (Svanborg-Eden *et al.*, 1982).

A promising anti-adhesive strategy for intervention in the adherence of harmful bacteria to the gastrointestinal tract may lie in the use of synthetic or naturally occurring receptor analogues or specific lectins which mimic bacterial adhesions in the diet (Gibbons and Dankers, 1983). These receptor analogues or lectins would respectively bind to potential pathogens and prevent their interaction with the mucosa or compete with potential bacterial pathogens for receptor sites on mucosal epithelial cells. Excessive intake of receptor analogues or lectins, however, may not be advisable, as the ecology of the indigenous flora of the digestive and respiratory tracts might be upset. It is probable that a controlled and well-balanced diet plays a central role in maintaining a proper balance between the indigenous bacteria and potential pathogens on the mucosal surface.

4.3 MICROBIAL INTERFERENCE

This approach involves the phenomenon of competitive exclusion or bacterial interference where normal bacterial flora with the same receptor specificity as the pathogen is introduced to a mucosal surface to prevent the colonization by the pathogen. *Lactobacillus* indigenous to the urogenital tract have been known to produce an inhibitory effect against the growth of uropathogens (Shahani *et al.*, 1977). In addition, these organisms have been shown to adhere to uroepithelial cells *in vitro* and to competitively exclude adherence of strains of *E. coli*, *K. pneumoniae*, and *P. aeruginosa* (Reid *et al.*, 1986). An *in vivo* model pretreating rat bladders with *Lactobacillus casei* prevented subsequent infection with uropathogens (Reid *et al.*, 1986). These studies suggest that, by supplementing normal flora with certain non-pathogens exhibiting the same binding specificity as certain known pathogens, it may be possible to prevent establishment of infection.

4.4 Fortification of Mucosal Surfaces

A novel approach to blocking bacterial attachment is to modify the surface of mucosal cells by the use of chemicals. This novel approach was taken by Parsons *et al.* (1979–1980), who noted that the urinary bladder is coated by a natural glycoaminoglycan layer and that this functions by sterically hindering the attachment of bacteria to specific receptors on underlying mucosal cells. He demonstrated that by stripping this coat, bacterial adherence was increased dramatically (Parsons *et al.*, 1979; Parsons, Pollen *et al.*, 1980; Parsons *et al.*, 1980). The glycoaminoglycan presumably acts by altering the electrostatic charge on the surface. Coating the surface once again with heparin or a synthetic glycoaminoglycan protected it from bacterial adherence. Heparin is a negatively charged compound, and its presence probably repels negatively charged bacteria.

5. *Concluding Remarks*

Studies on bacterial adhesins, particularly those relating to type 1 and P fimbriae of *E. coli*, have revealed the highly ordered nature of their biosynthesis and assembly as well as their structural complexity. Many cell-surface fimbriae protrude radially from the bacteria; and because of their length and rigidity are possibly the first and perhaps the only bacterial surface components directly encountered by host cells. The fine binding specificity of the adhesins together with the expression, density, and accessibility of their complementary receptors on host tissue generally dictates the site and level of bacterial colonization and infection in the host.

Increasingly, it is becoming recognized that bacterial adhesins play a crucial role not only in the initial attachment phase of the infectious process but also at the subsequent stages of infection including invasion and dissemination. Certain bacteria have the capability of utilizing their adhesins to modulate bacterial interactions with phagocytes and cells involved in the inflammatory response of the host to promote their survival *in vivo*. In addition to interacting with host cells and glycoconjugates (soluble or immobilized), certain adhesins can increase their range of habitats by promoting bacterial attachment to medical devices placed in the host and to indigenous microflora colonizing certain niches in the host. In recognition of the multiple roles played by bacterial adhesins in initiating and sustaining bacterial infections, several therapeutic modalities directed at interrupting bacterial adherence have been examined as a means to preventing or reducing bacterial infections in man and animals. Many of these modalities include immunological and non-immunological approaches and some have proven to be highly effective in animal studies. It is noteworthy that the choice of an anti-adhesive approach for a particular infection must take into consideration the population at risk. If the population is small and the risk is limited to finite periods of exposure such as in bacterial endocarditis, where host is transiently but predictably at risk, then anti-adhesive measures including the use of receptor analogues may be beneficial. If, on the other hand, the population is large and exposure to potential pathogens is continuous, immunoprophylaxis with antiadhesive vaccines may provide long-lasting protection against adhesion of pathogenic bacteria.

6. *Acknowledgements*

I am grateful to Drs Teruo Ikeda and Rupinder Tewari for help in the preparation of this manuscript. I would like to thank Vourdonna Knoeppel for her excellent secretarial assistance. My studies described here were supported by Public Health Service grant, AI-13550, from the National Institute of Health.

7. *References*

Abraham, S.N. and Beachey, E.H. (1985) . In "Advances in Host Defense Mechanisms", Vol. 4 (eds J.I. Gallin and A.S. Fauci), pp. 63–88. Raven Press, New York.

Abraham, S.N., Babu, J.P., Giampapa, C.S., Hasty, D.L., Simpson, W.A. and Beachey E.H. (1985). Protection against *Escherichia coli*-induced urinary tract infections with hybridoma antibodies directed against type 1 fimbriae or complementary D-mannose receptors. Infect. Immun. 48, 625–628.

Abraham, S.N., Beachey, E.H. and Simpson, W.A. (1983). Adherence of *Streptococcus pyogenes*, *Escherichia coli*, and *Pseudomonas aeruginosa* to fibronectin coated and uncoated epithelial cells. Infect. Immun. 41, 1261–1268.

Abraham, S.N., Hasty, D.L., Simpson, W.A. and Beachey, E.H. (1983). Anti-adhesive properties of a quaternary structure-specific hybridoma antibody against type 1 fimbriae of *Escherichia coli*. J. Exp. Med. 158, 1128–1144.

Abraham, S.N., Goguen, J.D., Sun, D., Klemm, P. and Beachey, E.H. (1987). Identification of two ancillary subunits of *Escherichia coli* type 1 fimbriae by using antibodies against synthetic oligopeptides of *fim* gene products. J. Bacterial. 169, 5530–5535.

Abraham, S.N., Goguen, J.D. and Beachey, E.H. (1988a). Hyperadhesive mutant of type 1 fimbriated *Escherichia coli* associated with the formation of FimH organelles (fimbriosomes). Infect. Immun. 56, 1023–1029.

Abraham, S.N., Sun, D., Dale, J.B. and Beachey, E.H. (1988b). Conservation of the D-mannose-adhesion protein among type 1 fimbriated members of the family *Enterobacteriaceae*. Nature (London) 336, 682–684.

Abramson, J.S. (1988). The pathogenesis of bacterial infections in infants and children: the role of viruses. Perspect. Biol. Med. 32, 63–72.

Adegbola, R.A. and Old, D.C. (1987). Antigenic relationships among type-1 fimbriae of Enterobacteriaceae revealed by immunoelectron-microscopy. J. Med. Microbiol. 24, 21–28.

Alkan, M.L. and Beachey, E.H. (1978). Excretion of lipoteichoic acid by group A streptococci: influence of penicillin on excretion and loss of ability to adhere to human oral epithelial cells. J. Clin. Invest. 61, 671–677.

Alkan, M.L., Ofek, I. and Beachey, E.H. (1977). Adherence of pharyngeal and skin strains of group A streptococci to human skin and oral epithelial cells. Infect. Immun. 18, 555–557.

Alkan, M.L., Wong, L., Silverblatt, F.J. (1986). Change in degree of type 1 piliation of *Escherichia coli* during experimental peritonitis in the mouse. Infect. Immun. 52, 549–554.

Andersson, B., Dahmen, J., Frejd, T., Leffler, H., Magnusson, G., Noori, G. and Svanborg-Eden, C. (1983). Identification of an active disaccharide unit of a glycoconjugate receptor for pneumococci attaching to human pharyngeal epithelial cells. J. Exp. Med. 158, 559–570.

Andersson, B., Porras O., Hanson, L.A., Lagergard, T. and Svanborg-Eden, C. (1986). Inhibition of attachment of *Streptococcus pneumoniae* and *Haemophilus influenzae* by human milk and receptor oligosaccharides. J. Infect. Dis. 153, 232–237.

Aronson, M., Medalia, O., Schori, L., Mirelman, D., Sharon, N. and Ofek, I. (1979). Prevention of colonization of the urinary tract of mice with *Escherichia coli* by blocking of bacterial adherence with methyl α-D-mannopyranoside. J. Infect. Dis. 139, 329–332.

Attridge, S., Hackett, J., Morona, R. and Whyte, P. (1988). Towards a live oral vaccine against enterotoxigenic *Escherichia coli* of swine. Vaccine 6, 387–389.

Babu, J.P., Simpson, W.A., Courtney, H.S. and Beachey, E.H. (1983). Interaction of human plasma fibronectin with cariogenic and noncariogenic oral streptococci. Infect. Immun. 41, 162–168.

Baddour, L.M., Christensen, G.D., Simpson, W.A. and Beachey, E.H. (1990). In "Principles and Practice of Infectious Diseases", (eds G.L. Mandell, R.G. Douglas, Jr. and J.E. Bennett), pp. 9–25. Churchill Livingstone, New York.

Baga, M., Goransson, M., Normark, S. and Uhlin, B.E. (1985). Transcriptional activation of a *Pap* pilus virulence operon from uropathogenic *Escherichia coli*. EMBO J. 4, 3887–3893.

Baga, M., Norgren, M. and Normark, S. (1987). Biogenesis of *E. Coli* Pap-pili: PapH, a minor pilin subunit involved in cell anchoring and length modulation. Cell 49, 241–251.

Baird, R.W., Courtney, H.S., Bronze, M.S. and Dale, J.B. (1990). Passive protection against group A streptococcal infection in mice by lipoteichoic acid. Clin. Res. 38, 353A.

Bar-Shavit, Z., Ofek, I., Goldman, R., Mirelman, D. and Sharon, N. (1977) Mannose residues on phagocytes as receptors for the attachment of *Escherichia coli* and *Salmonella typhi*. Biochem. Biophys. Res. Commun. 78, 455–460.

Beachey, E.H. (ed). (1980). "Bacterial Adherence, Receptor Recognition", Series B, Vol. 6. Chapman & Hall, London.

Beachey, E.H. (1981). Bacterial adherence: adhesion receptor interactions mediating the attachment of bacteria to mucosal surfaces. J. Infect. Dis. 143, 325–345.

Beachey, E.H. and Ofek, I. (1976). Epithelial cell binding of group A streptococci by lipoteichoic acid on fimbriae denuded of M protein. J. Exp. Med. 143, 759–771.

Beachey, E.H., Eisenstein, B.I. and Ofek, I. (1982). In "Bacterial adherence in infectious diseases". (Current concepts series). Upjohn, Kalamazoo.

Bjork, S., Breimer, M.E, Hansson, G.C., Karlsson, K.-A. and Leffler, H. (1987). J. Biol. Chem. 262, 6758–6765.

Brennan, M.J., Hannah, J.H. and Leininger, E. (1991). Adhesion of *Bordetella pertussis* to sulfatide and to the GalNAcβ4Gal sequence found in glycospingolipids. J. Biol. Chem. 266, 18 827–18 831.

Brinton, C.C. Jr. (1959). Non-flagellar appendages of bacteria. Nature 183, 782–786.

Brinton, C.C. (1965). The structure, function, synthesis and DNA and RNA transport in gram-negative bacteria. Ann. NY Acad. Sci. 27, 1003–1054.

Brinton, C.C., Wood, S.W. and Brown, A. (1982). In "Bacterial Vaccines, Seminars in Infectious Disease", Vol. IV (eds L. Weinstein and B.N. Fields), pp. 140–159. Thieme-Stratton, New York.

Buhler, T., Hoschutzky, H. and Jann, K. (1991). Analysis of colonization factor antigen I, an adhesin of entertoxigenic *Escherichia coli* 0.78:H11: fimbrial morphology and location of the receptor-binding site. Infect. Immun. 59, 3876–3882.

Ceri, H. and Westra, Y. (1988). Host binding proteins and bacterial adhesion: ecology and binding model. Biochem. Cell Biol. 66, 541–548.

Cheung, A., Koomey, J.M., Lee, S., Jaffe, E.A. and Fischetti, V.A. (1991). Recombinant human tumor necrosis factor alpha promotes adherence of *Staphylococcus aureus* to cultured human endothelial cells. Infect. Immun. 59, 3827–3831.

Chick, S., Harber, M.J., MacKenzie, R. and Asscher, A.W. (1981). Modified method for studying bacterial adhesion to isolated uro-epithelial cells and uromucoid. Infect. Immun. 34, 256–261.

Christensen, G.D., Simpson, W.A., Bisno, A.L. *et al.* (1982). Adherence to slime-producing strains of *Streptococcus epidermidis* to smooth surfaces. Infect Immun. 37, 318–326.

Clegg, S. and Gerlach, G.F. (1987). Enterobacterial fimbriae. J. Bacterial. 169, 934–938.

Clegg, S., Hull, S., Hull, R. and Pruckler, J. (1985). Construction and comparison of recombinant plasmids encoding type 1 fimbriae of members of the family *Enterobacteriaceae*. Infect. Immun. 48, 275–279.

Clegg, S., Pruckler J. and Purcell, B.K. (1985). Complementation analyses of recombinant plasmids encoding type 1 fimbriae of members of the family *Enterobacteriaceae*. Infect. Immun. 50, 338–340.

Courtney, H.S., Ofek, I., Simpson, W.A. and Beachey, E.H. (1981). Characterization of lipoteichoic acid binding to polymorphonuclear leucocytes of human blood. Infect. Immun. 32, 625–631.

Courtney, H.S., Simpson, W.A. and Beachey, E.H. (1983). Binding of streptococcal lipoteichoic acid to fatty acid binding sites on human plasma fibronectin. J. Bacteriol. 153, 763–770.

Courtney, H.S., Hasty, D.L. and Ofek, I. (1990). In "Microbial Cell Surface Hydrophobicity", (eds R.J. Doyle and M. Rosenberg), pp. 361–386. American Society for Microbiology, Washington, D.C.

Courtney, H.S., von Hunolstein, C., Dale, J.B., Bronze, M.S., Beachey, E.H. and Hasty, D.L. (1992). Lipoteichoic acid and M protein: dual adhesins of group A streptococci. Microb. Pathog. (In press).

Dahmen, J., Frejd, T., Magnusson, G., Noori, G. and Carlstrom, A.S. (1984). 2-Bromoethyl glycosides in glycoide synthesis: preparation of glycoproteins containing α-L-Fuc-(1 → 2)-D-Gal and β-D-Gal-(1 → 4)D-GlcNAc. Carbohydr. Res. 125, 237–245.

Dean, F.A. (1990). Comparison of receptors for 987P pili of enterotoxigenic *Escherichia coli* in the small intestines of neonatal and older pig. Infect. Immun. 58, 4030–4035.

Dell'Olio, F., de Kanter, F.J.J., van den Eijnden, D.H. and Serafini-Cessi, F. (1988). Structural analysis of the preponderant high-mannose oligosaccharide of human Tamm-Horsfall glycoprotein. Carbohydr. Res. 178, 327–332.

Denich, K., Blyn, L.B., Craiu, A. Braaten, B.A., Hardy, J., Low, D.A. and O'Hanley, P.D. (1991). DNA sequences of three papA genes from uropathogenic *Escherichia coli* strains: evidence of structural and serological conservation. Infect. Immun. 59, 3849–3858.

Diaz-Mitoma, F., Harding, G.K.M., Hoban, D.J. *et al.* (1987). Clinical significance of a test for slime production in ventriculoperitoneal shunt infections caused by coagulase-negative staphylococci. J. Infect. Dis. 156, 555–560.

Dodd, D.C. and Eisenstein, B.I. (1982). Antigenic quantitation inhibition assay of type 1 fimbriae on the surface of *Escherichia coli* cells by an enzyme linked immunosorbent inhibition assay. Infect Immun. 38, 764–769.

Duckworth, M., Jackson, D., Zak, K. and Heckels, J.E. (1983) Structural variation in pili expressed during gonococcal infections. J. Gen. Microbiol. 129, 1593–1596.

Duguid, J.P., Clegg, S. and Wilson, M.I. (1979). The fimbrial and non-fimbrial haemagglutinins of *Escherichia coli*. J. Med. Microbiol. 12, 213–227.

Duguid, J.P. and Old, D.C. (1980). In "Bacterial Adherence, Receptor and Recognition", Series B, Vol. 6 (ed E.H. Beachey), pp. 184–217. Chapman & Hall, London.

Eisenstein, B.I. (1990). In "Enterobacteriaceae in Principles and Practice of Infectious Diseases", (eds G.L. Mandell, R. Gordon Danglers and J.E. Bennet), pp. 1658–1673. Churchill Livingston, New York.

Ellen, R.P. and Gibbons, R.J. (1974). Parameters affecting the adherence and tissue tropisms of *Streptococcus pyogenes*. Infect. Immun. 9, 85–91.

Eshdat, Y., Silverblatt, F.J. and Sharon, N. (1981). Dissociation and reassembly of *Escherichia coli* type 1 pili. J. Bacteriol. 148, 308–314.

Fader, R.C. and Davis, C.P. (1980) Effect of piliation on *Klebsiella pneumoniae* infection in rat bladders. Infect. Immun. 30, 554–561.

Fainstein, V., Musher, D.M. and Cate, T.R. (1980). Bacterial adherence to pharyngeal cells during viral infection. J. Infect. Dis. 141, 172–176.

Finlay, B.B. and Falkow, S. (1989). Common themes in microbial pathogenicity. Microbial. Rev. 53, 210–230.

Firon, N., Ashkenazis, S., Mirelman, D., Ofek, I. and Sharon, N. (1987). Aromatic α-glycosides of mannose are powerful inhibitors of the adherence of type 1 fimbriated *Escherichia coli* to yeast and intestinal cells. Infect. Immun. 55, 472–476.

Firon, N., Ofek, I. and Sharon, N. (1982). Interaction of mannose containing oligosaccharides with the fimbrial lectin of *Escherichia coli*. Biochem. Biophys. Res. Commun. 105, 1426–1432.

Firon, N., Ofek, I. and Sharon, N. (1983). Carbohydrate specificity of the surface lectins of *Escherichia coli*, *Klebsiella pneumoniae* and *Salmonella typhimurium*. Carbohydr. Res. 120, 235–249.

Firon, N., Ofek, I. and Sharon, N. (1984). Carbohydrate-binding sites of the mannose-specific fimbrial lectins of Enterobacteria. Infect. Immun. 43, 1088–1090.

Gaastra, W., de Graaf, F.K. (1982). Host-specific fimbrial adhesins of noninvasive enterotoxigenic *Escherichia coli* strains. Microbial. Rev. 46, 129–161.

Garegg, P.J. and Oscarsson, S. (1985). Synthesis of 6- and 6′-deoxy derivatives of methyl 4-O-α-D-galatopyranosyl-β-D-galactopyranoside for studies of inhibition of pyelonephritogenic fimbriated *E. coli* adhesion to urinary epithelium-cell surfaces. Carbohydr. Res. 137, 270–275.

Gbarah, A., Gahmberg, C.G., Ofek, I., Jacobi, U. and Sharon, N. (1991). Identification of the leukocyte adhesion molecules CD11/CD18 as receptors for type 1 fimbriated (mannose specific) *Escherichia coli*. Infect. Immun. 59, 4524–4530.

Gerlach, G.D., Clegg, S., Ness, N.J., Swenson, D.L., Allen, B.L. and Nichols, W.D. (1989). Expression of type 1 fimbriae and mannose-sensitive hemagglutinin by recombinant plasmids. Infect. Immun. 57, 764–770.

Giampapa, C.S., Abraham, S.N., Chiang, T.M. and Beachey, E.H. (1988). Isolation and characterization of a receptor for type 1 fimbriae of *Escherichia coli* from guinea pig erythrocytes. J. Biol. Chem. 263, 5362–5367.

Gibbons, R.J. and Dankers, I. (1983). Association of food lectins with human oral epithelial cells *in vivo*. Arch. Oral. Biol. 28, 561–566.

Gibbons, R.J. and Van Houte, J. (1975). Bacterial adherence in oral microbiology. Annu. Rev. Microbiol. 29, 19–44.

Gibbons, R.J. and Van Houte, J. (1980). In "Bacterial Adherence, Receptor and Recognition", Series B, Vol. 6 (ed E.H. Beachey), pp. 60–104. Chapman & Hall, New York.

Goetz, M.B., Kuriyama, S.M. and Silverblatt, F.J. (1987). Phagolysosome formation by polymorphonuclear neutrophilic leukocytes after ingestion of *Escherichia coli* that express type 1 pili. J. Infect. Dis. 156, 229–233.

Goetz, M.B. and Silverblatt, F.J. (1987). Stimulation of human polymorphonuclear leukocyte oxidative metabolism by type 1 pili from *Escherichia coli*. Infect. Immun. 55, 534–540.

Gong, M., Batliwala, H. and Makowski, J. (1992). FASEB J. 6, 4471.

Gristina, A.G., Naylor, P.T. and Myrvik, Q.N. (1990). Musculoskeletal infection, microbial adhesion, and antibiotic resistance. Infect. Dis. Clin. North Am. 4, 391–408.

Hagberg, L., Engberg, I., Feter, R., Lam, J., Olling, S. and Svanborg-Eden, C. (1983) Ascending unobstructed urinary tract infection in mice caused by pyelonephritogenic *Escherichia coli* of human origin. Infect. Immun. 40, 273–283.

Hagberg, L., Hull, R., Hull S., Falkow, S., Freter, R. and Svanborg-Eden, C. (1983). Contribution of adhesion to bacterial persistence in the mouse urinary tract. Infect. Immun. 40, 265–272.

Hansen, M.S. and Brinton, C.C. (1988). Identification and characterization of *E. coli* type 1 pilus tip adhesion protein. Nature (London) 332, 265–268.

Hanski, E. and Caparon, M. (1992). Protein F, a fibronectin-binding protein, is an adhesin of the group A streptococcus (*Streptococcus pyogenes*). P.N.A.S. (In press).

Hansson, G.C., Karlsson, K.-A., Larson, G., Stromberg, N. and Thurin, J. (1985). Anal. Biochem. 146, 158–163.

Hasty, D.L., Beachey, E.H., Courtney, H.S. and Simpson, W.A. (1989). In "Fibronectin in Health and Disease", (ed S.E. Carsons), pp. 89–112. CRC Press, Inc., Boca Raton, Florida.

Hasty, D.L., Ofek, I., Courtney, H.S. and Doyle, R.J. (1992). Multiple adhesins of streptococci. Infect. Immun. (In press).

Hogt, A.H., Feijen, J., Dankert, J. *et al.* (1982). In "Proceedings of International Conference of Biomedical Polymers", pp. 39–47. Biological Engineers Society, London.

Holgersson, J., Karlsson, K.-A., Karlsson, P., Norrby, E., Orvell, C., Stromberg, N. (1985). In "World's Debt to Pasteur", (eds H. Koprowski and S.A. Plotkin), pp. 273–301. Alan R. Liss, New York.

Holgersson, J., Stromberg, N. and Breimer, M.E. (1988). Biochimie (In press).

Hoschutzky, H., Lottspeich, F. and Jann, K. (1989). Isolation and characterization of the α-galactosyl-1,4-β galactosyl-specific adhesin (P Adhesin) from fimbriated *Escherichia coli*. Infect. Immun. 57, 76–81.

Hull, R.A., Gill, R.E., Hsu, P., Minshaw, B.H. and Falkow, S. (1981). Construction and expression of recombinant plasmids encoding type 1 and D-mannose-resistant pili from a urinary tract infection *Escherichia coli* isolate. Infect. Immun. 33, 933–938.

Hull, R.A., Hull, S.I. and Falkow, S. (1984). Frequency of gene sequences necessary for pyelonephritis-associated pili expression among isolates of *Enterobacteriaceae* from human extra-intestinal infections. Infect. Immun. 43, 1064–1067.

Hultgren, S.J., Porter, T.N., Schaeffer, A.J. and Duncan, J.L. (1985). Role of type 1 pili and effects of phase variation on lower urinary tract infections produced by *Escherichia coli*. Infect. Immun. 50, 370–377.

Hultgren, S.J., Lindberg, F., Magnusson, G., Kihlberg, J., Tennent, J.M. and Normark, S. (1989). The PapG adhesin of uropathogenic *Escherichia coli* contains separate regions for receptor binding and for the incorporation into the pilus. Proc. Natl Acad. Sci. USA 86, 4357–4361.

Hultgren, S.J., Normark, S. and Abraham, S.N. (1991). In "Annual Review of Microbiology", (ed L.N. Ornston), pp. 383–415. Annual Reviews Inc., Palo Alto.

Isaacson, R.E., Dean, E.A., Morgan, R.L. and Moon, H.W. (1980). Immunization of suckling pigs against enterotoxigenic *Escherichia coli* induced diarrheal disease by vaccinating dams with purified K99 or 987 pili: antibody production in response to vaccination. Infect. Immun. 29, 824–826.

Isberg, R.R. (1991). Discrimination between intracellular uptake and surface adhesion of bacterial pathogens. Science 252, 934–938.

Iwahi, T., Abe, Y., Nakao, M., Imada, A. and Tsuchiya, K. (1983). Role of type 1 fimbriae in the pathogenesis of ascending urinary tract infection induced by *Escherichia coli* in mice. Infect. Immun. 39, 1307 –1315.

Jayappa, H.G., Goodnow, R.A., Geary, S.J. (1985). Role of *Escherichia coli* type 1 pilus in colonization of porcine-ileum and its protective nature as a vaccine antigen in controlling colibacillosis. Infect. Immun. 48, 350–354.

Kaack, M.B., Robert, J.A., Baskin, G. and Patterson, G.M. (1988). Maternal immunization with P fimbriae for the prevention of neonatal pyelonephritis. Infect. Immun. 56, 1–6.

Kabat, E.A. (1978). Dimensions and specificities of recognition sites on lectins and antibodies. J. Supramol. Struct. 8, 79–88.

Kallenius, G., Mollby, R., Svenson, S.B., Windberg, J., Lundblud, A. *et al.* (1980). The P^k antigen as receptor for the haemagglutinin of pyelonephritogenic *Escherichia coli*. F.E.M.S. Microbial. Lett. 7, 297–302.

Kallenius, G., Svenson, S.B., Mollby, R. and Winberg, J. (1981). Structure of carbohydrate part of receptor on human uroepithelial cells for pyelonephritogenic *Escherichia coli*. Lancet 2, 604–606.

Karlan, M.S., Skobel, A., Grizzard, M., *et al.* (1980). Myringotomy tube materials: Bacterial adhesion and infection. Otolaryngol. Head Neck Surg. 88, 783–794.

Karlsson, K.-A. (1988). In "Molecular Mechanisms of Microbial Adhesion", (eds L. Switalski, M. Hood and E.H. Beachey), pp. 77–96. Springer-Verlag, Berlin.

Karlsson, K.-A. (1989). In "Annu. Rev. Biochem", pp. 309–350. Annual Reviews Inc., Palo Alto.

Karlsson, K.-A. and Stromberg, N. (1987). Methods Enzymol. 138, 220–232.

Karr, J.F., Nowicki, B., Truong, L.D., Hull, R.A. and Hull, S.I. (1989). Purified P fimbriae from two cloned gene clusters of a single pyelonephritogenic strain adhere to unique structures in the human kidney. Infect. Immune. 57, 3594–3600.

Keith, B.R., Harris, S.L., Russell, P.W. and Orndorff, P.E. (1990). Effect of type 1 piliation on *in vitro* killing of *Escherichia coli* by mouse peritoneal macrophages. Infect. Immun. 58, 3448–3454.

Klemm, P. (1985). Fimbrial adhesins of *Escherichia coli*. Review Infect. Dis. 7, 321–339.

Klemm, P. and Christensen, G. (1987). Three *fim* genes required for the regulation of length and mediation of adhesion of *Escherichia coli* isolate. Infect. Immun. 33, 933–938.

Kihlberg, J., Frejd, T., Jansson, K., Sundin, A. and Magnusson, G. (1988). Synthetic receptor analogues: Preparation and calculated conformations of the 2-deoxy, 6.O-methyl, 6-deoxy and 6-deoxy-6-fluoro derivations of methyl 4-O-α-D-galacto-pyranoside (methyl β-D-galabioside). Carbohydr. Res. 176, 271–286.

Konig, W., Scheffer, J., Bremm, K.D., Hacker, J. and Goebel, W. (1985). Role of bacterial adherence and toxin production from *Escherichia coli* on leukotriene generation from human polymorphonuclear granulocytes. Int. Arch. Allergy Appl. Immunol. 77, 118–120.

Konig, B., Konig, W., Scheffer, J., Hacker, J. and Goebel, W. (1986). Role of *Escherichia coli* alpha-hemolysin and bacterial adherence in infection: requirement for release of inflammatory mediators from granulocytes and mast cells. Infect. Immun. 54, 886–892.

Konig, W., Konig, B., Scheffer, J., Hacker, J. and Goebel, W. (1989). Role for cloned virulence factors (bacterial adhesins, hemolysin) from uropathogenic *Escherichia coli* strains for the release of inflammatory mediators from granulocytes and mast cells. Immunology 67, 401–407.

Korhonen, T.K., Vaisanen, V., Saxin, H., Hultberg, H. and Svensson, S.B. (1982). P antigen recognizing fimbriae from human uropathogenic *Escherichia coli* strains. Infect. Immun. 37, 286–291.

Krivan, H.C., Ginsburg, V., Roberts, D.D. (1988). *Pseudomonas aeruginosa* and *Pseudomonas cepacia* isolated from cystic fibrosis patients bind specifically to gangliotetraosylceramide (asialo

GM1) and gangliotriaosylceramide (asialo GM2). Arch. Biochem. Biophys. 260, 493–496.

Krivan, H.C., Roberts, D.D., Ginsburg, V. (1988). Many pulmonary pathogenic bacteria bind specifically to the carbohydrate sequence GalNAcβ1-4Gal found in some glycolipids. Proc. Natl Acad. Sci. USA 85, 6157–6161.

Krogfelt, K.A., Bergmans, H. and Klemm, P. (1990). Direct evidence that the FimH protein is the mannose specific adhesin of *Escherichia coli* type 1 fimbriae. Infect. Immun. 58, 1995–1999.

Kuehn, M.J., Heuser, J., Normark, S. and Hultgren, S. (1992). The distal ends of P pili fibers are distinct fibrillar adhesive spikes in uropathogenic *E. coli*. Nature, 356, 252–255.

Kuriyama, S.M. and Silverblatt, F.J. (1986). Effect of Tamm-Horsfall urinary glycoprotein on phagocytosis and killing of type 1-fimbriated *Escherichia coli*. Infect. Immun. 51, 193–198.

Lancefield, R. (1962). Current knowledge of type-specific M antigens of group A streptococci. J. Immunol. 89, 307–313.

Lecce, J.G., Balsbaugh, R.K., Clare, D.A. and King, M.W. (1982). Rotavirus and hemolytic enteropathogenic *Escherichia coli* in weanling diarrhea of pigs. J. Clin. Microbiol. 16, 715–723.

Leffler, H. and Svanborg-Eden, C. (1980). Chemical identification of a glycosphingolipid receptor for *Escherichia coli* attaching to human urinary tract epithelial cells and agglutinating human erythrocytes. FEMS Microbiol. Lett. 8, 127–134.

Leffler, H. and Svanborg-Eden, C. (1986). In "Microbial Lectins and Agglutinins", (ed D. Mirelman), pp. 83–111. John Wiley & Sons, Inc., New York.

Levine, M.M., Blak, R.E., Brinton, C.C., Clements, M.L. *et al.* (1983). Reactogenicity immunogenicity and efficacy of *Escherichia coli* type 1 somatic pili parenteral vaccine in man. Scand. J. Infect. Dis. [Suppl.] 33, 83–95.

Lindberg, F.P., Lund, B. and Normark, S. (1984). Genes of pyelonephritogenic *E. coli* required for digalactoside-specific agglutination of human cells. EMBO J. 3, 1167–1173.

Lundberg, F.P., Lund, B., Johansson, L. and Normark, S. (1987) Localization of the receptor-binding protein at the tip of the bacterial pilus. Nature 328, 84–87.

Linder, H., Engberg, I., Jann, K. and Svanborg-Eden, C. (1988). Induction of inflammation by *Escherichia coli* on the mucosal level: requirement of adherence and endotoxin. Infect. Immun. 56, 1309–1313.

Linder, H., Engberg, I., Mattsby Baltzer, I., Jann, K. and Svanborg-Eden, C. (1988). Induction of inflammation by *Escherichia coli* on the mucosal level: requirement for adherence and endotoxin. Infect. Immun. 56, 1309–1313.

Lindstedt, R., Larson, G., Falk, P., Jodal, U., Leffler, H. and Svanborg, C. (1991). The receptor repertoire defines the host range for attaching *Escherichia coli* strains that recognize globo-A. Infect. Immun. 59, 1086–1092.

Locci, R., Peters, G. and Pulverer, G. (1981). Microbial colonization of prosthetic devices. III. Adhesion of staphylococci to lumina of intravenous catheters perfused with bacterial suspensions. Zentralbl. Bakteriol. Mikrobiol. Hyg. 172, 300–307.

Loomes, L.M., Uemura, K.-I., Childs, R.A., Paulson, J.C. and Rogers, G.N. (1984). Nature 307, 560–563.

Lund, B., Lindberg, F.P., Baga, M. and Normark, S. (1985). Globoside-specific adhesins of uropathogenic *Escherichia coli* are encoded by similar transcomplementable gene clusters. J. Bacteriol. 162, 1293–1301.

Lund, B., Lindberg, F.P., Marklund, B.I. and Normark, S. (1987). The PapG protein is the α-D-galactopyranosyl-(1–4)-β-D-galactopyranose-binding adhesin of uropathogenic *Escherichia coli*. Proc. Natl Acad. Sci. USA 84, 5898–5902.

Lund, B., Marklund, B.I., Stromberg, N., Lindberg, F., Karlsson, K.A. and Normark, S. (1988) Uropathogenic *Escherichia coli* express serologically identical pili with different receptor binding specificities. Mol. Microbiol. 2, 255–263.

Marre, R. and Hacker, J. (1987). Role of S- and common-type 1-fimbriae of *Escherichia coli* in experimental upper and lower urinary tract infection. Microb. Pathogen 2, 223–226.

Marshall, J.S. and Bienenstock, J. (1990). Mast cells. Springer Semin Immunopathol. 12, 191–202.

Maurer, L. and Orndorff, P.E. (1985). A new locus, *pilE*, required for the binding of type 1 piliated *Escherichia coli* to erythrocytes. FEMS Microbiol. Lett. 30, 59–66.

Maurer, L. and Orndorff, P.E. (1987). Identification and characterization of genes determining receptor binding and pilus length of *Escherichia coli* type 1 pili. J. Bacteriol. 169, 640–645.

McIntire, F.C., Crosby, L.K. and Vatter, A.E. (1982). Inhibitors of co-aggregation between *Actinomyces viscosus* T14V and *Streptococcus sanguis* 34: beta-galactosides, related sugars, and anionic amphipathic compounds. Infect. Immun. 36, 371–378.

McMichael, J.C. and Ou, J.T. (1979). Structure of common pili from *Escherichia coli*. J. Bacteriol. 138, 969–975.

Mergenhagen, S.E., Sandberg, A.L., Chassy, B.M., Brennan, M.J., Yeung, M.K., Donkersloot, J.A. and Cisar, J.O. (1987). Molecular basis of bacterial adhesion in the oral cavity. Rev. Infec. Dis. 9, Suppl. 5, S467–S470.

Meyer, T.F., Gibbs, C.P. and Haas, R. (1990). Variation and control of protein expression in *Neisseria*. Annu. Rev. Microbiol. 44, 451–477.

Minion, F.C., Abraham, S.N., Beachey, E.H. and Goguen, J.D. (1986). The genetic determinant of adhesive function in type 1 fimbriae of *Escherichia coli* is distinct from the gene encoding the fimbrial subunit. J. Bacteriol. 165, 1033–1036.

Miorner, H., Johansson, G. and Kronvall, G. (1983). Lipoteichoic acid is the major cell wall component responsible for surface hydrophobicity of group A streptococci. Infect. Immun. 39, 336–343.

Mirelman, D. (editor). (1986). "Microbial Lectins and Agglutinins: properties and Biological Activity", pp. 84–110. John Wiley & Sons, Inc., New York.

Moch, T., Hoschutzky, H., Hacker, J., Kroncke, K.-D. and Jann, K. (1987). Isolation and characterization of the α-sialyl-β-2,3-galactosyl-specific adhesin from fimbriated *Escherichia coli*. Proc. Natl Acad. Sci. USA 84, 3462–3466.

Nagy, L.K., MacKenzie, T. and Painter, K.R. (1985). Protection of the nursing pig against experimentally induced enteric colibacillosis by vaccination of dam with fimbrial antigens of *E. coli* (K88, K99 and 987P). The Veterinary Record 117, 408–414.

Nealon, T.J. and Mattingly, S.J. (1984). Role of cellular lipoteichoic acids in mediating adherence of serotype III strains of group B streptococci to human embryonic, fetal, and adult epithelial cells. Infect. Immun. 43, 523–530.

Neeser, J-R., Koellreutter, B and Wuersch, P. (1986). Oligomannoside-type glycopeptides inhibiting adhesion of

Escherichia coli strains mediated by type 1 pili: preparation of potent inhibitors from plant glycoproteins. Infect. Immun. 52, 428–436.

Newton, S.M.C., Jacob, C.O. and Stocker, B.A.D. (1989). Immune response to cholera toxin epitope inserted in *Salmonella* flagellin. Science 244, 70–72.

Normark, S., Lark, D., Hull, R., Norgren, M. and Baba, M. (1983). Genetics of digalactoside-binding adhesin from a uropathogenic *Escherichia coli* strain. Infect. Immun. 41, 942–949.

Normark, S., Baga, M., Goransson, M., Lindberg, F.P., Lund, B., *et al.* (1986). In "Microbial Lectins and Agglutinins: properties and Biological Activity", (ed D. Mirelman), pp. 113–143. Wiley Intersci., New York.

Ofek, I. and Beachey, E.H. (1980); In "Bacterial Adherence, Receptor and Recognition", Series B, Vol. 6 (ed E.H. Beachey), pp. 1–29. Chapman & Hall, London.

Ofek, I. and Sharon, N. (1990). Adhesins as lectins: specificity and role in infection. Curr. Top. Microbiol. Immunol. 151, 91–113.

Ofek, I., Beachey, E.H., Eyal, F. and Morrison, J.C. (1977). Postnatal development of binding of streptococci and lipoteichoic acid by oral mucosal cells of humans. J. Infect. Dis. 135, 267–274.

Ofek, I., Simpson, W.A. and Beachey, E.H. (1982). Formation of molecular complexes between a structurally defined M protein and acylated or deacylated lipoteichoic acid of *Streptococcus pyogenes*. J. Bacteriol. 149, 426–433.

O'Hanley, P., Lark, D., Normark, S., Falkow, S. and Schoolnik, G.K. (1983). Mannose-sensitive and Gal-Gal binding *Escherichia coli* pili from recombinant strains. Chemical, functional, and serologic properties. J. Exp. Med. 158, 1713–1719.

O'Henley, P., Lark, D. and Falkow, S. (1985). Molecular basis of *Escherichia coli* colonization of the upper urinary tract in BALB/c mice. J. Clin. Invest. 75, 347–360.

Orndorff, P.E. and Falkow, S. (1984). Organization and expression of genes responsible for type 1 piliation in *Escherichia coli*. J. Bacteriol. 159, 736–744.

Orndorff, P.E. and Bloch, C.A. (1990). The role of type 1 pili in the pathogenesis of *Escherichia coli* infections: a short review and some new ideas. Microb. Pathog. 9, 75–90.

Orskov, I., Orskov, F. and Birch-Andersen, A. (1980). Comparison of *Escherichia coli* fimbrial antigen F7 with type 1 fimbriae. Infect. Immun. 27, 657–666.

Parry, S.H., Abraham, S.N. and Sussman, M. (1983). In "Immunologic Aspects of Urinary Tract Infection in Children", (ed H. Schulte-Wisserman), pp. 113–126. Thieme, Stuttgart.

Parsons, C.L., Mulholland, S.G. and Anwar, H. (1979). Antibacterial activity of bladder surface mucin duplicated by exogenous glycosaminoglycan (heparin). Infect. Immun. 24, 552–557.

Parsons, C.L., Pollen, J.J., Anwar, H., *et al.* (1980). Antibacterial activity of bladder surface mucin duplicated in the rabbit bladder by exogenous glycosaminoglycan (sodium pantosanpolysulphate). Infect. Immun. 27, 876–881.

Parsons, C.L. Stauffer, C. and Schmidt, J.D. (1980). Bladder-surface glycosaminoglycans: An efficient mechanism of environmental adaptation. Science 208, 605–607.

Patel, P., Marrs, C.F., Mattick, J.S., Ruehl, W.W., Taylor, R.K. and Koomey, M. (1991 Shared antigenicity and immunogenicity of type 4 pilins expressed by *Pseudomonas aeruginosa*, *Moraxella bovis*, *Neisseria gonorrhoeae*, *Dichelobacter nodosus*, and *Vibrio cholerae*. Infect. Immun. 59, 4674–4676.

Pecha, B., Low, D. and O'Hanley, P. (1989). J. Clin. Invest. 83, 2102–2108.

Peterson, P.K., Schmeling, D., Cleary, P.P., Wilkinson, B.J., Kim, Y. and Quie, P.G. (1979). Inhibition of alternative complement pathway opsonization by group A streptococcal M protein. J. Infect. Dis. 139, 575–585.

Ponniah, S., Endres, R.O., Hasty, D.L. and Abraham, S.N. (1991). Fragmentation of *Escherichia coli* type 1 fimbriae exposes crytic D-mannose-binding sites. J. Bact. 173, 4195–4202.

Ramphal, R. and Pyle, M. (1983). Evidence for mucins and sialic acid as receptors for *Pseudomonas aeruginosa* in the lower respiratory tract. Infect. Immun. 41, 339–344.

Reid, G. and Sobel, J.D. (1987). Bacterial adherence in the pathogenesis of urinary tract infection: a review. Rev. Infect. Dis. 9, 470–487.

Reid, G., Chan, R.C.Y., Bruce, A.W. and Costerton, J.W. (1989). Prevention of urinary tract infection in rats with an indigenous *Lactobacillus casei* strain. Infect. Immun. 49, 320–324.

Reid, G., Hagberg, L., Svanborg-Eden, C., Cook, R.L., Bruce, A.W. and Costerton, J.W. (1986). In "Abstracts of the annual meeting of the American Society for Microbiology", American Society for Microbiology, Washington, D.C.

Rhen, M., Makela, P.H. and Korhonen, T.K. (1983). P-fimbriae of *Escherichia coli* are subject to phase variation. FEMS Microbiol. Lett. 19, 267–271.

Rhen, M., Tenhunen, J., Vaisanen-Rhen, U., Pere, A., Baga, M. and Korhonen, T. (1986). Fimbriation and P-antigen recognition of *Escherichia coli* strains harboring mutated recombinant plasmids encoding fimbrial adhesins of the uropathogenic *E. coli* strain KS71. J. Gen. Microbiol. 132, 71–77.

Roberts, J.A., Hardaway, K., Kaack, B., Fussel, E.H. and Baskin, G. (1984). Prevention against pyelonephritis by immunization with p-fimbriae. J. Urol. 131, 602–612.

Roberts, J.A., Kaack, B., Kallenius, G., Mollby, R., Winberg, J. and Svenson, S.B. (1984). Receptors for pyelonephritogenic *Escherichia coli* in primates. J. Urol. 131, 163–168.

Rothe, M.J., Nowak, M. and Kerdel, F.A. (1990). The mast cell in health and disease. J. Am. Acad. Dermatol. 23, 615–624.

Rugo, H.S., O'Hanley, P., Bishop, A.G., Pearce, M.K., Abrams, J.S., Howard, M. and O'Garra, A. (1992). Local cytokine production in a murine model of *Escherichi coli* pyelonephritis. J. Clin. Invest. 89, 1032–1039.

Rutter, J.M. and Jones, G.W. (1973). Protection against enteric disease caused by *Escherichia coli*. A model for vaccination with a virulence determinant. Nature 242, 531–532.

Sandberg, A.L., Mudrick, L.L., Cisar, J.O., Brennan, M.J., Mergenhagen, S.E. and Vatter, A.E. (1986). Type 2 fimbrial lectin-mediated phagocytosis of oral *Actinomyces* spp. by polymorphonuclear leukocytes. Infect. Immun. 54, 472–476.

Sandberg, A.L., Mudrick, L.L., Cisar, J.O., Metcalf, J.A. and Malech, H.L. (1988). Stimulation of superoxide and lactoferrin release from polymorphonuclear leukocytes by the type 2 fimbrial lectin *Actinomyces riscosus* T14V. Infect. Immun. 56, 267–269.

Sanford, B.A., Shelokov, A. and Ramsey, M.A. (1978). Bacterial adherence to virus-infected cells: a cell culture model of bacteria superinfection. J. Infect. Dis. 137, 176–181.

Sanford, B.A., Smith, N., Shelokov, A. and Ramsey, M.A. (1980). Adherence of influenza A viruses to group B streptococci. J. Infect. Dis. 141, 496–506.

Savage, D.C. (1972). In "Microbial Pathogenicity in Man and Animals, Symposia of the Society of General Microbiology", No. XXII, (eds. H. Smith and J.H. Pearce) pp. 25–57. Soc. General Microbiol. Cambridge University Press, London.

Schadow, K.H., Simpson, W.A. and Christensen, G.D. (1988). Characteristics of adherence to plastic tissue culture plates of coagulase-negative staphylococci exposed to subinhibitory concentrations of antimicrobial agents. J. Infect. Dis. 157, 71–77.

Scheffer, J., Konig, W., Hacker, J. and Goebel, W. (1985). Bacterial adherence and hemolysin production from *Escherichia coli* induces histamine and leukotriene release from various cells. Infect. Immun. 50, 271–278.

Scheffer, J., Vosbeck, K. and Konig, W. (1986). Induction of inflammatory mediators from human polymorphonuclear granulocytes and rat mast cells by haemolysin-positive and -negative *E. coli* strains with different adhesins. Immunology 59, 541–548.

Schifferli, D.M., Abraham, S.N. and Beachey, E.H. (1987). Use of monoclonal antibodies to probe subunit- and polymer-specific epitopes of 987P fimbriae of *Escherichia coli*. Infect. Immun. 55, 923–930.

Schmidt, G., Hacker, J., Wood, G. and Marre, R. (1989). Oral vaccination of rats with live avirulent *Salmonella* derivatives expressing adhesive fimbrial antigens of uropathogenic *Escherichia coli*. FEMS Microbiol. Immunol. 47, 229–236.

Schmidt, M.A., O'Hanley, P. and Schoolnik, G.K. (1984). Gal-Gal pyelonephritis *Escherichia coli* pili linear immunogenic and antigenic epitopes. J. Exp. Med. 161, 705–717.

Schmidt, M.A., O'Hanley, P., Lark, D. and Schoolnik, G.K. (1988). Synthetic peptides corresponding to protective epitopes of *Escherichia coli* digalactoside-binding pilin prevent infection in a murine pyelonephritis model. Proc. Natl Acad. Sci. USA 85, 1247–1251.

Schoolnick, G.K., Tai, J.Y. and Gotchlich, E. (1982). In "Bacterial Vaccines, Seminars in Infection Disease", Vol. IV, (eds L. Weinstein and B.N. Fields), pp. 172–180. Thieme-Stratton, New York.

Selinger, D.S., Julie, N., Reed, W.P. *et al.* (1978). Adherence of group A streptococci to pharyngeal cells: a role in the pathogenesis of rheumatic fever. Science 201, 455–457.

Sellwood, R., Gibbons, R.J. Jones, G.W. and Rutter, J.M. (1975). Adhesion of enteropathogenic *Escherichia coli* to pig intestinal brush borders: The existence of two pig phenotypes. J. Med. Microbiol. 8, 405–411.

Serafini Cessi, F., Dall'Olio, F. and Malagolini, N. (1984). High-mannose oligosaccharides from human Tamm-Horsfall glycoprotein. Biosci. Rep. 4, 269–274.

Shahani, K.M., Vakil, J.R. and Kilara, A. (1977). Natural antibiotic activity of *Lactobacillus acidophilus* and *bulgaricus*. II. Isolation of acidophilin from *L. acidophilus*. Cultured Dairy Products Journal 12, 8 –11.

Siegel, M., Olsen, D., Critchlow, C. and Buchanan, T.M. (1982). Gonococcal pili: safety and immunogenicity in humans and antibody function *in vitro*. J. Infect. Dis. 145, 300–310.

Silverblatt, F.J., Weinstein, R. and Rene, P. (1982). Protection against experimental pyelonephritis by antibody to pili. Scand. J. Infect. Dis. [Suppl] 33, 79–82.

Simpson, W.A. and Beachey, E.H. (1983). Adherence of group A streptococci to fibronectin on oral epithelial cells. Infect. Immun. 39, 275–279.

Simpson, W.A., Ofek, I. and Sarasohn, C. (1980). Characteristics of the binding of streptococcal lipoteichoic acid to human oral epithelial cells. J. Infect. Dis. 141, 457–462.

Simpson, W.A., Courtney, H. and Beachey, E.H. (1982). In "Microbiology", (ed D. Schlessinger), pp. 346–347. American Society for Microbiology, Washington, D.C.

Smith, A.L. (1987). Pathogenesis of *Haemoophilus influenzae* meningitis. Pediatr. Infect. Dis. J. 6, 783–786.

Stromberg, N., Deal, C., Nyberg, G., Normark, S., So, M. and Karlsson, K-A. (1988). Identification of carbohydrate structures that are possible receptors for *Neisseria gonorrhoeae*. Proc. Natl Acad. Sci. USA 85, 4902–4906.

Stromberg, N., Marklund, B.I., Lund, B., Ilver, D., Hamers, A. and Gaastra, W. (1990). Host-specificity of uropathogenic *Escherichia coli* depends on differences in binding specificity to Galα1-4Galβ-containing isoreceptors. EMBO J. 9, 2001–2010.

Stromberg, N., Hultgren, S., Russell, D.G. and Normark, S. (1992). In "Encyclopedia of Microbiology", Vol. 3, pp. 1–16. Academic Press, Inc., New York.

Sugarman, B. (1982). Adherence of bacteria to urinary catheters. Urol. Res. 10, 37–40.

Sugarman, B. (1982). *In vitro* adherence of bacteria to prosthetic vascular grafts. Infection 10, 9–14.

Sugarman, B. and Musher, D. (1981). Adherence of bacteria to suture materials (41141). Proc. Soc. Exp. Biol. Sci. 167, 156–160.

Svanborg-Eden, C., Hagberg, L., Hanson, L.A. *et al.* (1981). In "Adhesion and Microorganism Pathogenicity, CIBA Foundation Symp. 80", pp. 161–187. Pitman Medical, London.

Svanborg-Eden, C., Freter, R., Hagberg, L., Hull, R., Hull, S., Leffler, H. and Schoolnik, G. (1982). Inhibition of experimental ascending urinary tract infection by an epithelial cell-surface receptor analogue. Nature 298, 560–562.

Svanborg-Eden, C., Andersson, B., Hagberg, L. *et al.* (1983). Receptor analogues and antipilus antibodies as inhibitors of bacterial attachment *in vivo* and *in vitro*. Ann. NY Acad. Sci. 409, 580–591.

Svanborg-Eden, C., Bjursten, L-M., Hull, R., Hull, S., Magnusson, K-E., Moldovano, Z. and Leffler, H. (1984). Influence of adhesins on the interaction of *Escherichia coli* with human phagocytes. Infect. Immun. 44, 672–680.

Tacket, C.O., Hone, D.M., Curtiss, III, R., Kelly, S.M., Losonsky, G., Guers, L., Harris, A.M., Edelman, R. and Levine, M.M. (1992). Comparison of the safety and immunogenicity of ΔaroC ΔaroD and Δcya Δcrp *Salmonella typhi* strains in adult volunteers. Infect. Immun. 60, 536–541.

Tennent, J.M., Hultgren, S., Forsman, K., Goransson, M., Marklund, B.I. *et al.* (1990). In "The Bacteria: Molecular Basis of Pathogenesis" (eds B.H. Iglewski and V.C. Clark), pp. 79–110. Academic, San Diego.

Teti, G., Chiofalo, M.S., Tomasello, F., Fava, C. and Mastroeni, F. (1987). Mediation of *Staphylococcus saprophyticus* adherence to uroepithelial cells by lipoteichoic acid. Infect. Immun. 55, 839–842.

Tewari, R., Ikeda, T., Little, J.R., MacGregor, J.I., Hultgren, S.J. and Abraham, S.N. (1994). PapG fibrillum component of P-pili creates a negatively charged cloud around P-piliated *Escherichia coli* which repels human neutrophils. (Submitted).

Tuomanen, E.I. and Hendley, J.O. (1983). Adherence of *Bordetella pertussis* to human respiratory epithelial cells. J. Infect. Dis. 148, 125–130.

Tylewska, S.K., Fischetti, V.A. and Gibbons, R.J. (1988). Binding selectivity of *Streptococcus pyogenes* and M-protein to epithelial cells differs from that of lipoteichoic acid. Curr. Microbiol. 16, 209–216.

van Alpen, L., Geelen-van den Broek, L., Blaas, L., van Ham, M. and Dankert, J. (1991). Blocking of fimbria-mediated adherence of *Haemophilus influenzae* by sialyl gangliosides. Infect. Immun. 59, 4473–4477.

van Die, I., Spierings, G., van Meyen, I., Zuidweg, E., Hoekstra, W. and Bergman, H. (1985). Cloning and genetic organization of the gene cluster encoding F7, fimbriae of a uropathogenic *Escherichia coli* and comparison with $F7_2$ gene cluster. FEMS Microbiol. Lett. 28, 329–334.

van Houte, J. (1982). Bacterial adherence and dental plaque formation. Infection 10, 252–260.

Ventur, Y., Scheffer, J., Hacker, J., Goebel, W. and Konig, W. (1990). Effects of adhesins from mannose-resistant *Escherichia coli* on mediator release from human lymphocytes, monocytes, and basophils and from polymorphonuclear granulocytes. Infect. Immun. 58, 1500–1508.

Wasserman, S.I. (1989). Mast cell-mediated inflammation in asthma. Ann. Allergy 63, 546–550.

Watts, T.H., Scraba, D.G., Paranchych, W. (1982). Formation of 9-nm filaments from pilin monomers obtained by octyl-glucoside dissociation of *Pseudomonas aeruginosa* pili. J. Bacteriol. 151, 1508–1513.

Whitnack, E., Poirier, T.P. and Beachey, E.H. (1988). In "Bacteria-Host Cell Interaction", (ed M.A. Horowitz), pp. 111–119. Alan R. Liss, Inc., New York.

Williams, J.D., Robin J.-L., Lewis, R.A., Lee, T.H. and Austen, K.F. (1986). Generation of leukotrienes by human monocytes pretreated with cytochalsin B and stimulated with formyl-methionyl-leukyl-phenylalanine. J. Immunol. 136, 642–648.

Woods, D.E., Straus, D.C., Johansson, W.G. and Bass, J.A. (1981). Role of fibronectin in the prevention of adherence of *Pseudomonas aeruginosa* to buccal cells. J. Infect. Dis. 143, 784–799.

Woods, D.E., Straus, D.E., Johansson, W.G. and Bass, J.A. (1981). Role of salivary protease activity in adherence of gram-negative bacilli to mammalian buccal epithelial cells *in vivo*. J. Clin. Invest. 68, 1435–1440.

Zafriri, D., Oron, Y., Eisenstein, B.I. and Ofek, I. (1987). Growth advantage and enhanced toxicity of *Escherichia coli* adherent to tissue culture cells due to restricted diffusion of products secreted by the cells. J. Clin. Invest. 79, 1210–1216.

Zak, K., Diaz, J.L., Jackson, D. and Heckels, J.E. (1984). Antigenic variation during infection with *Neisseria gonorrhoeae*: detection of antibodies to surface proteins in sera. J. Infect. Dis. 149, 166–174.

13. Cell Adhesion Molecules in Host-Parasite Interactions

Christian F. Ockenhouse

1. Introduction

Parasitic diseases cause enormous mortality and morbidity for hundreds of millions of people in tropical and subtropical regions of the world perpetuating economic underdevelopment and rampant poverty. With increasing international travel and world-wide military operations, parasites infect not only indigenous populations but also travellers and military personnel. Despite past success with anti-parasitic chemotherapy, recent increases in infections and deaths attributed to protozoan parasites has focused attention on new and innovative means of combating the diseases caused by these organisms. The rational approach to the design of novel anti-parasitic therapies and preventive vaccines is predicated on an understanding of the molecular basis by which parasites establish infection, replicate, and provoke pathological processes.

The inception of disease by parasites requires a complex interplay between parasite surface ligands and complementary host cell surface receptors. The susceptibility to disease is influenced by both parasite and genetically-determined host factors. Many protozoan parasites have evolved elaborate and intricate life cycles in invertebrate

Adhesion Molecules
ISBN 0-12-741440-1

and vertebrate hosts, requiring interactions of multiple developmentally-regulated parasite surface ligands with assorted host cell surface receptors, and secreted extra-cellular matrix molecules which link the parasite to its cell receptor. The specificity of the molecular interaction between parasite and host is extraordinary. For example, the resistance to infection by the malaria parasite *Plasmodium vivax* is due to the absence of the Duffy blood group antigen on the surface of human red blood cells which is required for invasion by the parasite (Miller *et al.*, 1976). Invasion of erythrocytes by *P. falciparum* parasites however, does not require the Duffy blood group antigen but utilizes sialic acid and glycophorin residues on red cells to infect its human host. Genotypic selection against the Duffy antigen by a majority of Black Africans demonstrates the importance parasites have in exerting selective pressure on populations exposed to this mosquito-transmitted disease.

The fate of the organism after it enters the host may be classified into three categories: (1) the colonization of host epithelial surfaces by parasites; (2) host cellular invasion by obligate intracellular parasites; and (3) the evasion from immune selection by parasite subversion of host cell receptors (Table 13.1). Factors which influence the development of parasitic disease include the portal of entry, parasite virulence and accessibility to cells and tissues, cellular polarity, and inflammatory mediators and cytokines which regulate the expression of cell surface receptors for parasite ligands.

Prior to the development of disease in a host, many parasites must colonize epithelial surfaces of particular tissues. Parasites such as *Entamoeba histolytica*, *Giardia*, and *Trichomonas* adhere to specific epithelial surfaces in colon, small intestine, and vagina, respectively. Tissue invasion by parasites such as *Entamoeba histolytica* follows binding to epithelial surface receptors and entails the sequential adhesion and de-adhesion to cellular receptors as the parasite metastasizes to distant sites. Many protozoan organisms are obligate intracellular parasites which depend upon the host cell for intracellular development, replication, and protection from immune surveillance. The host cell required for parasite invasion may be limited to one cell type, such as the invasion of mononuclear phagocytes by *Leishmania* promastigotes, or may be non-specific as represented by the penetration of *Toxoplasma* tachyzoites or *Trypanosoma cruzi* trypomastigotes into multiple cell types both *in vitro* and *in vivo*. Parasites also use host adhesion molecules to evade immune selective pressure. Erythrocytes infected with the malaria parasite, *Plasmodium falciparum*, sequester in certain tissues and organs by adhering to capillary and post-capillary endothelial cell adhesion molecules. This phenomenon known as malaria parasite sequestration presumably protects the parasite from the phagocytic and filtering functions of the spleen as well as cell-mediated immune mechanisms.

The molecular characterization of many cell surface receptors has encouraged a network of investigators to identify and characterize host cell adhesion receptors subverted by parasites. The design of effective inhibitors of parasite adhesion and the construction of subunit vaccines which direct the immune response against

Table 13.1 Parasite adhesion to host cells

	Parasite	Parasite Stage	Host cell
I. Colonization and Tissue Metastasis			
	E. histolytica	trophozoite	colonic epithelium
	Giardia	trophozoite	small intestine epithelium
	Trichomonas	trichomonad	vaginal epithelium
	Cryptosporidium	sporozoite	intestinal epithelium
II. Host Cell Invasion			
	Plasmodium	merozoite, PRBC*	red blood cell
		sporozoite	hepatocyte
		zygote, oocyst	mosquito epithelial cell
	Babesia	merozoite	erythrocyte
	Leishmania	promastigote	mononuclear phagocyte
	Trichinella	larvae	intestinal epithelium muscle
	T. cruzi	trypomastigote	fibroblasts, macrophages muscle, myocardial cells
	Toxoplasma	trophozoites	most nucleated cells
III. Immune Evasion			
	P. falciparum	PRBC	endothelium
	Babesia	PRBC	endothelium

* PRBC, malaria parasitized red blood cell

parasite-encoded counter-receptors depends upon the knowledge of the molecular basis of parasite adhesion. This review will focus on specific examples of host-parasite interactions and immunological and pharmacological interventions which block adhesion.

2. *Host-Parasite Interactions in Human Malaria*

Malaria remains one of the major infectious diseases today in the tropics. It is caused by four species of the protozoan parasite, *Plasmodium*: *P. falciparum*, *P. vivax*, *P. malariae*, and *P. ovale*. Infection with each species of the parasite causes disease, however it is infection with *P. falciparum* which is responsible for the morbidity and mortality resulting in excess of one million deaths per year. Malaria parasites are a superb model to examine their interaction with host cell adhesion molecules. Counter-receptors on the parasite surface or encoded by the parasite and expressed on the surface of parasitized cells which bind multiple adhesion molecules in the mosquito vector and in the human host, depending upon the parasite's developmental stage.

2.1 LIFE CYCLE

Familiarity with a parasite's life cycle is essential in order to design pharmacological or immunological interventions which interrupt necessary developmental stages. Malaria sporozoites are the infective developmental stage for man. Sporozoites develop within the female *Anopheles* midgut and migrate anteriorly to its salivary glands prior to inoculation into the mammalian host during a blood meal. The sporozoites rapidly enter liver hepatocytes where they develop into exo-erythrocytic forms. The intra-hepatocytic parasites multiply and are released by the liver cell into the circulation where they bind and invade human erythrocytes. Depending upon malaria species, intra-erythrocytic asexual parasites multiply synchronously for 48–96 h to form mature trophozoites and schizonts. The parasitized red cells rupture releasing merozoites to launch another round of erythrocyte invasion and proliferation. Coincident with asexual parasite development, sexual stage gametocyte parasites are ingested by Anopheline mosquitoes, fertilize to form a zygote, and invade mosquito epithelial cells.

The design of anti-malarial treatments and preventive vaccines depends upon a critical understanding of obligatory parasite-host interactions which occur at each step of the parasite's life cycle. Host cell receptors and parasite counter-receptors important in malaria infections are summarized in Table 13.2 and discussed in greater detail below.

2.2 SPOROZOITE INVASION OF HEPATOCYTES

The developmentally-regulated circumsporozoite protein (CSP) is the major immunodominant surface protein on the sporozoite. The CS protein consists of tandem repeats of amino acids with highly conserved amino acid sequences flanking the central repeat region (Dame *et al.*, 1984). Monoclonal antibodies (mAbs) and protective polyclonal antibodies which recognize epitopes within the CS repeat region inhibit sporozoite invasion of hepatocyte and hepatoma cell lines (Hollingdale, *et al.*,

Table 13.2 Host receptor interactions with human malaria parasites and malaria-infected erythrocytes

Developmental Stage	*Host Cell*	*Receptor*	*Parasite Ligand*	*Reference*
P. falciparum				
zygote/ookinete	mosquito epithelial cell	?	Pfs25	Kaslow *et al.*, 1988
sporozoite	hepatocyte	35/55 kDa	CSP*	Hollingdale *et al.*, 1984; Aley *et al.*, 1986
merozoite	erythrocyte (invasion)	glycophorin	GBP*	Perkins, 1984; Pasvol *et al.*, 1982a,b
		sialic acid	EBA*	Sim *et al.*, 1990
trophozoite/schizont (PRBC)*	endothelium	CD36	sequestrin	Ockenhouse *et al.*, 1989, 1991
		ICAM-1	?	Berendt *et al.*, 1989
		thrombospondin	?	Roberts *et al.*, 1991
	erythrocyte (rosette formation)	CD36?	?	Kaul *et al.*, 1991
P. vivax				
merozoite	erythrocyte (invasion)	Duffy antigen (Fy^aFy^b	135 kDa	Miller *et al.*, 1976; Fang *et al.*, 1991

* CSP, circumsporozoite protein; GBP, glycophorin-binding protein; EBA, erythrocyte-binding antigen; PRBC, parasitized red blood cell

1984). These observations suggest that the CS protein may be important for parasite adhesion and invasion into liver cells. Initially, effort was directed toward the synthesis of recombinant or synthetic peptide vaccines composed entirely of the tandemly-repeated amino acid sequences within the CS protein. Sporozoites are obligate intracellular parasites and invasion into liver cells likely occurs via one or more hepatocyte surface receptors. While mAbs to the CSP repeat-domain block sporozoite invasion into hepatoma cells, this region alone is not responsible for sporozoite-hepatocyte recognition, since the amino acid repeat region vary in composition, size and number depending upon the species. Two regions of conserved amino acid sequences (regions I and II) flank the central tandem repeats suggesting that they may have important biological function. Aley *et al.* (1986) proposed that two hepatoma cell proteins bind sporozoites. A conserved ten amino acid peptide sequence in region located amino-terminal to the central repeat region co-precipitated two hepatoma cell proteins of 35 and 55 kDa.

2.2.1 Cell-Adhesive Motifs in the Sporozoite CS Protein

Region II located carboxy-terminal to the CS repeat region is also highly conserved among malarial sporozoite CS proteins and contains a novel four amino acid cell-adhesive motif, Val-Thr-Cys-Gly (VTCG). Peptides which contain the VTCG amino acid sequence bind T lymphocytic and myeloid cell lines (Rich *et al.*, 1990) and melanoma cells (Prater *et al.*, 1991). Other proteins which contain the VTCG cell adhesion motif and participate in cell-cell and cell-matrix adhesion include the type 1 repeats of thrombospondin (Lawler and Hynes, 1986), the complement regulatory protein properdin (Goundis and Reid, 1988) and two additional *P. falciparum* proteins, TRAP (thrombospondin-related anonymous protein) (Robson *et al.*, 1988) and Pfs25 (Kaslowet *et al.*, 1988) (Fig. 13.1). Rich *et al.* (1990) surmised that the CS protein region II containing the VTCG sequence is part of a recognition site for a liver cell surface receptor for sporozoite attachment and invasion. It is quite interesting that this adhesive motif is present in three *P. falciparum* proteins from the infective stage sporozoite (CSP), the asexual erythrocytic stage (TRAP) and the sexual stage (Pfs25). In fact, Kaslow *et al.* (1988) postulated that Pfs 25 on malarial ookinetes may bind to receptors on mosquito epithelial cells. Whether the VTCG adhesive motif is important in mediating sporozoite-hepatocyte recognition awaits experimental confirmation.

2.3 ERYTHROCYTE RECEPTORS FOR *P. FALCIPARUM MEROZOITES*

Merozoites released by hepatocytes into the circulation attach to and invade erythrocytes by specific receptors

Proteins with Val-Thr-Cys-Gly Cell-Adhesive Motif

	Residue no.			
CS protein	Pf 340-387	E W S P C S	V T C G	N G
	Pv 312-323	E W T P C S	V T C G	V G
	Pk 297-308	E W T P C S	V T C G	N G
Thrombospondin	425-436	P W S S C S	V T C G	D G
	482-493	P W D I C S	V T C G	G G
Properdin	58-69	L W G P C S	V T C S	E G
TRAP	249-260	E W S P C S	V T C G	K G

Figure 13.1 Protein homology to the Val-Thr-Cys-Gly cell adhesive motif. The VTCG (single amino acid designation) sequence has been implicated in mediating cell-cell and cell-matrix interactions (Rich *et al.*, 1990). The circumsporozoite protein (CSP), thrombospondin related anomalous protein (TRAP), and Pfs25 are malaria proteins synthesized by the sporozoite, asexual, and sexual stage parasites respectively.

(reviewed by Hadley *et al.*, 1986). The adaptation of *P. falciparum* malaria parasites to continuous *in vitro* cultivation advanced the study of the cellular and molecular requirements critical for merozoite binding and invasion into human erythrocytes. The invasion of *P. falciparum* merozoites into human red blood cells is specific and involves the recognition of clusters of carbohydrate and sialic acid on glycophorin A, B, and C (Pasvo *et al.*, 1982a, 1982b, 1984; Jungery *et al.*, 1983). Erythrocytes deficient in glycophorin show significant reduction in merozoite invasion (Miller *et al.*, 1977; Pasvol *et al.*, 1982b, 1984), and antibodies to glycophorin and soluble glycophorin inhibit invasion (Perkins, 1981; Jungery *et al.*, 1983). Proteolytic treatment of human erythrocytes with trypsin which cleaves glycophorin A and C, or the removal of N-acetyl neuraminic acid (sialic acid) by neuraminidase abolishes invasion of laboratory-adapted parasites (Breuer *et al.*, 1983). Although these findings provide compelling evidence that sialic acid is a receptor, or part of the receptor, on the red cell membrane, Hadley (1988) demonstrated that some *P. falciparum* merozoites invade erythrocytes lacking glycophorin A and B in a sialic acid-independent manner.

2.4 MEROZOITE COUNTER-RECEPTORS

Several merozoite surface proteins have been postulated to bind red cell glycophorin and/or sialic acid. Jungery *et al.* (1983) showed that metabolically-labelled *P. falciparum* parasite proteins of 210, 140, and 35 kDa attach to glycophorin and that the 140 and 35 kDa proteins were specifically eluded from a glycophorin column by N-acetyl-glucosamine. Likewise, Perkins (1984) described two glycophorin binding-proteins of 155 and 130 kDa. The 130 kDa protein gene was subsequently isolated, cloned, and sequenced (Ravetch *et al.*, 1985).

After the initial non-specific attachment of the merozoite to the red cell, the parasite reorients itself so that the apical end of the parasite lies in direct contact with the erythrocyte surface. The apical region of merozoite contains organelles (micronemes and rhoptry) which release soluble parasite-derived proteins. The erythrocyte binding-antigen (EBA) described by Camus and Hadley (1988) was originally isolated from the culture medium supernatants of rupturing late-stage schizont stage *P. falciparum* parasites. EBA secreted from the apical end of the parasite binds to sialic acid residues on glycophorin O-linked tetrasaccharides of human and murine erythrocytes but not to red cells treated with neuraminidase. The gene encoding EBA has recently been cloned and sequenced, and codes for 1435 amino acids (Sim *et al.*, 1990). The predicted size of the protein based upon its primary structure is consistent with the size of authentic culture medium-derived EBA which migrates in SDS-PAGE as a 175 kDa non-glycosylated protein. Polyclonal antibodies to a synthetic peptide encompassing residues 1062–1103 inhibit merozoite invasion by 80% and block binding of authentic EBA to red cells. This peptide sequence is conserved in laboratory-adapted and wild-type *P. falciparum* strains is a good candidate for inclusion into a multivalent malaria vaccine.

2.5 ERYTHROCYTE RECEPTORS AND COUNTER-RECEPTORS FOR *P: VIVAX*

Though *P. vivax* has not been adapted to continuous culture *in vitro*, the molecular analysis of parasite invasion into red cells has relied on the *in vivo* adaptation of *P. vivax* parasites in *Saimiri* monkeys, and by analogy, the study of invasion of erythrocytes by the closely related monkey parasite, *P. knowlesi*. Miller *et al.* (1976) first demonstrated that individuals with a genotype found predominantly in African and American Blacks who are Duffy blood group negative (*FyFy*) are resistant to invasion by *P. vivax*. Human erythrocytes which lack the blood groups antigens Fy^a and Fy^b are resistant to invasion by *P. vivax* and *P. knowlesi* merozoites. The Fy^a antigen comprises an epitope on an erythrocyte membrane protein migrating in SDS-PAGE at 35–43 kDa (Hadley *et al.*, 1984).

The *P. vivax* merozoite binding ligand for the Duffy blood group determinant has recently been identified (Fang *et al.*, 1991). This protein is stored within the microneme organelles at the apical end of the parasite and facilitates parasite invasion into erythrocytes bearing the Duffy blood group antigen.

2.6 CYTO-ADHERENCE OF *P. FALCIPARUM*-INFECTED RED BLOOD CELLS

The asexual erythrocytic stage of *P. falciparum* causes the symptoms and pathology of malaria. Ring stage intra-erythrocytic parasites and sexual stage gametocytes circulate unobstructed in the vasculature, and detection of these circulating parasites by blood smear provides the basis for the diagnosis of infection. The morbidity and mortality resulting from infection with *P. falciparum* occurs as a result of the obstruction of capillaries and venules by erythrocytes parasitized (PRBC) with the mature trophozoite and schizont stage of the parasite (Fig. 13.2). Sequestration of parasitized red blood cells

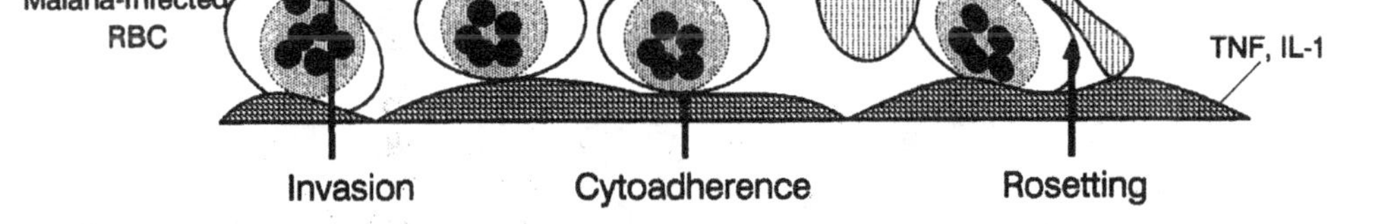

Figure 13.2 Sequestration by red cells infected with *Plasmodium falciparum*. Three adhesion pathways used by malaria parasites include red cell invasion by merozoite parasites, cyto-adherence of trophozoite/schizont intra-erythrocytic parasites to endothelium, and rosetting of uninfected red cells to parasitized erythrocytes. Cyto-adherence and rosette formation contribute to the microvascular obstruction by malaria-infected erythrocytes in the brain resulting in cerebral malaria. Adhesion of parasitized red cells to endothelium is regulated by circulating inflammatory cytokines such as TNF-α and interleukin-1.

(PRBC) within the brain leads to the most fatal form of the disease, cerebral malaria. Hence, the molecular basis of parasite sequestration has been the subject of intensive investigation. Two separate adhesion mechanisms contribute to vascular obstruction by PRBC. First, PRBC regardless of strain or geographic origin bind to human vascular endothelial cell receptors and secondly, aggregates of uninfected red cells bind to an infected erythrocyte forming rosettes. Sequestration impedes cerebral blood flow causing headache, altered mental consciousness, coma, and ultimately death associated with cerebral malaria.

Of the four human malaria species, only *P. falciparum* has developed a mechanism to avoid splenic clearance of infected erythrocytes and to maximize parasite survival. Several factors such as tissue preference, disease severity, elaboration of cytokines, and receptor expression contribute to the adhesion of PRBC to endothelium. Three adhesion molecules CD36, ICAM-1, and thrombospondin bind PRBC *in vitro*, and evidence to support their role in the pathogenesis of malaria sequestration is presented below.

2.6.1 CD36 as a Cyto-adherence Receptor

Listed below are several criteria to validate a putative endothelial cell sequestration receptor for erythrocytes infected with *P. falciparum*.

1. Anti-receptor antibodies should block PRBC adhesion
2. Lack of PRBC binding to cells not expressing the receptor
3. Correlation of receptor defects with abnormal binding function
4. A solid substrate coated with purified receptor supports PRBC adhesion
5. Purified receptor competes with membrane-bound receptor for PRBC adhesion
6. Binding of PRBC isolates from naturally-acquired malaria infections to receptor
7. Expression of receptor on cerebral microvasculature *in vivo*
8. Reversal of adhesion *in vivo* with purified receptor or receptor-like analogues

Initially, we had observed that the adhesion of PRBC to human monocytes (Fig. 13.3) was similar to that described for PRBC adhesion to human umbilical vein endothelial cells and C32 melanoma cells *in vitro*. MAbs OKM5 and OKM8 raised against human monocytes immunoprecipitate an 88 kDa trypsin-resistant glycoprotein identified as CD36 (Talle *et al.*, 1983). CD36 cross-reacts with microvascular endothelial cells, C32 melanoma cells and platelets. Anti-CD36 antibodies block PRBC binding by ⩾95% to cells expressing the CD36 antigen (Barnwell *et al.*, 1985; Ockenhouse *et al.*, 1989a). The surface expression of CD36 on melanoma cell lines correlates with the ability of these cells to bind PRBC (Panton *et al.*, 1987). Differentiated U937 myelomonocytic cells but not undifferentiated cells bind PRBC and acquire the CD36 surface antigen when incubated with phorbol esters (Ockenhouse and Chulay, 1988). Finally, human platelets and monocytes of the Nak^a-negative phenotype which lack detectable CD36 fail to bind malaria-infected erythrocytes (Tandon *et al.*, 1992).

CD36, an integral membrane glycoprotein of 471 amino acids, 10 N-linked glycosylation sites, and two membrane spanning regions (Oquendo *et al.*, 1989) serves as a natural receptor for the extracellular matrix molecules collagen (Tandon *et al.*, 1989) and thrombospondin (Asch *et al.*, 1987). Cells transfected with CD36 cDNA bind erythrocytes infected with *P. falciparum* (Berendt *et al.*, 1989; Oquendo *et al.*, 1989). CD36 shares extensive homology to a rat lysosomal integral membrane protein, LIMPII which may comprise

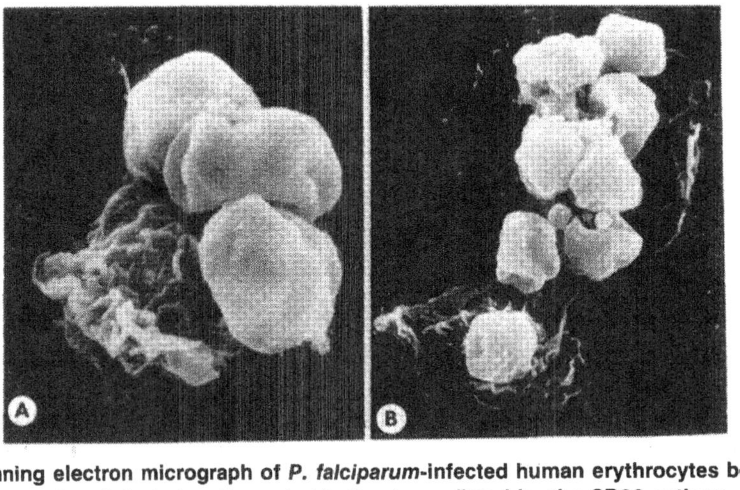

Figure 13.3 Scanning electron micrograph of *P. falciparum*-infected human erythrocytes bound to human monocytes. Infected erythrocyte binding is mediated by the CD36 antigen.

a new superfamily of proteins with unique adhesive properties (Vega *et al.*, 1992).

A solid substrate coated with CD36 purified from human platelets (platelet GPIV) supports the binding of human or monkey red blood cells infected with *P. falciparum* parasites but does not support adhesion of uninfected erythrocytes nor erythrocytes infected with immature ring stage parasites (Ockenhouse *et al.*, 1989a). Purified ^{125}I-labelled CD36 binding to PRBC is inhibited by excess unlabelled CD36 and anti-CD36 mAbs. All naturally-acquired *P. falciparum* isolates examined to date retain the ability to bind CD36, although binding does not vary with clinical severity of disease (Hasler *et al.*, 1990; Ockenhouse *et al.*, 1991a). The CD36 receptor is expressed on the cerebral microvasculature of brain tissue from malaria-infected and uninfected accident victims (Aikawa *et al.*, 1990). Six of the seven proposed criteria to validate an *in vivo* role for an endothelial receptor are met by CD36 and provide compelling evidence that CD36 is the principal receptor *in vivo* to which PRBC bind.

In addition to its role as a cell surface receptor, CD36 transduces cellular activation signals upon ligation with appropriate natural or artificial ligands. MAbs, collagen, and *P. falciparum*-infected erythrocytes have been shown to activate platelets and monocytes through CD36 (Ockenhouse *et al.*, 1989b). The mechanism for cellular activation is not precisely known, but *src*-related protein tyrosine kinases Fyn, Lyn, and Yes are coprecipitated with CD36 (Huang *et al.*, 1991). It is tempting to speculate that the activation of monocytes, platelets, and endothelial cells by CD36-dependent binding of *P. falciparum*-infected erythrocytes causes not only mechanical obstruction of post-capillary blood flow but contributes to host pathology by the release of inflammatory cytokines and destructive vasoactive mediators.

We have recently identified a *P. falciparum* PRBC counter-receptor, sequestrin, which binds to CD36 (Ockenhouse *et al.*, 1991b). Monospecific polyclonal anti-idiotypic antibodies raised against the antigen-binding site of an inhibitory mAb OKM8, contain the internal image of the receptor (Ab2) and recognize a complementary CD36 binding site on the surface of PRBC. Anti-idiotype antibodies react with the surface of PRBC by indirect immunofluorescence (Color Plate II), inhibit the binding of PRBC to membrane-bound CD36, and immunoprecipitate a 270 kDa ^{125}I-surface-labelled and H^3-isoleucine metabolically-labelled parasite protein which can be pre-cleared with CD36-Sepharose. These results indicate that a conserved and antigenically-invariant parasite-derived surface protein and not an altered erythrocyte membrane protein binds CD36. Sequestrin is a candidate malaria vaccine antigen and research is now focusing on the molecular cloning of this functional parasite protein.

2.6.2 ICAM-1 as a Cyto-adherence Receptor

Large vein endothelial cells such as those derived from human umbilical veins express low levels of CD36 compared to microvascular endothelium. In an attempt to identify alternative endothelial cell receptors, Berendt *et al.* (1989) selected parasitized erythrocytes which bind umbilical vein endothelial cells. Cells transfected with cDNA of intercellular adhesion molecule-1 (ICAM-1) bind PRBC from some laboratory-adapted strains of *P. falciparum*. PRBC binding correlates with the expression of ICAM-1 on tumor necrosis factor-α (TNF-α)-stimulated endothelial cells and is inhibited by anti-ICAM-1 mAbs (Berendt *et al.*, 1989). Tissue culture plates coated with purified ICAM-1 bind laboratory-adapted PRBC and wild-type isolates from patients with both uncomplicated and cerebral malaria (Ockenhouse *et al.*, 1991a).

ICAM-1, a member of the immunoglobulin supergene family of proteins, consists of five extracellular domains, a transmembrane domain, and a cytoplasmic domain (Simmons *et al.*, 1988; Staunton *et al.*, 1988). ICAM-1 is an endothelial cell adhesion receptor regulated by inflammatory cytokines that is critical for adhesive functions in the immune system. The leucocyte integrins LFA-1 and Mac-1 bind to sites located in the first and third NH_2-terminal domains of ICAM-1 (Staunton *et al.*, 1990; Diamond *et al.*, 1991). The human major type rhinovirus subverts ICAM-1 by binding to specific residues within the first NH_2-terminal domain of ICAM-1 (Greve *et al.*, 1989; Staunton *et al.*, 1990; McClelland *et al*, 1991). Using domain deletion mutations and human-mouse chimeric ICAM-1 molecules we have shown that *P. falciparum*-infected red cells bind to the first amino-terminal Ig-like domain (Fig. 13.4) (Ockenhouse *et al.*, 1992). Characterization of single and multiple amino acid substitution mutants revealed that the binding sites for PRBC are distinct from LFA-1 and human rhinovirus (Fig. 13.5). Since cells transfected with mouse ICAM-1 do not bind PRBC, we generated single amino acid mutants which substitute mouse for human residues. A double amino acid ($Gly^{15}Ser^{16}$ to $Ser^{15}Ala^{16}$) and a single amino acid substitution (Leu^{18} to Gln^{18}) abolished PRBC-binding to transfected cells without disrupting overall conformation or affecting LFA-1 or rhinovirus binding (Fig. 13.5). Overlapping synthetic hexapeptides spanning the critical contact site at Leu^{18} inhibited PRBC adhesion to ICAM-1-coated tissue culture plates while hexapeptides flanking this region had no effect (Ockenhouse *et al.*, 1992). The inhibition of PRBC binding to ICAM-1 by peptides was approximately three orders of magnitude less than that observed using soluble recombinant ICAM-1. Amino acid substitutions which disrupt the overall structure of the ICAM-1 molecule abolish PRBC binding to transfected cells indicating that the conformation of the sequence encompassing the binding site may be relevant.

ICAM-1 expression on endothelium *in vitro* and *in vivo* is strongly up-regulated by TNFα and interleukin-1. Interestingly in several clinical studies, levels of plasma

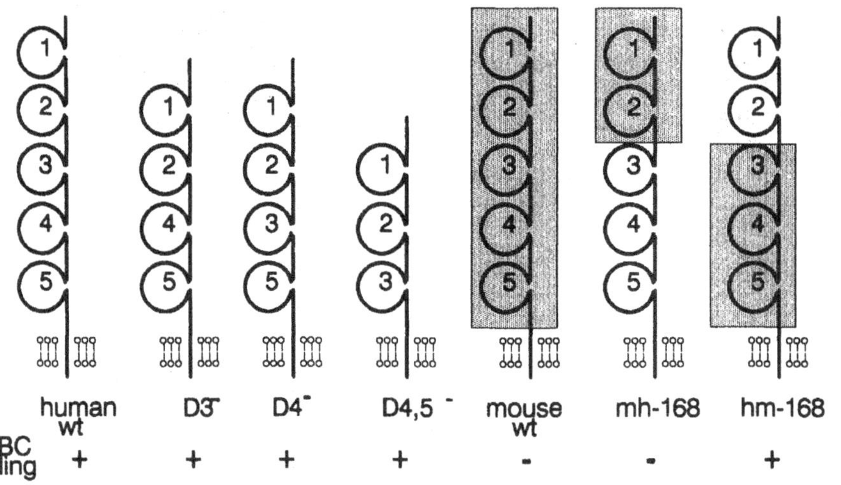

Figure 13.4 ***P. falciparum*-infected erythrocyte adhesion to transfected cells expressing domain deletion or human-mouse chimeric ICAM-1 molecules. PRBC bind human wild type (wt) but not murine ICAM-1 (murine ICAM-1 domains are shaded). Human-mouse chimeric ICAM-1 exchanges at residue 168 within domain 2 localize the PRBC binding sites within the first two amino-terminal domains. PRBC, parasitized red blood cell.**

TNFα concentration among children with severe and complicated *P. falciparum* malaria were significantly higher than in comparably ill aparasitemic children (Grau *et al.*, 1989; Shaffer *et al.*, 1991). While TNFα has no effect on parasite multiplication *in vitro*, TNFα may increase the density of ICAM-1 receptors in the cerebral vasculature, increasing the affintiy of PRBC for the vessel wall.

2.6.3 Thrombospondin as a Cyto-adherence Receptor

Thrombospondin (thrombin-sensitive glycoprotein, TSP) is a 420 kDa molecule composed of three identical disulfide linked chains. Thrombospondin is an extracellular matrix glycoprotein synthesized by a wide variety of cells and has been localized to the subendothelial region of the vessel wall. TSP appears to mediate essential roles in hemostasis and tumor cell adhesion. Roberts and coworkers (1985) showed that *P. falciparum* infected erythrocytes bind to tissue culture plates coated with Ca^{2+}-replete TSP but not to plates coated with fibronectin, laminin, or vitronectin. Polyclonal anti-TSP antibodies and purified TSP inhibited PRBC binding to TSP-coated plates, to melanoma cells (Roberts *et al.*, 1985), and to endothelium in an *ex vivo* rat meso-appendix model of parasitized erythrocyte sequestration (Rock *et al.*, 1988). Wild-type Falciparum malaria PRBC bind to TSP and correlates with PRBC binding to CD36 (Sherwood *et al.*, 1987; Hauler *et al.*, 1990). The correlation observed between PRBC adhesion to CD36 and thrombospondin, and experimental data suggesting that CD36 may be a cell surface receptor for TSP (Asch *et al.*, 1987) has provoked speculation that PRBC bind to CD36 via a TSP molecular bridge.

Several studies do not support the hypothesis that TSP alone is a PRBC receptor *in vivo*. In the absence of any detectable TSP, PRBC bind directly to CD36 (Barnwell *et al.*, 1989). Although purified TSP and anti-TSP antiserum block IRBC binding to TSP adsorbed to a solid substrate, no inhibition was observed to cells (Ockenhouse and Chulay, 1988). PRBC cyto-adherence to several cell lines correlates with CD36 but not TSP expression (Panton *et al.*, 1987; Sherwood *et al.*, 1989). TSP is not required as a molecular link between infected red cells and CD36, since cells transfected with cDNA encoding CD36 do not bind thrombospondin but readily bind large numbers of PRBC (Oquendo *et al.*, 1989). Lastly, Nak[a]-platelets defective in CD36

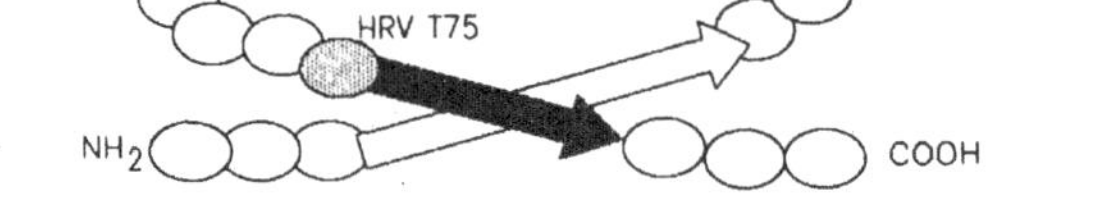

Figure 13.5 Model of tertiary structure of ICAM-1 domain 1 with locations of mutations affecting *P. falciparum*-infected erythrocyte binding (closed circles). The predicted β strands (arrows) and β turns (circles) were adapted from Staunton *et al.*, 1990. Mutations within the first turn and extending into β strand B, abolish PRBC binding and are distinct from mutations which affect binding of human rhinovirus (Ockenhouse *et al.*, 1992).

expression bind TSP to the same degree as normal platelets but do not bind PRBC (Tandon *et al.*, 1992).

2.6.4 Malaria Rosette Formation

Rosette formation in which aggregates of 10–20 uninfected red cells bind to an erythrocyte infected with the mature stage of the parasite is another cyto-adherent property. Similar to other adhesion pathways, rosette formation is divalent cation-dependent and heparin-sensitive (Udomsangpetch *et al.*, 1989; Carlson *et al.*, (1990a, 1990b). Heparin and antibodies to the histidine-rich malaria protein, PfHRP1, disrupt rosette formation (Carlson *et al.*, 1990a). Studies which have associated rosette formation with cerebral malaria (Carlson *et al.*, 1990b) have not been confirmed (Hasler *et al.*, 1990; Ho *et al.*, 1991). Under conditions of flow using a rat mesocecum *ex vivo* preparation, a higher peripheral resistance in venules was attributed to rosette-forming PRBC, but not to PRBC which do not spontaneously form rosettes (Kaul *et al.*, 1991). Microvascular blockage was abolished in venules when PRBC were incubated with the CD36 mAb OKM8 but not with the cross-reactive mAb OKM5. The contribution of CD36 to rosette formation is unclear but PRBC could bind to uninfected immature erythrocytes expressing low levels of surface CD36 which disappear as erythrocytes age (Ockenhouse, unpublished observations). Children infected with *P. falciparum* die not only from cerebral malaria but also from severe anemia resulting from red cell destruction and ineffective erythropoiesis. The release of large numbers of marrow-derived reticulocytes expressing CD36 into the peripheral circulation may augment rosette formation. The characterization of an additional heparin-sensitive uninfected red cell receptor for PRBC awaits further study.

Malaria parasite sequestration is much more complex than originally envisioned. The multiplicity of cell adhesion molecules which bind PRBC *in vitro* has raised several questions regarding the *in vivo* relevance of endothelial receptors and the severity of infection. Why is cerebral malaria observed in only 1–2% of children infected with the parasite although parasite sequestration occurs in all infections? Does the expression of several endothelial receptors concurrently on the cerebral vasculature direct PRBC migration to where adhesion is the strongest? Does rosetting of uninfected erythrocytes around a parasitized red cell account for the difference in uncomplicated versus severe malaria infections? Finally, do subpopulations of naturally-occuring PRBC which bind one or multiple cell adhesion molecules account for differences in clinical outcomes?

3. *Host-parasite Interactions in Human Leishmaniasis*

Human leishmaniasis resulting from infection with various species of the obligate intracellular protozoan parasite, *Leishmania*, causes a spectrum of clinical entities ranging from a self-healing superficial skin ulcer to a fatal visceral infection. The infective metacyclic promastigote stage parasites which are transmitted to humans by the bite of a sandfly, attaches to and invades cells of the host mononuclear phagocyte system. The transformation of the promastigote into the intracellular amastigote stage protects the parasite from macrophage oxidative-dependent killing. Several monocyte/macrophage cell surface receptors are subverted by *Leishmania* promastigotes (Table 13.3), and the predilection for one specific receptor depends upon the following factors: (1) Are infective or non-infective promastigotes used in invasion assays?; (2) Does the attachment to and invasion of macrophages by promastigotes occur in the absence or

Table 13.3 ***Leishmania*** **- Mononuclear phagocyte interactions**

Macrophage receptor	*Leishmania ligand*	*Serum requirement (C′ component)*	*Reference*
Mannosyl/Fucosyl Receptor (MFR)	LPG*	serum-independent	Channon *et al.*, 1984; Wilson and Pearson, 1986 Handman and Goding, 1985
CD18 (CR3 and p150,95)	LPG	serum-independent	Talamas-Rohana *et al.*, 1990
CR3	LPG	serum-dependent (C3)	Puentes *et al.*, 1988
CR3	gp63	serum-independent	Russell and Wilhelm, 1986; Wozencraft and Blackwell, 1987
CR3	gp63	serum-dependent (iC3b)	Blackwell *et al.*, 1985; Russell, 1987
CR1	gp63	serum-dependent (C3b)	da Silva *et al.*, 1989
Fibronectin	gp63	serum-dependent	Wyler *et al.*, 1985; Rizui *et al.*, 1988
Advanced glycosylation end product	?	serum-independent	Mosser *et al.*, 1987

* LPG, lipophosphoglycan

presence of serum?; (3) Which *Leishmania* species is used to investigate parasite-host interactions? The binding of leishmanial promastigotes to macrophages probably consist of concurrent receptor interactions which modulate the effects of each other.

3.1 MANNOSYL/FUCOSYL RECEPTOR (MFR)

The MFR receptor recognizes ligands containing D-mannose, L-fucose, or N-acetyl-D-glucosamine terminal residues of glycoconjugates. Such residues are present on the surface of *Leishmania* promastigotes (Dwyer and Gottlieb, 1983; Turco *et al.*, 1984). Inhibitors of MFR block the attachment and ingestion by murine resident peritoneal macrophages by *L. donovani* promastigotes (Channon *et al.*, 1984). Yeast mannan, a polysaccharide with high mannose linkages in the terminal sugars inhibits in a dose-dependent manner promastigote attachment to macrophages (Channon *et al.*, 1984), and depletion of membrane MFR by plating macrophages on surfaces coated with mannan abolishes promastigote attachment and ingestion (Blackwell *et al.*, 1985). Antibodies to lipophosphoglycan (LPG) block attachment of *Leishmania major* promastigotes to macrophages. LPG is a unique lipid-containing glycoconjugate consisting of repeating phosphorylated disaccharide units linked together via a carbohydrate core to a novel lysoalkyl-phosphatidyl-inositol lipid anchor. It has been suggested that LPG binds to the macrophage lectin-like binding sites (Handman and Goding, 1985).

3.2 PROMASTIGOTE BINDING TO CR3

Leishmania promastigotes incubated in normal human serum fix complement leading to high affinity interactions with the host cell surface type 3 (CR3) and type 1 (CR1) complement receptors. Alternatively, promastigotes incubated in the absence of normal human serum bind to complement components synthesized and expressed on the macrophage surface (Wozencraft and Blackwell, 1987). Invasion of macrophages by CR3 or CR1 depends upon the infectivity of the promastigote, the *Leishmania* species, and the inclusion of normal human serum in *in vitro* assays. Both serum and non-serum-incubated *L. donovani* promastigotes bind to and invade macrophages through CR3 (CD18) (Blackwell *et al.*, 1985; Mosser and Edelson, 1985). CR3 expresses binding sites for the ligands iC3b and bacterial LPS. Anti-CR3 mAbs M1/70 and OKM1 inhibit promastigote-macrophage binding, while mAb OKM10, which inhibits iC3b binding to the receptors has no effect on promastigote attachment (Blackwell *et al.*, 1985; da Silva *et al.*, 1989). This suggests that promastigote binding to CR3 is not via the iC3b binding site but to a lectin-binding site recognized by mAb OKM1. By contrast, Wilson and Pearson (1988) demonstrated that CR3 on human monocytes facilitates binding and ingestion of *L. major* promastigotes coated with iC3b. The inconsistency of these results is not easily resolved unless separate promastigote surface molecules bind two distinct CR3 binding domains.

3.3 LIPOPHOSPHOGLYCAN AND GP63 COUNTER-RECEPTORS FOR CR3

The lipophosphoglycan moiety on the surface of Leishmania promastigotes binds directly to CR3 (Talamas-Rohana *et al.*, 1990) or indirectly via complement components deposited on the promastigote surface (Puentes *et al.*, 1988). LPG reconstituted onto the surface of hydrophobic beads binds macrophages expressing CR3 but not to macrophages from individuals genetically-deficient in the leucocyte integrin complex CD18 which includes CR3, p150,95, and LFA-1 (Talamas-Rohana *et al.*, 1990). Competitive binding and antibody-inhibition assays have established that the LPS-binding site on CR3 is subverted by promastigote surface LPG.

The major promastigote surface protein, gp63 binds macrophages directly and is the major C3-acceptor site (Russell, 1987; Russell and Wilhelm, 1986). Russell (1987) showed that ^{125}I-labelled C3 binds liposomes containing gp63 and factors B and D, and postulated that the binding of complement to promastigote surface occurs by the alternative pathway. Alternatively, purified gp63 reconstituted onto beads binds directly to macrophage CR3 in the absence of complement at the Arg-Gly-Asp (RGD) binding site (Russell and Wright, 1988). Synthetic peptides which include the RGD sequence and antibodies which block the RGD recognition domain inhibit gp63 binding to CR3. The amino acid sequence for gp63 does not include the RGD cell adhesion motif, nevertheless, the CR3 binding site for *Leishmania* gp63 probably lies close to or coincident with the RGD binding site.

3.4 PROMASTIGOTE BINDING TO CR1

Promastigotes in the stationary phase of growth differ from parasites in the log phase in their infectivity of mice *in vivo* (Sacks *et al.*, 1985). The metacyclic promastigote stage transmitted by the sandfly into the host's circulation is infective for man. Da Silva and colleagues (1989) observed that infective promastigotes incubated in normal human serum invade macrophages by the type 1 (CR1) complement receptor. Complement deposited on the infective metacyclic parasite's surface is cleaved to C3b. C3b-coated promastigote binding to CR1 is inhibited by anti-CR1 antibodies. Blocking CR3 with anti-CR3 antibodies has no effect on infective stage attachment but does block binding of non-infective promastigotes (da Silva *et al.*, 1989). Unlike infective promastigotes, non-infective parasites bind both CR1 and CR3 independently of serum.

3.5 PROMASTIGOTE BINDING TO FIBRONECTIN

Fibronectin, the high molecular weight extracellular matrix glycoprotein may mediate the invasion of parasites into macrophages. Anti-fibronectin antibodies immuno-precipitate gp63 from surface-labelled and metabolically-labelled promastigote parasites and inhibit parasite attachment to cells (Wyler *et al.*, 1985; Rizui *et al.*, 1988). Antibodies to a fibronectin peptide fragment containing the RGDS amino acid sequence in addition to several anti-RGDS antibodies recognize gp63. One possible interpretation of these findings is that fibronectin-coated promastigote parasites bind directly to the RGDS binding site on CR3.

4. *Host-parasite Interactions in American Trypanosomiasis*

Chagas disease or American trypanosomiasis results from infection with the obligate intracellular parasite, *Trypanosma cruzi*. Transmission of the infective metacyclic trypomastigote stage parasites by reduvid bugs to human hosts occurs by parasite penetration through openings in the skin. *T. cruzi* trypomastigotes invade a variety of host cell types including macrophages, fibroblasts, and muscle cells. Invasion of cells by trypomastigotes is influenced by cellular polarity (Schenkman *et al.*, 1988). Confluent Madin-Darby canine kidney epithelial cells resist parasite invasion while non-confluent cells are readily invaded. The accessibility of parasites to the basolateral domain is critical for parasite invasion, whereas exposure to the cell's apical surface in confluent cell monolayers has no effect on parasite internalization.

4.1 FIBRONECTIN AND FC RECEPTORS FOR *T. CRUZI*

Adhesion and internalization of trypomastigotes by phagocytic and non-phagocytic means is enhanced by incubation of parasites with normal serum. ^{125}I-fibronectin binding to *T. cruzi* trypomastigotes is specific, saturable, and reversible with excess unlabelled fibronectin (Ouaissi *et al.*, 1984, 1986). Adhesion to cells by *T. cruzi* trypomastigotes is inhibited by N-acetyl-glucosamine which binds a lectin-like parasite surface protein with specificity for N-acetyl-glucosamine (Crane and Dvorak, 1982). Fibronectin contains a N-acetyl-lactosamine type carbohydrate moiety which comprises N-acetyl-glucosamine residues. Peptides containing the RGD cell adhesion-motif block parasite invasion of cells, implicating fibronectin in parasite recognition.

Fc receptors on the surface of *T. cruzi* trypomastigotes increase infectivity for rat heart myoblasts (Rodriguez de Cuna *et al.*, 1991). Incubation of parasites with IgG1 or the addition of purified Fc receptors *in vitro* results in increased parasite infectivity. Saturable ^{125}I-labelled Fc binding to *T. cruzi* is inhibited by unlabelled Fc (Rodriguez de Cuna *et al.*, 1991).

4.2 C1q RECEPTORS FOR *T. CRUZI*

MAbs to type 1 and type 3 complement receptors (CR1, CR3) do not inhibit cellular attachment and invasion of trypomastigotes (Joiner *et al.*, 1985). Serum-incubated trypomastigotes produce an inhibitor of the complement cascade, preventing C3 but not C1q deposition on the parasite surface. In contrast, non-infective *T. cruzi* epimastigotes fix large numbers of C3 molecules on their surface and bind to cell surface complement receptors by a 72 kDa parasite surface protein (Joiner *et al.*, 1985). Rimoldi and colleagues (1988, 1989) speculated that infective trypomastigotes incubated in human serum bind to C1q receptors by C1q deposition on the parasite surface. Heat-inactivated C1q, or removal of C1q, impedes parasite adhesion to human foreskin fibroblasts, while the addition of purified C1q back to C1q-depleted serum restores attachment and internalization.

5. *Host-parasite Interactions in Human Amoebiasis*

The protozoan parasite, *Entamoeba histolytica*, responsible for human amebiasis colonizes and invades human colonic epithelium, causing amoebic colitis with symptoms ranging from watery diarrhea to fulminant bloody dysentery. Trophozoite parasites, found at the borders of liver abscesses, may occasionally migrate to more distant sites such as the lung, pericardium, and nervous system. The pathogenesis of the disease is incompletely understood, but involves parasite virulence factors, mucosal barriers, host immune defences, and cell surface receptors which influence parasite adhesion and invasion.

5.1 POLYSACCHARIDE ADHESION RECEPTORS

Colonization of colonic epithelium by trophozoite parasites requires a galactose or N-acetyl-galactosamine inhibitable lectin-like interaction between parasite and host. Simple sugars such as galactose (Gal) and N-acetyl-D-galactosamine (GalNAc) or asialoglycoproteins with multiple terminal Gal residues inhibit amebic adhesion to Chinese hamster ovary (CHO) cells, whereas mannose, sialic acid, and N-acetyl-glucosamine had no effect

(Ravdin and Guerrant, 1981). Trophozoite binding to cells is inhibited when lactosamine-containing N-linked carbohydrates, or Gal and GalNAc residues, on N- and O-linked carbohydrates are lacking, whereas binding is increased on mutant cells expressing terminal Gal residues or neuraminidase-treated cells lacking sialic acid residues (Ravdin *et al.*, 1985, 1989). Human colonic mucins bind to a Gal-inhibitable adherence lectin and inhibit binding to CHO cells (Chadee *et al.*, 1987).

5.2 TROPHOZOITE LECTIN-LIKE COUNTER-RECEPTORS

The trophozoite surface lectin implicated in binding to intestinal epithelium is a heterodimeric molecule of 170 and 35 kDa linked by disulfide bonds (Petri *et al.*, 1987, 1989). Purified lectin competitively inhibits parasite adhesion to cells *in vitro* (Petri *et al.*, 1987). MAbs to the 170 kDa subunit inhibit and stimulate trophozoite adhesion, providing a model to study the intricate mechanisms implicated in parasite adhesion and de-adhesion as the organism invades intestinal epithelium and metastasizes to distant organs (Petri *et al.*, 1990). MAbs which inhibit parasite adhesion to red blood cells and epithelial cells immunoprecipitate a 112 kDa amoeba surface protein from invasive but not adhesion-deficient parasite mutants (Arroyo and Orozco, 1987). The binding and ligation of a specific cell surface receptor by one or more trophozite complementary ligands triggers the release of a membrane ion-channel, forming protein which induces host cells to undergo cytolysis, facilitating parasite invasion.

6. *Conclusions*

The characterization of parasite binding sites on adhesion molecules has stimulated investigation into pharmacological and immunological inhibitors of parasite-host adhesion pathways. Research is currently directed towards the development of soluble receptor-like analogues which compete with the membrane-bound receptor(s) for parasite adhesion. Recombinant soluble ICAM-1 is an effective inhibitor of malaria parasitized erythrocyte binding to both purified ICAM-1 and cytokine-activated endothelium (Ockenhouse *et al.*, 1992). Receptor-based synthetic peptides and anti-idiotypic antibodies which molecularly mimic the parasite-binding domain of the receptor, block binding of parasites to purified adhesion molecules. Although these approaches would inhibit or reverse adhesion *in vivo*, killing of the parasite is dependent upon innate immune mechanisms to control parasitemia. The construction of recombinant immuno-adhesins is a promising strategy designed to be therapeutically effective through at least two mechanisms (Staunton *et al.*, 1992). First, a chimeric immuno-adhesin constructed by the molecular fusion of a receptor-like domain to the constant domains of the human IgG1 heavy chain antibody molecule, would bind the parasite surface and block adhesion to its membrane-bound receptor. Secondly, the immuno-adhesin would sensitize the parasite for immune-mediated elimination by phagocytosis, complement-mediated lysis, or antibody-dependent cell-mediated killing (ADCC) (Byrn *et al.*, 1990).

The molecular characterization of several cell surface receptors involved in cell-cell and cell-matrix binding has greatly influenced our understanding of the key mechanisms by which parasites initiate infectious disease. The assortment of functionally and immunologically distinct receptors subverted by parasites illustrates that adhesion is of such significance as to require redundant and diverse receptor-mediated pathways. Finally, the highly co-ordinated actions between parasite-encoded surface ligands and multiple surface adhesion receptors re-emphasizes the critical importance adhesion molecules exert on parasite survival, and engenders the hope that effective therapeutic interventions may soon control and eliminate these scourages of mankind.

7. *References*

Aikawa, M., Iseki, M., Barnwell, J.W., Taylor, D., Oo, M.M. and Howard, R. (1990). The pathology of human cerebral malaria. Am. J. Trop. Med. Hyg. 43, 30–37.

Aley, S.B., Bates, M.D., Tam, J.P. and Hollingdale, M.R. (1986). Synthetic peptides from the circumsporozoite proteins of *Plasmodium falciparum* and *Plasmodium knowlesi* recognize the human hepatoma cell line Hep G2A16 *in vitro*. J. Exp. Med. 164, 1915–1922.

Arroyo, R. and Orozco, E. (1987). Localization and identification of an *Entamoeba histolytica* adhesin. Mol. Biochem. Parasitol. 23, 151–158.

Asch, A., Barnwell, J.W., Silverstein, R. and Nachman, R.L. (1987). Isolation of the thrombospondin membrane receptor. J. Clin. Invest. 79, 1054–1061.

Barnwell, J.W., Ockenhouse, C.F. and Knowles, D.M. II, (1985). Monoclonal antibody OKM5 inhibits the *in vitro* binding of *Plasmodium falciparum*-infected erythrocytes to monocytes, endothelial, and C32 melanoma cells. J. Immunol. 135, 3494–3497.

Barnwell, J.W., Asch, A.S., Nachman, R.L., Yamaya, M., Aikawa, M. and Ingravallo, P. (1989). A human 88 kDa membrane glycoprotein (CD36) functions *in vitro* as a receptor for a cytoadherence ligand on *Plasmodium falciparum*-infected erythrocytes. J. Clin. Invest. 84, 765–772.

Berendt, A.R., Simmons, D.L., Tansey, J., Newbold, C.I. and Marsh, K. (1989). Intercellular adhesion molecule-1 is an endothelial cell adhesion receptor for *Plasmodium falciparum*. Nature (Lond.) 341, 57–59.

Blackwell, J.M., Ezekowitz, R.A.B., Roberts, M.B., Channon, J.Y., Sim, R.B. and Gordon, S. (1985). Macrophage complement and lectin-like receptors bind *Leishmania* in the absence of serum. J. Exp. Med. 162, 324–331.

Breuer, W.V., Ginsburg, H. and Cabantchik, Z.I. (1983). An assay of malaria parasite invasion into human erythrocytes. The effects of chemical and enzymatic modification of erythrocyte membrane. Biochem. Biophys. Acta. 755, 263–271.

Byrn, R.A., Mordenti, J., Lucas, C., Smith, D., Marsters, S.A., Johnson, J.S., Cossum, P., Chamow, S.M., Wurm, F.M., Gregory, T., Groopman, J.E. and Capon, D.J. (1990). Biological properties of a CD4 immunoadhesin. Nature 344, 667–670.

Camus, D. and Hadley, T.J. (1988). A *Plasmodium falciparum* antigen that binds to host erythrocytes and merozoites. Science 230, 553–556.

Carlson, J., Helmby, H., Hill, A.V.S., Brewster, D., Greenwood, B.M. and Wahlgren, M. (1990a). Human cerebral malaria: association with erythrocyte rosetting and lack of anti-rosetting antibodies. Lancet 336, 1457–1460.

Carlson. J., Holmquist, G., Taylor, D.W., Perlmann, P. and Wahlgren, M. (1990b). Antibodies to a histidine-rich protein (PfHRP1) disrupt spontaneously formed *Plasmodium falciparum* erythrocyte rosettes. Proc. Natl Acad. Sci. USA 87, 2511–2515.

Chadee, K., Petri, W.A., Jr., Innes, D.J. and Ravdin, J.I. (1987). Rat and human colonic mucins bind to and inhibit adherence lectin of *Entamoeba histolytica*. J. Clin. Invest. 80, 1245–1254.

Channon, J.Y., Roberts, M.B. and Blackwell, J.M. (1984). A study of the differential respiratory burst activity elicited by promastigotes and amastigotes of *Leishmania donovani* in murine resident peritoneal macrophages. Immunol. 53, 345–355.

Dame, J.B., Williams, J.L., McCutchan, T.F., Weber, J.C., Wirtz, R.A., Hockmeyer, W.T., Maloy, W.L., Haynes, J.D., Schneider, I., Roberts, D., Sanders, G.S., Premkumar Reddy, E., Diggs, C.L. and Miller, L.H. (1984). Structure of the gene encoding the immunodominant surface antigen on the sporozoite of the human malaria parasite, *Plasmodium falciparum*. Science 225, 593–599.

DaSilva, R.P., Hall, B.F., Joiner, K.A. and Sacks, D.C. (1989). CRl; the C3b receptor, mediates binding of infective *Leishmania major* metacyclic promastigotes to human macrophages. J. Immunol. 143, 617–622.

Diamond, M.S., Staunton, D.E., Marlin, S.D. and Springer, T.A. (1991). Binding of the integrin Mac-1 (CD11b/CD18) to the third immunoglobulin-like domain of ICAM-1 (CD54) and its regulation by glycosylation. Cell 65, 961–971.

Dwyer, D.M. and Gottlieb, M. (1983). The surface membrane chemistry of *Leishmania*: its possible role in parasite sequestration and survival. J. Cell. Biochem. 23, 345–355.

Fang, X., Kaslow, D.C., Adams, J.H. and Miller, L.H. (1991). Cloning of the *Plasmodium vivax* Duffy receptor. Mol. Biochem. Parasitol. 44, 125–132.

Goundis, D. and Reid, K.B.M. (1988). Properdin, the terminal complement components, thrombospondin and the circumsporozoite protein of malaria parasites contain similar sequence motifs. Nature 335, 82–85.

Grau, G.E., Taylor, T.E., Molyneux, M.E., (1989). Tumor necrosis factor and disease severity in children with falciparum malaria. N. Eng. J. Med. 320, 1586–1591.

Greve, J.M., Davis, G., Meyer, A.M., Forte, C.P., Yost, S.C., Marlor, C.W., Kamarck, M.E. and McClelland, A. (1989). The major human rhinovirus receptor is ICAM-1. Cell 56, 839–847.

Hadley, T.J., David, P.H., McGinnis, M. and Miller, L.H. (1984). Identification of an erythrocyte component carrying the Duffy Blood group Fy^a antigen. Science 223, 597–599.

Hadley, T.J., Klotz, F.W. and Miller, L.H. (1986). Invasion of erythrocytes by malaria parasites: a cellular and molecular overview. Annu. Rev. Microbiol. 40, 451–477.

Hadley, T.J., Klotz, F.W., Pasvol, G., Haynes, J.D., McGinnis, M.H., Okubo, Y. and Miller, L.H. (1988). Malaria parasites invade erythrocytes that lack glycophorins A and B (M^kM^k): strain differences indicate receptor heterogeneity and two pathways for invasion. J. Clin. Invest. 80, 1190–1193.

Handman, E. and Goding, J.W. (1985). *The Leishmania* receptors for macrophages is a lipid-containing glycoconjugate. EMBO J. 4, 329–336.

Hasler, T., Handunnetti, S.M., Aguiar, J.C., van Schravendijk, M.R., Greenwood, B.M., Lallinger, G., Cegielski, P. and Howard, R.J. (1990). *In vitro* rosetting, cytoadherence, and micro-agglutination properties of Plasmodium falciparum-infected erythrocytes from Gambian and Tanzanian patients. Blood 76, 1845–1852.

Ho, M., Davis, T.M.E., Silamut, K., Bunnag, D. and White, N.J. (1991). *In vitro* rosette formation of *P. falciparum*-infected erythrocytes in acute malaria. Infect Immun. 59, 2135–2139.

Hollingdale, M.R., Nardin, E.H., Tharavanij, S., Schwartz, A.L. and Nussenzweig, R.S. (1984). Inhibition of entry of *Plasmodium falciparum* and *Plasmodium vivax* sporozoites into cultured cells; an *in vitro* assay of protective antibodies. J. Immunol. 132, 909–913.

Horley, K.J., Carpenito, C., Baker, B. and Takei, F. (1989). Molecular cloning of murine intercellular adhesion molecule (ICAM-1). EMBO J. 8, 2889–2896.

Huang, M., Bolen, J.B., Barnwell, J.W., Shattil, S.J. and Brugge, J.S. (1991). Membrane glycoprotein IV (CD36) is physically associated with the Fyn, Lyn, and Yes protein-tyrosine kinases in human platelets. Proc. Natl Acad, Sci. USA 88, 7844–7848.

Joiner, K.A., Hieny, S., Kirchhoff, L.V. and Sher, A. (1985). gp72, the 72 kDa glycoprotein is the membrane acceptor site for C3 on *Trypanosoma cruzi*. J. Exp. Med. 161, 1196–1212.

Jungery, M., Boyle, D., Patel, T., Pasvol, G. and Weatherall, D.J. (1983). Lectin-like polypeptides of *P. falciparum* bind to red cell sialoglycoproteins. Nature. 301, 704–705.

Kaslow, D.C., Quakyi, I.A., Syin, C., Raum, M.G., Keister, D.B., Coligan, J.E., McCutchan, T.F. and Miller, L.H. (1988). A vaccine candidate from the sexual stage of human malaria that contains EGF-like domains. Nature 333, 74–76.

Kaul, D.K., Roth, E.F. Jr., Nagel, R.L., Howard, R.J. and Handunnetti, S.M. (1991). Rosetting of *Plasmodium falciparum*-infected red blood cells with uninfected red blood cells enhances microvascular obstruction under flow conditions. Blood 78, 812–819.

Lawler, J. and Hynes, R.O. (1986). The structure of human thrombospondin, an adhesive glycoprotein with multiple calcium binding sites and homologies with several different proteins. J. Cell. Biol. 103, 1635–1648.

Lehker, M.W., Arroyo, R. and Alderette, J.F. (1991). The regulation by iron of the synthesis of adhesions and cytoadherence levels in the protozoan *Trichomonas vaginalis*. J. Exp. Med. 174, 311–318.

McClelland, A., Debar, J., Connolly Yost, S., Meyer, A.M., Marlor, C.W. and Greve, J.M. (1991). Identification of monoclonal antibody epitopes and critical residues for rhinovirus binding in domain 1 of intercellular adhesion molecule 1. Proc. Natl Acad. Sci. USA 88, 7993–7997.

Miller, L.H., Mason, S.J., Clyde, D.F. and McGinnis, M.H. (1976). The resistance factor of *Plasmodium vivax* in Blacks; the Duffy blood group genotype FyFy. New Engl. J. Med. 295, 302–304.

Miller, L.H., Haynes, J.D., McAuliffe, F.M., Shiroishi, T., Durocher, J.R. and McGinnis, M.H. (1977). Evidence for differences in erythrocyte surface receptors for the malaria parasite, *Plasmodium falciparum* and *Plasmodium knowlesi*. J. Exp. Med. 146, 277–281.

Mosser, D.M. and Edelson, P.J. (1985). The mouse macrophage receptor for C3bi (CR3) is a major mechanism in the phagoctosis of leishmania promastigotes. J. Immunol. 135, 2785–2789.

Mosser, D.M., Vlassara, H., Edelson, P.J. and Cerami, A. (1987). *Leishmania* promastigotes are recognized by the macrophage receptor for advanced glycosylation endproducts. J. Exp. Med. 165,140–145.

Ockenhouse, C.F. and Chulay, J.D. (1988). *Plasmodium falciparum* sequestration: evidence that OKM5 antigen (CD36) mediates cytoadherence of parasitized erythrocytes to a myelomonocytic cell line. J. Infect. Dis. 157, 584–588.

Ockenhouse, C.F., Tandon, N.N., Magowan, C., Jamieson G.A., and Chulay, J.D. (1989a). Identification of a platelet membrane glycoprotein as a falciparum malaria sequestration receptor. Science (Wash.D.C.) 243, 1469–1471.

Ockenhouse, C.F., Magowan, C. and Chulay, J.D. (1989b). Activation of monocytes and platelets by monoclonal antibodies or malaria-infected erythrocytes binding to the CD36 surface receptor *in vitro*. J. Clin. Invest. 84, 468–475.

Ockenhouse, C.F., Ho, M., Tandon N.N., VanSeventer, G.A., Shaw, S., White, N., Jamieson, G.A., Chulay, J.D. and Webster, H.K. (1991a). Molecular basis of sequestration in severe and uncomplicated *Plasmodium falciparum* malaria: differential adhesion of infected erythrocytes to CD36 and ICAM-1. J. Infect Dis. 164, 163–169.

Ockenhouse, C.F., Klotz, F.W., Tandon, N.N. and Jamieson, G.A. (1991b). Sequestrin, a CD36 recognition protein on *Plasmodium falciparum*-infected erythrocytes identified by anti-idiotype antibodies. Proc. Natl Acad. Sci. USA 88, 3175–3179.

Ockenhouse, C.F., Betageri, R., Springer, T.A. and Staunton, D.E. (1992). *Plasmodium falciparum*-infected erythrocytes bind ICAM-1 at a site distinct from LFA-1, Mac-1, and human rhinovirus. Cell 68, 63–69.

Oquendo, P., Hundt, E., Lawler, J. and Seed, B. (1989) CD36 directly mediates cytoadherence of *Plasmodium falciparum* parasitized erythrocytes. Cell 58, 95–101.

Ouaissi, M.A., Afchain, D., Capron, A. and Grimaud, J.A. (1984). Fibronectin receptors on *Trypanosoma cruzi* trypomastigotes and their biological function. Nature (Lond.) 308, 380–382.

Ouaissi, M.A., Cornette, J., Afchain, D., Capron, A., Gras-Masse, H. and Tartar, A. (1986). *Trypanosoma cruzi* infection inhibited by peptides modelled from a fibronectin cell attachment domain. Science 234, 603–607.

Panton, L.J., Leech, J.H., Miller, L.H. and Howard, R.J. (1987). Cyto-adherence of *Plasmodium falciparum*-infected erythrocytes to human melanoma cell lines correlates with surface OKM5 antigen. Infect. Immun. 55, 2754–2758.

Pasvol, G., Jungery, M., Weatherall, D.J., Parson, S.F., Anstee, D.J. and Tanner. M.J.A. (1982a). Glycophorin as a possible receptor for *Plasmodium falciparum*. Lancet 2, 947–950.

Pasvol, G., Wainscoat, J.S. and Weatherall, D.J. (1982b). Erythrocytes deficient in glycophorin resist invasion by the malarial parasite, *Plasmodium falciparum*. Nature (Lond.) 297, 64–66.

Pasvol, G., Anstee, D and Tanner, M.J.A. (1984). Glycophorin C and the invasion of red blood cells by *Plasmodium falciparum*. Lancet, i, 907–908.

Perkins, M. (1981). Inhibitory effects of erythrocyte membrane proteins on the *in vitro* invasion of the human malarial parasite (*Plasmodium falciparum*) into its host cell. J. Cell Biol. 90, 563–567.

Perkins, M.E. (1984). Surface proteins of *Plasmodium falciparum* merozoites binding to the erythrocyte receptor, glycophorin. J. Exp. Med. 160, 788–798.

Petri, W.A., Jr., Smith, R.D., Schlesinger, P.H., Murphy, C.F. and Ravdin, J.I. (1987). Isolation of the galactose-binding lectin that mediates the in vitro adherence of *Entamoeba histolytica*. J. Clin. Invest. 80, 1238–1244.

Petri, W.A., Jr., Chapman, M.D., Snodgrass, T., Mann, B.J., Broman, J. and Ravdin, J.I. (1989). Subunit structure of the galactose and N-acetyl-D-galactosamine inhibitable lectin of *Entamoeba histolytica*. J. Biol. Chem. 264, 3007–3012.

Petri, W.A., Jr., Snodgrass, T.L., Jackson, T., Gathiram, V., Simjee, A.E., Chadee, K. and Chapman, M.D. (1990). Monoclonal antibodies directed against the galactose-binding lectin of *Entamoeba histolytica* enhance adherence. J. Immunol. 144, 4803–4809.

Prater, C.A., Plotkin, J., Jaye, D. and Frazier, W.A. (1991). The properdin-like type I repeats of human thrombospondin contain a cell attachment site. J. Cell Biol. 112, 1031–1040.

Puentes, S.M., Sacks, D., da Silva, R.P. and Joiner, K.A. (1988). Complement binding by two developmental stages of *Leishmania major* promastigotes varying in expression of a surface lipophosphoglycan. J. Exp. Med. 167, 887–902.

Ravdin, J.I. and Guerrant, R.L. (1981). Role of adherence in cytopathogenic mechanisms of *Entamoeba histolytica*. Study with mammalian tissue culture cells and human erythrocytes. J. Clin. Invest. 68, 1305–1313.

Ravdin, J.I., John, J.E., Johnston, L.I., Innes, D.J. and Guerrant, R.L. (1985). Adherence of *Entamoeba histolytica* trophozoites to rat and human colonic mucosa. Infect. Immun. 48, 292–297.

Ravdin, J.I., Stanley, P., Murphy, C.F. and Petri, W.A., Jr. (1989). Characterization of cell surface carbohydrate receptors for *Entamoeba histolytica* adherence lectin. Infect. Immun. 57, 2179–2186.

Ravetch, J.V., Kochan, J. and Perkins, M. (1985). Isolation of the gene for a glycophorin-binding protein implicated in erythrocyte invasion by a malaria parasite. Science 227, 1593–1597.

Rich, K.A., George, F.W., IV., Law, J.L. and Martin, W.J. (1990). Cell-adhesive motif in region II of malarial circumsporozoite protein. Science 249, 1574–1577.

Rimoldi, M.T., Sher, A., Hieny, S., Lituchy, A., Hammer, C.H. and Joiner, K. (1988). Developmentally regulated expression by *Trypanosoma cruzi* of molecules which accelerate the decay of complement C3 convertases. Proc. Natl Acad. Sci. USA 85, 193–197.

Rimoldi, M.T., Tenney, A.J., Bobak, D.A. and Joiner, K.A. (1989). Complement component C1q enhances invasion of human mononuclear phagocytes and fibroblasts by *Trypanosoma cruzi* trypomastigotes. J. Clin. Invest. 84, 1982–1989.

Rizui, F.S., Ouassi, M.A., Marty, B., Santoro, F. and Capron, A. (1988). The major surface protein of Leishmania promastigotes is a fibronectin-like molecule. Eur. J. Immunol. 18, 473–479.

Roberts, D.D., Sherwood, J.A., Spitalnik, S.L., Panton, L.J., Howard, R.J., Dixit, V.M., Frazier, W.A., Miller, L.H., Ginsburg, V. (1985). Thrombospondin binds falciparum malaria parasitized erythrocytes and may mediate cyto-adherence. Nature 318, 64–66.

Robson, K.J.H., Hall, J.R.S., Jennings, M.W., Harris, T.J.R., Marsh, K., Newbold, C.I., Tate, V.E. and Weatherall, D.J. (1988). A highly conserved amino-acid sequence in thrombospondin, properdin, and in proteins from sporozoites and blood stages of a human malaria parasite. Nature 335, 79–82.

Rock, E.P., Roth, E.F., Rojas-Corona, R.R., Sherwood, J.A., Nagel, R.L., Howard, R.J. and Kaul, D.K. (1988) Thrombospondin mediates the cyto-adherence of *Plasmodium falciparum*-infected red cells to vascular endothelium in shear flow conditions. Blood 71, 71–75.

Rodriguez De Cuna, C., Kierszenbaum, F. and Wirth, J.J. (1991). Binding of the specific ligand to Fc receptors on *Trypanosoma cruzi* increases the infective capacity of the parasite. Immunol. 72, 114–120.

Russell, D.G. (1987). The macrophage-attachment glycoprotein gp63 is the predominant C3-acceptor site on *Leishmania mexicana* promastigotes. Eur. J. Biochem. 164, 213–222.

Russell, D.G. and Wilhelm, H. (1986). The involvement of the major surface glycoprotein (gp63) of *Leishmania* promastigotes in attachment to macrophages. J. Immunol. 136, 2613–2620.

Russell, D.G. and Wright, S.D. (1988). Complement receptor type 3 (CR3) binds to a Arg-Gly-Asp-containing region of the major surface glycoprotein, gp63, of *Leishmania* promastigotes. J. Exp. Med. 168, 279–292.

Sacks, D.L., Hieny, S. and Sher, A. (1985). Identification of cell surface carbohydrate and antigenic changes between non-infective and infective developmental stages of *Leishmania major* promastigotes. J. Immunol. 135, 564–569.

Schenkman, S., Andrews, N.W., Nussenzweig, V. and Robbins, E.S. (1988) *Trypanosoma cruzi* invade a mammalian epithelial cell in a polarized manner. Cell 55, 157–165.

Shaffer, N., Grau, G.E., Hedberg, K., Davachi, F., Lyamba, B., Hightower, A.W., Breman, J.G. and Nguyen-Dinh, P. (1991). Tumor necrosis factor and severe malaria. J. Infect. Dis. 163, 96–101.

Sherwood, J.A., Roberts, D.D., Marsh, K., Harvey, E.B., Spitalnik, S.L., Miller, L.H. and Howard, R.J. (1987). Thrombospondin binding by parasitized erythrocyte isolates in falciparum malaria. Am. J. Trop. Med. Hyg. 36, 228–233.

Sherwood, J.A., Roberts, D.D., Spitalnik, S.L., Marsh, K., Harvey, E.B., Miller, L.H. and Howard, R.J. (1989). Studies of the receptors on melanoma cells for *Plasmodium falciparum* infected erythrocytes. Am. J. Trop. Med. Hyg. 40, 119–127.

Sim, B.K.L., Orlandi, P.A., Haynes, J.D., Klotz, F.W., Carter, J.M., Camus, D., Zegans, M.E. and Chulay, J.D. (1990). Primary structure of the 175 K *Plasmodium falciparum* erythrocyte binding antigen and identification of a peptide which elicits antibodies that inhibit malaria merozoite invasion. J. Cell Biol. 111, 1877–1884.

Simmons, D., Makgoba, M.W. and Seed, B. (1988). ICAM, an adhesion ligand for LFA-1, is homologous to the neural cell adhesion molecule NCAM. Nature (Lond.). 331, 624–627.

Staunton, D.E., Marlin, S.D., Stratowa, C., Dustin, M.L. and Springer, T.A. (1988). Primary structure of ICAM-1 demonstrates interaction between members of the immunoglobulin and integrin supergene families. Cell. 52, 925–933.

Staunton, D.E., Merluzzi, V.J., Rothlein, R., Barton, R., Marlin, S.D. and Springer, T.A. (1989). A cell adhesion molecule, ICAM-1, is the major surface receptor for rhinoviruses. Cell. 56, 849–853.

Staunton, D.E., Dustin, M.L., Erickson, H.P. and Springer, T.A. (1990). The arrangement of the immunoglobulin-like domains of ICAM-1 and the binding sites for LFA-1 and rhinovirus. Cell. 61, 243–254.

Staunton, D.E., Ockenhouse, C.F. and Springer, T.A. (1992). Soluble intercellular adhesion molecule I-immunoglobin G1 immunoadhesin mediates phagocytosis of malaria-infected erythrocytes. J. Exp. Med. 176, 1471–1476.

Talamas-Rohana, P., Wright, S.., Lennartz, M.R. and Russell, D.G. (1990). Lipophosphoglycan from *Leishmania mexicana* promastigotes bind to members of the CR3, p150,95, and LFA-1 family of leukocyte integrins. J. Immunol. 144, 4817–4824.

Talle, M.A., Rao, P.E., Westberg, E., Allegai, N., Makowski, M., Mittler, R.S. and Goldstein, G. (1983). Patterns of antigenic expression on human monocytes as defined by monoclonal antibodies. Cell. Immunol 78, 83–99.

Tandon, N.N., Kralisz, U. and Jamieson, G.A. (1989). Identification of GPIV (CD36) as a primary receptor for platelet-collagen adhesion. J. Biol. Chem. 264, 7576–7583.

Tandon, N.N., Ockenhouse, C.F., Greco, N.J. and Jamieson, G.A. (1992). Adhesive functions of platelets lacking GPIV (CD36). Blood (In press).

Turco, S.J., Wilkerson, M.A. and Clawson, D.R. (1984). Expression of an unusual acidic glycoconjugate in *Leishmania donovani*. J. Biol. Chem. 259, 3883–3889.

Udomsangpetch, R., Wahlin, B., Carlson, J., Berzins, K., Torii, M., Aikawa, M., Perlmann, P. and Wahlgren, M. (1989). *Plasmodium falciparum*-infected erythrocytes form spontaneous erythrocyte rosettes. J. Exp. Med. 169, 1835–1840.

Vega, M.A., Segui-Real, B., Garcia, J.A., Cales, C., Rodriguez, F., Vanderkerckhove, J., and Sandoval, I. (1991). Cloning, sequencing, and expression of a cDNA encoding rat LIMPII, a novel 74 kDa lysosomal membrane protein related to the surface adhesion protein CD36. J. Biol. Chem. 266, 16 818–16 824.

Wilson, M.E. and Pearson, R.D. (1988). Roles of CR3 and mannose receptors in the attachment and ingestion of *Leishmania donovani* by human mononuclear phagocytes. Infect. Immun. 56, 363–369.

Wozencraft, A.O. and Blackwell, J.M. (1987). Increased infectivity of stationary-phase promastigotes of *Leishmania donovani*: correlation with enhanced C3 binding capacity and CR3-mediated attachment to host macrophages. Immunol. 60, 559–566.

Wyler, D.J., Sypek, J.P. and McDonald, J.A. (1985). *In vitro* parasite-monocyte interactions in human leishmaniasis: possible role of fibronectin in parasite attachment. Infect. Immun. 49, 305–311.

Glossary

α_1, α_2 receptors Adrenoceptor subtypes
α_1-ACT α1-Antichymotrypsin
α1AT α_1-Antiprotease inhibitor *also known as* α_1-antitrypsin
$\alpha_{1\text{-PI}}$ α_1-Proteinase inhibitor
α_2-M α_2-macroglobulin
A Absorbance
AI, AII Angiotensin I, II
Å Angstrom
AA Arachidonic acid
aa Amino acids
AAb Autoantibody
Ab Antibody
Ab1 Idiotype antibody
Ab2 Anti-idiotype antibody
Ab2α Anti-idiotype antibody which binds outside the antigen binding region
Ab2Lβ Anti-idiotype antibody which binds to the antigen binding region
Ab3 Anti-anti-idiotype antibody
Abcc Antibody dependent cellular cytotoxicity
ABA-L-GAT Arsanilic acid conjugated with the synthetic polypeptide L-GAT
AC Adenylate cyclase
ACAT Acyl-co-enzyme-A acyltransferase
ACAID Anterior chamber-associated immune deviation
ACE Angiotensin-converting enzyme
ACh Acetylcholine
ACTH Adrenocorticotrophin hormone
Ado Adenosine
ADP Adenosine diphosphate
AES Anti-eosinophil serum
Ag Antigen
AGE Advanced glycosylation endproduct
AGEPC 1-*O*-alkyl-2-acetyl-*sn*-glyceryl-3-phosphocholine; *also known as* PAF and APRL
AH Acetylhydrolase
AID Autoimmune disease
AIDS Acquired immune deficiency syndrome
A/J A Jackson inbred mouse strain
ALP Anti-leukoprotease
cAMP cyclic adenosine monophosphate *also known as* adenosine 3′, 5′-phosphate)
AM Alveolar macrophage
AML Acute myelogenous leukaemia
AMP Adenosine monophosphate
ANAb Anti-nuclear antibodies
ANCA Anti-neutrophil cytoplasmic autoantibodies
cANCA Cytoplasmic ANCA
pANCA Perinuclear ANCA
AND Anaphylactic degranulation
ANF Atrial natriuretic factor
ANP Atrial natriuretic peptide
Anti-I-A, Anti-I-E Antibody against class II MHC molecule encoded by I-A locus, I-E locus
anti-Ig Antibody against an immunoglobulin
anti-RTE Anti-tubular epithelium
AP-1 Activator protein-1
APA B-azaprostanoic acid
APAS Antiplatelet antiserum
APC Antigenpresenting cell
APD Action potential duration
ApO-B Apolipoprotein B
APRL Anti-hypertensive polar renal lipid *also known as* PAF
APUD Amine precursor uptake and decarboxylation
AR-CGD Autosomal recessive form of chronic granulomatous disease
ARDS Adult respiratory distress syndrome
AS Ankylosing spondylitis
ASA Acetylsalicylic acid *also known as* aspirin
4-ASA, 5-ASA 4-, 5-aminosalicylic acid
ATHERO-ELAM A monocyte adhesion molecule
ATL Adult T cell leukaemia
ATP Adenosine triphosphate
ATPγs Adenosine 3′ thiotriphosphate
AUC Area under curve
AVP Arginine vasopressin

β_1, β_2 receptors Adrenoceptor subtypes
β_2 (CD18) A leucocyte integrin
β_2M β_2-Microglobulin
β-TG β-Thromboglobulin
B_7/BB_1 *Known to be* expressed on B cell blasts and immunostimulatory dendritic cells
BAF Basophilactivating factor
BAL Bronchoalveolar lavage
BALF Bronchoalveolar lavage fluid
BALT Bronchus-associated lymphoid tissue
B cell Bone marrow-derived lymphocyte
BCF Basophil chemotactic factor
B-CFC Basophil colony-forming cell
BCG Bacillus Calmette-Guérin
bFGF Basic fibroblast growth factor
Bg Birbeck granules
BHR Bronchial hyperresponsiveness
b.i.d. *Bis in die* (twice a day)
Bk Bradykinin
Bk_1, Bk_2 receptors Bradykinin receptor subtypes *also known as* B_1 and B_2 receptors
Bk_2 receptor Bradykinin receptor subtype
Bl-CFC Blast colony-forming cells
B-lymphocyte Bursa-derived lymphocyte
BM Bone marrow
BMCMC Bone marrow cultured mast cell
BMMC Bone marrow mast cell
BOC-FMLP Butoxycarbonyl-FMLP
bp Base pair
BPB Para-bromophenacyl bromide
BPI Bacterial permeability-increasing protein
BSA Bovine serum albumin

C1, C2...C9 The 9 main components of complement
C1 inhibitor A serine protease inhibitor which inactivates C1r/C1s
C1q Complement fragment 1q (anaphylatoxin)
C1qR Receptor for C1w; facilitates attachment of immune complexes to mononuclear leucocytes and endothelium
C3a Complement fragment 3a (anaphylatoxin)
$C3a_{72-77}$ A synthetic carboxyterminal peptide C3a analogue

C3aR Receptor for anaphylatoxins, C3a, C4a, C5a
C3b Complement fragment 3b (anaphylatoxin)
C3bi Inactivated form of C3b fragment of complement
C4b Complement fragment 4b (anaphylatoxin)
C4BP C4 binding protein; plasma protein which acts as cofactor to factor I inactivate C3 convertase
C5a Complement fragment 5a (anaphylatoxin)
C5aR Receptor for anaphylatoxins C3a, C4a and C5a
C5b Complement fragment 5b (anaphylatoxin)
$C_\epsilon 2$, $C_\epsilon 3$, $C_\epsilon 4$ Heavy chain of immunoglobulin E: domain 2, 3 and 4
Ca *The chemical symbol* for calcium
$[Ca^{2+}]_i$ Intracellular free calcium concentration
CAH Chronic active hepatitis
CALLA Common lymphoblastic leukaemia antigen
CALT Conjunctival associated lymphoid tissue
cAMP cyclic adenosine monophosphate *also known as* adenosine 3′, 5′-phosphate)
CAM Cell adhesion molecule
CAP57 Cationic protein from neutrophils
CatG Cathepsin G
CB Cytochalasin B
CBH Cutaneous basophil hypersensitivity
CBP Cromolyn-binding protein
CCK Cholecystokinin
CCR Creatinine clearance rate
CD Cluster of differentiation (a system of nomenclature for surface molecules on cells of the immune system); cluster determinant
CD1 Cluster of differentiation 1 *also known as* MHC class I-like surface glycoprotein
CD1a Isoform a *also known as* non-classical MHC class I-like surface antigen; present on thymocytes and dendritic cells
CD1b *Known to be* present on thymocytes and dendritic cells
CD1c Isoform c *also known as* non-classical MHC class I-like surface antigen; present on thymocytes
CD2 Defines T cells involved in antigen nonspecific cell activation
CD3 *Also known as* T cell receptor-associated surface glycoprotein on T cells
CD4 Defines MHC class II-restricted T cell subsets
CD5 *Known to be* present on T cells and a subset of B cells; *also known as* Lyt 1 in mouse
CD7 Cluster of differentiation 7; present on most T cells and NK cells
CD8 Defines MHC class I-restricted T cell subset; present on NK cells
CD10 *Known to be* common acute leukaemia antigen
CD11a *Known to be* an α chain of LFA-1 (leucocyte function antigen-1) present on several types of leucocyte and which mediates adhesion
CD11c *Known to be* a complement receptor 4 α chain.
CD13 Aminopeptidase N; present on myeloid cells
CD14 *Known to be* a lipid-anchored glycoprotein; present on monocytes
CD15 *Known to be* Lewis X, fucosyl N acetyllactosamine
CD16 *Known to be* Fcγ receptor III
CD16-1, CD16-2 Isoforms of CD16
CD19 Recognizes B cells and follicular dendritic cells
CD20 *Known to be* a pan B cell
CD21 C3d receptor
CD23 Low affinity FcεR
CD25 Low affinity receptor for interleukin-2
CD27 Present on T cells and plasma cells
CD28 Present on resting and activated T cells and plasma cells
CD30 Present on activated B and T cells
CD31 *Known to be* on platelets, monocytes, macrophages, granulocytes, B-cells and endothelial cells; *also known as* PECAM
CD32 Fcγ receptor II
CD33⁺ *Known to be* a monocyte and stem cell marker
CD34[1] *Known to be* a stem cell marker
CD35 C3b receptor
CD36 *Known to be* a macrophage thrombospondin receptor
CD40 Present on B cells and follicular dendritic cells
CD41 *Known to be* a platelet glycoprotein
CD44 *Known to be* a leucocyte adhesion molecule; *also known as* hyaluronic acid cell adhesion molecule (H-CAM), Hermes antigen, extracellular matrix receptor III (ECMIII); present on polymorphonuclear leucocytes
CD45 *Known to be* a pan leucocyte marker
CD45RO *Known to be* the isoform of leukosialin present on memory T cells
CD46 *Known to be* a membrane cofactor protein
CD49 Cluster of differentiation 49
CD51 *Known to be* vitronectin receptor alpha chain
CD54 *Known to be* Intercellular adhesion molecule-1 *also known as* ICAM-1
CD57 Present on T cells and NK subsets
CD58 A leucocyte function-associated antigen-3, *also known to be* a member of the β-2 integrin family of cell adhesion molecules
CD59 *Known to be* a low molecular weight HRf present to many haematopoetic and nonhaematopoetic cells
CD62 *Known to be present on* activated platelets and endothelial cells; *also known as* P-selectin
CD64 *Known to be* Fcγ receptor I
CD65 *Known to be* fucoganglioside
CD68 Present on macrophages
CD69 *Known to be* an activation inducer molecule; present on activated lymphocytes
CD72 Present on B-lineage cells
CD74 An invariant chain of class II B cells
CDC Complement-dependent cytotoxicity
cDNA Complementary DNA
CDP Choline diphosphate
CDR Complementary-determining region
CD_{xx} Common determinant *xx*
CEA Carcinoembryonic antigen
CETAF Corneal epithelial T cell activating factor
CF Cystic fibrosis
Cf Cationized ferritin
CFA Complete Freund's adjuvant
CFC Colony-forming cell
CFU Colony-forming unit
CFU-S Colonyforming unit, spleen
CGD Chronic granulomatous disease
cGMP cyclic guanosine monophosphate also known as guanosine 3′, 5′-phosphate
CGRP Calcitonin gene-related peptide
CH2 Hinge region of human immunoglobulin
CHO Chinese hamster ovary
CI Chemical ionization
CIBD Chronic inflammatory bowel disease
CK Creatine phosphokinase
CKMB The myocardial-specific isoenzyme of creatine phosphokinase
Cl *The chemical symbol for* chloride
CL Chemiluminescent
CLA Cutaneous lymphocyte antigen
CL18/6 Anti-ICAM-1 monoclonal antibody
CLC Charcot–Leyden crystal

CMC Critical micellar concentration
CMI Cell mediated immunity
CML Chronic myeloid leukaemia
CMV Cytomegalovirus
CNS Central nervous system
CO Cyclooxygenase
CoA Coenzyme A
CoA-IT Coenzyme A-independent transacylase
Con A Concanavalin A
COPD Chronic obstructive pulmonary disease
COS Fibroblast-like kidney cell line established from simian cells
CoVF Cobra venom
CP Creatine phosphate
c.p.m. Counts per minute
CPJ Cartilage/pannus junction
Cr *The chemical symbol for* chromium
CR Complement receptor
CR1, CR2 & CR4c Complement receptor types 1, 2 and 4
CR3-α Complement receptor type 3-α
CRF Corticotrophin-releasing factor
CRH Corticotrophin-releasing hormone
CRI Cross-reactive idiotype
CRP Creactive protein
CSA Cyclosporin A
CSF Colony-stimulating factor
CSS Churg-Strauss syndrome
CT Computed tomography
CTAP-III Connective tissue-activating peptide
CTD Connective tissue diseases
C terminus Carboxy terminus of peptide
CThp Cytotoxic T lymphocyte precursors
CTL Cytotoxic T lymphocyte
CTLA-4 *Known to be* co-expressed with CD20 on activated T cells
CTMC Connective tissue mast cell
CVF Cobra venom factor

Da Dalton (the unit of relative molecular mass)
DAF Decayaccelerating factor
DAG Diacylglycerol
DAO Diamine oxidase
D-Arg D-Argininecs
DArg-[Hyp^3, DPhe^7]-BK A bradykinin B_2 receptor antagonist. Peptide derivative of bradykinin
DArg-[Hyp^3, Thi^5, DTic^7, Tic^8]-BK A bradykinin B_2 receptor antagonist. Peptide derivative of bradykinin
DC Dendritic cell
DCF Oxidized DCFH
DCFH 2′, 7′-dichlorofluorescin
DEC Diethylcarbamazine
desArg^9-BK Carboxypeptidase N product of bradykinin
desArg^{10}KD Carboxypeptidase N product of kallidin
DFMO α-Difluoromethyl ornithine
DFP Diisopropyl fluorophosphate
DGLA Dihomo-γ-linolenic acid
DH Delayed hypersensitivity
DHR Delayed hypersensitivity reaction
DIC Disseminated intravascular coagulation
DL-CFU Dendritic cell/Langerhans cell colony forming
DLE Discoid lupus erythematosus
DMARD Diseasemodifying antirheumatic drug
DMF *N*, *N*-dimethylformamide
DMSO Dimethyl sulfoxide
DNA Deoxyribonucleic acid
D-NAME D-Nitroarginine methyl ester
DNase Deoxyribonuclease
DNCB Dinitrochlorobenzene
DNP Dinitrophenol
Dpt4 *Dermatophagoides pteronyssinus* allergen 4
DGW2, DR3, DR7 HLA phenotypes
DREG-56 (Antigen) L-selectin
DREG-200 A monoclonal antibody against L-selectin
ds Double-stranded
DSCG Disodium cromoglycate
DST Donor-specific transfusion
DTH Delayed-type hypersensitivity
DTPA Diethylenetriamine pentaacetate
DTT Dithiothreitol
dv/dt Rate of change of voltage within time

ε Molar absorption coefficient
EA Egg albumin
EAE Experimental autoimmune encephalomyelitis
EAF Eosinophil-activating factor
EAR Early phase asthmatic reaction
EAT Experimental autoimmune thyroiditis
EBV Epstein-Barr virus
EC Endothelial cell
ECD Electron capture detector
ECE Endothelin-converting enzyme
E-CEF Eosinophil cytotoxicity enhancing factor
ECF-A Eosinophil chemotactic factor of anaphylaxis
ECG Electrocardiogram
ECGF Endothelial cell growth factor
ECGS Endothelial cell growth supplement
E. coli *Escherichia coli*
ECP Eosinophil cationic protein
ED_{35} Effective dose producing 35% maximum response
ED_{50} Effective dose producing 50% maximum response
EDF Eosinophil differentiation factor
EDN Eosinophil-derived neurotoxin
EDRF Endothelium-derived relaxant factor
EDTA Ethylenediamine tetraacetic acid *also known as* etidronic acid
EE Eosinophilic eosinophils
EEG Electroencephalogram
EET Epoxyeicosatrienoic acid
EFA Essential fatty acid
EFS Electrical field stimulation
EG1 Monoclonal antibody specific for the cleaved form of eosinophil cationic peptide
EGF Epidermal growth factor
EGTA Ethylene glycol-bis(β-aminoethyl ether) *N*, *N*, *N*, ′*N*′-tetraacetic acid
EI Electron impact
eIF-2 Subunit of protein synthesis initiation factor
ELF Respiratory epithelium lung fluid
ELISA Enzyme-linked immunosorbent assay
EMS Eosinophiliamyalgia syndrome
ENS Enteric nervous system
EO Eosinophil
EO-CFC Eosinophil colonyforming cell
EOR Early onset reaction also known as EAR
EPA Eicosapentaenoic acid
EpDIF Epithelial-derived inhibitory factor *also known as* epithelium-derived relaxant factor
EPO Eosinophil peroxidase
EPOR Erythropoietin receptor
EPR Effector cell protease
EPX Eosinophil protein X
ER Endoplasmic reticulum
E-selectin Endothelial selectin *formerly known as* endothelial leucocyte adhesion molecule-1 (ELAM-1)
ESP Eosinophil stimulation promoter
ESR Erythrocyte sedimentation rate
ET, ET-1 Endothelin, -1
ETYA Eicosatetraynoic acid
FA Fatty acid
FAB Fast-electron bombardment
Fab Antigen binding fragment
F(ab′)2 Fragment of an immunoglobulin produced pepsin treatment
FACS Flow activated cell sorter
factor B Serine protease in the C3 converting enzyme of the alternative pathway
factor D Serine protease which cleaves factor B

factor H Plasma protein which acts as a co-factor to factor I
factor I Hydrolyses C3 converting enzymes with the help of factor H
FAD Flavine adenine dinucleotide
FBR Fluorescence photobleaching recovery
Fc Crystallizable fraction of immunoglobulin molecule
Fcγ Receptor for Fc portion of IgG
FcγRI Ig Fc receptor I *also known as* CD64
FcγRII Ig Fc receptor II *also known as* CD32
FcγRIII Ig Fc receptor III *also known as* CD16
$Fc_ε$RI High affinity receptor for IgE
$Fc_ε$RII Low affinity receptor for IgE
FcR Receptor for Fc region of antibody
FCS Foetal calf (bovine) serum
FEV_1 Forced expiratory volume in 1 second
FGF Fibroblast growth factor
FID Flame ionization detector
FITC Fluorescein isothiocyanate
FKBP FK506-binding protein
FLAP 5-Lipoxygenase-activating protein
FMLP *N*-Formyl-methionyl-leucyl-phenylalanine
FNLP Formyl-norleucyl-leucyl-phenylalanine
FOC Follicular dendritic cell
FPR Formyl peptide receptor
FS cell Folliculostellate cell
FSG Focal sequential glomerulosclerosis
FSH Follicle stimulating hormone
5-FU 5-fluorouracil

Ga G-protein
G6PD Glucose 6-phosphate dehydrogenase
GABA γ-aminobutyric acid
GAG Glycosaminoglycan
GALT Gut-associated lymphoid tissue
GAP GTPase-activating protein
GBM Glomerular basement membrane
GC Guanylate cyclase
GC-MS Gas chromatography mass spectroscopy
G-CSF Granulocyte colony-stimulating factor
GDP Guanosine 5′-diphosphate
GEC Glomerular epithelial cell
GF-1 An insulin-like growth factor
GFR Glomerular filtration rate
GH Growth hormone
GH-RF Growth hormone-releasing factor
Gi Family of pertussis toxin sensitive G-proteins
GI Gastrointestinal
GIP Granulocyte inhibitory protein
GlyCam-1 Glycosylation-dependent cell adhesion molecule-1
GMC Gastric mast cell
GM-CFC Granulocyte-macrophage colony-forming cell
GM-CSF Granulocyte-macrophage colony-stimulating factor
GMP Guanosine monophosphate (guanosine 5′-phosphate)
Go Family of pertussis toxin sensitive G-proteins
GP Glycoprotein
gp45-70 Membrane co-factor protein
$gp90^{MEL}$ 90 kd glycoprotein recognised by monoclonal antibody MEL-14; *also known as* L-selectin
GPIIb-IIIa Glycoprotein IIb-IIIa *known to be* a platelet membrane antigen
$GppCH_2P$ Guanyl-methylene diphosphanate *also known as* a stable GTP analogue
GppNHp Guanylyl-imidiodiphosphate *also known as* a stable GTP analogue
cGRGDSP Glycine-arginine-glycine-aspartic acid serine-proline
Gro Growth-related oncogene
GRP Gastrin-related peptide
Gs Stimulatory G protein
GSH Glutathione (reduced)
GSSG Glutathione (oxidized)
GTP Guanosine triphosphate
GTP-γ-S Guanarine 5′*O*-(3-thiotriphosphate)
GTPase Guanidine triphosphatase
GVHD Graft-versus-host-disease
GVHR Graft-versus-host-reaction

H Histamine
H_1, H_2, H_3 Histamine receptor types 1, 2 and 3
H_2O_2 *The chemical symbol for* hydrogen peroxide
Hag Haemagglutinin
Hag-1, Hag-2 Cleaved haemagglutinin subunits-1, 2
H & E Haematoxylin and eosin
hIL Human interleukin
Hb Haemoglobin
HBBS Hank's balanced salt solution
H-CAM Hyaluronic acid cell adhesion molecule
HDC Histidine decarboxylase
HDL High-density lipoprotein
HEL Hen egg white lysozyme
HEPE Hydroxyeicosapentanoic acid
HEPES *N*-2-Hydroxylethylpiperazine-*N*′-2-ethane sulphonic acid
HES Hypereosinophilic syndrome
HETE 5, 8, 9, 11 and 15 Hydroxyeicosatetraenoic acid
5(S)HETE A stereo isomer of 5-HETE
HETrE Hydroxyeicosatrienoic acid
HEV High endothelial venule
HFN Human fibronectin
HGF Hepatocyte growth factor
HHTrE 12(*S*)-Hydroxy-5, 8, 10-heptadecatrienoic acid
HIV Human immunodeficiency virus
HL60 Human promyelocytic leukaemia cell line
HLA Human leucocyte antigen
HLA-DR2 Human histocompatability antigen class II
HMG CoA Hydroxylmethylglutaryl coenzyme A
HMW High molecular weight
HMT Histidine methyltransferase
HMVEC Human microvascular endothelial cell
HNC Human neutrophil collagenase (MMP-8)
HNE Human neutrophil elastase
HNG Human neutrophil gelatinase (MMP9)
HODE Hydroxyoctadecanoic acid
HPETE, 5-HPETE & 15-HPETE 5 and 15 Hydroperoxyeicosatetraenoic acid
HPETrE Hydroperoxytrienoic acid
HPODE Hydroperoxyoctadecanoic acid
HPLC High performance liquid chromatography
HRA Histamine-releasing activity
HRAN Neutrophil-derived histamine-releasing activity
HRf Homologous-restriction factor
HRF Histamine-releasing factor
HRP Horseradish peroxidase
HSA Human serum albumin
HSP Heatshock protein
HS-PG Heparan sulphate proteoglycan
HSV, HSV1 Herpes simplex virus, -1
^{3}HTdR Tritiated thymidine
5-HT 5-Hydroxytryptamine *also known as* Serotonin
HTLV-1 Human T-cell leukaemia virus-1
HUVEC Human umbilical vein endothelial cell
[Hyp^3]BK Hydroxyproline derivative of bradykinin
[Hyp^4]-KD Hydroxyproline derivative of kallidin

I_{sc} Short-circuit current
Ia Immune reactionassociated antigen
Ia+ Murine class II major histocompatibility complex antigen
IB4 Anti-CD18 monoclonal antibody

IBD Inflammatory bowel disease
IBMX Isobutylmethylxanthine
IBS Inflammatory bowel syndrome
iC3 Inactivated C3
iC4 Inactivated C4
IC_{50} Concentration producing 50% inhibition
ICAM Intercellular adhesion molecules
ICAM-1, ICAM-2, ICAM-3 Intercellular adhesion molecules-1, -2, -3
cICAM-1 Circulating form of ICAM-1
ICE IL-1β-converting enzyme
i.d. Intradermal
IDC Interdigitating cell
IDD Insulin-dependent (type 1) diabetes
IEL Intraepithelial leucocyte
IELym Intraepithelial lymphocytes
IFA Incomplete Freund's adjuvant
IFN Interferon
IFNαc, IFNβ, IFNγ Interferons α, β, γ
Ig Immunoglobulin
IgA, IgE, IgG, IgM Immunoglobulins A, E, G, M
IgG1 Immunoglobulin G class 1
IgG_{2a} Immunoglobulin G class 2a
IGF-1 Insulin-like growth factor
Ig-SF Immunoglobulin supergene family
IGSS Immuno-gold silver stain
IHC Immunohistochemistry
IHES Idiopathic hypereosinophilic syndrome
IκB NFκB inhibitor protein
IL Interleukin
IL-1, Il-2...IL-8 Interleukins-1, -2...-8
IL-1α, IL-1β Interleukin-1α, 1β
ILR Interleukin receptor
IL-1R, IL-2R; IL-3R – IL-6R Interleukin-1 – 6 receptors
IL-1Ra Interleukin-1 receptor antagonist
IL-2Rβ Interleukin-2 receptor β
IMF Integrin modulating factor
IMMC Intestinal mucosal mast cell
i.p. Intraperitoneally
IP_3 Inositol triphosphate
IP_4 Inositol tetrakisphosphate
IPF Idiopathic pulmonary fibrosis
IPO Intestinal peroxidase
IpOCOCq Isopropylidene OCOCq
IRAP IL-1 receptor antagonist protein
IRF-1 Interferon regulatory factor 1
ISCOM Immune-stimulating complexes
ISGF3 Interferon-stimulated gene Factor 3
ISGF3α, ISGFγ α, γ subunits of ISGF3
IT Immunotherapy
ITP Idiopathic thrombocytopenic purpura
i.v. Intravenous

K *The chemical symbol for* potassium
K_a Association constant
kb Kilobase
20KDHRF A homologous restriction factor; binds to C8
65KDHRF A homologous restriction factor, *also known as* C8 binding protein; interferes with cell membrane poreformation by C5b-C8 complex
Kcat Catalytic constant; a measure of the catalytic potential of an enzyme
K_d Equilibrium dissociation constant
kD Kilodalton
K_D Dissociation constant
KD Kallidin
Ki Antagonist binding affinity
Ki67 Nuclear membrane antigen
KLH Keyhole limpet haemocyanin
KOS KOS strain of herpes simplex virus

λ_{max} Wavelength of maximum absorbance
LAD Leucocyte adhesion deficiency
LAK Lymphocyte-activated killer (cell)
LAM, LAM-1 Leucocyte adhesion molecule, -1
LAR Late-phase asthmatic reaction
L-Arg L-Arginine
LBP LPS binding protein
LC Langerhans cell
LCF Lymphocyte chemoattractant factor
LCR Locus control region
LDH Lactate dehydrogenase
LDL Low-density lipoprotein
LDV Laser Doppler velocimetry
Le^x(Lewis X) Leucocyte ligand for selectin
LFA Leucocyte function-associated antigen
LFA-1 Leucocyte functionassociated antigen1; *also known to be* a member of the β-2 integrin family of cell adhesion molecules
LG β-Lactoglobulin
LGL Large granular lymphocyte
LH Luteinizing hormone
LHRH Luteinizing hormone-releasing hormone
LI Labelling index
LIS Lateral intercellular spaces
LMP Low molecular mass polypeptide
LMW Low molecular weight
L-NOARG L-Nitroarginine
LO Lipoxygenase
5-LO, 12-LO, 15-LO 5-, 12-, 15-Lipoxygenases
LP(a) Lipoprotein a
LPS Lipopolysaccharide
L-selectin Leucocte selectin, formerly known as monoclonal antibody that recognizes murine L-selectin (MEL-14 antigen), leucocyte cell adhesion molecule-1 (LeuCAM-1), lectin cell adhesion molecule-1 (LeCAM-1 or LecCAM-1), leucocyte adhesion molecule-1 (LAM-1)
LT Leukotriene
LTA_4, LTB_4, LTC_4, LTD_4, LTE_4 Leukotrienes A4, B_4, C_4, D_4 and E_4
L_y-1^+ (Cell line)
LX Lipoxin
LXA_4, LXB4, LXB_4, LXC_4, LXD_4, LXE_4 Lipoxins A_4, B_4, C_4, D_4 and E_4

M Monocyte
M3 Receptor Muscarinic receptor subtype 3
M-540 Merocyanine-540
mAb Monoclonal antibody
mAb IB4, mAb PB1.3, mAb R 3.1, mAb R 3.3, mAb 6.5, mAb 60.3 Monoclonal antibodies IB4, PB1.3, R 3.1, R 3.3, 6.5, 60.3
MAC Membrane attack molecule
Mac Macrophage (also abbreviated to MΦ)
Mac-1 Macrophage-1 antigen; a member of the β-2 integrin family of cell adhesion molecules (also abbreviated to MΦ1), *also known as* monocyte antigen-1 (M-1), complement receptor-3 (CR3), CD11b/CD18
MAF Macrophageactivating factor
MAO Monoamine oxidase
MAP Monophasic action potential
MAPTAM An intracellular Ca^{2+} chelator
MARCKS Myristolated, alanine-rich C kinase substrate; specific protein kinase C substrate
MBP Major basic protein
MBSA Methylated bovine serum albumin
MC Mesangial cells
M cell Microfold or membranous cell of Peyer's patch epithelium
MCP Membrane cofactor protein
M-CSF Monocyte/macrophage colony-stimulating factor c
MC_T Tryptase-containing mast cell
MC_{TC} Tryptase and chymase-containing mast cell
MDA Malondialdehyde
MDGF Macrophage-derived growth factor
MDP Muramyl dipeptide
MEA Mast cell growth enhancing activity

MEL Metabolic equivalent level
MEM Minimal essential medium
MG Myasthenia gravis
MGSA Melanoma-growth-stimulatory activity
MHC Major histocompatibility complex
MI Myocardial ischaemia
MIF Migration inhibition factor
mIL Mouse interleukin
MI/R Myocardial ischaemia/reperfusion
MIRL Membrane inhibitor of reactive lysis
mixCFC Colony-forming cell mix
Mk Megakaryocyte
MLC Mixed lymphocyte culture
MLymR Mixed lymphocyte reaction
MLR Mixed leucocyte reaction
MMC Mucosal mast cell
MMCP Mouse mast cell protease
MMP, MMP1 Matrix metalloproteinase, -1
MNA 6-Methoxy-2-napthylacetic acid
MNC Mononuclear cells
MΦ Macrophage (also abbreviated to Mac)
MPO Myeloperoxidase
MRI Magnetic resonance imaging
mRNA Messenger ribonucleic acid
MS Mass spectrometry
MSS Methylprednisoline sodium succinate
MT Malignant tumour
MW Molecular weight

Na *The chemical symbol for* sodium
NA Noradrenaline *also known as* norepinephrine
NAAb Natural autoantibody
NAb Natural antibody
NADH Reduced nicotinamide adenine dinucleotide
NADP Nicotinamide adenine diphosphate
NADPH Reduced nicotinamide adenine dinucleotide phosphate
NAF Neutrophil activating factor
LNAME L-Nitroarginine methyl ester
NANC Non-adrenergic, non-cholinergic
NAP Neutrophil-activating peptide
NAP-1, NAP-2 Neutrophil-activating peptides -1 and -2
NBT Nitroblue tetrazolium
NC1 Noncollagen 1
N-CAM Neural cell adhesion molecule
NCEH Neutral cholesteryl ester hydrolase
NCF Neutrophil chemotactic factor
NDGA Nordihydroguaretic acid
NDP Nucleoside diphosphate
Neca 5′-(*N*-ethyl carboxamido)-adenosine
NED Nedocromil sodium
NEP Neutral endopeptidase (EC 3.4.24.11)
NF-AT Nuclear factor of activated T lymphocytes
NF-κB Nuclear factor-κB
NgCAM Neural-glial cell adhesion molecule
NGF Nerve growth factor
NGPS Normal guinea-pig serum
NIH 3T3 (fibroblasts) National Institute of Health 3T3-Swiss albino mouse fibroblast
NIMA Non-inherited maternal antigens
Nk Neurokinin
NK Natural killer
Nk-1, Nk2, NK-3 Neurokinin receptor subtypes 1, 2 and 3
NkA Neurokinin A
NkB Neurokinin B
LNMMA L-Nitromonomethyl arginine
NMR Nuclear magnetic resonancec
NO *The chemical symbol for* nitric oxide
NPK Neuropeptide K
NPY Neuropeptide Y
NRS Normal rabbit serum
NSAID Non-steroidal anti-inflammatory drug
NSE Nerve-specific enolase
NT Neurotensin
N terminus Amino terminus of peptide

$\cdot O_2^-$ Oxygen free radical
OA Osteoarthritis
OAG Oleoyl acetyl glycerol
OD Optical density
ODC Ornithine decarboxylase
ODS Octadecylsilyl
•OH *The chemical symbol for* hydroxyl radical
OT Oxytocin
OVA Ovalbumin
ox-LDL Oxidized low-density lipoprotein

Ψa Apical membrane potential
P Probability
P Phosphate
P_aO_2 Arterial oxygen pressure
P_i Inorganic phosphate
p150, 95 A member of the β-2-integrin family of cell adhesion molecules; *also known as* CD11c
PA Phosphatidic acid
pA_2 Negative logarithm of the antagonist dissociation constant
PAF Platelet-activating factor *also known as* APRL and AGEPC
PAGE Polyacrylamide gel electrophoresis
PAI Plasminogen activator inhibitor
PAM Pulmonary alveolar macrophages
PAS Periodic acid-Schiff reagent
PBA Polyclonal B cell activators
PBC Primary biliary cirrhosis
PBL Peripheral blood lymphocytes
PBMC Peripheral blood mononuclear cells
PBS Phosphate-buffered saline
PC Phosphatidylcholine
PCA Passive cutaneous anaphylaxis
pCDM8 Eukaryotic expression vector
PCNA Proliferating cell nuclear antigen
PCR Polymerase chain reaction
p.d. Potential difference
PDBu 4α-phorbol 12, 13-dibutyrate
PDE Phosphodiesterase
PDGF Platelet-derived growth factor
PDGFR Platelet-derived growth factor receptor
PE Phosphatidylethanolamine
PECAM-1 Platelet endothelial cell adhesion molecule-1; *also known as* CD31
PEG Polyethylene glycol
PET Positron emission tomography
PEt Phosphatidylethanolamine
PF_4 Platelet factor 4
PG Prostaglandin
PGAS Polyglandular autoimmune syndrome
PGD_2 Prostaglandin D_2
PGE1, PGE_2, PGF_2, $PGF_{2\alpha}$, PGG_2, PGH^2 Prostaglandins E_1, E_2, F_2, $F_{2\alpha}$, G_2, H_2
PGF, PGH Prostaglandins F and H
PGI_2 Prostaglandin I_2 *also known as* prostacyclin
P_aO_2 Arterial oxygen pressure
PGP Protein generelated peptide
Ph^1 Philadelphia (chromosome)
PHA Phytohaemagglutinin
PHD PHD[8(1-hydroxy-3-oxo-propyl)-9, 12-dihydroxy-5, 10 heptadecadienic acid]
PHI Peptide histidine isoleucine
PHM Peptide histidine methionine
P_i Inorganic phosphate
PI Phosphatidylinositol
PI-3, 4-P2 Phosphatidylinositol 3, 4-biphosphate
PI-3, 4, 5-P3 Phosphatidylinositol 3, 4, 5-trisphosphate
PI-3-kinase Phosphatidylinositol-3-kinase
PI-4-kinase Phosphatidylinositol-4-kinase
PI-3-P Phosphatidylinositol-3-phosphate
PI-4-P Phosphatidylinositol-4-phosphate
PI-4, 5-P2 Phosphatidylinositol 4, 5-biphosphate
PIP Phosphatidylinositol monophosphate

PIP_2 Phosphatidylinositol biphosphate
PK Protein kinase
PKA, PKC Protein kinases A and C
PL Phospholipase
PLA, PLA_2, PLC, PLD Phospholipases A, A_2, C and D
PLN Peripheral lymph node
PLNHEV Peripheral lymph node HEV
PLP Proteolipid protein
PLT Primed lymphocyte typing
PMA Phorbol myristate acetate
PMC Peritoneal mast cell
PMN Polymorphonuclear neutrophil
PMSF Phenylmethylsulphonyl fluoride
PNAd Peripheral lymph node vascular addressin
PNH Paroxysmal nocturnal hemoglobinuria
PNU Protein nitrogen unit
p.o. *Per os* (by mouth)
PPD Purified protein derivative
PPME Polymeric polysaccharide rich in mannose-6-phosphate moieties
PRA Percentage reactive activity
PRD, PRDII Positive regulatory domain, -II
PR3 Proteinase-3
PRBC Parasitized red blood cell
proET-1 Proendothelin-1
PRL Prolactin
PRP Platelet-rich plasma
PS Phosphatidylserine
P-selectin Platelet selectin *formerly known as* platelet adctivation-dependent granule external membrane protein (PADGEM), granule membrane protein of MW 140 kD (GMP140)
PT Pertussis toxin
PTCA Percutaneous transluminal coronary angioplasty
PTCR Percutaneous transluminal coronary recanalization
Pte-H_4 Tetrahydropteridine
PUFA Polyunsaturated fatty acid
PUMP-1 Punctuated metalloproteinase
PWM Pokeweed mitogen

q.i.d. *Quater in die* (four times a day)
QRS Segment of electrocardiogram

•R Free radical
R15.7 Anti-CD18 monoclonal antibody
RA Rheumatoid arthritis
RANTES A member of the IL8 supergene family (*R*egulated on *a*ctivation, *n*ormal *T* *e*xpressed and *s*ecreted)
RAST Radioallergosorbent test
RBC Red blood cell
RBF Renal blood flow
RBL Rat basophilic leukaemia
RE RE strain of herpes simplex virus type 1
REA Reactive arthritis
REM Relative electrophoretic mobility
RER Rough endoplasmic reticulum
RF Rheumatoid factor
RFL-6 Rat foetal lung-6
RFLP Restriction fragment length polymorphism
RGD Arginine-glycine-asparagine
rh- Recombinant human – (prefix usually referring to peptides)
RIA Radioimmunoassay
RMCP, RMCPII Rat mast cell protease, -II
RNA Ribonucleic acid
RNase Ribonuclease
RNHCl *N*-Chloramine
RNL Regional lymph nodes
ROM Reactive oxygen metabolite
ROS Reactive oxygen species
R-PIA *R*-(1-methyl-1-phenyltheyl)-adenosine
cRPMI 1640 Roswell Park Memorial Institute 1640 medium
RS Reiter's syndrome
RSV Rous sarcoma virus
RTE Rabbit tubular epithelium
RTE-a-5 Rat tubular epithelium antigen a-5
r-tPA Recombinant tissue-type plasminogen activator
RW Ragweed

S Svedberg (unit of sedimentation density)
SALT Skin-associated lymphoid tissue
SAZ Sulphasalazine
SC Secretory component
SCF Stem cell factor
SCFA Short-chain fatty acid
SCG Sodium cromoglycate *also known as* DSCG
SCID Severe combined immunodeficiency syndrome
sCR1 Soluble type-1 complement receptors
SCW Streptococcal cell wall
SD Standard deviation
SDS Sodium dodecyl sulphate
SDS-PAGE Sodium dodecyl sulphate-polyacrylamide gel electrophoresis
SEM Standard error of the mean
SGAW Specific airway conductance
SHR Spontaneously hypertensive rat
SIRS Soluble immune response suppressor
SIV Simian immunodeficiency virus
SK Streptokinase
SLE Systemic lupus erythematosuscs
SLe^x Sialyl Lewis X antigen
SLO Streptolysin-O
SLPI Secretory leucocyte protease inhibitor
SM Sphingomyelin
SNAP *S*-Nitroso-*N*-acetylpenicillamine
SNP Sodium nitroprusside
SOD Superoxide dismutase
SOM Somatostatin *also known as* somatotrophin release-inhibiting factor
SOZ Serum-opsonized zymosan
SP Sulphapyridine
SR Systemic reaction
SRBC Sheep red blood cells
SRS Slowreacting substance
SRS-A Slow-reacting substance of anaphylaxis
Sub P Substance P

T Thymus-derived
$t_{1/2}$ Half-life
T84 Human intestinal epithelial cell line
TauNHCl Taurine monochloramine
TBM Tubular basement membrane
TCA Trichloroacetic acid
T cell Thymus-derived lymphocyte
TCR T cell receptor α/β or γ/δ heterodimeric forms
TDI Toluene diisocyanate
TEC Tubular epithelial cell
TF Tissue factor
Tg Thyroglobulin
TGF Transforming growth factor
TGFα, TGFβ, TGFβ_1 Transforming growth factors α, β and β_1
T_H T helper cell
$T_{H}o$ T Helper o
T_{Hp} T helper precursor
TH0, T_H1, T_H2 Subsets of helper T cells
THP-1 Human monocytic leukaemia
Thy 1+ Murine T cell antigen
t.i.d. *Ter in die* (three times a day)
TIL Tumour-infiltrating lymphocytes
TIMP Tissue inhibitors of metalloproteinase
TIMP-1, TIMP-2 Tissue inhibitors of metalloproteinases 1 and 2
Tla Thymus leukaemia antigen
TLC Thin-layer chromatography
TLCK Tosyl-lysyl-CH_2Cl
TLP Tumour-like proliferation
Tm T memory
TNF, TNF-α Tumour necrosis factor, -α
tPA Tissue-type plasminogen activator
TPA 12-*O*-tetradeconylphorbol-13-acetate
TPCK Tosyl-phenyl-CH_2Cl
TPK Tyrosine protein kinases

TPP Transpulmonary pressure
TRAP Thrombospondin related anomalous protein
Tris Tris(hydroxymethyl) aminomethane
TSH Thyroid-stimulating hormone
TSP Thrombospondin
TTX Tetrodotoxin
TX Thromboxane
TXA_2, TXB_2 Thromboxane A_2, B_2
Tyk2 Tyrosine kinase

U937 (cells) Histiocytic lymphoma, human
UC Ulcerative colitis
UDP Uridine diphosphate
UPA Urokinase-type plasminogen activator
UTP Uridine triphosphate
UV Ultraviolet

VAP Viral attachment protein
VC Veiled cells
VCAM, VCAM-1 Vascular cell adhesion molecule, -1, *also known as* inducible cell adhesion molecule MW 110 kD (INCAM-110)
VF Ventricular fibrillation
VIP Vasoactive intestinal peptide
VLA Very late activation antigen beta chain; *also known as* CD29
VLA α2 Very late activation antigen alpha 2 chain; *also known as* CD49b
VLA α4 Very late activation antigen alpha 4 chain; *also known as* CD49d
VLA α6 Very late activation antigen alpha 6 chain; *also known as* CD49f
VLDL Very low-density lipoprotein
***V* max** Maximal velocity
***V* min** Minimal velocity
VN Vitronectin
vp Viral protein
VP Vasopressin
VPB Ventricular premature beat
VT Ventricular tachycardia

W Murine dominant white spotting mutation
WBC White blood cell
WGA Wheat germ agglutinin

XO Xanthine oxidase

Y1/82A A monoclonal antibody detecting a cytoplasmic antigen in human macrophages

ZA Zonulae adherens
ZAS Zymosan-activated serum
zLYCK Carboxybenzyl-Leu-Tyr-CH_2Cl
ZO Zonulae occludentes

Key to Illustrations

Helper lymphocyte

Suppressor lymphocyte

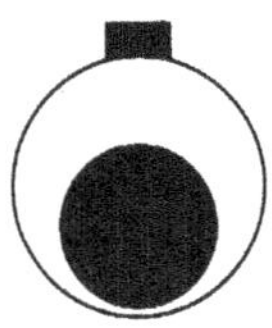

Killer lymphocyte

Plasma cell

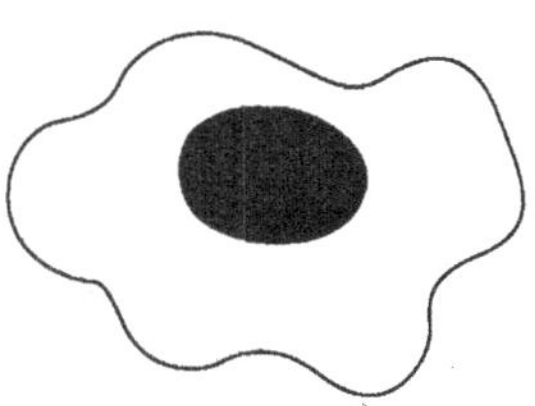

Bacterial or Tumour cell

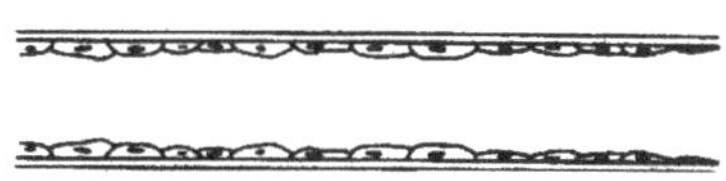

Blood vessel lumen

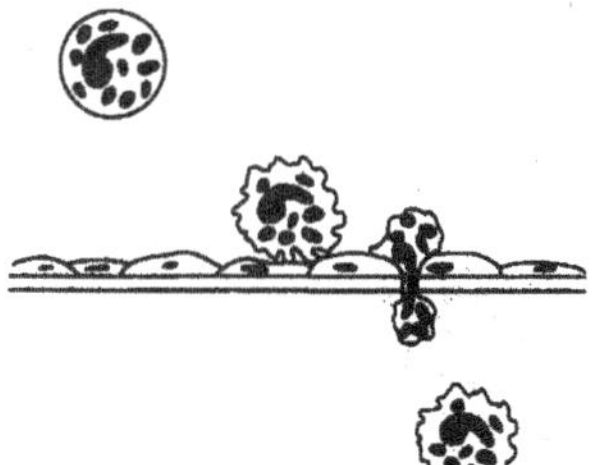

Eosinophil passing through vessel wall

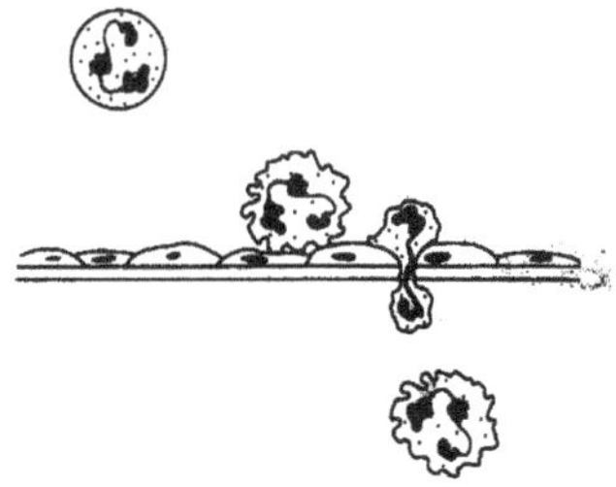

Neutrophil passing through vessel wall

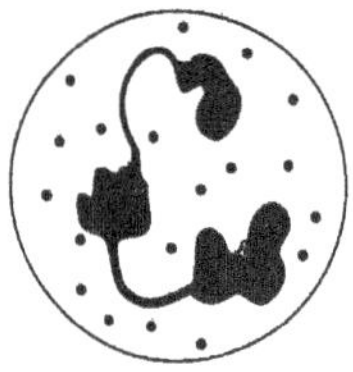

Resting neutrophil

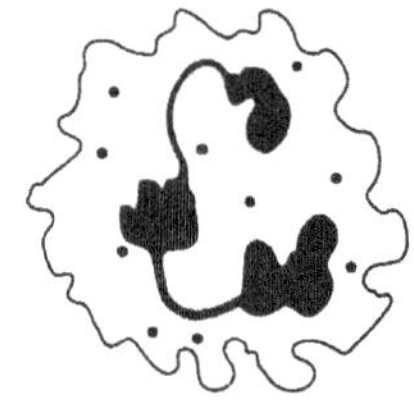

Activated neutrophil

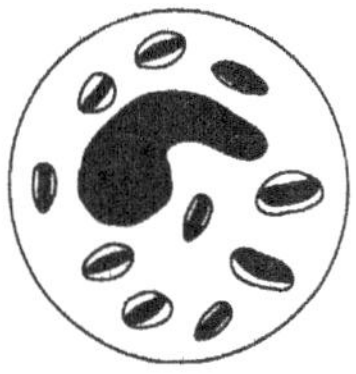

Resting eosinophil

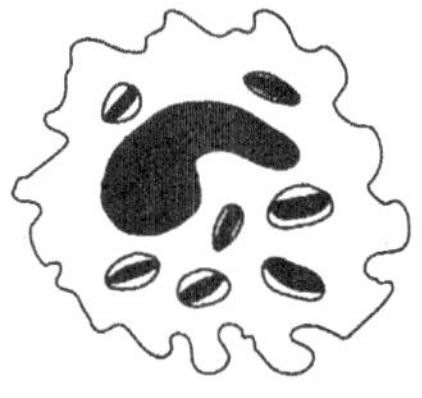

Activated eosinophil

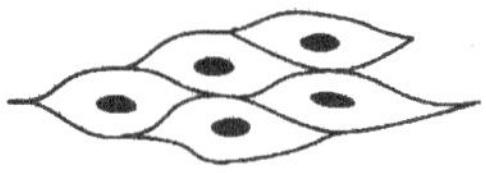

Smooth muscle

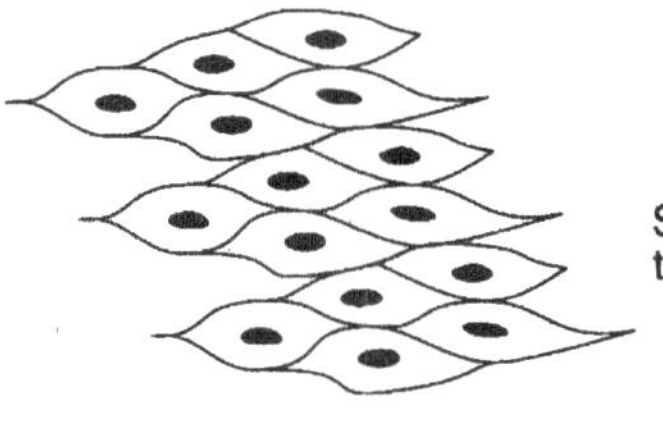

Smooth muscle thickening

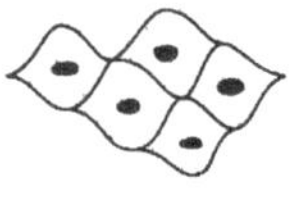

Smooth muscle contraction

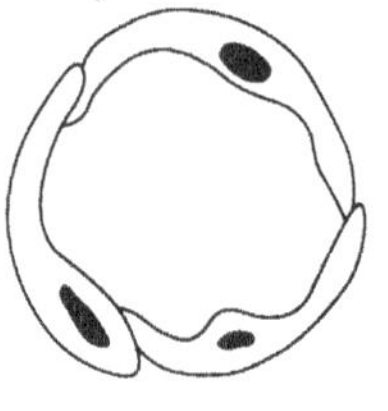

Normal blood vessel

Endothelial cell permeability

Resting macrophage

Activated macrophage

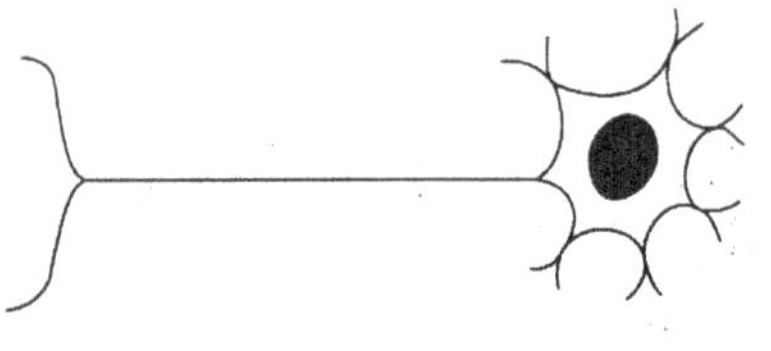

Nerve

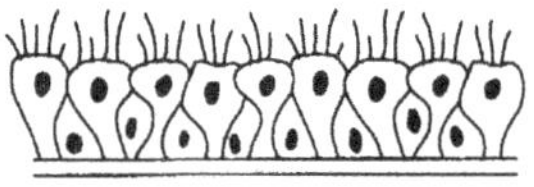

Intact epithelium

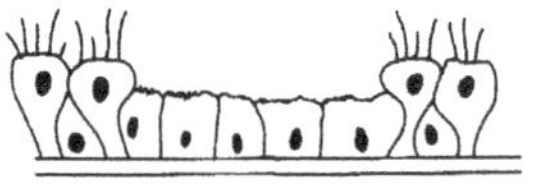

Damaged epithelium

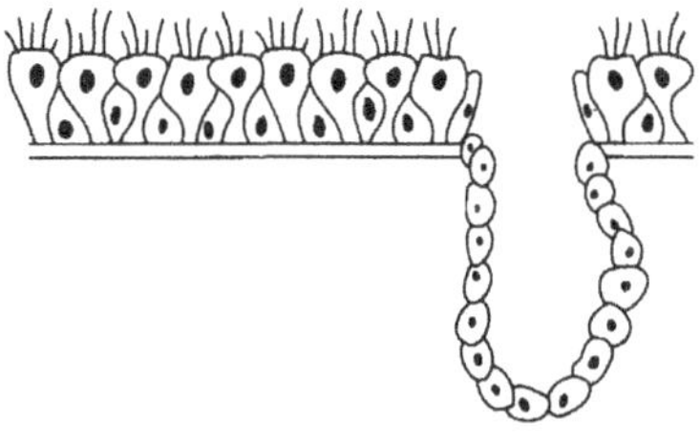

Intact epithelium with submucosal gland

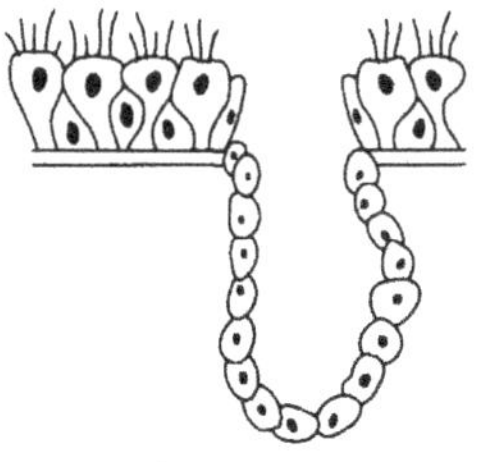

Normal submucosal gland

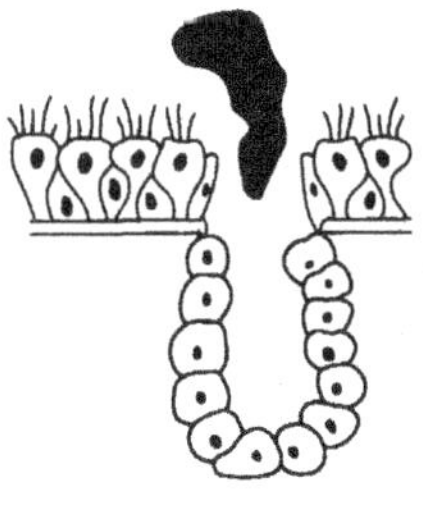

Hypersecreting submucosal gland

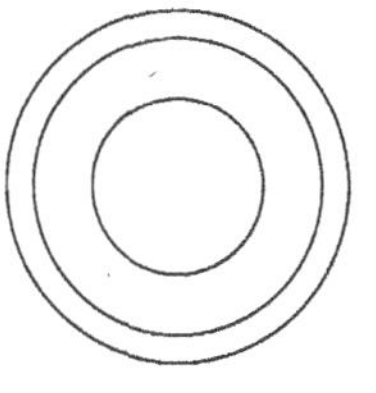

Normal airway

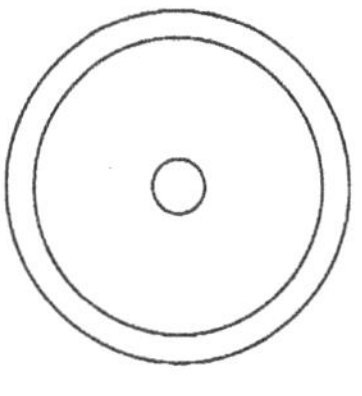

Oedema

Bronchospasm

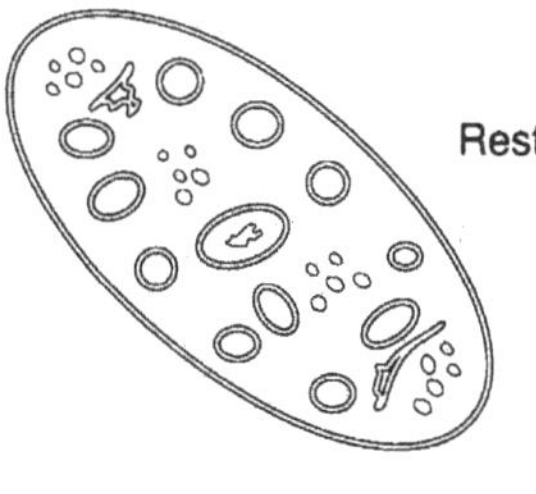

Resting platelet

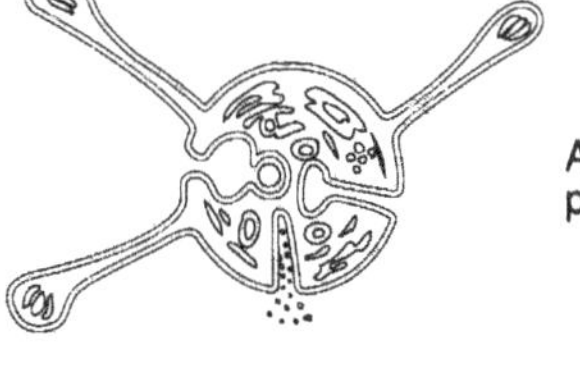

Activated platelet

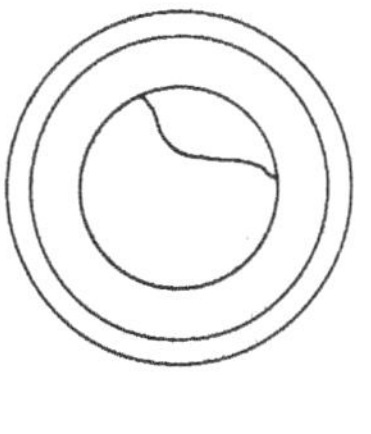

Airway hypersecreting mucus

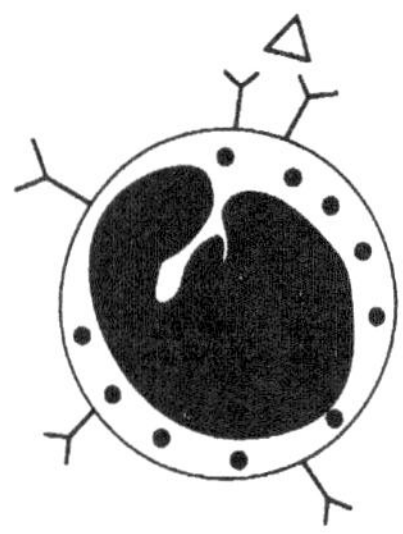

Resting basophil

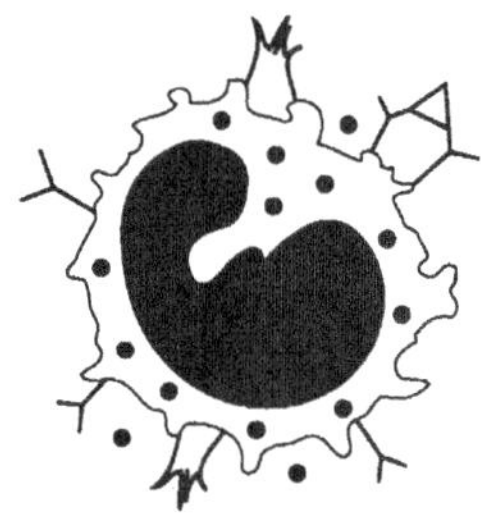

Activated basophil

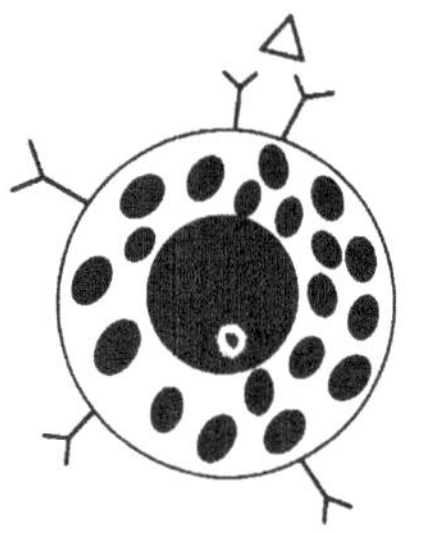

Resting mast cell

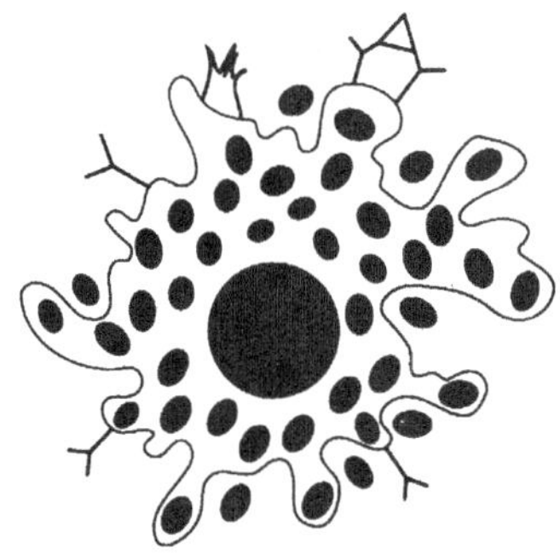

Activated mast cell

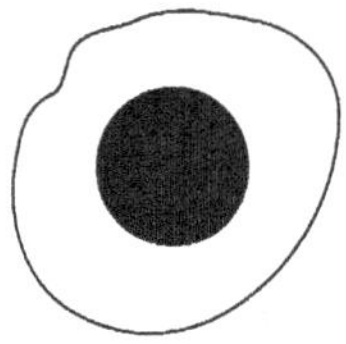

Resting chondrocyte

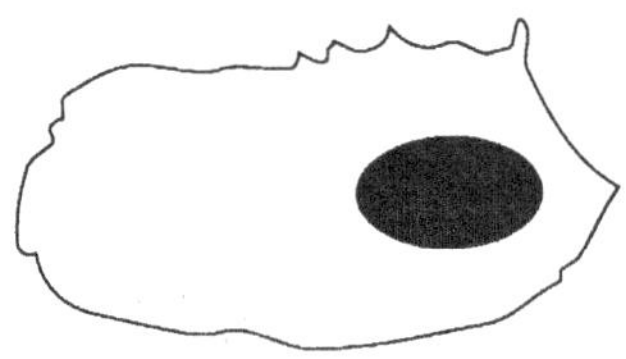

Activated chondrocyte

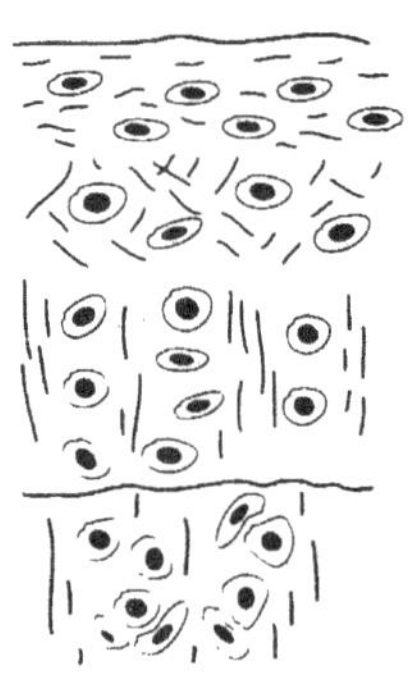

Cartilage

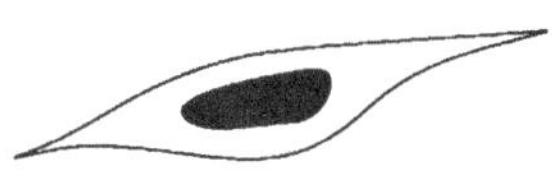

Fibroblast

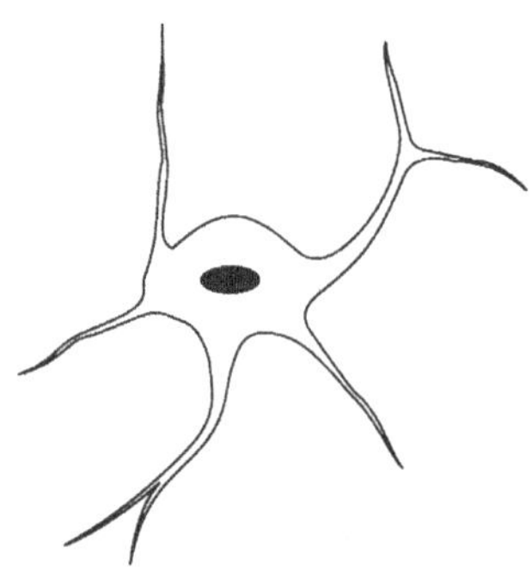

Dendritic cell/ Langerhans cell

Arteriole

Venule

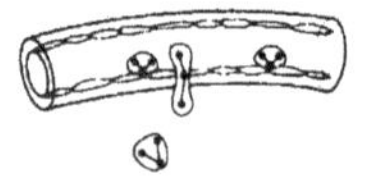

Inflamed venule

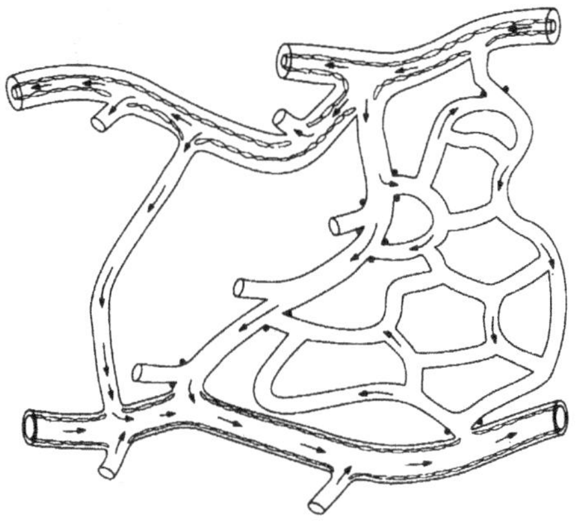

Microcirculatory system

Index

Edwards Brothers Malloy
Ann Arbor MI. USA
July 1, 2015